PRENTICE HALL

mygeologyplace™

Where geology comes to you.

W9-BTB-078

Students with new copies of Keller, *Introduction to Environmental Geology, Fourth Edition* have full access to the book's MyGeologyPlace—a 24/7 study tool with quizzes, web destinations, and other features designed to help you make the most of your study time.

Just follow the easy website registration steps listed below...

Your Access Code is:

If there is no metallic coating covering the access code above, the code may no longer be valid and you will need to purchase online access using a major credit card to use the website. To do so, go to www.prenhall.com/keller, click the cover for Keller, *Introduction to Environmental Geology, Fourth Edition*, then click the "Get Access" button, and follow the instructions under "Students" to Purchase Access Now.

Registration Instructions for MyGeologyPlace

1. Go to *www.prenhall.com/keller*
2. Click the cover for Keller, *Introduction to Environmental Geology, Fourth Edition*.
3. Click "Register".
4. Using a coin (not a knife) scratch off the metallic coating above to reveal your Access Code.
5. Complete the online registration form, choosing your own personal Login Name and Password.
6. Enter your pre-assigned Access Code exactly as it appears above.
7. Complete the online registration form by entering your School Location information.
8. After your personal Login Name and Password are confirmed by e-mail, go back to *www.prenhall.com/keller*, click your book's cover, enter your new Login Name and Password, and click "Log In".

Minimum system requirements
PC Operating Systems:

Windows 2000/XP
Pentium II 233 MHz processor. 64 MB RAM
In addition to the minimum memory required by your OS.
Internet Explorer(TM) 6.0 or,
Internet Explorer(TM) 7.0

Macintosh Operating Systems:
Macintosh Power PC with OS X (10.4)
In addition to the RAM required by your OS, this product requires 64 MB RAM, with 40MB Free RAM, with Virtual Memory enabled
Safari 2.0

Macromedia Shockwave(TM)
8.50 release 326 plugin
Macromedia Flash Player 6.0.79 & 7.0
Acrobat Reader 6.0.1
800 x 600 pixel screen resolution

TECHNICAL SUPPORT Call 1-800-677-6337. Phone support is available Monday–Friday, 8am to 8pm and Sunday 5pm to 12am, Eastern time. Visit our support site at http://247.prenhall.com. E-mail support is available 24/7.

Introduction to
Environmental Geology

FOURTH EDITION

Introduction to
Environmental Geology

Edward A. Keller

University of California, Santa Barbara

PEARSON

Prentice
Hall

Upper Saddle River, NJ 07458

Library of Congress Cataloging-in-Publication Data

Keller, Edward A., 1942-
    Introduction to environmental geology/Edward A. Keller.—4th ed.
        p. cm.
Includes bibliographical references and index.
ISBN 0-13-225150-7
  1. Environmental geology.   I. Title.

QE38.K46 2007
550–dc22   2006037684

Publisher: Daniel Kaveney
Editor-in-Chief, Science: Nicole Folchetti
Assistant Editor: Sean Hale
Executive Managing Editor: Kathleen Schiaparelli
Assistant Managing Editor: Beth Sweeten
Marketing Manager: Amy Porubsky
Media Editor: Andrew Sobel
Creative Director: Juan López
Art Director: Heather Scott
Interior and Cover Design: Michael Fruhbeis
Manufacturing Manager: Alexis Heydt-Long
Senior Managing Editor, Art Production and Management: Patricia Burns
Manager, Production Technologies: Matthew Haas
Managing Editor, Art Management: Abigail Bass
Illustrations: GEX, Inc.
Manufacturing Buyer: Ilene Kahn
Editorial Assistants: John DeSantis, Jessica Neumann
Director, Image Resource Center: Melinda Patelli
Manager, Rights and Permissions: Zina Arabia
Manager, Interior Research: Beth Brenzel
Manager, Visual Research and Permissions: Karen Sanatar
Image Permission Coordinator: Debbie Hewitson
Photo Researcher: Truitt and Marshall
Composition: ICC Macmillan Inc.
Cover Image: Dave Martin/AP Wide World Photos

© 2008, 2005, 2002, 1999 Pearson Education, Inc.
Pearson Prentice Hall
Pearson Education, Inc.
Upper Saddle River, NJ 07458

Pearson Prentice Hall™ is a trademark of Pearson Education, Inc.

Printed in the United States of America
10 9 8 7 6 5 4 3 2 1

ISBN-13: 978-0-13-225150-1
ISBN-10:   0-13-225150-7

Pearson Education Ltd., *London*
Pearson Education Australia, PTY. Limited, *Sydney*
Pearson Education Singapore, Pte. Ltd
Pearson Education North Asia Ltd, *Hong Kong*
Pearson Education Canada, Ltd., *Toronto*
Pearson Educacion de Mexico, S.A. de C.V.
Pearson Education—Japan, *Tokyo*
Pearson Education Malaysia, Pte. Ltd

BRIEF CONTENTS

- A chapter summary reinforces the major points of the chapter to help you refocus on the important subjects.
- The foundations of environmental geology are presented in Chapters 1 through 4, and Chapters 5 through 20 contain a discussion that revisits the five fundamental principles in terms of the material presented in the chapter.
- Detailed references are supplied at the end of the text to provide additional readings and to give credit to the scholars who did the research reported in the chapter.
- Key terms are presented at the end of the chapters. These will help you identify the important concepts and terminology necessary to better understand the chapter.
- Review questions help with your review of important subject matter.
- Critical thinking questions stimulate you to think about some of the important issues in the chapters and try to relate these to your life and society.

The appendixes in *Introduction to Environmental Geology,* Fourth Edition, are intended to add additional information useful in helping you understand some of the more applied aspects of environmental geology. This information may be most useful in supplementing laboratory exercises and field exercises in which you may participate. Specific topics include:

- Identification of rocks and minerals with accompanying tables and suggestions.
- Introduction to topographic and geologic maps with specific information concerning how to read topographic maps, construct topographic profiles, and understand geologic maps.
- Introduction to Digital Elevation Models (DEMs) and Global Positioning System instrumentation (GPS).
- Discussion of how geologists determine and interpret geologic time.
- Darcy's Law, with example of how we use it to solve groundwater problems.
- A glossary of terms used in the field of environmental geology.

The New Instructional Package

Prentice Hall has assembled a greatly improved resource package for *Introduction to Environmental Geology,* Fourth Edition.

For the Instructor

- *Instructor Resource Center on CD-ROM, with PowerPoint Presentations and Animations* **(0-13-225153-1):** Included on the Instructor Resource CD are three PowerPoint® presentations for each chapter: (1) **Lecture outline**— "plug-and-play" lecture presentations based on the outline of the text to get you up and running as quickly as possible; (2) **Art only**—every illustration and most of the photos in the text, in order, pre-loaded onto PowerPoint slides; (3) **Animations**—high-quality animations of key geologic processes (see below for more details). Also included are all illustrations and a selection of photos for the text in 16-bit, low-compression JPEG files. All images are manually adjusted for color, brightness, and contrast.
- *PH Geoscience Animations* **(0-13-600377-X):** The Prentice Hall Geoscience Animation Library resulted from a survey in which we asked instructors to identify the concepts most difficult to teach using traditional, static resources. Then we animated them. Created through a unique collaboration among five of Prentice Hall's leading geoscience authors, these animations represent a

BRIEF CONTENTS

Hazard City: Assignments in Applied Geology CD-ROM, Third Edition

This CD is included inside every new copy of this textbook. In these 11 assignments, students investigate the fictional town of Hazard City and play the role of a practicing geologist.

NEW Map Reading: Students learn basic map reading skills and apply them by planning a route to collect stream water specimens in a case of suspected contamination.

NEW Tsunami/Storm Surge: Students research the causes, controls, and impacts of tsunamis and storm surge. They then use maps and basic math to determine maximum warning time, number of residents needing evacuation, and the percentage of a town to be rebuilt after a tsunami/storm surge of known run up.

Ground Water Contamination: Students use field and laboratory data to prepare a countour map of the water table, determine the direction of ground water flow and map a contaminated area.

Volcanic Hazard Assessment: Researching volcanic hazards, collecting field information, and decision-making are all used to determine the potential impact of a volcanic eruption on different parts of Hazard City.

Landslide Hazard Assessment: Students research the factors that determine landslide hazard at five construction sites and make recommendations for development.

Earthquake Damage Assessment: Students research the effects of earthquakes on buildings, explore Hazard City, and determine the number of people needing emergency housing given an earthquake of specific intensity.

Flood Insurance Rate Maps: Flood insurance premiums are estimated using a flood insurance rate map, insurance tables and site characteristics.

Snowpack Monitoring: Students utilize climatic data to estimate variables that are key to flood control and water supply management.

Coal Property Evaluation: The potential value of a mineral property is estimated by learning about mining and property evaluation, then applying that knowledge in a resource calculation.

Landfill Siting: Students use maps and geological data to determine if any of five proposed sites meet the requirements of the State Administrative Code for landfill siting.

Shoreline Property Assessment: Students visit four related waterfront building sites—some developed and some not—and analyze the risk each faces due to shoreline erosion processes.

CONTENTS

PART THREE
Resources and Pollution   379

PREFACE

The main objective of *Introduction to Environmental Geology,* Fourth Edition, is to help equip students—particularly those who intend to take only a single science course—with an understanding of the interactions between geologic processes, ecological processes, and society. During the first half of the twenty-first century, as the human population increases and the use of resources grows, many decisions concerning our use of those resources, such as water, soil, air, minerals, energy, and space to live, will determine our standard of living and the quality of our environment. Scientific knowledge combined with our values will dictate those decisions. Your charge, as a future leader or informed citizen, is to choose paths of development that are good for people and the environment, that larger community that includes plants, animals, water, and air—in other words, the environment consisting of ecosystems that we and all living things depend upon for our well-being.

Earth's dynamic and changing environment constitutes one of the most compelling and exciting areas of study. Environmental geology is the application of geologic information to the entire spectrum of interactions between people and the physical environment. During a course in environmental geology, you will develop an understanding of how geology interacts with major environmental problems facing people and society. This is the essence of *Introduction to Environmental Geology*, Fourth Edition. Our strategy with this text is to:

- Introduce you to the basic concepts and principles of physical and environmental geology, focusing on Earth materials and processes.

- Provide you with sufficient information concerning natural hazards and the geologic environment so that you will be a more informed citizen. You will be better prepared to make decisions concerning where you live and how society responds to natural hazards and catastrophes such as earthquakes, volcanic eruptions, and flooding.

- Help you develop an understanding of relationships between natural resources and pollution. We seek, find, and use resources and, as a result, may pollute our environment. Thus, it is important to know how we might minimize pollution problems.

- Help you understand the basic concepts of environmental management as they relate to the geologic environment in areas such as waste management, environmental health, global change, and environmental assessment.

After finishing your course in environmental geology, you will be better prepared to make decisions concerning where you build or buy a home, what resources you choose to utilize, and appropriate environmental actions relevant to society and Earth's ecosystems from a local to a global scale.

Five Fundamental Concepts

To this end, this book introduces a device we call the "Fundamental Concepts of Environmental Geology." These five concepts are designed to provide a memorable, transportable framework of understanding that you can carry away from the class and use throughout life to make informed choices about your interaction with and effect upon geologic processes:

- *Human population growth:* Population growth is the number one environmental problem. As population increases, so do our effects and demands on the environment.

- *Sustainability:* Sustainability is the long-term environmental objective of providing for the future of humans and other living things who share the planet.
- *Earth as a system:* The activities of human beings can have important effects on any or all of Earth's systems, often affecting the global environment.
- *Hazardous Earth processes, risk assessment, and perception:* Earth's hazardous processes have always occurred and will always occur. Human beings need to recognize the threat of hazards, assess the risk to life and property, and either avoid them or plan accordingly.
- *Scientific knowledge and values:* Scientific inquiries often provide a variety of potential solutions to environmental problems. The solution we choose is a direct reflection of our value system.

These concepts are introduced in the first chapter and then highlighted throughout the text (look for the "Revisiting the Fundamental Concepts" section at the end of every chapter). By tying the content to these five principles, the text provides a framework for understanding that will extend far beyond the confines of this course and into your everyday life.

Organization

Introduction to Environmental Geology, Fourth Edition, is well suited to your study of environmental geology, whether you are a geology major or are taking this class as a science elective. I have organized *Introduction to Environmental Geology*, Fourth Edition, to flow naturally from the introduction of fundamental principles of environmental science and geology, to more specific information concerning how Earth works, to natural processes and hazards, to understanding natural resources and their management, with the objective of minimizing environmental degradation. We end with a detailed discussion of global change, focusing on climate and some important interactions between society and the geologic environment.

Introduction to Environmental Geology, Fourth Edition, consists of 20 chapters arranged in four parts:

- Part 1 introduces philosophy and fundamental concepts, the structure of Earth and plate tectonics, and the origin and significance of minerals and rocks. Thus, Part 1 presents fundamentals of physical geology with important environmental information necessary to understand the remainder of the text. Chapter 1 introduces five fundamental concepts of environmental science, with an emphasis on the geologic environment. Chapter 2 discusses the structure of Earth and the important subject of plate tectonics and how our planet works from a geologic perspective. Chapter 3 presents geologic information concerning rocks and minerals necessary for understanding environmental geology problems and solutions to those problems. In Chapter 3, we also introduce some of the fundamental principles of geology, including the law of original horizontality, the law of cross-cutting relationships, the concept of the depositional environment, the concept of the rock cycle, and the principle of magmatic differentiation. Chapter 4 presents basics of ecology with links to geology. An ecosystem includes a community of life and it's non-living environment (rock, soil, etc.). The new emerging study of geology linked to ecology is exciting and offers many opportunities.
- Part 2 addresses natural hazards, including an introduction to natural hazards (Chapter 5), earthquakes (Chapter 6), volcanic activity (Chapter 7), rivers and flooding (Chapter 8), landslides (Chapter 9), coastal processes (Chapter 10), and impacts of extraterrestrial objects (Chapter 11). The intent is not to provide copious amounts of detailed information concerning these processes but to focus on the basics involved and the environmental concerns of Earth processes and natural hazards.

- Part 3 presents the major resources associated with the geologic environment and the subject of pollution. Important topics include water resources (Chapter 12), water pollution (Chapter 13), mineral resources (Chapter 14), energy resources (Chapter 15), soils (Chapter 16), waste management (Chapter 17), and air pollution (Chapter 18). The focus is to present the basic principles concerning natural resources and to identify potential environmental problems and solutions.

- Part 4 is concerned with the important topics of global change, environmental management, and relationships between environment and society. Chapter 19 discusses global change with a focus on global warming and stratospheric ozone depletion. Finally, in Chapter 20, which is a "capstone," we discuss relationships between environment and society with topics such as environmental health, land-use planning, environmental law, environmental impact analysis, how we may achieve the goal of obtaining environmental sustainability, and what that environment might entail.

Major New Material in the Fourth Edition

The fourth edition benefited greatly from feedback from instructors using the text—most of the changes reflect their thoughtful reviews. One new chapter—Ecology and Geology (Chapter 4) recognizes important links between physical and biological processes in the study of the environment. In addition to this new chapter, new or extensively revised discussions include:

- Why solving environmental problems is difficult (Chapter 1)
- Environmental policy and the precautionary principle (Chapter 1)
- Tsunamis (Chapter 6)
- Evaluation of flooding (Chapter 8)
- La Conchita landslide (Chapter 9)
- Hurricanes (Chapter 10)
- Water resources (Chapter 13)
- Peak oil (Chapter 15)
- E-waste (Chapter 17)
- Air toxics (Chapter 18)
- Global warming (Chapter 19)
- What a sustainable environment involves—a vision for the future (Chapter 20)

In addition to the above mentioned new chapter and new or revised discussions, the discussion of many topics has been updated. At the request of reviewers, many figures have been revised to more clearly illustrate the topics under discussion.

Features of the Text

This book is sensitive to the study needs of students. Each chapter is clearly structured to help you understand the material and effectively review the major concepts. To help you use the material from the book, each chapter is organized with the following study aids:

- Learning objectives that state clearly what you should be able to do upon completing the chapter.
- Selected features, called *Case History* or *A Closer Look*, are added where appropriate to help you relate topics in the text to the world around you.

- A chapter summary reinforces the major points of the chapter to help you refocus on the important subjects.

- The foundations of environmental geology are presented in Chapters 1 through 4, and Chapters 5 through 20 contain a discussion that revisits the five fundamental principles in terms of the material presented in the chapter.

- Detailed references are supplied at the end of the text to provide additional readings and to give credit to the scholars who did the research reported in the chapter.

- Key terms are presented at the end of the chapters. These will help you identify the important concepts and terminology necessary to better understand the chapter.

- Review questions help with your review of important subject matter.

- Critical thinking questions stimulate you to think about some of the important issues in the chapters and try to relate these to your life and society.

The appendixes in *Introduction to Environmental Geology,* Fourth Edition, are intended to add additional information useful in helping you understand some of the more applied aspects of environmental geology. This information may be most useful in supplementing laboratory exercises and field exercises in which you may participate. Specific topics include:

- Identification of rocks and minerals with accompanying tables and suggestions.

- Introduction to topographic and geologic maps with specific information concerning how to read topographic maps, construct topographic profiles, and understand geologic maps.

- Introduction to Digital Elevation Models (DEMs) and Global Positioning System instrumentation (GPS).

- Discussion of how geologists determine and interpret geologic time.

- Darcy's Law, with example of how we use it to solve groundwater problems.

- A glossary of terms used in the field of environmental geology.

The New Instructional Package

Prentice Hall has assembled a greatly improved resource package for *Introduction to Environmental Geology,* Fourth Edition.

For the Instructor

- *Instructor Resource Center on CD-ROM, with PowerPoint Presentations and Animations* **(0-13-225153-1):** Included on the Instructor Resource CD are three PowerPoint® presentations for each chapter: (1) **Lecture outline**— "plug-and-play" lecture presentations based on the outline of the text to get you up and running as quickly as possible; (2) **Art only**—every illustration and most of the photos in the text, in order, pre-loaded onto PowerPoint slides; (3) **Animations**—high-quality animations of key geologic processes (see below for more details). Also included are all illustrations and a selection of photos for the text in 16-bit, low-compression JPEG files. All images are manually adjusted for color, brightness, and contrast.

- *PH Geoscience Animations* **(0-13-600377-X):** The Prentice Hall Geoscience Animation Library resulted from a survey in which we asked instructors to identify the concepts most difficult to teach using traditional, static resources. Then we animated them. Created through a unique collaboration among five of Prentice Hall's leading geoscience authors, these animations represent a significant leap forward in lecture presentation. Available on the Instructor Resource Center on CD, the animations are provided as Flash files and, for your convenience, pre-loaded into PowerPoint slides. The list of animations includes:

- Convergent Margins
- Seafloor Spreading
- Faults
- Transform Faults
- Foliation
- Folding
- P & S Waves
- Stream Processes
- Angular Unconformity and Nonconformity
- Global Warming
- Beach Drift
- Seismograph Operations
- Breakup of Pangaea
- Nebular Hypothesis
- Oxbow Lake Formation
- Crater Lake
- Igneous Features
- Hydrologic Cycle
- Tidal Cycle
- Glacial Processes—Ice Budget
- Relative Dating
- Tectonic Settings of Volcanic Activity
- Glacial Processes—Plucking and Moraines
- Water Phases
- Wave Motion
- Coastal Processes—Jetties, Groins, Breakwaters
- Ocean Circulation
- Accretion of Terranes
- Global Atmospheric Circulation
- Cyclones and Anticyclones

- *Transparencies* (0-13-600575-6): Full-color transparencies containing a carefully selected, vibrant assembly of figures, tables, and graphs taken from the text, manually adjusted for color, brightness, and contrast for optimal projection.

- *Instructor's Manual with Test Item File* (0-13-255152-7): Authored by Glenn D. Thackray of Idaho State University, the Instructor's Manual provides chapter outlines and objectives, classroom discussion topics, and answers to the end-of-chapter questions in the text. The test item file includes nearly 1,000 multiple choice, true/false, and short-answer test questions based on the text. The test item file is also available in both **WebCT** and **Blackboard formats** for easy import into your course management system.

For the Student

- MyGeologyPlace: **www.prenhall.com/keller** Organized by chapter for easy integration into the course, this website offers numerous review exercises from which students can get automatic feedback, updated Internet links for further exploration, and critical thinking questions. In addition, the Regional Updates section links students to USGS Fact Sheets applicable to their local area.

- *Hazard City: Assignments in Applied Geology, 3e:* Included inside every new copy of the text, *Hazard City* provides meaningful, easy-to-assign, and easy-to-grade assignments. Based on the idealized town of Hazard City, the assignments put students in the role of a practicing geologist—gathering and analyzing real data, evaluating risk, and making assessments and recommendations. The third edition of this widely used CD-ROM contains two new modules: Map Reading and Tsunami/Storm Surge. In Map Reading, students apply basic map reading skills by planning a route to collect stream water specimens in a case of suspected contamination. In Tsunami/Storm Surge, students use maps and basic math to determine the potential impact of a tsunami or storm surge on a coastal community.

Acknowledgments

Successful completion of this book was greatly facilitated by the assistance of many individuals, companies, and agencies. In particular, I am indebted to the U.S. Geological Survey and their excellent environmental programs and publications. To the Internet as a tremendous tool to quickly contact people and organizations doing environmental work. To authors of papers cited in this book, I offer my thanks and appreciation for their contributions. Without their work, this book could not have been written. I must also thank the thoughtful people who dedicated valuable time completing reviews of chapters or the entire book. Their efforts have greatly contributed to this work. I wish to thank Scott Brame, Clemson University; John Bratton, Stonehill College; Eleanor J. Camann, Georgia Southern University; Elizabeth Catlos, Oklahoma State University; Raymond M. Coveney, Jr., University of Missouri; William M. Harris, University of St. Thomas; Michael Krol, Bridgewater State College; Dan Leavell, Ohio State University—Newark; J. Barry Maynard, University of Cincinnati; Stephen R. Newkirk, University of Memphis; Michael Phillips, Illinois Valley Community College; Hongbing Sun, Rider University; and Cynthia Venn, Bloomsburg University.

Special thanks go to Tanya Atwater, William Wise, and Frank Spera for their assistance in preparing the chapters on plate tectonics, minerals and rocks, and impacts, respectively. I greatly appreciate the review of the new chapter on geology and ecology by Carla D'Antonio, who provided important information and advice on basic principles of ecology and ecological restoration.

I am particularly indebted to my editors at Prentice Hall. Special thanks go to Geosciences Publisher Daniel Kaveney, whose enthusiasm, intelligence, encouragement, ideas, and creativity made this book possible. I greatly appreciate the assistance of Amy Porubsky, marketing manager. I also appreciate the efforts of Brittney Corrigan-McElroy, production editor, and Jerry Marshall, photo acquisitions. Art was rendered by MapQuest and Imagineering. Thanks to Heather Scott, art director, for her work in updating the interior and cover design for this new edition. I appreciate the encouragement and support from my wife, Valery, who assisted by pointing out ways to improve the content and presentation.

Edward A. Keller
Santa Barbara, California

For the people of the Gulf Coast and the City of New Orleans who lost so much. Hopefully, the Federal Emergency Management Administration, the U.S. Army Corps of Engineers, and local and national elected officials have learned from Hurricane Katrina to better plan for disasters and catastrophes. There will be more, even stronger hurricanes . . . how we choose to be prepared reflects our values and hopefully elevates our human spirit and compassion.

About the Author

Edward A. Keller

Ed Keller is a professor, researcher, writer, and, most importantly, mentor and teacher to undergraduate and graduate students. Currently, Dr. Keller's students are working on earthquake hazards, how waves of sediment move through a river system following disturbance, and geologic controls on habitat to endangered southern steelhead trout. Born and raised in California (Bachelor's degrees in Geology and Mathematics from California State University at Fresno, Master's degree in Geology from the University of California at Davis), it was while pursuing his Ph.D. in Geology from Purdue University in 1973 that Ed wrote the first edition of *Environmental Geology*. The text soon became a foundation of the environmental geology curriculum. Ed joined the faculty of the University of California at Santa Barbara in 1976 and has been there since, serving multiple times as the chair of both the Environmental Studies and Hydrologic Science programs. In that time he has been an author on more than 100 articles, including seminal works on fluvial processes and tectonic geomorphology. Ed's academic honors include the Don J. Easterbrook Distinguished Scientist Award, Geological Society of America (2004), the Quatercentenary Fellowship from Cambridge University, England (2000), two Outstanding Alumnus Awards from Purdue University (1994, 1996), a Distinguished Alumnus Award from California State University at Fresno (1998), and the Outstanding Outreach Award from the Southern California Earthquake Center (1999).

Ed and his wife Valery, who brings clarity to his writing, love walks on the beach at sunset and when the night herons guard moonlight sand at Arroyo Burro Beach in Santa Barbara.

PART ONE

Foundations
of Environmental Geology

The objective of Part 1 is to present the five fundamental principles of environmental geology and the important information necessary to understand the rest of the text. Of particular importance are (1) the fundamental concepts of environmental science, emphasizing the geologic environment; (2) the structure of Earth and, from a plate tectonics perspective, how our planet works; (3) geologic information concerning rocks and minerals necessary to understand environmental geology problems and solutions to those problems; and (4) linkages between geologic processes and the living world.

Chapter 1 opens with a definition and discussion of environmental geology, followed by a short history of the universe and the origin of Earth. Of particular importance is the concept of geologic time, which is critical in evaluating the role of geologic processes and human interaction in the environment. Five fundamental concepts are introduced: human population growth, sustainability, Earth as a system, hazardous Earth processes, and scientific knowledge and values. These are revisited throughout the text. Chapter 2 presents a brief discussion of the internal structure of Earth and a rather lengthy treatment of plate tectonics. Over periods of several tens of millions of years, the positions of the continents and the development of mountain ranges and ocean basins have dramatically changed our global environment. The patterns of ocean currents, global climate, and the distribution of living things on Earth are all, in part, a function of the processes that have constructed and maintained continents and ocean basins over geologic time.

Minerals and rocks and how they form in geologic environments are the subjects of Chapter 3. Minerals and rocks provide basic resources that our society depends on for materials to construct our homes, factories, and other structures; to manufacture airplanes, trains, cars, buses, and trucks that move people and goods around the globe; and to maintain our industrial economy, including everything from computers to eating utensils. The study of minerals and rocks aids in our general understanding of Earth processes at local, regional, and global levels. This knowledge is particularly important in understanding hazardous processes, including landslides and volcanic eruptions, in which properties of the rocks are intimately related to the processes and potential effects on human society.

Geology and ecology and the many links between the two are presented in Chapter 4. An ecosystem includes the non-living environment, which is the geologic environment. In addition, the living part of an ecosystem (community of organisms) has many important feedback cycles and links to important landscape and geologic processes. Chapter 4 presents some basics of ecology for geologists and emphasizes their relationship to environmental geology.

ONE

Philosophy and Fundamental Concepts

Learning Objectives

In this chapter we discuss and define geology and environmental geology, focusing on aspects of culture and society that are particularly significant to environmental awareness. We present some basic concepts of environmental science that provide the philosophical framework of this book. After reading this chapter, you should be prepared to discuss the following:

- Geology and environmental geology as a science
- Increasing human population as the number one environmental problem
- The concept of sustainability and important factors related to the "environmental crisis"
- Earth as a system and changes in systems
- The concepts of environmental unity and uniformitarianism and why they are important to environmental geology
- Hazardous Earth processes
- Scientific knowledge and values
- The scientific method
- Geologic time and its significance
- The precautionary principle
- Why solving environmental problems can be difficult

Easter Island
A story of rise and fall of a society that overused its resources.
(Tom Till/Getty Images Inc.)

CASE HISTORY | Easter Island: Are We on the Same Path at a Global Scale?

Easter Island, at 172 km², is a small, triangular-shaped, volcanic island located several thousand kilometers west of South America, with a subtropical climate. Polynesian people first reached the island approximately 1,500 years ago. When the Polynesians first arrived, they were greeted by a green island covered with forest, including large plam trees. By the sixteenth century, 15,000 to 30,000 people were living there. They had established a complex society spread among small villages and they raised crops and chickens to supplement the fish, marine mammals, and seabirds that sustained their diet. For religious reasons, they carved massive statues (called moai) from volcanic rock. The statues have the form of a human torso with stone headdress. Most are about 7 m high (21 ft), but some were higher than 20 m. The statues were moved into place at various locations on the island using ropes with tree trunks as rollers.

When Europeans reached Easter Island in the seventeenth century, only about 2,000 people were living on the island. The main symbols of the once-vibrant civilization were the statues, most of which had been toppled and damaged. No trees were growing on the island and the people were living in a degraded environment.

Why Did the Society Collapse? Evidently, the society collapsed in just a few decades, probably the result of degradation of the island's limited resource base. As the human population of the island increased, more and more land was cleared for agriculture while remaining trees were used for fuel and for moving the statues into place. Previously, the soils were protected beneath the forest cover and held water in the subtropical environment. Soil nutrients were probably supplied by dust from thousands of kilometers away that reached the island on the winds. Once the forest was cleared, the soils eroded and the agricultural base of the society was diminished. Loss of the forest also resulted in loss of forest products necessary for building homes and boats, and, as a result, the people were forced to live in caves. Without boats, they could no longer rely on fish as a source of protein. As population pressure increased, wars between villages became common, as did slavery and even cannibalism, in attempts to survive in an environment depleted of its resource base.

Lessons Learned. The story of Easter Island is a dark one that vividly points to what can happen when an isolated area is deprived of its resources through human activity: Limited resources cannot support an ever-growing human population.

Although the people of Easter Island did deplete their resources, the failure had some factors they could not understand or recognize. Easter Island has a naturally fragile environment[1] compared to many other islands the Polynesians colonized.

- The island is small and very isolated. The inhabitants couldn't expect help in hard times from neighboring islands.

- Volcanic soils were originally fertile, but agricultural erosion was a problem and soil-forming processes on the island were slow compared to more tropical islands. Nutrient input to soils from atmospheric dust from Asia was not significant.

- The island's three volcanoes are not active, so no fresh volcanic ash added nutrients to the soils. The topography is low with gentle slopes. Steep high mountains generate clouds, rain, and runoff that nourishes lowlands.

- With a subtropical climate with annual rainfall of 80 cm (50 in), there was sufficient rainfall, but the water quickly infiltrated through the soil into porous volcanic rock.

- There are no coral reefs at Easter Island to provide abundant marine resources.

There is fear today that our planet, an isolated island in space, may be reaching the same threshold faced by the people of Easter Island in the sixteenth century. In the twenty-first century, we are facing limitations of our resources in a variety of areas, including soils, fresh water, forests, rangelands, and ocean fisheries. The primary question from both an environmental perspective and for the history of humans on Earth is: Will we recognize the limits of Earth's resources before it is too late to avoid the collapse of human society on a global scale? Today there are no more frontiers on Earth, and we have a nearly fully integrated global economy. With our modern technology, we have the ability to extract resources and transform our environment at rates much faster than any people before us. The major lesson from Easter Island is clear: Develop a sustainable global economy that ensures the survival of our resource base and other living things on Earth, or suffer the consequences.[1]

Some aspects of the history of Easter Island have recently been challenged as being only part of the story. Deforestation certainly played a role in the loss of the trees, and rats that arrived with the Polynesians were evidently responsible for eating seeds of the palm trees, not allowing regeneration. The alternative explanation is that the Polynesian people on Easter Island at the time of European contact in 1722 numbered about 3,000 persons. This population may have been close to the maximum reached in about the year 1350. Following contact, introduced diseases and enslavement, resulted in reduction of the population to about 100 by the late 1870s.[2] As more of the story of Easter Island emerges from scientific and social studies, the effects of human resource exploitation, invasive rats, and European contact will become clearer. The environmental lessons of the collapse will lead to a better understanding of how we can sustain our global human culture.

1.1 Introduction to Environmental Geology

Everything has a beginning and an end. Our Earth began about 4.6 billion years ago when a cloud of interstellar gas known as a solar nebula collapsed, forming protostars and planetary systems (see A Closer Look: Earth's Place in Space). Life on Earth began about 3.5 billion years ago, and since then multitudes of diverse organisms have emerged, prospered, and died out, leaving only fossils to mark their place in Earth's history. Just a few million years ago, our ancestors set the stage for the present dominance of the human species. As certainly as our Sun will die, we too will eventually disappear. Viewed in terms of billions of years, our role in Earth's history may be insignificant, but for those of us now living and for our children and theirs, our impact on the environment is significant indeed.

A CLOSER LOOK | Earth's Place in Space

The famous geologist Preston Cloud wrote:

> Born from the wreckage of stars, compressed to a solid state by the force of its own gravity, mobilized by the heat of gravity and radioactivity, clothed in its filmy garments of air and water by the hot breath of volcanoes, shaped and mineralized by 4.6 billion years of crustal evolution, warmed and peopled by the Sun, this resilient but finite globe is all our species has to sustain it forever.[3]

In this short, eloquent statement, Cloud takes us from the origin of Earth to the concept of sustainability that today is at the forefront of thinking about the environment and our future.

We Have a Right to Be Here. The place of humanity in the universe is stated well in the *Desiderata:* "You are a child of the universe, no less than the trees and the stars; you have the right to be here. And whether or not it is clear to you, no doubt the universe is unfolding as it should."[4] To some this might sound a little out of place in science but, as emphasized further by Cloud, people can never escape the fact that we are one piece of the biosphere, and, although we stand high in it, we are not above it.[3]

Origin of the Universe. Figure 1.A presents an idealized view of the history of the universe with an emphasis on the origin of our solar system and Earth. Scientists studying the stars and the origin of the universe believe that about 12 billion years ago, there was a giant explosion known as the big bang. This explosion produced the atomic particles that later formed galaxies, stars, and planets. It is believed that about 7 billion years ago, one of the first generations of giant stars experienced a tremendous explosion known as a *supernova.* This released huge amounts of energy, producing a solar nebula, which is thought to be a spinning cloud of dust and gas. The solar nebula condensed as a result of gravitational processes, and our Sun formed at the center, but some of the particles may have been trapped in solar orbits as rings, similar to those we observe around the planet Saturn. The density of particles in individual rings was evidently not constant, so

gravitational attraction from the largest density of particles in the rings attracted others until they collapsed into the planetary system we have today. Thus, the early history of planet Earth, as well that of the other planets in our solar system, was characterized by intense bombardment of meteorites. This bombardment was associated with accretionary processes—that is, the amalgamation of various sized particles, from dust to meteorites, stony asteroids, and ice-rich comets many kilometers in diameter—that resulted in the formation of Earth about 4.6 billion years ago.[3,5] This is the part of Earth's history that Cloud refers to when he states that Earth was born from the wreckage of stars and compressed to a solid state by the force of its own gravity. Heat generated deep within Earth, along with gravitational settling of heavier elements such as iron, helped differentiate the planet into the layered structure we see today (see Chapter 2).

Origin of Atmosphere and Water on Earth. Water from ice-cored comets and outgassing, or the release of gases such as carbon dioxide and water vapor, from volcanoes and other processes, produced Earth's early atmosphere and water. About 3.5 billion years ago the first primitive life-forms appeared on Earth in an oxygen-deficient environment. Some of these primitive organisms began producing oxygen through photosynthesis, which profoundly affected Earth's atmosphere. Early primitive, oxygen-producing life probably lived in the ocean, protected from the Sun's ultraviolet radiation. However, as the atmosphere evolved and oxygen increased, an ozone layer was produced in the atmosphere that shielded Earth from harmful radiation. Plants evolved that colonized the land surface, producing forests, meadows, fields, and other environments that made the evolution of animal life on the land possible.[3]

The spiral of life generalized in Figure 1.A delineates evolution as life changed from simple to complex over several billion years of Earth's history. The names of the eras, periods, and epochs that geologists use to divide **geologic time** are labeled with their range in millions or billions of years from the present (Table 1.1). If you go on to study geology, they will

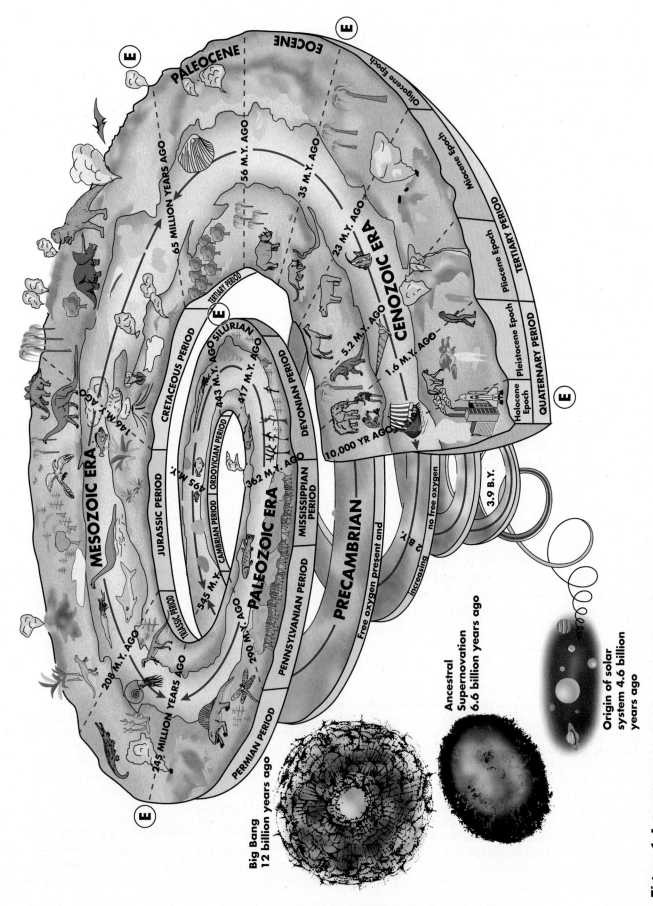

Figure 1.A **Earth history** Idealized diagram of the history of the universe and Earth, with emphasis on the biological evolution of Earth from simple life-forms of the Precambrian to humans today. Precambrian = 4.6 billion years ago to 545 million years ago. Red arrows are boundaries for eras (Table 1.1). (E) is time of mass extinction event. *(Modified after U.S. Geological Survey; and Cloud, P. 1978. Cosmos, Earth and man. New Haven, CT: Yale University Press)*

TABLE 1.1 Geologic Time with Important Events

Era	Period	Epoch	Million Years before Present	Life	Events	Earth	Million Years before Present	True Scale (Million Years before Present)
Cenozoic	Quaternary	Holocene	0.01	• Extinction event • Modern humans	◄► Ice Age	Formation of Transverse Ranges, CA	1.65	Cenozoic
Cenozoic	Quaternary	Pleistocene	1.65	• Early humans				
Cenozoic	Tertiary	Pliocene	5.2					
Cenozoic	Tertiary	Miocene	23	• Grasses • Whales		Formation of Andes Mountains		
Cenozoic	Tertiary	Oligocene	35	• Extinction event		Collision of India with Asia forming Himalayan Mountains and Tibetan Plateau		
Cenozoic	Tertiary	Eocene	56	• Mammals expand				
Cenozoic	Tertiary	Paleocene	65	• Dinosaur extinction,[1] extinction event	◄►	Rocky Mountains form	65	Mesozoic
Mesozoic	Cretaceous		146	• Flowering plants		Emplacement of Sierra Nevada Granites (Yosemite National Park)		
Mesozoic	Jurassic		208	• Birds	◄►	• Supercontinent Pangaea begins to break up		
Mesozoic	Triassic		245	• Mammals • Dinosaurs			245	Paleozoic
Paleozoic	Permian		290	• Extinction event • Reptiles	• Ice Age			
Paleozoic	Carboniferous		363	• Trees (coal swamps) • Extinction event	◄►	Appalachian Mountains form		
Paleozoic	Devonian		417	• Land plants • Extinction event				
Paleozoic	Silurian		443					
Paleozoic	Ordovician		495	• Fish				
Paleozoic	Cambrian		545	• Explosion of organisms with shells	• Ice Age		545	Precambrian
Precambrian			2500	• Multicelled organisms	• Ice Age			
Precambrian			3500	• Free oxygen in atmosphere and ozone layer in stratosphere				
Precambrian			4000	• Primitive life (first fossils)		• Oldest rocks		
Precambrian			4600			• Age of Earth	4600	4600

[1] Some scientists believe that not all dinosaurs became extinct but that some dinosaurs evolved to birds.

become as familiar to you as the months of the year. The boundaries between eras, periods, and epochs are based on both the study of what was living at the particular time and on important global geologic events in Earth's history. Relative ages of rocks are based on the assemblage of fossils—that is, evidence for past life such as shells, bones, teeth, leaves, seeds—that are found in rocks or sediments. A general principle of geology, known as the **law of faunal assemblages,** states that rocks with similar fossils are most likely of a similar geologic age. For example, if we find bones of dinosaurs in a rock, we know the rocks are Mesozoic in age. Fossils provide relative ages of rocks; numerical, or absolute, dates depend upon a variety of sophisticated chemical age-dating techniques. These age-dating techniques allow geologists to often pinpoint the geologic age of rocks containing fossils to within a few million years or better.

Evolution as a Process. The evolutionary process as deduced from the fossil record has not been a smooth continuous one but instead has been punctuated by explosions of new species at some times and extinction of many species at other times. Five mass extinction events are shown in Figure 1.A.

Evolution and extinction of species are natural processes, but for those times when many species became extinct at approximately the same time, we use the term *mass extinction*. For example, the dinosaurs became extinct approximately 65 million years ago. Some geologists believe this mass extinction resulted from climatic and environmental changes that naturally occurred on Earth; others believe the planet was struck by a "death star," an asteroid of about 10 km (6 mi) in diameter, that crashed into what is today the Yucatan Peninsula in Mexico. It is believed that another such impact would produce firestorms and huge dust clouds that would circle Earth in the atmosphere for a prolonged period of time, blocking out sunlight, greatly reducing or stopping photosynthesis, and eventually leading to mass extinction of both the species that eat plants and the predators that feed on the plant eaters.[5]

It is speculated that asteroids of the size that may have caused the dinosaurs to become extinct are not unique, and such catastrophic impacts have occurred at other times during Earth history. Such an event is the ultimate geologic hazard, the effects of which might result in another mass extinction, perhaps including humans! (See Chapter 11.) Fortunately, the probability of such an occurrence is very small during the next few thousand years. In addition, we are developing the technology to identify and possibly deflect asteroids before they strike Earth. The history of our solar system and Earth, briefly outlined here, is an incredible story of planetary and biological evolution. What will the future bring? We do not know, of course, but certainly it will be punctuated by a change, and as the evolutionary processes continue, we too will evolve, perhaps to a new species. Through the processes of pollution, agriculture, urbanization, industrialization, and the land clearing of tropical forest, humans appear to be causing an acceleration of the rate of extinction of plant and animal species. These human activities are significantly reducing Earth's biodiversity—the number and variability of species over time and space (area)—and are thought to be a major environmental problem because many living things, including humans, on Earth depend on the environment with its diversity of life-forms for their existence.

Geologically speaking, we have been here for a very short time. Dinosaurs, for example, ruled the land for more than 100 million years. Although we do not know how long our own reign will be, the fossil record suggests that all species eventually become extinct. How will the history of our own species unfold, and who will write it? Our hope is to leave something more than some fossils that mark a brief time when *Homo sapiens* flourished. Hopefully, as we evolve we will continue to become more environmentally aware and find ways to live in harmony with our planet.

Geology is the science of processes related to the composition, structure, and history of Earth and its life. Geology is an interdisciplinary science, relying on aspects of chemistry (composition of Earth's materials), physics (natural laws), and biology (understanding of life-forms).

Environmental geology is applied geology. Specifically, it is the use of geologic information to help us solve conflicts in land use, to minimize environmental degradation, and to maximize the beneficial results of using our natural and modified environments. The application of geology to these problems includes the study of the following (Figure 1.1).

1. Earth materials, such as minerals, rocks, and soils, to determine how they form, their potential use as resources or waste disposal sites, and their effects on human health

2. Natural hazards, such as floods, landslides, earthquakes, and volcanic activity, in order to minimize loss of life and property

3. Land for site selection, land-use planning, and environmental impact analysis

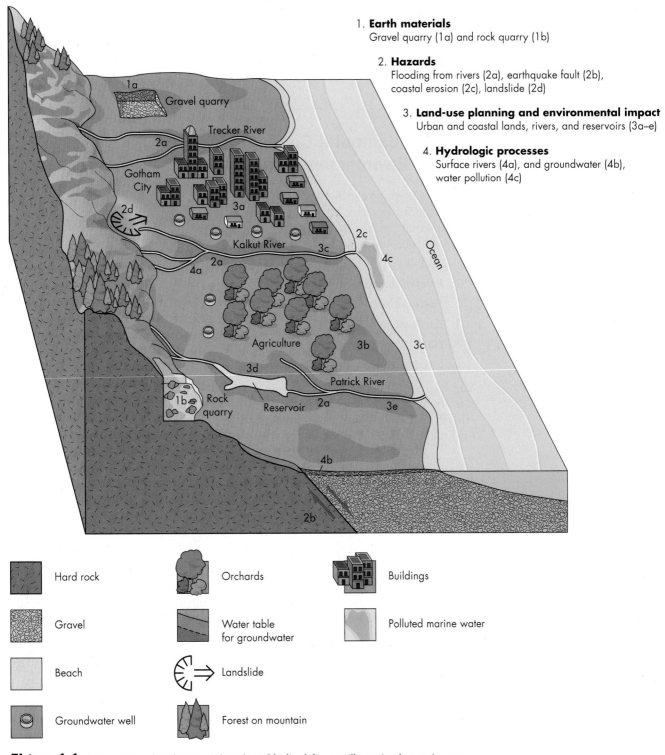

1. **Earth materials**
   Gravel quarry (1a) and rock quarry (1b)

2. **Hazards**
   Flooding from rivers (2a), earthquake fault (2b), coastal erosion (2c), landslide (2d)

3. **Land-use planning and environmental impact**
   Urban and coastal lands, rivers, and reservoirs (3a–e)

4. **Hydrologic processes**
   Surface rivers (4a), and groundwater (4b), water pollution (4c)

Figure 1.1 Components of environmental geology Idealized diagram illustrating four main areas of study for environmental geology. Geologic processes encompass all of the four areas. These offer employment opportunities for geologists, engineers, and hydrologists.

4. Hydrologic processes of groundwater and surface water to evaluate water resources and water pollution problems

5. Geologic processes, such as deposition of sediment on the ocean floor, the formation of mountains, and the movement of water on and below the surface of Earth, to evaluate local, regional, and global change

Considering the breadth of its applications, we can further define environmental geology as the branch of Earth science that studies the entire spectrum of human interactions with the physical environment. In this context, environmental geology is a branch of *environmental science*, the science of linkages between physical, biological, and social processes in the study of the environment.

1.2 Fundamental Concepts of Environmental Geology

Before we begin to explore the many facets of environmental geology presented in this textbook, there are some basic concepts that need to be introduced. These five fundamental concepts serve as a conceptual framework upon which the rest of the textbook will build. As you read through *Introduction to Environmental Geology*, you will notice that these concepts are revisited throughout the text.

1. *Human population growth*
2. *Sustainability*
3. *Earth as a system*
4. *Hazardous Earth processes*
5. *Scientific knowledge and values*

The five concepts presented here do not constitute a list of all concepts that are important to environmental geologists, and they are not meant to be memorized. However, a general understanding of each concept will help you comprehend and evaluate the material presented in the rest of the text.

Concept One: Human Population Growth
The number one environmental problem is the increase in human population

The number one environmental problem is the ever-growing human population. For most of human history our numbers were small as was our input on Earth. With the advent of agriculture, sanitation, modern medicine, and, especially, inexpensive energy sources such as oil, we have proliferated to the point where our numbers are a problem. The total environmental impact from people is estimated by the impact per person times the total number of people. Therefore, as population increases, the total impact must also increase. As population increases, more resources are needed and, given our present technology, greater environmental disruption results. When local population density increases as a result of political upheaval and wars, famine may result (Figure 1.2).

Exponential Growth

What Is the Population Bomb? Overpopulation has been a problem in some areas of the world for at least several hundred years, but it is now apparent that it is a global problem. From 1830 to 1930, the world's population doubled from 1 to 2 billion people. By 1970 it had nearly doubled again, and by the year 2000 there were about 6 billion people on Earth. The problem is sometimes called the population bomb, because the exponential growth of the human population results in the explosive increase in the number of people (Figure 1.3). **Exponential growth** for increase in humans means that the number of people added to the population each year is not constant; rather, a constant percentage of the current population is added each year. As an analogy, consider a high-yield savings account that pays

Figure 1.2 Famine Korem Camp, Ethiopia, in 1984. Hungry people are forced to flee their homes as a result of political and military activity and gather in camps such as these. Surrounding lands may be devastated by overgrazing from stock animals, gathering of firewood, and just too many people in a confined area. The result may be famine. *(David Burnett/Contact Press Images, Inc.)*

interest of 7 percent per year. If you start with $100, at the end of the first year you have $107, and you earned $7 in interest. At the end of the second year, 7 percent of $107 is $7.49, and your balance is $107 plus $7.49, or $114.49. Interest in the third year is 7 percent of 114.49, or $8.01, and your account has $122.51. In 30 years you will have saved about $800.00. Read on to find out how I know this.

There are two important aspects of exponential growth:

- The **growth rate,** measured as a percentage
- The **doubling time,** or the time it takes for whatever is growing to double

Figure 1.4 illustrates two examples of exponential growth. In each case, the object being considered (student pay or world population) grows quite slowly at first,

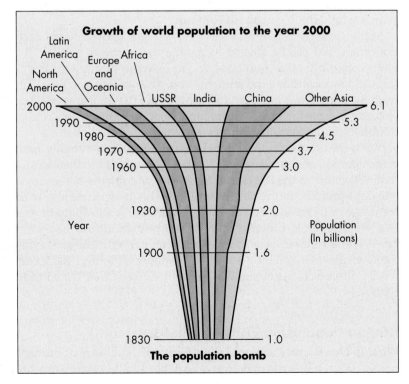

The population bomb

Figure 1.3 **The population bomb** The population in 2006 is 6.6 billion and growing *(Modified after U.S. Department of State)*

Figure 1.4 **Exponential growth**
(a) Example of a student's pay, beginning at 1 cent for the first day of work
and doubling daily for 31 days.
(b) World population. Notice that both
curves have the characteristic *J* shape,
with a slow initial increase followed by
a rapid increase. The actual shape of
the curve depends on the scale at
which the data are plotted. It often
looks like the tip of a skateboard.
*(Population data from U.S. Department
of State)*

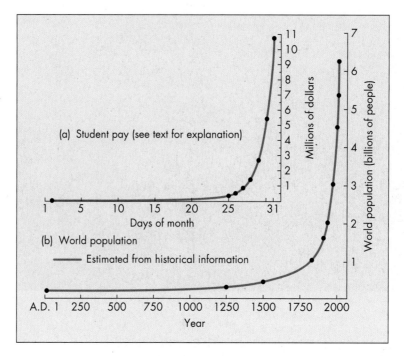

begins to increase more rapidly, and then continues at a very rapid rate. Even very modest rates of growth eventually produce very large increases in whatever is growing.

How Fast Does Population Double? A general rule is that doubling time (D) is roughly equal to 70 divided by the growth rate (G):

$$D = 70/G$$

Using this approximation, we find that a population with a 2 percent annual growth rate would double in about 35 years. If it were growing at 1 percent per year, it would double in about 70 years.

Many systems in nature display exponential growth some of the time, so it is important that we be able to recognize such growth because it can eventually yield incredibly large numbers. As an extreme example of exponential growth (Figure 1.4a), consider the student who, after taking a job for 1 month, requests from the employer a payment of 1 cent for the first day of work, 2 cents for the second day, 4 cents for the third day, and so on. In other words, the payment would double each day. What would be the total? It would take the student 8 days to earn a wage of more than $1 per day, and by the eleventh day, earnings would be more than $10 per day. Payment for the sixteenth day of the month would be more than $300, and on the last day of the 31-day month, the student's earnings for that one day would be more than $10 million! This is an extreme case because the constant rate of increase is 100 percent per day, but it shows that exponential growth is a very dynamic process. The human population increases at a much lower rate—1.4 percent per year today—but even this slower exponential growth eventually results in a dramatic increase in numbers (Figure 1.4b). Exponential growth will be discussed further under Concept Three, when we consider systems and change.

Human Population Through History

What Is Our History of Population Growth? The story of human population increase is put in historic perspective in Table 1.2. When we were hunter-gatherers,

TABLE 1.2   How We Became 6 Billion +

40,000–9,000 B.C.: Hunters and Gatherers

Population density about 1 person per 100 km² of habitable areas;* total population probably less than a few million; average annual growth rate less than 0.0001% (doubling time about 700,000 years)

9,000 B.C.–A.D. 1600: Preindustrial Agricultural

Population density about 1 person per 3 km² of habitable areas (about 300 times that of the hunter and gatherer period); total population about 500 million; average annual growth rate about 0.03% (doubling time about 2,300 years)

A.D. 1600–1800: Early Industrial

Population density about 7 persons per 1 km² of habitable areas; total population by 1800 about 1 billion; annual growth rate about 0.1% (doubling time about 700 years)

A.D. 1800–2000: Modern

Population density about 40 persons per 1 km²; total population in 2000 about 6.1 billion; annual growth rate at 2000 about 1.4% (doubling time about 50 years)

*Habitable area is assumed to be about 150 million square kilometers (58 million square miles). Modified after Botkin, D. B., and Keller, E. A. 2000. *Environmental science*, 3rd ed. New York: John Wiley and Sons.

our numbers were very small, and growth rates were very low. With agriculture, growth rates in human population increased by several hundred times owing to a stable food supply. During the early industrial period (A.D. 1600 to 1800) growth rates increased again by about 10 times. With the Industrial Revolution, with modern sanitation and medicine, the growth rates increased another 10 times. Human population reached 6 billion in 2000. By 2013 it will be 7 billion and by 2050 it will be about 9 billion. That is 1 billion new people in only 13 years and 3 billion (about one-half of today's population) in 50 years. By comparison, total human population had reached only 1 billion in about A.D. 1800, after over 40,000 years of human history! Less developed countries have death rates similar to those of more developed countries, but their birth rates are twice those of developed countries. India will likely have the greatest population of all countries by 2050, with about 18 percent of the total world population, followed by China with 15 percent. Together, these two countries will then have about one-third of the total world population by 2050.[6]

Population Growth and the Future

How Many People Can Earth Comfortably Support? Because Earth's population is increasing exponentially, many scientists are concerned that in the twenty-first century it will be impossible to supply resources and a high-quality environment for the billions of people who may be added to the world population. Three billion more people by 2050, with almost all of the growth in the developing countries, is cause for concern. Increasing population at local, regional, and global levels compounds nearly all environmental geology problems, including pollution of ground and surface waters; production and management of hazardous waste; and exposure of people and human structures to natural processes (hazards) such as floods, landslides, volcanic eruptions, and earthquakes.

There is no easy answer to the population problem. In the future we may be able to mass-produce enough food from a nearly landless agriculture, or use artificial growing situations, to support our ever-growing numbers. However, the ability to feed people does not solve the problems of limited space available to people and maintenance or improvement of their quality of life. Some studies suggest that the present population is already above a comfortable **carrying capacity** for the planet. Carrying capacity is the maximum number of people Earth can hold without causing environmental degradation that reduces the ability of the planet to support the population. The role of education is paramount in the population problem. As people (particularly women) become more educated, the population growth rate tends to decrease. As the rate of literacy increases, population growth is reduced. Given the variety of cultures, values, and norms in the world today, it appears that our greatest hope for population control is, in fact, through education.[7]

The Earth Is Our Only Suitable Habitat. The Earth is now and for the foreseeable future the only suitable habitat we have, and its resources are limited. Some resources, such as water, are renewable, but many, such as fuels and minerals, are not. Other planets in our solar system, such as Mars, cannot currently be considered a solution to our resource and population problems. We may eventually have a colony of people on Mars, but it would be a harsh environment, with people living in bubbles.

When resource and other environmental data are combined with population growth data, the conclusion is clear: It is impossible, in the long run, to support exponential population growth with a finite resource base. Therefore, one of the primary goals of environmental work is to ensure that we can defuse the population bomb. Some scientists believe that population growth will take care of itself through disease and other catastrophes, such as famine. Other scientists are optimistic that we will find better ways to control the population of the world within the limits of our available resources, space, and other environmental needs.

Good News on Human Population Growth. It is not all bad news regarding human population growth; for the first time since the mid-1900s, the rate of increase in human population is decreasing. Figure 1.5 shows that the number of people added to the total population of Earth peaked in the late 1980s and has generally decreased since then. This is a milestone in human population growth and it is encouraging.[8] From an optimistic point of view, it is possible that our global population of 6 billion persons in 2000 may not double again. Although

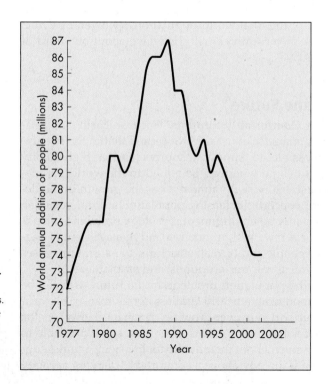

Figure 1.5 **Good news on population growth** World annual increase in population peaked in the late 1980s. Today it is at a level comparable to the late 1970s. This increase is like adding two Californias each year. *(Data from the U.S. Bureau of the Census and Worldwatch Institute)*

population growth is difficult to estimate because of variables such as agriculture, sanitation, medicine, culture, and education, it is estimated that by the year 2050 human population will be between 7.3 and 10.7 billion, with 8.9 billion being most likely. Population reduction is most likely related to the education of women, the decision to marry later in life, and the availability of modern birth control methods. Until the growth rate is zero, however, population will continue to grow. About 20 countries, mostly in Western Europe but including China, have achieved a total fertility rate (number of children per woman) less than 2.1, which is the level necessary for replacement.

Concept Two: Sustainability
Sustainability is the environmental objective

What is sustainability? **Sustainability** is something that we are struggling to define. One definition is that sustainability is development that ensures that future generations will have equal access to the resources that our planet offers. Sustainability also refers to types of development that are economically viable, do not harm the environment, and are socially just.[7] Sustainability is a long-term concept, something that happens over decades or even over hundreds of years. It is important to acknowledge that sustainability with respect to use of resources is possible for renewable resources such as air and water. Sustainable development with respect to nonrenewable resources such as fossil fuels and minerals is possible by, first, extending their availability through conservation and recycling; and second, rather than focusing on when a particular nonrenewable resource is depleted, focusing on how that mineral is used and develop substitutes for those uses.

There is little doubt that we are using living environmental resources such as forests, fish, and wildlife faster than they can be naturally replenished. We have extracted minerals, oil, and groundwater without concern for their limits or for the need to recycle them. As a result, there are shortages of some resources. We must learn how to sustain our environmental resources so that they continue to provide benefits for people and other living things on the planet.

We stated in Concept One, with respect to humans and resources, that Earth is the only place to live that is now accessible to us, and our resources are limited. To meet future resource demands and to sustain our resources, we will need large-scale recycling of many materials. Most materials can theoretically be recycled. The challenge is to find ways to do it that do not harm the environment, that increase the quality of life, and that are economically viable. A large part of our solid and liquid waste disposal problems could be alleviated if these wastes were reused or recycled. In other words, many wastes that are now considered pollutants can be turned into resources. Land is also an important resource for people, plants, and animals as well as for manufacturing, mining, and energy production; transportation; deposition of waste products; and aesthetics. Owing in part to human population increases that demand more land for urban and agricultural purposes, human-induced change to Earth is increasing at a rapid rate. A recent study of human activity and the ability to move soil and rock concluded that human activity (agriculture, mining, urbanization, and so on) moves as much or more soil and rock on an annual basis than any other Earth process (Figure 1.6), including mountain building or river transport of sediment. These activities and their associated visual changes to Earth (for example, leveling hills) suggest that human activity is the most significant process shaping the surface of Earth.[9] (See A Closer Look: Human Landscape Modification: Ducktown, Tennessee.) We'll discuss land-use planning in Chapter 20.

Figure 1.6 Mining A giant excavating machine in this mine can move Earth materials at a rate that could bury one of the Egyptian Pyramids in a short time. *(Joseph J. Scherschel/NGS Image Collection)*

A CLOSER LOOK | Human Landscape Modification: Ducktown, Tennessee

A Man-Made Desert in Tennessee? The land surrounding Ducktown once looked more like the Painted Desert of Arizona than the lush vegetation of the Blue Ridge Mountains of the southeastern United States (Figure 1.B, part a).[10] The story of Ducktown starts in 1843 when what was thought to be a gold rush turned out to be a rush for copper. By 1855,

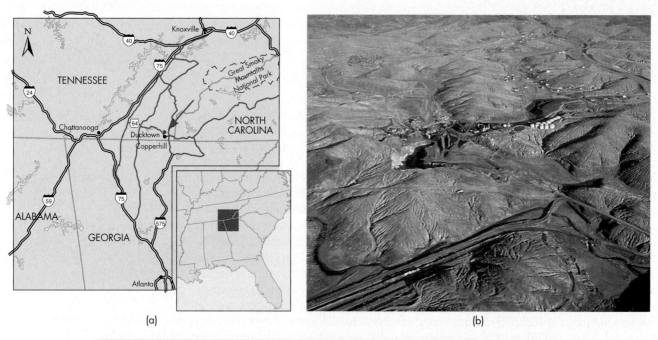

(a)   (b)

(c)

Figure 1.B **The lasting effects of land abuse** (a) Location of Ducktown, Tennessee. (b) The human-made desert resulting from mining activities around Ducktown more than 100 years ago. Extensive soil erosion and loss of vegetation have occurred, and complete recovery will probably take more than 100 years. *(Kristoff, Emory/NGS Image Sales)* (c) Ducktown area in recent years, showing the process of recovery. *(Tennessee Valley Authority)*

30 companies were transporting copper ore by mule over the mountains to a site called Copper Basin and to Ducktown. Huge ovens—open pits 200 m (656 ft) long and 30 m (98 ft) deep—were constructed to separate the copper from zinc, iron, and sulfur. The local hardwood forest was cut to fuel these ovens, and the tree stumps were pulled and turned into charcoal. Eventually, every tree over an area of about 130 km^2 (50 mi^2), or an area equal to approximately four times that of Manhattan Island, was removed. The ovens produced great clouds of noxious gas that were reportedly so thick that mules wore bells to keep from colliding with people and each other. The sulfur dioxide gas and particulates produced acid rain and acid dust that killed the remaining vegetation. This loss of vegetation led to extensive soil erosion, leaving behind a hard mineralized rock cover resembling a desert. The scarred landscape is so large that it is one of the few human landmarks visible from space (Figure 1.B, part b).

People Are Basically Optimistic About Their Future. The devastation resulting from the Ducktown mining activity also produced adverse economic and social change. Nevertheless, people in Ducktown remain optimistic. A sign at the entry to the town states, "Copper made us famous. Our people made us great." The revegetation process started in the 1930s, and most of the area is now covered with some vegetation (Figure 1.B, part c). However, it will probably take hundreds of years for the land to completely recover. The lessons learned from the restoration of the Copper Basin will provide useful information for other areas in the world where human-made deserts occur, such as the area around the smelters in Sudbury, Ontario (Figure 1.C). However, there is still concern for mining areas, particularly in developing countries, where landscape destruction similar to that at Copper Basin is still ongoing.[12]

Figure 1.C **Air pollution** Area around Sudbury, Ontario, devoid of vegetation because of air pollution from smelters, smokestacks in background. *(Bill Brooks/Masterfile Corporation)*

Are We in an Environmental Crisis? Demands made on diminishing resources by a growing human population and the ever-increasing production of human waste have produced what is popularly referred to as the **environmental crisis.** This crisis in the United States and throughout the world is a result of overpopulation, urbanization, and industrialization, combined with too little ethical regard for our land and inadequate institutions to cope with environmental stress.[10] The rapid use of resources continues to cause environmental problems on a global scale, including

- Deforestation and accompanying soil erosion and water and air pollution occur on many continents (Figure 1.7a).
- Mining of resources such as metals, coal, and petroleum wherever they occur produces a variety of environmental problems (Figure 1.7b).
- Development of both groundwater and surface-water resources results in loss of and damage to many environments on a global scale (see Case History: The Aral Sea: The Death of a Sea).

On a positive note, we have learned a great deal from the environmental crisis, particularly concerning the relationship between environmental degradation and resource utilization. Innovative plans for sustainable development of resources, including water and energy, are being developed to lessen a wide variety of environmental problems associated with using resources.

Figure 1.7a **Logging** Clear-cut timber harvesting exposes soils, compacting them and generally contributing to an increase in soil erosion and other environmental problems. *(Edward A. Keller)*

Figure 1.7b **Mining** Large open pit mines such as this one east of Silver City, New Mexico, are necessary if we are to obtain resources. However, they do cause disturbance to the surface of the land, and reclamation may be difficult or nearly impossible in some instances. *(Michael Collier)*

Do We Need to Save Earth or Ourselves? The environmental slogan of the 1990s was "save our planet." Is Earth's very survival really in danger? In the long view of planetary evolution, it seems highly likely that Earth will outlive the human race. Our Sun is likely to last another several billion years at least, and even if all humans became extinct in the next few years, life would still flourish on our planet. The environmental degradation we have imposed on the landscape, atmosphere, and waters might last for a few hundreds or thousands of years, but they would eventually be cleansed by natural processes. Therefore, our major concern is the quality of the human environment, which depends on sustaining our larger support systems, including air, water, soil, and other life.

Concept Three: Earth as a System
Understanding Earth's systems and their changes
is critical to solving environmental problems.

A **system** is any defined part of the universe that we select for study. Examples of systems are a planet, a volcano, an ocean basin, or a river (Figure 1.8). Most systems contain several component parts that mutually adjust to function as a whole, with changes in one component bringing about changes in other components. For example, the components of our global system are water, land, atmosphere, and life. These components mutually adjust, helping to keep the entire Earth system operating.[11]

Figure 1.8 **River as a system** Image of part of the Amazon River system (blue) and its confluence with the Rio Negro (black). The blue water of the Amazon is heavily laden with sediment, whereas the water of the Rio Negro is nearly clear. Note that as the two large rivers join, the waters do not mix initially but remain separate for some distance past the confluence. The Rio Negro is in flood stage. The red is the Amazon rain forest, and the white lines are areas of human-caused disturbances such as roads. *(Earth Satellite Corporation/Science Photo Library/Photo Researchers, Inc.)*

CASE HISTORY | The Aral Sea: The Death of a Sea

The Aral Sea, located between Kazakhstan and Uzbekistan, formerly part of the Union of Soviet Socialist Republics, was a prosperous tourist vacation spot in 1960. Water diversion for agriculture nearly eliminated the Aral Sea in a period of only 30 years. It is now a dying sea surrounded by thousands of square kilometers of salt flats, and the change is permanently damaging the economic base of the region.

In 1960, the area of the Aral Sea was about 67,000 km^2 (around 26,200 mi^2). Diversion of the two main rivers that fed the sea has resulted in a drop in surface elevation of more than 20 m (66 ft) and loss of about 28,000 km^2 (10,800 mi^2) of surface area (Figure 1.D). Towns that were once fishing centers on the shore are today about 30 km (19 mi) inland. Loss of the sea's moderating effect on weather is changing the regional climate; the winters are now colder, and the sum-

mers warmer. Windstorms pick up salty dust and spread it over a vast area, damaging the land and polluting the air.

The lesson to be learned from the Aral Sea is how quickly environmental damage can bring about regional change. Environmentalists, including geologists, worry that what people have done to the Aral region is symptomatic of what we are doing on many fronts on a global scale.[13] Today an ambitious restoration project is underway to save the northern, smaller part of the lake. A low dam has been constructed across the lake just south of where the Syr Darya flows into the lake (see Figure 1.D). With water conservation of the river water, more water is flowing in the lake and the dam keeps the water in the northern part of the lake bed. Water levels there are rising and some fishing has returned. This is a promising sign, but much more needs to be done.

Figure 1.D Dying sea (a) The Aral Sea is a dying sea, surrounded by thousands of square kilometers of salt flats. *(Courtesy of Philip P. Micklin)* (b) Water diversion for agriculture has nearly eliminated the sea. The two ships shown here are stranded high and dry along the shoreline, which contains extensive salt flats formed as the Aral Sea has evaporated. *(David Turnley/CORBIS)*

Input-Output Analysis

Input-output analysis is an important method for analyzing change in open systems. Figure 1.9 identifies three types of change in a pool or stock of materials; in each case the net change depends on the relative rates of the input and output. Where the input into the system is equal to the output (Figure 1.9a), a rough steady state is established and no net change occurs. The example shown is a university in which students enter as freshmen and graduate four years later at a constant rate. Thus, the pool of university students remains a constant size. At the global scale, our planet is a roughly steady-state system with respect to energy:

Figure 1.9 Change in systems
Major ways in which a pool or stock of some material may change. *(Modified after Ehrlich, P. R., Ehrlich, A. H., and Holdren, J. P. 1977. Ecoscience: Population, resources, environment, 3rd ed. San Francisco: W. H. Freeman)*

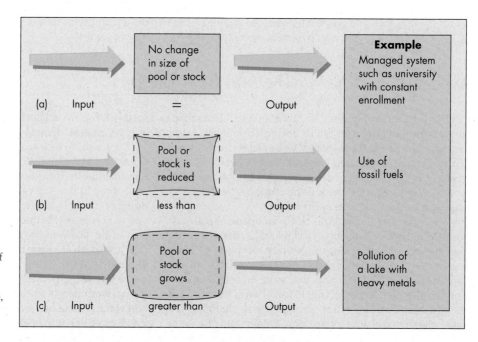

Incoming solar radiation is roughly balanced by outgoing radiation from Earth. In the second type of change, the input into the system is less than the output (Figure 1.9b). Examples include the use of resources such as fossil fuels or groundwater and the harvest of certain plants or animals. If the input is much less than the output, then the fuel or water source may be completely used up, or the plants or animals may become extinct. In a system in which input exceeds output (Figure 1.9c), the stock of whatever is being measured will increase. Examples include the buildup of heavy metals in lakes from industrial pollution or the pollution of soil and water.

How Can We Evaluate Change? By evaluating rates of change or the input and output of a system, we can derive an **average residence time** for a particular material, such as a resource. The average residence time is a measure of the time it takes for the total stock or supply of the material to be cycled through a system. To compute the average residence time (T; assuming constant size of the system and constant rate of transfer), we take the total size of the stock (S) and divide it by the average rate of transfer (F) through the system:

$$T = S/F$$

For example, if a reservoir holds 100 million cubic meters of water, and both the average input from streams entering the reservoir and the average output over the spillway are 1 cubic meter per second, then the average residence time for a cubic meter of water in the reservoir is 100 million seconds, or about 3.2 years (Figure 1.10). We can also calculate average residence time for systems that vary in size and rates of transfer, but the mathematics is more difficult. It is often possible to compute a residence time for a particular resource and then to apply the information to help understand and solve environmental problems. For example, the average residence time of water in rivers is about 2 weeks compared with thousands of years for some groundwater. Thus, strategies to treat a one-time pollution event of oil spilled in a river will be much different from those for removing oil floating on groundwater that resulted from a rupture of an underground pipeline. The oil in the river is a relatively accessible, straightforward, short-term problem,

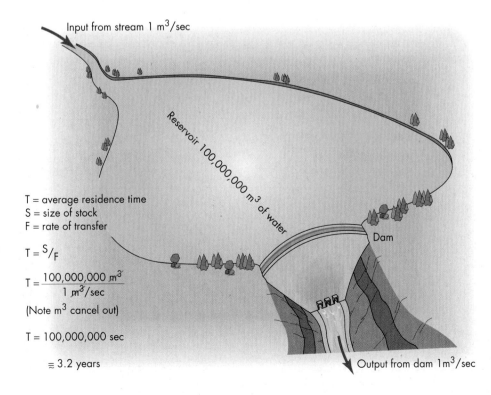

Input from stream 1 m³/sec

Reservoir 100,000,000 m³ of water

Dam

T = average residence time
S = size of stock
F = rate of transfer

$T = \dfrac{S}{F}$

$T = \dfrac{100{,}000{,}000 \ m^3}{1 \ m^3/sec}$

(Note m³ cancel out)

$T = 100{,}000{,}000 \ sec$

$\equiv 3.2$ years

Output from dam 1m³/sec

Figure 1.10 **Average residence time** Calculation of the average residence time for a cubic meter of water in a reservoir where input = output = 1 m³ per second and the size of the reservoir is constant at 100,000,000 m³ of water.

whereas polluted groundwater is a more difficult problem because it moves slowly and has a long average residence time. Because it may take from several to hundreds of years for pollution of groundwater to be naturally removed, groundwater pollution is difficult to treat.

Predicting Changes in the Earth System

The idea that "the present is the key to the past," called **uniformitarianism,** was popularized by James Hutton, referred to by some scholars as the father of geology, in 1785 and is heralded today as a fundamental concept of Earth sciences. As the name suggests, uniformitarianism holds that processes we observe today also operated in the past (flow of water in rivers, formation and movement of glaciers, landslides, waves on beaches, uplift of the land from earthquakes, and so on). Uniformitarianism does not demand or even suggest that the magnitude (amount of energy expended) and frequency (how often a particular process occurs) of natural processes remain constant with time. We can infer that, for as long as Earth has had an atmosphere, oceans, and continents similar to those of today, the present processes were operating.

Present Human Activity Is Part of the Key to Understanding the Future. In making inferences about geologic events, we must consider the effects of human activity on the Earth system and what effect these changes to the system as a whole may have on natural Earth processes. For example, rivers flood regardless of human activities, but human activities, such as paving the ground in cities, increase runoff and the magnitude and frequency of flooding. That is, after the paving, floods of a particular size are more frequent, and a particular rainstorm can produce a larger flood than before the paving. Therefore, to predict the long-range effects of flooding, we must be able to determine how future human activities will change the size and frequency of floods. In this case, *the present is the key to the future.* For example, when environmental geologists examine recent landslide deposits (Figure 1.11) in an area designated to become a housing development, they must use uniformitarianism to infer where there will be future landslides as well as to predict what effects urbanization will have on the magnitude and frequency of future landslides. We will now consider linkages between processes.

Figure 1.11 Urban development The presence of a landslide on this slope suggests that the slope is not stable and further movement may occur in the future. This is a "red flag" for future development in the area.
(Edward A. Keller)

Environmental Unity

The principle of **environmental unity**, which states that one action causes others in a chain of actions, is an important principle in the prediction of changes in the Earth system. For example, if we constructed a dam on a river, a number of changes would occur. Sediment that moved down the river to the ocean before construction of the dam would be trapped in the reservoir. Consequently, beaches would be deprived of the sediment from the river, and the result of that deprivation may be increased coastal erosion. There being less sediment on the beach may also affect coastal animals such as sand crabs and clams that use the sand. Thus, building the dam would set off a chain or series of effects that would change the coastal environment and what lived there. The dam would also change the hydrology of the river and would block fish from migrating upstream. We will now consider global linkages.[11]

Earth Systems Science

Earth systems science is the study of the entire planet as a system in terms of its components (see A Closer Look: The Gaia Hypothesis). It asks how component

A CLOSER LOOK | The Gaia Hypothesis

Is Earth Analogous to an Organism? In 1785 at a meeting of the prestigious Royal Society of Edinburgh, James Hutton, the father of geology, said he believed that planet Earth is a superorganism (Figure 1.E). He compared the circulation of Earth's water, with its contained sediments and nutrients, to the circulation of blood in an animal. In Hutton's metaphor, the oceans are the heart of Earth's global system, and the forests are the lungs.[15] Two hundred years later, British scientist and professor James Lovelock introduced the **Gaia hypothesis,** reviving the idea of a living Earth. The hypothesis is named for Gaia, the Greek goddess Mother Earth.

The Gaia hypothesis is best stated as a series of hypotheses:

- *Life significantly affects the planetary environment.* Very few scientists would disagree with this concept.

- *Life affects the environment for the betterment of life.* This hypothesis is supported by some studies showing that life on Earth plays an important role in regulating planetary climate so that it is neither too hot nor too cold for life to survive. For example, it is believed that single-cell plants floating near the surface of the ocean partially control the carbon dioxide content of the atmosphere and thereby global climate.[15]

- *Life deliberately or consciously controls the global environment.* There are very few scientists who accept this third hypothesis. Interactions and the linking of processes that operate in the atmosphere, on the surface of Earth, and in the oceans are probably sufficient to explain most of the mechanisms by which life affects the environment. In contrast, humans are beginning to make decisions concerning the global environment, so the idea that humans can consciously influence the future of Earth is not an extreme view. Some people have interpreted this idea as support for the broader Gaia hypothesis.

Gaia Thinking Fosters Interdisciplinary Thinking. The real value of the Gaia hypothesis is that it has stimulated a lot of interdisciplinary research to understand how our planet works. As interpreted by most scientists, the hypothesis does not suggest foresight or planning on the part of life but rather that natural processes are operating.

Figure 1.E **Home** Image of Earth centering on the North Atlantic Ocean, North America, and the polar ice sheets. Given this perspective of our planet, it is not difficult to conceive it as a single large system. *(Earth Imaging/Getty Images Inc.)*

systems (subsystems of the Earth system) such as the atmosphere (air), hydrosphere (water), biosphere (life), and lithosphere (rocks) are linked and have formed, evolved, and been maintained; how these components function; and how they will continue to evolve over periods ranging from a decade to a century and longer.[14] Because these systems are linked, it is also important to understand and be able to predict the impacts of a change in one component on the others. The challenge is to learn to predict changes likely to be important to society and then to develop management strategies to minimize adverse environmental impacts. For example, the study of atmospheric chemistry suggests that our atmosphere has changed over millenia. Trace gases such as carbon dioxide have increased by about 100 percent since 1850. Chlorofluorocarbons (CFCs), used as refrigerants and aerosol-can propellants, released at the surface have migrated to the stratosphere, where they react with energy from the Sun, causing destruction of the ozone layer that protects Earth from harmful ultraviolet radiation. The important topics of global change and Earth systems science will be discussed in Chapter 19, following topics such as Earth materials, natural hazards, and energy resources.

Concept Four: Hazardous Earth Processes

There have always been Earth processes that are hazardous to people. These natural hazards must be recognized and avoided when possible, and their threat to human life and property must be minimized.

We humans, like all animals, have to contend with natural processes such as storms, floods, earthquakes, landslides, and volcanic eruptions that periodically damage property and kill us. During the past 20 years, natural hazards on Earth have killed several million people. The annual loss was about 150,000 people, and financial damages were about $20 billion.

Natural Hazards That Produce Disasters Are Becoming Superdisasters Called Catastrophes. Early in human history, our struggle with natural Earth processes was mostly a day-to-day experience. Our numbers were neither great nor concentrated, so losses from hazardous Earth processes were not significant. As people learned to produce and maintain a larger and, in most years, more abundant food supply, the population increased and became more concentrated locally. The concentration of population and resources also increased the impact that periodic earthquakes, floods, and other natural disasters had on humans. This trend has continued, so that many people today live in areas likely to be damaged by hazardous Earth processes or susceptible to the adverse impact of such processes in adjacent areas. An emerging principle concerning natural hazards is that as a result of human activity (population increase and changing the land through agriculture, logging, mining, and urbanization) what were formerly disasters are becoming catastrophes. For example,

- Human population increase has forced more people to live in hazardous areas such as floodplains, steep slopes (where landslides are more likely), and near volcanoes.

- Land-use transformations including urbanization and deforestation increase runoff and flood hazard and may weaken slopes, making landslides more likely.

- Burning vast amounts of oil, gas, and coal has increased the concentration of carbon dioxide in the atmosphere, contributing to warming the atmosphere and oceans. As a result, more energy is fed into hurricanes. The number of hurricanes has not increased, but the intensity and size of the storms have increased.

We can recognize many natural processes and predict their effects by considering climatic, biological, and geologic conditions. After Earth scientists have identified potentially hazardous processes, they have the obligation to make the information available to planners and decision makers, who can then consider ways of avoiding or minimizing the threat to human life or property. Put concisely, this process consists of assessing the risk of a certain hazard in a given area and basing planning decisions on that risk assessment. Public perception of hazards also plays a role in the determination of risk from a hazard. For example, although they probably understand that the earthquake hazard in southern California is real, the residents who have never experienced an earthquake first hand may have less appreciation for the seriousness of the risk of loss of property and life than do persons who have experienced an earthquake.

Concept Five: Scientific Knowledge and Values
The results of scientific inquiry to solve a particular environmental problem often provide a series of potential solutions consistent with the scientific findings. The chosen solution is a reflection of our value system.

What Is Science? To understand our discussion of scientific knowledge and values, let us first gain an appreciation for the conventions of scientific inquiry. Most scientists are motivated by a basic curiosity about how things work. Geologists are excited by the thrill of discovering something previously unknown about how the world works. These discoveries drive them to continue their work. Given that we know little about internal and external processes that form and maintain our world, how do we go about studying it? The creativity and insight that may result from scientific breakthroughs often begin with asking the right question pertinent to some problem of interest to the investigators. If little is known about the topic or process being studied, they will first try to conceptually understand what is going on by making careful observations in the field or, perhaps, in a laboratory. On the basis of his or her observations, the scientist may then develop a question or a series of questions about those observations. Next the investigator will suggest an answer or several possible answer to the question. The possible answer is a **hypothesis** to be tested. The best hypotheses can be tested by designing an experiment that involves data collection, organization, and analysis. After collection and analysis of the data, the scientist interprets the data and draws a conclusion. The conclusion is then compared with the hypothesis, and the hypothesis may be rejected or tentatively accepted. Often, a series of questions or multiple hypotheses are developed and tested. If all hypotheses suggested to answer a particular question are rejected, then a new set of hypotheses must be developed. This method is sometimes referred to as the **scientific method**. The steps of the scientific method are shown in Figure 1.12. The first step of the scientific method is the formation of a question—in this case, "Where does beach sand come from?" In order to explore this question, the scientist spends some time at the beach. She notices some small streams that flow into the ocean; she knows that the streams originate in the nearby mountains. She then refines her question to ask specifically, "Does beach sand come from the mountains to the beach by way of streams?" This question is the basis for the scientist's hypothesis: Beach sand originates in the mountains. To test this hypothesis, she collects some sand from the beach and from the streams and some rock samples from the mountains. She then compares their mineral content. She finds that the mineral content of all three is roughly the same. She draws a conclusion that the beach sand does come from the mountains, and so accepts her hypothesis. If her hypothesis had proved to be wrong, she would have had to formulate a new hypothesis. In complex geologic

Figure 1.13 **Eroding a valley**
Idealized diagram of progressive incision of a river into a sequence of horizontal rocks. The side slope is steep where rocks are hard and resistant to incision, and the rate of incision is generally less than about 0.01 mm per year (about 0.0004 in. per year). For softer rocks, where the side slope is gentle, the rate of incision may exceed 1 mm per year (0.039 in. per year). If the canyon incised about 1 km (0.62 mi) in 1 million years, the average rate is 1 mm per year (0.039 in. per year). *(Modified after King, P. B., and Schumm, S. A., 1980. The physical geography of William Morris Davis. Norwich, England: Geo Books)*

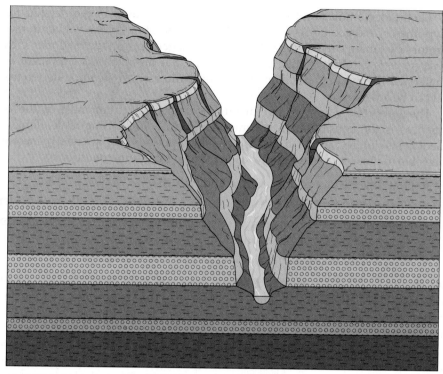

(a) Incision at about 250,000 yrs

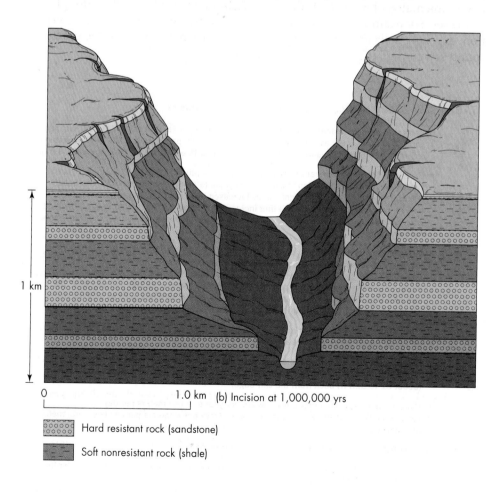

1 km

0   1.0 km   (b) Incision at 1,000,000 yrs

Hard resistant rock (sandstone)

Soft nonresistant rock (shale)

We can recognize many natural processes and predict their effects by considering climatic, biological, and geologic conditions. After Earth scientists have identified potentially hazardous processes, they have the obligation to make the information available to planners and decision makers, who can then consider ways of avoiding or minimizing the threat to human life or property. Put concisely, this process consists of assessing the risk of a certain hazard in a given area and basing planning decisions on that risk assessment. Public perception of hazards also plays a role in the determination of risk from a hazard. For example, although they probably understand that the earthquake hazard in southern California is real, the residents who have never experienced an earthquake first hand may have less appreciation for the seriousness of the risk of loss of property and life than do persons who have experienced an earthquake.

Concept Five: Scientific Knowledge and Values

The results of scientific inquiry to solve a particular environmental problem often provide a series of potential solutions consistent with the scientific findings. The chosen solution is a reflection of our value system.

What Is Science? To understand our discussion of scientific knowledge and values, let us first gain an appreciation for the conventions of scientific inquiry. Most scientists are motivated by a basic curiosity about how things work. Geologists are excited by the thrill of discovering something previously unknown about how the world works. These discoveries drive them to continue their work. Given that we know little about internal and external processes that form and maintain our world, how do we go about studying it? The creativity and insight that may result from scientific breakthroughs often begin with asking the right question pertinent to some problem of interest to the investigators. If little is known about the topic or process being studied, they will first try to conceptually understand what is going on by making careful observations in the field or, perhaps, in a laboratory. On the basis of his or her observations, the scientist may then develop a question or a series of questions about those observations. Next the investigator will suggest an answer or several possible answer to the question. The possible answer is a **hypothesis** to be tested. The best hypotheses can be tested by designing an experiment that involves data collection, organization, and analysis. After collection and analysis of the data, the scientist interprets the data and draws a conclusion. The conclusion is then compared with the hypothesis, and the hypothesis may be rejected or tentatively accepted. Often, a series of questions or multiple hypotheses are developed and tested. If all hypotheses suggested to answer a particular question are rejected, then a new set of hypotheses must be developed. This method is sometimes referred to as the **scientific method.** The steps of the scientific method are shown in Figure 1.12. The first step of the scientific method is the formation of a question—in this case, "Where does beach sand come from?" In order to explore this question, the scientist spends some time at the beach. She notices some small streams that flow into the ocean; she knows that the streams originate in the nearby mountains. She then refines her question to ask specifically, "Does beach sand come from the mountains to the beach by way of streams?" This question is the basis for the scientist's hypothesis: Beach sand originates in the mountains. To test this hypothesis, she collects some sand from the beach and from the streams and some rock samples from the mountains. She then compares their mineral content. She finds that the mineral content of all three is roughly the same. She draws a conclusion that the beach sand does come from the mountains, and so accepts her hypothesis. If her hypothesis had proved to be wrong, she would have had to formulate a new hypothesis. In complex geologic

Figure 1.12 **Science** The steps in the scientific method.

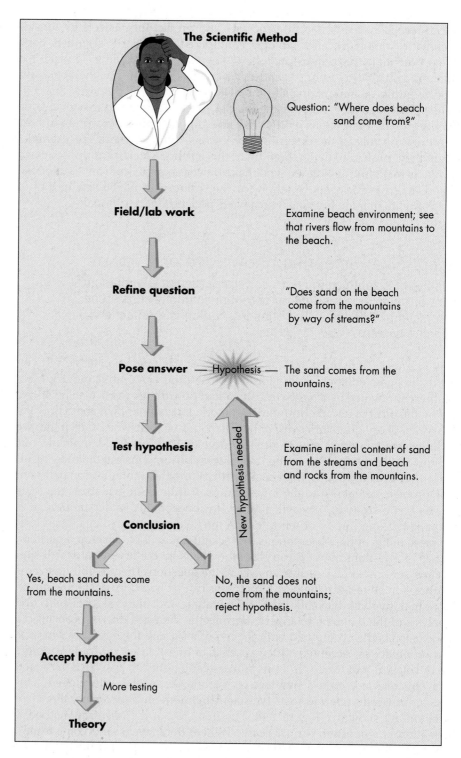

problems, multiple hypotheses may be formulated and each tested. This is the method of multiple working hypotheses. If a hypothesis withstands the testing of a sufficient number of experiments, it may be accepted as a **theory.** A theory is a strong scientific statement that the hypothesis supporting the theory is likely to be true but has not been proved conclusively. New evidence often disproves existing hypotheses or scientific theory; absolute proof of scientific theory is not possible. Thus, much of the work of science is to develop and test hypotheses, striving to reject current hypotheses and to develop better ones.

Laboratory studies and fieldwork are commonly used in partnership to test hypotheses, and geologists often begin their observations in the field or in the laboratory by taking careful notes. For example, a geologist in the field may create a *geologic map*, carefully noting and describing the distribution of different Earth materials. The map can be completed in the laboratory, where the collected material can be analyzed.

The important variable that distinguishes geology from most of the other sciences is the consideration of time (see the Geologic Time Scale, Table 1.1). Geologists' interest in Earth history over time periods that are nearly incomprehensible to most people naturally leads to some interesting questions:

- How fast are mountains uplifted and formed?
- How fast do processes of erosion reduce the average elevation of the land?
- How fast do rivers erode canyons to produce scenic valleys such as Yosemite Valley and the Grand Canyon (Figure 1.13)?
- How fast do floodwaters, glaciers, and lava flows move?

As shown in Table 1.3, rates of geologic processes vary from a fraction of a millimeter per year to several kilometers per second. The fastest rates are more than a trillion times the slowest. The most rapid rates, a few kilometers per second, are for events with durations of a few seconds. For example, uplift of 1 m (3.3 ft.) during an earthquake may seem like a lot, but when averaged over 1,000 years (the time between earthquakes), it is a long-term rate of 1 mm per year (0.039 in. per year), a typical uplift rate in forming mountains. Of particular importance to environmental geology is that human activities may accelerate the rates of some processes. For example, timber harvesting and urban construction remove vegetation, exposing soils and increasing the rate of erosion. Conversely, the practice of sound soil conservation may reduce rates.

TABLE 1.3   Some Typical Rates of Geologic Processes

Slow Rates	■ Uplift that produces mountains. Generally 0.5 to 2 mm per year (about 0.02 to 0.08 in. per year). Can be as great as 10 mm per year (about 0.39 in. per year). It takes (with no erosion) 1.5 million to 6 million years to produce mountains with elevations of 3 km (around 1.9 mi).
	■ Erosion of the land. Generally 0.01 to 1 mm per year (about 0.004 to 0.039 in. per year). It takes (with no uplift) 3 million to 300 million years to erode a landscape by 3 km (about 1.9 mi). Erosion rate may be significantly increased by human activity such as timber harvesting or agricultural activities that increase the amount of water that runs off the land, causing erosion. Rates of uplift generally exceed rates of erosion, explaining why land above sea level persists.
	■ Incision of rivers into bedrock, producing canyons such as the Grand Canyon in Arizona. Incision is different from erosion, which is the material removed over a region. Rates are generally 0.005 to 10 mm per year (about 0.0002 to 0.39 in. per year). Therefore, to produce a canyon 3 km (around 1.9 mi) deep would take 300 thousand to 600 million years. The rate of incision may be increased several times by human activities such as building dams because increased downcutting of the river channel occurs directly below a dam.
Intermediate Rates	■ Movement of soil and rock downslope by creeping in response to the pull of gravity. Rate is generally 0.5 to 1.2 mm per year (about 0.02 to 0.05 in. per year).
	■ Coastal erosion by waves. Generally 0.25 to 1.0 m per year (0.82 to 3.28 ft per year). Thus, to provide 100 years' protection from erosion, a structure should be built about 25 to 100 m (about 82 to 328 ft) back from the cliff edge.
Fast Rates	■ Glacier movement. Generally a few meters per year to a few meters per day. ■ Lava flows. Depends on type of lava and slope. From a few meters per day to several meters per second.
	■ River flow in floods. Generally a few meters per second.
	■ Debris avalanche, or flow of saturated earth, soil, and rocks downslope. Can be greater than 100 km (62 mi) per hour.
	■ Earthquake rupture. Several kilometers per second.

Figure 1.13 Eroding a valley
Idealized diagram of progressive incision of a river into a sequence of horizontal rocks. The side slope is steep where rocks are hard and resistant to incision, and the rate of incision is generally less than about 0.01 mm per year (about 0.0004 in. per year). For softer rocks, where the side slope is gentle, the rate of incision may exceed 1 mm per year (0.039 in. per year). If the canyon incised about 1 km (0.62 mi) in 1 million years, the average rate is 1 mm per year (0.039 in. per year). *(Modified after King, P. B., and Schumm, S. A., 1980. The physical geography of William Morris Davis. Norwich, England: Geo Books)*

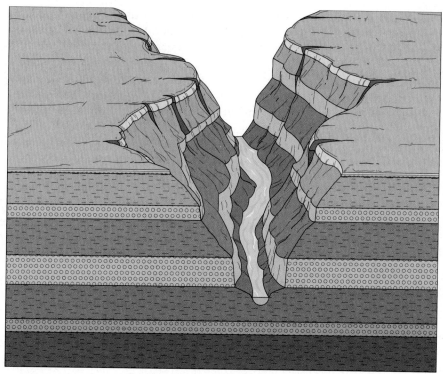

(a) Incision at about 250,000 yrs

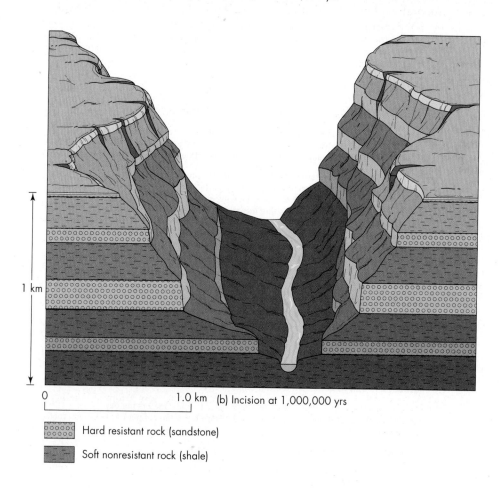

(b) Incision at 1,000,000 yrs

Hard resistant rock (sandstone)

Soft nonresistant rock (shale)

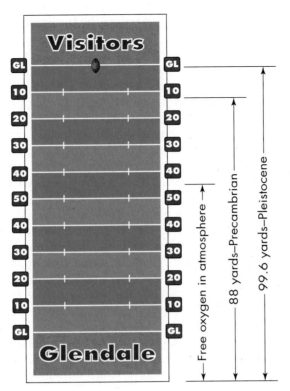

Figure 1.14 Time Geologic time as represented by a football field. See the text for further explanation.

Humans evolved during the Pleistocene epoch (the last 1.65 million years), which is a very small percentage of the age of Earth. To help you conceptualize the geologic time scale, Figure 1.14 illustrates all of geologic time as analogous to yards on a football field. Think back to your high school days, when your star kick-off return player took it deep into your end zone. Assume that the 100 yard field represents the age of Earth (4.6 billion years), making each yard equal to 45 million years. As your star zigs and zags and reaches the 50 yard line, the crowd cheers. But in Earth history he has traveled only 2,250 million years and is still in a primitive oxygen-deficient environment. At the opponent's 45 yard line, free oxygen in the atmosphere begins to support life. As our runner crosses the 12 yard line, the Precambrian period comes to an end and life becomes much more diversified. At less than half a yard from the goal line, our star runner reaches the beginning of the Pleistocene, the most recent 1.65 million years of Earth history, when humans evolved. As he leaps over the 1 inch line and in for the touchdown, the corresponding period in Earth history is 100,000 years ago, and modern humans were living in Europe. Another way to visualize geologic time is to imagine that one calendar year is equal to the age of Earth, 4.6 billion years. In this case, Earth formed on January 1; the first oxygen in the atmosphere did not occur until July; and mammals did not make their appearance until December 18. The first human being arrived on the scene on December 31 at 6 P.M.; and recorded history began only 48 seconds before midnight on December 31!

In answering environmental geology questions, we are often interested in the latest Pleistocene (the last 18,000 years), but we are most interested in the last few thousand or few hundred years of the Holocene epoch, which started approximately 10,000 years ago (see Appendix D, How Geologists Determine Time). Thus, in geologic study, geologists often design hypotheses to answer questions integrated through time. For example, we may wish to test the hypothesis that burning fossil fuels such as coal and oil, which we know releases carbon dioxide into the atmosphere, is causing global warming by trapping heat in the lower atmosphere. We term this phenomenon the greenhouse effect, which is discussed in detail in Chapter 19. One way to test this hypothesis would be to show that before the Industrial Revolution, when we started burning a lot of coal and, later, oil to power the new machinery of the time period, the mean global temperature was significantly lower than it is now. We would be particularly interested in the last few hundred to few thousand years before temperature measurements were recorded at various spots around the planet as they are today. To test the hypothesis that global warming is occurring, the investigator could examine prehistoric Earth materials that might provide indicators of global temperature. This examination might involve studying glacial ice or sediments from the bottoms of the oceans or lakes to estimate past levels of carbon dioxide in the atmosphere. Properly completed, studies can provide conclusions that enable us to accept or reject the hypothesis that global warming is occurring.

Our discussion about what science is emphasizes that science is a process. As such it is a way of knowing that constitutes a current set of beliefs based on the application of the scientific method. Science is not the only way a set of beliefs are established. Some beliefs are based on faith, but these, while valid, shouldn't be confused with science. The famous Roman philosopher Cicero once concluded that divine providence, or as we call it now, *intelligent design*, was responsible for the organization of nature and harmony that maintained the environment for all people. As modern science emerged with the process of science, other explanations emerged. This has included explanations for biological evolution by biologists, the understanding of space and time by physicists, and the explanation that continents and ocean basins form through plate tectonics by geologists.

Culture and Environmental Awareness

Environmental awareness involves the entire way of life that we have transmitted from one generation to another. To uncover the roots of our present condition, we must look to the past to see how our culture and our political, economic, ethical, religious, and aesthetic institutions affect the way we perceive and respond to our physical environment.

An ethical approach to maintaining the environment is the most recent development in the long history of human ethical evolution. A change in the concept of property rights has provided a fundamental transformation in our ethical evolution. In earlier times, human beings were often held as property, and their masters had the unquestioned right to dispose of them as they pleased. Slaveholding societies certainly had codes of ethics, but these codes did not include the idea that people cannot be property. Similarly, until very recently, few people in the industrialized world questioned the right of landowners to dispose of land as they please. Only within this century has the relationship between civilization and its physical environment begun to emerge as a relationship involving ethical considerations.

Environmental (including ecological and land) ethics involves limitations on social as well as individual freedom of action in the struggle for existence in our stressed environment. A **land ethic** assumes that we are responsible not only to other individuals and society, but also to the total environment, the larger community consisting of plants, animals, soil, rocks, atmosphere, and water. According to this ethic, we are the land's citizens and protectors, not its conquerors. This role change requires us to revere, love, and protect our land rather than allow economics to determine land use.[16] The creation of national parks and forests is an example of protective action based on a land ethic. Yellowstone National Park, in Wyoming and Montana, was the first national park in the United States, established in March 1872. Yellowstone led to the creation of other national parks, monuments, and forests, preserving some of the country's most valued aesthetic resources. Trees, plants, animals, and rocks are protected within the bounds of a national park or forest. In addition, rivers flow free and clean, lakes are not overfished or polluted, and mineral resources are protected. Last, the ethic that led to the protection of such lands allows us the privilege of enjoying these natural areas and ensures that future generations will have the same opportunity. We will now change focus to discuss why solving environmental problems tends to be difficult and introduce the emerging environmental policy tool known as the precautionary principle.

Why Is Solving Environmental Problems So Difficult?

Many environmental problems tend to be complex and multifaceted. They may involve issues related to physical, biological, and human processes. Some of the problems are highly charged from an emotional standpoint and potential solutions are often vigorously debated.

There are three main reasons that solving environmental problems may be difficult:

- Expediential growth is often encountered. Expediential growth means that the amount of change may be happening quickly whether we are talking about an increase or decrease.

- There are often lag times between when a change occurs and when it is recognized as a problem. If the lag time is long, it may be very difficult to even recognize a particular problem.

- An environmental problem involves the possibility of irreversible change. If a species becomes extinct, it is gone forever.

Environmental policy links to environmental economics are in their infancy. That is, the policy framework to solve environmental problems is a relatively new arena. We are developing policies such as the precautionary principle and finding ways to evaluate the economics of gains and losses from environmental change. For example, how do you put a dollar amount on aesthetics or living in a quality environment? What the analysis often comes down to is an exercise in values clarification. Science can provide a number of potential solutions to problems but which solution we pick will depend upon our values.

Precautionary Principle

What Is the Precautionary Principle? Science has the role of trying to understand physical and biological processes associated with environmental problems such as global warming, exposure to toxic materials, and depletion of resources, among others. However, all science is preliminary and it is difficult to prove relationships between physical and biological processes and link them to human processes. Partly for this reason, in 1992, the Rio Earth Summit on sustainable development supported the **precautionary principle.** The idea behind the principle is that when there exists a potentially serious environmental problem, scientific certainty is not required to take a precautionary approach. That is, better safe than sorry. The precautionary principle thus contributes to the critical thinking on a variety of environmental concerns, for example, manufacture and use of toxic chemicals or burning huge amounts of coal as oil becomes scarcer. It is considered one of the most influential ideas for obtaining an intellectual, environmentally just policy framework for environmental problems.[17]

The precautionary principle recognizes that scientific proof is not possible in most instances, and management practices are needed to reduce or eliminate environmental problems believed to result from human activities. In other words, in spite of the fact that full scientific certainty is not available, we should still take cost-effective action to solve environmental problems.

The Precautionary Principle May Be Difficult to Apply. One of the difficulties in applying the precautionary principle is the decision concerning how much scientific evidence is needed before action on a particular problem should be taken. This is a significant and often controversial question. An issue being considered has to have some preliminary data and conclusions but awaits more scientific data and analysis. For example, when considering environmental health issues related to burning coal, there may be an abundance of scientific data about air, water, and land pollution, but with gaps, inconsistencies, and other scientific uncertainties. Those in favor of continuing or increasing the use of coal may argue that there is not sufficient proof to warrant restricting its use. Others would argue that absolute proof of safety is necessary before a big increase in burning of coal is allowed. The precautionary principle, applied to this case, would be that lack of full scientific certainty concerning the use of coal should not be used as a reason for not taking, or postponing, cost-effective measures to reduce or prevent environmental degradation or heath problems. This raises the question of what constitutes a cost-effective measure. Determination of benefits and costs of burning more coal compared to burning less or treating coal more to clean up the fuel should be done, but other economic analysis may also be appropriate.[17,18]

There will be arguments over what is sufficient scientific knowledge for decision making. The precautionary principle may be difficult to apply, but it is becoming a common part of the process of environmental analysis and policy when applied to environmental protection and environmental health issues. The European Union has been applying the principle for over a decade, and the City and County of San Francisco in 2003 became the first government in the United States to make the precautionary principle the basis for its environmental policy.

Applying the precautionary principle requires us to use the principle of environmental unity and predict potential consequences of activities before they occur. Therefore, the precautionary principle has the potential to become a proactive, rather than reactive, tool in reducing or eliminating environmental degradation resulting from human activity. The principle moves the burden of proof of no harm from the public to those proposing a particular action. Those who develop new chemicals or actions are often, but not always, against the precautionary principle. The opponents often argue that applying the principle is too expensive and will stall progress. It seems unlikely that the principle will be soon applied across the board in the United States to potential environmental problems. Nevertheless, it will likely be invoked more often in the future. When the precautionary principle is applied, it must be an honest debate between all informed and potentially affected parties. The entire range of alternative actions should be considered, including taking no action.

Science and Values

We Are Creatures of the Pleistocene. There is no arguing that we are a very successful species that until recently has lived in harmony with both our planet and other forms of life for over 100 thousand years. We think of ourselves as modern people, and certainly our grasp of science and technology has grown tremendously in the past several hundred years. However, we cannot forget that our genetic roots are in the Pleistocene. In reality our deepest beliefs and values are probably not far distant from those of our ancestors who sustained themselves in small communities, moving from location to location and hunting and gathering what they needed. At first thought this statement seems inconceivable and not possible to substantiate considering the differences between our current way of life and that of our Pleistocene ancestors. It has been argued that studying our Pleistocene ancestors, with whom we share nearly identical genetic information, may help us understand ourselves better.[19] That is, much of our human nature and in fact our very humanity may be found in the lives of the early hunters and gatherers, explaining some of our current attitudes toward the natural world. We are more comfortable with natural sounds and smells like the movement of grass where game is moving or the smell of ripe fruit than the shril noise of horns and jackhammers and smell of air pollution in the city. Many of us enjoy sitting around a campfire roasting marshmallows and telling stories about bears and rattlesnakes. We may find a campfire comforting even if smoke stings our eyes because our Pleistocene ancestors knew fire protected them from predators such as bears, wolves, and lions. If you want to liven up a campfire talk, start telling grizzly bear stories!

Solutions we choose to solve environmental problems depend upon how we value people and the environment. For example, if we believe that human population growth is a problem, then conscious decisions to reduce human population growth reflects a value decision that we as a society choose to endorse and implement. As another example, consider flooding of small urban streams. Flooding is a hazard experienced by many communities. Study of rivers and their natural processes leads to a number of potential solutions for a given flood hazard. We may choose to place the stream in a concrete box—a remedy that can significantly reduce the flood hazard. Alternatively, we may choose to restore our urban streams and their floodplains, the flat land adjacent to the river that periodically floods, as greenbelts. This choice will reduce damage from flooding while providing habitat for a variety of animals including raccoons, foxes, beavers, and muskrats that use the stream environment; resident and migratory birds that nest, feed, and rest close to a river; and a variety of fish that live in the river system. We will also be more comfortable when interacting with the river. That is why river parks are so popular.

The coastal environment, where the coastline and associated erosional processes come into conflict with development, provides another example for science and values. Solutions to coastal erosion may involve defending the coast, along with its urban development, at all cost by constructing "hard structures" such as seawalls. Science tells us that consequences from the hard solution generally include reduction or elimination of the beach environment in favor of protecting development. Science also tells us that using appropriate setbacks from the erosion zone of coastal processes provides a buffer zone from the erosion, while maintaining a higher quality coastal environment that includes features such as beaches and adjacent seacliffs or dune lines. The solution we pick depends upon how we value the coastal zone. If we value the development more than the beach, then we may choose to protect development at all cost. If we value the beach environment, we may choose more flexible options that allow for erosion to take place naturally within a buffer zone between the coast and development.

By the year 2050, the human population on our planet will likely increase to about 9 billion people, about 3 billion more than today. Thus, it appears that during the next 50 years crucial decisions must be made concerning how we will deal with the increased population associated with increased demands on resources including land, water, minerals, and air. The choices we make will inevitably reflect our values.

SUMMARY

The immediate causes of the environmental crisis are overpopulation, urbanization, and industrialization, which have occurred with too little ethical regard for our land and inadequate institutions to cope with environmental stress. Solving environmental problems involves both scientific understanding and the fostering of social, economic, and ethical behavior that allows solutions to be implemented. Beyond this, complex environmental problems can be difficult to solve due to the possibility of exponential growth, lag times between cause and effect, and irreversible consequences. A new emerging policy tool is the precautionary principle. The idea behind the principle is that when a potentially serious environmental problem exists, scientific certainty is not required to take a precautionary approach and find a cost-efficient solution. Some environmental problems are sufficiently serious that it is better to be safe than sorry.

Five fundamental concepts establish a philosophical framework for our investigation of environmental geology:

1. The increasing world population is the number one environmental problem.

2. Sustainability is the preferred solution to many environmental problems.

3. An understanding of the Earth system and rates of change in systems is critical to solving environmental problems.

4. Earth processes that are hazardous to people have always existed. These natural hazards must be recognized and avoided when possible, and their threat to human life and property minimized.

5. Results of scientific inquiry to solve a particular environmental problem often result in a series of potential solutions consistent with the scientific findings. Which solution we choose reflects our value system.

Key Terms

average residence time (p. 21)

carrying capacity (p. 13)

doubling time (p. 11)

Earth systems science (p. 23)

environmental crisis (p. 17)

environmental geology (p. 8)

environmental unity (p. 23)

exponential growth (p. 10)

Gaia hypothesis (p. 23)

geologic time (p. 5)

geology (p. 8)

growth rate (p. 11)

hypothesis (p. 25)

input-output analysis (p. 20)

land ethic (p. 30)

law of faunal assemblages (p. 8)

precautionary principle (p. 31)

scientific method (p. 25)

sustainability (p. 15)

system (p. 18)

theory (p. 26)

uniformitarianism (p. 22)

Review Questions

1. What is environmental geology?

2. Define the components of the scientific method.

3. What are the roots of the so-called environmental crisis?

4. Why are we so concerned about the increase in human population?

5. What is sustainability?

6. Define the principle of environmental unity, and provide a good example.

7. What is exponential growth?

8. What is Earth systems science, and why is it important?

9. What do we mean by average residence time?

10. How can the principle of uniformitarianism be applied to environmental geology?

11. What is the Gaia hypothesis?

12. What is the precautionary principle and why is it important?

13. Why is solving complex environmental problems often difficult?

Critical Thinking Questions

1. Assuming that there is an environmental crisis today, what possible solutions are available to alleviate the crisis? How will solutions in developing countries differ from those in highly industrialized societies? Will religion or political systems have a bearing on potential solutions? If so, how will they affect the solutions?

2. It has been argued that we must control human population because otherwise we will not be able to feed everyone. Assuming that we could feed 10 billion to 15 billion people on Earth, would we still want to have a smaller population than that? Why?

3. We state that sustainability is the environmental objective. Construct an argument to support this statement. Are the ideas of sustainability and building a sustainable economy different in developing, poor countries than in countries that are affluent and have a high standard of living? How are they different, and why?

4. The concept of environmental unity is an important one today. Consider some major development being planned for your region and outline how the principle of environmental unity could help in determining the project's potential environmental impact. In other words, consider a development and then a series of consequences resulting from it. Some of the impacts may be positive and some may be negative in your estimation.

5. Do you believe we have a real connection to our Pleistocene ancestors? Could such a connection explain our childlike love of baby animals or the storytelling around a camp fire? Is the human race's long history of hunting and gathering, during which our genetic evolution occurred, reflected in our values?

6. Is the Gaia hypothesis science? How could you test the main parts? Which would be hard to test? Why?

7. Defend or criticize the notion that increase in human population is *the* environmental problem and that sustainability is the solution.

8. Do you think the precautionary principle should be applied to the problem of controlling the growth of the human population? If you do, how could it be applied?

TWO

Internal Structure of Earth and Plate Tectonics

Written with the assistance of Tanya Atwater

Learning Objectives

The surface of Earth would be much different—relatively smooth, with monotonous topography—if not for the active tectonic processes within Earth that produce earthquakes, volcanoes, mountain chains, continents, and ocean basins.[1] In this chapter we focus directly on the interior of Earth, with the following learning objectives:

- Understand the basic internal structure and processes of Earth
- Know the basic ideas behind and evidence for the theory of plate tectonics
- Understand the mechanisms of plate tectonics
- Understand the relationship of plate tectonics to environmental geology

The San Andreas fault in southern California is the major boundary between the Pacific and North American plates. Here in the Indio Hills, the fault is delineated by lines of native palm trees. *(Edward A. Keller)*

CASE HISTORY | Two Cities on a Plate Boundary

California straddles the boundary between two tectonic plates, which are discussed in detail in this chapter. That boundary between the North American and Pacific plates is the notorious San Andreas Fault (Figure 2.1). A fault is a fracture along which one side has moved relative to the other, and the San Andreas Fault is a huge fracture zone, hundreds of kilometers long. Two major cities, Los Angeles to the south and San Francisco to the north, are located on opposite sides of this fault. San Francisco was nearly destroyed by a major earthquake in 1906, which led to the identification of the fault. Many of the moderate to large earthquakes in the Los Angeles area are on faults related to the San Andreas fault system. Most of the beautiful mountain topography in coastal California near both Los Angeles and San Francisco is a direct result of processes related to movement on the San Andreas Fault. However, this beautiful topography comes at a high cost to society. Since 1906, earthquakes on the San Andreas fault system or on nearby faults, undoubtedly influenced by the plate boundary, have cost hundreds of lives and many billions of dollars in property damage. Construction of buildings, bridges, and other structures in California is more expensive than elsewhere because they must be designed to withstand ground shaking caused by earthquakes. Older structures have to be *retrofitted*, or have changes made to their structure, to withstand the shaking, and many people purchase earthquake insurance in an attempt to protect themselves from the "big one."

Los Angeles is on the Pacific plate and is slowly moving toward San Francisco, which is on the North American plate.

Figure 2.1 San Andreas Fault
Map showing the San Andreas Fault and topography in California. Arrows show relative motion on either side of the fault. *(R. E. Wallace/National Earthquake Information Center. U.S.G.S.)*

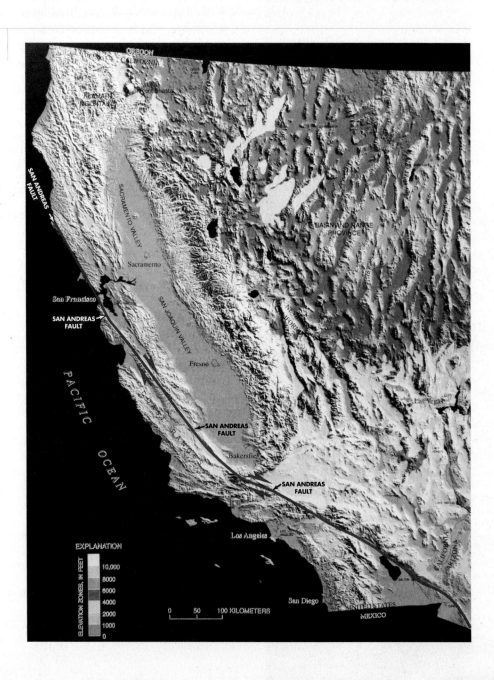

In about 20 million years the cities will be side by side. If people are present, they might be arguing over which is a suburb of the other. Of course, there will still be a plate boundary between the Pacific and North American plates 20 million years from now, because large plates have long geologic lives, on the order of 100 million years. However, the boundary may not be the San Andreas Fault. The plate boundary will probably have moved eastward, and the topography of what is now California may be somewhat different. In fact, some recent earthquake activity in California, such as the large 1992 Landers earthquake, east of the San Andreas fault, may be the beginning of a shift in the plate boundary.

2.1 Internal Structure of Earth

You may be familiar with the situation comedy *Third Rock from the Sun*, a phrase that refers to our planet Earth. Far from being a barren rock, Earth is a complex dynamic planet that in some ways resembles a chocolate-covered cherry. That is, Earth has a rigid outer shell, a solid center, and a thick layer of liquid that moves around as a result of dynamic internal processes. The internal processes are incredibly important in affecting the surface of Earth. They are responsible for the largest landforms on the surface: continents and ocean basins. The configuration of the continents and ocean basins in part controls the oceans' currents and the distribution of heat carried by seawater in a global system that affects climate, weather, and the distribution of plant and animal life on Earth. Finally, Earth's internal processes are also responsible for regional landforms including mountain chains, chains of active volcanoes, and large areas of elevated topography, such as the Tibetan Plateau and the Rocky Mountains. The high topography that includes mountains and plateaus significantly affects both global circulation patterns of air in the lower atmosphere and climate, thereby directly influencing all life on Earth. Thus, our understanding of the internal processes of Earth is of much more than simply academic interest. These processes are at the heart of producing the multitude of environments shared by all living things on Earth.

The Earth Is Layered and Dynamic. Earth (Figure 2.2a) has a radius of about 6,300 km (4,000 mi) (Figure 2.2b). Information regarding the internal layers of the Earth is shown in Figure 2.2b. We can consider the internal structure of Earth in two fundamental ways:

- by composition and density (heavy or light).
- by physical properties (for example, solid or liquid, weak or strong).

Our discussion will explore the two ways of looking at the interior of our planet. Some of the components of the basic structure of Earth[1] are

- A solid inner **core** with a thickness of more than 1,300 km (808 mi) that is roughly the size of the moon but with a temperature about as high as the temperature of the surface of the Sun.[2] The inner core is believed to be primarily metallic, composed mostly of iron (about 90 percent by weight), with minor amounts of elements such as sulfur, oxygen, and nickel.

- A liquid outer core with a thickness of just over 2,000 km (1,243 mi) with a composition similar to that of the inner core. The outer core is very fluid, more similar to water than to honey. The average density of the inner and outer core is approximately 10.7 grams per cubic centimeter (0.39 pounds per cubic inch). The maximum near the center of Earth is about 13 g/cm^3 (0.47 lb/in^3). By comparison, the density of water is 1 g/cm^3 (0.04 lb/in^3) and the average density of Earth is approximately 5.5 g/cm^3 (0.2 lb/in^3).

- The **mantle**, nearly 3,000 km (1,864 mi) thick, surrounds the outer core and is mostly solid, with an average density of approximately 4.5 g/cm^3 (0.16 lb/in^3). Rocks in the mantle are primarily iron- and magnesium-rich silicates. Interestingly, the density difference between the outer core and the

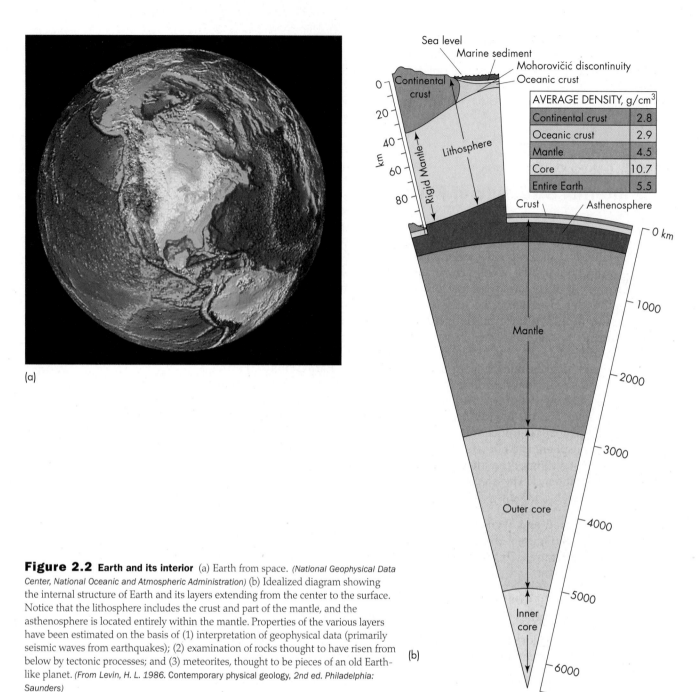

AVERAGE DENSITY, g/cm³	
Continental crust	2.8
Oceanic crust	2.9
Mantle	4.5
Core	10.7
Entire Earth	5.5

Figure 2.2 Earth and its interior (a) Earth from space. *(National Geophysical Data Center, National Oceanic and Atmospheric Administration)* (b) Idealized diagram showing the internal structure of Earth and its layers extending from the center to the surface. Notice that the lithosphere includes the crust and part of the mantle, and the asthenosphere is located entirely within the mantle. Properties of the various layers have been estimated on the basis of (1) interpretation of geophysical data (primarily seismic waves from earthquakes); (2) examination of rocks thought to have risen from below by tectonic processes; and (3) meteorites, thought to be pieces of an old Earth-like planet. *(From Levin, H. L. 1986. Contemporary physical geology, 2nd ed. Philadelphia: Saunders)*

overlying mantle is greater than that between the rocks at the surface of Earth and the overlying atmosphere! In the case of the outer core and mantle, the more fluid phase of the outer core is beneath the solid phase of the mantle. This is just the opposite of the case of the rock-atmosphere relationship, where the fluid atmosphere overlies the solid lithosphere. Because it is liquid, the outer core is dynamic and greatly influences the overlying mantle and, thus, the surface of Earth.

- The **crust,** with variable thickness, is the outer rock layer of the Earth. The boundary between the mantle and crust is known as the Mohorovičić discontinuity (also called the **Moho**). It separates the lighter rocks of the crust with an average density of approximately 2.8 g/cm³ (0.10 lb/in³) from the denser rocks of the mantle below.

Continents and Ocean Basins Have Significantly Different Properties and History. Within the uppermost portion of the mantle, near the surface of Earth our terminology becomes more complicated. For example, the cool, strong outermost layer of Earth is also called the **lithosphere** (*lithos* means "rock"). It is much stronger and more rigid than the material underlying it, the **asthenosphere** (*asthenos* means "without strength"), which is a hot and slowly flowing layer of relatively weak rock. The lithosphere averages about 100 km (62 mi) in thickness, ranging from a few kilometers (1 to 2 mi) thick beneath the crests of mid-ocean ridges to about 120 km (75 mi) beneath ocean basins and 20 to 400 km (13 to 250 mi) beneath the continents. The crust is embedded in the top of the lithosphere. Crustal rocks are less dense than the mantle rocks below, and oceanic crust is slightly denser than continental crust. Oceanic crust is also thinner: The ocean floor has a uniform crustal thickness of about 6 to 7 km (3.7 to 4.4 mi), whereas the crustal thickness of continents averages about 35 km (22 mi) and may be up to 70 km (44 mi) thick beneath mountainous regions. Thus, the average crustal thickness is less than 1 percent of the total radius of Earth and can be compared to the thin skin of a tangerine. Yet it is this layer that is of particular interest to us because we live at the surface of the continental crust.

In addition to differences in density and thickness, continental and oceanic crust have very different geologic histories. Oceanic crust of the present ocean basins is less than approximately 200 million years old, whereas continental crust may be several billion years old. Three thousand kilometers (1,865 mi) below us, at the core-mantle boundary, processes may be occurring that significantly affect our planet at the surface. It has been speculated that gigantic cycles of convection occur within Earth's mantle, rising from as deep as the core-mantle boundary up to the surface and then falling back again. The concept of **convection** is illustrated by heating a pan of hot water on a stove (Figure 2.3). Heating the water at the bottom of the pan causes the water to become less dense and more unstable, so it rises to the top. The rising water displaces denser, cooler water, which moves laterally and sinks to the bottom of the pan. It is suggested that Earth layers contain convection cells and operate in a similar fashion.

A complete cycle in the mantle may take as long as 500 million years.[1] Mantle convection is fueled at the core-mantle boundary both by heat supplied from the molten outer core of Earth and by radioactive decay of elements (such as uranium) scattered throughout the mantle. Let us now examine some of the observations and evidence that reveal the internal structure of Earth.

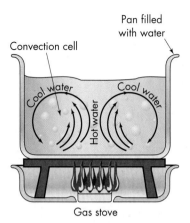

Figure 2.3 Convection
Idealized diagram showing the concept of convection. As the pan of water is heated, the less dense hot water rises from the bottom to displace the denser cooler water at the top, which then sinks down to the bottom. This process of mass transport is called convection, and each circle of rising and falling water is a convection *cell*.

2.2 How We Know about the Internal Structure of Earth

What We Have Learned about Earth from Earthquakes. Our knowledge concerning the structure of Earth's interior arises primarily from our study of **seismology**. Seismology is the study of earthquakes and the passage of seismic waves through Earth.[3] When a large earthquake occurs, seismic energy is released and seismic waves move both through Earth and along its surface. The properties of these waves are discussed in detail in Chapter 6 with earthquake hazards.

Some waves move through solid and liquid materials while others move through solid, but not liquid materials. The rates at which seismic waves propagate are on the order of a few kilometers per second (1 or 2 miles per second). Their actual velocity varies with the properties of the materials through which the waves are propagating (moving). When the seismic waves encounter a boundary, such as the mantle-core boundary, some of them are *reflected* back. Others cross the boundary and are *refracted* (change the direction of propagation). Still others fail to propagate through the liquid outer core. Thousands of seismographs (instruments that record seismic waves) are stationed around the world. When an earthquake occurs, the reflected and refracted waves are recorded when they emerge at the surface. Study of these waves has been a powerful tool for deducing the layering of the interior of Earth and the properties of the materials found there.

In summary, the boundaries that delineate the internal structure of Earth are determined by studying seismic waves generated by earthquakes and recorded on seismographs around Earth. As seismology has become more sophisticated, we have learned more and more about the internal structure of Earth and are finding that the structure can be quite variable and complex. For example, we have been able to recognize

- where magma, which is molten rock material beneath Earth's surface, is generated in the asthenosphere

- the existence of slabs of lithosphere that have apparently sunk deep into the mantle

- the extreme variability of lithospheric thickness, reflecting its age and history

2.3 Plate Tectonics

The term *tectonics* refers to the large-scale geologic processes that deform Earth's lithosphere, producing landforms such as ocean basins, continents, and mountains. Tectonic processes are driven by forces within the Earth. These processes are part of the tectonic system, an important subsystem of the Earth system.

Movement of the Lithospheric Plates

What Is Plate Tectonics? The lithosphere is broken into large pieces called *lithospheric plates* that move relative to one another (Figure 2.4a).[4] Processes associated with the creation, movement, and destruction of these plates are collectively known as **plate tectonics.**

Locations of Earthquakes and Volcanoes Define Plate Boundaries. A lithospheric plate may include both a continent and part of an ocean basin or an ocean region alone. Some plates are very large and some are relatively small, though they are significant on a regional scale. For example, the Juan de Fuca plate off the Pacific Northwest coast of the United States, which is relatively small, is responsible for many of the earthquakes in northern California. The boundaries between lithospheric plates are geologically active areas. Most earthquakes and many volcanoes are associated with these boundaries. In fact, plate boundaries are defined by the areas in which concentrated seismic activity occurs (Figure 2.4b). Over geologic time, plates are formed and destroyed, cycling materials from the interior of Earth to the surface and back again at these boundaries (Figure 2.5). The continuous recycling of tectonic processes is collectively called the *tectonic cycle.*

Seafloor Spreading Is the Mechanism for Plate Tectonics. As the lithospheric plates move over the asthenosphere, they carry the continents embedded within them.[5] The idea that continents move is not new; it was first suggested by German scientist Alfred Wegener in 1915. The evidence he presented for **continental drift** was based on the congruity of the shape of continents, particularly those across the Atlantic Ocean, and on the similarity in fossils found in South America and Africa. Wegener's hypothesis was not taken seriously because there was no known mechanism that could explain the movement of continents around Earth. The explanation came in the late 1960s, when **seafloor spreading** was discovered. In seafloor regions called **mid-oceanic ridges,** or **spreading centers,** new crust is

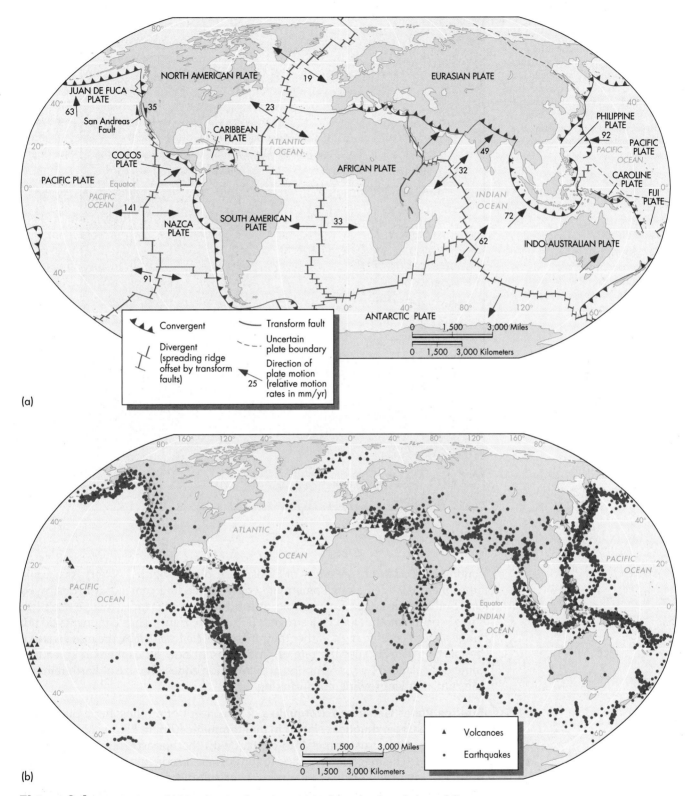

(a)

(b)

Figure 2.4 Earth's plates (a) Map showing the major tectonic plates, plate boundaries, and direction of plate movement. *(Modified from Christopherson, R. W. 1994.* Geosystems, *2nd ed. Englewood Cliffs, NJ: Macmillan)* (b) Volcanoes and earthquakes: Map showing location of volcanoes and earthquakes. Notice the correspondence between this map and the plate boundaries. *(Modified after Hamblin, W. K. 1992.* Earth's dynamic systems, *6th ed. New York: Macmillan)*

Figure 2.5 **Model of plate tectonics** Diagram of the model of plate tectonics. New oceanic lithosphere is being produced at the spreading ridge (divergent plate boundary). Elsewhere, oceanic lithosphere returns to the interior of Earth at a convergent plate boundary (subduction zone). *(Modified from Lutgens, F., and Tarbuck, E. 1992. Essentials of geology. New York: Macmillan)*

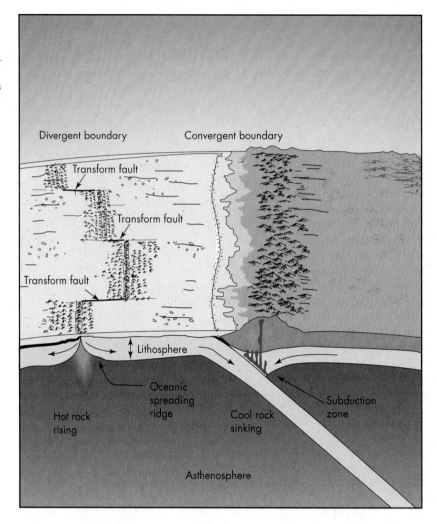

continuously added to the edges of lithospheric plates (Figure 2.5, left). As oceanic lithosphere is added along some plate edges (spreading centers), it is destroyed along other plate edges, for example, at **subduction zones** (areas where one plate sinks beneath another and is destroyed) (Figure 2.5, right). Thus continents do not move *through* oceanic crust; rather they are *carried along with it* by the movement of the plates. Also, because the rate of production of new lithosphere at spreading centers is balanced by consumption at subduction zones, the size of Earth remains constant, neither growing nor shrinking.

Sinking Plates Generate Earthquakes. The concept of a lithospheric plate sinking into the upper mantle is shown in diagrammatic form in Figure 2.5. When the wet, cold oceanic crust comes into contact with the hot asthenosphere, magma is generated. The magma rises back to the surface, producing volcanoes, such as those that ring the Pacific Ocean basin, over subduction zones. The path of the descending plate (or *slab*, as it sometimes is called) into the upper mantle is clearly marked by earthquakes. As the oceanic plate subducts, earthquakes are produced both between it and the overriding plate and within the interior of the subducting plate. The earthquakes occur because the sinking lithospheric plate is relatively cooler and stronger than the surrounding asthenosphere; this difference causes rocks to break and seismic energy to be released.[6]

The paths of descending plates at subduction zones may vary from a shallow dip to nearly vertical, as traced by the earthquakes in the slabs. These dipping planes of earthquakes are called **Wadati-Benioff zones** (Figure 2.6). The very

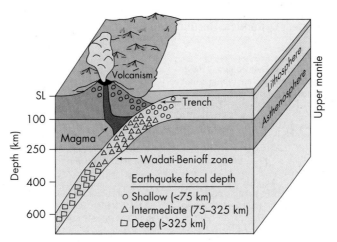

Figure 2.6 Subduction zone
Idealized diagram of a subduction zone showing the Wadati-Benioff zone, which is an array of earthquake foci from shallow to deep that delineate the subduction zone and the descending lithospheric plate.

existence of Wadati-Benioff zones is strong evidence that subduction of rigid "breakable" lithosphere is occurring.[6]

Plate Tectonics Is a Unifying Theory. The theory of plate tectonics is to geology what Darwin's origin of species is to biology: a unifying concept that explains an enormous variety of phenomena. Biologists now have an understanding of evolutionary change. In geology, we are still seeking the exact mechanism that drives plate tectonics, but we think it is most likely convection within Earth's mantle. As rocks are heated deep in Earth, they become less dense and rise. Hot materials, including magma, leak out, and are added to the surfaces of plates at spreading centers. As the rocks move laterally, they cool, eventually becoming dense enough to sink back into the mantle at subduction zones. This circulation is known as convection, which was introduced in Section 2.1. Figure 2.7 illustrates the cycles of convection that may drive plate tectonics.

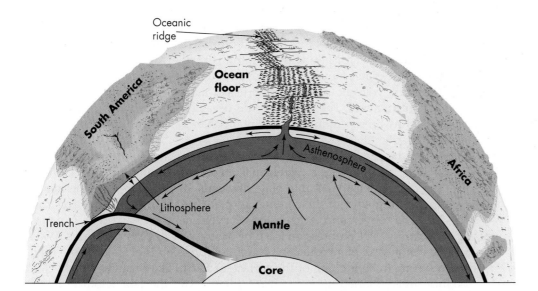

Figure 2.7 **Plate movement** Model of plate movement and mantle. The outer layer (or lithosphere) is approximately 100 km (approximately 62 mi) thick and is stronger and more rigid than the deeper asthenosphere, which is a hot and slowly flowing layer of relatively low-strength rock. The oceanic ridge is a spreading center where plates pull apart, drawing hot, buoyant material into the gap. After these plates cool and become dense, they descend at oceanic trenches (subduction zones), completing the convection system. This process of spreading produces ocean basins, and mountain ranges often form where plates converge at subduction zones. A schematic diagram of Earth's layers is shown in Figure 2.2b. (*Grand, S. P. 1994. Mantle shear structure beneath the Americas and surrounding oceans. Journal of Geophysical Research 99:11591–621. Modified after Hamblin, W. K. 1992. Earth's dynamic systems, 6th ed. New York: Macmillan*)

TABLE 2.1 Types of Plate Boundaries: Dynamics, Results, and Examples

Plate Boundary	Plates Involved	Dynamics	Results	Example
Divergent	Usually oceanic	Spreading. The two plates move away from one another and molten rock rises up to fill the gap.	Mid-ocean ridge forms and new material is added to each plate.	African and North American plate boundary (Figure 2.4a) **Mid-Atlantic Ridge**
Convergent	Ocean-continent	Oceanic plate sinks beneath continental plate.	Mountain ranges and a subduction zone are formed with a deep trench. Earthquakes and volcanic activity are found here.	Nazca and South American plate boundary (Figure 2.4a) **Andes Mountains** **Peru-Chile Trench**
Convergent	Ocean-ocean	Older, denser, oceanic plate sinks beneath the younger, less dense oceanic plate.	A subduction zone is formed with a deep trench. Earthquakes and volcanic activity are found here.	Fiji plate (Figure 2.4a) **Fiji Islands**
Convergent	Continent-continent	Neither plate is dense enough to sink into the asthenosphere; compression results.	A large, high mountain chain is formed, and earthquakes are common.	Indo-Australian and Eurasian plate boundary (on land) (Figure 2.4a) **Himalaya Mountains**
Transform	Ocean-ocean or continent-continent	The plates slide past one another.	Earthquakes common. May result in some topography.	North American and Pacific plate boundary (Figure 2.10) **San Andreas fault**

Types of Plate Boundaries

There are three basic types of plate boundaries: divergent, convergent, and transform, shown in Figures 2.4 and 2.5 and Table 2.1. These boundaries are not narrow cracks as shown on maps and diagrams but are zones that range from a few to hundreds of kilometers across. Plate boundary zones are narrower in ocean crust and broader in continental crust.

Divergent boundaries occur where new lithosphere is being produced and neighboring parts of plates are moving away from each other. Typically this process occurs at mid-ocean ridges, and the process is called seafloor spreading (Figure 2.5). Mid-ocean ridges form when hot material from the mantle rises up to form a broad ridge typically with a central rift valley. It is called a rift valley, or rift, because the plates moving apart are pulling the crust apart and splitting, or rifting, it. Molten volcanic rock that is erupted along this rift valley cools and forms new plate material. The system of mid-oceanic ridges along divergent plate boundaries forms linear submarine mountain chains that are found in virtually every ocean basin on Earth.

Convergent boundaries occur where plates collide. If one of the converging plates is oceanic and the other is continental, an oceanic-continental plate collision results. The higher-density oceanic plate descends, or subducts, into the mantle beneath the leading edge of the continental plate, producing a subduction zone (Figure 2.5). The convergence or collision of a continent with an ocean plate can result in compression. Compression is a type of stress, or force per unit area. When an oceanic-continental plate collision occurs, compression is exerted on the lithosphere, resulting in shortening of the surface of Earth, like pushing a table cloth to produce folds. Shortening can cause folding, as in the table cloth example, and faulting, or displacement of rocks along fractures to thicken the lithosphere (Figure 2.8a). This process of deformation produces major mountain chains and volcanoes such as the Andes in South America and the Cascade Mountains in the Pacific Northwest of the United States (see A Closer Look: The Wonder of Mountains). If two oceanic lithospheric plates collide (oceanic-to-oceanic plate collision), one plate subducts beneath the other, and a subduction zone and arc-shaped chain of volcanoes known as an *island arc* are formed (Figure 2.8b) as, for example, the Aleutian Islands of the North Pacific. A **submarine trench,** relatively

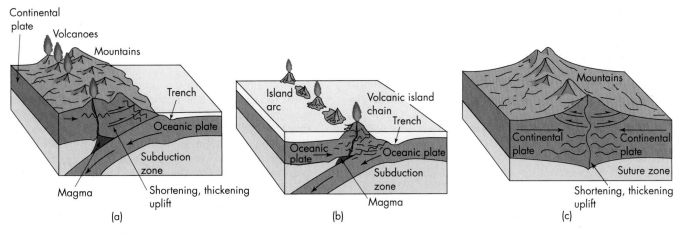

Figure 2.8 **Convergent plate boundaries** Idealized diagram illustrating characteristics of convergent plate boundaries: (a) continental-oceanic plate collision, (b) oceanic-oceanic plate collision, and (c) continental-continental plate collision.

narrow, usually several thousand km long and several km deep depression on the ocean floor, is often formed as the result of the convergence of two colliding plates with subduction of one. A trench is often located seaward of a subduction zone associated with an oceanic-continental plate or oceanic-oceanic plate collision. Submarine trenches are sites of some of the deepest oceanic waters on Earth. For example, the Marianas trench at the center edge of the Philippine plate is 11 km (7 mi) deep. Other major trenches include the Aleutian trench south of Alaska and the Peru-Chile trench west of South America. If the leading edges of both plates contain relatively light, buoyant continental crust, subduction into the mantle of one of the plates is difficult. In this case a continent-to-continent plate collision occurs, in which the edges of the plates collide, causing shortening and lithospheric thickening due to folding and faulting. (Figure 2.8c). Where the two plates join is known as a *suture zone*. Continent-to-continent collision has produced some of the highest mountain systems on Earth, such as the Alpine and Himalayan mountain belts (Figure 2.9). Many older mountain belts were formed in a similar way; for example, the Appalachians formed during an ancient continent-to-continent plate collision 250 to 350 million years ago.

Figure 2.9 **Mountains in Italy** Mountain peaks (the Dolomites) in southern Italy are part of the Alpine mountain system formed from the collision between Africa and Europe.
(Edward A. Keller)

A CLOSER LOOK | The Wonder of Mountains

Mountains have long fascinated people with their awesome presence. We are now discovering a fascinating story concerning their origin. The story removes some of the mystery as to how mountains form, but it has not removed the wonder. The new realization that mountains are systems (see Chapter 1) resulting from the interaction between tectonic activity (that leads to crustal thickening), the climate of the mountain, and Earth surface processes (particularly erosion) has greatly expanded our knowledge of how mountains develop.[7,8] Specifically, we have learned the following:

- Tectonic processes at convergent plate boundaries lead to crustal thickening and initial development of mountains. The mean (or average) elevation that a mountain range attains is a function of the uplift rate, which varies from less than 1 mm to about 10 mm per year (0.04 to 0.4 in. per year). The greater the rate of uplift, the higher the point to which the mean elevation of a mountain range is likely to rise during its evolution.

- As a mountain range develops and gains in elevation, it begins to modify the local and regional climate by blocking storm paths and producing a "rain shadow" in which the mountain slopes on the rain-shadow side receive much less rainfall than does the other side of the mountain. As a result, rates of runoff and erosion on the side of the rain shadow are less than for the other side. Nevertheless, the rate of erosion increases as the elevation of the mountain range increases, and eventually the rate of erosion matches the rate of uplift. When the two match, the mountain reaches its maximum mean elevation, which is a dynamic balance between the uplift and erosion. At this point, no amount of additional uplift will increase the mean elevation of the mountains above the dynamic maximum. However, if the uplift rate increases, then a higher equilibrium mean elevation of the range may be reached. Furthermore, when the uplift ceases or there is a reduction in the rate of uplift, the mean elevation of the mountain range will decrease.[7] Strangely, the elevations of individual peaks may still increase!

- Despite erosion, the elevation of a mountain peak in a range may actually increase. This statement seems counterintuitive until we examine in detail some of the physical processes resulting from erosion. The uplift that results from the erosion is known as isostatic uplift. **Isostasy** is the principle whereby thicker, more buoyant crust stands topographically higher than crust that is thinner and denser. The principle governing how erosion can result in uplift is illustrated in Figure 2.A. The fictitious Admiral Frost has been marooned on an iceberg and is uncomfortable being far above the surface of the water. He attempts to remove the ice that is above the water line. Were it not for isostatic (buoyant) uplift, he would have reached his goal to be close to the water line. Unfortunately for Admiral Frost, this is not the way the world works; continuous isostatic uplift of the block as ice is removed always keeps one-tenth of the iceberg above the water. So, after removing the ice above the water line, he still stands almost as much above the water line as before.[7]

- Mountains, of course, are not icebergs, but the rocks of which they are composed are less dense than the rocks of the mantle beneath. Thus, they tend to "float" on top of the denser mantle. Also, in mountains, erosion is not uniform but is generally confined to valley walls and bottoms. Thus, as erosion continues and the mass of the mountain range is reduced, isostatic compensation occurs and the entire mountain range rises in response. As a result of the erosion, the maximum elevation of mountain peaks actually may increase, while the mean elevation of the entire mountain block decreases. As a general rule, as the equivalent of 1 km (0.6 mi) of erosion across the entire mountain block occurs, the mean elevation of mountains will rise approximately five-sixths of a kilometer (one-half mile).

In summary, research concerning the origin of mountains suggests that they result in part from tectonic processes that cause the uplift, but they also are intimately related to climatic and erosional processes that contribute to the mountain building process. Erosion occurs during and after tectonic uplift, and isostatic compensation to that erosion occurs for millions of years. This is one reason it is difficult to remove mountain systems from the landscape. For example, mountain systems such as the Appalachian Mountains in the southeastern United States were originally produced by tectonic uplift several hundred million years ago when Europe collided with North America. There has been sufficient erosion of the original Appalachian Mountains to have removed them as topographic features many times over were it not for continued isostatic uplift in response to the erosion.

Transform boundaries, or transform faults, occur where the edges of two plates slide past one another, as shown in Figure 2.5. If you examine Figures 2.4a and 2.5, you will see that a spreading zone is not a single, continuous rift but a series of rifts that are offset from one another along connecting transform faults. Although the most common locations for transform plate boundaries are within oceanic crust, some occur within continents. A well-known continental transform boundary is the San Andreas Fault in California, where the rim of the Pacific plate is sliding horizontally past the rim of the North American plate (see Figures 2.4a and 2.10).

Locations where three plates border one another are known as **triple junctions.** Figure 2.10 shows several such junctions: Two examples are the meeting point of the Juan de Fuca, North American, and Pacific plates on the West Coast of North America (this is known as the Mendocino triple junction) and the junction of the spreading ridges associated with the Pacific, Cocos, and Nazca plates west of South America.

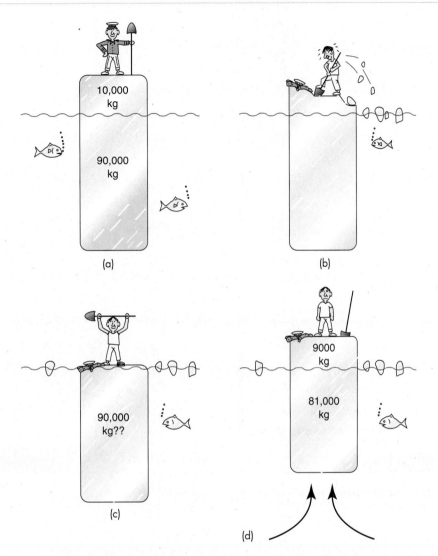

Figure 2.A Isostasy Idealized diagram or cartoon showing the principle of isostatic uplift. Admiral Frost is left adrift on an iceberg and is uncomfortable being so far above the surface of the water (a). He decides to remove the 10,000 kg (22,046 lb) of ice that is above the water line on the iceberg on which he is standing (b). Were it not for isostatic (buoyant) uplift, Admiral Frost would reach his goal (c). However, in a world with isostasy, uplift results from removal of the ice, and there is always one-tenth of the iceberg above the water (d). What would have happened if Admiral Frost had elected to remove 10,000 kg of ice from only one-half of the area of ice exposed above the sea? Answer: The maximum elevation of the iceberg above the water would have actually increased. Similarly, as mountains erode, isostatic adjustments also occur, and the maximum elevation of mountain peaks may actually increase as a result of the erosion alone! *(From Keller, E. A., and Pinter, N. 1996.* Active tectonics. *Upper Saddle River, NJ: Prentice Hall)*

Rates of Plate Motion

Plate Motion Is a Fast Geologic Process. The directions in which plates move are shown on Figure 2.4a. In general, plates move a few centimeters per year, about as fast as some people's fingernails or hair grows. The Pacific plate moves past the North American plate along the San Andreas Fault about 3.5 cm per year (1.4 in. per year), so that features such as rock units or streams are gradually displaced over time where they cross the fault (Figure 2.11). During the past 5 million years, there has been about 175 km (about 110 miles) of displacement, a distance equivalent to driving two hours at 55 mph on a highway along the San Andreas Fault. Although the central portions of the plates move along at a steady slow rate, plates interact at their boundaries, where collision or subduction or both occur, and movement may not be smooth or steady. The plates often get stuck together. Movement is analogous to sliding one rough wood board over another. Movement

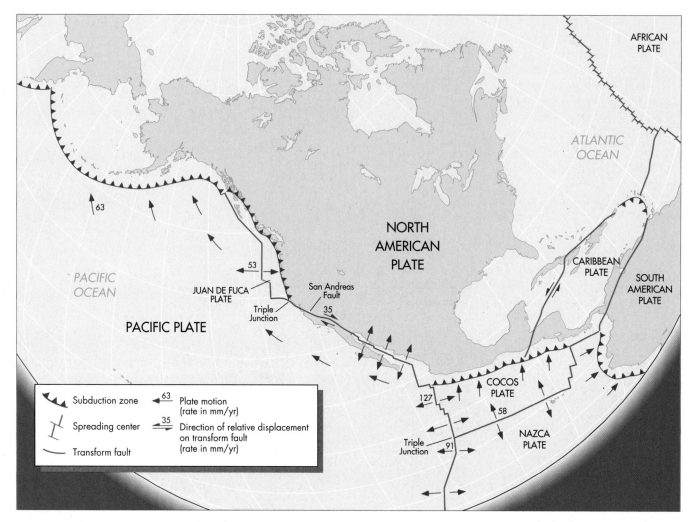

Figure 2.10 **North American plate boundary** Detail of boundary between the North American and Pacific plates. *(Courtesy of Tanya Atwater)*

Figure 2.11 **The San Andreas Fault** The fault is visible from the lower left to upper right diagonally across the photograph, as if a gigantic plow had been dragged across the landscape. *(James Balog/Getty Images Inc.)*

occurs when the splinters of the boards break off and the boards move quickly by one another. When rough edges along the plate move quickly, an earthquake is produced. Along the San Andreas Fault, which is a transform plate boundary, the displacement is horizontal and can amount to several meters during a great earthquake. During an earthquake in 1857 on the San Andreas Fault a horse corral across the fault was reportedly changed from a circle to an "S" shape. Fortunately, such an event generally occurs at any given location only once every 100 years or so. Over long time periods, rapid displacement from periodic earthquakes and more continuous slow "creeping" displacements add together to produce the rate of several centimeters of movement per year along the San Andreas Fault.

2.4 A Detailed Look at Seafloor Spreading

When Alfred Wegener proposed the idea of continental drift in 1915, he had no solid evidence of a mechanism that could move continents. The global extent of mid-oceanic ridges was discovered in the 1950s, and in 1962 geologist Harry H. Hess published a paper suggesting that continental drift was the result of the process of seafloor spreading along those ridges. The fundamentals of seafloor spreading are shown in Figure 2.5. New oceanic lithosphere is produced at the spreading ridge (divergent plate boundary). The lithospheric plate then moves laterally, carrying along the embedded continents in the tops of moving plates. These ideas produced a new major paradigm that greatly changed our ideas about how Earth works.[3,6,9]

The validity of seafloor spreading was established from three sources: (1) identification and mapping of oceanic ridges, (2) dating of volcanic rocks on the floor of the ocean, and (3) understanding and mapping of the paleomagnetic history of ocean basins.

Paleomagnetism

We introduce and discuss Earth's magnetic field and paleomagnetic history in some detail in order to understand how seafloor spreading and plate tectonics were discovered. Earth has had a magnetic field for at least the past 3 billion years[2] (Figure 2.12a). The field can be represented by a dipole magnetic field with lines of magnetic force extending from the South Pole to the North Pole. A dipole magnetic field is one that has equal and opposite charges at either end. Convection occurs in the iron-rich, fluid, hot outer core of Earth because of compositional

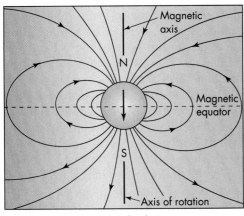
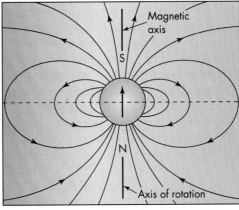

(a) Normal polarity   (b) Reversed polarity

Figure 2.12 **Magnetic reversal** Idealized diagram showing the magnetic field of Earth under (a) normal polarity and (b) reversed polarity. *(From Kennett, J. 1982. Marine geology. Englewood Cliffs, NJ: Prentice Hall)*

changes and heat at the inner-outer core boundary. As more buoyant material in the outer core rises, it starts the convection (Figures 2.3 and 2.7). The convection in the outer core, along with the rotation of Earth that causes rotation of the outer core, initiates a flow of electric current in the core. This flow of current within the core produces and sustains Earth's magnetic field.[2,3]

Earth's magnetic field is sufficient to permanently magnetize some surface rocks. For example, volcanic rock that erupts and cools at mid-oceanic ridges becomes magnetized at the time it passes through a critical temperature. At that critical temperature, known as the *Curie point*, iron-bearing minerals (such as magnetite) in the volcanic rock orient themselves parallel to the magnetic field. This is a permanent magnetization known as *thermoremnant magnetization*.[3] The term **paleomagnetism** refers to the study of the magnetism of rocks at the time their magnetic signature formed. It is used to determine the magnetic history of Earth.

The magnetic field, based on the size and conductivity of Earth's core, must be continuously generated or it would decay away in about 20,000 years. It would decay because the temperature of the core is too high to sustain permanent magnetization.[2]

Earth's Magnetic Field Periodically Reverses. Before the discovery of plate tectonics, geologists working on land had already discovered that some volcanic rocks were magnetized in a direction opposite to the present-day field, suggesting that the polarity of Earth's magnetic field was reversed at the time the volcanoes erupted and the rocks cooled (Figure 2.12b). The rocks were examined for whether their magnetic field was normal, as it is today, or reversed relative to that of today, for certain time intervals of the Earth's history. A chronology for the last few million years was constructed on the basis of the dating of the "reversed" rocks. You can verify the current magnetic field of the Earth by using a compass; at this point in Earth's history the needle points to the north magnetic pole. During a period of reversed polarity, the needle would point south! The cause of **magnetic reversals** is not well known, but it is related to changes in the convective movement of the liquid material in the outer core and processes occurring in the inner core. Reversals in Earth's magnetic field are random, occurring on average every few hundred thousand years. The change in polarity of Earth's magnetic field takes a few thousand years to occur, which in geologic terms is a very short time.

What Produces Magnetic Stripes? To further explore the Earth's magnetic field, geologists towed magnetometers, instruments that measure magnetic properties of rocks, from ships and completed magnetic surveys. The paleomagnetic record of the ocean floor is easy to read because of the fortuitous occurrence of the volcanic rock basalt (see Chapter 3) that is produced at spreading centers and forms the floors of the ocean basins of Earth. The rock is fine-grained and contains sufficient iron-bearing minerals to produce a good magnetic record. The marine geologists' discoveries were not expected. The rocks on the floor of the ocean were found to have irregularities in the magnetic field. These irregular magnetic patterns were called anomalies or perturbations of Earth's magnetic field caused by local fields of magnetized rocks on the seafloor. The anomalies can be represented as stripes on maps. When mapped, the stripes form quasi-linear patterns parallel to oceanic ridges. The marine geologists found that their sequences of stripe width patterns matched the sequences established by land geologists for polarity reversals in land volcanic rocks. Magnetic survey data for an area southwest of Iceland are shown on Figure 2.13. The black stripes represent normally magnetized rocks and the intervening white stripes represent reversed magnetized rocks.[10] Notice that the stripes are not evenly spaced but have patterns that are symmetrical on opposite sides of the Mid-Atlantic Ridge (Figure 2.13).

Why Is the Seafloor No Older than 200 Million Years? The discovery of patterns of magnetic stripes at various locations in ocean basins allowed geologists to infer

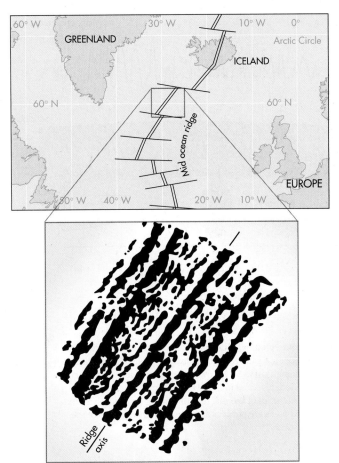

Figure 2.13 Magnetic anomalies on the seafloor Map showing a magnetic survey southwest of Iceland along the Mid-Atlantic Ridge. Positive magnetic anomalies are black (normal) and negative magnetic anomalies are white (reversed). Note that the pattern is symmetrical on the two sides of the mid-oceanic ridge. *(From Heirtzler, J. R., Le Pichon, X., and Baron, J. G. 1966. Magnetic anomalies over the Reykjanes Ridge. Deep-Sea Research 13:427–43)*

numerical dates for the volcanic rocks. Merging the magnetic anomalies with the numerical ages of the rocks produced the record of seafloor spreading. The spreading of the ocean floor, beginning at a mid-oceanic ridge, could explain the magnetic stripe patterns.[11] Figure 2.14 is an idealized diagram showing how seafloor spreading may produce the patterns of magnetic anomalies (stripes). The pattern shown is for the past several million years, which includes several periods of normal and reversed magnetization of the volcanic rocks. Black stripes represent normally magnetized rocks, and brown stripes are rocks with a reversed magnetic signature. Notice that the most recent magnetic reversal occurred approximately 0.7 million years ago. The basic idea illustrated by Figure 2.14 is that rising magma at the oceanic ridge is extruded, or pushed out onto the surface, through volcanic activity, and the cooling rocks become normally magnetized. When the field is reversed, the cooling rocks preserve a reverse magnetic signature, and a brown stripe (Figure 2.14) is preserved. Notice that the patterns of magnetic anomalies in rocks on both sides of the ridge are mirror images of one another. The only way such a pattern might result is through the process of seafloor spreading. Thus, the pattern of magnetic reversals found on rocks of the ocean floor is strong evidence that the process of spreading is happening. Mapping of magnetic anomalies, when combined with age-dating of the magnetic reversals in land rocks creates a database that suggests exciting inferences; Figure 2.15 shows the age of the ocean floor as determined from this database. The pattern, showing that the youngest volcanic rocks are found along active mid-oceanic ridges, is consistent with the theory of seafloor spreading. As distance from these ridges increases, the age of the ocean floor also increases, to a maximum of about 200 million years, during the early Jurassic period (see Table 1.1). Thus, it appears that the present ocean floors of the world are no older than

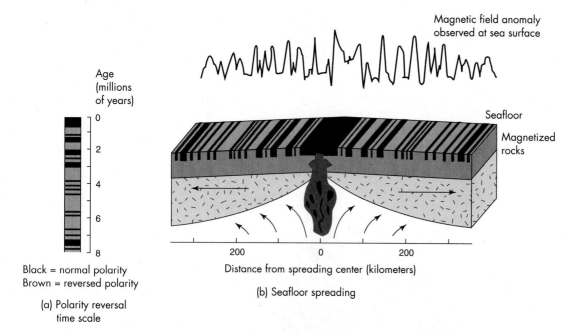

Figure 2.14 **Magnetic reversals and seafloor spreading** Idealized diagram showing an oceanic ridge and the rising of magma, in response to seafloor spreading. As the volcanic rocks cool, they become magnetized. The black stripes represent normal magnetization; the brown stripes are reversed magnetization. The record shown here was formed over a period of several million years. Magnetic anomalies (stripes) are a mirror image of each other on opposite sides of the mid-oceanic ridge. Thus, the symmetrical bands of the normally and reversely magnetized rocks are produced by the combined effects of the reversals and seafloor spreading. *(Courtesy of Tanya Atwater)*

200 million years. In contrast, rocks on continents are often much older than Jurassic, going back about 4 billion years, almost 20 times older than the ocean floors! We conclude that the thick continental crust, by virtue of its buoyancy, is more stable at Earth's surface than are rocks of the crust of the ocean basins. Continents form by the processes of accretion of sediments, addition of volcanic materials, and collisions of tectonic plates carrying continental landmasses. We will continue this discussion when we consider the movement of continents during the past 200 million years. However, it is important to recognize that it is the pattern of magnetic stripes that allows us to reconstruct how the plates and the continents embedded in them have moved throughout history.

Hot Spots

What Are Hot Spots? There are a number of places on Earth called **hot spots,** characterized by volcanic centers resulting from hot materials produced deep in the mantle, perhaps near the core-mantle boundary. The partly molten materials are hot and buoyant enough to move up through mantle and overlying moving tectonic plates.[3,6] An example of a continental hot spot is the volcanic region of Yellowstone National Park. Hot spots are also found in both the Atlantic and Pacific Oceans. If the hot spot is anchored in the slow-moving deep mantle, then, as the plate moves over a hot spot, a chain of volcanoes is produced. Perhaps the best example of this type of hot spot is the line of volcanoes forming the Hawaiian-Emperor Chain in the Pacific Ocean (Figure 2.16a). Along this chain, volcanic eruptions range in age from present-day activity on the big island of Hawaii (in the southeast) to more than 78 million years ago near the northern end of the Emperor Chain. With the exception of the Hawaiian Islands and some coral atolls (ringlike coral islands such as Midway Island), the chain consists of submarine volcanoes known as *seamounts*. Seamounts are islands that were eroded by waves and submarine landslides and subsequently sank beneath the ocean surface. As seamounts move farther off the hot spot, the volcanic rocks

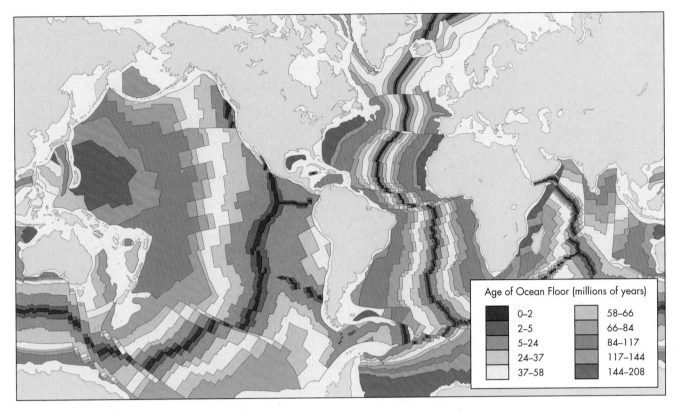

Figure 2.15 **Age of the ocean floor** Age of the seafloor is determined from magnetic anomalies and other methods. The youngest ocean floor (red) is located along oceanic ridge systems, and older rocks are generally farther away from the ridges. The oldest ocean floor rocks are approximately 180 million years old. *(From Scotese, C. R., Gahagan, L. M., and Larson, R. L. 1988. Plate tectonic reconstruction of the Cretaceous and Cenozoic ocean basins.* Tectonophysics 155:27–48)

the islands are composed of cool and the oceanic crust they are on becomes denser, and sinks.

Seamounts constitute impressive submarine volcanic mountains. In the Hawaiian Chain the youngest volcano is Mount Loihi, which is still a submarine volcano, presumably directly over a hot spot, as idealized on Figure 2.16b. The ages of the Hawaiian Islands increase to the northwest, with the oldest being Kauai, about 6 million years old. Notice in Figure 2.16a that the line of seamounts makes a sharp bend at the junction of the Hawaiian and Emperor Chains. The age of the volcanic rocks at the bend is about 43 million years, and the bend is interpreted to represent a time when plate motions changed.[12] If we assume that the hot spots are fixed deep in the mantle, then the chains of volcanic islands and submarine volcanoes along the floor of the Pacific Ocean that get older farther away from the hot spot provide additional evidence to support the movement of the Pacific plate. In other words, the ages of the volcanic islands and submarine volcanoes could systematically change as they do only if the plate is moving over the hot spot.

2.5 **Pangaea and Present Continents**

Plate Tectonics Shapes Continents and Dictates the Location of Mountain Ranges. Movement of the lithospheric plates is responsible for the present shapes and locations of the continents. There is good evidence that the most recent global episode of continental drift, driven by seafloor spreading, started about 180 million years ago, with the breakup of a supercontinent called Pangaea (this name, meaning "all lands," was first proposed by Wegener). Pangaea (pronounced pan-jee-ah) was enormous, extending from pole to pole and over halfway around Earth near the equator (Figure 2.17). Pangaea had two parts

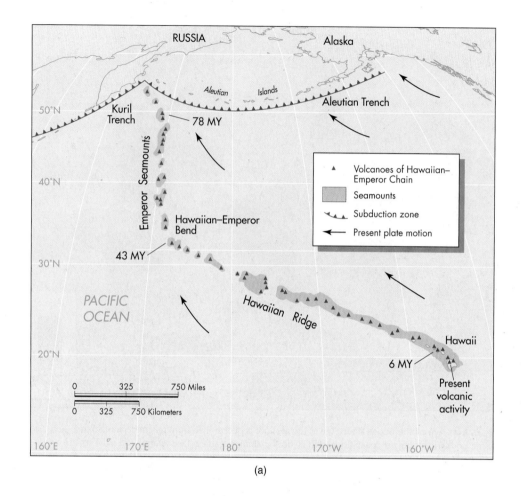

(a)

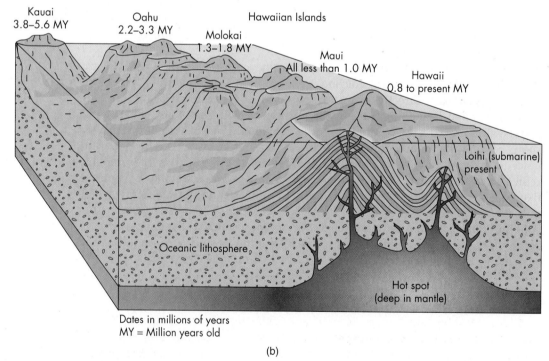

Dates in millions of years
MY = Million years old

(b)

Figure 2.16 Hawaiian hot spot (a) Map showing the Hawaiian-Emperor Chain of volcanic islands and seamounts. Actually, the only islands are Midway Island and the Hawaiian Islands at the end of the chain, where present volcanic activity is occurring. (*Modified after Claque, D. A., Dalrymple, G. B., and Moberly, R. 1975. Petrography and K-Ar ages of dredged volcanic rocks from the western Hawaiian Ridge and southern Emperor Seamount chain. Geological Society of America Bulletin 86:991–98*) (b) Sketch map showing the Hawaiian Islands, which range in age from present volcanic activity to about 6 million years old on the island of Kauai. (*From Thurman, Oceanography, 5th ed. Columbus, OH: Merrill, plate 2*)

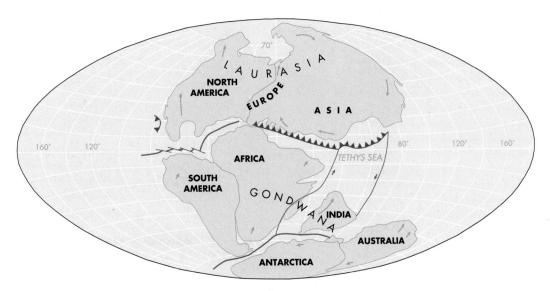

(a) 180 million years ago

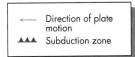
←——— Direction of plate motion
▲▲▲ Subduction zone

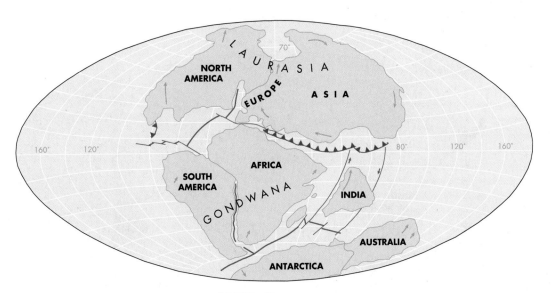

(b) 135 million years ago

Figure 2.17 **Two hundred million years of plate tectonics** (a) The proposed positions of the continents at 180 million years ago; (b) 135 million years ago; (c) 65 million years ago; and (d) at present. Arrows show directions of plate motion. See text for further explanation of the closing of the Tethys Sea, the collision of India with China, and the formation of mountain ranges. *(From Dietz, R. S., and Holden, J. C. 1970. Reconstruction of Pangaea: breakup and dispersion of continents, Permian to present. Journal of Geophysical Research 75(26):4939–56. Copyright by the American Geophysical Union. Modifications and block diagrams from Christopherson, R. W. 1994. Geosystems, 2nd ed. Englewood Cliffs, NJ: Macmillan)*

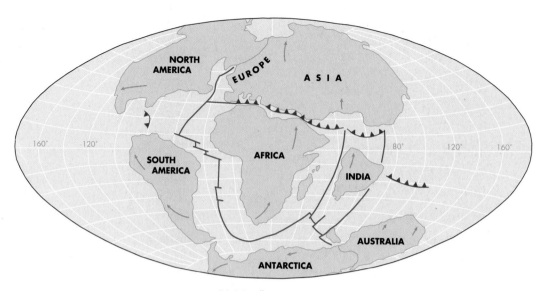

(c) 65 million years ago

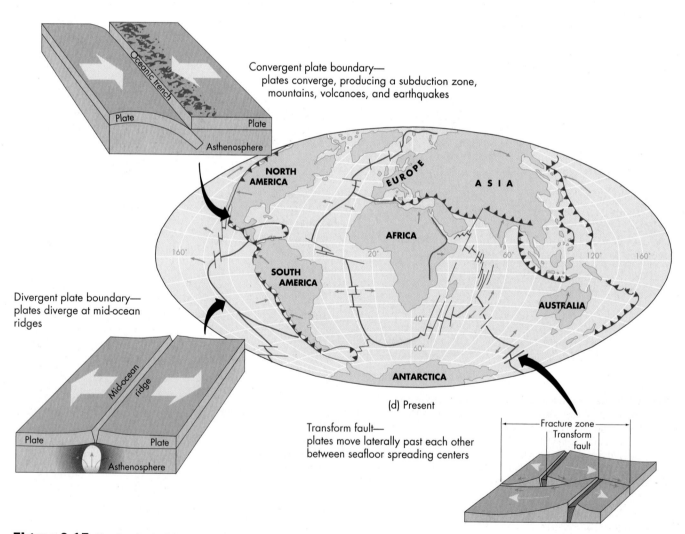

Convergent plate boundary—
plates converge, producing a subduction zone,
mountains, volcanoes, and earthquakes

Divergent plate boundary—
plates diverge at mid-ocean
ridges

(d) Present

Transform fault—
plates move laterally past each other
between seafloor spreading centers

Figure 2.17 Two hundred million years of plate tectonics (*Continued*)

(Laurasia to the North and Gondwana to the South) and was constructed during earlier continental collisions. Figure 2.17a shows Pangaea as it was nearly 200 million years ago. Seafloor spreading over the past 200 million years separated Eurasia and North America from the southern land mass; Eurasia from North America; and the southern continents (South America, Africa, India, Antarctica, and Australia) from one another (Figure 2.17b–d). The Tethys Sea, between Africa and Europe-Asia (Figure 2.17a–c), closed, as part of the activity that produced the Alps in Europe. A small part of this once much larger sea remains today as the Mediterranean Sea (Figure 2.17d). About 50 million years ago India crashed into China. That collision, which has caused India to forcefully intrude into China a distance comparable from New York to Miami, is still happening today, producing the Himalayan Mountains (the highest mountains in the world) and the Tibetan Plateau.

Understanding Plate Tectonics Solves Long-Standing Geologic Problems. Reconstruction of what the supercontinent Pangaea looked like before the most recent episode of continental drift has cleared up two interesting geologic problems:

- Occurrence of the same fossil plants and animals on different continents that would be difficult to explain if they had not been joined in the past (see Figure 2.18).

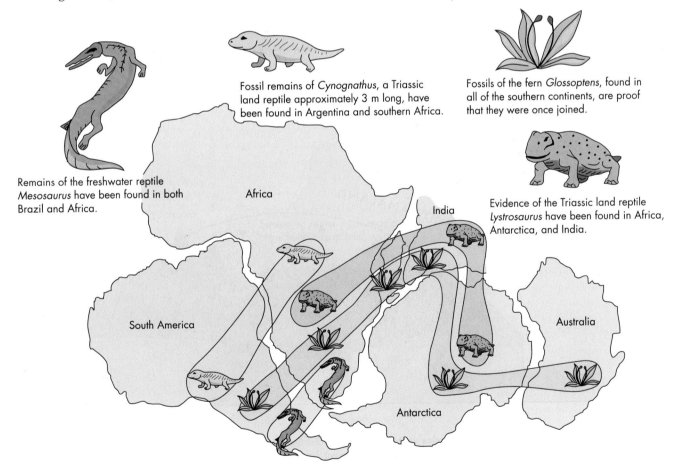

Figure 2.18 **Paleontological evidence for plate tectonics** This map shows some of the paleontological (fossil) evidence that supports continental drift. It is believed that these animals and plants could not have been found on all of these continents were they not once much closer together than they are today. Major ocean basins would have been physical barriers to their distribution. *(From Hamblin, W. K. 1992. Earth's dynamic systems, 6th ed. New York: Macmillan)*

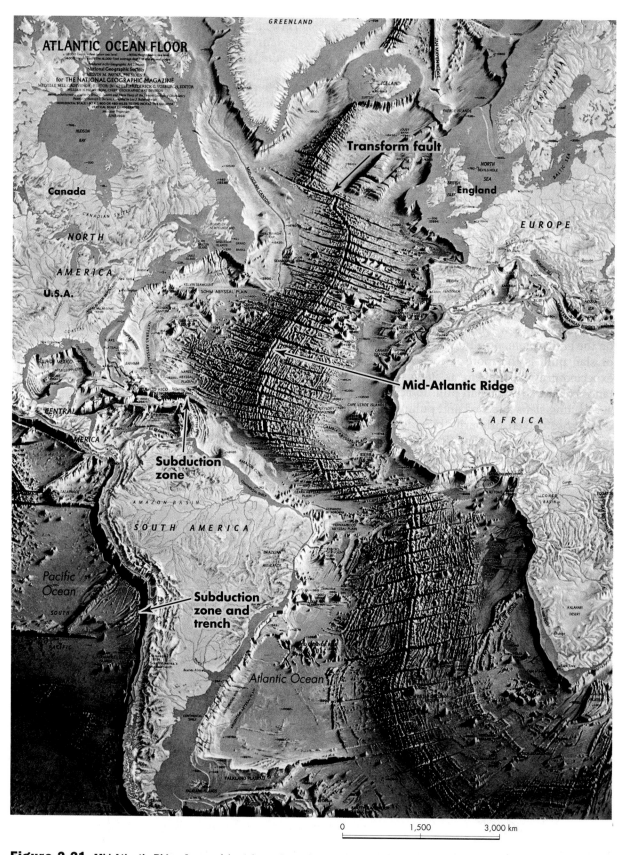

Figure 2.21 **Mid Atlantic Ridge** Image of the Atlantic Ocean basin showing details of the seafloor. Notice that the width of the Mid-Atlantic Ridge is about one-half the width of the ocean basin. *(Heinrich C. Berann/NGS Image Collection)*

(Laurasia to the North and Gondwana to the South) and was constructed during earlier continental collisions. Figure 2.17a shows Pangaea as it was nearly 200 million years ago. Seafloor spreading over the past 200 million years separated Eurasia and North America from the southern land mass; Eurasia from North America; and the southern continents (South America, Africa, India, Antarctica, and Australia) from one another (Figure 2.17b–d). The Tethys Sea, between Africa and Europe-Asia (Figure 2.17a–c), closed, as part of the activity that produced the Alps in Europe. A small part of this once much larger sea remains today as the Mediterranean Sea (Figure 2.17d). About 50 million years ago India crashed into China. That collision, which has caused India to forcefully intrude into China a distance comparable from New York to Miami, is still happening today, producing the Himalayan Mountains (the highest mountains in the world) and the Tibetan Plateau.

Understanding Plate Tectonics Solves Long-Standing Geologic Problems. Reconstruction of what the supercontinent Pangaea looked like before the most recent episode of continental drift has cleared up two interesting geologic problems:

- Occurrence of the same fossil plants and animals on different continents that would be difficult to explain if they had not been joined in the past (see Figure 2.18).

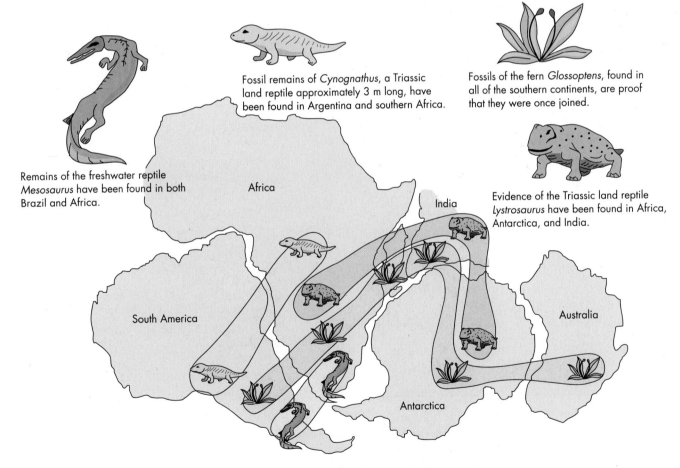

Fossil remains of *Cynognathus*, a Triassic land reptile approximately 3 m long, have been found in Argentina and southern Africa.

Fossils of the fern *Glossoptens*, found in all of the southern continents, are proof that they were once joined.

Remains of the freshwater reptile *Mesosaurus* have been found in both Brazil and Africa.

Evidence of the Triassic land reptile *Lystrosaurus* have been found in Africa, Antarctica, and India.

Africa

India

South America

Australia

Antarctica

Figure 2.18 Paleontological evidence for plate tectonics This map shows some of the paleontological (fossil) evidence that supports continental drift. It is believed that these animals and plants could not have been found on all of these continents were they not once much closer together than they are today. Major ocean basins would have been physical barriers to their distribution. *(From Hamblin, W. K. 1992. Earth's dynamic systems, 6th ed. New York: Macmillan)*

■ Evidence of ancient glaciation on several continents, with inferred directions of ice flow, that makes sense only if the continents are placed back within Gondwanaland (southern Pangaea) as it was before splitting apart (see Figure 2.19).

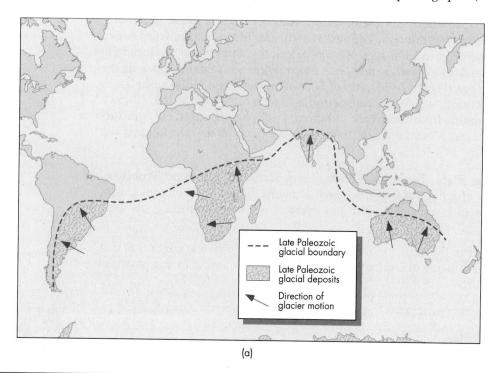

(a)

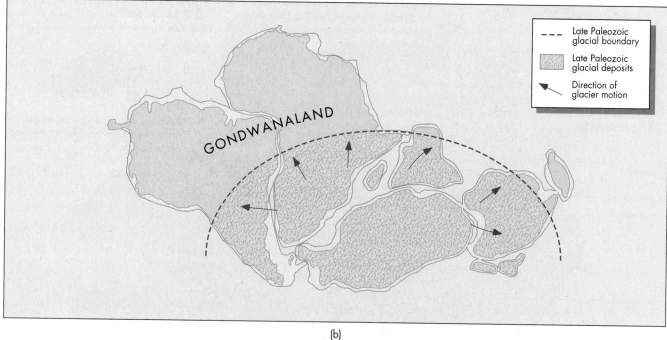

(b)

Figure 2.19 **Glacial evidence for plate tectonics** (a) Map showing the distribution of evidence for late Paleozoic glaciations. The arrows indicate the direction of ice movement. Notice that the arrows are all pointing away from ocean sources. Also these areas are close to the tropics today, where glaciation would have been very unlikely in the past. These Paleozoic glacial deposits were formed when Pangaea was a supercontinent, before fragmentation by continental drift. (b) The continents are restored (it is thought that continents drifted north away from the South Pole). Notice that the arrows now point outward as if moving away from a central area where glacial ice was accumulating. Thus, restoring the position of the continents produces a pattern of glacial deposits that makes much more sense. *(Modified after Hamblin, W. K. 1992. Earth's dynamic systems, 6th ed. New York: Macmillan)*

2.6 How Plate Tectonics Works: Putting It Together

Driving Mechanisms That Move Plates. Now that we have presented the concept that new oceanic lithosphere is produced at mid-oceanic ridges because of seafloor spreading and that old, cooler plates sink into the mantle at subduction zones, let us evaluate the forces that cause the lithospheric plates to actually move and subduct. Figure 2.20 is an idealized diagram illustrating the two most likely driving forces, ridge push and slab pull.

The mid-oceanic ridges or spreading centers stand at elevations of 1 to 3 km (3,000 to 9,000 ft) above the ocean floor as linear, gently arched uplifts (submarine mountain ranges; see Figure 2.21) with widths greater than the distance from Florida to Canada. The total length of mid-oceanic ridges on Earth is about twice the circumference of Earth. Ridge push is a gravitational push, like a gigantic landslide, away from the ridge crest toward the subduction zone (the lithosphere slides on the asthenosphere). Slab pull results because as the lithospheric plate moves farther from the ridge, it cools, gradually becoming denser than the asthenosphere beneath it. At a subduction zone, the plate sinks through lighter, hotter mantle below the lithosphere, and the weight of this descending plate pulls on the entire plate, resulting in slab pull. Which of the two processes, ridge push or slab pull, is the more influential of the driving forces? Calculations of the expected gravitational effects suggest that ridge push is of relatively low importance compared with slab pull. In addition, it is observed that plates with large subducting slabs attached and pulling on them tend to move much more rapidly than those driven primarily by ridge push alone (for example, the subduction zones surrounding the Pacific Basin). Thus, slab pull may be more influential in moving plates than ridge push.

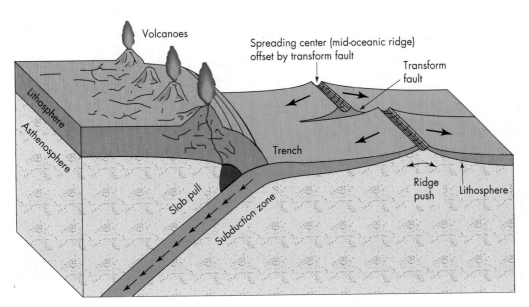

Figure 2.20 Push and pull in moving plates Idealized diagram showing concepts of ridge push and slab pull that facilitate the movement of lithospheric plates from spreading ridges to subduction zones. Both are gravity driven. The heavy lithosphere falls down the mid-oceanic ridge slope and subducts down through the lighter, hotter mantle. *(Modified after Cox, A., and Hart, R. B. 1986. Plate tectonics. Boston: Blackwell Scientific Publications)*

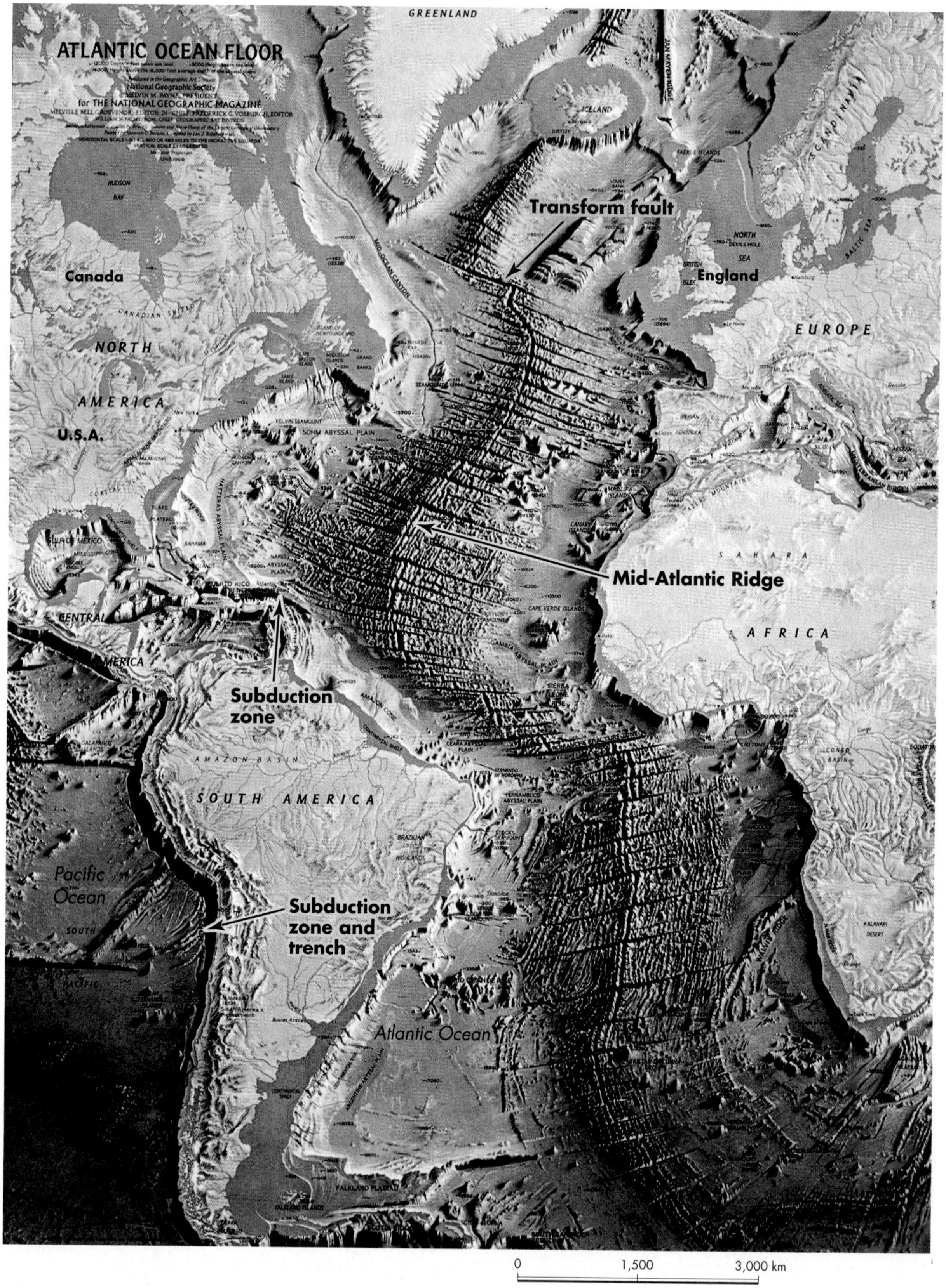

Figure 2.21 **Mid Atlantic Ridge** Image of the Atlantic Ocean basin showing details of the seafloor. Notice that the width of the Mid-Atlantic Ridge is about one-half the width of the ocean basin. *(Heinrich C. Berann/NGS Image Collection)*

2.7 **Plate Tectonics and Environmental Geology**

Plate Tectonics Affects Us All. The importance of the tectonic cycle to environmental geology cannot be overstated. Everything living on Earth is affected by plate tectonics. As the plates slowly move a few centimeters each year, so do the continents and ocean basins, producing zones of resources (oil, gas, and minerals), as well as earthquakes and volcanoes (Figure 2.4b). The tectonic processes occurring at plate boundaries largely determine the types and properties of the rocks upon which we depend for our land, our mineral and rock resources, and the soils on which our food is grown. For example, large urban areas, including New York and Los Angeles, are developed on very different landscapes, but both have favorable conditions for urban development. New York (Figure 2.22a) is sited on the "trailing edge" of the North American plate, and the properties of the coastline are directly related to the lack of collisions between plates in the area. The divergent plate boundary at the Mid-Atlantic Ridge between North America and Africa is several thousand kilometers (over 1,500 miles) to the east. The collision boundaries between the North American and Caribbean plates and between the North American and Pacific plates are several thousands of kilometers (over 1,500 miles) to the south and west, respectively (see Figure 2.4a). The passive processes of sedimentation from rivers, glaciers, and coastal processes, depositing sediments on rifted and thinned continental crust, instead of the more active crustal deformation that produces mountains, have shaped the coastline of the eastern United States north of Florida. The breakup of Pangaea about 200 million years ago (Figure 2.17) produced the Atlantic Ocean, which, with a variety of geologic processes, including erosion, deposition, and glaciation over millions

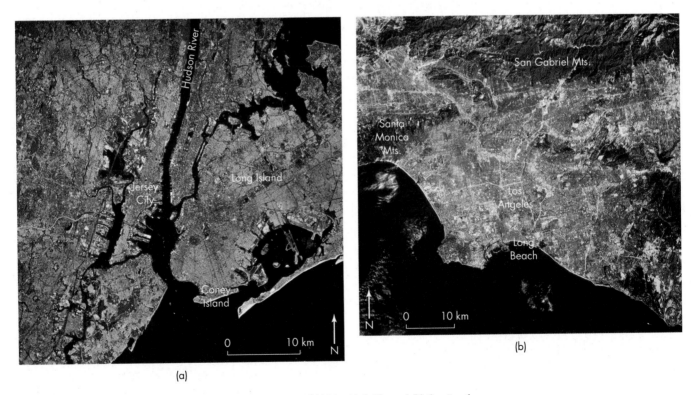

(a)

(b)

Figure 2.22 Los Angeles and New York Satellite images of (a) New York City and (b) the city of Los Angeles. Both are coastal cities; however, Los Angeles is surrounded by mountains, whereas New York is sited in a relatively low-relief area characteristic of much of the Atlantic coastal environment. For these images, healthy vegetation is red, urban development is blue, beaches are off-white, and water is black. *(Science Source/Photo Researchers, Inc.)*

of years, eventually led to the development of the beautiful but subdued topography of the coastal New York area. In contrast, the Los Angeles metropolitan area is near the "leading edge" of the boundary between the North American and Pacific plates (Figure 2.10), characterized by active, vigorous crustal deformation (uplift; subsidence, or sinking of the ground's surface; and faulting) associated with the San Andreas fault, a transform boundary. The deformation has produced the Los Angeles Basin, rimmed by rugged mountains and uplifted coastline (Figure 2.22b).

Plate motion over millions of years can change or modify flow patterns in the oceans and the atmosphere, influencing or changing global climate as well as regional variation in precipitation. These changes affect the productivity of the land and its desirability as a place to live. Plate tectonics also determines, in part, what types of minerals and rocks are found in a particular region. We will explore how rocks and minerals are influenced by plate tectonics in Chapter 3.

SUMMARY

Our knowledge concerning the structure of Earth's interior is based on the study of seismology. Thus we are able to define the major layers of Earth, including the inner core, outer core, mantle, and crust. The uppermost layer of Earth is known as the lithosphere, which is relatively strong and rigid compared with the soft asthenosphere found below it. The lithosphere is broken into large pieces called plates that move relative to one another. As these plates move, they carry along the continents embedded within them. This process of plate tectonics produces large landforms, including continents, ocean basins, mountain ranges, and large plateaus. Oceanic basins are formed by the process of seafloor spreading and are destroyed by the process of subduction, both of which result from convection within the mantle.

The three types of plate boundaries are divergent (mid-oceanic ridges, spreading centers), convergent (subduction zones and continental collisions), and transform faults. At some locations, three plates meet in areas known as triple junctions. Rates of plate movement are generally a few centimeters per year.

Evidence supporting seafloor spreading includes paleomagnetic data, the configurations of hot spots and chains of volcanoes, and reconstructions of past continental positions.

The driving forces in plate tectonics are ridge push and slab pull. At present we believe the process of slab pull is more significant than ridge push for moving tectonic plates from spreading centers to subduction zones.

Plate tectonics is very important in environmental geology because everything living on Earth is affected by it.

Key Terms

asthenosphere (p. 41)

continental drift (p. 42)

convection (p. 41)

convergent boundary (p. 46)

core (p. 39)

crust (p. 40)

divergent boundary (p. 46)

hot spot (p. 54)

isostasy (p. 48)

lithosphere (p. 41)

magnetic reversal (p. 52)

mantle (p. 39)

mid-oceanic ridge (p. 42)

Moho (p. 40)

paleomagnetism (p. 52)

plate tectonics (p. 42)

seafloor spreading (p. 42)

seismology (p. 41)

spreading center (p. 42)

subduction zone (p. 44)

submarine trench (p. 46)

transform boundary (p. 48)

triple junction (p. 48)

Wadati-Benioff zone (p. 44)

Review Questions

1. What are the major differences between the inner and outer cores of Earth?

2. How are the major properties of the lithosphere different from those of the asthenosphere?

3. What are the three major types of plate boundaries?

4. What is the major process that is thought to produce Earth's magnetic field?

5. Why has the study of paleomagnetism and magnetic reversals been important in understanding plate tectonics?

6. What are hot spots?

7. What is the difference between ridge push and slab pull in the explanation of plate motion?

Critical Thinking Question

1. Assume that the supercontinent Pangaea (Figure 2.17) never broke up. Now deduce how Earth processes, landforms, and environments might be different than they are today with the continents spread all over the globe.

Hint: Think about what the breakup of the continents did in terms of building mountain ranges and producing ocean basins that affect climate and so forth.

THREE

Minerals and Rocks

Written with the assistance of William Wise

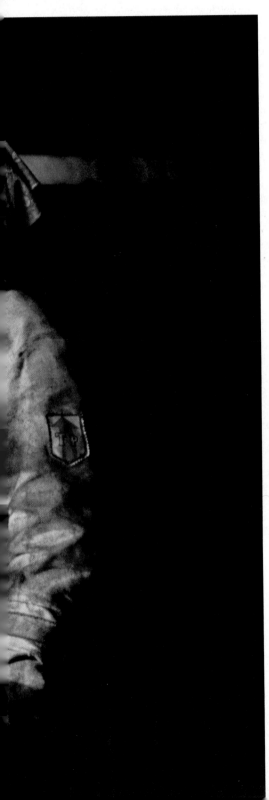

Learning Objectives

Minerals and rocks are part of our inheritance from the geologic past. They are the basic building blocks of the solid Earth and are particularly important in environmental geology for the following reasons: (1) Minerals and rocks are among the basic resources we depend upon for building our homes, driving our automobiles, and manufacturing our computers, among many other things; (2) minerals and rocks play an important role in many Earth processes, such as landsliding, coastal erosion, and volcanic activity; (3) study of minerals and rocks provides important information concerning the history of Earth; and (4) knowledge of mineral and rock processes and properties is an integral part of understanding how Earth works and how to best manage Earth's resources. Mineral resources are discussed in detail in Chapter 14.

In this chapter we focus on the following learning objectives:

- Understand minerals in terms of their chemistry and internal structure
- Know the major groups of important rock-forming minerals and their environmental significance
- Understand the rock cycle and how it interacts with plate tectonics
- Know the three rock laws
- Know the basic rock types and their environmental significance
- Know basic rock structures

Workers in this factory using asbestos were not aware of the potential health hazards to which they were exposed. *(David & Peter Turnley/CORBIS)*

CASE HISTORY | The Asbestos Controversy

For many people, the word *asbestos* is a red flag that signals a hazard to human health because asbestos may cause fatal lung disease. In response to the perceived hazard, people in the United States have spent large sums of money to remove asbestos from old buildings, such as schools and other public buildings, where it was used in ceiling and floor tiles and for insulation (Figure 3.1). The asbestos controversy is concerned with when and where asbestos should be removed, but first let us define asbestos and discuss the hazards it causes.

Asbestos is broadly defined as small, elongated mineral fibers that are present in certain silicate minerals and rocks. Silicate minerals and rocks are those that contain silicon (Si) and oxygen (O) in their chemical composition. Asbestos has proved to be a useful material, particularly for its fire-retardant properties. It is also used in brake linings and in a variety of insulations. However, human exposure to asbestos has caused lung disease, including cancer. Realization of the health hazard led to efforts to reduce and remove asbestos or to ban it outright.

The mineral quartz, which is not asbestos but may be present in crushed rock as small mineral fragments or grains, is also considered a probable carcinogenic material to humans. As a result, any natural material in the United States containing one-tenth of 1 percent of free silica, including quartz, must display hazardous warning signs. Theoretically, trucks that transport crushed rocks need to carry warning signs; a truck driver transporting crushed stones in Delaware was issued a citation for not displaying such signs on his truck! A local Delaware paper reported facetiously that beaches composed of silica sand might also present a public health hazard, so warnings should be posted on the beaches. The author of the article went on to state that much misunderstanding has resulted from well-meaning efforts to extrapolate data from environmental toxicity studies without understanding common natural characteristics of minerals.[1,2] The author's point was that, although sand is composed of quartz, it is in the form of hard, rounded particles, not potentially hazardous fibrous dust.

The different types of asbestos are not equally hazardous. In fact, exposure to "white asbestos," or chrysotile, the variety most commonly used in the United States, is evidently not very harmful. On the other hand, "blue asbestos," the mineral crocidolite, is known to cause lung disease.[2]

A good deal of fear is associated with nonoccupational exposure to asbestos, particularly exposure of children. Identifying potentially hazardous blue asbestos in schools, public buildings, and homes and then removing or covering it to avoid human exposure is an important goal. Funds spent to remove hazardous blue asbestos constitute a good investment in environmental health. Large amounts of money have been spent to remove white asbestos, which is not very hazardous and may not always require removal. Deciding whether to remove asbestos requires the careful study of fibrous minerals and rocks in the context in which they are used; only then can we understand their potential toxicity to people and other living things. In the larger context, the asbestos controversy emphasizes the need to know more about the minerals we use in modern society and how they are linked to the environment.

(a)

(b)

Figure 3.1 Asbestos (a) Woman wearing protective clothing while inspecting asbestos. *(Rene Sheret/Getty Images Inc.)* (b) Sample of white asbestos, showing the fluffy, fibrous nature of this commercial-grade asbestos. The scale along the bottom is in millimeters. *(From Skinner, H. C. W., Ross, M., and Frondel, C. 1988. Asbestos and other fibrous materials: Mineralogy, crystal chemistry, and health effects. New York: Oxford University Press)*

3.1 Minerals

What Is a Mineral? A mineral is a naturally occurring, solid Earth material that has formed by geologic processes. A more precise definition will follow after we introduce some basic chemistry necessary for deeper understanding of minerals. For now, think of minerals as the building blocks of rocks that form the solid framework of Earth's crust.

Atoms and Elements

The study of atoms and elements of importance to geologic processes is a part of *geochemistry*, the study of the chemistry of Earth. Geochemistry includes the study of the natural distribution of chemical elements in minerals, rocks, soils, water, and the atmosphere, along with the transfer of atoms through the environment. Let us start with some definitions.

All matter, including rocks, minerals, water, and you, is composed of *atoms*. An atom is the smallest part of a chemical element that can take part in a chemical reaction or combine with another *atom*. An element is a chemical substance composed of identical atoms that may not be separated into different substances by ordinary chemical means. An atom of a particular element is denoted by its atomic symbol, often the abbreviation of the first letter or first and second letters of the English or Latin name of that element.[3] For example, the atomic symbol for an atom of carbon is C, and oxygen's atomic symbol is O. The Periodic Table (Table 3.1) includes a list of known elements.

Conceptual Model of an Atom

Idealized diagrams of an atom are shown in Figure 3.2. These models show the core, or nucleus, of an atom and three subatomic particles: *protons*, *neutrons*, and *electrons*. The nucleus of the atom is composed of protons, which carry a positive charge, and neutrons, which carry no charge. Electrons carry a negative charge and are found outside of the nucleus. The number of protons in the nucleus of an atom for a particular element is unique and defines its *atomic number* as listed in Table 3.1. For example, hydrogen has 1 proton in its nucleus, oxygen has 8 protons, silicon has 14, gold has 79, uranium has 92, and so on. Also shown in Table 3.1 are those elements that are relatively abundant in the minerals and rocks of Earth's crust, for example silicon, oxygen, aluminum, and iron; those of major importance to life, including

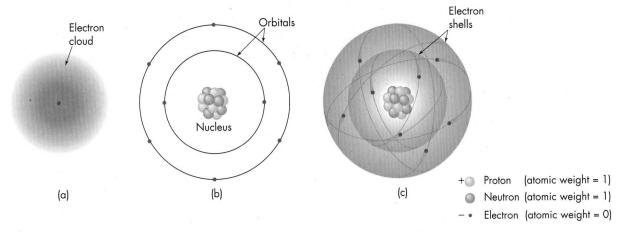

Figure 3.2 The atom (a) Idealized diagram showing the basic structure of an atom as a nucleus surrounded by an electron cloud. (b) This conceptualized view is only an approximation or model showing the nucleus of the atom surrounded by orbiting electrons. (c) A more realistic view of an atom consists of electron shells surrounding the nucleus (shown here is oxygen; the size of the nucleus relative to electron shells has been greatly exaggerated in order to be viewed). *(After Press, F., and Siever, R. 1994. Understanding Earth. New York: W. H. Freeman)*

TABLE 3.1   Periodic Table of the Elements Showing Elements That Are Relatively Abundant in Earth's Crust and Some Environmentally Important Trace Elements

carbon, oxygen, and nitrogen; and those that have environmental importance in very small amounts, referred to as *trace elements*, such as cobalt, zinc, lead and uranium.

An atom may be conceptualized as a nucleus surrounded by an electron cloud (Figure 3.2a). Electrons can also be depicted as revolving in orbits around the nucleus, much as Earth and our neighboring planets revolve around the Sun (Figure 3.2b). Although this is a satisfactory model for conceptualizing what an atom looks like, it is too simple and inaccurate from a mathematical standpoint.[3] Electrons are actually arranged in energy levels, or shells, around the nucleus (Figure 3.2c). Because the electrons are negatively charged, those closest to the positively charged nucleus are held more tightly than those in the outer shells.

Although electrons carry a negative charge, they have a very small mass compared with protons and neutrons. As a result, almost the entire mass of an atom is concentrated in the nucleus. The sum of the number of neutrons and protons in the nucleus of an atom is known as the *atomic mass number*. Because neutrons carry no charge, an atom is electrically balanced when the number of protons is equal to the number of electrons. However, atoms may gain or lose loosely bound electrons that are in the outer shells. An atom that has lost or gained electrons is known as an ion. If an atom loses electrons, it becomes positively charged and is known as a *cation*. For example, potassium, atomic symbol K, becomes K^+ after losing one electron. If an atom gains electrons, it becomes negatively charged and is known as an *anion*. Oxygen, atomic symbol O, becomes O^{2-} after gaining two electrons.

Isotopes

What Is an Isotope? Atoms of the same element always have the same atomic number. However, they may have a variable number of neutrons and therefore a variable atomic mass number. Two atoms of the same element with a different number of neutrons in the nucleus are known as *isotopes*. Two isotopes of the element carbon, which is the building block of the living world, are $^{12}_6C$ and $^{13}_6C$. Atoms of both of these isotopes have an atomic number of 6, but their atomic mass numbers are 12 and 13, respectively. These isotopes are sometimes written as C-12 and C-13. Of the two, C-12 is the common isotope.

Why Are Isotopes Important in Studying the Environment? It is a triumph of science that so much has been learned from studying isotopes. For example, some isotopes are called radioisotopes because they are unstable and undergo nuclear decay (spontaneously change and emit nuclear radiation). These are discussed in Chapter 15 with nuclear energy. Other elements have stable isotopes that do not undergo nuclear decay. For example, oxygen has two stable isotopes, $^{16}_8O$ and $^{18}_8O$. Study of the abundance of these two oxygen isotopes in sea water, glacial ice (frozen water), and very small marine organisms known as forminifera has provided exciting evidence about global climate change and how much glacial ice has been present at different times in Earth history (Figure 3.3). $^{16}_8O$ is lighter than $^{18}_8O$ so is preferentially evaporated with water (H_2O) from the ocean. During times when lots

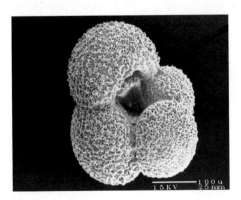

Figure 3.3 **Forminifera** A type of planktonic animal useful for extracting oxygen isotope information. Specimen taken from ocean floor sediments. Distance across is about 0.1 mm. *(Courtesy of Tessa Hill)*

of glacial ice is on the land, the ocean is relatively depleted in $^{16}_8O$, and forminifera then have less $^{16}_8O$ in the mineral material of their shells. Forminifera occur as fossils with sediment that accumulates on the ocean floor. When fossil forminifera are collected, dated, and the relative amounts of the two oxygen isotopes determined, we can infer how much ice was on land, and if the climate was warm or cold.

Mineral Chemistry: Compounds

A **mineral** is formally defined as an element or chemical compound that must[4]

1. Be naturally formed. This requirement excludes human-made substances such as human-made diamonds.
2. Normally be a solid. This requirement excludes most fluids and all gases.
3. Have a characteristic chemical formula.
4. Have a characteristic crystalline structure, in most cases. Although some geologists require that the substance have a crystalline structure, the definition of a mineral was revised in the mid-1990s to include some noncrystalline solids and even some liquids such as mercury.

Minerals can be either elements or compounds. We discussed elements earlier. A compound is a substance composed of two or more elements that can be represented by a chemical formula, such as PbS, or lead sulfide. Naturally occurring PbS is the mineral *galena* that we mine to obtain lead. Thus, minerals may be composed of either a single element, such as the mineral diamond, which is composed of carbon (C), or several elements in a compound such as the mineral galena, lead sulfide (PbS).

Minerals and Chemical Bonding

The atoms that constitute a mineral are held together by *chemical bonding*. Bonding results from attractive forces between atoms, sharing of electrons, or both. Types of bonds determine some of the primary physical properties of minerals; they explain in part, for example, why diamonds and graphite have the same chemical formula of pure carbon (C) but are so different.

There are four main types of chemical bonds in minerals: covalent, ionic, Van der Waals, and metallic. *Covalent bonds* form when atoms share electrons. For example, diamonds, one of the hardest substances on Earth, are composed of covalently bonded carbon atoms (Figure 3.4). Gem-quality diamonds are clear, hard crystals that jewelers may cut to form beautiful gemstones of high value. Covalent bonds are stronger than *ionic bonds*, which form because of an attraction between negatively

(a)

(b)

Figure 3.4 **Diamond** (a) Diamond crystal. *(J & L Weber/Peter Arnold, Inc.)* (b) Idealized diagram of internal structure of diamond, with balls representing carbon atoms joined by rods representing strong covalent bonds. *(Charles D. Winters/Photo Researchers, Inc.)*

and positively charged ions. An example of an ionic bond is the attraction of Na^+ and Cl^- ions, forming sodium chloride, the mineral halite. Compounds with ionic bonds are much more soluble and easily dissolved in water than are those with co-valent bonds. Therefore, minerals with ionic bonds tend to be chemically active and mobile in the environment. *Van der Waals bonds* involve a weak attraction between chains or sheets of ions that themselves are likely to be bonded by stronger covalent and ionic bonds. As previously mentioned, the mineral graphite, like diamond, is composed of pure carbon. However, the similarity ends there. Graphite is black and consists of soft sheets of carbon atoms that easily *part*, or break from one another (a mineral property called *cleavage*, see Appendix A). Graphite is the "lead" used in pencils. It is also a good dry lubricant, sprayed as a dust into door locks to help the parts of the lock move freely. *Metallic bonds* form between metal atoms. Gold con-tains metallic bonds. Gold's properties, such as its ability to conduct an electrical current, and malleability, the ability to form paper-thin sheets, are both due to metal-lic bonds. In a metallic bond, electrons are shared by all atoms of the solid rather than by only specific atoms, as in covalent bonds. The electrons can flow, making gold both an excellent conductor of electricity and easy to pound into a thin sheet.[5] As a result, gold is in high demand for electronic processing.

When learning about chemical bonding of minerals, keep in mind that you are dealing with a complex subject: Bonding in minerals may not be strictly of a par-ticular type but may have properties of the various types. Thus, for a particular mineral, more than one type of bond is often present.[5]

Crystalline Structure of Minerals

Now that we know that chemical bonds hold the atoms of minerals together, let's consider the term *crystalline*. *Crystalline* refers to the orderly regular repeating geometric patterns of atoms found in most minerals. The smallest unit of this geo-metric pattern in a crystal is called the *unit cell*. A crystal is composed of stacking unit cells.[3] The internal structure for a given mineral typically contains a particu-lar symmetry that determines the external form of the crystal. Figure 3.5 shows some of the common shapes of crystals for selected minerals.

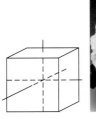
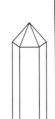

(a) Cube
   4-sided crystal
   mineral examples:
   halite, pyrite

(b) Hexagon
   6-sided crystal
   mineral example:
   quartz

(c) Octahedron
   8-sided crystal
   mineral examples:
   diamond, fluorite

(d) Dodecahedron
   12-sided crystal
   mineral example:
   garnet

Figure 3.5 **Shapes of crystals**
Some common shapes of crystals for selected minerals. *([a] Alfred Pasieka/ SPL/Photo Researchers, Inc.; [b] Charles D. Winters/Timeframe Photography Inc./ Photo Researchers, Inc.; [c] Charles D. Winters/Photo Researchers, Inc.; [d] E.R. Degginger/ Photo Researchers, Inc.)*

Figure 3.6 Crystal structure
(a) The structure of sodium chloride, showing the arrangement of six sodium (Na$^+$) ions around one chloride (Cl$^-$) ion. (b) The same structure with ions drawn to their relative sizes. *(After Gribble, C. D., ed. 1988. Rutley's elements of mineralogy, 27th ed. Boston: Unwin Hyman)*

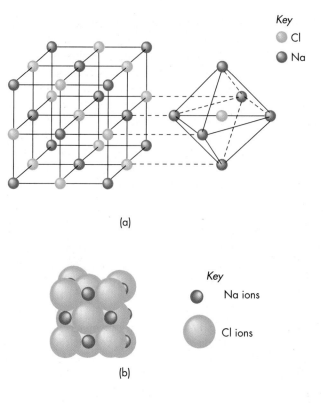

The internal crystal structure or framework for halite is shown in Figure 3.6. The structure is a cube, with sodium ions at the corners and chlorine ions at intermediate positions in the *crystal lattice*, the framework that defines the regular geometric pattern of atoms in a crystal. Notice that in Figure 3.6b, which is drawn to relative scale, the sodium ions are approximately half the size of the chloride ions. The length along one of the edges is approximately 0.56 nanometers (1 nm = 0.000000001 m). This distance is so short that it needs to be magnified approximately a million times to be the same size as a grain of sand or the head of a pin.

3.2 Important Rock-Forming Minerals

Although there are more than 4,000 minerals, only a few dozen are common constituents of Earth materials on or near Earth's surface. These few minerals are often important to many environmental concerns and are necessary in identifying most Earth materials we know as rocks, which are aggregates of one or more minerals. Some selected minerals are shown in Figure 3.7.

Determining a mineral's chemical composition and crystalline structure requires the use of sophisticated instruments. When we identify minerals from hand specimens, samples that you can hold in your hand, we also use appearance and some physical properties to assist us. Mineral properties, with their chemical compositions, are defined in Appendix A, with a table of some of the common minerals likely to be encountered in environmental geology work. Finally, **weathering,** the way that a mineral breaks down by physical and chemical processes at or near the surface of Earth, is important in forming sediments and soil (see A Closer Look: Weathering).

Although it is not often stated, geologists learn to identify the common minerals in hand specimens or under a microscope by the mental process of pattern recognition, based on clues of color, specific gravity, crystal form, cleavage, and so on. After careful examination of a number of samples, you learn to recognize a particular mineral, and the identification can be verified by a simple test or two. Recognition of the important rock-forming minerals is an important skill for a geologist.

Figure 3.7 **Some common minerals** (a) Cluster of quartz crystals from Brazil. Some are colorless and some are rose colored. Quartz is a very hard, common rock-forming mineral. *(Arnold Fisher/Science Photo Library/Photo Researchers Inc.)* (b) One of the several varieties of feldspar, the most common rock-forming mineral in Earth's crust. *(Stuart Cohen/Comstock)* (c) Yellow and pink clay minerals on the wall of Paint Mines, Colorado. These are only two examples of the many clay minerals that, when present in soils, may exhibit undesirable properties such as low strength, high water content, poor drainage, and high shrink-swell potential. *(Mark D. Phillips/Photo Researchers, Inc.)* (d) The dark mineral in this hand specimen of a rock is the black mica called biotite. It is a common mineral in granitic rocks as well as some metamorphic rocks. *(Barry L. Runk/Grant Heilman Photography, Inc.)* (e) Calcite, the abundant mineral in limestone and marble. Limestone terrain is associated with caverns, sinkholes, subsidence, and potential water pollution and construction problems. *(Betty Crowell/Faraway Places)* (f) Pyrite (fool's gold) is iron sulfide, a common mineral associated with ore deposits and coal that reacts with water and oxygen to form sulfuric acid. *(Betty Crowell/Faraway Places)* (g) Fragment of native copper. *(Comstock)*

The important groups of rock-forming minerals are summarized in Table 3.2 (on page 77). Each group is based primarily upon chemistry, and this abbreviated list serves as a summary for our discussion of types of minerals. We will now discuss each of the major mineral groups.

Silicates

Roughly 72 percent of the weight of Earth's crust is composed of oxygen (about 45 percent) and silicon (about 27 percent). These two elements, in combination with a few others including aluminum (8 percent), iron (6 percent), calcium (5 percent), magnesium (3 percent), sodium (2 percent), and potassium (2 percent), account for the chemical composition of minerals that make up about 98 percent by weight of Earth's crust. Astoundingly, about 94 percent by volume of the crust is oxygen! Geologists commonly report the abundance of a material by either mass (weight) or volume. Although mass is normally measured in kilograms, the correct unit for weight is Newtons. Newtons are calculated as kilograms multiplied by gravity; since gravity is a constant on Earth, we often ignore it in common uses of everyday

A CLOSER LOOK | Weathering

What Is Weathering? Briefly defined, weathering is the breaking apart or chemical decomposition of minerals and rocks at or near the surface of Earth by physical, biological, and chemical processes. Weathering processes are important for various reasons:

- They are some of the primary processes involved in the formation of soil (see Chapter 16).

- They prepare, that is, reduce in size and weaken, Earth material for easier transport by running water or slope processes such as landslides.

- Through chemical decomposition of some rock types, such as limestone, discussed in Section 3.6, they produce characteristic landforms, such as surface pits or sinkholes, and subsurface voids, such as caverns (see Chapter 12).

- They are responsible for the transformation of minerals, such as transforming feldspars to clay minerals.

As stated above, the processes involved in weathering are physical, chemical, and biological, or some combination of these. For example, physical weathering or breaking of rocks may be caused by frost action; water in rock fractures freezes and expands, breaking the mineral grains apart. Examples of biologically induced weathering include burrowing of animals and microbial digestion of rocks by a variety of organisms. The term *chemical weathering* refers to the partial dissolution of rocks and their minerals by chemical reactions,

usually in the presence of natural acidic solutions such as carbonic acid, commonly found in soils. Even rainwater is slightly acidic because carbon dioxide in the atmosphere combines with the water to produce a weak carbonic acid ($CO_2 + H_2O \rightarrow H_2CO_3$). Stronger acids may be formed as a result of water's mixing with sulfide minerals to form sulfuric acid. The net result from weathering is that near the surface of Earth, rocks are broken down or dissolved. This breakdown prepares the sediment that commonly constitutes the load that rivers transport to a lake or the ocean. In environmental geology, when describing Earth materials such as minerals, rocks, and soils at a site, one must carefully evaluate how those materials have weathered and where the weathered products such as clay minerals are concentrated.

You are probably familiar with the fight to control rust on your car, bicycle, or tools. Rust results from the oxidation, a chemical weathering process, of iron. Rust is composed of a group of minerals collectively named *limonite*, often a soft earthy material with a yellow-brown color. Rust weakens metals, reducing their strength and usefulness. Similarly, the chemical weathering of an otherwise strong rock by the "rusting" of iron-bearing minerals or formation of clay along fractures in the rock may weaken the rock (Figure 3.A). The weathered rock will not be as strong or as desirable for a foundation material for a structure such as a concrete dam and may be likely to fail by a landslide.

Figure 3.A Chemical weathering Weathered clay (iron stained orange) and thin lighter layer of calcite and gypsum minerals that form along weathered layers can weaken rock. *(Edward A. Keller)*

life. Therefore, in Europe cheese is sold by the kilogram, and in the United States it is sold by the pound. Whatever the units, we all know that a cup filled with Styrofoam weighs less than a cup of sand. *Volume* refers to the amount of space a material takes up. One cubic meter of Styrofoam and one cubic meter of sand have the same volume, but the sand is much heavier than the Styrofoam.

Minerals that include the elements silicon (Si) and oxygen (O) in their chemical composition are called *silicates*; these are the most abundant of the rock-forming minerals. The basic building block of all silicate minerals is the pyramid-shaped silicon–oxygen tetrahedron composed of relatively large oxygen ions at the corners, with the small silicon ion in the center (Figure 3.8). The tetrahedron may be present in a particular mineral framework as isolated tetrahedra, single or double chains, sheets, or complex networks (Figure 3.9).[3] The silicon–oxygen

TABLE 3.2   Rock-Forming Minerals by Groups, Based Mostly on Chemistry

Mineral Group	Examples	Chemical Formula	Comments
Silicates	Quartz	SiO_2	Common
	Plagioclase feldspars	$(Na, Ca)Al(Si, Al)_2O_8$	Very common
	Pyroxene	$(Ca, Mg, Fe)_2Si_2O_6$	Ferromagnesian mineral
Carbonates	Calcite	$CaCO_3$	Main minerals in limestone and marbles
	Dolomite	$(Ca, Mg)CO_3$	
Oxides	Hematite	Fe_2O_3	Primary ore of iron
	Bauxite	Hydrous aluminum oxides	Primary ore of aluminum
Sulfides	Pyrite	FeS_2	Major constituent of acid mine drainage
	Galena	PbS	Primary ore of lead
Native elements	Gold	Au	Precious metal, industrial uses
	Diamond	C	Jewelry, industrial uses
	Sulfur	S	Used to produce sulfuric acid

Note: A more extensive list is in Appendix A.

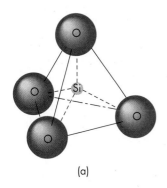

(a)

(b)

Figure 3.8 **Silicon–Oxygen tetrahedron** (a) Idealized diagram of the silicon–oxygen tetrahedron. This view is expanded to show the relatively large oxygen ions at the corners of the tetrahedron with the small silicon ion in the center. Chemical bonds are represented by the dashed lines between silicon and oxygen ions. (b) Diagram of the tetrahedron more as it is found in nature, with the oxygen ions touching each other. The silicon ion is the dashed circle in the central part of the tetrahedron. *(After Dietrich, R. V., and Skinner, V. J. 1979.* Rocks and rock minerals. *New York: John Wiley)*

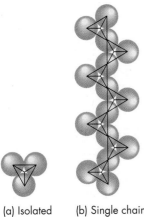

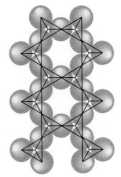

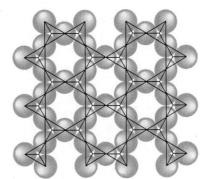

Too complex to show in two-dimensional drawing

(a) Isolated
Olivine
$(Mg, Fe)_2SiO_2$

(b) Single chain
Pyroxene
$(Ca, Mg, Fe)_2SiO_2$

(c) Double chain
Amphibole
$(Na, Ca)_2(Mg, Al, Fe)_5Si_8O_{22}(OH)_2$

(d) Sheet
Muscovite
$KAl_2(AlSi_3)O_{10}(OH)_2$

(e) Complex network
Quartz
SiO_2

Figure 3.9 **Linking silicon–oxygen tetrahedron** Idealized diagram showing several ways silicon–oxygen tetrahedra may be linked. *(After Press, F., and Siever, R. 1994.* Understanding Earth. *New York: W. H. Freeman)*

Figure 3.10 Conchoidal fractures This type of fracture looks like arcs of circles in natural volcanic glass (obsidian). Quartz also fractures conchoidally. The bar is 2.5 cm (1 in.) long. *(Edward A. Keller)*

tetrahedra combine with other elements such as calcium, magnesium, sodium, and aluminum to form the various silicate minerals. The arrangement of the silicon–oxygen tetrahedra in their various forms determines the properties of minerals (see Figures 3.8 and 3.9). The most important rock-forming silicate minerals or mineral groups are quartz, the feldspar group, micas, and the ferromagnesian group.

Quartz. *Quartz*, a form of silicon dioxide (SiO_2) with a network structure of silicon–oxygen tetrahedra, is one of the most abundant minerals in the crust of Earth. It can usually be recognized by its hardness, which is greater than glass, and by the characteristic way it fractures—conchoidally (Figure 3.10). Pure quartz is colorless, that is, clear, but most quartz crystals contain impurities that can make them white, rose, purple, or smoky black, among other colors (Figure 3.7a). Some of these colored varieties are semiprecious gemstones, such as purple amethyst. Large, six-sided, clear, pointed crystals of quartz are abundant in nature and are often sold as crystals. Because it is very resistant to the natural weathering processes and the processes that transport mineral grains, such as in rivers, quartz is a common mineral in river and beach sands.

Feldspars. *Feldspars* are aluminosilicates, containing silicon (Si), oxygen (O), and aluminum (Al) in combination with potassium (K), sodium (Na), or calcium (Ca) in a network structure of silicon–oxygen tetrahedra. The most abundant group of rock-forming minerals, constituting 60 percent of Earth's crust, feldspars are commercially important in the ceramics and glass industries. They are generally gray or pink and are fairly hard (Figure 3.7b).

There are two major types of feldspar: (1) alkali feldspars, $(Na, K)AlSi_3O_8$, which represent several feldspar minerals rich in potassium (k-feldspar) or sodium; and (2) plagioclase feldspars, $(Na, Ca)Al(Si, Al)Si_2O_8$, which are a series of minerals ranging from sodium-rich to calcium-rich. In the chemical formulas, the elements in parentheses, as for example the (Na, Ca) in plagioclase, represent elements that can substitute for one another.

Feldspars may weather chemically to form clay minerals (Figure 3.7c) with important environmental implications (see both A Closer Look: Weathering and A Closer Look: Clay).

Mica. *Mica* is a name for a group of important rock-forming minerals formed from sheets of silicon–oxygen tetrahedra, including the colorless mica, muscovite, $KAl_2(AlSi_3)O_{10}(OH)_2$, and the ferromagnesian mica, biotite, defined below. The micas are distinguished by a perfect *basal cleavage*, that is, they cleave parallel to the base of the crystal and peel into sheets. The mineral muscovite was used in

A CLOSER LOOK | Clay

The term *clay* is an important term in environmental geology and can be defined in several ways. We use the term *clay-sized* particles to describe those that are less than 0.004 mm in diameter, very small indeed. We also refer to clay as very fine mineral fragments defined in terms of chemical composition and internal crystal structure. The numerous clay minerals may be classified into several groups based on chemical characteristics and atomic structure. Some clays will take in a lot of water and expand upon wetting, a topic discussed in Chapter 16.

People have been using clays for thousands of years because they may be molded to form anything from crude pots to fine china to building blocks, including brick or sun-dried adobe.

Because some clay minerals have a high capacity for absorbing liquids, they have a variety of uses for medicine and industry. For example, some "kitty litter" is composed primarily of clays. Clay minerals are used as a filler for paper, and clay is also used to coat paper, particularly when a glossy finish is desired.

Clays are so often used in industrial processes that most of them have been synthesized from basic ingredients. When found in nature, clays are generally a warning to the environmental geologist. For example, clays may fill in fractures of rocks, thereby weakening the rocks. They also may be present in particular soil layers, where they may present problems in stabilizing a slope or preparing for the foundation of a building.

pioneer times as window material and later in doors of ovens, enabling one to see what was cooking.

Ferromagnesian Minerals. *Ferromagnesian minerals* are a group of silicates in which the silicon and oxygen combine with iron (Fe) and magnesium (Mg). These are the dark (black, brown, or green) minerals in most rocks. Black mica, biotite, $K(Mg, Fe)_3AlSi_3O_{10}(OH)_2$, is formed from sheets of the silicon–oxygen tetrahedra (Figure 3.7d). Three other important groups of ferromagnesian minerals used to identify igneous rocks, or rocks formed from the solidification of magma, are *olivine*, $(Mg, Fe)_2SiO_4$, a group of minerals formed from individual silicon–oxygen tetrahedra, with magnesium (Mg) and iron (Fe) substituting for each other in variable amounts from pure magnesium to pure iron; *pyroxene*, $(Ca, Mg, Fe)_2Si_2O_6$, a single-chain silicon–oxygen tetrahedra group of minerals; and *amphibole*, $(Na, Ca)_2(Mg, Al, Fe)_5Si_8O_{22}(OH)_2$, a double-chain silicon–oxygen tetrahedra group of minerals.

Because they are not very resistant to weathering and erosional processes, ferromagnesian minerals tend to be altered or removed from their location relatively quickly. They weather readily and combine with oxygen during the chemical process known as *oxidation*. Oxides are mineral compounds that link oxygen with metallic elements, as, for example, hematite and limonite, hydrous iron oxide (rust). Ferromagnesian minerals combine readily with other elements to form clays and soluble salts. Ferromagnesian minerals when abundant, may produce weak rocks, and builders must be cautious when evaluating a construction site that contains rocks high in ferromagnesian minerals. Caution is especially important for highway, tunnel, or reservoir planning.

Other Important Rock-Forming Minerals

Oxides. Earth materials containing useful minerals, especially metals, that can be extracted at a profit are called *ores* (ores are discussed further in Chapter 14). Iron and aluminum, probably the most important metals in our industrial society, are extracted from ores containing iron and aluminum oxides. The most important iron ore is hematite (an iron oxide, Fe_2O_3) and the most important aluminum ore is bauxite (a mixture of several aluminum oxides). Magnetite (Fe_3O_4, also an iron oxide, but economically less important than hematite) is common in many rocks. Magnetite, also known as lodestone, is a natural magnet that attracts and holds iron particles. Where particles of magnetite or other dark minerals are abundant, they may produce layers of black sand in streams or beach deposits.

Figure 3.11 **Limestone caverns**
Lechuguilla Cave, Carlsbad Caverns
National Park, New Mexico. Spelunker
(person exploring cave) Neeld Messler
is working his way down into the cave.
*(Michael Nichols/National Geographic
Image Collection)*

Carbonate Minerals. Environmentally, the most important *carbonate mineral* is calcite (calcium carbonate, $CaCO_3$), shown in Figure 3.7e. This mineral is the major constituent of limestone and marble, two very important rock types. Many marine organisms from oysters and clams to forminifera (discussed earlier with oxygen isotopes) have shells composed of carbonate minerals. Chemical weathering of such rocks by water dissolves the calcite, often producing caverns, sinkholes, or surface pits (Figure 3.11).

Sulfide Minerals. Pyrite (iron sulfide, FeS_2), also known as fool's gold, is a *sulfide mineral* and is shown in Figure 3.7f. Sulfides can also be associated with environmental degradation, which typically occurs when roads, tunnels, or mines cut through coal-bearing rocks that contain sulfide minerals. The exposed sulfides oxidize, or combine with oxygen, in the presence of water to form compounds such as sulfuric acid that may enter and pollute streams and other environments (Figure 3.12). This is a major problem in the coal regions of Appalachia and many other regions of the world where sulfur-rich coal and sulfide minerals are mined (see Chapter 15).

Native Elements. Minerals formed of a single element are called *native elements*; gold, silver, copper (Figure 3.7g), and diamonds are examples of native elements. They have long been sought as valuable minerals. They occur in rather small

Figure 3.12 **Acid mine
drainage** This stream in Jackson
County, Ohio, is polluted by acid
mine drainage from a strip mine.
(Matt Meadows/Peter Arnold, Inc.)

accumulations but are occasionally found in sufficient quantities to justify mining. As we mine these minerals in ever lower-grade deposits, the environmental impact will increase because the larger the mine the greater the environmental impact. (The environmental effects of mining are discussed in Chapter 14.)

3.3 Rock Cycle

What Is a Rock? A **rock** is an aggregate of one or more minerals. That is, some rocks are formed from a single mineral, and others are composed of several minerals. Although rocks vary greatly in their composition and properties, they can be classified into three general types, or families, according to their mineralogy—or mineral composition, chemical composition, and *texture* (size, shape, and arrangement of grains) (see Appendix B)—and how they were formed during the **rock cycle** (Figure 3.13). We may consider this cycle a worldwide rock-recycling system linking subsurface processes driven by Earth's internal heat, which melts or changes rocks in the tectonic cycle, to surface processes driven by solar energy. The rock cycle produces three general families of rock: igneous rocks, sedimentary rocks, and metamorphic rocks. Crystallization of molten rock produces **igneous rocks** both beneath and on Earth's surface. Rocks at or near the surface break down chemically and physically by weathering, forming sediments that are transported by wind, water, and ice to depositional basins, such as the ocean (see A Closer Look: Weathering). The accumulated layers of sediment eventually undergo chemical and physical changes, forming **sedimentary rocks.** Deeply buried sedimentary rocks may be metamorphosed, or altered in form by heat, pressure, or chemically active fluids, to produce **metamorphic rocks.** These metamorphic rocks may be buried still more deeply and melt, beginning the cycle again. Some variations of this idealized sequence are shown in Figure 3.13. For example, an igneous or metamorphic rock may be altered into a new metamorphic rock without undergoing weathering or erosion. To summarize, the type of rock formed in the rock cycle depends on the rock's environment.

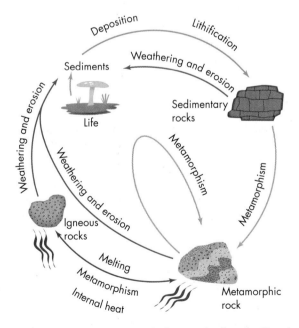

Figure 3.13 The rock cycle Idealized rock cycle showing the three families of rocks and important processes that form them. Life is part of the cycle, adding carbon and other elements to sediments that contribute to sedimentary rocks including coal and other fossil fuels. Linkages between life and geology are discussed in Chapter 4.

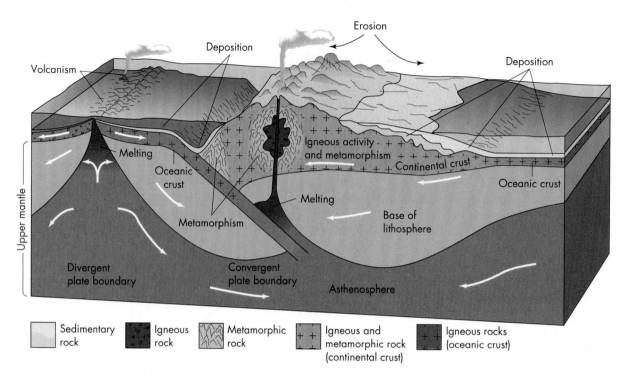

Figure 3.14 **Rock environments** Idealized diagram showing some of the environments in which sedimentary, igneous, and metamorphic rocks form. *(Modified after Judson, S., Kauffman, M. E., and Leet, L. D. 1987. Physical geology, 7th ed. Englewood Cliffs, NJ: Prentice Hall)*

Rock Cycle and Plate Tectonics

Plate tectonics provides several environments for rock formation, with specific rock-forming processes occurring at each type of plate boundary (Figure 3.14). When we consider the rock cycle alone, we are concerned mainly with the recycling of rock and mineral materials. However, the tectonic processes that drive and maintain the rock cycle are essential in determining the properties of the resulting rocks. Therefore, our interest in plate tectonics is more than academic; we build our homes, industries, and roads and grow our crops on these Earth materials.

3.4 Three Rock Laws

Understanding Earth history through geologic time requires knowing some fundamental laws. Three of the most important are

1. The **law of crosscutting relationships** states that a rock is younger than any other rock that it cuts.
2. The **law of original horizontality** states that when originally deposited, sedimentary layers are nearly horizontal.
3. The **law of superposition** states that if a series of layered sediments have not been overturned, the oldest layers are on the bottom and the youngest are on top.

We will return to these laws in our discussion of rocks, their geologic history, and their environmental significance. We will begin our discussion with igneous rocks and then will consider sedimentary rocks and metamorphic rocks.

3.5 Igneous Rocks

Igneous rocks have crystallized from *magma*, a mobile mass of hot, quasi-liquid Earth material consisting of a mixture of melted and solid materials. Magma is often generated in the upper asthenosphere or the lithosphere (Figure 3.14). You

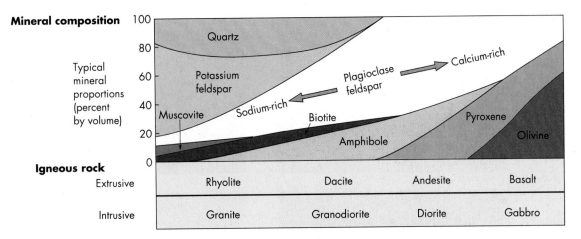

Figure 3.15 Igneous rocks Types of extrusive and intrusive igneous rocks and their characteristic mineral composition. Note that every extrusive rock has an intrusive counterpart. For example, rhyolite and granite are composed of the same minerals, but rhyolite cooled at the surface of Earth whereas granite cooled beneath the surface of Earth. *(Modified after Davidson, J. P., Reed, W. E., and Davis, P. M. 1997. Exploring Earth. Upper Saddle River, NJ: Prentice Hall)*

can visualize magma as hot, slushy, cherry pie filing consisting of liquid filling mixed with solid cherries.

Intrusive Igneous Rocks

If magma cools slowly and crystallizes well below the surface of Earth, the result is *intrusive* igneous rock. Individual mineral grains can be seen with the naked eye. Crystals in intrusive igneous rocks that are larger than the surrounding crystals, well formed, and surrounded by relatively finer-grained crystals are known as *phenocrysts*. When intrusive igneous rocks are exposed on Earth's surface, we can assume that erosion has removed the surface material originally covering them. As the molten magma rises toward the surface, it displaces the rock it intrudes, often breaking off portions of the intruded rock and incorporating them as it crystallizes. These foreign blocks, known as *inclusions*, are evidence of forcible intrusion.

Common types of intrusive igneous rocks such as granite, granodiorite, diorite, and gabbro and their general mineralogy is shown in Figure 3.15. For example, granite is composed mostly of potassium-rich alkali feldspar and quartz with minor amounts of sodium-rich feldspar, the micas muscovite and biotite, and amphibole (Figure 3.16a).

Batholiths and Plutons

Magma that forms a body of intrusive igneous rock emplaced in the crust in a particular region may be truly gigantic, often exceeding thousands of cubic kilometers. The largest intrusions are called *batholiths*, which are composed of a series of smaller intrusions known as plutons. *Plutons* are variable in size but may be as small as a few kilometers in width. Some intrusions, such as dikes and sills, are tabular in form, relatively long and narrow (Figure 3.17). Dikes intrude through and cut across existing rocks, whereas sills intrude parallel to the rock layers. Most dikes we observe in the field are relatively small, less than 1 m to a few meters in width. Dikes can form complex patterns that either cut across one another or are in a radial pattern; these patterns are known as dike swarms. Most batholiths are formed of granitic rocks, but dikes and sills may be basaltic. Sometimes it is possible to distinguish younger from older plutons by examining them in the field where they are exposed by erosion. For example, the light gray pluton in Figure 3.17 contains inclusions, or pieces of the dark gray pluton, within it. Inclusions allow geologists to determine the relative ages of the plutons. A pluton that contains inclusions of another pluton within it must be the younger of the two. Therefore, the light gray pluton is younger than the dark gray pluton. The youngest intrusion in Figure 3.17

Figure 3.16 Some common rock types (a) A specimen of granite. The pink fine-grained mineral is feldspar, the dark mineral is ferromagnesian, and the very light-colored mineral is quartz. The large crystal in the center is known as a phenocryst, which is a coarse crystal in a finer matrix of other crystals. The phenocryst in this case is feldspar. *(Betty Crowell/Faraway Places)* (b) Granite dome in Yosemite Valley, California. (c) Layered rocks here are basalt flows along the Palouse River in Washington. *(Betty Crowell/Faraway Places)* (d) Rocks exposed here in the foreground are sandstone. The parent material was a sand dune that has been cemented and turned to stone, Zion National Park, Utah. *(George Hunter/Getty Images Inc.)* (e) The rock shown here is a conglomerate found in Death Valley, California. Individual particles that are cemented together are easily seen; the vertical lines are fractures in the rock. *(Betty Crowell/Faraway Places)* (f) The rock exposed here is a highly weathered limestone in the Yunnan Province of China. The pinnacles are formed by chemical weathering of the limestone and removal of cover soil material. *(Betty Crowell/Faraway Places)* (g) Schist, a foliated metamorphic rock in which parallel alignment of mineral grains, in this case mica, produces the foliation and reflects the light. *(Marvin B. Winter/Photo Researchers, Inc.)* (h) Gneiss, a foliated metamorphic rock with minerals segregated into white bands of feldspar and dark bands of ferromagnesian minerals. *(John Buitenkant/Photo Researchers, Inc.)*

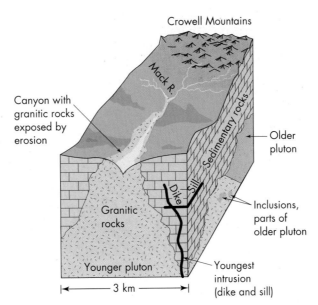

Figure 3.17 Igneous intrusions
Idealized block diagram showing several types of igneous intrusions. The older pluton is known to be older; it is cut by the younger pluton and inclusions of the older pluton are found in the younger. The youngest intrusions in the diagram are the dike and sill. Notice that the younger pluton is exposed at the surface in a canyon similar to the exposure of granitic rocks in Yosemite shown in Figure 3.16b. Assuming that the pluton was emplaced at a depth of more than 10 km (about 6.2 mi), one finds an appreciation for the tremendous amount of erosion necessary to expose granitic rocks in canyons such as Yosemite Valley.

is the dike and sill that cuts across the younger pluton. This analysis demonstrates the *law of crosscutting relationships*, which states that a rock is younger than any other rock that it cuts.

Why Magma Rises and Intrudes Other Rocks

One commonly asked question is, "Why do masses of partially melted rock or magma rise or intrude the surrounding rocks?" A probable explanation is that once the internal heat of Earth has formed a mass of magma, it is hotter and less dense than the surrounding rocks, and therefore rises much as a helium-filled balloon rises in the atmosphere or as a scuba diver's air bubbles rise through the water. Presumably, once the density differences are equalized, the mass of magma or intrusion ceases its upward journey toward the surface of Earth, and the magma completes the crystallization of minerals to form intrusive igneous rocks. Often, before crystallization to form intrusive igneous rocks occurs, however, some magma reaches the surface of Earth to cause volcanic eruptions and form extrusive igneous rocks.

Extrusive Igneous Rocks

Igneous rocks that crystallize at the surface of Earth are called *extrusive*. Extrusive igneous rocks often form from lava, or molten magma flowing from volcanoes. They can also form from pyroclastic debris, composed of fragmented magma that has rapidly solidified as it was blown out of a volcano.

Extrusive igneous rocks are fine-grained owing to rapid cooling and crystallization, and individual crystals or grains usually cannot be seen with the naked eye. However, extrusive igneous rocks may have previously formed phenocrysts of minerals that crystallized when the magma was deeper within the Earth and crystallization was slower. The texture of such a rock, which has relatively few, often well-formed, phenocrysts surrounded by a mass of fine-grained crystals, is called *porphyritic*. There are different types of extrusive igneous rock, such as rhyolite, dacite, andesite, and basalt, just as there are different types of intrusive igneous rock (Figure 3.15). The specific type is based on differences in the rocks' chemical composition. Notice on Figure 3.15 that the minerals in rhyolite are the same as in granite, as is the case with dacite and granodiorite, andesite and diorite, and basalt and gabbro. It is the size of the crystals formed by the process of crystallization, either at the surface or deep beneath the surface of Earth, that differentiates extrusive from intrusive igneous rocks. We often identify intrusive igneous rocks from hand specimens by examining the phenocrysts present. For example, andesite may

Figure 3.18 Andesite Hand specimen of andesite with white feldspar phenocrysts. Bar is 2.5 cm (1 in.). *(Edward A. Keller)*

have phenocrysts of feldspar (Figure 3.18), whereas basalt is more likely to have phenocrysts of pyroxene or olivine. We will discuss extrusive rocks and their differences in more detail in Chapter 7 when we discuss volcanoes.

Extrusive rocks form from lava or magma that crystallizes at the surface. A lava flow may be mixed with cemented fragments of broken lava and ash, called *volcanic breccia*, or with thick vesicular, or cavity-filled, zones produced as gas escapes from cooling lava.

Pyroclastic debris, also known as tephra, ejected from a volcano produces a variety of extrusive rocks (Figure 3.19). Debris consisting of rock and glass fragments less than 4 mm (about 0.16 in.) in diameter are called volcanic ash; when this ash is compacted, cemented, or welded together, it is called *tuff*. Pyroclastic activity also produces larger fragments that, when mixed with ash and cemented together, form the rock called volcanic breccia.

Extrusive igneous rocks may develop into piles of layered volcanic rocks surrounding volcanoes that form entire islands such as the Hawaiian Islands in the Pacific. They may also form basaltic plateaus that cover many thousands of square kilometers with a series of basaltic flows of great thickness (see Figure 3.16c, part of the Snake River Basaltic Plateau in Washington).

Figure 3.19 Pyroclastic debris Pyroclastic deposits, fine-grained volcanic ash, pumice (light-colored layers), and coarser pyroclastic fragments (marble sized) mixed with ash; north flank of Tenerife, Canary Islands. *(Edward A. Keller)*

Igneous Rocks and the Environment

Intrusive and extrusive igneous rocks have a wide variety of properties, and generalizations are difficult. We can, however, make three environmental points associated with these rocks:

- Intrusive igneous rocks, especially granite, are generally strong rocks that make a good foundation for many structures such as dams and large buildings. Blocks of these rocks are often resistant to weathering and are used for a variety of construction purposes.[6,7]

- Lava flows that have cooled and solidified often exhibit extensive *columnar jointing* (Figure 3.20). Columnar jointing is a type of fracturing that occurs during cooling that may lower the strength of the rock. Solidified lava flows may also have subterranean voids known as lava tubes. Lava tubes can collapse from the weight of the overlying material or can carry large amounts of groundwater. Both features can cause problems during the planning, design, or construction phases of a project.

- Tuff is generally a soft, weak rock that may have very low strength.[6] The strength of a tuff rock depends upon the degree to which it has become cemented or welded. Some tuff may be altered through weathering into a type of clay known as *bentonite*, an extremely unstable material. When bentonite is wet, it expands to many times its original volume and is unstable.

It is difficult to make generalizations about the suitability of extrusive igneous rocks for specific uses. Careful field examination is always necessary before large structures are built on such rocks.[6] Planning, designing, and constructing engineering projects on extrusive rocks, especially pyroclastic debris, can be complicated and risky.[7] This fact was tragically emphasized on June 5, 1976, when the Teton Dam in Idaho failed, killing 14 people and inflicting approximately $1 billion in property damage. Just before the failure, a whirlpool several meters across was seen in the reservoir, strongly suggesting that a tunnel of free-flowing water had developed beneath the dam. In fact, the dam was built upon highly fractured volcanic rocks. Water began moving under the foundation as the reservoir filled. When the subsurface moving water came into contact with the dam, it quickly eroded a tunnel through the base of the dam, causing the whirlpool. The dam collapsed just minutes later, and a wall of water up to 20 m (66 ft) high rushed downstream, destroying homes, farms, equipment, animals, and crops for 160 km (100 mi) downstream from the dam failure.

Figure 3.20 Columnar jointing or fracturing These joints or fractures in basalt form because of contraction during cooling. This photo was taken at Devil's Postpile National Monument, Sierra Nevada, California. *(A. G. Sylvester)*

3.6 Sedimentary Rocks

Sedimentary rocks constitute about 75 percent of all rocks exposed at the surface of Earth (Figure 3.21). Their common environments are shown in Figure 3.14. There are two types of sedimentary rock: *detrital*, or clastic, sedimentary rocks, which form from broken parts of previously existing rocks; and *chemical*, or nonclastic, sedimentary rocks, which are deposited when chemical or biochemical processes cause minerals to form from substances dissolved in water.

Sedimentary rocks form when sediments are transported, deposited, and then transformed into rock, by natural cementation or compaction or both. After deposition, both physical and chemical changes in sediments occur in response to increased pressure and temperature. The increase in pressure and temperature is not sufficient to produce metamorphic rocks. These changes in the sediments are a result of their being buried and of fluids migrating through them. These processes are collectively referred to as *diagenesis*. Some of these processes are shown in idealized form in Figure 3.22.

- Sediment from a river is delivered to a sedimentary basin, in this case, an ocean. The basin is only one of many types of *sedimentary environments* on Earth. Other sedimentary environments include lakes, river floodplains, sand dunes, and glacial environments.

- As the river enters the ocean at point A, the coarser sediment (sand) is deposited first at point B, and finer silt and clay are deposited farther from the shoreline where the transport processes are weakest, at point C. The sediment is deposited in beds, or layers. Individual beds are nearly horizontal when deposited, as stated in the *law of original horizontality*. The sequence of beds are known as *strata*. The top bed, and others, may have been produced by a single large flood event from the river. The different layers are separated by *bedding planes* (Figure 3.23) that denote the top and bottom of a particular sedimentary bed. The layering of sedimentary rocks therefore results from changes in deposition associated with grain size, difference in composition of the sediments, or a series of events such as large floods that periodically bring a greater amount of sediment into the sedimentary basin. A particular sedimentary bed may change or grade laterally from one type of sediment to another—for example, the sand, silt, and clay illustrated in Figure 3.22. The lateral

Figure 3.21 **Sequence of sedimentary rock** The Grand Canyon of the Colorado River exposes a spectacular section of sedimentary rocks. In this view looking toward the south rim at sunset, the top sedimentary unit is the Kaibab limestone, below which is a series of sandstones, shales, and the famous Red Wall limestone. *(Jack W. Dykinga)*

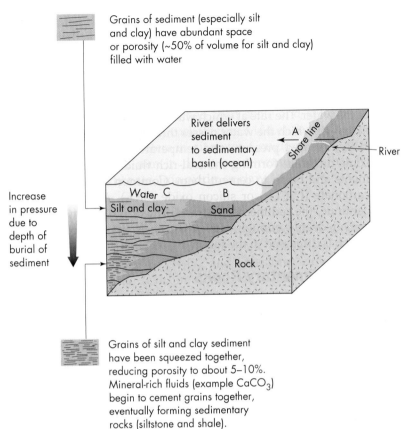

Grains of sediment (especially silt and clay) have abundant space or porosity (~50% of volume for silt and clay) filled with water

Increase in pressure due to depth of burial of sediment

Grains of silt and clay sediment have been squeezed together, reducing porosity to about 5–10%. Mineral-rich fluids (example CaCO₃) begin to cement grains together, eventually forming sedimentary rocks (siltstone and shale).

Figure 3.22 **Diagenesis** Idealized block diagram illustrating some of the processes of diagenesis that transform sediments into sedimentary rocks. Points A, B, and C correspond to locations discussed in the text.

change in sediment is referred to as a *facies* change; in this case, a sandy facies has changed into a silt and clay facies in the same sedimentary unit.

- As the sedimentary basin sinks or sea level rises, sediments (strata) up to several kilometers in thickness may be deposited. Recall the spectacular section of sedimentary rocks in the Grand Canyon shown in Figure 3.21.

- When the sediment was originally deposited, there was open space between the grains. This space between the grains, known as porosity, is filled with water; therefore, for each cubic meter (about 35 ft³) of the sediment, half of the space is occupied by water between the grains. The oldest sedimentary beds are located at the bottom of the sedimentary basin. As stated in the *law of*

Figure 3.23 **Tilted sedimentary rocks** Lines generally running from lower left to upper right are bedding planes of these sedimentary rocks. The layers are inclined about 30 degrees. The law of original horizontality states that these bedding planes were close to horizontal when the sediment grains forming the rocks were deposited. We assume they are at 30 degrees today because they have been uplifted and tilted by tectonic processes.
(Edward A. Keller)

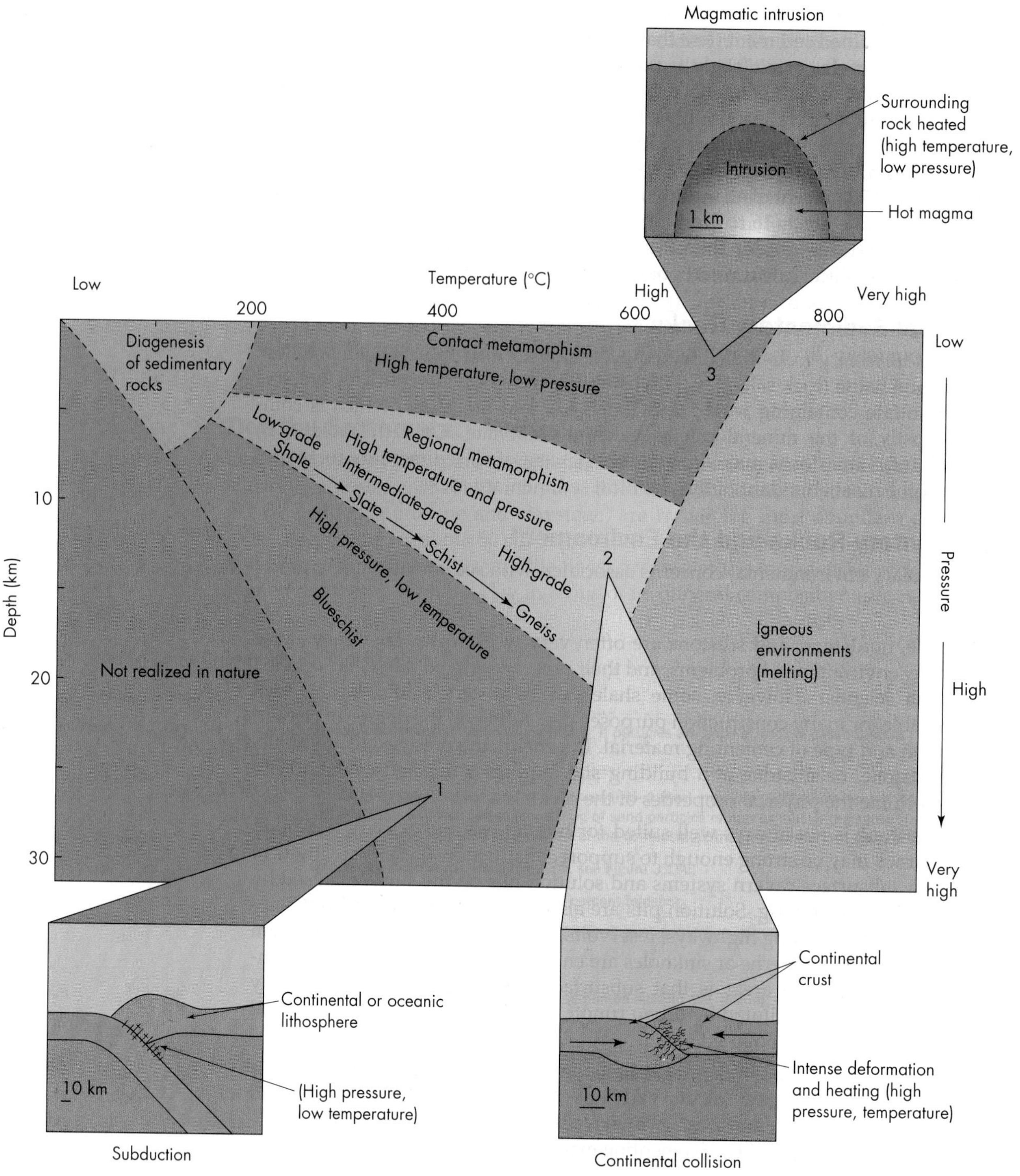

Magmatic intrusion

Surrounding rock heated (high temperature, low pressure)

Intrusion

1 km

Hot magma

Temperature (°C)

Low   200   400   High 600   Very high   800

Low

Diagenesis of sedimentary rocks

Contact metamorphism
High temperature, low pressure

3

Low-grade
Shale

High temperature and pressure

Regional metamorphism

Intermediate-grade

Slate

Schist

High-grade

Gneiss

High pressure, low temperature

2

Blueschist

Pressure

Not realized in nature

Igneous environments (melting)

High

Depth (km)

10

20

30

1

Very high

Continental or oceanic lithosphere

10 km

(High pressure, low temperature)

Subduction

Continental crust

10 km

Intense deformation and heating (high pressure, temperature)

Continental collision

Figure 3.24 **Geologic environments of metamorphism** Highly idealized diagram showing temperature and pressure conditions for diagenesis of sediments and the main types of metamorphism resulting from different pressure-temperature conditions and geologic environments. Subduction zones are characterized by high-pressure metamorphism (1) in which pressure increases as the oceanic lithosphere, with its load of marine sediments, is subducted. Regional metamorphism (2) results from intense deformation, with high temperatures and pressures resulting from continental collisions. Rocks surrounding a body of cooling magma (3) experience high-temperature contact metamorphism at relatively low pressure. *(Modified after Davidson, J. P., Reed, W. E., and Davis, P. M. 1997. Exploring Earth. Upper Saddle River, NJ: Prentice Hall)*

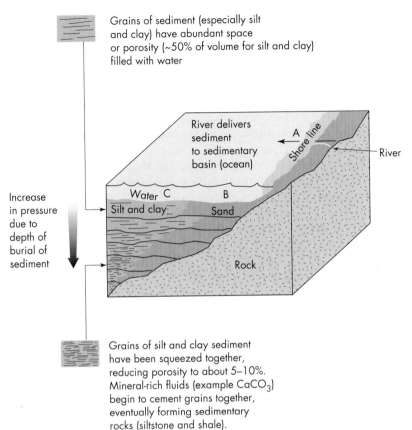

Grains of sediment (especially silt and clay) have abundant space or porosity (~50% of volume for silt and clay) filled with water

Increase in pressure due to depth of burial of sediment

Grains of silt and clay sediment have been squeezed together, reducing porosity to about 5–10%. Mineral-rich fluids (example $CaCO_3$) begin to cement grains together, eventually forming sedimentary rocks (siltstone and shale).

Figure 3.22 Diagenesis
Idealized block diagram illustrating some of the processes of diagenesis that transform sediments into sedimentary rocks. Points A, B, and C correspond to locations discussed in the text.

change in sediment is referred to as a *facies* change; in this case, a sandy facies has changed into a silt and clay facies in the same sedimentary unit.

- As the sedimentary basin sinks or sea level rises, sediments (strata) up to several kilometers in thickness may be deposited. Recall the spectacular section of sedimentary rocks in the Grand Canyon shown in Figure 3.21.

- When the sediment was originally deposited, there was open space between the grains. This space between the grains, known as porosity, is filled with water; therefore, for each cubic meter (about 35 ft³) of the sediment, half of the space is occupied by water between the grains. The oldest sedimentary beds are located at the bottom of the sedimentary basin. As stated in the *law of*

Figure 3.23 Tilted sedimentary rocks Lines generally running from lower left to upper right are bedding planes of these sedimentary rocks. The layers are inclined about 30 degrees. The law of original horizontality states that these bedding planes were close to horizontal when the sediment grains forming the rocks were deposited. We assume they are at 30 degrees today because they have been uplifted and tilted by tectonic processes.
(Edward A. Keller)

superposition, if a series of sedimentary beds have not been overturned, then the youngest beds are on the top and the oldest on the bottom. The sedimentary deposits near the bottom are subjected to both an increase in pressure due to the weight of the overlying sediments and an increase in temperature resulting from natural heat flow from the interior of Earth. The increase in pressure causes the grains to be squeezed tightly together, forcing out some of the water. The rate at which water is forced out is dependent in part on the ease with which the water passes through the pore spaces between the particles. Increased pressure and temperature may facilitate the dissolving of some minerals, forming mineral-rich fluids that migrate through the sediments and begin to cement them. Common cementing materials might be carbonate ($CaCO_3$) or silicon oxide (SiO_2). As these diagenetic processes continue, the sediments are transformed into sedimentary rock. If they are buried deeply enough, then diagenesis may proceed to metamorphism, which is discussed in the following section.

Detrital Sedimentary Rocks

Detrital sedimentary rocks are classified according to their grain size as either shale, siltstone, sandstone, or conglomerate (Table 3.3A). Shale and siltstone, along with poorly bedded mudstone and claystone, are by far the most abundant detrital rocks, accounting for about 50 percent of all sedimentary rocks. Composed of clay- and silt-sized particles, they are the finest grained of the four detrital rock types. Remember that we defined clay in terms of its mineral composition (see A Closer

TABLE 3.3   Detrital (A) and Chemical (B) Sedimentary Rocks

	Size	Sediment	Rock
(A) Detrital (Clastic) Sedimentary Rocks	Greater than 2 mm	Gravel	**Conglomerate:** often has a sandy matrix. If particles are angular, rock is called breccia (Figure 3.17d).
	$\frac{1}{16}$–2 mm	Sand	**Sandstone:** generally designated as coarse (0.5–1 mm), medium (0.25–0.49 mm), or fine (0.13–0.24 mm) if well-sorted (consisting of sand particles of approximately the same size). Have also been subdivided on the basis of the composition; the more important types are:
			Quartzose sandstone (mainly quartz, see Figure 3.17c)
			Arkosic sandstone (arkose): over 20 percent feldspar
			Graywacke: poorly sorted (consisting of sand particles of many sizes mixed together); contains rock fragments with a clay matrix
	$\frac{1}{256}$–$\frac{1}{16}$ mm	Silt	**Siltstone** (mudstone): compacted or cemented silt and clay lacking fine lamination
	Less than $\frac{1}{256}$ mm	Clay	**Shale:** compacted or cemented silt and/or clay with fine laminations along which rock easily splits (fissility). Rocks composed of clay that lack fissility are called **claystone. Mudstone** is an unlaminated mixture of silt and clay.
	Principal Composition	**Principal Texture**	**Rock**
(B) Chemical (Nonclastic) Sedimentary Rocks	Calcite, $CaCO_3$	Fine[1]	**Limestone:** often of biologic origin and may contain fossils. *Coquina* is limestone composed mainly of fossils or fossil fragments. Effervesces in diluted hydrochloric acid.
	Calcite, $CaCO_3$	Fine	**Chalk:** soft, white limestone formed by the accumulation of microscopic shells. Effervesces with dilute hydrochloric acid.
	Silica, SiO_2	Fine	**Chert:** hardness 6 or 7[3], often white, flint is black or dark gray.
	Gypsum, $CaSO_4 \cdot 2H_2O$	Fine to coarse[2]	**Gypsum:** hardness 2. One good cleavage and two poorer cleavages.
	Halite, NaCl	Fine to coarse	**Rock salt:** cubic crystals and cleavage may be visible. Salty taste.
	Silica, SiO_2	Fine	**Diatomite:** soft, white rock formed by the accumulation of microscopic shells composed of silica. Distinguished from chalk by lack of effervescence in diluted hydrochloric acid.

[1]Cannot see grains with naked eye; generally grain size is less than 1/16 mm.
[2]Can see grains with naked eye; generally grain size is greater than 2 mm.
[3]See Appendix A for definition of hardness.
Modified after Foster, R. J. 1991. *Geology*, 6th ed. Upper Saddle River, NJ: Prentice Hall.

Look: Clay) and stated that geologists also use the term in a textural sense to mean the very fine-grained sediment (less than 1/256 mm in diameter).

When layers of clay or silt (slightly coarser-grained sediment, 1/256 to 1/16 millimeters in diameter) are compacted or cemented, they form shale or siltstone. Siltstone is more massive, generally lacking bedding planes, whereas shale characteristically has closely spaced bedding planes.

Sandstone and *conglomerate* are coarser grained than shale and siltstone and make up about 25 percent of all sedimentary rock (Figure 3.16d,e). Sand-sized particles of sandstone are 1/16 to 2 millimeters in diameter. Conglomerate contains gravel-sized particles greater than 2 mm in diameter, cemented together with material such as silica, calcium carbonate, iron oxide, or clay.

Chemical Sedimentary Rocks

Chemical sedimentary rocks are classified according to their mineral composition; they include halite (rock salt, NaCl), gypsum (hydrated calcium sulfate, which is calcium sulfate containing water, $CaSO_4 \cdot 2H_2O$), and limestone, which is composed mostly of the mineral calcite (calcium carbonate, $CaCO_3$) (Table 3.3B) (Figure 3.16f). Limestones make up about 25 percent of all sedimentary rocks and are by far the most abundant of the chemical sedimentary rocks.

Sedimentary Rocks and the Environment

Three primary environmental concerns associated with sedimentary rocks are as follows.

- Shale, mudstone, and siltstone are often very weak rocks. They may cause many environmental problems, and their presence is a red flag to the applied Earth scientist. However, some shale can be a very stable, strong rock suitable for many construction purposes depending on the degree of cementation and type of cementing material. In general, the presence of any shale, mudstone, or siltstone at a building site requires a detailed evaluation to determine the physical properties of the rock.

- Limestone is not always well suited for human use and activity. Although this rock may be strong enough to support construction, it easily weathers to form subsurface cavern systems and solution pits on the surface caused by chemical weathering. Solution pits are also known as sinkholes (see Chapter 12). Constructing highways, reservoirs, and other structures is a problem in areas where caverns or sinkholes are encountered. Another problem common in limestone areas is that subsurface water in cavern systems may quickly become polluted by urban runoff, with little or no chance of natural purification of the polluted water.

- Cementing materials in detrital sedimentary rocks may be troublesome. Silica cement is the strongest; calcium carbonate tends to dissolve in weak acid; and clay may be unstable and wash away. It is always advisable to carefully evaluate the strength and stability of cementing materials in detrital sedimentary rocks.

3.7 Metamorphic Rocks

Igneous and metamorphic rocks together account for more than 90 percent of all rocks in Earth's crust. Metamorphic rocks are changed rocks; that is, heat, pressure, and chemically active fluids produced by the tectonic cycle or resulting from the presence of an intrusion may change the mineralogy and texture of preexisting rocks, in effect producing new rocks. Generalized geologic environments along with pressure temperature conditions that result in metamorphism are idealized in Figure 3.24. The three major types of metamorphism are

1. High-pressure, low-temperature metamorphism; characteristic of subduction zones.

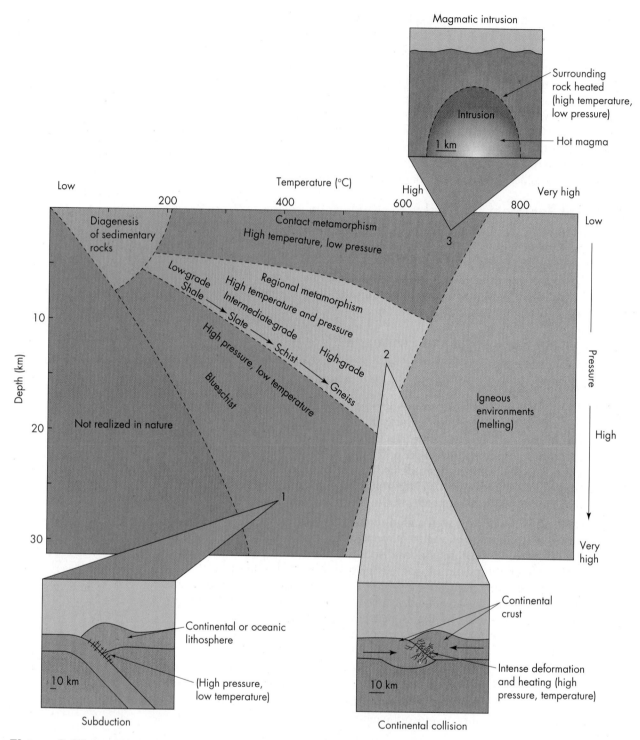

Figure 3.24 Geologic environments of metamorphism Highly idealized diagram showing
temperature and pressure conditions for diagenesis of sediments and the main types of metamorphism
resulting from different pressure-temperature conditions and geologic environments. Subduction zones
are characterized by high-pressure metamorphism (1) in which pressure increases as the oceanic
lithosphere, with its load of marine sediments, is subducted. Regional metamorphism (2) results from
intense deformation, with high temperatures and pressures resulting from continental collisions. Rocks
surrounding a body of cooling magma (3) experience high-temperature contact metamorphism at
relatively low pressure. *(Modified after Davidson, J. P., Reed, W. E., and Davis, P. M. 1997. Exploring Earth.
Upper Saddle River, NJ: Prentice Hall)*

2. High-temperature, high-pressure metamorphism; characteristic of *regional metamorphism* that might be produced during continental collision. Notice that in this environment rock metamorphism changes low-grade sedimentary rocks to high-grade metamorphic rocks with increasing temperature and pressure. The rocks progressively change from sedimentary rock to low-grade metamorphic rock known as slate, to an intermediate-grade metamorphic rock known as schist, and finally to a high-grade metamorphic rock known as gneiss. The original shale, a sedimentary rock, is thus transformed to slate, then schist, then gneiss, through progressive regional metamorphism as a result of increasing temperature and pressure.

3. High-temperature, low-pressure metamorphism, or *contact metamorphism*. When magma intrudes the upper crust, it heats nearby surrounding rocks, but because it is close to the surface, the pressure is relatively low. This increased temperature causes the formation of particular types of metamorphic minerals and rocks in the contact zone.

Foliated Metamorphic Rocks

Slate is a relatively low-grade metamorphic rock resulting from the metamorphism of the sedimentary rock shale as shown in Figure 3.25 and outlined in Table 3.4. As illustrated in Figure 3.25, tectonic stress applied to shale under relatively low temperatures and pressures produces a new rock. *Slate* is referred to as a *foliated* metamorphic rock, in which the mineral grains such as mica are aligned either in parallel, layering, or banding structure, producing a rock cleavage or *foliation*. If the slate is subjected to higher temperatures and pressures, it may change to a higher-grade metamorphic rock known as *schist*. Processes of recrystallization intensify the parallel alignment of mineral grains and crystal size increases, often rendering the crystals relatively easy to identify (Figure 3.16g). If our original shale is subjected to even higher temperatures and pressures, then *gneiss*, with a foliated texture characterized by banding of light and dark minerals, may form (Figure 3.16h).

Nonfoliated Metamorphic Rocks

Not all metamorphic rocks are foliated. *Marble*, consisting mostly of the mineral calcite ($CaCO_3$), is an example of a nonfoliated metamorphic rock. It results from

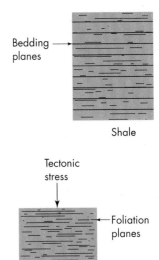

Figure 3.25 Transformation of shale to slate Idealized diagram showing the transformation from the sedimentary rock shale to the metamorphic rock slate. The foliation planes in the slate develop as a result of tectonic stress, with recrystallization of the minerals as a result of relatively low-grade metamorphism. Note that the foliation planes form at right angles to the orientation of the tectonic stress. In a gross sense, the initial grains in the rock look something like a house of cards. When you place your palm on the house of cards and apply pressure, it collapses and the cards lie flat and stacked on the table, much like the foliation planes in the slate.

TABLE 3.4 **Metamorphism of Shale**

Sedimentary Rock	Low-Grade Metamorphism	Intermediate-Grade Metamorphism	High-Grade Metamorphism
Shale →	Slate →	Schist →	Gneiss
Clay	Very fine clay particles begin to be transformed into very fine-grained mica crystals. Foliation forms.	Mica crystals become large and rock develops strong foliation.	Mica is transformed mostly to feldspar, producing banding (foliation) of light and dark minerals, such as light feldspar with dark amphibole.

Figure 3.26 Mountainside view of metamorphic rocks Metamorphic rocks exposed at a mountain range scale, Kejser Franz Joseph Fiord, East Greenland. The east-dipping (center) white unit is marble. This complex exposure is 800 m (2,625 ft) high and includes several normal or extensional faults that merge downward to a very large extensional fault known as a detachment fault. Metamorphic rock types exposed include gneisses as well as the beautiful white marble. *(Ebbe Hartz, University of Oslo)*

the metamorphism of limestone, usually due to regional metamorphism or, in some instances, contact metamorphism (Figure 3.26). The metamorphic processes cause the calcite to recrystallize into larger crystals. If the parent rock is relatively pure limestone, then a beautiful white marble that Michelangelo might have chosen for his work is produced. *Quartzite* is another important nonfoliated metamorphic rock. Quartzite forms from the regional metamorphism of quartz-rich sandstone.

Metamorphic Rocks and the Environment

There are several points to be made concerning metamorphic rocks and the environment.

Figure 3.27 Slate as roofing material Slate roofs on homes and a pub in southern England. *(Edward A. Keller)*

- *Slate* is generally an excellent foundation material. It has also been used for constructing chalkboards, beds for pool tables, roofing material, and decorative stone counters (Figure 3.27). *Schist*, a coarse-grained metamorphic rock, is composed of soft minerals, making it a poor foundation material for large structures (Figure 3.16g). *Gneiss*, a coarse-grained, banded metamorphic rock, is usually a hard, tough rock similar in most respects to granite and suitable for most engineering purposes (Figure 3.16h).

- Foliation planes of metamorphic rocks are potential planes of weakness. The strength of the rock, its potential to slide, and the movement of water through the rock all vary with the orientation of the foliation. Consider, for example, the construction of road cuts and dams in terrain where foliated metamorphic rocks are common. For road cuts in metamorphic rocks, the preferred orientation is parallel to the foliation, or rock cleavage, with planes dipping away from the road cut. Foliation planes inclined toward the road result in unstable blocks that might fall or slide toward the road (Figure 3.28). And, since groundwater tends to flow down foliation, foliation planes inclined toward the road would cause a drainage problem. In the construction of dams, the preferred orientation is nearly vertical foliation planes parallel to the horizontal axis of the structure[6] (Figure 3.29). This position minimizes the chance of leaks and unstable blocks, which are a serious risk when the foliation planes are parallel to the reservoir walls (see Case History: St. Francis Dam).

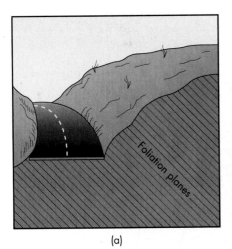

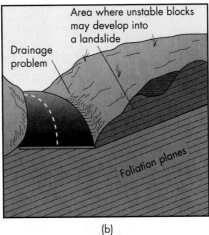

Figure 3.28 Foliation and highway stability Two possible orientations of foliation in metamorphic rock and the effect on highway stability. (a) Where the foliation is inclined away from the road, there is less likelihood that unstable blocks will fall on the roadway. (b) Where the foliation is inclined toward the road, unstable blocks of rock above the road may produce a landslide hazard.

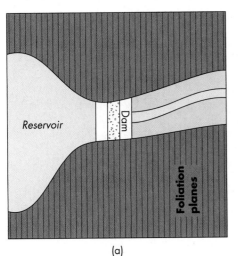

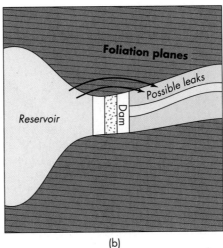

Figure 3.29 Foliation and dam stability Two possible orientations of foliation in metamorphic rocks at a dam and reservoir site. (a) The most favorable orientation is for the foliation to be parallel to the axis of the dam. (b) The least favorable orientation is for the foliation planes to be perpendicular to the axis of the dam. This results because water may flow along the foliation, causing water to leak from the reservoir.

CASE HISTORY | St. Francis Dam

On the night of March 12, 1928, more than 500 lives were lost and $10 million in property damage was sustained as ravaging floodwaters raced down the San Francisquito Canyon near Saugus, California. The 63 m (207 ft) high St. Francis Dam, with a main section 214 m (702 ft) long and holding 47 million cubic meters (almost 1.7 billion cubic feet) of water, had failed (Figure 3.B).

Why Did the Dam Fail? The causes of the failure were clearly geologic. The east canyon wall was metamorphic rock with foliation planes parallel to the wall. Before the failure, both recent and ancient landslides indicated the instability of the rock. The west wall was sedimentary rock with prominent high topographic ridges that suggested the rock was strong and resistant. Unfortunately, the rocks disintegrated when

they became wet. This characteristic was not discovered and tested until after the dam had failed. The contact between the two rock types is a fault with an approximately 1.5 m-thick zone (about 5 ft) of crushed and altered rock. The fault was shown on California's 1922 fault map but either was not recognized or was ignored. Three processes combined to cause the tragedy: slipping of the metamorphic rock, the primary cause of failure; disintegration and sliding of the sedimentary rock; and leakage of water along the fault zone, which washed out the crushed rock. Together, these processes destroyed the bond between the concrete and the rock and precipitated the failure.[7,8] This disaster did serve to focus public attention on the need for geologic investigation as part of siting reservoirs. Such investigations are now standard procedure.

(a)

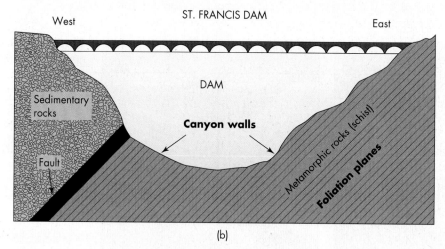

ST. FRANCIS DAM

West   East

Sedimentary rocks

DAM

Canyon walls

Fault

Metamorphic rocks (schist)

Foliation planes

(b)

(c)

Figure 3.B Failure of the St. Francis Dam (a) Before failure. (b) Geology along the axis of the dam. (c) After failure. *(Courtesy of Los Angeles Department of Water and Power)*

3.8 Rock Strength and Deformation

The strength of Earth materials, which is generally defined as resistance to failure such as fracturing, sliding, or flowing, varies with composition, texture, and location. Weak rocks, such as those containing many altered ferromagnesian minerals, may slowly flow under certain conditions and be very difficult to tunnel through. In contrast, some rocks are very strong and need little or no support. However, even strong rocks, when buried deep within Earth, may also flow and be deformed (for example: rupture, rotation, or change in shape such as shortening, thickening, or thinning). Thus, the strength of a rock may be quite different in different environments.

Deformation of Earth materials may be *elastic, plastic, ductile,* or *brittle.* Elastic deformation occurs when the deformed material returns to its original shape after the stress is removed. Examples include compressing a rubber ball, stretching a rubber band, or drawing and releasing an archer's bow. Plastic deformation, on the other hand, is characterized by permanent change. That is, the material does not return to its original shape after the stress has been removed. Examples of plastic deformation include compressing snow into a snowball or stretching chewing gum. If a material ruptures before plastic deformation, it is called *brittle*; materials that rupture after considerable plastic deformation are known as *ductile.* The type of deformation that occurs and the rate of deformation depend upon a variety of circumstances, including the type of rock, the magnitude of the stress it is experiencing, how quickly or slowly the stresses are applied, and the temperature of the material. For example, glass at room temperature is hard and behaves like a brittle substance, but glass heated to a high enough temperature is soft and ductile, deforming plastically before rupturing. That is why glassblowers can make such beautiful objects.

The strength of rocks is often recorded as the compressive strength, tensile strength, or shear strength, referring to the magnitude of a particular stress necessary to cause deformation (Figure 3.30). Before an earthquake, rocks undergo

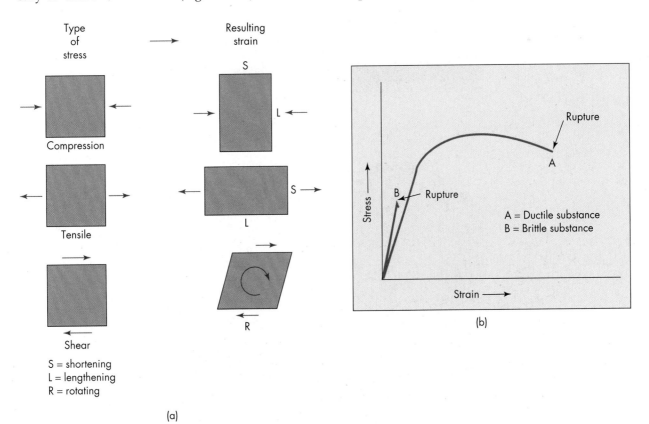

Figure 3.30 Stress and strain (a) Types of stress and resulting strain. (b) Relationships between stress and strain in rocks. A brittle rock has no plastic deformation (the curved part of line A).

stress, followed by strain (deformation). Rupture of the rocks produces an earthquake. Explaining this phenomenon further is beyond the scope of our discussion here, but you will encounter it if you take structural geology or a course in earthquake geology.

3.9 Rock Structures

There are many types of rock structures. Common structures include fractures, folds, and unconformities. We will discuss each in turn.

Fractures

Two common types of rock fracture are *joints*, along which no displacement occurs, and *faults*, along which displacement has occurred (see Chapter 6). Several rock fractures, in this case joints, are shown in Figure 3.31. A fault that displaces

(a)

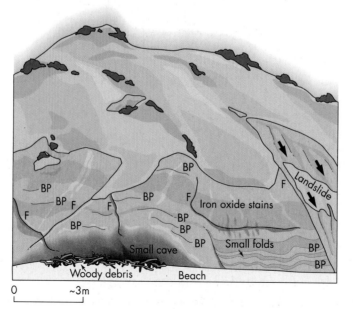

(b)

Figure 3.31 Fractures and folds (a) Light brown shale along a southern California beach. (b) Drawing showing several properties of the rock: fractures (F), folds, iron oxide stains, bedding planes (BP), and small landslide. *(Edward A. Keller)*

Figure 3.32 **Fault outcrop (natural exposure)** Fault on the eastern coast of Santa Cruz Island, California. Arrows show direction of relative displacement. *(Edward A. Keller)*

rock on Santa Cruz Island in Southern California is shown in Figure 3.32. Fractures in rocks, which vary from small hairline breaks to large fault zones up to a kilometer or so wide and hundreds of kilometers long, have environmental significance because

- They are conduits for fluids, most often water, possibly including pollutants.

- They are zones of weakness in rocks. Fractures generally lower the strength of a rock and thus its suitability as a foundation for everything from buildings to bridges, dams, and runways for aircraft.

- Once a fracture has developed it is subject to weathering, which widens it and may produce unstable clay minerals that may easily wash out or facilitate landslides. When large displacements of a meter (3.3 ft) or so occur on fractures, large earthquakes are produced (see Chapter 6).

Folds

When rocks are shortened by lateral compression, a series of folds may develop. As a demonstration, push horizontally on a tablecloth and observe the series of arch-shaped convex folds, called *anticlines*, and bowl-shaped concave folds, called *synclines*, that form. These folds are shown on Figure 3.33a and will be discussed in greater detail with earthquakes in Chapter 6. If active folding is presently occurring in an area, a series of linear anticlinal ridges separated by synclinal valleys may be present as shown on Figure 3.33a. A series of folds is a *fold belt*, and the basic structure that produces linear valleys and ridges can remain as part of the landscape for several hundred million years. The long-lasting nature of a fold belt is illustrated by the eroded fold belt that part of the Appalachian Mountains formed in the southeastern United States (Figure 3.33c; also see Chapter 2). The Appalachian Mountains formed when Europe collided with America over 200 million years ago. The harder rocks form linear ridges and the softer rocks form linear valleys of the Appalachian landscape we observe today. These linear ridges and valleys played an important part in our Civil War; armies might have been only a ridge apart without necessarily knowing each other's position.

Unconformities

An *unconformity* is a significant break or gap in the geologic record. It is a time when erosion rather than deposition occurred. Unconformities are important in understanding the geologic history of a region; they are also important because they are often a natural boundary between two rock types with different characteristics.

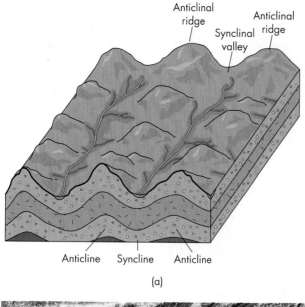

(a)

(b)

(c)

Figure 3.33 Anticlines and synclines (a) Block diagram of anticlines and synclines. (b) In areas with surface expression of active (present) folding, anticlines often form elongated hills and anticlinal ridges such as Sheep Mountain anticline, Wyoming. *(John Shelton)* (c) High-altitude image of the fold belt of the Ridge and Valley Province of the Appalachian Mountains of the southeastern United States. *(Earth Satellite Corporation/Science Photo Library/Photo Researchers, Inc.)*

Springs and oil deposits are sometimes associated with unconformities where downward-migrating water or upward-migrating oil is pushed to the surface.

Unconformities represent a time period when erosion was occurring, and an *erosion surface* may have developed if the land was above sea level. Today, all land on Earth that is above sea level is part of an erosion surface. For example, the Piedmont province in the southeastern United States is a developing erosion surface. If sea level were to rise and cover the land near Charlotte, North Carolina, the igneous and metamorphic rocks that are millions of years old would be buried by young marine sediments. The then buried erosion surface would be recognized as an unconformity in the rock record. There are three types of unconformity:

- *Nonconformity:* forms between older igneous or metamorphic rocks and younger sedimentary rocks (Figure 3.34).
- *Angular unconformity:* younger sedimentary rocks are located upon an erosion surface, below which older sedimentary rocks are tilted or folded (Figures 3.34 and 3.35).
- *Disconformity:* formed as an erosion surface between parallel layers of sedimentary rock. They may be hard to recognize as there may be little visible discontinuity compared with nonconformity and angular unconformity.

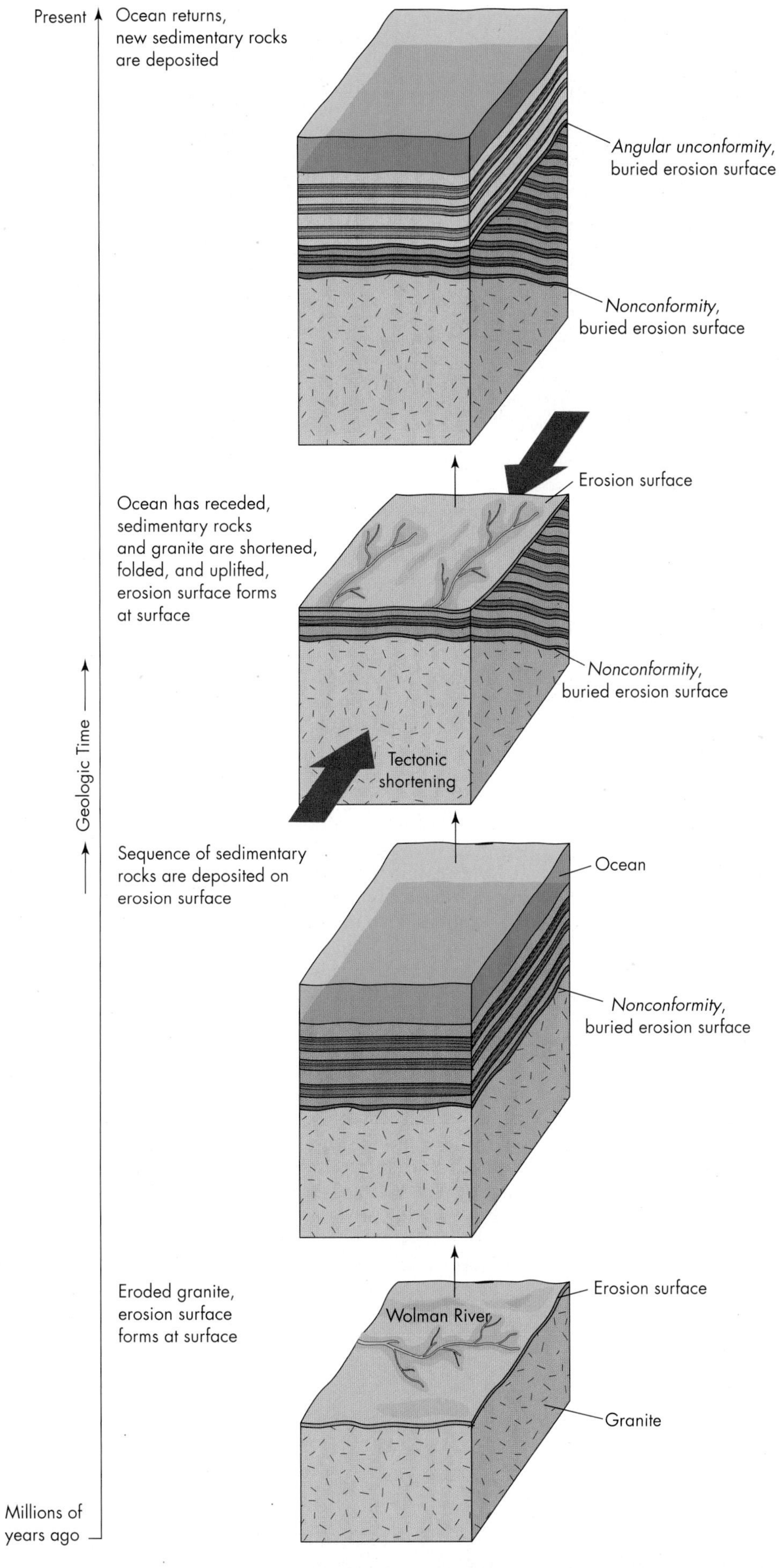

Present

Ocean returns,
new sedimentary rocks
are deposited

Angular unconformity,
buried erosion surface

Nonconformity,
buried erosion surface

Erosion surface

Ocean has receded,
sedimentary rocks
and granite are shortened,
folded, and uplifted,
erosion surface forms
at surface

Nonconformity,
buried erosion surface

Tectonic
shortening

Geologic Time

Sequence of sedimentary
rocks are deposited on
erosion surface

Ocean

Nonconformity,
buried erosion surface

Eroded granite,
erosion surface
forms at surface

Erosion surface

Wolman River

Granite

Millions of
years ago

Figure 3.34 **How unconformities may form** Block diagrams showing how a nonconformity and angular unconformity may form.

Angular unconformity

Figure 3.35 Angular unconformity Angular unconformity between older, more tilted sedimentary layers and younger sedimentary layers. *(A. G. Sylvester)*

SUMMARY

Minerals and rocks are the basic building blocks of the solid Earth and form some of our most basic resources on which we rely heavily for our modern civilization. Minerals and rocks also play an important role in many Earth surface processes, such as landslides, earthquakes, and volcanic activity. Finally, the study of minerals and rocks provides important information concerning the history of Earth.

A mineral is a naturally occurring, normally crystalline element or compound formed by geologic processes. Although there are more than 4,000 minerals, we need to know only a few of them to identify most rocks and help evaluate environmental problems. The major groups of minerals of interest to environmental geology are silicates, oxides, carbonates, sulfides, and native elements.

Rocks are aggregates of one or more minerals that are classified into three general types or families according to how they were formed in the rock cycle. These are igneous, sedimentary, and metamorphic. These three rock types are constantly being created and destroyed as part of the rock cycle. The rock cycle is intimately related to plate tectonics.

Three important rock laws fundamental to understanding Earth history are (1) the law of crosscutting relationships, (2) the law of original horizontality, and (3) the law of superposition.

Igneous rocks are those that crystallized from magma. They can be either extrusive, meaning they cooled at the surface of Earth, or intrusive, meaning they cooled beneath Earth's surface.

Sedimentary rocks form from parts of other rocks or by chemical processes. They can be either detrital, formed from the lithification of rock fragments, or chemical, produced when chemical or biochemical processes cause solid materials to form from substances dissolved in water.

Metamorphic rocks are rocks changed by heat, pressure, chemically active fluids, or some combination of those factors. They can be foliated, having layered or banded alignment of mineral grains, or nonfoliated, with mineral grains lacking layering or banding.

Minerals and rocks have physical and chemical properties that may have environmental importance. Examples include the relationship between pyrite and acid mine drainage, calcite/limestone and cavern formation, basalt and lava tubes, and schist/foliation and landslides.

The strength of a rock depends upon several factors, including composition, texture, structure, and where it is in or on Earth.

Common rock structures include fractures, faults, folds, and unconformities. These structures may have important environmental consequences.

Weathering is the physical breaking apart or chemical decomposition of minerals and rocks at or near the surface of Earth by physical, biological, and chemical processes. Weathering is most important in the formation of soils, preparation of Earth materials for transport by surficial processes, and transformation of feldspar minerals to clay minerals.

Key Terms

igneous rocks (p. 81)

law of crosscutting relationships (p. 82)

law of original horizontality (p. 82)

law of superposition (p. 82)

metamorphic rocks (p. 81)

mineral (p. 72)

rock (p. 81)

rock cycle (p. 81)

sedimentary rocks (p. 81)

weathering (p. 74)

Review Questions

1. What is a mineral?

2. Define *weathering*.

3. Define *clay*.

4. Define *rock*.

5. Differentiate between igneous, sedimentary, and metamorphic rocks.

6. What are the major components of the rock cycle?

7. What is the law of crosscutting relationships?

8. State the laws of original horizontality and superposition.

9. Define *batholith* and *pluton*.

10. How might a porphyritic texture be produced?

11. What is the difference between a detrital and a chemical sedimentary rock?

12. What factors determine the strength of a rock?

13. What factors contributed to the failure of the St. Francis Dam?

14. What are the main rock structures?

15. How might an angular unconformity be produced?

Critical Thinking Question

1. Consider the case history concerning the use of asbestos and the need to remove white asbestos, which is not thought to cause environmental health problems. Suppose you are the superintendent of schools in a large inner-city school system with many old buildings that contain asbestos: pipes wrapped in asbestos, asbestos ceiling and floor tiles, and fire-retardant asbestos used in the auditorium. What steps would you take to determine if there was a hazard, and how would you communicate with parents who are worried that their children are being exposed to harmful materials? Outline a plan of action.

FOUR

Ecology and Geology

Learning Objectives

Ecology and geology are linked in many fascinating and important ways. These linkages and their utility in restoring environments such as rivers, wetlands, or beaches are emphasized in this chapter. Important learning objectives are

- Know some of the basic concepts of ecology and linkages to geology
- Understand the importance of relationships between geology and biodiversity
- Know what factors increase or decrease biodiversity
- Know what human domination of ecosystems is and how we can reduce the human footprint on the environment
- Know why we need an appropriate environmental ethic on a time scale relevant to people today
- Know what ecological restoration and the processes of restoration are

This stream in the mountains of southern California has deep cool pools that form in part because of the large boulders that converge flow, causing scour. The processes of scour produce the pool which is an important habitat for trout.

CASE HISTORY | Endangered Steelhead Trout in Southern California: It's All About Geology

Coastal streams that emerge from southern California mountains and flow into the ocean are not commonly thought of as trout habitat. Steelhead trout are born in mountain streams and travel to the ocean where they remain for several years before returning to spawn. These fish are more commonly associated with streams of the Pacific Northwest. Nevertheless, populations of steelhead trout exist from San Diego in furthermost southern California north to south of San Francisco where they merge with their more northern relatives.

Southern California has a semiarid climate and stream flow is extremely variable. Lower parts of streams often dry up in the summer and much of the entire stream may dry up during drought years that occur periodically. In wet years, especially following wildfires, floods with large amounts of sediment are common. Headwater streams following wildfire or landslides may become choked with gravel that in subsequent years spreads through the system, providing important habitat for fish and other aquatic species.

Summer low flow is particularly important to southern steelhead, which are an endangered species. Adults enter the streams from the ocean during winter months to spawn and may return to the ocean to spawn again in future years. The eggs hatch in the gravel of the stream and young fish reside in the stream for a period of months to a year or so before the urge to go to the ocean moves them to migrate. As part of a study to evaluate the steelhead habitat in the Santa Monica Mountains near Los Angeles, several stream systems were observed during the summer low-flow months. One of the goals as part of a plan to recover endangered southern steelhead was to identify which streams in the Santa Monica Mountains were most capable of supporting steelhead trout. Several of them, including Malibu Creek and Topanga Creek, were known to have steelhead. Geology of the Santa Monica Mountains was found to be an important factor in enhancing the summer low flow that is the major limiting factor for steelhead survival in southern California. Where aquifers are present and groundwater is forced to the surface due to rock fractures or faults, seeps and springs are more common. It was found that on the scale of the Santa Monica Mountains the eastern portion of the range offers higher potential for summer low flow due to favorable geology. As a series of offshore faults come on land and cross the streams, more abundant summer low flow occurs. The faults form a barrier to groundwater, forcing the water toward the surface where it emerges as seeps and springs. During the late fall of 2005 a number of pools were observed in Topanga Canyon (Figure 4.1) and no fish were observed in most of the pools. In two or three pools where fish were observed, seeps and springs from fractures and faults in the sedimentary rocks were clearly providing a source of cold water. The pools were at places where rock banks and large boulders were in the channel. The rocks and boulders constrict the channel, producing a zone of fast water at high flow that scours a pool, providing a low-flow habitat for fish.

The study of the Santa Monica Mountains suggests that it is important to consider geologic factors in streams when assessing fish habitat. In southern California it is clear that the geology and groundwater are important in understanding fish habitat.

4.1 Ecology for Geologists: Basic Terms

Ecology is the study of controls over the distribution and abundance of living things. More generally ecology is the study of living things (organisms) and their interactions and linkages to each other and to the nonliving environment. The nonliving environment is largely controlled by physical and chemical processes related to the geologic cycle (see Figure 3.13). The complex interactions between life and the broader environment are responsible for the creation and maintenance of our living world. Geologic processes from the global to the regional down to the smallest scales, such as a rock under which a lizard lives, greatly affect life processes.

Discussion of relationships between ecology and geology starts by defining a few terms and principles. These include species, population, ecological community, habitat, and niche. Following these definitions, we will turn our discussion to ecosystems where geology plays a full role in partnership with life. A *species* is a group of individuals capable of interbreeding. Population may be defined as a group of individuals of the same species living in the same area. The *ecological community* is a group of populations of different species living in the same area with varying degrees of interaction with each other. We use the term *habitat* to denote where a particular species lives; how it makes a living is known as its *niche*. For example, think about a mountain lion in the mountains of Montana. We say

Figure 4.1 **Fish habitat: It is about geology** This pool in Topanga Canyon is at a site where fractures and faults cause springs that introduce cold water into the stream system. Without this cold water there would be much less habitat for the endangered southern steelhead. *(Edward A. Keller)*

the habitat is the mountains, but the niche of the mountain lion is eating deer and other large mammals.[1]

Before leaving our discussion of some of the basic terms of ecology we will briefly consider types of species. Some species are considered to be indigenous in that they are found in the area where they evolved. Others are exotic species brought into an area or region by humans for a variety of purposes or as accidentals. For example, acacia trees were brought to the United States and planted as wind breaks in arid regions. Two varieties of eucalyptus trees from Australia have been imported and widely planted, as have been numerous other plants and animals from around the Earth. Most exotic species when introduced do not cause problems, but some do. Sometimes we refer to problem exotic species as invasive species. These species will compete with indigenes species and may displace them. Some invasive species are brought in accidentally due to transporting material around the world, whereas others are brought in intentionally. In either case, negative aspects of invasive species are often not anticipated. Introduction of invasive species is one of the major reasons for the extinction of plants and animals around the globe.

Two other terms that are useful are biosphere and biota. The *biosphere* is the part of Earth where life exists and *biota* refers to all organisms living in an area or region up to and including the entire Earth. With these basic definitions behind us, we will consider what an ecosystem is, types of ecosystems, relations between geology and biodiversity, and human domination of ecosystems.

What Is an Ecosystem and How Does It Work?

An **ecosystem** is an ecological community and its nonliving environment in which energy flows and chemicals (such as nutrients and water) cycle. Thus the ecosystem is geology, chemistry, and hydrology and functional linkages with life are many and complex. The basics of this are shown in Figure 4.2, and it's important to remember that energy flows through ecosystems where chemicals are recycled and used numerous times. Sometimes we refer to "ecosystem function," which is rates of flow of energy and cycling of nutrients or other chemicals through an ecosystem. In addition to ecosystem function, other characteristics are structure, process, and change. Ecosystem structure includes two parts: the community of organisms and the nonliving (geologic) environment. The two main processes of ecosystems are energy flow and chemical cycling. Finally, succession is an orderly and sometimes not so orderly change of species as an ecosystem evolves following a disturbance such as volcanic eruption, flood, or wildfire. If the disturbance results in a new land surface such as new land added by volcanic eruption to an island, the succession is called primary. More commonly disturbance involves reestablishment of existing ecosystems following disturbance and is called secondary succession. Secondary succession often involves plants that are called pioneers because they can do well with a lot of light and grow quickly. With time, as more nutrients are cycled and the system develops, the middle stage of succession occurs; this is characterized by the greatest number of species and their ability to use energy and cycle material. The later stages of succession are dominated by fewer species, and in a forest we say these are the old growth. The popular idea of a "balance of nature" with orderly succession to a climax condition where little changes and the system is in equilibrium is an imaginary condition and a concept

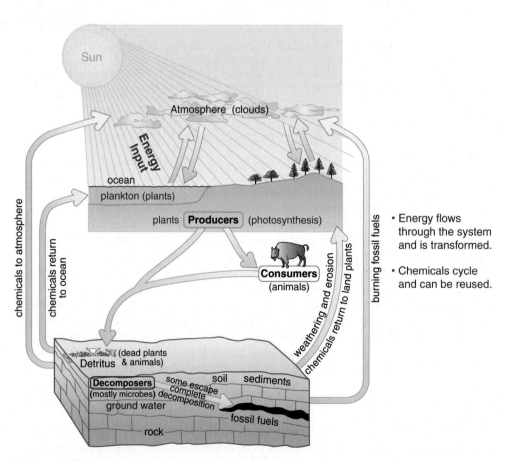

Figure 4.2 Ecosystem basics Idealized diagram showing the basics of an ecosystem in terms of energy flowing and chemicals cycling. *(Edward A. Keller)*

largely rejected by ecologists. Disturbance and change on a variety of scales of time and space are the norm.[1]

There are several types of ecosystems, including indigenous, natural, human modified, and human made or constructed. Completely natural indigenous ecosystems on land are hard to find because human activity has been so pervasive or invasive that almost all ecosystems have been modified by human use and interest. For example, some of the waste of our society, such as lead that is emitted into the atmosphere, is transported around the planet, affecting all ecosystems it comes into contact with, often far from human populations.

Some ecosystems, over a wide range of sites and purposes, are constructed by humans. For example, we may construct shallow ponds or a series of canals known as bioswales that collect runoff of surface water. Marsh plants, such as cattails, when planted in ponds or canals, use and remove nutrients in water that's delivered to them as a waste or pollutant, helping clean the water. Specially designed wetland ecosystems have been constructed where bacteria and plants process mine wastewater and help remove toxins from water. Other large-scale ecosystems are constructed to partially treat urban wastewater. Human-constructed ecosystems are part of what is known as biological engineering.

Natural Service Functions of Ecosystems

Earth is a suitable place to live because the environment produces the necessary resources that living things need to survive. As one of the many species on Earth, we extract resources and receive the benefits of natural service functions from ecosystems (also called ecosystem services). By natural service functions we mean those processes of ecosystems that are responsible for producing clean water and air as well as the mixtures of plants and animals that are necessary for our survival. For example, ecosystem services help cycle elements through the environment, provide nutrients to plants, remove pollutants from water, and through soil fertility allow for increased crop production.

Natural service functions may include buffering functions such as protection from natural hazards such as landsliding and flooding. For example, plants on steep slopes contribute to soil stability through the interaction of growing roots in the soil that increases the strength of slope materials and provides protection from failure by landslide. The roots, especially those with a diameter about the size of a pencil's, bind the soil together much like steel bars in concrete. Similarly, plants on the banks of a stream provide a root mass that stabilizes the soil and helps retard stream-bank erosion. Fresh or saltwater marshes provide a buffer that absorbs wave energy and energy from winds. Coastal marshes protect against coastal flooding and also helps reduce coastal erosion.

We have grown accustomed to the natural service functions that the ecosystems of Earth provide. On the other hand, we sometimes change the land and reduce or eliminate these service functions. For example, when we remove coastal marshes, we are more vulnerable to coastal erosion and flooding from storms such as hurricanes that occasionally strike the coastline. When we drain marshes and wetlands along rivers, we reduce their ability to store water and as a result increase the flood hazard.

4.2 Geology and Biodiversity

Biodiversity is an important concept in environmental science. Most commonly, when we think about biodiversity, we are discussing the number or abundance of species in an ecosystem or ecological community. The number of species is often referred to as species richness whereas the relative proportion of species in an ecosystem is referred to as its species evenness. Another term sometimes encountered is dominant species, which is the species or multiple species that would be most commonly observed in an ecosystem. Plant ecologists use the concept of importance value of a species or multiple species to an ecosystem. The important principle with respect to biodiversity is that geology influences biodiversity from the smallest scale on a hill slope to continental-scale features such as a mountain range.

Biodiversity of Trees in North America and Europe

A fascinating and interesting relationship between geology and biodiversity at the continental scale is the distribution and number of native tree species in North America versus Europe. North America has many more native species of trees than are found in Europe and the hypothesized reason is related to linkages between the ice ages, glaciers, and plate tectonics that determine the orientation of mountain ranges. In North America, the major young mountain ranges run generally north–south and include the Rocky Mountains and the mountains of the West Coast. In Europe the dominant young mountain ranges resulted from the collision between the African and European plates that produced the Alpine range in Spain east through the Himalayas (Figure 4.3). So how might this be related to the number of species of trees in North America and Europe?

The Last Glacial Maxima about 20,000 years ago, when glacial ice covered about 30 percent of the continental area, was a harsh environment for trees. As the continental glaciers grew in North America and Europe, the trees had to migrate in front of the advancing ice. In North America, they found corridors to migrate to the south, but in Europe they were blocked by the east–west Alps that were glaciated. In Europe the trees became trapped between the glaciers and many more species became extinct than in North America. Thus, we see the biodiversity of native trees in North America and Europe was significantly affected by continental-scale mountain building related to plate tectonics.[2]

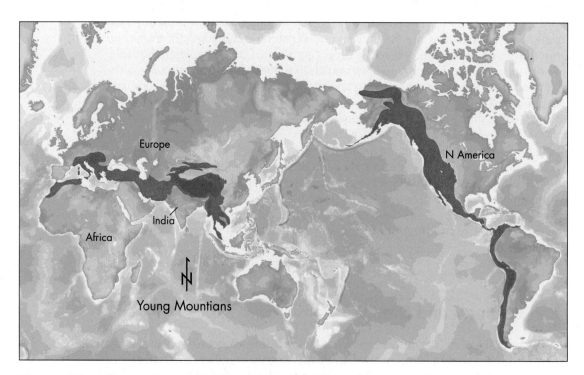

Figure 4.3 **Major young mountains on Earth** A map showing the major young mountain ranges. Notice mountain ranges in North America are roughly north–south and the major mountain ranges in Europe, the Alps and Himalayas, are much more east–west. *(Edward A. Keller)*

Community Effects and Keystone Species: How Are These Concepts Related to Geology?

Two or more organisms may interact in complex ways to affect other organisms in an ecosystem. These organisms are further linked to the nonliving environment and combined ecosystem functions (rates of flow of energy and cycling of nutrients) in ways that help maintain the system. Some individual species have a strong community effect with an influence disproportionate to their abundance. We call these species *keystone species*, and there are many examples of relationships between geology and keystone species that help maintain ecosystems and biodiversity. We will discuss two of these: stream processes in Yellowstone National Park and the kelp forest of southern California.

Stream Processes and Ecology: Story of Mountain Streams, Elk, and Wolves in Yellowstone National Park

Mountain streams are robust systems consisting of the stream channel, bed, and banks composed of silt, sand, gravel, and bedrock. Cool clean water is supplied by the geologic environment through snowmelt and rain that infiltrate the rocks and soil to emerge in the stream as a series of seeps or springs. This water provides the summer low flow. The water itself supports life in the stream, including fish and other organisms, as well as streamside vegetation in the riparian environment, which refers to the environment adjacent to stream channels that is different from the adjacent uplands. Stream bank vegetation also helps retard erosion of the banks and the introduction of an adverse amount of sediment into the stream system. Some of the riparian vegetation, such as cottonwood and willow trees, is a prime food for animals such as a deer and elk that browse on the plants. If there is too much browsing, the abundance of trees will be dramatically reduced; this damages the stream environment by reducing the shade to the stream and also increases bank erosion and introduces harmful fine sediment into the water. Fine sediment (such as silt) is a type of water pollution because it reduces soil resources (it is a product of soil erosion), degrades water quality, and can directly clog gills of aquatic animals. In the stream bed fine sediment may fill the spaces between gravel particles or seal the bed with mud, damaging fish and aquatic insect habitat.

The above scenario was played out in Yellowstone National Park between the 1960s and 1990s, during a period when wolves were not present and elk were not culled. The willows and other streamside vegetation were greatly denuded by browsing elk. Then in the mid-1990s wolves were reintroduced to the Yellowstone National Park ecosystem. Observations by wildlife scientists reported that wolves were more successful in hunting elk in areas where the elk had to negotiate more complex, changing topography, common along the streams where the elk were feeding. The elk soon responded and avoided the stream environments except when absolutely necessary. Over a four-year period, from 1998 to 2002, the percentage of willows eaten by elk dropped from about 90 percent to nearly zero. It is expected that as the streamside vegetation contributes to recovery, the stream channel and banks will also recover to become more as they were prior to the 1920s, when the last wolves in Yellowstone were killed (Figure 4.4). Although the reintroduction of large predators such as wolves is controversial, the wolves are playing a role in the ecosystems in which they are present. Wolves, by hunting elk and scaring them away from the streams, exert a positive force that improves the stream banks and water quality (Figure 4.5). Thus, wolves are a keystone species and affect the broader community of organisms. Wolves are not interested in protecting the streamside environment, but through hunting elk (Figure 4.6) they help maintain a higher quality stream environment.[3]

Figure 4.4 Wolves and streams
(a) Willows along Blacktale Creek in the spring of 1996 are nearly absent because of heavy browsing by elk. *(Yellowstone National Park)* (b) Following reintroduction of wolves only six years earlier, thicker stands of willows are clearly evident. *(William J. Ripple)*

(a)

(b)

	Without Wolves	With Wolves
Predators ↓	Wolves not present (1926–1995) ↘	Wolves restored (post-1995) ↘
Prey ↓	Elk browse on woody species (aspen, cottonwood, and willow) on stream banks ↘	Elk adjust to predation risk ↘
Plants ↓	Decreased abundance of stream side species (aspen, cottonwood, willows, and others) ↙ ↓ ↘	Increased recruitment of woody browse species ↙ ↓ ↘
Other ecosystem responses ↓	Loss of riparian functions Loss of beaver Loss of food web support for other stream side plants and animals ↘ ↙ ↙ Channel erosion and widening, loss of wetlands, loss of connection between streams and adjacent floodplains	Recovery of stream functions Recovery of beaver Recovery of food web for other stream side plants and animals ↘ ↓ ↙ Channels stabilize, recovery of wetlands and stream hydrology

Figure 4.5 Ecosystem processes with and without wolves Diagram showing processes that occur with and without wolves for streams in Yellowstone National Park. Without wolves, extensive browsing by elk degraded this stream environment which has recovered since wolves have been reintroduced. *(Modified after Ripple, J. W., and Beschta, R. L. 2004. Wolves and the ecology of fear; can predation risk structure ecosystems? BioScience 54(8): 755–766)*

Figure 4.6 **Wolves are natural predators of elk** in Yellowstone National Park and their reintroduction has benefited stream ecosystems in unexpected ways. *(Tom McHugh/Photo Researchers, Inc.)*

Coastal Geology, Kelp, Urchins, and Sea Otters

The kelp forests in southern California (Figure 4.7) are a remarkable ecosystem based in part on the local geology. Kelp are a type of large marine alga that grows incredibly fast. The giant kelp of southern California can grow up to more than 25 cm per day, reaching a height of over 10 m above the sea floor. The three parts of the kelp plant include the rootlike holdfast, stem (stipe), and system of blades (leaves). They also have a number of flotation devices near the blades so they will remain in an upright position in the water column. At the surface, particularly at low tide, the sea surface looks like a nearly solid mat of kelp. Below the surface the forest consists of stipes, which are attached to the bottom by the holdfasts. The holdfast is attached to boulders or the rocky bottom. Some of the rocky environments to which they may become attached are wave-cut platforms found offshore in fairly shallow water and rock reefs, which may have water depths of up to several tens of meters. The wave-cut platform has variable relief depending upon the local resistance of the particular rock shelf. Rock reefs often result from geologic uplift from faulting or the presence of more resistant (hard) rock. From Santa Barbara south to Los Angeles these structures are parallel to shore and roughly east–west. Nearly all the offshore reefs that support kelp forests owe their existence to an active geologic environment.

Holdfasts are attached to the rocks with what looks like a root system but it does not take up nutrients. Rather, the function of the holdfast is to hold the plant in place. If the holdfast breaks loose or is destroyed by some process, the kelp will float free and drift to the shore. During storms the kelp moving near the surface will apply stress to the lower plant, causing the holdfast to come loose. The holdfast, with bits of rock, may be transported to the beach to become part of the sediment carried in the near-shore environment. This process can move rock

Figure 4.7 **The kelp forest of southern California** is a remarkably diverse ecosystem. Geology plays an important role in the development of the kelp forest, in which the plants are attached to the ocean bottom in relatively shallow water of wave-cut platforms or uplifted rock reefs. *(Flip Nicklin/Minden Pictures)*

particles from offshore to the beach, while distributing organic material from offshore to the beach.

The kelp forest flourishes in areas of upwelling nutrients and relatively cold water. Growth rates are greatly reduced in years when the marine water warms up, as, for example, during El Niño years when warmer water is present. There is a variety of organisms near the bottom of the kelp forest, including urchins, which are spiny animals that feed on the holdfast of kelp. When urchins feed and break the holdfast, the kelp will float free and die. There are several predators of sea urchins, including humans, who collect them for their eggs, or roe, which is valuable, and sea otters, who eat the urchins (Figure 4.8).

Studies conducted in Alaska suggest that those areas that no longer have many otters are impoverished. In some areas, the sea urchins are so plentiful that kelp scarcely exists. Where sea otters have been restored to their former range after they were nearly exterminated for their valuable fur, the kelp returns and

Figure 4.8 **Sea otters and the kelp forest** Sea otters are a keystone species of the kelp forest. They feed on urchins and other shellfish. *(FRANS LANTING/MINDEN PICTURES)*

flourishes.[1] Linkages between the geologic environment, kelp forest, urchins, and sea otters are idealized in Figure 4.9. Notice that when sea otters are present there are fewer urchins and more abundant kelp.

Sea otters are a keystone species with a strong community effect because when they eat sea urchins, they make it more likely to have a healthy kelp forest. The kelp forest and exposed rock provides the structure and food resources for other animals including fish. Sea otters have no interest in protecting kelp and they don't hang around at the base of the kelp keeping the urchins off. However, through their feeding on urchins, sea otters help ensure a healthy, productive kelp forest.[1]

There is currently a controversy about whether sea otters should continue to expand into their previous habitat in southern California south of Pont Conception, northwest of Santa Barbara. The conflict results because people harvest urchins and fishermen are afraid that if sea otters become abundant, the number of urchins will be reduced to the point that it will not be profitable to harvest them. The sea otters, of course, are unaware of this controversy and are simply moving back into their historic habitat. If the sea otters are allowed to migrate south, their reestablishment will be a slow process and in fact may not occur for a variety of reasons. First and foremost, the coastal waters in some areas of southern California are polluted and pollutants may sicken the sea otters. Also, the climate is changing and ocean waters are warming, which may make the southern regions of the habitat less viable for kelp forests and the urchins that feed on kelp holdfasts. In addition, sea otters are sensitive to water temperature. They have high fat density that provides good insolution in cold water, but can result in overheating in warm water.

Our discussion of wolves in Yellowstone and sea otters in the kelp forest and their relations to the hydrologic and geologic environment is fairly straightforward.

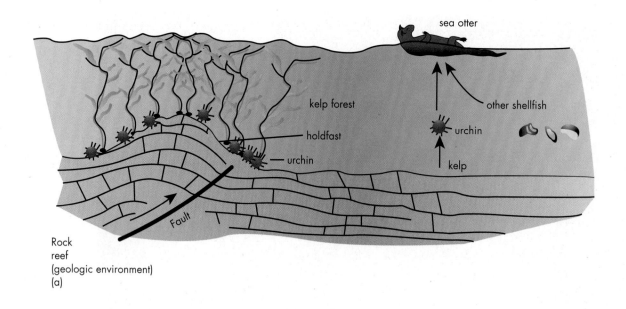

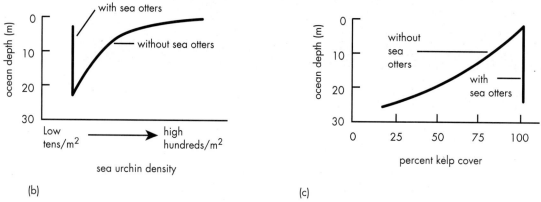

Figure 4.9 **See otters affect kelp forest** Idealized diagrams showing the effects of sea otters on the kelp forest and in particular the abundance of kelp. (a) The geologic environment with uplifted rock reef, kelp, urchins, and sea otters. (b) Without sea otters the density of urchins is high. (c) With sea otters the percent of kelp cover is high. Notice when sea otters are absent, there is a large reduction in the kelp.

There are numerous other examples of how keystone species affect ecosystems. For example, American bison through their grazing practices helped maintain biodiversity of the tall grass prairie and the soil where they roamed (Figure 4.10).

Some species that are not large individuals or highly visible in their ecosystems have large importance from an ecologic perspective. For example, reef-building coral and algae are species that create physical habitat in the coral reef environments of the world. The coral and algae provide the framework upon which other reef life exists (Figure 4.11). Due to warming of the oceans, pollution, and overfishing, coral reefs around the globe are in decline in many places. Finally, ancient coral reefs are present as limestone rock that forms the foundation of many areas where people live, including much of Florida and the Great Lakes area.

Factors That Increase or Decrease Biodiversity

Following our discussion of ecological communities, ecosystems, and keystone species, we can more directly address what sorts of processes are likely to either

Figure 4.10 American bison greatly influenced the tall grass prairie that they roamed across as they grazed. Their grazing habit helped maintain the biodiversity of the prairie grasses and the entire landscape, including soils and wetlands known as buffalo wallows. *(Lowell Georgia/Corbis)*

increase or decrease biodiversity. We are primarily concerned with species richness, that is, the number of species, but biodiversity also relates to genetic diversity, which is the number of genes found in the population that may not always be expressed by the morphology or function of the particular organism.

What Factors Increase Biodiversity?

Biodiversity maybe increased by several factors[1] including

- Presence of diverse habitat with many potential niches. For example, a river with variable depth, turbulence, velocity, and amount of large woody debris (stems and rootwads of trees) will support more species.

Figure 4.11 Coral reefs are composed of a number of organisms including algae and corals, which build up the basic reef structure that greatly affects biodiversity of the marine environment. *(Cousteau Society/ The Image Bank/Getty Images)*

- Moderate amounts of disturbance, such as wildfires, violent storms, or volcanic activity, that provides new or renewed habitat.
- Presence of harsh environments, such as hot springs or nutrient poor rocks and soil may have specialized species that increase diversity at the regional scale.
- Relatively constant environmental factors such as temperature, precipitation, and elevation. This is one of the reasons why there are so many species near the equator. There is a relatively constant supply of energy from the sun near the equator and conditions such as temperature and humidity are relatively constant, leading to greater diversity of organisms.
- Evolution, or generation of biodiversity, which generally refers to slow changes over geologic time in organisms. However, sometimes evolution can occur rapidly with some species.
- An environment that is highly modified by life, as, for example, rich organic soil. At regional scales more biologically productive areas tend to be more diverse.
- Geology, which affects ecosystem function and process from small- to continental-scale environments. See, for example, the earlier discussion of biodiversity and orientation of mountain ranges.

What Factors Reduce Biological Diversity?

Biological diversity may be reduced[1] by

- Presence of extreme environments, as, for example, hot springs or tar seeps that locally provide a much more limited set of habitats and niches for life. From above we see that extreme environments may increase biodiversity at the regional scale. Thus the effect changes with scale.
- Extreme disturbance, or very frequent disturbance such as regional-scale fires, storms, or volcanic activity that catastrophically disrupt ecosystems.
- Transformation of the land, which fragments ecosystems. For examples, above and below a dam on a river that block migration of fish, construction of a large reservoir that leaves a series of small islands isolated from the mainland, and urbanization of a region with few corridors between rural lands for migration of plants and animals.
- Presence of environmental stresses, such as pollution.
- Habitat simplification, for example, agriculture that reduces habitat and construction of engineering structures to control flooding or erosion (see A Closer Look: Seawalls and Biodiversity).
- Introduction of exotic, intrusive species that compete with indigenous species or cause predation or disease in indigenous species.
- Presence of mountain ranges that block or restrict migration of plants and animals.

Human Domination of Ecosystems

It is apparent that we dominate almost all ecosystems on Earth. Study of this domination has led to the general conclusion that the domination has not yet produced a global disaster. However, in some areas disastrous conditions have occurred. What is apparent is that there are many factors that are linked in complex ways to human population increase as well as transformation of the land for human use and global change in climate and the biogeochemical cycles. These processes are resulting in the reduction of biological diversity, including the loss of entire ecosystems and the extinction of many species. We are apparently in a large extinction event that in the last 2,000 years has resulted in a reduction of biodiversity, particularly for birds, mammals, and fish as well as many insects and other species. The rates of extinction during the last two millennia are much larger than would be expected by geologic background rates.

The most significant factor that has led to human domination of Earth's ecosystems is land transformations. These transformations occur as we develop our agriculture, industry, recreation, and urban centers.

A CLOSER LOOK | Seawalls and Biodiversity

A seawall is a structure made of concrete, large boulders, or sometimes wood constructed parallel to the shore with the objective of stopping coastal erosion (Figure 4.A). Waves break on the wall (not the exposed beach), reducing coastal erosion. Seawalls cause an initial narrowing of the beach because they physically intrude on the beach. The seawall causes waves and their energy to be reflected. The reflected waves erode sand near the wall, which further narrows the beach. Unless there is a lot of sand available to replace the eroded sand, the beach over a period of years to decades will become narrower and narrower. In some cases the beach may disappear, except at the lowest tides.

A variety of species use a beach ecosystem. Clams and other shellfish, worms, and sand crabs live below the surface.

Animals including birds and fish hunt and eat what is in the sand or washed out by wave action. Kelp, driftwood, and other plant debris can be found on the surface along with sand fleas and insects that feed on the beach debris. When a beach is narrowed by a seawall, all the life is squeezed into a narrower zone. There are fewer animals in the sand and fewer birds to feed and rest on the beach. There are also less driftwood and stranded kelp on the beach which provide habitat for insects. As a result, biodiversity of the beach decreases.[4] An idealized diagram of the changes is shown in Figure 4.B. The big idea is that as we reduce habitat of species in an ecosystem, their diversity changes. Some species are reduced in number, and others may disappear. This is especially likely if a beach nearly disappears following construction of a seawall.

Figure 4.A Seawall narrows of beach This seawall in southern California is affecting the beach environment by narrowing the sandy beach. *(Edward A. Keller)*

Effects of a seawall over time

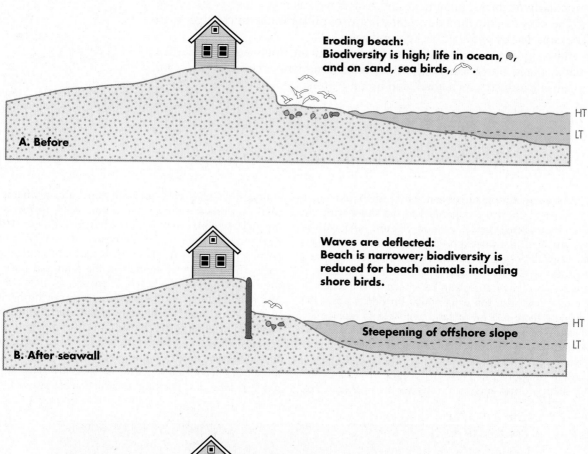

Eroding beach:
Biodiversity is high; life in ocean, ⦿,
and on sand, sea birds, ⌒.

HT
LT

A. Before

Waves are deflected:
Beach is narrower; biodiversity is
reduced for beach animals including
shore birds.

Steepening of offshore slope

HT
LT

B. After seawall

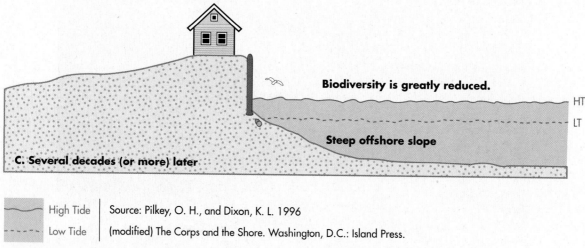

Biodiversity is greatly reduced.

HT
LT

Steep offshore slope

C. Several decades (or more) later

| | High Tide | Source: Pilkey, O. H., and Dixon, K. L. 1996 |
| | Low Tide | (modified) The Corps and the Shore. Washington, D.C.: Island Press. |

Figure 4.B **Seawalls and ecology** Idealized diagram showing the effect of seawalls on a sandy beach environment over a period of decades. As the beach slowly narrows, the biodiversity is reduced for beach animals, including shore birds. In addition, narrower beaches also have less organic debris, including driftwood, that provides habitat for organisms living on the beach.

The Golden Rule of the Environment: A Geologic Perspective All About Timing

Stephen Jay Gould, a famous geologist and ecologist, stated that a proper scientific analysis of the environmental crisis facing people today requires the use of an appropriate scale of time and space. Gould argues that there is "Earth time"(deep time), which is basically geologic, and then there is "human time," which is very much shorter and of interest to people on Earth today. He stated that modern human beings are only one of millions of species that have been on Earth and that each species is unique and has its own value. Gould argues that there is an appropriate environmental ethic based upon the issue of the human time scale versus the "majesty but irrelevance" of geologic time.

Gould's basic conclusion is that we need to make a "pact" with Earth (our home) because She holds all the important playing cards and as such has a power over us. We are in great need of developing a more compatible relationship with our planet on our time. Earth does not need an agreement with us. Such an agreement would be a gift from Mother Earth to us, and is a variation of the *golden rule:* Do unto others as you would have them do unto you. Gould argues that we should cut a deal while Earth is still able to enter into such an agreement on our time scale. He states, "If we scratch her, she will bleed, kick us out, bandage up, and go about her own business at her own time scale."[5] In other words, if we continue to treat our planet with disrespect, we will eventually degrade the environment for humans and many other living things. As a result, we might become extinct sooner, along with many other species. If we sustain ecosystems and resources, we might avoid extinction a bit longer. From a deep time perspective, over hundreds of millions to billions of years, our species will have little effect on Earth's history. However, on a time scale of interest to us we would like to hang around as long as possible. We are not talking about saving Earth. Earth will go on at least a few billion more years with or without us! We are talking about sustaining Earth systems that we depend upon for our health and well-being.

What Can We Do to Reduce the Human Footprint on the Environment?

The human footprint on the environment is the impact we have on our planet, including its resources and ecosystems. Just recognizing that the human footprint is growing is a step (poor pun) in the right direction. We recognize that our total footprint is the product of the footprint per person times the total number of persons. In order to reduce this impact we need to either reduce the number of people or reduce the impact per person. The ways most often mentioned are to reduce human population, use resources more efficiently, and learn to manage our waste better. It is also important to gain a better scientific understanding of ecosystems. We need to know more about how ecosystems are linked to human-induced change and how our social-cultural environment is linked to those changes. When we do this, we recognize the importance of human-dominated ecosystems and their linkages to the broader world. Finally, recognizing that we have a significant footprint, we have a moral responsibility to manage ecosystems better. If we wish to maintain biological diversity of "wild ecosystems" with their "wild species," then we will have to make conscious decisions about how we manage the environment. Particularly important will be more active management, because we recognize that the human population increase is unprecedented in Earth's history. Never before has one species so clearly dominated the world. In some cases, changes are increasing and accelerating, and the more we know about these the better prepared we will be for active management of our planet.[6]

4.3 Ecological Restoration

Restoration ecology is the science behind the process of ecological restoration. Our discussion focuses on the application of the science (restoration ecology) to restoration projects. A simple definition of **ecological restoration** is that it is the process of altering a site or area with the objective of reestablishing indigenous, historical ecosystems. There are many potential restoration projects, including river restoration for fish and other wildlife habitat; dam removal to reunite fragmented river ecosystems (see A Closer Look: Restoration of the Kissimmee River);

Restoration of the Kissimmee River in central and southern Florida is the most ambitious river restoration project in the United States. Prior to channelization, the river was approximately 160 km (100 mi) long, meandering through a flood plain that was several kilometers wide (Figure 4.C). The river had an unusual hydrology because it inundated its floodplain for prolonged periods of time. As a result, the floodplain and river supported a biologically diverse ecosystem consisting of wetland plants, wading birds, waterfowl, fish, and other wildlife.

Prior to about 1940 few people lived in the Kissimmee basin and the land use was primarily farming and cattle ranching.

However, rapid development and growth occurred following World War II, and in 1947, widespread flooding occurred as a hurricane moved through the basin. As a result, the state of Florida requested the federal government to design a flood-control plan for central and southern Florida. The channelization of the Kissimmee River was planned from about 1954 to 1960, and between 1962 and 1971 the river was channelized. Approximately two-thirds of the floodplain was drained and a canal was excavated, turning the meandering river into a straight canal. As a result of the channelization, ecosystem function was degraded. Wetlands with populations of birds and fish were drastically reduced. Outrage over the

Figure 4.C Restoration of the Kissimmee River (a) The Kissimmee River in central and southern Florida prior to channelization. (*South Florida Water Management District*) (b) The Kissimmee following restoration which straightened the channel, essentially producing a river as a ditch. (*South Florida Water Management District*)

(a)

(b)

loss of the river went on years before the current restoration efforts were discussed and planned. The purpose of the restoration is to return a portion of the river to its historic meandering riverbed and wide floodplain. Specific objectives[9] are

- Restore historic biological diversity and ecosystem function.
- Re-create the historic pattern of wetland plant communities.
- Reestablish the historic hydrologic conditions with prolonged flooding of the floodplain.
- Re-create the historic river floodplain environment and its connection to the main river.

The restoration project was authorized by the U.S. Congress in 1992 in partnership with the South Florida Water Management District and the U.S. Army Corps of Engineers. The general plan for the restoration is shown in Figure 4.D.

The restoration of the Kissimmee River is an ongoing project that began several years ago. By 2001, approximately 12 km of the nearly straight channel have been restored to a meandering channel with floodplain wetlands about 24 km long. This returned the ecosystems to a more natural state as water again flowed through a meandering channel and onto the floodplain. As a result, wetland vegetation was reestablished and birds and other wildlife are returning. The potential flood hazard is being addressed as the restoration plans allow the river to meander through its floodplain while maintaining flood protection. Retaining flood protection is the reason the entire river will not be returned to what it was prior to channelization; that is, some of the structures that control the floodwaters will be removed and others will be maintained.

The cost of the restoration for the Kissimmee River will be several times greater than it was to channelize it. However, the project reflects our values in maintaining biological diversity and providing for recreational activities in a more natural environment.

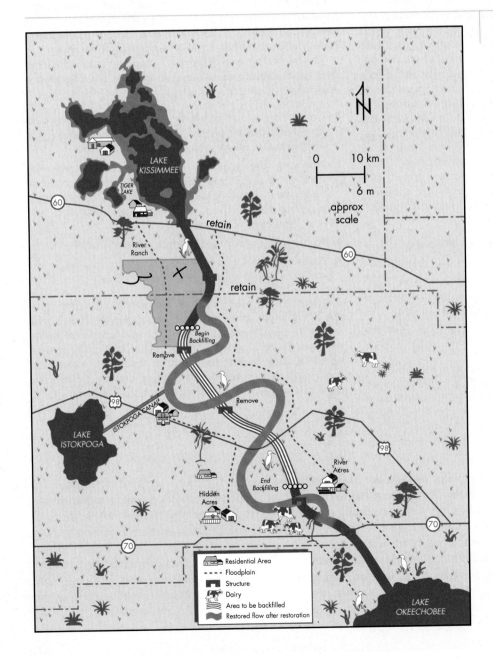

Figure 4.D Kissimmee River restoration plan General plan for restoration of the Kissimmee River. One of the major objectives is to restore biological diversity in ecosystem function of the river as well as re-create the historic river floodplain environment that is connected to the main river. *(South Florida Water Management District)*

A CLOSER LOOK | Restoration of the Florida Everglades

The Florida Everglades is considered to be one of the nation's most valuable ecological treasures. The Everglades' ecosystem stretches from a series of small lakes near Orlando, Florida, southward to Florida Bay for a length of several hundred kilometers (Figure 4.E). That portion of the ecosystem north of Lake Okeechobee comprises the drainage area feeding into the lake, while the area south of the lake is a long system of wetlands that may be hydrologically described as a shallow, wide body of slowly moving water. Tourists from around the world have visited the Everglades for generations. They come to see the unique landscape as well as the wildlife. The Everglades is home to more than 11,000 species of plants, as well as several hundred bird species and numerous species of fish and marine mammals including the endangered Florida manatee. The wetlands and surrounding areas are the last remaining habitat for nearly 70 threatened or endangered species, including the Florida panther and the American crocodile. Since about 1900, much of the Everglades has been drained for agriculture and urban developments, and only 50 percent of the original wetlands remain. A complex system of canals and levees controls much of the flow of water for a variety of purposes, including flood control, water supply, and land drainage.

Over many decades the draining of the wetlands and the encroachment of urban areas have degraded the Everglades' ecosystem to the point where the resource was likely to be lost without a restoration program. That restoration is now underway and it is the largest environmental project in a wetland in the world. The program is a 30-year endeavor that will cost upwards of $10 billion. Restoration goals[10] include

- Restoration of more natural hydrologic processes.
- Enhancement and recovery of native and particularly endangered species.
- Improvement of water quality, especially control of nutrients from agricultural and urban areas.
- Restoration of habitat for all wildlife that uses the Everglades.

The restoration plan is an aggressive one that involves federal, state, and tribal partners as well as numerous other groups interested in the Everglades. The progress to date is notable, as pollution of water flowing into the Everglades from agricultural activities has been reduced by about one-half. As a result, the water entering the system is cleaner than it has been for years. In addition, thousands of hectares have been treated to remove invasive, exotic species such as Brazilian pepper trees and tilapia (a fish from Africa) with the objective of improving and conserving habitat for a variety of endangered and other species.

The plan to restore the Everglades is a long-term one and involves many years of scientific research yet to be completed. The program is complicated by the fact that over 5 million people live in south Florida. The area has a rapidly growing economy, and many urban issues related to water quality and land use need to be addressed. Perhaps the biggest issue is the water and the plan will need to carefully consider restoration that delivers the water in the proper amount, quantity, and place to support ecosystems in the Everglades. This will be a challenge because millions of people in South Florida are also competing for the water.

The overriding goal of the restoration is to ensure the long-term sustainability of water resources in the Everglades' ecosystem.[10] Doing this includes

- Controlling human population in south Florida, access to the Everglades by people, and human development that encroaches on the Everglades.
- Applying the principle of environmental unity (see Chapter 1; everything affects everything else) to better understand and anticipate possible consequences resulting from changes to the geologic, hydrologic, and ecological parts of the system.
- Applying the precautionary principle (see Chapter 1).
- Analyzing rates and changes of systems related to the water, land, and wildlife as linked to people.

It is clear that the people of south Florida and the nation value the Everglades' ecosystem and are applying science with values to implement a far-reaching ecological restoration program. Hopefully it will be successful and future generations will look back to this point in time as a "watershed" (poor pun) that started the ball rolling to protect one of the most valuable ecological resources in North America.

floodplain restoration to improve ecological function; wetland restoration of fresh or saltwater marshes for a variety of purposes from flood control to improving wildlife habitat to providing a buffer to coastal erosion (see A Closer Look: Restoration of the Florida Everglades); beach and coastal sand dune restoration to help manage beach erosion and provide wildlife habitat (see A Closer Look: Coastal Sand Dune Restoration at Pocket Beaches: University of California, Santa Barbara); restoration of habitat for endangered species; reshaping the land, drainage, and vegetation following surface mining; and restoring habitat for native species and wildlife impacted by clearcut logging or for improved fire management.

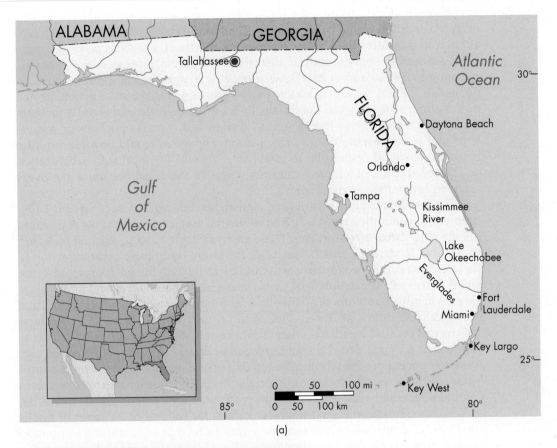

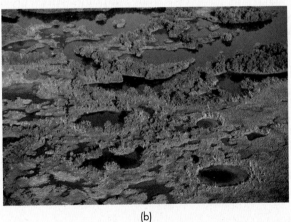

(b)

Figure 4.E Florida Everglades (a) Map of Florida showing the Everglades in southernmost Florida. The Everglades is a wide, shallow, flowing riverlike system extending from Lake Okeechobee south to the Atlantic Ocean. (b) The Florida Everglades is a diverse ecosystem. (*Farrell Grehan/CORBIS*)

A common objective of the restoration process is to change a degraded ecosystem so that it resembles a less human-disturbed ecosystem and contains the structure, function, diversity, and processes of the desired ecosystem. In an attempt to add a human and social component to ecological restoration, it might also be defined as the process of deliberately modifying a site or area to compensate for environmental degradation caused by humans. The purpose of the restoration is to reestablish sustainable ecosystems and develop a new relationship between the natural environment and the human-modified environment.[7] When we try to apply these ideas and principles, we come to the realization that many human-altered environments are nothing like natural systems. Furthermore, it is nearly impossible to restore many sites to their initial conditions because they have been irreversibly changed. For example, many coastal wetlands and river wetlands have been drained, filled in, and developed for human use. Returning them to their original condition would be impossible. Therefore, a more practical approach is to attempt to transform the present ecosystems and landscapes into ones that more closely resemble ecosystems less disturbed by human processes. For example, planning to restore a river that flows through a city might to include removal of nonnative species, replanting native plants, and allowing the river to behave in a more natural state within its floodplain. In doing so, the restoration would produce a greenbelt along a river that is well vegetated and contains portions of ecosystems that were present prior to urbanization. The result would be a compromise between ecological restoration to establish what was there prior to urbanization and the production of a more natural river system that functions more like rivers found in a nonurban environment.[8]

It is possible to restore a variety of landscapes, but regardless of what is being restored, consideration must be given to several major factors that we sometimes call the "Big Three" in restoration. These are hydrologic process, soil and rock, and vegetation. Hydrologic process includes the entire hydrologic cycle, referring to surface waters and groundwater. Soils and rocks include the microorganisms that reside in the soil. Vegetation is the cover material on land and in wetland ecosystems. When we apply the Big Three in a restoration project we soon realize that ecological restoration, as it is applied today, is part art and part applied science. This results because we do not often know enough about particular ecosystems to define their function, structure, and process accurately. Ecological restoration also has a component of societal contribution that goes along with the scientific contribution. In this respect restoration is a social activity. The contribution from people in communities where restoration will take place, or as they are sometimes called, the stakeholders, is important in establishing goals for restoration. A series of public meetings for comment and recommendations are held prior to starting restoration projects. There is feedback between the goals and scientific endpoints, which are a consequence of the goals. For example, if we are restoring a salt marsh and want the plants in the marsh to reduce the nutrient loading to a particular level, then that level is a scientific endpoint. If it turns out that the endpoint may not be reached, then further consideration of goals may be in order. The scientific measures refer to the measurements and monitoring that happen along the way toward achievement of scientific endpoints and the goals of the restoration. The contribution of science is important in developing the endpoints and measures.

Ecological restoration may vary from very simple procedures, such as removing exotic species of plants and planting desired species, with a focus on the restoration of composition and structure, to reconstruction of the entire landscape with a focus on process. Reconstruction of a landscape and its ecosystems requires intimate knowledge of geologic and hydrologic processes. Table 4.1 lists details of the processes of restoration.

Regardless of the nature and extent of the restoration being done, it is a negotiated process between people interested in the project, those doing the restoration, and scientists making recommendations, gathering data, and analyzing results. It is also important to recognize that evaluation of the restoration project starts long before construction. This evaluation involves identifying environmentally sensitive aspects of the project and how they might interact with important historic or cultural values. Prior to construction a careful evaluation of the geologic, hydrologic, and vegetation conditions at the site is necessary to match the environment to the restoration procedures. In other words, we evaluate the hydrology, geology, and vegetation in terms of the overall goals of the project and what is being restored. During restoration, monitoring is often necessary to help evaluate whether the objectives are being achieved. Also, during the entire project, including the development of goals, those doing the restoration

TABLE 4.1 Steps and Procedures in Planning and Initiating an Ecological Restoration Project

1. Develop an ecological description of the area to be restored.

2. Provide a clear understanding of the need for the restoration.

3. Define the objectives and goals of the project.

4. Specifically state the procedures that will be used to achieve the restoration.

5. Clearly know the reference ecosystem that the restoration is attempting to reach.

6. Determine how the restored ecosystem will be self-sustaining; that is, provide for flow of energy and cycling of chemicals to ensure long-term self-maintenance of the restored ecosystem and stable linkages to other ecosystems.

7. State the standards of performance during restoration and monitoring following completion.

8. Work with all people (stakeholders) interested in the project from initiation through completion and postproject monitoring.

9. Examine what the potential consequences of the project are likely to be; that is, apply the principle of environmental unity, that everything affects everything else and anticipate what primary, secondary, and tertiary effects may be.

Source: Modified after Society for Ecological Restoration, 2004. The SER international primer on ecological restoration, www.SER.org.

should work closely with the community people involved, including environmentalists, property owners, and other interest groups, to maximize opportunities to cooperate in the restoration project. Finally, it's important to recognize that ecological restoration, while often successful, is likely to have mixed success. This is particularly likely in the first years of a restoration project, when unexpected difficulties often occur, making changes in the restoration plan necessary. There may also be natural events such as droughts or severe storms that will interfere with the restoration activities.

Another concept related to ecological restoration is biological engineering, which is the use of vegetation in engineering projects to achieve specific goals such as protecting stream banks from erosion.[7,8] On a broader scale, ecological engineering includes designing and constructing ecosystems.

A CLOSER LOOK | Coastal Sand Dune Restoration at Pocket Beaches: University of California, Santa Barbara

The University of California, Santa Barbara, has several kilometers of beach on its campus. Two small pocket beaches exist at locations where the sea cliff is interrupted or cut through by prehistoric channels. At these locations some sand blows from the beaches inland for several tens of meters and a series of low coastal sand dunes have developed. Over the years, the dunes were colonized by a South African species of ice plant. The ice plant came to cover the entire dune, inhibiting normal dune function by allowing no other plants to grow and inhibiting natural sand movement. As a result, the biodiversity of the dune ecosystem was greatly reduced. Restoration of the dunes has involved solarizing the ice plant. This is done by covering the plants with black plastic for several months until they die and turn into mulch. The dune is then planted with native dune species (Figure 4.F). As native vegetation returns to the dunes, the ecosystem is restored. The restoration of the dunes was simple and straightforward, with limited objectives of removing exotic, invasive species and replacing them with native species. When this was done the function of the ecosystem recovered and biological diversity increased. In this case the restoration of composition (plants) promotes the restoration of process (allows the sand in the dunes to move). However, in the Everglades and Kissimmee River examples hydrologic processes need to be restored prior to restoring composition.

Figure 4.F Sand dune restoration Sand dunes at the University of California, Santa Barbara, following restoration. The invasive species has been removed and native dunes species have been planted. *(Edward A. Keller)*

SUMMARY

Ecology is the study of living things and their interactions and linkages to each other and their nonliving environment. It can also be viewed as the study of factors influencing the distribution and abundance of species. An ecosystem is a community of organisms and their interactions with the nonliving environment in which energy flows and chemicals cycle. Therefore, an understanding of ecosystems involves understanding the geologic, biogeochemical, and hydrologic environment.

Geology has many linkages to biodiversity through its influence on the abundance and distribution of plants and animals. Examples include trout habitat in southern California; the forests of North America and Europe; wolves and elk in Yellowstone National Park and their relationship to stream processes; and the kelp forest in California related to the interaction of sea otters and the urchins they eat.

Human activity and interest dominate almost all ecosystems on Earth. Our processes (influences) are reducing biodiversity, including the loss of entire ecosystems and the extinction of many species. The processes of most concern are land transformation, as, for example, transformation of the land from a natural environment to agriculture or urban uses; introduction of invasive species; processes that cause global change; and change to biogeochemical cycles. Reducing the human footprint on the ecosystems of Earth is an important objective and will include, among other activities, controlling human population, managing Earth's resources for sustainability, and managing our waste more efficiently.

We are in need of an appropriate environmental ethic based on a time scale of interest to people. If we choose to sustain resources and ecosystems, we will have a more compatible relationship with our home (Earth) and our species will be able to remain on the planet a bit longer.

Ecological restoration is emerging as an important process with a variety of goals and objectives. In general, ecological restoration has the purpose of reestablishing sustainable ecosystem structure, process, and function.

Revisiting Fundamental Concepts

Human Population Growth

Controlling human population growth is an important objective in sustaining our environment and reducing the human footprint on the environment. If we are unable to control human growth, we will have little success in achieving sustainable management of our resources.

Sustainability

Sustainability is the environmental objective. Sustaining Earth's ecosystems is necessary if we hope to sustain our human system. We share our planet with a vast number of other living things and many of these are necessary for our own well-being. Therefore, when we consider geology and ecosystems, we need to consider those linkages that will lead to sustainable development in its broadest context.

Earth as a System

Ecosystems are an important component of Earth systems that link living organisms with their nonliving environment. Geologic and hydrologic systems have simple to complex relationships with communities of organisms. These organisms have important linkages to the diversity of living organisms in ecosystems. Maintaining living systems in a desired and sustainable state must involve understanding the role of geologic and hydrologic processes and maintaining these processes in a way that supports the desired biological conditions.

Hazardous Earth Processes, Risk Assessment, and Perception

Ecosystems have natural service functions that are linked to hazards such as landslides and flooding. Bank vegetation on stream banks increases the stability of stream banks and their resistance to erosion. Wetlands moderate floodwater by storing and releasing it slowly, thus reducing flood hazards. Wetlands provide a buffer to storm waves from the ocean and thus help protect coastal areas from erosion and flooding.

Scientific Knowledge and Values

The science of ecology has provided a lot of information about ecosystems. How we use that knowledge to restore and maintain ecosystems will reflect our values.

Key Terms

ecology (p. 106)
ecosystem (p. 108)

biodiversity (p. 109)
ecological restoration (p. 121)

Review Questions

1. Define an ecosystem.
2. What is the relationship between a species and ecological community?
3. What is meant by biodiversity?
4. What factors may increase or decrease biodiversity?
5. What was learned by the case histories of wolves in Yellowstone National Park and sea otters in the kelp forest of southern California?
6. How do seawalls reduce biodiversity?
7. What is meant by the golden rule of the environment?
8. What is ecological restoration?
9. What is the difference between ecological restoration, naturalization, and biological engineering?
10. What are some important linkages between geology and ecology?

Critical Thinking Questions

1. An ecosystem consists of an ecological community as well as its nonliving environment. Which of the two do you think is more important and why? In other words, do you think the physical environment comes before what lives there or does what lives there affect the physical environment to a greater extent than the role of geology? Or are both things equally important?

2. Visit a local stream. Carefully examine the stream and surrounding environment and try to determine the amount of human domination. Is the stream you are looking at a candidate for restoration? If so, what could be done? If not, why not?

3. Why may some geologists and hydrologists not be particularly aware of ecosystem function in terms of the biological environment and why are some biologists unaware of the details of the physical and hydrological conditions of the ecosystems they work with? What can be done to narrow the gap between those working on living systems and those working on geological or hydrological systems?

PART TWO

Earth Processes and Natural Hazards

Our focus in Part 2 turns to the major natural hazards: an introduction to hazardous processes (Chapter 5), earthquakes, including tsunamis (Chapter 6), volcanic activity (Chapter 7), flooding (Chapter 8), landslides (Chapter 9), coastal processes, including hurricanes (Chapter 10) , and impacts of asteroids or comets (Chapter 11). The purpose is not to provide extensive amounts of detailed information concerning these natural processes that we define as hazards but to focus on the basics involved and the environmental concerns resulting from interactions between people and natural processes and hazards. The major principles presented are: 1) Earth is a dynamic environment, and change resulting from natural processes is the norm rather than an exception; 2) we must strive to

learn all we can about natural processes and hazards so that effects on human society may be minimized; and 3) human population increase and changing land use are greatly increasing the threat of loss of life and property to natural hazards. Of particular importance is the recognition of locations where hazardous processes are going to occur and their natural or human-induced return period (time between events). We will learn that the environmentally preferred adjustment to natural hazards is environmental planning to avoid those locations where the hazards are most likely to occur and to zone the land appropriately. Environmental planning involves detailed study of natural processes and mapping of those processes to produce environmental maps useful in the planning process.

FIVE

Introduction to Natural Hazards

Learning Objectives

Natural hazards are naturally occurring processes that may be dangerous to human life and structures. Volcanic eruptions, earthquakes, floods, and hurricanes are all examples of natural hazards. Human population continues to increase, and there is a need to develop environmentally sound strategies to minimize the loss of life and property damage from hazards, especially in urban areas. The study of hazardous processes, therefore, constitutes one of the main activities of environmental geology. The learning objectives for this chapter are

- Understand why increasing population and changing land use increase the threat of loss of life and property from a natural disaster to the level of a catastrophe

- Know the conditions that make some natural Earth processes hazardous to people

- Understand how a natural process that gives rise to disasters may also be beneficial to people

- Understand the various natural processes that constitute hazards to people and property

- Know why history, linkages between processes, prediction, and risk assessment are important in determining the threat from natural hazards

- Know how people perceive and adjust to potential natural hazards

- Know the stages of recovery following natural disasters and catastrophes

Hurricane Katrina, a giant storm Hurricane Katrina approaching New Orleans region—aerial image. *(NOAA)*

CASE HISTORY	Hurricane Katrina, Most Serious Natural Catastrophe in U.S. History

Hurricane Katrina (see opening photograph) made landfall in the early evening of August 29, 2005, about 45 km (30 miles) to the east of New Orleans. Katrina was a huge storm that caused serious damage up to 160 km (100 mi) from its center. The storm produced a storm surge (a mound of water pushed onshore by the storm) of 3 to 6 meters (9 to 20 feet). Much of the coastline of Louisiana and Mississippi was devastated as coastal barrier islands and beaches were eroded and homes destroyed. Property damage from Hurricane Katrina and costs to rehabilitate or rebuild the area may exceed $100 billion, making it the most costly hurricane in the history of the United States. The number of human deaths will never be known for sure, as many bodies may have been washed out to sea or buried too deep to be found. The official number of deaths is 1,836. The hurricane and subsequent flooding set into motion a series of events that caused significant environmental consequences. Initial loss of life and property from wind damage and storm surge was immense. Entire coastal communities disappeared with their fishing industry.

At first it was thought the city of New Orleans had been spared, as the hurricane did not make a direct hit. The situation turned into a catastrophe when water from Lake Pontchartrain, north of the city and connected to the Gulf, flooded the city. Levees capped with walls, constructed to keep the water in the lake and protect low-lying parts of the city, collapsed in two locations and water poured in. Another levee failed on the Gulf side of the city and contributed to the flooding. Approximately 80 percent of New Orleans was under water from knee deep to rooftop or greater depths (Figure 5.1). People who could have evacuated but didn't, and those who couldn't because they lacked transportation, took the brunt of the storm. A part of the city that wasn't flooded was in the French Quarter (Old Town), which is the area of New Orleans famous for music and Mardi Gras. The people who built New Orleans over two hundred years ago realized that much of the area was low in elevation and built on the natural levees of the Mississippi River. The natural levees formed by periodic overbank deposition of sediment

from the river over thousands of years. They are parallel to the river channel and are higher than the adjacent land, providing natural flood protection.

As marshes and swamps were drained in lower areas, the city expanded into low areas with a much greater flood hazard. Much of the city is in a natural bowl and parts are a meter or so (3 to 9 feet) below sea level (Figure 5.2).

It has been known for a long time that if a large hurricane were to make a direct or near-direct hit on the city, extensive flooding and losses would result. The warnings were not completely ignored, but sufficient funds were not forthcoming to maintain the levee and system of floodwalls to protect low-lying areas of the city from a large hurricane.

The fact that the region is subsiding at highly variable rates from 1 to 4 m (3 to 12 feet) per 100 years contributed to the flood hazard. Over short periods, 50 percent to 75 percent of the subsidence in some areas is natural, resulting from geologic processes (movement along faults) that formed the Gulf of Mexico and the Mississippi River delta.[1] As much as several meters (more than 10 feet) of subsidence has occurred in the last 100 years, and during that period sea level has risen about 20 cm (8 in.). The rise is due in part to global warming. As the Gulf water and ocean water warm, they expand, raising sea level. The subsidence results in part from a number of human processes including extraction of groundwater, oil, and gas, as well as loss of freshwater wetlands that compact and sink when denied sediment from the Mississippi River. Because the Mississippi River has artificial levees (embankments constructed by humans), it no longer delivers sediment to the wetlands. Therefore wetlands have stopped building up from sediment accumulation. Before the levees were constructed in the Mississippi River delta, floodwater with its sediment spread across the delta helping maintain wetland soils and plants. The freshwater wetlands near New Orleans were largely removed during past decades. As wetlands are removed, they are replaced by saltwater ecosystems as sea level rises and the land continues to subside. The freshwater wetlands are a better buffer to winds and storm waves than are saltwater wetlands. Tall trees such as cypress and other plants of freshwater wetlands (Figure 5.3) provide roughness that slows down water from high waves or storm surge moving inland. It's well known that one of the natural service functions of both saltwater and freshwater coastal wetlands is to provide protection to inland areas from storms.

New Orleans Will Be Rebuilt.   The people of New Orleans are resilient and will move back into this famous historic city. We have learned from this event and better measures are being taken to ensure that this kind of catastrophe is less likely to occur in the future. For example, in some areas where flood waters, even with future levee failure, will be less than one story high, reconstruction includes flood-proof buildings by constructing living areas on the second floor.

The U.S. Army Corps of Engineers, which is largely responsible for the flood protection of New Orleans, produced a draft report in June 2006 concerning the hurricane and flood protection system. They acknowledged that the system had evolved piecemeal over a number of decades and

Figure 5.1 Katrina floods New Orleans New Orleans was flooded when flood defenses failed during Hurricane Katrina. *(N. Smiley/Pool/ Dallas Morning News/Corbis)*

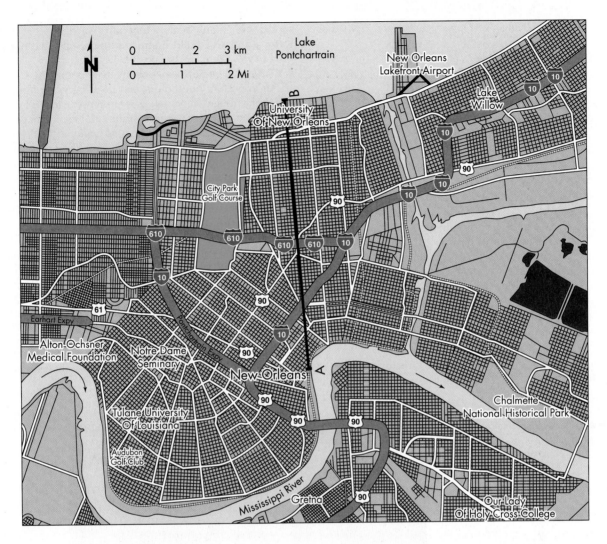

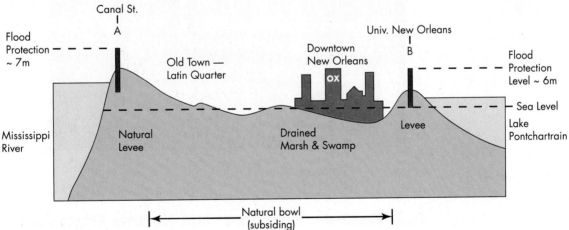

Figure 5.2 **Map and cross section of New Orleans** Much of the city is below sea level between the Mississippi River and Lake Pontchartrain. *(Edward A. Keller)*

that it was a system in name only. Nevertheless, the flood protection structures were constructed to protect the city and their failure was responsible for a majority of the flooding that resulted from Hurricane Katrina in 2005.

Some conclusions of the report[2] are

■ There were no fallback redundancies (second-tier protection) to the flood protection if the primary flood control

structures failed. Pumping stations designed to remove floodwaters were the only example of a redundant system and these were not designed to function in a major hurricane with extensive flooding.

■ Hurricane Katrina exceed the design criteria of the flood protection structures. Some levee and floodwall failures resulted from being overtopped by floodwaters. Others failed, without being overtopped, by erosion from their front side.

- Regional subsidence has been faster than was appreciated. The heights of the flood protection structures were not adjusted for subsidence, and some of the floodwalls and levees were as much as 1m (3 ft) below the elevation they were designed at.

- Although scientific knowledge about hurricanes and storm surges has increased, this did not lead to updating the flood control plan. What adjustment did occur was fragmented rather than consistent and uniform.

- Consequences of flooding were also concentrated. More than 75 percent of the people who died were over 60 years old and were located in areas with greatest depth of flooding. The larger number of deaths of the elderly occurred because poor, elderly people and disabled people were the least able to evacuate without assistance.

- Parts of the flood and hurricane protection system have been repaired at a cost of about $800 million since Katrina, and these are the strongest parts of the flood protection. At present, the protection level remains the same as before Katrina.

In fairness to the Corps, flood control structures are often underfunded, and as with New Orleans, construction is spread over many years. Hopefully, we have learned from Katrina and will build a stronger, more effective hurricane protection system for the city of New Orleans, as well as other U.S. cities where hurricanes are likely to occur.

An important question is, can flooding occur again even if higher, stronger flood defenses are constructed? Of course it can—when a bigger storm strikes, future damage is inevitable. If freshwater marshes are restored and the river waters of the Mississippi allowed to flow through them again, they will help provide a buffer from winds and waves. As previously mentioned, freshwater marshes have trees that provide a roughness to the land that slows wind and retards the advance of waves from the gulf. Every 1.5 km (1 mile) of marsh land can reduce waves by about 25 cm (1 ft).

Tremendous amounts of money will need to be spent to make New Orleans more resistant to future storms. In light of the many billions of dollars in damages from catastrophes, it seems prudent to spend the money in a proactive way to protect important resources, particularly in our major cities.

In the remainder of this chapter, we will discuss some of the principles of natural processes we know as hazards and how they produce disasters and catastrophes. You will learn how poor land uses and changing land uses, coupled with population increase, greatly increase the risk of some hazards.

Figure 5.3 Cypress tree freshwater wetlands along the coast of Louisiana. *(Tim Fitzharris/Minden Pictures)*

5.1 Hazards, Disasters, and Natural Processes

Natural Disasters: Loss of Life and Property Damages

Natural disasters, which are events that cause great loss of life or property damage, or both, such as earthquakes, floods, cyclones (hurricanes), have in the past few decades killed several million people, with an average worldwide annual loss of life of about 150,000 people. The financial losses resulting from natural disasters now exceed $50 billion per year and do not include social impacts such as loss of employment, mental anguish, and reduced productivity. Three individual disasters, a cyclone accompanied by flooding in Bangladesh in 1970, an earthquake in China in 1976, and a tsunami in the Indian Ocean in 2004, each claimed over 250,000 lives. These terrible disasters (catastrophes) were caused by natural hazards that have always existed—atmospheric disturbance and tectonic movement—but their extent was affected by human population density and land-use patterns.

Why Natural Processes Are Sometimes Hazards

Natural hazards are basically natural processes. These processes become hazardous when people live or work in areas where they occur. Natural processes can also become hazards when land-use changes, such as urbanization or deforestation, affect natural processes, causing flooding or landsliding. It is the environmental geologist's role to identify potentially hazardous processes and make this information available to planners and decision makers so that they can formulate various alternatives to avoid or minimize the threat to human life or property. However, the naturalness of hazards is a philosophical barrier that we encounter whenever we try to minimize their adverse effects. For example, we try to educate people that a river and floodplains, the flat land adjacent to the river, are part of the same natural system and that we should expect floods on floodplains as the name suggests. Minimizing the flood hazard may be as simple as not building on floodplains! However, this seemingly logical solution is difficult to get across to people who see floodplains as flat land on which to build houses.

Magnitude and Frequency

The *impact* of a disastrous event is in part a function of its *magnitude*, or amount of energy released, and *frequency*, or recurrence interval; however, it is influenced by many other factors, including climate, geology, vegetation, population, and land use. In general, the frequency of such an event is inversely related to the magnitude. Small earthquakes, for example, occur more often than do large ones (see A Closer Look: The Magnitude-Frequency Concept).

Benefits of Natural Hazards

It is ironic that the same natural events that take human life and destroy property also provide us with important benefits or natural service functions. For example, periodic flooding of the Mississippi River supplies nutrients to the floodplain, on which form the fertile soils used for farming. Flooding, which causes erosion on mountain slopes, also delivers river sediment to beaches (Figure 5.4) and flushes

(a)   (b)

Figure 5.4 **Dams and beaches** (a) Sediment from the Ventura River in southern California is delivering some sand to beaches in the region; however, an old upstream dam (b) is storing sand that might otherwise nourish the beach. The dam is scheduled to be removed. *([a] Pacific Western; [b] Edward A. Keller)*

A CLOSER LOOK | The Magnitude-Frequency Concept

The **magnitude-frequency concept** states that there is generally an inverse relationship between the magnitude of an event and its frequency. For example, the larger the flood, the less frequently such a flood occurs. The concept also includes the idea that much of the work of forming Earth's surface occurs through events of moderate magnitude and frequency rather than by common natural processes of low magnitude and high frequency or by extreme events of high magnitude and low frequency.

As an analogy to the magnitude-frequency concept, consider the work of reducing the extent of a forest by resident termites, human loggers, and elephants (Figure 5.A). The termites are numerous and work quite steadily, but they are so small that they can never do enough work to destroy all the trees. The people are fewer and work less often, but, being stronger than termites, they can accomplish more work in a given time. Unlike the termites, the people can eventually fell most of the trees. The elephants are stronger still and can knock down many trees in a short time, but there are only a few of them and they rarely visit the forest. In the long run the elephants do less work than the people and bring about less change.

In our analogy it is humans who, with a moderate expenditure of energy and time, do the most work and change the forest most drastically. Similarly, natural events with moderate energy expenditure and moderate frequency are often the most important shapers of the landscape. For example,

Figure 5.A Human scale of change Human beings with our high technology are able to down even the largest trees in our old-growth forests. The lumberjack shown here is working in a national forest in the Pacific Northwest. *(William Campbell/ Sygma Photo News)*

pollutants from estuaries in the coastal environment. Landslides may bring benefits to people when landslide debris forms dams, creating lakes in mountainous areas (Figure 5.5). Although some landslide-created dams will collapse and cause hazardous downstream flooding, dams that remain stable can provide valuable water storage and are an important aesthetic resource.

Volcanic eruptions have the potential to produce catastrophes; however, they also provide us with numerous benefits. They often create new land, as in the case of the Hawaiian Islands, which are completely volcanic in origin (Figure 5.6). Nutrient-rich volcanic ash may settle on existing soils and quickly become incorporated, creating soil suitable for wild plants and crops. Earthquakes can also provide us with valuable services. When rocks are pulverized during an earthquake, they may form an impervious clay zone known as a **fault gouge** along the fault. In many places, fault gouge has formed groundwater barriers upslope from a fault, producing natural subsurface dams and water resources. Along some parts of the San Andreas fault in the arid Coachella Valley near Indio, California, this process has produced oases, in which pools of water are surrounded by native palm trees in an otherwise desert environment (Figure 5.7). In addition, earthquakes are also important in mountain building and thus are directly responsible for many of the scenic resources of the western United States.

most of the sediment carried by rivers in regions within a subhumid climate (most of the eastern United States) is transported by flows of moderate magnitude and frequency. However, there are many exceptions. In arid regions, for example, much of the sediment in normally dry channels may be transported by rare high-magnitude flows produced by intense but infrequent rainstorms. Along the barrier-island coasts of the eastern United States, high-magnitude storms often cut inlets that cause major changes in the pattern and flow of sediment (Figure 5.B).

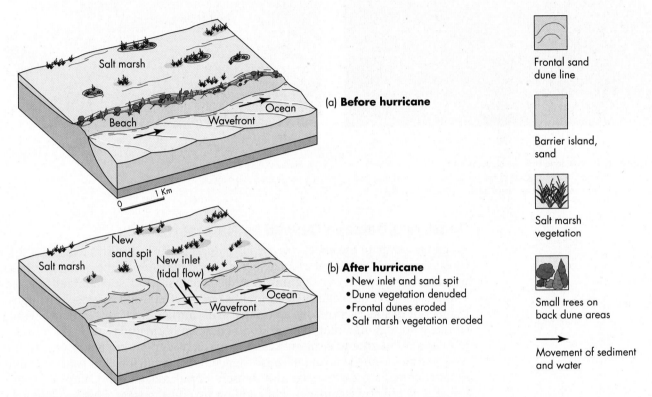

(a) **Before hurricane**

(b) **After hurricane**
- New inlet and sand spit
- Dune vegetation denuded
- Frontal dunes eroded
- Salt marsh vegetation eroded

Frontal sand dune line

Barrier island, sand

Salt marsh vegetation

Small trees on back dune areas

→ Movement of sediment and water

Figure 5.B Hurricanes change coasts Idealized diagram showing the formation of an inlet through a barrier island resulting from erosion during a hurricane. (a) Before and (b) after hurricane.

Figure 5.5 Landslide dam
Landslide dam forming a lake in Utah.
(Michael Collier)

(a)

(b)

Figure 5.6 New land from volcanic eruption New land being added to the island of Hawaii. (a) The plume of smoke in the central part of the photograph is where hot lava is entering the sea. (b) Closeup of an advancing lava front near the smoke plume. *(Edward A. Keller)*

Death and Damage Caused by Natural Hazards

When we compare the effects of various natural hazards, we find that those that cause the greatest loss of human life are not necessarily the same as those that cause the most extensive property damage. Table 5.1 summarizes selected information about the effects of natural hazards in the United States. The largest number of deaths each year is associated with tornadoes (Figure 5.8) and windstorms, although lightning (Figure 5.9), floods, and hurricanes also take a heavy toll. Loss of life due to earthquakes can vary considerably from one year to the next, as a single great quake can cause tremendous human loss. It is estimated that a large, damaging earthquake in a densely populated part of California could inflict $100 billion in damages while killing several thousand people.[3] The 1994 Northridge earthquake in the Los Angeles area killed approximately 60 people and caused more than $30 billion in property damage. In fact, property damage from individual hazards is considerable. Floods, landslides, frost, and expansive soils each cause mean annual damages in the United States in excess of $1.5 billion. Surprisingly, expansive soils, clay-rich soils that expand and contract with wetting and drying, are one of the most costly hazards, causing over $3 billion in damages

(a)

(b)

Figure 5.7 Oases and faults (a) Native palm trees along the San Andreas fault, Coachella Valley, California. The fault dams groundwater that the trees use. (b) In some cases the water forms surface pools and an oasis. *(Edward A. Keller)*

TABLE 5.1   Effects of Selected Hazards in the United States

Hazard	No. of Deaths per Year	Occurrence Influenced by Human Use	Catastrophe Potential[2]
Flood	86	Yes	H
Earthquake[1]	50 + ?	Yes	H
Landslide	25	Yes	M
Volcano[1]	<1	No	H
Coastal erosion	0	Yes	L
Expansive soils	0	No	L
Hurricane	55	Perhaps	H
Tornado and windstorm	218	Perhaps	H
Lightning	120	Perhaps	L
Drought	0	Perhaps	M
Frost and freeze	0	Yes	L

[1]Estimate based on recent or predicted loss over 150-year period. Actual loss of life and/or property could be much greater.

[2]Catastrophe potential: high (H), medium (M), low (L).

Source: Modified after White, G. F., and Haas, J. E. 1975. *Assessment of research on natural hazards*. Cambridge, MA: MIT Press.

annually to building foundations, sidewalks (Figure 5.10), driveways, and swimming pools.

An important aspect of all natural hazards is their potential to produce a **catastrophe.** A catastrophe is any situation in which the damages to people, property, or society in general are sufficient such that recovery or rehabilitation is a

(a)   (b)

Figure 5.8 **Tornado hazard** (a) Tornado in Tampa Bay, Florida, on July 12, 1995. *(Brian Baer/ St. Petersburg Times/AP/Wide World Photos)* (b) Mobile homes destroyed by tornado that struck Benton, Louisiana, on April 4, 1999. *(Eric Gay/AP/Wide World Photos)*

Figure 5.9 Lightning strike Lightning is responsible for more than 100 deaths each year in the United States. Shown here are lightning strikes near Walton, Nebraska. *(Joel Sartore/NGS Image Collection)*

long, involved process.[4] Table 5.1 shows the catastrophe potential for the hazards considered. The events most likely to produce a catastrophe are floods, hurricanes, tornadoes, earthquakes, volcanic eruptions, and large wildfires (not included in Table 5.1). Landslides, which generally cover a smaller area, have only a moderate catastrophe potential. The catastrophe potential of drought is also moderate: though a drought may cover a wide area with high financial losses, there is usually plenty of warning time before its worst effects are experienced. Hazards with a low catastrophe potential include coastal erosion, frost, lightning, and expansive soils.

The effects of natural hazards change with time. Changes in land-use patterns that influence people to develop on marginal lands, urbanization that changes the physical properties of Earth materials, and increasing population all alter the effects of natural hazards. Although damage from most hazards in the United States is increasing, the number of deaths from many hazards is decreasing because of better hazard forecasting and warning to the public.

Figure 5.10 Soil hazard
Organic-rich expansive soils are cracking the walls of this building in Spain.
(Edward A. Keller)

5.2 Evaluating Hazards: History, Linkages, Disaster Prediction, and Risk Assessment

Fundamental Principles Concerning Natural Hazards

The understanding of natural hazards and how we might minimize their impact on people and the environment is facilitated through the recognition of five principles:

1. *Hazards are known from scientific evaluation.* Natural hazards, such as earthquakes, volcanic eruptions, landslides, and floods, are natural processes that can be identified and studied using the scientific method. Most hazardous events and processes can be monitored and mapped, and their future activity can be evaluated based on the frequency of past events, patterns, and types of precursor events.

2. *Risk analysis is an important component in understanding impacts resulting from hazardous processes.* Hazardous processes are amenable to risk analysis based on the probability of an event occurring and the consequences resulting from that event; for example, if we were to estimate that in any given year in Los Angeles or Seattle there is a 5 percent chance of a moderate earthquake occurring. If we know the consequence of that earthquake in terms of loss of life and damage, then we can calculate the risk to society of that earthquake actually happening.

3. *Hazards are linked.* Hazardous processes are linked in many ways, from simple to complex. For example, earthquakes can produce landslides and giant sea waves called tsunamis, and hurricanes often cause flooding and erosion.

4. *Hazardous events that previously produced disasters are often now producing catastrophes.* The size of the natural hazardous event as well as its frequency is influenced by human activity. As a result of increasing human population and poor land-use practices, what used to be disasters are often now catastrophes.

5. *Consequences of hazards can be minimized.* Minimizing the potential adverse consequences and effects of natural hazards requires an integrated approach that includes: scientific understanding; land-use planning (regulation and engineering); and proactive disaster preparedness.

Role of History in Understanding Hazards

A fundamental principle of understanding natural hazards is that they are repetitive events, and therefore studying their history provides much needed information in any hazard reduction plan. Whether we are studying flood events, landslides, volcanic eruptions, or earthquakes, the historical and recent geologic history of an area is a primary data set. For example, if we wish to evaluate the flooding history of a particular river, one of the first tasks is to study the previous floods of that river system. This study should include detailed evaluation of aerial photographs and maps reaching as far back as the record allows. For prehistoric events, we can study the geologic environment for evidence of past floods, such as the sequence of flood deposits on a floodplain. Often, these contain organic material that may be dated to provide a history of prehistoric flood events. This history is then linked with the documented historical record of high flows, providing a perspective on flooding of the river system being evaluated. Similarly, if we are investigating landslides in a particular river valley, studying the documented historical occurrence of these events and linking that information to prehistoric landslides will provide basic data necessary to better predict landslides.

The hydrologists' role in flood analysis is to evaluate stream flow records taken from sites, known as gauging stations, where stream flow recorders have been established (Figure 5.11). Unfortunately, except for larger rivers, the records are usually relatively short, covering only a few years. Most small streams have no

Figure 5.11 **Monitoring stream flow** Stream gauging station on the Merced River in Yosemite National Park continuously monitors the flow of water in the river. Solar cells provide power. This is not a "run of the mill" station. It is designed to educate park visitors on how stream flow is recorded. *(Edward A. Keller)*

gauging station at all. Geologists have the observation skills, tools, and training to "read the landscape." They can evaluate prehistoric evidence for natural hazards and link this information with the modern record to provide the perspective of time on a particular process. Environmental geologists also have the ability to recognize landforms associated with hazardous processes. In addition, they recognize that the nature and extent of the hazard varies as the assemblage of landforms varies. For example, flooding that occurs in a river valley with a flat adjacent floodplain is very different from flooding on a delta. The river and floodplain constitute a relatively simple system consisting of a single channel bordered by the floodplain (Figure 5.12). Deltas, however, are more complex landforms, produced when a river enters a lake or an ocean (Figure 5.13). Deltas often have multiple channels that receive floodwaters at various times and places, varying the position

Figure 5.12 **Floodplain** Mission Creek, California (left), and floodplain (right) with an urban park on it. The location of the park is an example of good use of a floodplain. *(Edward A. Keller)*

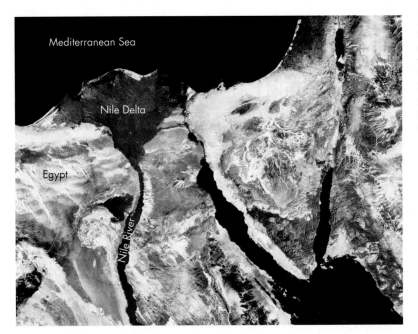

Figure 5.13 **Delta** Infrared image of the Nile Delta (upper left) and surrounding region. Healthy vegetation is red. The white strips at the delta edge are sandy islands that have a serious erosion problem since the construction of the Aswan Dam (not shown) in 1964. *(Earth Satellite Corporation/SPL/Photo Researchers, Inc.)*

of the channel and thus the energy of a flood. Processes on deltas are discussed in detail in Chapter 8, but the general principle of instability of channels associated with different types of landforms is the idea we wish to emphasize here.

In summary, before we can truly understand the nature and extent of a natural hazard, for example, flooding at a particular site, we must study in detail the history of the site, especially the occurrence, location, and effects of past floods. Understanding this history provides a perspective on the hazard that allows for the "big picture" to be better understood and appreciated. Integrating historical information with both present conditions and land-use change of the recent past, such as deforestation and urbanization, allows for better understanding of the hazard. This results because land-use changes can increase the impact of hazards such as landslides and floods. Studying the record also enables more reliable prediction of future events.

Linkages between Hazardous Events

Linkages between natural processes that are hazardous to people generally fall into two categories. First, many of the hazards themselves are linked. For example, hurricanes are often associated with flooding, and intense precipitation associated with hurricanes causes coastal erosion and landslides on inland slopes. Natural hazards and the characteristics of Earth materials provide a second type of linkage. For example, the sedimentary rock known as shale is composed of loosely cemented or compacted tiny sediments that are prone to landslides. Granite provides another example of the linkage between natural hazards and Earth material characteristics. Although generally strong and durable, granite is prone to sliding along fractures within the rock.

Disaster Forecast, Prediction, and Warning

A **prediction** of a hazardous event such as an earthquake involves specifying the date, time, and size of the event. This is different from predicting where or how often a particular event such as a flood will occur. A **forecast,** on the other hand, has ranges of certainty. The weather forecast for tomorrow may state there is a 40 percent chance of showers. Learning how to predict or forecast disasters in order to minimize loss of life and property damage is an important endeavor. For each particular hazard, we have a certain amount of information; in some cases, this information allows us to predict or forecast events accurately. When insufficient information is available, the best we can do is to locate areas where disastrous events have occurred and infer where and when similar future events might take place. If we know both the probability and the possible consequences of an event's occurring at a particular location, we can assess the risk the event poses to people and property, even if we cannot accurately predict when it will occur.

The effects of a specific disaster can be reduced if we can forecast or predict the event and issue a warning. In a given situation, most or all of the following elements are involved:

- Identifying the location where a hazardous event will likely occur
- Determining the probability that an event of a given magnitude will occur
- Observing precursor events
- Forecasting or predicting the event
- Warning the public

Location. For the most part, we know *where* a particular kind of event is likely to occur. On a global scale, the major zones for earthquakes and volcanic eruptions have been delineated by mapping earthquake foci and the locations of recent volcanic rocks and volcanoes. On a regional scale, we can predict from past eruptions which areas in the vicinity of certain volcanoes are most likely to be threatened by large mudflows or ash in the event of future eruptions. This risk has been delineated for most large volcanoes, including the Pacific Northwest's Cascade Range and volcanoes in Alaska, Japan, Italy, Mexico, Central and South America, Hawaii, and numerous other volcanic islands in the oceans of the world. On a local scale, detailed work with soils, rocks, and hydrology may identify slopes that are likely to fail and cause a landslide or where expansive soils exist. Certainly we can predict where flooding is likely to occur from the location of the floodplain and evidence from recent floods such as the location of flood debris and high-water line.

Probability of Occurrence. Determining the probability that a particular event will occur in a particular location within a particular time span is an essential goal of hazard evaluation. For many large rivers we have sufficient records of flow to develop probability models that can reasonably predict the average number of floods of a given magnitude that will occur in a given time period. Likewise, droughts may be assigned a probability on the basis of past occurrence of rainfall in the region. However, these probabilities are similar to the chances of throwing a particular number on a die or drawing an inside straight in poker; the element of chance is always present. For example, the 10 year flood may occur on the average of every 10 years, but it is possible for several floods of this magnitude to occur in any one year, just as it is possible to throw two straight sixes with a die.

Precursor Events. Many hazardous events are preceded by **precursor events.** For example, the surface of the ground may creep, or move slowly down a slope, for a period of time, days to months, before a landslide. Often the rate of creep increases up to when the landslide occurs. Volcanoes sometimes swell or bulge before an eruption, and often emissions of volcanic gases accompanied by seismic activity significantly increase in local areas surrounding the volcano. Foreshocks and anomalous, or unusual, uplift may precede earthquakes. Precursor events help predict when and where an event is likely to happen. For example, landslide creep or swelling of a volcano may result in the issuance of a warning, allowing people to evacuate a hazardous area.

Forecast. When a forecast of an event is issued, the certainty of the event is given, usually as the percent chance of something happening. When we hear a forecast of a hazardous event, it means we should be prepared for the event.

Prediction. It is sometimes possible to accurately *predict* when certain natural events will occur. Flooding of the Mississippi River, which occurs in the spring in

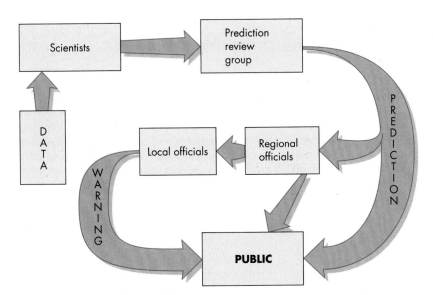

Figure 5.14 Hazard prediction or warning Possible flow path for issuance of a natural disaster prediction or warning.

response to snowmelt or very large regional storm systems, is fairly common, and we can often predict when the river will reach a particular flood stage, or water level. When hurricanes are spotted far out to sea and tracked toward the shore, we can predict when and where they will likely strike land. Tsunamis, or seismic sea waves, generated by disturbance of ocean waters by earthquakes or submarine volcanoes, may also be predicted. The tsunami warning system has been fairly successful in the Pacific Basin and can predict the arrival of the waves. A short time prediction of a hazardous event such as a hurricane motivates us to act now to reduce potential consequences before the event happens.

Warning. After a hazardous event has been predicted or a forecast has been made, the public must be warned. Information leading to the **warning** of a possible disaster such as a large earthquake or flood should move along a path similar to that shown in Figure 5.14. The public does not always welcome such warnings, however, especially when the predicted event does not come to pass. In 1982, when geologists advised that a volcanic eruption near Mammoth Lakes, California, was quite likely, the advisory caused a loss of tourist business and apprehension on the part of the residents. The eruption did not occur, and the advisory was eventually lifted. In July 1986, a series of earthquakes occurred over a 4 day period in the vicinity of Bishop, California, in the eastern Sierra Nevada. The initial earthquake was relatively small and was felt only locally; but a later, larger earthquake causing some damage also occurred. Investigators concluded there was a high probability an even larger quake would occur in the same area in the near future and issued a warning. Local business owners, who feared the loss of summer tourism, felt that the warning was irresponsible; in fact, the predicted quake never materialized.

Incidents of this kind have led some people to conclude that scientific predictions are worthless and that advisory warnings should not be issued. Part of the problem is poor communication between the investigating scientists and reporters for the media (see A Closer Look: Scientists, Hazards, and the Media). Newspaper, television, and radio reports may fail to explain the evidence or the probabilistic nature of disaster prediction. This failure leads the public to expect completely accurate statements as to what will happen. Although scientific predictions of volcanic eruptions and earthquakes are not always accurate, scientists have a responsibility to publicize their informed judgments. An informed public is better able to act responsibly than an uninformed public, even if the subject makes people uncomfortable. Ship captains, who depend on weather advisories and warnings of changing conditions, do not suggest that they would be better off not knowing about an impending storm, even though the storm might veer and miss the ship.

A CLOSER LOOK | Scientists, Hazards, and the Media

People today learn what is happening in the world by watching television, listening to the radio, surfing the Internet, or reading newspapers and magazines. Reporters for the media are generally more interested in the impact of a particular event on people than in its scientific aspects. Even major volcanic eruptions or earthquakes in unpopulated areas may receive little media attention, whereas moderate or even small events in populated areas are reported in great detail. The news media want to sell stories, and spectacular events that affect people and property "sell."[5]

Establishing good relations between scientists and the news media is a goal that may be difficult to always achieve. In general, scientists tend to be conservative, critical people who are afraid of being misquoted. They may perceive reporters as pushy and aggressive or as willing to present half-truths while emphasizing differences in scientific opinion to embellish a story. Reporters, on the other hand, may perceive scientists as an uncooperative and aloof group who speak in an impenetrable jargon and are unappreciative of the deadlines that reporters face.[5] These statements about scientists and media reporters are obviously stereotypic. In fact,

both groups have high ethical and professional standards; nevertheless, communication problems and conflicts of interest often occur, affecting the objectivity of both groups.

Because scientists have an obligation to provide the public with information about natural hazards, it is good policy for a research team to pick one spokesperson to interact with the media to ensure that information is presented as consistently as possible. Suppose, for example, that scientists are studying a swarm of earthquakes near Los Angeles and speculation exists among them regarding the significance of the swarm. Standard operating procedure for Earth scientists working on a problem is to develop several working hypotheses and future scenarios. However, when scientists are working with the news media on a topic that concerns people's lives and property, their reports should be conservative evaluations of the evidence at hand, presented with as little jargon as possible. Reporters, for their part, should strive to provide their readers, viewers, or listeners with accurate information that the scientists have verified. Embarrassing scientists by misquoting them will only lead to mistrust and poor communication between scientists and journalists.

Just as weather warnings have proved very useful for planning ships' routes, official warnings of hazards such as earthquakes, landslides, and floods are also useful to people making decisions about where they live, work, and travel.

Consider once more the prediction of a volcanic eruption in the Mammoth Lakes area of California. The seismic data suggested to scientists that molten rock was moving toward the surface. In view of the high probability that the volcano would erupt and the possible loss of life if it did, it would have been irresponsible for scientists not to issue an advisory. Although the eruption did not occur, the warning led to the development of evacuation routes and consideration of disaster preparedness. This planning may prove useful in the future; it is likely that a volcanic eruption will occur in the Mammoth Lakes area in the future. The most recent event occurred only 600 years ago! In the end, the result of the prediction is a better informed community that is better able to deal with an eruption when it does occur.

Risk Assessment

Before discussing and considering adjustments to hazards, people must have a good idea of the risk that they face under various scenarios. Risk assessment is a rapidly growing field in the analysis of hazards, and its use should probably be expanded.

Risk Determination. The **risk** of a particular event is defined as the product of the probability of that event's occurring multiplied by the consequences should it actually occur.[6] Consequences, such as damages to people, property, economic activity, and public service, may be expressed in a variety of scales. If, for example, we are considering the risk from earthquake damage to a nuclear reactor, we may evaluate the consequences in terms of radiation released, which can further be translated into damage to people and other living things. In any such assessment, it is important to calculate the risks of various possible events—in this example, earthquakes of various magnitudes. A large earthquake has a lower probability of occurring than does a small one, but its consequences are likely to be greater.

Acceptable Risk. Determining *acceptable risk* is more complicated. The risk that an individual is willing to endure is dependent upon the situation. Driving an automobile is fairly risky, but most of us accept that risk as part of living in a modern world. However, acceptable risk from a nuclear power plant is very low because we consider almost any risk of radiation poisoning unacceptable. Nuclear power plants are controversial because many people perceive them as high-risk facilities. Even though the probability of a nuclear accident due to a geologic hazard such as an earthquake may be quite low, the associated consequences could be high, resulting in a relatively high risk.

Institutions, such as the government and banks, approach the topic of acceptable risk from an economic point of view rather than a personal perception of the risk. For example, a bank will consider how much risk they can tolerate with respect to flooding. The federal government may require that any property that receives a loan from them not have a flood hazard that exceeds 1 percent per year, that is, protection up to and including the 100 year flood.

Problems and Opportunities for Risk Assessment. A frequent problem of risk analysis, with the exception of flooding on a river with a long record of past floods, is lack of reliable data available for analyzing the probability of an event. It can be difficult to assign probabilities to geologic events such as earthquakes and volcanic eruptions, because the known chronology of past events is often inadequate.[6] Similarly, it may be very difficult to determine the consequences of an event or series of events. For example, if we are concerned about the consequences of releasing radiation into the environment, local biological, geologic, hydrologic, and meteorological information must be gathered to evaluate the radiation's effects. This information may be complex and difficult to analyze. Despite these limitations, methods of determining the probability of earthquakes and volcanic eruptions are improving, as is our ability to estimate consequences of hazardous events. Certainly, risk assessment is a step in the right direction and should be expanded. As more is learned about determining the probability and consequences of a hazardous event, we will be able to provide more reliable forecasts and predictions necessary for decision making, such as when to issue a warning or evacuate people from harm's way.

5.3 The Human Response to Hazards

Often, the manner in which we deal with hazards is primarily *reactive*. After a disaster we engage in searching for and rescuing survivors, firefighting, and providing emergency food, water, and shelter. There is no denying that these activities reduce loss of life and property and need to be continued. However, the move to a higher level of hazard reduction will require increased efforts to *anticipate* disasters and their impact. Land-use planning to avoid hazardous locations, hazard-resistant construction, and hazard modification or control such as flood control channels are some of the adjustments that anticipate future disastrous events and may reduce our vulnerability to them.[4]

Reactive Response: Impact of and Recovery from Disasters

The impact of a disaster upon a population may be either direct or indirect. *Direct effects* include people killed, injured, displaced, or otherwise damaged by a particular event. *Indirect effects* generally include responses to the disaster such as emotional distress, donation of money or goods, and paying taxes to finance the recovery. Direct effects are felt only by those individuals immediately affected by the disaster, whereas indirect effects are felt by the populace in general.[7,8]

The stages of recovery following a disaster are emergency work, restoration of services and communication lines, and reconstruction. Figure 5.15 shows an idealized model of recovery. This model can be applied to actual recovery activities following events such as the 1994 Northridge earthquake in the Los Angeles area. Restoration began almost immediately after the earthquake. For example, in the first few weeks and months after the earthquake, roads were repaired and utilities were restored with the help of an influx of dollars from federal programs, insurance companies, and other sources. The damaged areas in Northridge moved quickly

Figure 5.15 **Recovery from disaster** Generalized model of recovery following a disaster. The first 2 weeks after a disaster are spent in a state of emergency, in which normal activities are ceased or changed. During the following 19 weeks, in the restoration phase, normal activities return and function, although perhaps not at predisaster level. During the reconstruction I phase, for almost 4 years after a disaster, the area is being rebuilt and normal activities return to predisaster levels. Finally, during reconstruction II, major construction and development are being done, and normal activities are improved and developed. *(From Kates, R. W., and Pijawka, D. 1977. Reconstruction following disaster. In From rubble to monument: The pace of reconstruction, ed. J. E. Haas, R. W. Kates, and M. J. Bowden. Cambridge, MA: MIT Press)*

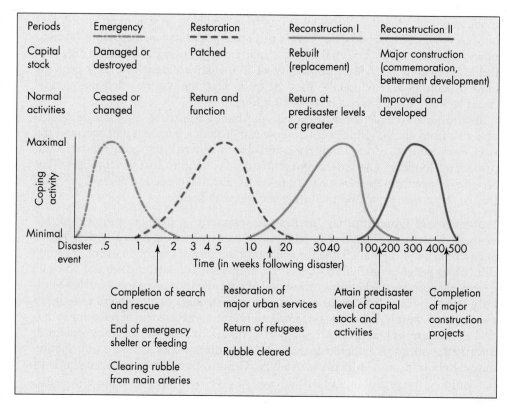

from the restoration phase to the reconstruction I stage. We are now well into the reconstruction II period following the Northridge earthquake, and it is important to remember lessons from two previous disasters: the 1964 earthquake that struck Anchorage, Alaska, and the flash flood that devastated Rapid City, South Dakota, in 1972. Anchorage began restoration approximately one month after the earthquake as money from federal programs, insurance companies, and other sources poured in. As a result, reconstruction was a hectic process, because everyone tried to obtain as much of the available funds as possible. In Rapid City, however, the restoration did not peak until approximately 10 weeks after the flood. The community took time to carefully think through the best alternatives to avoid future flooding problems. As a result, Rapid City today has an entirely different land use on the floodplain. The floodplain is now a greenbelt, with golf courses and other such activities, which has greatly reduced the flood hazard in the area (Figure 5.16). Conversely, the rapid restoration and reconstruction in Anchorage were accompanied by little land-use planning. Apartments and other buildings were hurriedly constructed across areas that had suffered ground rupture and had simply been filled in and regraded. By ignoring the potential benefits of careful land-use planning, Anchorage is vulnerable to the same type of earthquake damage that it suffered in 1964.[4,7,8]

In Northridge, the effects of the earthquake on highway overpasses and bridges, buildings, and other structures have been carefully evaluated to determine how improved engineering standards for construction of new structures or strengthening of older structures might be implemented during the reconstruction II period (Figure 5.15). Future earthquakes of moderate to large intensity are certain to occur again in the Los Angeles area. Therefore, we must continue efforts in the area of earthquake hazard reduction.

Anticipatory Response: Perceiving, Avoiding, and Adjusting to Hazards

The options we choose, individually or as a society, for avoiding or minimizing the impacts of disasters depend in part on our hazard perception. A good deal of work has been done in recent years to try to understand how people perceive various natural hazards. This work is important because the success of hazard reduction

Figure 5.16 Floodplain management Rapid City, South Dakota. A 1972 flood, prior to management, killed over 200 people and destroyed many homes in the floodplain. The floodplain of Rapid Creek downstream of Canyon Lake (center) is now the site of a golf course (lower left). *(Courtesy of Perry Rahn)*

programs depends on the attitudes of the people likely to be affected by the hazard. Although there is informed perception and understanding of a hazard at the institutional level among government agencies such as the U.S. Geological Survey and state or local flood control agencies, this may not filter down to the general population. This lack of understanding is particularly true for events that occur infrequently; people are more aware of situations such as large floods and brush or forest fires that may occur every few years or decades (Figure 5.17). There are often local ordinances to help avoid or minimize damages resulting from these events. For example, in some areas of southern California local regulations stipulate that homes must be roofed with shingles that do not burn readily. Other regulations

Figure 5.17 Killer wildfire Wildfire in October 1991 devastated this Oakland, California, neighborhood. This fire killed 25 people and destroyed over 3,000 homes, with damages of about $1.7 billion. *(Tom Benoit/Getty Images Inc.)*

include mandatory installation of sprinkler systems and clearing lots of brush. Such safety measures are often noticeable during the rebuilding phase following a fire.

One of the most environmentally sound adjustments to hazards involves **land-use planning.** That is, people can avoid building on floodplains, in areas where there are active landslides, or in places where coastal erosion is likely to occur. In many cities, floodplains have been delineated and zoned for a particular land use. With respect to landslides, legal requirements for soil engineering and engineering geology studies at building sites may greatly reduce potential damages. Damages from coastal erosion can be minimized by requiring adequate setback of buildings from the shoreline or seacliff. Although it may be possible to control physical processes in specific instances, land-use planning to accommodate natural processes is often preferable to a technology-based solution that may or may not work.

Insurance. Inurance is another option that people may exercise in dealing with natural hazards. Flood insurance and earthquake insurance are common in many areas. However, because of large losses following the 1994 Northridge earthquake, several insurance companies announced they would no longer offer earthquake insurance to residents of the area.

Evacuation. In the states along the Gulf of Mexico and along the eastern coast of the United States, evacuation is an important option or adjustment to the hurricane hazard. Often, there is sufficient time for people to evacuate provided they heed the predictions and warnings. However, if people do not react quickly and the affected area is a large urban region, then evacuation routes may be blocked by residents leaving in a last-minute panic. Successful evacuation from areas of volcanic eruptions is mentioned in Chapter 7.

Disaster Preparedness. Individuals, families, cities, states, or even entire nations can practice **disaster preparedness.** Training individuals and institutions to handle large numbers of injured people or people attempting to evacuate an area after a warning is issued is particularly important in disaster preparedness.

Artificial Control of Natural Processes

Attempts at *artificial control of natural processes* such as landslides, floods, and lava flows have had mixed success. Seawalls constructed to control coastal erosion may protect property to some extent but tend to narrow or even eliminate the beach. Even the best designed artificial structures cannot be expected to adequately defend against an extreme event, although retaining walls and other structures that defend slopes from landslides have generally been successful when well designed (Figure 5.18). Even the casual observer has probably noticed the variety of such structures along highways and urban land in hilly areas. These have limited impact

Figure 5.18 Protecting a slope Retaining wall being constructed along a canyon side in an area where landslides were formerly common. The wall protects the homes that were constructed too close to the edge of the steep canyon. If the homes were built with adequate setback from the canyon, the wall would not be necessary. *(Edward A. Keller)*

on the environment but are necessary where construction demands that artificial cuts be excavated or where unstable slopes impinge on human structures. Common methods of flood control are channelization and construction of dams and levees. Unfortunately, flood control projects tend to provide floodplain residents with a false sense of security; no method can be expected to absolutely protect people and their property from high-magnitude floods. We will return to this discussion in Chapter 8.

All too often people choose to simply bear the loss caused by a natural disaster. Many people are optimistic about their chances of making it through any sort of disaster and take little action in their own defense, particularly for hazards such as volcanic eruptions and earthquakes that occur only rarely in a particular area. Regardless of the strategy we choose either to minimize or to avoid hazards, it is imperative that we understand and anticipate hazards and their physical, biological, economic, and social impacts.

5.4 Global Climate and Hazards

Global and regional climatic change may significantly affect the incidence of hazardous natural events such as storms (floods and erosion), landslides, drought, and fires. Global warming associated with climatic change may also have an impact on natural hazards. (We will discuss global warming in detail in Chapter 19.)

How might a climatic change affect the magnitude and frequency of disastrous natural events? As a result of global warming, sea level will rise as glacial ice melts and warming ocean waters expand. This rise in sea level will lead to an increase in coastal erosion. With a change in climatic patterns, food production areas will shift as some receive more precipitation and others less than they do now. Deserts and semiarid areas would likely expand, and more northern latitudes could become more productive. Such changes could lead to global population shifts, which could precipitate wars or major social and political upheavals.

Global warming, with warming of the oceans of the world, will channel more energy from warmer ocean water into the atmosphere. Warming the atmosphere will likely increase the frequency and severity of hazardous weather-related processes including thunderstorms, tornadoes, and hurricanes. The year 1998 set a record for economic losses from weather-related disasters, causing about $100 billion in economic losses worldwide. Losses from storms, floods, fires, and droughts in 1998 (one of the warmest years on record) were greater than the losses for the entire decade of the 1980s (Figure 5.19). Worldwide, the impact on people from weather-related disasters in 1998 was catastrophic, killing approximately 32,000 people while displacing another 300 million people from their homes.[9] The trend is clear—the number of severe weather events (storms, floods, drought, heat waves, and cold) is increasing.

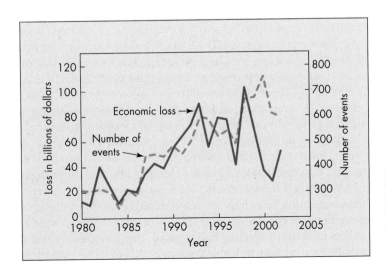

Figure 5.19 **Severe weather hazards increasing** Worldwide number of events and economic losses from weather-related natural hazards (1980–2002). *(Modified after Sawing, J. L. 2003. Vital Signs. New York: W. W. Norton. World Watch Institute)*

5.5 Population Increase, Land-Use Change, and Natural Hazards

Population Increase and Hazardous Events

Population growth throughout the world is a major environmental problem. And, as our population increases, the need for planning to minimize losses from natural disasters also increases. Specifically, an increase in population puts a greater number of people at risk from a natural event; it also forces more people to settle in hazardous areas, creating additional risks. The risks of both high population density and living in a danger zone are dramatically illustrated by the loss of thousands of lives in Colombia in 1985. (See A Closer Look: Nevado del Ruiz.)

Mexico City is another example of the risks associated with high population density coupled with living in a dangerous zone. Mexico City is the center of the world's most populous urban area. Approximately 23 million people are concentrated in an area of about 2,300 km^2 (890 mi^2) and about one-third of the families live in a single room. The city is built on ancient lake beds that accentuate earthquake shaking, and parts of the city have been sinking at the rate of a few centimeters per year owing in part to groundwater withdrawal. The subsidence has not been uniform, so the buildings tilt and are even more vulnerable to the shaking of earthquakes.[10] In September 1985, Mexico endured a large damaging earthquake that killed about 10,000 people in Mexico City alone.

Land-Use Change and Hazardous Events

During the 1990s there was a record number of great catastrophes worldwide. Although there are over 500 disasters from natural hazardous events each year, only a few are classified as a great catastrophe, which is one that results in deaths or losses so great that outside assistance is required.[11] During the past half-century, there has been a dramatic increase in great catastrophes, as illustrated in Figure 5.20. For these events, flooding is the major killer of people, followed by earthquakes, volcanic eruptions, and windstorms. From 1985 to 1995 over 550,000 people died as a result of natural hazards. The vast majority of deaths (96 percent) were in the developing countries. Asia suffered the greatest losses from 1985 to 1997, with 77 percent of the total deaths and 45 percent of the economic losses. It could have been even worse from 1985 to 1999, were it not for improvements in warning, disaster preparedness, and sanitation following disasters.[11] Nevertheless, economic losses have increased at a much faster rate than have the number of deaths.

Three of the deadly catastrophes resulting from natural hazards linked to changes in land use were Hurricane Mitch in 1998, which devastated Central America; the 1998 flooding of the Yangtze River in China; and Hurricane Katrina in 2005. Hurricane Mitch caused approximately 11,000 deaths, and the floods in the Yangtze River resulted in nearly 4,000 deaths. It has been speculated that damages from these events in Central America and China were particularly severe because of land-use changes that had occurred. For example, Honduras has already lost nearly half of its forests owing to timber harvesting, and a fire with an area of about 11,000 km^2 (4,250 mi^2) occurred in the region before the hurricane. As a result of the previous deforestation and the fire, hillsides were stripped of vegetation and washed away; along with them went farms, homes, roads, and bridges (Figure 5.21). The story is similar in central China; recently, the Yangtze River basin lost about 85 percent of its forest as a result of timber harvesting and conversion of land to agriculture. As a result of the land-use changes in China, flooding of the Yangtze River is probably much more common than it was previously.[9] Hurricane Katrina (see case history opening this chapter) emphasized several poor land-use choices, including the removal of coastal wetlands that provide a buffer to wind and storm waves.

The hazardous events that caused catastrophes in the late 1990s and early twenty-first century in Central America, China, the United States, and other parts of the world may be early warning signs of things to come. It is apparent that

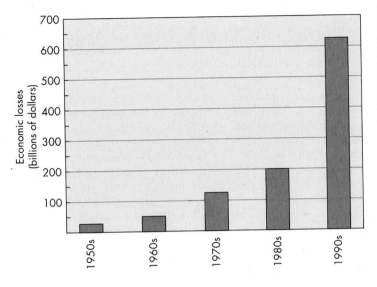

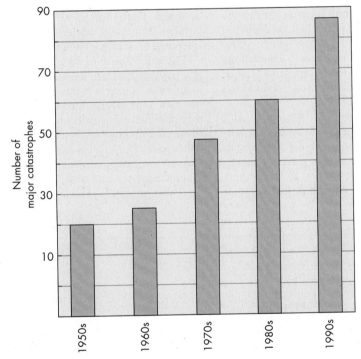

Figure 5.20 **Disasters are increasing** Major disasters from the 1950s to the 1990s. (a) Economic losses and (b) number of major catastrophes. *(Data from Abramovitz, J. N. 2001. Averting unnatural disasters. In Brown, L. R., et al. State of the world. 2001, pp. 123–42. Worldwatch Institute. New York: W. W. Norton)*

Figure 5.21 **Hurricane disaster** Homes destroyed by flooding and landslides in Honduras from Hurricane Mitch in 1998. *(Luis Elvir/AP/Wide World Photos)*

A CLOSER LOOK

Nevado del Ruiz: A Story of People, Land Use, and Volcanic Eruption

A fundamental principle of this chapter is that population increase, coupled with land-use change, has intensified impacts of natural hazards. In some cases, disasters from natural hazards have become catastrophes. For example, when the Colombian volcano Nevado del Ruiz erupted in 1845, a mudflow roared down the east slope of the mountain, killing approximately 1,000 people. Deposits from that event produced rich soils in the Lagunilla River valley, and an agricultural center developed there. The town that the area supported was known as Armero, and by 1985 it was a prosperous community with a growing population of about 23,000 people. On November 13, 1985, another mudflow associated with a volcanic eruption buried Armero, leaving about 21,000 people dead or missing while inflicting more than $200 million in property damage. It was population increase during the 140 years between the mudflows that multiplied the death toll by more than 20 times. The flat lands and rich soil produced by the previous volcanic eruption lured people into a hazardous area. Ironically, the area was decimated by the same type of event that had earlier produced productive soils, stimulating development and population growth.[12]

The volcanic eruption of November 13, 1985, followed a year of precursor activity, including earthquakes and hot-spring activity. Volcano monitoring began in July 1985, and in October a hazards map was completed that correctly identified the events that occurred on November 13. The report and accompanying map gave a 100 percent probability that

potentially damaging mudflows would be produced by an eruption, as had been the case with previous eruptions.

The November 13 event, as expected, started with an explosive eruption that produced pyroclastic flows of hot volcanic ash that scoured and melted glacial ice on the mountain. The ash and melting glacial ice produced water that generated mudflows that raced down river valleys. Figure 5.C shows the areas affected by magma that came into contact with water or ice, causing violent steam and ash explosions, called *base surges*, and pyroclastic flows. It also shows the location of glacial ice that contributed the water necessary to produce the mudflows. Of particular significance was the mudflow that raced down the River Lagunillas and destroyed part of the town of Armero, where most of the deaths occurred. Figure 5.D shows the volcano and the town of Armero. Mudflows buried the southern half of the town, sweeping buildings completely off their foundations.[13]

The real tragedy of the catastrophe was that the outcome was predicted; in fact, there were several attempts to warn the town and evacuate it. Hazard maps were circulated in October, but they were largely ignored. Figure 5.E shows the hazard map of events expected before the eruption and the events that occurred. This graphically illustrates the usefulness of volcanic risk maps.[14] Despite these warnings there was little response, and as a result approximately 21,000 people died. Early in 1986 a permanent volcano observatory center was established in Colombia to continue monitoring the Ruiz volcano as well as

Figure 5.C Eruption of Nevado del Ruiz Map of the volcano Nevado del Ruiz area, showing some of the features associated with the eruption of November 13, 1985. *(Modified after Herd, D. G. 1986. The Ruiz volcano disaster. EOS, Transactions of the American Geophysical Union, May 13: 457–60)*

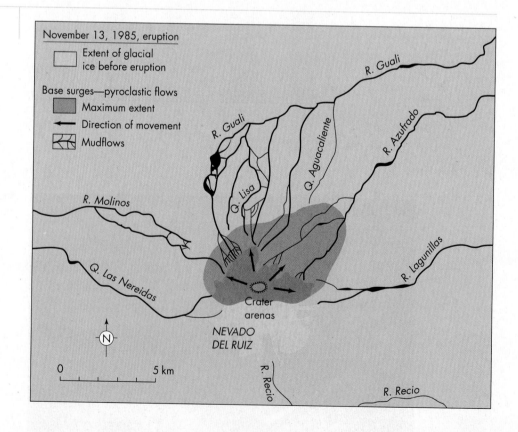

others in South America. South America should now be better prepared to deal with volcanic eruptions. Had there been better communication lines from civil defense headquarters to local towns and a better appreciation of potential volcanic hazards

even 40 km (about 25 mi) from the volcano, evacuation would have been possible for Armero. It is hoped that the lessons learned from this event will help minimize future loss of life associated with volcanic eruptions and other natural disasters.

(a)

(b)

Figure 5.D Nevado del Ruiz (a) The volcano before the eruption. *(Photo El Espectador/Corbis/Sygma)* (b) Catastrophic mudflow: The eruption generated a mudflow that nearly destroyed the town of Armero, killing 21,000 people. *(J. Langevin/Corbis/Sygma)*

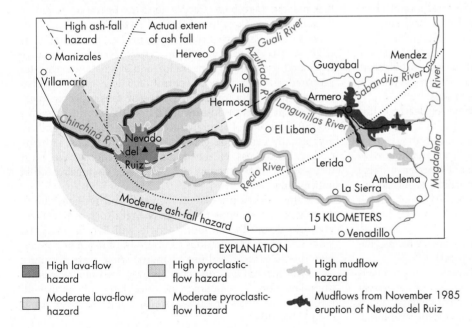

Figure 5.E Volcanic hazard map
Produced and circulated 1 month before the November 13, 1985, eruption of Nevado del Ruiz and mudflows that buried Armero, Colombia. The actual mudflow deposits are shown in red. *(Wright, T. L., and Pierson, T. C. 1992. U.S. Geological Survey Circular 1973)*

human activities are likely increasing the impacts of natural disasters. In recognition of the influence of human activities, China has banned timber harvesting in the upper Yangtze River basin, has prohibited imprudent floodplain land uses, and has allocated several billion dollars for reforestation. The lesson being learned is that if we wish to minimize damages from natural hazards in the future, we need to consider land rehabilitation. The goal must be to achieve sustainable development based on both restoration and maintenance of healthy ecosystems.[9] This goal will be difficult to reach given the pressures of human population growth in many parts of the world. The effects of population increase emphasize the need to control human population growth if we are to solve pressing environmental problems and reach our goal of sustaining our environment.

SUMMARY

Our discussion of natural processes suggests a view of nature as dynamic and changing. This understanding tells us that we cannot view our environment as fixed in time. A landscape without natural hazards would also have less variety; it would be safer but less interesting and probably less aesthetically pleasing. The jury is still out on how much we should try to control natural hazards and how much we should allow them to occur. However, we should remember that disturbance is natural and that management of natural resources must include management for and with disturbances such as fires, storms, and floods.

A fundamental principle of environmental geology is that there have always been Earth processes dangerous to people. These become hazards when people live close to the source of danger or when they modify a natural process or landscape in a way that makes it more dangerous. Natural events that will continue to cause deaths and property damage include flooding, landslides, earthquakes, volcanic activity, wind, expansive soils, drought, fire, and coastal erosion. The frequency of a hazardous event is generally inversely related to its magnitude; its impact on people depends on its frequency and magnitude as well as on such diverse factors as climate, geology, vegetation, and human use of the land. The same natural events that create disasters may also bring about benefits, as when river flooding or a volcanic eruption supplies nutrients to soils.

The events causing the greatest number of deaths in the United States are tornadoes and windstorms, lightning, floods, and hurricanes, although a single great earthquake can take a very large toll. Floods, landslides, frost, and expansive soils cause the greatest property damage. Events most likely to produce a catastrophe are floods, hurricanes, tornadoes, earthquakes, volcanic eruptions, and fires. Although in the United States land-use changes, urbanization, and population increase are causing damages from most hazards to increase, better predictions and warning systems are reducing the number of deaths from these hazardous processes.

Some disastrous events can be forecasted or predicted fairly accurately, including some river floods and the arrival of coastal hurricanes and tsunamis. Precursor events can warn experts about impending earthquakes and volcanic eruptions.

Once an event has been forecasted or predicted, this information must be made available to planners and decision makers in order to minimize the threat to human life and property. The manner in which a warning is issued and how scientists communicate with the media and public are particularly significant. For many hazards we cannot determine when a specific event will occur but only the probability of occurrence, based on the record of past occurrences. The risk associated with an event is the product of the event's probability of occurrence and the likely consequences should it actually occur.

The impact of a disaster upon a population includes direct effects—people killed, dislocated, or otherwise damaged—and indirect effects—emotional distress, donation of money or goods, and paying taxes to finance recovery. Recovery often has several stages, including emergency work, restoration of services and communication, and reconstruction.

The options that individuals or societies choose for avoiding or adjusting to natural hazards depend in part on how hazards are perceived. People tend to be more aware of hazards that are likely to occur often, such as the several hurricanes in the Atlantic that could strike the East and Gulf Coasts of North America each year. Options to adjust to hazards range from land-use planning, insurance, and evacuation to disaster preparedness and artificial control of natural processes. For hazards that occur rarely in particular areas, people often choose to just bear the loss incurred from the hazard. Attempts to artificially control natural processes have had mixed success and usually cannot be expected to defend against extreme events. Regardless of the approach we choose, we must increase our understanding of hazards and do a better job of anticipating them.

As the world's population increases and we continue to modify our environment through changes such as urbanization and deforestation, more people will live on marginal lands and in more hazardous locations. As a result of population pressure and land-use changes, what were formerly localized hazards and disasters are now becoming catastrophes. Therefore, as population increases, better planning at all levels will be necessary if we are to minimize losses from natural hazards.

Revisiting Fundamental Concepts

Human Population Growth

Increase in human population is forcing many people to live in areas where natural hazards are more likely, as for example on floodplains, steep slopes, and flanks of volcanoes. Population pressure has accelerated land-use changes as more land is urbanized, farmed, mined, and harvested for timber. As a result, hazardous events that were formerly disasters are becoming catastrophes.

Sustainability

Living in hazardous areas such as on floodplains is not a sustainable practice, as communities will continue to suffer property damage and loss of life. Ensuring that future generations inherit a quality environment requires that we minimize losses from natural hazards. We cannot sustain soil resources if we continue to log forests on steep slopes, thereby increasing the occurrence of landslides and flooding that erode soil.

Earth as a System

Dynamic Earth systems such as the atmosphere, hydrosphere, and lithosphere are responsible for producing processes that are hazardous to people. Change in one part of a system affects other parts, often in a complex way. For example, burning fossil fuels is releasing carbon dioxide into the atmosphere, causing global warming. One result is that some places will receive more rain and experience more violent storms. These increases will cause changes in the intensity of floods and landslide activity.

Scientific Knowledge and Values

Intensive study of hazardous Earth processes has greatly increased our knowledge of where these processes occur, how they work, the damages they do, and how best to minimize damages. People's values, coupled with scientific knowledge, will determine in part the choices we make to reduce the impact of hazardous events. For example, study of hospitals and schools in an earthquake-prone region may reveal that many are old and likely to fail during a large earthquake. Fixing the hospitals and schools to withstand shaking is a logical answer, but it may cost billions of dollars. Our value of human life compared with other values will determine whether we spend the money to do the necessary work to strengthen the hospitals and schools.

Key Terms

catastrophe (p. 141)

disaster preparedness (p. 152)

fault gouge (p. 138)

forecast (p. 145)

land-use planning (p. 152)

magnitude-frequency concept (p. 138)

precursor events (p. 146)

prediction (p. 145)

risk (p. 148)

warning (p. 147)

Review Questions

1. What were the two main lessons learned from the 1985 eruption of Nevado del Ruiz in Colombia?

2. What is a catastrophe, and how does it differ from a disaster?

3. What do we mean by the magnitude and frequency of a natural process?

4. What is the main conclusion from the magnitude-frequency concept?

5. What is the role of history in understanding natural hazards?

6. List some potential linkages between hazardous events.

7. What are some of the methods of predicting where a disaster is likely to occur?

8. What is the difference between a forecast and prediction?

9. Why do you think there sometimes are strained relationships between the media and scientists?

10. How may the risk of a particular event be defined?

11. What is the difference between a reactive response and an anticipatory response in hazard reduction?

12. What are some of the common adjustments that limit or reduce the effects of natural hazards?

13. What is the role of global climate in the occurrence of natural hazardous events?

14. How does human population increase result in disasters becoming catastrophes?

Critical Thinking Questions

1. List all the natural processes that are hazardous to people and property in the region where you live. What adjustments have you and the community made to lessen the impacts of these hazards? Could more be done? What? Which alternatives are environmentally preferable?

2. Assume that in the future we will be able to predict with a given probability when and where a large damaging earthquake will occur. If the probability that the earthquake will occur on a given date is quite low, say 10 percent, should the general public be informed of the forecast? Should we wait until a 50 percent confidence or even a 90 percent confidence is assured? Does the length of time between the forecast and the event have any bearing on your answers?

3. Find a friend, and one of you take the role of a scientist and another a news reporter. Assume that the news reporter is interviewing the scientist about the nature and extent of hazardous processes in your town. After the interview, jot down some of your thoughts concerning ways in which scientists communicate with newspeople. Are there any conflicts?

4. Develop a plan for your community to evaluate the risk of flooding. How would you go about determining an acceptable risk?

5. Do you agree or disagree that land-use change and population increase are increasing the risk from natural processes? Develop a hypothesis and discuss how it might be tested.

SIX

Earthquakes and Related Phenomena

Learning Objectives

The study of earthquakes is an exciting field with significant social consequences including potential catastrophic loss of life; damage or loss of homes, large buildings, and infrastructure such as roads, train tracks, airports, dams, and power plants; disruption of people's lives; and loss of income. In this chapter we focus on the following learning objectives:

- Understand the relationship of earthquakes to faulting

- Understand how the magnitude of an earthquake is determined

- Know the types of earthquake waves, their properties, and how strong ground motion is produced

- Understand how seismic risk is estimated

- Know the major effects of earthquakes

- Know how an earthquake may produce a tsunami

- Understand the components of the earthquake cycle

- Understand the methods that could potentially predict earthquakes

- Understand the processes of earthquake hazard reduction and how people adjust to and perceive the hazard

Los Angeles earthquake Collapse of a freeway in the Los Angeles area as the result of the 1994 Northridge earthquake. *(Les Stone/Corbis/Sygma)*

CASE HISTORY | Northridge, 1994

The 1994 Northridge earthquake that struck the Los Angeles area on January 17 was a painful wake-up call to southern Californians. The earthquake killed 57 people and caused about $40 billion in property damage. Several sections of freeways were heavily damaged, as were parking structures and more than 3,000 buildings (Figure 6.1). The Northridge earthquake is one in a series of moderate-sized earthquakes that have recently occurred in southern California.

The rupture of rocks that produced the Northridge earthquake was initiated on a steep fault at a depth of approximately 18 km (11 mi). The rupture quickly propagated upward (northward and westward) but did not reach the surface, stopping at a depth of several kilometers. At the same time, the rupture progressed laterally in a mostly westward direction, 20 km (12.5 mi). The geometry of the fault movement is shown in Figure 6.2. The movement produced uplift and folding of part of the Santa Susana Mountains, a few kilometers north of Northridge.[1]

The Northridge earthquake terrified people, especially children. The shaking, which lasted about 15 seconds, was intense; people were thrown out of bed, objects flew across rooms, chimneys tumbled, walls cracked, and Earth groaned and roared with each passing earthquake wave. When the shaking stopped, people had little time to recover before strong aftershocks started.

(a)

(b)

Figure 6.1 **Earthquake in Los Angeles urban region** Damage from the 1994 Northridge, California, earthquake. (a) A parking structure. *(R. Forrest Hopson)* (b) Damage to the Kaiser Permanente Building. *(A. G. Sylvester)*

6.1 Introduction to Earthquakes

There are approximately 1 million earthquakes a year that can be felt by people somewhere on Earth. However, only a small percentage of these can be felt very far from their source. Earthquakes can be compared with one another by the energy they release, their *magnitude,* or by their intensity of shaking, referred to as ground motion, and the resulting impact on people and society. Table 6.1 lists selected major earthquakes that have struck the United States since the early nineteenth century.

6.2 Earthquake Magnitude

When a news release is issued about an earthquake, it generally gives information about where the earthquake started, known as the epicenter. The **epicenter** is the location on the surface of Earth above the **focus,** which is the point at depth where the rocks ruptured to produce the earthquake (Figure 6.3). The news also reports **moment magnitude,** which is a measure of the energy released by the earthquake. The moment magnitude is based in part upon important physical characteristics,

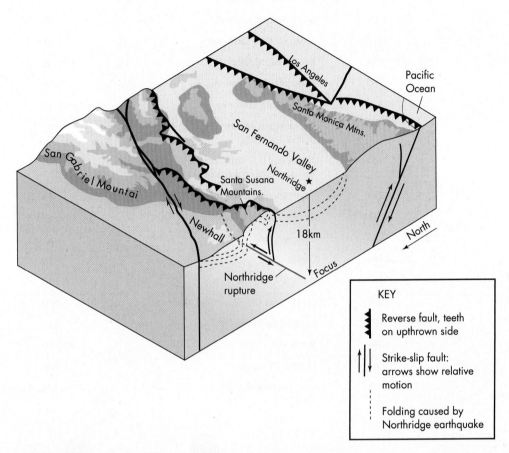

Figure 6.2 **Details of an earthquake** Block diagram showing the fault that produced the 1994 Northridge earthquake. During the earthquake, the Santa Susana Mountains were folded, uplifted 38 cm (15 in.), and moved 21 centimeters (8.2 in.) to the northwest. *(Courtesy of Pat Williams, Lawrence Livermore Laboratory)*

including the area that ruptured along a fault plane during an earthquake, the amount of movement or fault slip during an earthquake, and the rigidity of the rocks.

Before the use of moment magnitude, *Richter magnitude,* named after the famous seismologist Charles Richter, was used to describe the energy released by an earthquake. Richter magnitude is based upon the *amplitude,* or size, of the largest seismic wave produced during an earthquake. A *seismograph* is an instrument that records earthquake displacements; seismographs produce *seismographic records,* or *seismograms.* The amplitude recorded is converted to a magnitude on a logarithmic scale; that is, each integer increase in Richter magnitude represents a tenfold increase in amplitude. For example, a Richter magnitude 7 earthquake produces a displacement on the seismogram 10 times larger than does a magnitude 6. Although the Richter magnitude remains the best known earthquake scale to many people, earthquake scientists, known as *seismologists,* do not commonly use it. For large, damaging earthquakes the Richter magnitude is approximately equal to the moment magnitude, which is more commonly used today. In this book we will simply refer to the size of an earthquake as its magnitude, M, without designating a Richter or moment magnitude.

TABLE 6.1 Selected Major Earthquakes in the United States

Year	Locality	Damage (millions of dollars)	Number of Deaths
1811–1812	New Madrid, Missouri	Unknown	Unknown
1886	Charleston, South Carolina	23	60
1906	San Francisco, California	524	700
1925	Santa Barbara, California	8	13
1933	Long Beach, California	40	115
1940	Imperial Valley, California	6	9
1952	Kern County, California	60	14
1959	Hebgen Lake, Montana (damage to timber and roads)	11	28
1964	Alaska and U.S. West Coast (includes tsunami damage from earthquake near Anchorage)	500	131
1965	Puget Sound, Washington	13	7
1971	San Fernando, California	553	65
1983	Coalinga, California	31	0
1983	Central Idaho	15	2
1987	Whittier, California	358	8
1989	Loma Prieta (San Francisco), California	5,000	62
1992	Landers, California	271	1
1994	Northridge, California	40,000	57
2001	Seattle, Washington	2,000	1
2002	South-Central Alaska	(sparsely populated area)	0

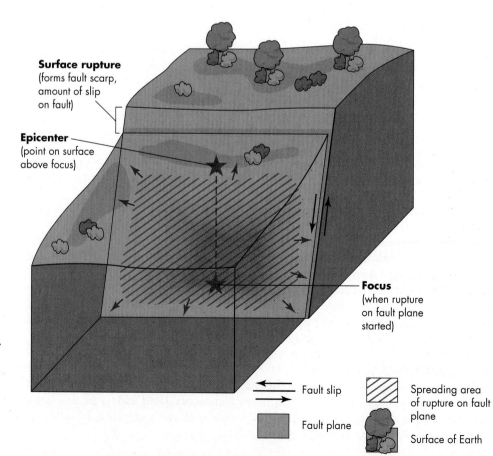

Surface rupture (forms fault scarp, amount of slip on fault)

Epicenter (point on surface above focus)

Focus (when rupture on fault plane started)

Fault slip

Fault plane

Spreading area of rupture on fault plane

Surface of Earth

Figure 6.3 Basic earthquake features Block diagram showing fault plane, amount of displacement, rupture area, focus, and epicenter. Rupture starts at the focus and propagates up, down, and laterally. During a major to giant earthquake, slip may be 2 to 20 meters along a fault length of 100 or more kilometers. Rupture area may be 1,000 square kilometers or more.

TABLE 6.2   Worldwide Magnitude and Frequency of Earthquakes by
Descriptor Classification

Descriptor	Magnitude	Average Annual No. of Events
Great	8 and Higher	1
Major	7–7.9	18
Strong	6–6.9	120
Moderate	5–5.9	800
Light	4–4.9	6,200 (estimated)
Minor	3–3.9	49,000 (estimated)
Very minor	<3.0	Magnitude 2–3 about 1,000 per day
		Magnitude 1–2 about 8,000 per day

U.S. Geological Survey. 2000. *Earthquakes, facts and statistics*. http://neic.usgs.gov. Accessed 1/3/00.

TABLE 6.3   Relationships between Magnitude, Displacement,
and Energy of Earthquakes

Magnitude Change	Ground Displacement Change[1]	Energy Change
1	10 times	About 32 times
0.5	3.2 times	About 5.5 times
0.3	2 times	About 3 times
0.1	1.3 times	About 1.4 times

[1]Displacement, vertical or horizontal, that is recorded on a standard seismograph.

Source: U.S. Geological Survey. 2000. *Earthquakes, facts and statistics*. neic.usgs.gov. Accessed 1/3/00.

The magnitude and frequency of earthquakes worldwide are shown in Table 6.2. An event of magnitude (M) 8 or above is considered a *great earthquake,* capable of causing widespread catastrophic damage. In any given year there is a good chance that one M 8 event will occur somewhere in the world. A M 7 event is a *major earthquake,* capable of causing widespread and serious damage. Magnitude 6 signifies a *strong earthquake* that can cause considerable damage, depending upon factors such as location, surface materials, and quality of construction. Ground motion can be recorded as the displacement or actual separation of rocks produced by an earthquake. Relationships between change in magnitude and change in displacement and energy are shown in Table 6.3. This table illustrates that the difference between a M 6 and a M 7 earthquake is considerable. A M 7 earthquake releases about 32 times more energy than a M 6 earthquake and the amount of displacement, or ground motion, is 10 times greater. If we compare a M 5 with a M 7 earthquake the differences are much greater. The energy released is about 1,000 times greater. The topic of earthquake magnitude is introduced here to help compare the severity of earthquakes. We will return to a more detailed discussion of this topic later in this chapter when earthquake processes are discussed.

Earthquake Catastrophes

Catastrophic, or great, earthquakes are devastating events that can destroy large cities and take thousands of lives in a matter of seconds. A sixteenth-century earthquake in China reportedly claimed 850,000 lives. More recently, a 1923 earthquake near Tokyo killed 143,000 people, and a 1976 earthquake in China killed several hundred thousand. In 1985, an earthquake originating beneath the Pacific

Figure 6.4 Earthquake damage
(a) This elevated road collapsed as a result of intense seismic shaking associated with the 1995 Kobe, Japan, earthquake. *(Naoto Hosaka/Getty Images Inc.)* (b) Collapsed buildings, Balakot, Pakistan, from M 7.6 earthquake in 2005. *(AP/Wide World Photos)*

(a)

(b)

Ocean off Mexico (M 8.1) caused 10,000 deaths in Mexico City, several hundred kilometers from the source. Exactly one year after Northridge, the January 17, 1995, Kobe, Japan, earthquake (M 7.2) killed more than 5,000 and injured 27,000 people while destroying 100,000 buildings and causing over $100 billion in property damages (Figure 6.4a). The January 26, 2001, Gujarat, India, earthquake (M 7.7) killed as many as 30,000 people, injured 166,000, damaged or destroyed about 1 million homes, and left 600,000 people homeless. An earthquake on October 8, 2005, of M 7.6, struck northern Pakistan. Although the epicenter was in Pakistan, extensive damage also occurred in Kashmir and India (Figure 6.4b). Over 80,000 people were killed and over 30,000 buildings collapsed. Entire villages were destroyed, some buried by landslides triggered by the violent shaking.[2,3]

6.3 Earthquake Intensity

A qualitative way of comparing earthquakes is to use the **Modified Mercalli Scale,** which describes 12 divisions of intensity based on observations concerning the severity of shaking during an earthquake (Table 6.4). Intensity reflects how people perceived the shaking and how structures responded to the shaking. Whereas a particular earthquake has only one magnitude, different levels of intensity may be assigned to the same earthquake at different locations, depending on proximity to the epicenter and local geologic conditions. Figure 6.5 is a map showing the spatial variability of intensity for the 1971 San Fernando earthquake (M 6.6). Such maps, produced from questionnaires sent to residents in the epicentral region after an earthquake, are a valuable, although crude, index of ground shaking.

One of the major challenges during a damaging earthquake is to quickly determine where the damage is most severe. An approach now being used is known as a **shake map** that shows the extent of potential damaging shaking following an earthquake. Data for a shake map are recorded from a dense network of high-quality seismograph stations. When seismic data are received at seismographic stations, the areas with the severest shaking are known within a minute or so after the shaking has ceased. This information is critical to direct an effective emergency response to those areas. The maps in Figure 6.6 show the shake map for the 1994 Northridge, California, earthquake (M 6.7), and the 2001 M 6.8 Seattle, Washington, earthquake. Notice that the magnitudes of the two earthquakes are very similar, but the intensity of shaking was greater for

TABLE 6.4   Modified Mercalli Intensity Scale (abridged)

Intensity	Effects
I	Felt by very few people.
II	Felt by only a few persons at rest, especially on upper floors of buildings. Delicately suspended objects may swing.
III	Felt quite noticeably indoors, especially on upper floors of buildings, but many people do not recognize it as an earthquake. Standing motor cars may rock slightly. Vibration feels like the passing of a truck.
IV	During the day felt indoors by many, outdoors by few. At night some awakened. Dishes, windows, doors disturbed; walls make cracking sound; sensation like heavy truck striking building; standing motor cars rock noticeably.
V	Felt by nearly everyone; many awakened. Some dishes, windows, and so on, broken; a few instances of cracked plaster; unstable objects overturned; disturbances of trees, poles, and other tall objects sometimes noticed. Pendulum clocks may stop.
VI	Felt by all; many frightened and run outdoors. Some heavy furniture moved; a few instances of fallen plaster or damaged chimneys. Damage is slight.
VII	Everybody runs outdoors. Damage negligible in buildings of good design and construction; slight to moderate in well-built ordinary structures; considerable in poorly built or badly designed structures; some chimneys broken. Noticed by persons driving cars.
VIII	Damage slight in specially designed structures; considerable in ordinary substantial buildings with partial collapse; great in poorly built structures; panel walls thrown out of frame structures; fall of chimneys, factory stacks, columns, monuments, walls; heavy furniture overturned; sand and mud ejected in small amounts; changes in well water; disturbs persons driving cars.
IX	Damage considerable in specially designed structures; well-designed frame structures thrown out of plumb; great in substantial buildings, with partial collapse. Buildings are shifted off foundations. Ground cracked conspicuously. Underground pipes are broken.
X	Some well-built wooden structures are destroyed; most masonry and frame structures with foundations destroyed; ground badly cracked. Rails bent. Landslides considerable from riverbanks and steep slopes. Shifted sand and mud. Water is splashed over banks.
XI	Few, if any (masonry) structures remain standing. Bridges are destroyed. Broad fissures are formed in ground. Underground pipelines are taken out of service. Earth slumps and land slips on soft ground occurs. Train rails are bent.
XII	Damage is total. Waves are seen on ground surfaces. Lines of sight and level distorted. Objects are thrown upward into the air.

Source: From Wood and Neuman, 1931, by U.S. Geological Survey, 1974, *Earthquake Information Bulletin* 6(5): 28.

Figure 6.5 Intensity of shaking Modified Mercalli Intensity Map for the 1971 San Fernando Valley, California, earthquake (M 6.6), determined after the earthquake. *(U.S. Geological Survey, 1974, Earthquake Information Bulletin 6[5])*

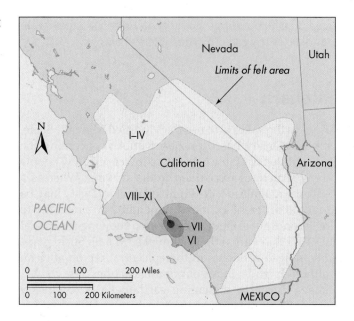

the Northridge earthquake. The technology to produce and distribute shake maps in the minutes following an earthquake was made available in 2002.[4] The cost of seismographs is small relative to damages from earthquake shaking, and the arrival of emergency personnel is critical in the first minutes and hours following an earthquake if people in collapsed buildings are to be rescued. The shake map is also very useful in helping locate areas where gas lines and other utilities are likely to be damaged. Clearly, the use of this technology, especially in our urban areas vulnerable to earthquakes, is a very desirable component of our preparedness for earthquakes.

6.4 Plate Boundary Earthquakes

California, which straddles two lithospheric plates that are moving past one another, experiences frequent damaging earthquakes. The 1989 Loma Prieta earthquake (M 7.1) on the San Andreas fault system south of San Francisco killed 62 people and caused $5 billion in property damage. Neither the Loma Prieta earthquake nor the Northridge earthquake (M 6.7) was considered a great earthquake. It has been estimated that a great earthquake occurring today in a densely populated part of southern California could inflict $100 billion in damage and kill several thousand people. Thus, the Northridge quake, as terrible as it was, was not the anticipated "big one." Given that earthquakes have the proven potential for producing a catastrophe, earthquake research is primarily dedicated to understanding earthquake processes. The more we know about the probable location, magnitude, and effects of an earthquake, the better we can estimate the damage that is likely to occur and make the necessary plans for minimizing loss of life and property.

Interplate earthquakes are those between two plates, initiated near plate boundaries and producing nearly continuous linear or curvilinear zones in which most seismic activity takes place (Figure 6.7). Most large U.S. earthquakes are interplate earthquakes in the West, particularly near the North American and Pacific plate boundaries (Table 6.1). However, large damaging *intraplate earthquakes*, located within a single plate, can occur far from plate boundaries.

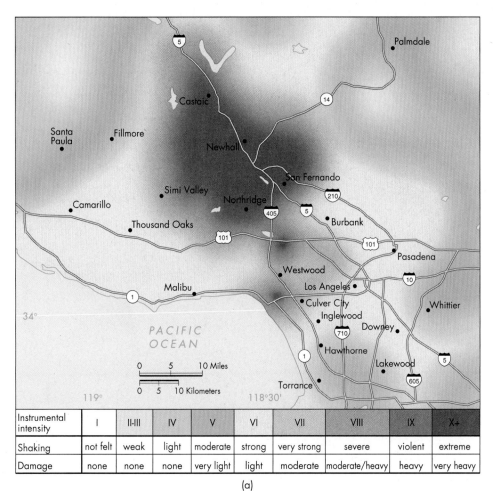

Instrumental intensity	I	II-III	IV	V	VI	VII	VIII	IX	X+
Shaking	not felt	weak	light	moderate	strong	very strong	severe	violent	extreme
Damage	none	none	none	very light	light	moderate	moderate/heavy	heavy	very heavy

(a)

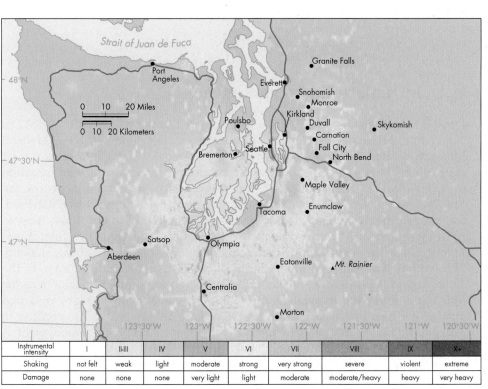

Instrumental intensity	I	II-III	IV	V	VI	VII	VIII	IX	X+
Shaking	not felt	weak	light	moderate	strong	very strong	severe	violent	extreme
Damage	none	none	none	very light	light	moderate	moderate/heavy	heavy	very heavy

(b)

Figure 6.6 **Real-time intensity of shaking** (a) Shake map for the 1994 Northridge, California, earthquake (M 6.7) determined after the earthquake occurred. *(U.S. Geological Survey, and courtesy of David Wald)* (b) The 2001 Seattle, Washingon, earthquake (M 6.8). *(Pacific Northwest Seismograph Network, University of Washington)*

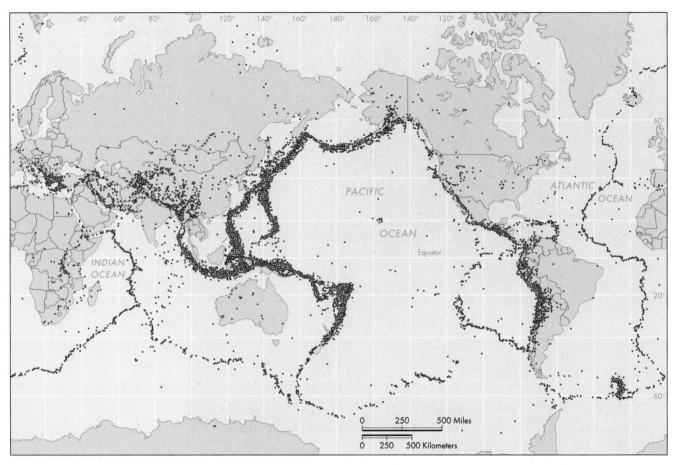

Figure 6.7 **Earthquakes at plate boundaries** Map of global seismicity (1963–1988), delineating plate boundaries and earthquake belts shown in Figure 6.9. Brown dots represent individual earthquakes. For the locations and names of Earth's tectonic plates refer to Figure 2.4. *(Courtesy of National Earthquake Information Center)*

6.5 Intraplate Earthquakes

Intraplate earthquakes with M 7.5+ that occurred in the winter of 1811–1812 in the central Mississippi Valley nearly destroyed the town of New Madrid, Missouri, while killing an unknown number of people. These earthquakes rang church bells in Boston! Seismic shaking produced intense surface deformation over a wide area from Memphis, Tennessee, north to the confluence of the Mississippi and Ohio Rivers. During the earthquakes, forests were flattened; fractures in the ground opened so wide that people had to cut down trees to cross them; and land sank several meters in some areas, causing flooding. It was reported that the Mississippi River actually reversed its flow during shaking. The earthquakes occurred along a seismically active structure known as the New Madrid seismic zone, which underlies the geologic structure known as the Mississippi River Embayment (Figure 6.8). The embayment is a downwarped area of Earth's crust where the lithosphere is relatively weak. The recurrence interval, or time between events, for major earthquakes in the embayment is estimated to be about 500 years.[5,6] Even in this "stable" interior of the North American plate, the possibility of future damage demands that the earthquake hazard in the area be considered when facilities such as power plants and dams are being designed and built. It is believed that the New Madrid seismic zone is a young, perhaps less than 10,000 year old, zone of deformation. The rate of

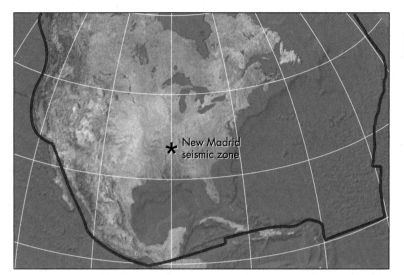

Figure 6.8a New Madrid seismic zone Location is thousands of kilometers from the nearest plate boundary (red lines).

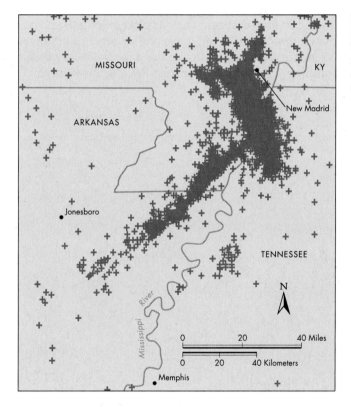

Figure 6.8b This zone is the most earthquake-prone region in the United States east of the Rocky Mountains. Locations of recorded minor earthquakes since 1974 are shown as crosses. *(U.S. Geological Survey)*

uplift is sufficient to produce significant topographic relief over a period of several hundred thousand years. The fact that topography of the uplifted region in the Mississippi River floodplain area is very minor supports the hypothesis that this is a very young fault system or one that has been recently reactivated. The New Madrid seismic zone is capable of producing great earthquakes and is the object of intensive research.

Another large damaging intraplate earthquake (M 7.5) occurred on August 31, 1886, near Charleston, South Carolina. The earthquake killed about 60 people and damaged or destroyed most buildings in Charleston. Effects of the earthquake were reported at distances exceeding 1,000 km (620 mi).

Intraplate earthquakes in the eastern United States are generally more damaging and felt over a much larger area than a similar magnitude earthquake in California. The reason is that the rocks in the eastern United States are generally stronger and less fractured and can more efficiently transmit earthquake waves than rocks in the west.

6.6 Earthquake Processes

Our discussion of global tectonics established that Earth is a dynamic, evolving system. Earthquakes are a natural consequence of the processes that form the ocean basins, continents, and mountain ranges of the world. Most earthquakes occur along the boundaries of lithospheric plates (Figures 6.7 and 6.9).

Faulting

The process of fault rupture, or *faulting,* can be compared to sliding two rough boards past one another. Friction along the boundary between the boards, analogous to a fault plane, may temporarily slow their motion, but rough edges break off and motion occurs at various places along the plane. For example, lithospheric plates that are moving past one another are slowed by friction along their boundaries. As a result, rocks along the boundary undergo strain, or deformation, resulting from stress produced by the movement. When stress on the rocks

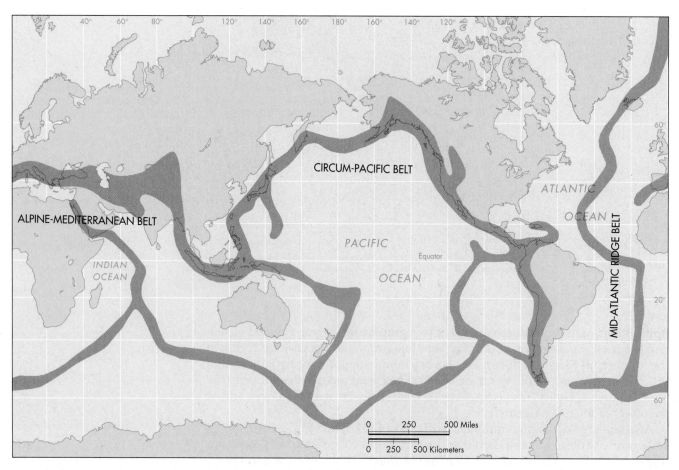

Figure 6.9 Earthquake belts Map of the world showing the major earthquake belts as shaded areas. *(National Oceanic and Atmospheric Administration)*

exceeds their strength, the rocks rupture, forming a fault and producing an **earthquake.** A **fault** is a fracture or fracture system along which rocks have been displaced; that is, one side of the fracture or fracture system has moved relative to the other side. The long-term rate of movement is known as the *slip rate* and is often recorded as millimeters per year (mm/yr) or meters per 1,000 years (m/ky). During a major to great earthquake, displacements of several meters may suddenly occur along a fault. When a rupture begins it starts at the focus and then grows or propagates up, down, and laterally along the fault plane during an earthquake. (See Figure 6.3.) The sudden rupture of the rocks produces shock waves called earthquake waves, or *seismic waves,* that shake the ground. In other words, the pent-up energy of the strained rocks is released in the form of an earthquake. Faults are therefore *seismic sources,* and identifying them is the first step in evaluating the risk of an earthquake, or seismic risk in a given area.

Fault Types

The major types of faults, based on the direction, or sense of the relative displacement, are shown on Figure 6.10. A *strike-slip fault* is one in which the sides of the fault are displaced horizontally; a strike-slip fault is called *right-lateral* if the right-hand side moves toward you as you sight, or look along the fault line, and *left-lateral* if the left-hand side moves toward you. A fault with vertical displacement is referred to as a *dip-slip fault.* A dip-slip fault may be a *reverse fault* or a *normal fault,* depending on the geometry of the displacement. Geologists use interesting terminology to distinguish reverse and normal faults. Notice in Figure 6.10 that there are two blocks separated by the fault plane. One way to remember the terminology for the two blocks is to imagine you are walking up the fault plane, like walking up a hill. The block you would be standing on is called the foot-wall, and the other block is called the hanging-wall. If the fault displacement is such that the hanging-wall moves up relative to the foot-wall, the fault is called a *reverse fault.* When the fault plane of a reverse fault has an angle of less than 45 degrees it is called a *thrust fault.* If the hanging-wall moves down relative to the foot-wall the fault is called a *normal fault.* Reverse and thrust fault displacement are associated with crustal shortening, whereas normal fault displacement is associated with crustal extension.

The faults shown in Figure 6.10 generally produce surface displacement or rupture. However, there are also *buried faults,* usually associated with folded rocks. Displacement and rupture of buried faults do not propagate to the surface even in large earthquakes, as was the case with the Northridge earthquake.

The relationship of a buried reverse fault to rock folding is shown in Figure 6.11. Shortening of a sequence of sedimentary rocks has produced folds called anticlines and synclines. *Anticlines* are arch-shaped folds; *synclines* are bowl-shaped folds. In this illustration, anticlines form ridges and synclines form basins at the surface of the ground. Notice that in the cores of two of the anticlines on the right, buried faulting has occurred. Faulting during earthquakes causes anticlinal mountains to be uplifted, whereas subsidence, or sinking of the ground surface, may occur in synclinal valleys.

Until recently it was thought that most active faults could be mapped because their most recent earthquake would cause surface rupture. Discovering that some faults are buried and that rupture does not always reach the surface has made it more difficult to evaluate the earthquake hazard in some areas.

Active Faults

Most geologists consider a particular fault to be an *active fault* if it can be demonstrated to have moved during the past 10,000 years, the Holocene epoch. The Quaternary period, spanning approximately the past 1.65 million years, is the most recent period of geologic time, and most of our landscape has been produced during that time (see Table 1.1). Fault displacement that has occurred during the Pleistocene

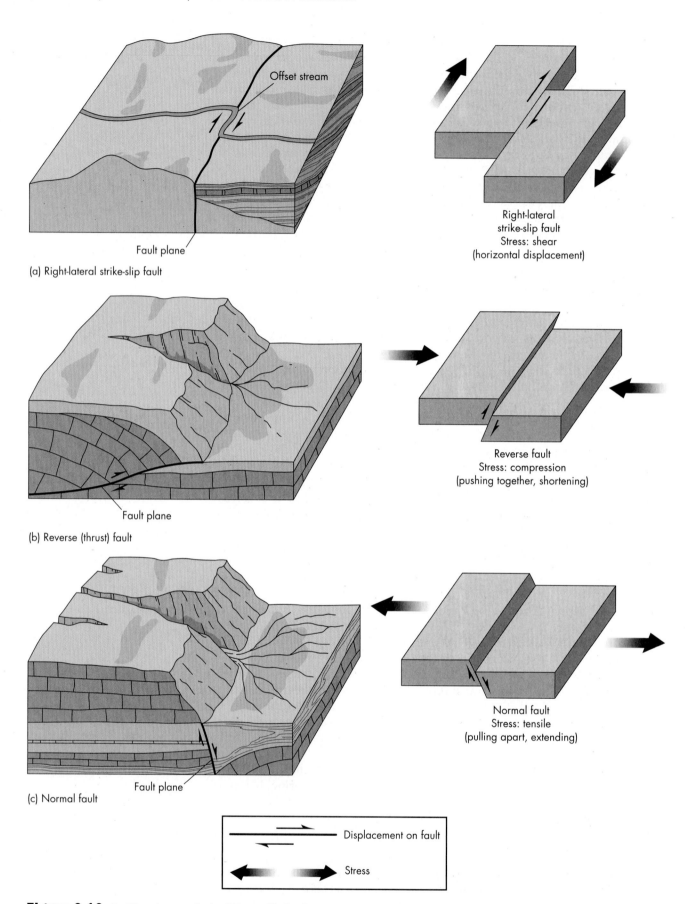

Offset stream

Fault plane

(a) Right-lateral strike-slip fault

Right-lateral
strike-slip fault
Stress: shear
(horizontal displacement)

Fault plane

(b) Reverse (thrust) fault

Reverse fault
Stress: compression
(pushing together, shortening)

Fault plane

(c) Normal fault

Normal fault
Stress: tensile
(pulling apart, extending)

Displacement on fault

Stress

Figure 6.10 Faulting changes the land Types of fault movement and effects on the landscape based on the sense of motion relative to the fault.

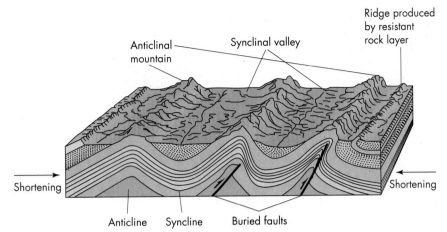

Anticlinal mountain · Synclinal valley · Ridge produced by resistant rock layer

Shortening → ← Shortening

Anticline · Syncline · Buried faults

Figure 6.11 Buried faults
Block diagram illustrating several types of common folds and buried reverse faults with possible surface expressions such as anticlinal mountains and synclinal valleys. *(Modified after Lutgens, F., and Tarbuck, E. 1992. Essentials of geology, 4th ed. New York: Macmillan)*

epoch of the Quaternary period, approximately 1.65 million to 10,000 years ago, but not in the Holocene, is classified as *potentially active* (Table 6.5).

Faults that have not moved during the past 1.65 million years are generally classified as *inactive*. However, we emphasize that it is often difficult to prove the activity of a fault in the absence of easily measured phenomena such as historical earthquakes. Demonstrating that a fault is active may require determining the past earthquake history, or *paleoseismicity*, on the basis of the geologic record. This determination involves identifying faulted Earth materials and determining when the most recent displacement occurred.

Tectonic Creep

Some active faults exhibit **tectonic creep,** that is, gradual displacement that is not accompanied by felt earthquakes. The process can slowly damage roads, sidewalks, building foundations, and other structures. Tectonic creep has damaged culverts under the football stadium of the University of California at Berkeley and periodic repairs have been necessary as the cracks developed. Movement of approximately 3.2 cm (1.3 in.) was measured in a period of 11 years. More rapid rates of tectonic creep have been recorded on the Calaveras fault zone, a segment of the San Andreas fault, near Hollister, California. At one location, a winery situated on the fault is slowly being pulled apart at about 1 cm (0.4 in.) per year (Figure 6.12). Damages resulting from tectonic creep generally occur along narrow fault zones subject to slow, continuous displacement.

TABLE 6.5   Terminology Related to Recovery of Fault Activity

Era	Period	Epoch	Years before Present	Fault Activity
Cenozoic	Quaternary	Historic Holocene	200	Active
			10,000	
		Pleistocene		Potentially active
			1,650,000	
	Tertiary	Pre-Pleistocene		Inactive
	Pre-Cenozoic time		65,000,000	
			4,500,000,000	
	Age of Earth			

Source: After California State Mining and Geology Board Classification, 1973.

Figure 6.12 Tectonic creep
This concrete culvert at the Almandea vineyards in Hollister, California, is being split by creep on the San Andreas fault. *(James A. Sugar/NGS Image Collection)*

Slow Earthquakes

Slow earthquakes are similar to other earthquakes in that they are produced by fault rupture. The big difference is that the rupture, rather than being nearly instantaneous, can last from days to months. The moment magnitude of slow earthquakes can be in the range of 6 to 7 because a large area of rupture is often involved, although the amount of slip is generally small (a centimeter or so). Slow earthquakes are a newly recognized fundamental Earth process. They are recognized through analysis of continuous geodetic measurement or GPS, similar to the devices that are used in automobiles to identify your location. These instruments can differentiate horizontal movement in the millimeter range and have been used to observe surface displacements from slow earthquakes. When slow earthquakes occur frequently, say every year or so, their total contribution to changing the surface of the earth to produce mountains over geologic time may be significant.[7]

6.7 Earthquake Shaking

Three important factors determine the shaking you will experience during an earthquake: (1) earthquake magnitude; (2) your distance from the epicenter; and (3) local soil and rock conditions. If an earthquake is of moderate magnitude (M 5 to 5.9) or larger, then strong motion or shaking may be expected. It is the strong motion from earthquakes that cracks the ground and makes Earth "rock and roll" to damage buildings and other structures.

Types of Seismic Waves

Some of the seismic waves generated by fault rupture travel within Earth and others travel along the surface. The two types of seismic waves that travel with a velocity of several kilometers per second through rocks are primary (P) and secondary (S) waves (see Figure 6.13a, b).

P waves, also called compressional waves, are the faster of the two and can travel through solid, liquid, and gaseous materials (see Figure 6.13a). The velocity of P waves through liquids is much slower. Interestingly, when P waves are propagated into the atmosphere, they are detectable to the human ear. This fact explains the observation that people sometimes *hear* an earthquake before they feel the shaking caused by the arrival of the slower surface waves.[8]

S waves, also called shear waves, can travel only through solid materials. Their speed through rocks such as granite is approximately one-half that of P waves. S waves produce an up-and-down or side-to-side motion at right angles to the direction of wave propagation, similar to the motion produced in a clothesline by pulling it down and letting go (see Figure 6.13b). Because liquids cannot spring

P wave: Direction of propagation

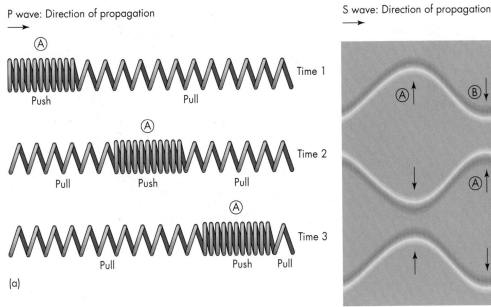

(a)

S wave: Direction of propagation

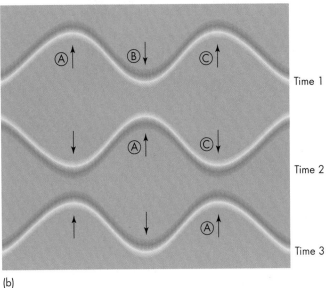

(b)

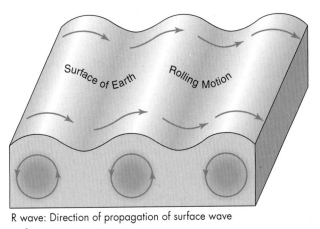

R wave: Direction of propagation of surface wave

(c)

Figure 6.13 **Seismic waves** Idealized diagram showing differences between P waves and S waves. (a) Visualize the P wave as dilation (pulling apart) and contraction (pushing together) the rings of a Slinky plastic spring. If you extend it about 3 m (about 10 ft) horizontally on a flat surface and then push from the left end, the zone of compression (contraction) will propagate to the right as shown. The rate of propagation of P waves through rocks such as granite is about 6 km/sec (3.7 mi/sec). Through liquids, P waves move much slower, about 1.5 km/sec (0.9 mi/sec) through water. (b) To visualize S waves, stretch a rope about 7 m (about 23 ft) between two chairs, or use a clothesline if you have one. Pull up the left end and then release it; the wave will propagate to the right, in this case up and down. Displacement for S waves is at right angles to the direction of wave propagation, just the opposite of P wave displacements that push and pull in the same direction as the wave propagates. Points A, B, and C are the positions of a specific part of the wave at different times. S waves travel through rocks such as granite approximately 3 km/sec (1.9 mi/sec). They cannot travel through liquids. (c) Surface or R waves. Notice that the vertical motion is at right angles to the direction of wave propagation and that the elliptical motion is opposite to the direction of propagation. Surface waves are often the most damaging of the seismic waves.

back when subjected to this type of motion, called sideways shear, S waves cannot move through liquids.[8]

When seismic waves reach the surface, complex **surface waves** (R waves) are produced. Surface waves, which move along Earth's surface, travel more slowly than either P or S waves and cause much of the earthquake damage to buildings and other structures. Damage occurs because surface waves have a complex horizontal and vertical ground movement or rolling motion that may crack walls and foundations of buildings, bridges, and roads. An important surface wave is the R wave, shown in Figure 6.13c. The rolling motion moves in the direction opposite to that of the wave, or propagation, and the vertical motion, or amplitude, is at right angles to the direction of propagation.

Seismograph

The *seismogram* is the written or digital record of an earthquake. In written form, it is a continuous line that shows vertical or horizontal Earth motions received at a seismic recording station and recorded by a seismograph. The components of a simple seismograph are shown in Figure 6.14a, and a photograph of a modern seismograph is shown in Figure 6.14b. An idealized written record, or seismogram,

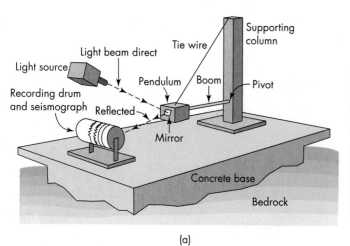

(a)

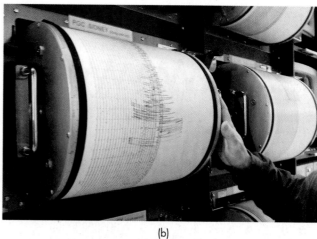

(b)

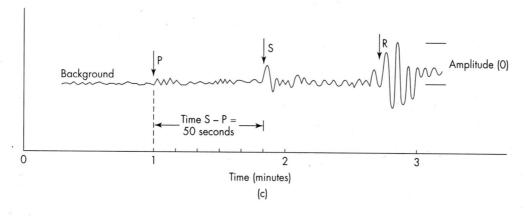

(c)

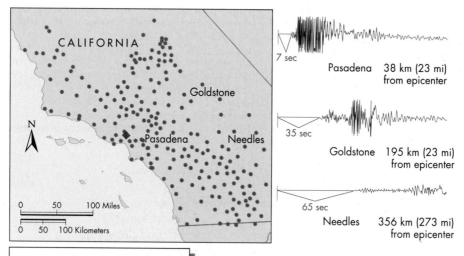

- Seismic stations

◆ Epicenter 1994
Northridge earthquake

No. of sec. Approximate length
of time between occurence
of earthquake and arrival of
P waves.

(d)

Pasadena   38 km (23 mi)
from epicenter

Goldstone   195 km (23 mi)
from epicenter

Needles   356 km (273 mi)
from epicenter

Figure 6.14 **Seismograph** (a) Simple seismograph showing how it works. (b) Modern seismograph at the Pacific Geoscience Centre in Canada. The earthquake is the February 28, 2001, Seattle earthquake (M 6.8). *(Ian McKain/AP Wide World Photos)* (c) Idealized seismogram for an earthquake. The amplitude of the R (surface) waves is greater than that of P and the S waves. The S–P time of 50 seconds tells us that the earthquake epicenter was about 420 km (261 mi) from the seismograph. (d) Differences in arrival time and amount of shaking at three seismic stations located from 38 to 356 km (24 to 221 mi) from the 1994 Northridge, California, earthquake. Notice that, with distance, the time of arrival of shaking increases and the amplitude of shaking decreases. *(Modified from Southern California Earthquake Center)*

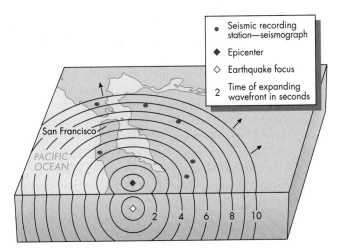

Figure 6.15 **Locating an earthquake** Idealized diagram of San Francisco, California, region with expanding wavefront of seismic waves from the focus of an earthquake. The arrival of waves at seismic recording stations may be used to mathematically calculate the location of the epicenter and its depth of focus. *(U.S. Geological Survey)*

Legend:
- Seismic recording station—seismograph
- ◆ Epicenter
- ◇ Earthquake focus
- 2 Time of expanding wavefront in seconds

is shown in Figure 6.14c. Notice that for Figure 6.14c the P waves arrive 50 seconds before the S waves and about 1 minute 40 seconds before the surface (R) waves. R waves have the largest amplitude and often cause the most damage to buildings.

The effect of distance on the seismogram is shown in Figure 6.14d for the 1994 M 6.7 Northridge, California, earthquake. Three seismographs from close to the epicenter at Pasadena (38 km or 24 mi) to far away at Needles (356 km or 221 mi) are shown. The shaking starts with the arrival of the P wave. Notice that the shaking arrives sooner and is more intense at Pasadena than at Needles.

The difference in arrival times at seismographs (S − P) can be used to locate the epicenter of an earthquake. Locating an epicenter requires at least three seismograms from three locations. For example, the S − P value of 50 seconds on Figure 6.14c suggests that the distance between the epicenter and the seismograph was about 420 km (261 mi). This is calculated algebraically using the S − P value and the velocity for P and S waves mentioned earlier; however, this calculation is beyond the scope of our discussion. Although we now know that the epicenter was about 420 km (261 mi) from the seismograph, this distance could be in any direction. When the records from seismographs from numerous seismic stations are analyzed using computer models, the epicenter for an earthquake can be located. The model is used to calculate the travel times of seismic waves to each seismograph from various depths of focus (Figure 6.15). Distances and depths to the focus are adjusted to provide the best fit of the data from all the seismographs. By a process of approximation, the location of the epicenter and depth to the focus are determined.

Frequency of Seismic Waves

Another important characteristic of earthquakes and their seismic waves is their frequency, measured in cycles per second, or hertz (Hz). To understand wave frequency, visualize water waves traveling across the surface of the ocean, analogous to seismic waves. If you are at the end of a pier and record the time when each wave peak moves by you toward shore, the average time period between wave peaks is the *wave period* and is usually measured in seconds. Consider the passing of each wave by you to be one cycle. If one wave passes the end of the pier on its way toward the shore every 10 seconds, the wave period is 10 seconds and there are 6 waves per minute. The frequency of the waves in cycles per second is 1 cycle divided by 10 seconds, or 0.1 Hz. Returning to earthquake waves, most P and S waves have frequencies of 0.5 to 20 Hz, or one-half to 20 cycles per second. Surface waves have lower frequencies than P and S waves, often less than 1 Hz.

Why is the wave frequency important? During an earthquake seismic waves with a wide range of frequencies are produced, and intense ground motion near the epicenter of a large earthquake is observed. High-frequency shaking causes low buildings to vibrate, and low-frequency shaking causes tall buildings to vibrate. It is the vibrating or shaking that damages buildings. Two points are important in understanding the shaking hazard: (1) Near the epicenter, both shorter buildings and taller buildings may be damaged by high- and low-frequency seismic waves. (2) With increasing distance from the epicenter the high-frequency waves are weakened or removed by a process called *attenuation*. Rapid shaking dies off quickly with distance. As a result, nearby earthquakes are often described as "jolting" and far-away earthquakes as "rolling." Low-frequency seismic waves of about 0.5 to 1.0 Hz can travel long distances without much attenuation. Therefore, they can damage tall buildings far from the epicenter. This fact has importance in planning to reduce earthquake damages and loss of life. Tall buildings need to be designed to withstand seismic shaking even if they are located hundreds of kilometers from large faults that are capable of producing strong to great earthquakes.

Material Amplification

Different Earth materials, such as bedrock, alluvium (sand and gravel), and silt and mud, respond differently to seismic shaking. For example, the intensity of shaking or strong ground motion of unconsolidated sediments may be much more severe than that of bedrock. Figure 6.16 shows how the *amplitude* of shaking, or the vertical movement, is greatly increased in unconsolidated sediments such as silt and clay deposits. This effect is called **material amplification.**

The Mexico City earthquake (M 8.1) demonstrated that buildings constructed on materials likely to accentuate and increase seismic shaking are extremely vulnerable to earthquakes, even if the event is centered several hundred kilometers away. Although seismic waves originating offshore initially contained many frequencies, those arriving at the city were low-frequency waves of about 0.5 to 1.0 Hz. It is speculated that when seismic waves struck the lake beds beneath Mexico City, the amplitude of shaking may have increased at the surface by a factor of 4 or 5. Figure 6.17 shows the geology of the city and the location of the worst damage. The intense regular shaking caused buildings to sway back and forth, and eventually many of them pancaked as upper stories collapsed onto lower ones.[9]

The potential for amplification of surface waves to cause damage was again demonstrated with tragic results during the 1989 Loma Prieta earthquake (M 7.1), which originated south of San Francisco. Figure 6.18 shows the epicenter and the areas that greatly magnified shaking. The collapse of a tiered freeway, which killed 41 people, occurred on a section of roadway constructed on bay fill and mud (Figure 6.19). Less shaking occurred where the freeway was constructed on older, stronger alluvium; in these areas the structure survived. Extensive damage

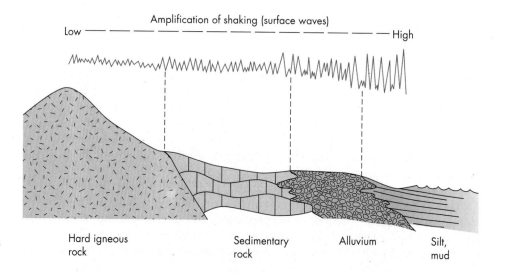

Figure 6.16 Amplification of shaking Generalized relationship between near-surface Earth material and amplification of shaking during a seismic event.

Amplification of shaking (surface waves)

Low ————————————— High

Hard igneous rock

Sedimentary rock

Alluvium

Silt, mud

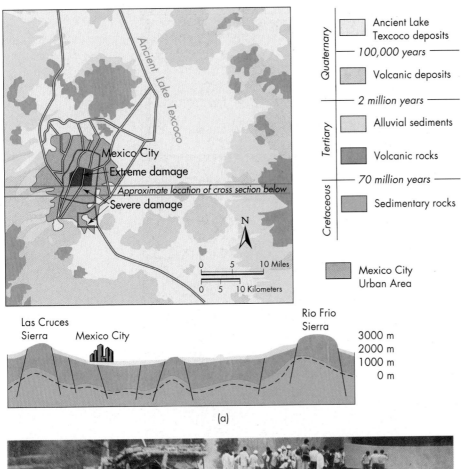

Figure 6.17 **Earthquake damage to Mexico City** (a) Generalized geologic map of Mexico City showing the ancient lake deposits where the greatest damage occurred. The zone of extreme damage is represented by the solid red area, and the severe damage zone is outlined in red. (b) One of many buildings that collapsed during the 1985 (M 8.1) earthquake. *(Both courtesy of T. C. Hanks and Darrell Herd/U.S. Geological Survey)*

(a)

(b)

was also recorded in the Marina district of San Francisco (Figure 6.20). This area was primarily constructed on bay fill and mud as well as debris dumped into the bay during the cleanup following the 1906 earthquake.[10]

Directivity

Rupture of rocks on a fault plane starts at a point and radiates, or propagates, from that point. The larger the area of rupture the larger the earthquake. **Directivity,** another amplification effect, results because the intensity of seismic shaking increases in the direction of the fault rupture. For example, fault rupture of the 1994

Figure 6.18 **Loma Prieta earthquake** San Francisco Bay region, showing the San Andreas fault and the epicenter of the 1989 earthquake, which had a magnitude of 7.1. The most severe shaking was on bay fill and mud, where a freeway collapsed and the Marina district was damaged. *(Modified after T. Hall. Data from U.S. Geological Survey)*

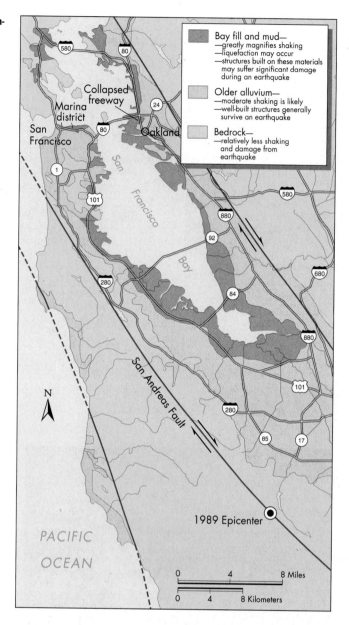

Northridge earthquake was up (north) and to the west, resulting in stronger ground motions from seismic shaking to the northwest (Figure 6.21). The stronger ground motion to the northwest is believed to have resulted from the movement of a block of Earth, the hanging-wall block, upward and to the northwest.[1]

Ground Acceleration During Earthquakes

Strong ground motion from earthquakes may be described in terms of the speed or velocity at which primary, secondary, and surface waves travel through rocks or across the surface of Earth. In general, these waves move a few kilometers per second. This velocity is analogous to the units used to describe the speed at which you drive a car, for example, 100 km per hour (62 mi per hour). Earthquake waves, however, travel much faster than cars, at a velocity of about 20,000 km (12,428 mi) per hour, similar to that of the U.S. Space Shuttle when it orbits Earth. Damage to structures from strong ground motion is related to both the amplitude of seismic surface waves and to the rate of velocity change of the seismic waves with time. The rate of change of velocity with time is referred to as acceleration. You may have learned in physical science that the acceleration of gravity is 9.8 meters

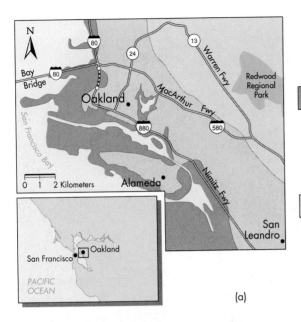

Collapse of two-tier section of Nimitz Freeway

Bay fill and mud
Greatly magnifies shaking—liquefaction may occur. Structures built on these materials may suffer significant damage during an earthquake.

Older alluvium
Moderate shaking is likely. Well-built structures generally survive an earthquake.

(a)

Figure 6.19 **Collapse of a freeway** (a) Generalized geologic map of part of the San Francisco Bay showing bay fill and mud and older alluvium. *(Modified after Hough, S. E., et al. 1990. Nature 344(6269): 853–55. Copyright © Macmillan Magazines Ltd., 1990. Used by permission of the author)* (b) Collapsed freeway as a result of the 1989 earthquake. *(Courtesy of Dennis Laduzinski)*

(b)

Figure 6.20 **Earthquake damage** Damage to buildings in the Marina district of San Francisco resulting from the 1989 M 7.1 earthquake. *(John K. Nakata/ U.S. Geological Survey)*

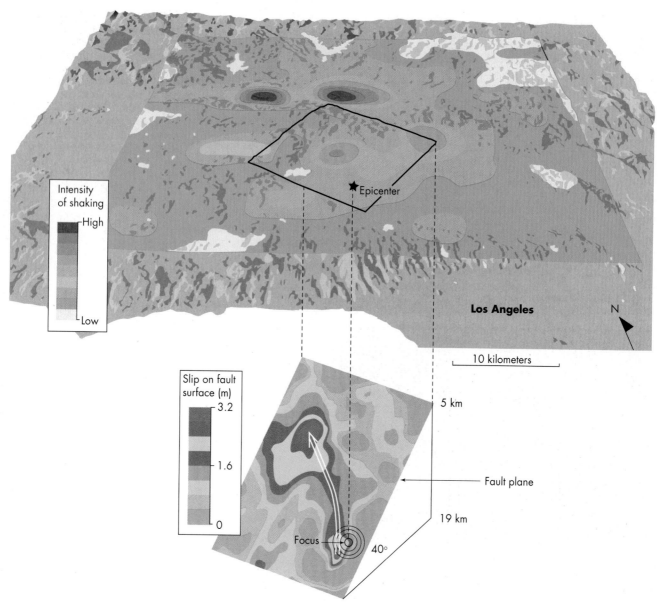

Figure 6.21 **Epicenter and rupture path** Aerial view of the Los Angeles region from the south showing the epicenter of the M 6.7 1994 Northridge earthquake, showing peak ground motion in centimeters per second and the fault plane in its subsurface position. The path of the rupture is shown, along with the amount of slip in meters along the fault plane. The area that ruptured along the fault plane is approximately 430 km² (166 mi²) and the fault plane is dipping at approximately 40 degrees to the south-southwest. Notice that the maximum slip and intensity of shaking both occur to the northwest of the epicenter. The fault rupture apparently began at the focus in the southeastern part of the fault plane and proceeded upward and to the northwest as shown by the arrow. *(U.S. Geological Survey. 1996. USGS response to an urban earthquake, Northridge '94. U.S. Geological Survey Open File Report 96–263)*

per second per second (written as 9.8 m/sec², or 32 ft/sec²). This acceleration is known as 1 *g*. If you parachute from an airplane and free fall for several seconds, you will experience the acceleration of gravity (1 *g*), and by the end of the first second of free fall, your velocity will be 9.8 m/sec. At the end of 2 seconds of free fall, your velocity will be about 20 m/sec, and it will be about 30 m/sec at the end of 3 seconds. This increase, or rate of change, in velocity with time is the acceleration. Earthquake waves cause the ground to accelerate both vertically and horizontally, just as your car accelerates horizontally when you step on the gas pedal. Specially designed instruments called accelometers measure and record the acceleration of the ground during earthquakes. The units of acceleration in earthquake studies

Figure 6.22 Buildings collapse in Russia Rows of prefabricated, unreinforced, five-story apartment buildings destroyed by the 1995 earthquake that struck the island of Sakhalin in Russia (M 7.5). *(Tanya Makeyeva/AP/Wide World Photos)*

are in terms of the acceleration of gravity. If the ground accelerates at 1 g, the value is 9.8 m/sec^2. If the acceleration of the ground is 0.5 g, the value is roughly 5 m/sec^2. One reason we are interested in acceleration of the ground from earthquakes is that when engineers design buildings to withstand seismic shaking, the building design criteria are often expressed in terms of a maximum acceleration of the ground, since it is the horizontal acceleration of the ground that causes most damage to buildings. For example, earthquakes with M 6.0 to 6.9, typical magnitudes for strong southern California earthquakes, will produce horizontal accelerations of about 0.3 to 0.7 g, with localized values near the epicenter exceeding 1.0 g. Horizontal accelerations in excess of about 0.3 g cause damage to some buildings, and at 0.7 g damage is widespread, unless buildings are designed and constructed to withstand strong ground motion. Therefore, if we wish to design buildings to withstand M 6 to 6.9 earthquakes, we need to conservatively design them to withstand ground accelerations of about 0.6 to 0.7 g. To put this in perspective, consider that homes constructed of adobe, which are common in rural Mexico, South America, and the Middle East, can collapse under a horizontal acceleration of 0.1 g.[8] Unreinforced, prefabricated concrete buildings are also very vulnerable, at 0.2 to 0.3 g, as was tragically illustrated in the 1995 Sakhalin, Russia, earthquake (M 7.5). Two thousand of the 3,000 people in the town of Neftegorsk were killed when 17 prefabricated apartment buildings collapsed into rubble (Figure 6.22).

Depth of Focus

Recall that the point or area within Earth where the earthquake rupture starts is called the focus (see Figure 6.3). The depth of focus of an earthquake varies from just a few kilometers deep to almost 700 km (435 mi) below the surface. The deepest earthquakes occur along subduction zones, where slabs of oceanic lithosphere sink to great depths. The 2001 Seattle, Washington, earthquake (M 6.8) is an example of a relatively deep earthquake. Although the Seattle event was a large quake, about the same size as the 1994 Northridge earthquake, the focus was deeper, at

(a)

Figure 6.23 **Ground rupture from faulting** (a) Cracking of the ground and damage to a building caused by the M 7.5 Landers earthquake, California. (b) Fence and dirt road offset to the right. *(Photos by Edward A. Keller)*

(b)

about 52 km (32 mi) beneath Earth's surface. The focus was within the subducting Juan de Fuca plate along a normal fault (see Figure 2.10). Seismic waves had to move up 52 km (32 mi) before reaching the surface, and they lost much of their energy in the journey. As a result, the earthquake caused relatively little damage for its magnitude. In contrast, most earthquakes in southern California have focal depths of about 10 to 15 km (6 to 9 mi), although deeper earthquakes have occurred. These relatively shallow earthquakes are more destructive than deeper earthquakes of comparable magnitude because they are deep enough to generate strong seismic shaking, yet sufficiently close to the surface to cause strong surface shaking. A M 7.5 earthquake on strike-slip faults in the Mojave Desert near Landers, California, in 1992 had a focal depth of less than 10 km (6 mi). This event caused extensive ground rupture for about 85 km (53 mi).[11] Local vertical displacement exceeded 2 m (6.5 ft), and extensive lateral displacements of about 5 m (16 ft) were measured (Figure 6.23). If the Landers event had occurred in the Los Angeles Basin, extensive damage and loss of life would have occurred.

6.8 Earthquake Cycle

Observations of the 1906 San Francisco earthquake (M 7.7) led to a hypothesis known as the **earthquake cycle.** The earthquake cycle hypothesis proposes that there is a drop in elastic strain after an earthquake and a reaccumulation of strain before the next event.

Strain is deformation resulting from stress. *Elastic strain* may be thought of as deformation that is not permanent, provided that the stress is released. When the stress is released, the deformed material returns to its original shape. If the stress continues to increase, the deformed material eventually ruptures, making the deformation permanent. For example, consider a stretched rubber band or a bent archery bow; continued stress will snap the rubber band or break the bow. When a stretched rubber band or bow breaks, it experiences a rebound, in which the broken ends snap back, releasing their pent-up energy. A similar effect, referred to as *elastic rebound,* occurs after an earthquake (Figure 6.24). At time 1, rocks on either side of a fault segment have no strain built up and show no deformation. At time 2, elastic strain begins to build, caused by the tectonic forces that pull the rocks in

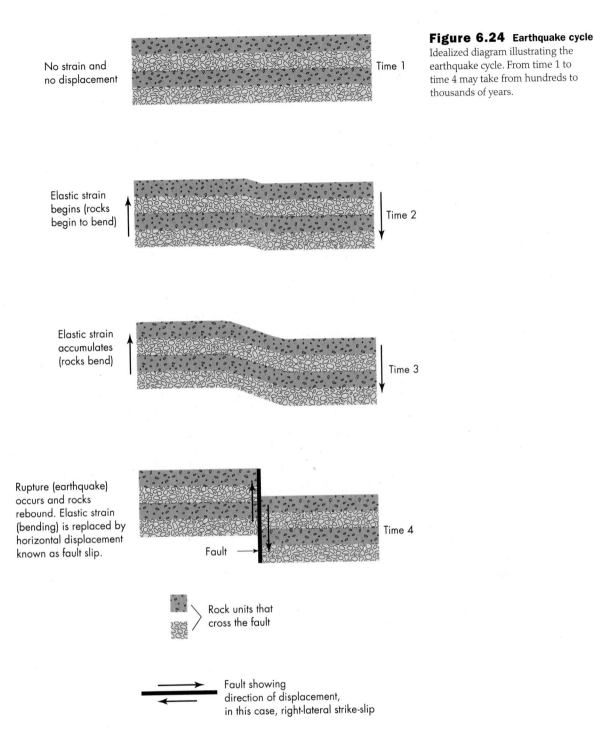

Figure 6.24 **Earthquake cycle**
Idealized diagram illustrating the earthquake cycle. From time 1 to time 4 may take from hundreds to thousands of years.

No strain and no displacement — Time 1

Elastic strain begins (rocks begin to bend) — Time 2

Elastic strain accumulates (rocks bend) — Time 3

Rupture (earthquake) occurs and rocks rebound. Elastic strain (bending) is replaced by horizontal displacement known as fault slip. — Fault — Time 4

Rock units that cross the fault

Fault showing direction of displacement, in this case, right-lateral strike-slip

opposite directions, referred to as shear stress. The rocks begin to bend. At time 3, elastic strain is accumulating, and the rocks have bent; however, they are still held together by friction. When the deformed rocks finally rupture, at time 4, the stress is released and the elastic strain suddenly decreases as the sides of the fault "snap" into their new, or permanently deformed, position. After the earthquake, it takes time for sufficient elastic strain to accumulate again to produce another rupture.[12]

Stages of the Earthquake Cycle

It is speculated that a typical earthquake cycle has three or four stages. The first is a long period of seismic inactivity following a major earthquake and associated *aftershocks*, earthquakes that occur anywhere from a few minutes to a year or so after the main event. This is followed by a second stage characterized by increased seismicity, as accumulated elastic strain approaches and locally exceeds the strength of the rock. The accumulated elastic strain initiates faulting that produces small earthquakes. The third stage of the cycle, which may occur only hours or days before the next large earthquake, consists of *foreshocks*. Foreshocks are small to moderate earthquakes that occur before the main event. For example, a M 6 earthquake may have foreshocks of about M 4. In some cases, this third stage may not occur. After the major earthquake, considered to be the fourth stage, the cycle starts over again.[12] Although the cycle is hypothetical and periods between major earthquakes are variable, the stages have been identified in the occurrence and reoccurrence of large earthquakes.

6.9 Earthquakes Caused by Human Activity

Several human activities are known to increase or cause earthquake activity. Damage from these earthquakes is regrettable, but the lessons learned may help to control or stop large catastrophic earthquakes in the future. Three ways that the actions of people have caused earthquakes are

- Loading the Earth's crust, as in building a dam and reservoir
- Disposing of waste deep into the ground through disposal wells
- Setting off underground nuclear explosions

Reservoir-Induced Seismicity

During the 10 years following the completion of Hoover Dam on the Colorado River in Arizona and Nevada, several hundred local tremors occurred. Most of these were very small, but one was M 5 and two were about M 4.[12] An earthquake in India, approximately M 6, killed about 200 people after the construction and filling of a reservoir. Evidently, fracture zones may be activated both by the increased load of water on the land and by increased water pressure in the rocks below the reservoir, resulting in faulting.

Deep Waste Disposal

From April 1962 to November 1965 several hundred earthquakes occurred in the Denver, Colorado, area. The largest earthquake was M 4.3 and caused sufficient shaking to knock bottles off shelves in stores. The source of the earthquakes was eventually traced to the Rocky Mountain Arsenal, which was manufacturing materials for chemical warfare. Liquid waste from the manufacturing process was being pumped down a deep disposal well to a depth of about 3,600 m (11,800 ft). The rock receiving the waste was highly fractured metamorphic rock, and injection of the new liquid facilitated slippage along fractures. Study of the earthquake activity revealed a high correlation between the rate of waste injection and the occurrence of earthquakes. When the injection of waste stopped, so did the earthquakes (Figure 6.25).[13] Fluid injection of waste as an earthquake trigger was an important occurrence because it directed attention to the fact that earthquakes and fluid pressure are related.

Nuclear Explosions

Numerous earthquakes with magnitudes as large as 5.0 to 6.3 have been triggered by underground nuclear explosions at the Nevada Test Site. Analysis of the aftershocks suggests that the explosions caused some release of natural tectonic strain. This led to discussions by scientists as to whether nuclear explosions might be used to prevent large earthquakes by releasing strain before it reached a critical point.

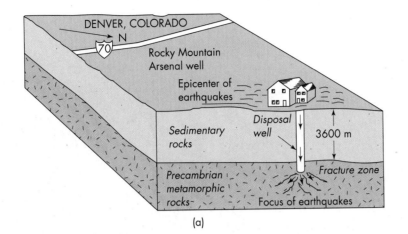

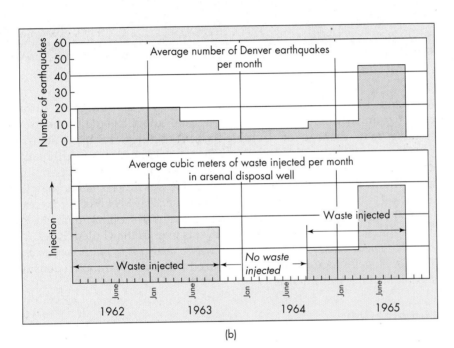

Figure 6.25 **Human-caused earthquakes** (a) Generalized block diagram showing the Rocky Mountain Arsenal well. (b) Graph showing the relationship between earthquake frequency and the rate of injection of liquid waste. *(After Evans, D. M. 1966. Geotimes 10. Reprinted by permission)*

6.10 **Effects of Earthquakes**

Shaking is not the only cause of death and damage in earthquakes: Catastrophic earthquakes have a wide variety of destructive effects. Primary effects, those caused directly by fault movement, include ground shaking and its effects on people and structures and surface rupture. Secondary effects induced by the faulting and shaking include liquefaction of the ground, landslides, fires, disease, tsunamis, and regional changes in land elevation.

Shaking and Ground Rupture

The immediate effects of a catastrophic earthquake can include violent ground shaking accompanied by widespread surface rupture and displacements. Surface rupture with a vertical component produces a *fault scarp*. A fault scarp is a linear, steep slope that looks something like a curb on a street. Fault scarps are often approximately 1 m (3.3 ft) or more high and extend for a variable distance, from a few tens of meters to a few kilometers along a fault (Figure 6.26). The 1906 San Francisco earthquake (M 7.7) produced 6.5 m (21.3 ft) of horizontal displacement, or fault-slip, along the San Andreas fault north of San Francisco and reached a

Figure 6.26 **Fault scarp**
Produced by the 1992 M 7.5
Landers earthquake in California.
(Edward A. Keller)

maximum Modified Mercalli Intensity of XI.[8] At this intensity, surface accelerations can snap and uproot large trees and knock people to the ground. The shaking may damage or collapse large buildings, bridges, dams, tunnels, pipelines, and other rigid structures. The great 1964 Alaskan earthquake (M 8.3) caused extensive damage to transportation systems, railroads, airports, and buildings. The 1989 Loma Prieta earthquake (M 7.1) was much smaller than the Alaska event, yet it caused about $5 billion of damage. The 1994 Northridge earthquake (M 6.7) caused 57 deaths while inflicting about $40 billion damage, making it one of the most expensive hazardous events ever in the United States. *The Northridge earthquake caused so much damage because there was so much there to be damaged.* The Los Angeles region is highly urbanized with high population density, and the seismic shaking was intense.

Liquefaction

Liquefaction is the transformation of water-saturated granular material, or sediments, from a solid to a liquid state. During earthquakes liquefaction may result from compaction of sediments during intense shaking. Liquefaction of near-surface water-saturated silts and sand causes the materials to lose their strength and to flow. As a result, buildings may tilt or sink into the liquefied sediments while tanks or pipelines buried in the ground may rise buoyantly.[14]

Landslides

Earthquake shaking often triggers many landslides, a comprehensive term for several types of hillslope failure, in hilly and mountainous areas. These can be extremely destructive and can cause great loss of life, as demonstrated by the 1970 Peru earthquake. In that event, more than 70,000 people died, and of those, 20,000 people were killed by a giant landslide that buried the cities of Yungan and Ranrahirca. Both the 1964 Alaskan earthquake and the 1989 Loma Prieta earthquake caused extensive landslide damage to buildings, roads, and other structures. The 1994 Northridge earthquake and aftershocks triggered thousands of landslides. A giant landslide from the side of a mountain was triggered by the 2002 Alaska earthquake (Figure 6.27). Thousands of other landslides were also triggered by the earthquake. Most were on the steep slopes of the Alaska Range.

A large landslide associated with the January 13, 2001, El Salvador earthquake (M 7.6) buried the community of Las Colinas, killing hundreds of people. Tragically, the landslide could probably have been avoided if the slope that failed had not

Figure 6.27 Earthquake-triggered landslides This giant landslide was one of thousands triggered by the 2002 (M 7.9) Alaskan earthquake. Landslide deposits cover part of the Black Rapids Glacier *(U.S. Geological Survey)*

previously been cleared of vegetation for the construction of luxury homes. (We will discuss landslides in greater detail in Chapter 9.)

Fires

Fire is a major hazard associated with earthquakes. Shaking of the ground and surface displacements can break electrical power and gas lines, thus starting fires. In individual homes and other buildings, appliances such as gas heaters may be knocked over, causing gas leaks that could be ignited. The threat from fire is intensified because firefighting equipment may become damaged and essential water mains may be broken during an earthquake. Earthquakes in both Japan and the United States have been accompanied by devastating fires (Figure 6.28). The San Francisco earthquake of 1906 has been repeatedly referred to as the

Figure 6.28 Earthquake and fire Fires associated with the 1995 Kobe, Japan, earthquake caused extensive damage to the city. *(CORBIS)*

"San Francisco Fire," and in fact 80 percent of the damage from that event was caused by firestorms that ravaged the city for several days. The 1989 Loma Prieta earthquake also caused large fires in the Marina district of San Francisco.

Disease

Landslides from the 1994 Northridge earthquake raised large volumes of dust, some of which contained fungi spores that cause valley fever. Winds carried the dust and spores to urban areas including Simi Valley, where an outbreak of valley fever occurred during an 8-week period after the earthquake. Two hundred cases were diagnosed, which is 16 times the normal infection rate; of these, 50 people were hospitalized and three died.[1] Earthquakes can rupture sewer and water lines, causing water to become polluted by disease-causing organisms. The death of animals and people buried in earthquake debris also produces potential sanitation problems that may result in an outbreak of disease.

Regional Changes in Land Elevation

Vertical deformation, including both uplift and subsidence, is another effect of some large earthquakes. Such deformation can cause regional changes in groundwater levels. The great 1964 Alaskan earthquake (M 8.3) caused regional uplift and subsidence.[15] The uplift, which was as much as 10 m (33 ft), and the subsidence, which was as much as 2.4 m (7.8 ft), caused effects ranging from severely disturbing or killing coastal marine life to changes in groundwater levels. In areas of uplift, canneries and fishermen's homes were displaced above the high-tide line, rendering docks and other facilities inoperable. The subsidence resulted in flooding of some communities. In 1992 a major earthquake (M 7.1) near Cape Mendocino in northwestern California produced approximately 1 m (3.3 ft) of uplift at the shoreline, resulting in the death of marine organisms exposed by the uplift.[16]

6.11 Tsunami

Tsunamis (the Japanese word for "large harbor waves") are produced by the sudden vertical displacement of ocean water.[17] Tsunamis are a serious natural hazard that can cause a catastrophe thousands of kilometers from where they originate. They may be triggered by several types of events such as large earthquakes that cause rapid uplift or subsidence of the seafloor; underwater landslides that may be triggered by an earthquake; collapse of part of a volcano that slides into the sea; submarine volcanic explosion; and impact in the ocean of an extraterrestrial object such as an asteroid or comet. Asteroid impact can produce a "Mega" tsunami about 100 times as high as the largest tsunami produced by an earthquake and put hundreds of millions of people at risk.[17] Fortunately, the probability of a large impact is very low. Of the above potential causes, tsunamis produced by earthquakes are by far the most common.

Damaging tsunamis in historic time have been relatively frequent and until recently confined to the Pacific Basin. Recent examples include

- The 1960 (M 9+) Chile earthquake that triggered a tsunami killing 61 people in Hawaii after traveling for 15 hours across the Pacific Ocean.
- The 1964 (M 8.3) Alaskan earthquake that generated a deadly tsunami that killed about 130 people in Alaska and California.
- The 1993 (M 7.8) earthquake in the sea of Japan that generated a tsunami that killed 120 people on the island of Okushiri, Japan.
- The 1998 (M 7.1) Papua New Guinea earthquake and submarine landslide that triggered a tsunami that killed more than 2,100 people.
- The 2004 (M 9) Sumatran earthquake that generated a tsunami that killed about 250,000 people (see A Closer Look: Indonesian Tsunami).

A CLOSER LOOK | Indonesian Tsunami

The Indonesian tsunami of 2004 made us aware of these giant waves and the destruction they can cause. Within a span of only a few hours, about 250,000 people were killed and millions were displaced as coastal area after coastal area around the Indian Ocean was struck by the series of tsunami waves.

The largest earthquake on Earth in the past 4 decades struck on the morning of Sunday, December 26, 2004, just off the Indonesian island of Sumatra (Figure 6.A). That earthquake caused the most damaging tsunami in recorded history. The magnitude of the earthquake was 9.0, and it

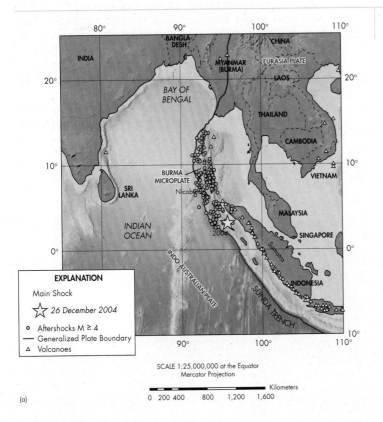

(a)

Figure 6.A Geologic setting
(a) of the Indonesian tsunami.
(Modified after U.S. Geological Survey)
(b) Idealized cross section of where a subduction zone earthquake might occur. *(Modified from United Kingdom Hydrographic Office)* Also see Figure 2.5.

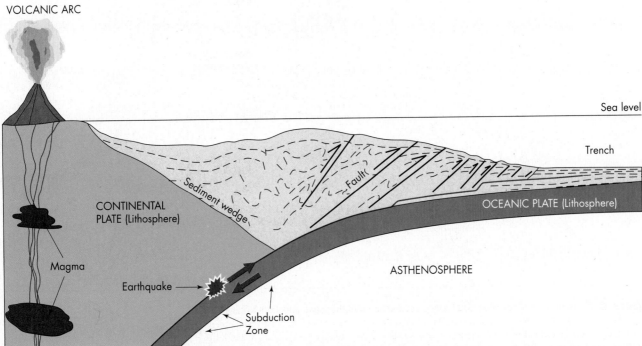

(b)

occurred on the plate boundary where the giant Indo-Australian Plate is being subducted beneath the smaller Burma Plate near Sumatra. The fault rupture was what is known as a "megathrust event" on the subduction zone, which dips gently (about 10°) to the northeast. Shaking from the earthquake lasted several minutes and displacement on the rupture surfaces (faults) between the plates was about 15 meters (50 feet). The total length of the rupture was about 1,200 kilometers (750 miles). The bottom of the sea was displaced by the earthquake to the west-southwest about 15 meters horizontally and several meters vertically. Displacement of the sea bottom disturbed and displaced the overlying waters of the Indian Ocean and a series of tsunami waves were generated. The effect is like throwing a giant boulder in your bathtub and watching the rings or waves of water spread out. In this case, though, the boulder came up from the bottom, but the result was the same and waves radiated outward moving at high speed across the Indian Ocean.

There was no warning system for tsunamis in the Indian Ocean as there are for the Pacific, and people were by and large caught by surprise. Warning by United States scientists, who recognized that a very large earthquake that could potentially produce a tsunami had occurred, was issued to the United States government as well as some governments in the region where the earthquake occurred. Unfortunately, the warnings were not timely enough for countries to take action, and warnings were complicated by the fact that there are insufficient linkages between the United States and other governments of the region. Had it been possible for warnings to be issued, thousands of lives could have been saved because the tsunami waves in some instances took several hours to reach the coastlines where people died. Even a warning of a half an hour or so would have been sufficient to move many people from low-lying coastal areas and saved many, many lives.[17] The first sensor on the seafloor to detect a tsunami for the Indian Ocean was in place by the end of 2006 and others will follow. Sirens are now in place along some coastal areas around the Indian Ocean to warn people if a tsunami is coming.

Over three-quarters of the deaths were in Indonesia, which suffered from both intense shaking from the earthquake that caused the tsunami and the tsunami itself. The site of the magnitude 9 earthquake and the generation of the tsunami and its travel over several hours with the number of deaths are shown on Figure 6.B. Notice that the time from generation

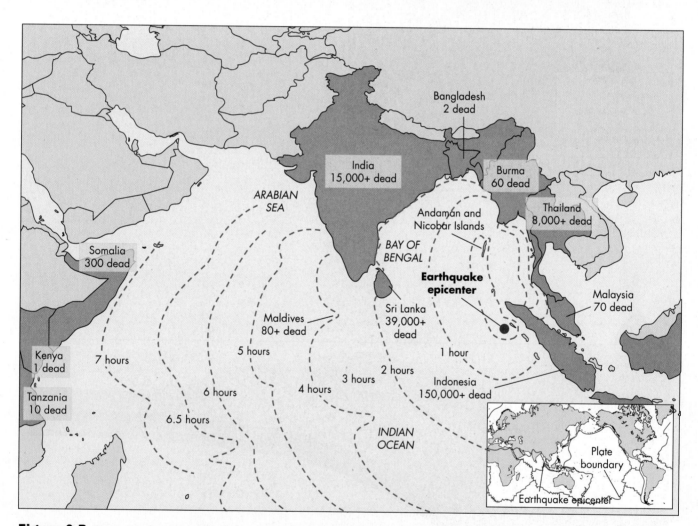

Figure 6.B Tsunami in December 2004 killed several hundred thousand people in Indonesia
This map shows the epicenter of the magnitude 9 earthquake that produced the Indonesian tsunami. Shown are the movement out of tsunami waves that devastated many areas in the Indian Ocean. Notice that the waves took approximately 7 hours to reach Somalia, where almost 300 people were killed. Most of the deaths were in Indonesia, where the waves arrived only about 1 hour after the earthquake. *(NOAA)*

of the tsunami to its arrival in Somalia was about 7 hours. In India, it was 2 hours, and earlier and later in other places depending on distance from the earthquake.

The coastal area of Banda Aceh, Indonesia, was nearly destroyed, as shown in before and after photographs (Figure 6.C). The destruction was a result of earthquake shaking, tsunami, and subsidence of the land from earthquake deformation.

Tourist areas in the region were hard hit, especially in Thailand, where several thousand tourists were killed. Figure 6.D shows the approaching wave in Phuket, Thailand. Some people seem to be mesmerized by the approaching waves while others are running in panic. On at least two beaches, everything wasn't quite as bleak. About 100 tourists and other people were saved when a 10 year-old British girl

sounded the warning in time for a beach in Thailand to be evacuated. She had previously reviewed a lesson in school about plate tectonics, earthquakes, and tsunamis. As part of that lesson, she learned that sometimes the sea recedes prior to the arrival of a tsunami. That is precisely what she observed and she warned her mother, eventually screaming, to get off the beach, that they were in danger. She convinced her mother as well as others on the beach and people in the hotel where they stayed. The beach was successfully evacuated. Her mother later stated she did not even know what a tsunami was, but that her daughter's school lesson had saved the day, for them at least. In a second case, a scientist staying at a Sri Lanka hotel witnessed a small wave rise up and inundate the swimming pool. This was followed by a 7 meter drop in sea

(a)

(b)

Figure 6.C Nearly complete destruction associated with Indonesian tsunami of 2004 (a) Banda Aceh before December 23, 2004. *(Digital Globe)* (b) Two days after the tsunami that struck the Indonesian provincial capital of Banda Aceh on the northern end of the island of Sumatra. Nearly all of the development has been damaged or destroyed. Notice along the top of the photograph the beach with the extensive erosion leaving what appears to be a number of small islands where a more continuous coast was formerly. Subsidence caused by the earthquake as well as the tsunami caused the destruction and changes in land elevation. *(Digital Globe)*

level. The scientist recognized that a big wave was coming and sounded the alarm. The hotel manager used a megaphone to warn people to get off the beach. Many people had gone down to the beach out of curiosity to see the exposed seafloor. When the big 7 meter wave arrived, most people were near stairs and escaped to higher floors. No hotel staff or guests were killed, but several people on the ground floor were swept out and survived by clinging to palm trees in the hotel garden.[18]

Some native people in Indonesia have a collective memory of tsunamis. When the earthquake occurred, some applied that knowledge and moved to high ground, saving entire small tribes on some islands.

Elephants saved a small group of about 12 tourists in Thailand from the 2004 tsunami.[19] Elephants that give tourists rides started trumpeting about the time the M 9 earthquake struck off the island of Sumatra. They became agitated again about 1 hour later. Those elephants that were not taking

tourists for rides broke loose from their strong chains and headed inland. Elephants that had tourists aboard for a ride didn't respond to their handlers and climbed a hill behind a beach resort where about 4,000 people were killed by the tsunami. When handlers recognized the advancing tsunami, additional tourists were lifted onto the elephants, which used their trunks to place the people on their backs (tourists usually mount elephants from a wooden platform), and moved inland. Tsunami waves surged about 1 km (0.5 mi) inland up from the beach. The elephants stopped just beyond where the waves ended their destructive path.

The question is, did the elephants know something that people did not? Animals have sensory ability that differs from humans. It's possible they heard the earthquake as it generated sound waves with low tones called infrasonic sound generated by the earthquake. Some people are also able to sense sound waves but don't generally perceive them

Figure 6.D **Tourists running for their lives** Man in foreground is looking back at the tsunami rushing toward him that is higher than the building. The location is Phuket, Thailand. Many of the people living there, as well as the tourists, did not initially think the wave would inundate the area where they were and when the waves arrived they thought they would be able to outrun the rising water. In some cases people did escape, but all too often people were drowned. (*ZUMA Press*)

as a hazard. The elephants may also have sensed the motion as the land vibrated from the earthquake. They fled inland, which was the only way they could go. This is apparently what happened, but the linkages between the elephants' sensory ability and their behavior is speculative. Nevertheless, the end result was a few lives were saved.[19]

Tropical ecology played a role in determining tsunami damage. Where the maximum intensity of the tsunami impact occurred, massive destruction of the coast was inevitable. In other areas, the damage was not uniform. Some coastal villages were destroyed while others were much less damaged. Those villages that were spared destruction were generally protected from the energy of tsunami waves by mangrove forest or several rows of plantation trees that shielded the villages, lessening the impact of the waves.[20]

The role of education of people living along shorelines that are susceptible to tsunamis is important. Lives would have been saved had more people recognized from Earth's behavior that a tsunami was likely. Those people experiencing a large earthquake would have known that a tsunami might be coming and move to higher ground. Thousands of miles away where the earthquake waves would not be felt, the Earth still provides signals of what may happen. The example of the schoolgirl clearly shows this: Telling people that if the water suddenly recedes, quickly exposing the sea bottom, you might expect it to come back as a tsunami wave. This would be their signal to move to higher ground. People should also be informed that tsunamis are seldom one wave but are in fact a series of waves with later ones sometimes being more damaging than earlier ones. The education of people close to where a tsunami may originate is particularly important as waves may arrive in 10–15 minutes following an earthquake. Geologists have warned that it is likely that another large tsunami will be generated by earthquakes off shore of Indonesia in the next few decades.

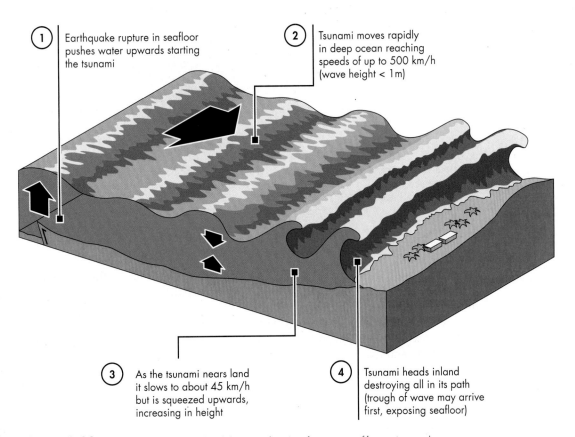

① Earthquake rupture in seafloor pushes water upwards starting the tsunami

② Tsunami moves rapidly in deep ocean reaching speeds of up to 500 km/h (wave height < 1m)

③ As the tsunami nears land it slows to about 45 km/h but is squeezed upwards, increasing in height

④ Tsunami heads inland destroying all in its path (trough of wave may arrive first, exposing seafloor)

Figure 6.29 **Tsunami damage** Idealized diagram showing the process of how a tsunami is produced by an earthquake. *(Modified after United Kingdom Hydrographic Office)*

How Do Earthquakes Cause a Tsunami? The processes that generate a tsunami from an earthquake are generalized in Figure 6.29. Faulting that ruptures the seafloor and displaces the overlying water starts the process. Generally, a M 7.5 or greater earthquake is necessary to generate a damaging tsunami.

Tsunami waves are relatively low in open ocean, being less than a meter high, but they travel at speeds of jet aircraft (750 km/hr, 500 mi/hr). When a tsunami wave strikes the coast, the energy of the wave is compressed in the shallow waters

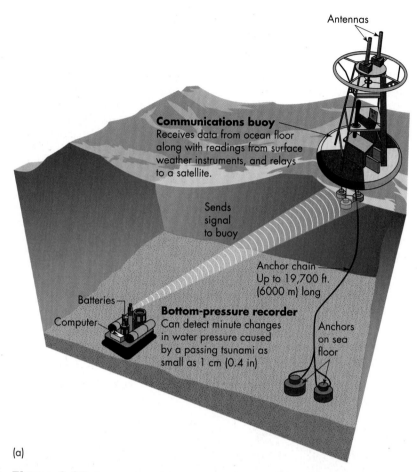

Antennas

Communications buoy
Receives data from ocean floor
along with readings from surface
weather instruments, and relays
to a satellite.

Sends
signal
to buoy

Anchor chain
Up to 19,700 ft.
(6000 m) long

Batteries

Computer

Bottom-pressure recorder
Can detect minute changes
in water pressure caused
by a passing tsunami as
small as 1 cm (0.4 in)

Anchors
on sea
floor

(a)

Figure 6.30 **Tsunami warning system** (a) A surface buoy and bottom sensor to detect a tsunami.
(b) Travel time (each band in one hour) for a tsunami generated in Hawaii. The wave arrives in Los Angeles
in about 5 hours. It takes about 12 hours for the waves to reach South America. Locations of six existing
tsunami instruments are shown. Others are being planned for the Atlantic and Caribbean. *(NOAA)*

and the height of the wave increases dramatically. Tsunami waves come in as a
series and later waves may be higher than earlier ones. When the water from
a wave retreats from the land, flowing back to sea, the return flow can be as
dangerous as an incoming wave.[17]

After an earthquake that produces a tsunami, the arrival time of waves can be
estimated to within plus or minus 1.5 minutes per hour of travel time. This information has been used to produce a tsunami warning system in the Pacific Ocean.
A surface buoy with a bottom sensor (Figure 6.30a) to detect a tsunami helps
provide data to make a map showing arrival times around the Pacific Ocean. The
map in Figure 6.30b is for a tsunami traveling from Hawaii.

6.12 Earthquake Risk and Earthquake Prediction

The great damage and loss of life associated with earthquakes are due in part to
the fact that they often strike without warning. A great deal of research is being
devoted to anticipating earthquakes. The best we can do at present is to use probabilistic methods to determine the risk associated with a particular area or with a
particular segment of a fault. Such determinations of risk are a form of long-term
prediction: We can say that an earthquake of a given magnitude or intensity has a
high probability of occurring in a given area or fault segment within a specified
number of years. These predictions assist planners who are considering seismic
safety measures or people who are deciding where to live. However, long-term
prediction does not help residents of a seismically active area to anticipate and

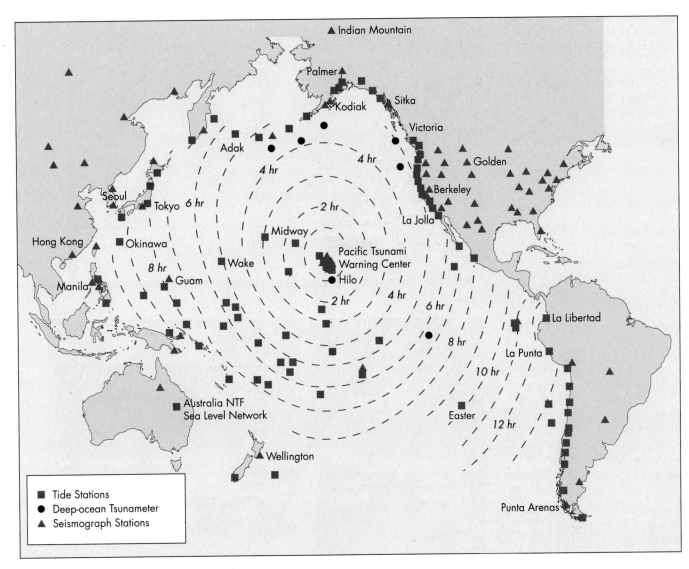

(b)

Figure 6.30 *(Continued)*

prepare for a specific earthquake over a period of days or weeks before an event. Short-term prediction specifying the time and place of the earthquake would be much more useful, but the ability to make such predictions has eluded us. Predicting imminent earthquakes depends to a large extent on observation of precursory phenomena, or changes preceding the event.

Estimation of Seismic Risk

The earthquake risk associated with particular areas is shown on seismic hazard maps, which are prepared by scientists. Some of these maps show relative hazard, that is, where earthquakes of a specified magnitude have occurred. However, a preferable method of assessing seismic risk is to calculate the probability of either a particular event or of the amount of shaking likely to occur. Figure 6.31a shows an earthquake hazard map for the United States. It was prepared on the basis of the probability of horizontal ground motion. The darkest areas on the map represent the regions of greatest seismic hazard, because it is those areas that will have the highest probability of experiencing the greatest seismic shaking. Regional earthquake hazard maps are valuable; however, considerably more data are necessary to evaluate hazardous areas in order to assist in the development of building codes and the determination of insurance rates. Figure 6.31b shows the

Figure 6.31 Earthquake hazard map (a) A probabilistic approach to the seismic hazard in the United States. Colors indicate level of hazard. (b) Probability of at least one M 6–7 or greater earthquake occurring before 2032 on a particular fault and entire region for the San Francisco Bay region. *(From U.S. Geological Survey Fact Sheets FS-131-02, 2002, and 039-03, 2004)*

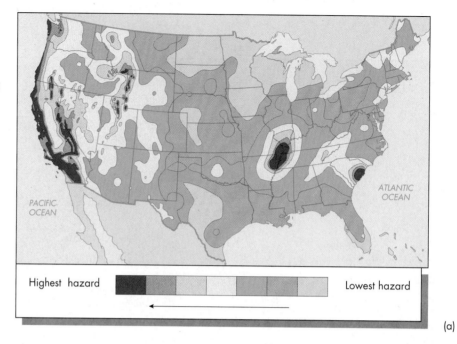

Highest hazard   Lowest hazard

(a)

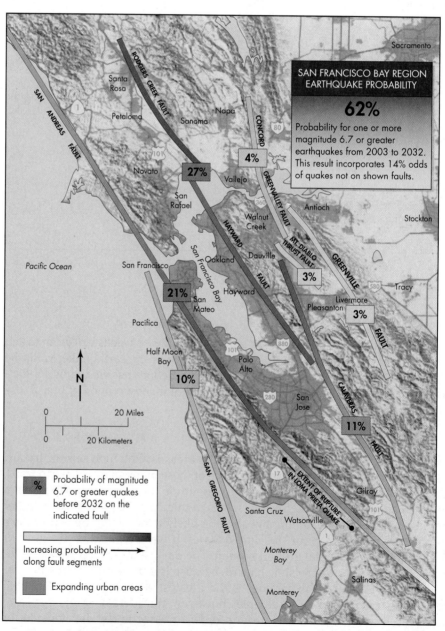

(b)

TABLE 6.6   Fault (Seismic Source) Type in California

Based on slip rate and potential earthquake a fault could produce. Type B collectively produce most of the damage in California because there are many more Type B faults than Type A, and Type C faults, while common, produce smaller earthquakes.

Fault Type	Description	DEFINITION[1]	
		Maximum Moment Magnitude, M[3]	Slip Rate, SR (mm/yr)[3]
A	Faults that are capable of producing large-magnitude events and that have a high rate of seismic activity[2]	M ≥ 7.0	SR ≥ 5
B	All faults other than Types A and C	M ≥ 7.0 M < 7.0 M ≥ 6.5	SR < 5 SR > 2 SR < 2
C	Faults that are not capable of producing large-magnitude earthquakes and that have a relatively low rate of seismic activity	M < 6.5	SR ≤ 2

[1]Both maximum moment magnitude and slip rate conditions must be satisfied concurrently when determining the seismic source type.

[2]Subduction sources shall be evaluated on a site-specific basis.

[3]≥ means greater than or equal to; < means less than.

Source: From *1997 California Uniform Building Code,* Table 16-U.

probability of a M 6.7 or larger earthquake from 2003 to 2032 occurring on faults in the San Francisco Bay region. There is a combined 62 percent probability of at least one M 6 to 7 or greater earthquake occurring before 2032.[21] In addition, the state of California now classifies faults on the basis of the slip rate of the fault and the maximum moment magnitude earthquake the fault can produce (Table 6.6). An example of a type A fault would be the San Andreas fault. The majority of faults in California that produce damaging earthquakes are type B. This results because there are many type B faults in California and they can produce strong major earthquakes (Table 6.2). This system of classification is thought to be more important than simply demonstrating whether a fault is active. Most known active faults in California have been classified by type.

Short-Term Prediction

The short-term prediction, or *forecast,* of earthquakes is an active area of research. Similar to a weather forecast, an earthquake forecast specifies a relatively short time period in which the event is likely to occur and assigns it a probability of occurring. The basic procedure for predicting earthquakes was once thought to be as easy as "one-two-three."[22] First, deploy instruments to detect potential precursors of a future earthquake; second, detect and recognize the precursors in terms of when an earthquake will occur and how big it will be; and third, after review of your data, publicly predict the earthquake. Unfortunately, earthquake prediction is much more complex than first thought.[22]

The Japanese made the first attempts at earthquake prediction with some success, using the frequency of microearthquakes (M less than 2), repetitive surveys of land levels, and a change in the local magnetic field of Earth. They found that earthquakes in the areas they studied were nearly always accompanied by swarms of microearthquakes that occurred several months before the major shocks. Furthermore, ground tilt was correlated strongly with earthquake activity.

Chinese scientists made the first successful prediction of a major earthquake in 1975. The February 4, 1975, Haicheng earthquake (M 7.3) destroyed or damaged about 90 percent of buildings in the city, which had a population of 9,000. The short-term prediction was based primarily on a series of foreshocks that began 4 days before the main event. On February 1 and 2, several shocks with a magnitude of less than 1 occurred. On February 3, less than 24 hours before the main shock, a foreshock of M 2.4 occurred, and in the next 17 hours, eight shocks with a magnitude greater than 3 occurred. Then, as suddenly as it began, the foreshock activity became relatively quiet for 6 hours until the main earthquake occurred.[23] Haicheng's population of 9,000 was saved by the massive evacuation from potentially unsafe housing just before the earthquake.

Unfortunately, foreshocks do not always precede large earthquakes. In 1976, one of the deadliest earthquakes in recorded history struck near the mining town of Tanshan, China, killing several hundred thousand people. There were no foreshocks. Earthquake prediction is still a complex problem, and it will probably be many years before dependable short-range prediction is possible. Such predictions most likely will be based upon precursory phenomena such as

- Patterns and frequency of earthquakes, such as the foreshocks used in the Haicheng prediction
- Preseismic deformation of the ground surface
- Emission of radon gas
- Seismic gaps along faults
- Anomalous animal behavior (?)

Preseismic Deformation of the Ground Surface

Rates of uplift and subsidence, especially when they are rapid or anomalous, may be significant in predicting earthquakes. For more than 10 years before the 1964 earthquake near Niigata, Japan (M 7.5), there was a broad uplift of Earth's crust of several centimeters near the Sea of Japan coast. Similarly, broad slow uplift of several centimeters occurred over a 5 year period before the 1983 Sea of Japan earthquake (M 7.7).[24]

Preinstrument uplifts of 1 to 2 m (3.3 to 6.6 ft) preceded large Japanese earthquakes in 1793, 1802, 1872, and 1927. Although these uplifts are not well understood, they could have been indicators of the impending earthquakes. These uplifts were recognized by sudden withdrawals of the sea from the land, often as much as several hundred meters in harbors. For example, on the morning of the 1802 earthquake, the sea suddenly withdrew about 300 m (984 ft) from a harbor in response to a preseismic uplift of about 1 m (3.3 ft). Four hours later the earthquake struck, destroying many houses and uplifting the land another meter, causing the sea to withdraw an even greater distance.[25]

Emission of Radon Gas

The levels of radon, a radioactive gas, have been observed to increase significantly before some earthquakes. It is believed that before an earthquake rocks expand, fracture, and experience an influx of water. Radon gas, which is naturally present in rocks, moves with the water. There was a significant increase in radon gas measured in water wells a month or so before the 1995 Kobe, Japan, earthquake (M 7.2).[26]

Seismic Gaps

Seismic gaps are defined as areas along active fault zones or within regions that are likely to produce large earthquakes but have not caused one recently. The lack of earthquakes is believed to be temporary; furthermore, these areas are thought to store tectonic strain and are thus candidates for future large earthquakes.[27] Seismic gaps have been useful in medium-range earthquake prediction. At least 10 large plate boundary earthquakes have been successfully forecast from seismic gaps since 1965, including one in Alaska, three in Mexico, one in South America, three in Japan, and one in Indonesia. In the United States, seismic gaps along the San Andreas fault include one near Fort Tejon, California, that last ruptured in 1857 and one along the Coachella Valley, a segment that has not produced a great earthquake for several hundred years. Both gaps are likely candidates to produce a great earthquake in the next few decades.[12,27]

Anomalous Animal Behavior (?)

Anomalous animal behavior has often been reported before large earthquakes. Reports have included dogs barking unusually, chickens refusing to lay eggs, horses or cattle running in circles, rats perching on power lines, and snakes crawling out in the winter and freezing. Anomalous behavior of animals was evidently common before the Haicheng earthquake.[23] Years ago it was suggested that sometime in the future we might have animals such as ground squirrels or snakes in cages along faults, and that somehow they would tell us when an earthquake was likely to occur in the near future. Undoubtedly, some animals are much more sensitive to Earth movements and possible changes in Earth before an earthquake or landslide than are people. However, the significance and reliability of animal behavior are very difficult to evaluate. There is little research going on to further explore anomalous animal behavior in the United States, but it remains one of the interesting mysteries surrounding earthquakes.

6.13 Toward Earthquake Prediction

We are still a long way from a working, practical methodology to reliably predict earthquakes.[28] However, a good deal of information is currently being gathered concerning possible precursor phenomena associated with earthquakes. To date the most useful precursor phenomena have been patterns of earthquakes, particularly foreshocks, and seismic gaps. Optimistic scientists around the world today believe that we will eventually be able to make consistent long-range forecasts (tens to a few thousand years), medium-range predictions (a few years to a few months), and short-range predictions (a few days or hours) for the general locations and magnitudes of large, damaging earthquakes.

Although progress on short-range (days to months) earthquake prediction has not matched expectations, medium- to long-range forecast (years to decades) based on probability of an earthquake occurring on a particular fault has progressed faster than expected. The October 28, 1983, Borah Peak earthquake in central Idaho (M 7.3) has been lauded as a success story for medium-range earthquake hazard evaluation. Previous evaluation of the Lost River fault suggested that the fault was active.[29] The earthquake killed two people and caused approximately $15 million in damages. Fault scarps up to several meters high and numerous ground fractures along the 36 km (22 mi) rupture zone of the fault were produced as a result of the earthquake. The important fact was that the scarp and faults produced during the earthquake were superimposed on previously existing fault scarps, validating the usefulness of careful mapping of scarps produced from prehistoric earthquakes. Remember; where the ground has broken before, it is likely to break again!

6.14 Sequence of Earthquakes in Turkey: Can One Earthquake Set Up Another?

There is considerable controversy regarding patterns of repetitive earthquakes. That is, do earthquakes on a given fault have return periods that are relatively constant with relatively constant-magnitude earthquakes, or do they tend to occur in clusters over a period of several hundred to several thousands of years? Some faults evidently do produce earthquakes of similar magnitude over relatively

Figure 6.32 **Sequence of earthquakes** Earthquakes (M greater than 6.7) on the North Anatolian fault in the twentieth century. Events of 1992 and 1951 are not shown. Year 1999 shows the rupture length of two events combined (Izmit, August; and Duzce, November). *(Modified after Reilinger, R., Toksot, N., McClusky, S., and Barka, A. 2000. 1999 Izmit, Turkey earthquake was no surprise. GSA Today 10(1):1–5)*

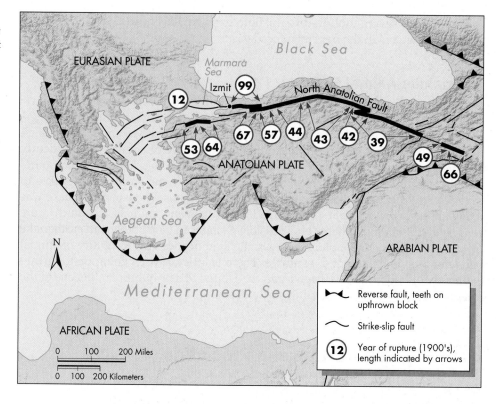

constant return periods. However, as we learn more about individual faults, we learn that there is a great deal of variability concerning magnitudes of earthquakes and their return periods for a given fault system. For example, in the twentieth century a remarkable series of earthquakes with magnitude greater than 6.7 generally occurred from east to west on the north Anatolian fault in Turkey, resulting in surface ruptures along a 1,000 km (621 mi) section of the fault (Figure 6.32). Two events in 1999—the Izmit (M 7.4) and the Duzce (M 7.1) earthquakes—were particularly severe, causing billions of dollars in property damage and thousands of deaths. The sequence of earthquakes has been described as a "falling-domino scenario," in which one earthquake sets up the next, eventually rupturing nearly the entire length of the fault in a cluster of events.[30,31] Clusters of earthquakes that form a progressive sequence of events in a relatively short period of time are apparently separated by a period of no earthquakes for several hundred years. A similar sequence or clustering may be occurring along the fault bordering the Sumatra Plate boundary that produced the 2004 tsunami. Three earlier large earthquakes from 1600 to 1833, discovered from the geologic record, preceded the 2004 event, and a M 8.7 event occurred in 2005. That earthquake did not produce a tsunami, but collapsed buildings and took about 1,000 lives.[3] Understanding processes related to clustering of earthquakes along a particular fault is very important if we are to plan for future seismic events in a given region. The fact that there may not have been a large damaging earthquake for several hundred years may not be as reassuring as we once thought.

6.15 The Response to Earthquake Hazards

Responses to seismic hazards in earthquake-prone areas include development of hazard-reduction programs, careful siting of critical facilities, engineering and land-use adjustments to earthquake activity, and development of a warning system. The extent to which these responses occur depends in part on people's perception of the hazard.

Earthquake Hazard-Reduction Programs

In the United States, the U.S. Geological Survey (USGS) as well as university and other scientists are developing a National Earthquake Hazard Reduction Program. The major goals[28] of the program are to

- *Develop an understanding of the earthquake source.* This requires an understanding of the physical properties and mechanical behavior of faults as well as development of quantitative models of the physics of the earthquake process.

- *Determine earthquake potential.* This determination involves characterizing seismically active regions, including determining the rates of crustal deformation, identifying active faults, determining characteristics of paleoseismicity, calculating long-term probabilistic forecasts, and finally, developing methods of intermediate- and short-term prediction of earthquakes.

- *Predict effects of earthquakes.* Predicting effects includes gathering of the data necessary for predicting ground rupture and shaking and for predicting the response of structures that we build in earthquake-prone areas and evaluating the losses associated with the earthquake hazard. (See A Closer Look: The Alaska Earthquake of 2002 and the Value of Estimating Potential Ground Rupture.)

- *Apply research results.* The program is interested in transferring knowledge about earthquake hazards to people, communities, states, and the nation. This knowledge concerns what can be done to plan better for earthquakes and reduce potential losses of life and property.

Adjustments to Earthquake Activity

The mechanism of earthquakes is still poorly understood; therefore, such adjustments as warning systems and earthquake prevention are not yet reliable alternatives. There are, however, reliable protective measures we can take:

- *Structural protection,* including the construction of large buildings and other structures such as dams, power plants, and pipelines able to accommodate moderate shaking or surface rupture. This measure has been relatively successful in the United States. (See A Closer Look: The Alaska Earthquake of 2002 and the Value of Estimating Potential Ground Rupture.) The 1988 Armenia earthquake (M 6.8) was somewhat larger than the 1994 Northridge event (M 6.7), but the loss of life and destruction in Armenia were staggering. At least 45,000 people were killed, compared with 57 in California, and near-total destruction occurred in some towns near the epicenter. Most buildings in Armenia were constructed of unreinforced concrete and instantly crumbled into rubble, crushing or trapping their occupants. The 2005 Pakistan M 7.6 earthquake killed over 80,000 people. Many of the deaths occurred as apartment buildings with little or no steel reinforcement collapsed to resemble a stack of pancakes.[3] This is not to say that the Northridge earthquake was not a catastrophe. It certainly was; the Northridge earthquake left 25,000 people homeless, caused the collapse of several freeway overpasses, injured approximately 8,000 people, and inflicted many billions of dollars in damages to structures and buildings (Figure 6.1). However, since most buildings in the Los Angeles Basin are constructed with wood frames or reinforced concrete, thousands of deaths were avoided in Northridge.

- *Land-use planning,* including the siting of important structures such as schools, hospitals, and police stations in areas away from active faults or sensitive Earth materials that are likely to increase seismic shaking. This planning involves zoning the ground's response to seismic shaking on a block-by-block basis. Zoning for earthquakes in land-use planning is necessary

A CLOSER LOOK | The Alaska Earthquake of 2002 and the Value of Estimating Potential Ground Rupture

On November 3, 2002, a magnitude 7.9 earthquake occurred in early afternoon on the Denali fault in south-central Alaska. That event produced approximately 340 km of surface rupture with maximum right-slip displacement of just over 8 meters (Figure 6.E) The earthquake struck a remote part of Alaska where few people live, and although it caused thousands of landslides and numerous examples of liquefaction and intense shaking, little structural damage and no deaths were recorded.[32]

The Denali fault earthquake demonstrated the value of seismic hazard and earthquake hazard evaluation. The fault was studied in the early 1970s as part of an evaluation of the Trans-Alaskan pipeline that today supplies approximately 17 percent of the domestic oil supply for the United States. Where the pipeline crosses the Denali fault, geologists determined that the fault zone was several hundred meters wide and might experience a 6 meter horizontal displacement from a magnitude 8 earthquake. These estimates were used in design of the pipeline, which included long horizontal steel beams with teflon shoes which allow the pipeline to slide horizontally approximately 6 meters. This was associated with a zigzag pattern of the pipeline that allows, along with the steel beams, the horizontal movement (Figure 6.E). The 2002 earthquake occurred within the mapped zone of the fault and sustained about 4.3 meters (14 feet) of right-lateral strike-slip. As a result the pipeline suffered little damage and there was no oil spill. The cost in 1970 of the engineering and construction that allowed the fault slip was about $3 million. Today, that looks very cost effective considering that approximately $25 million worth of oil per day is transported via the pipeline. If the pipeline had been ruptured, then the cost of repair and cleanup of the environment might have cost several hundred million dollars.[33]

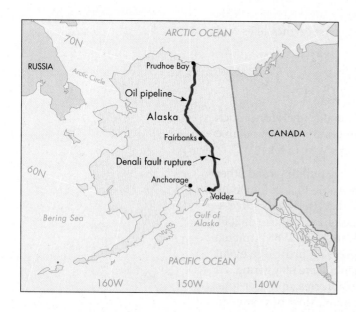

Figure 6.E **A Trans-Alaska oil pipeline survives (M 7.9) earthquake** The Alaskan pipeline was designed to withstand several meters of horizontal displacement on the Denali fault. The 2002 earthquake caused a rupture of 4.3 m (14 ft) beneath the pipeline. The built-in bends on slider beams with teflon shoes accommodated the rupture as designed. This was strong confirmation of the value of estimating potential ground rupture and taking action to remove the potential threat and damage. *[Alyeska Pipeline Service Company (ASPC)]*

because ground conditions can change quickly in response to shaking. In urban areas, where property values may be as high as millions to billions of dollars per block, we need to produce detailed maps of ground response to accomplish zonation. These maps will assist engineers when designing buildings and other structures that can better withstand seismic shaking. Clearly, zonation requires a significant investment of time and money; however, the first step is to develop methods that adequately predict the ground motion from an earthquake at a specific site.

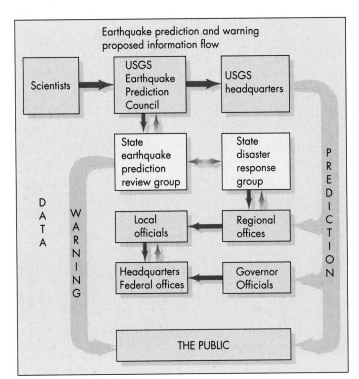

Figure 6.33 Issuing a prediction or warning A federal plan for issuance of earthquake predictions and warning: the flow of information. *(From McKelvey, V. E. 1976. Earthquake prediction—Opportunity to avert disaster. U.S. Geological Survey Circular 729)*

- *Increased insurance and relief measures* to help adjustments after earthquakes. After the 1994 Northridge earthquake, total insurance claims were very large, and some insurance companies terminated earthquake insurance.

We hope eventually to be able to predict earthquakes. The federal plan for issuing prediction and warning is shown on Figure 6.33. Notice how it is related to the general flow path for issuance of a disaster prediction shown in Figure 5.14. The general flow of information moves from scientists to a prediction council for verification. Once verified, a prediction that a damaging earthquake of a specific magnitude would occur at a particular location during a specified time would be issued to state and local officials. These officials are responsible for issuing a warning to the public to take defensive action that, one hopes, has been planned in advance. Potential response to a prediction depends upon lead time, but even a few days would be sufficient to mobilize emergency service, shut down important machinery, and evacuate particularly hazardous areas.

Earthquake Warning Systems

Technically, it is feasible to develop an earthquake warning system that would provide up to about 1 minute's warning to the Los Angeles area before the arrival of damaging earthquake waves from an event several hundred kilometers away. This type of system is based on the principle that the warning sent by a radio signal via satellite relay travels much faster than seismic waves. The Japanese have had a system for nearly 20 years that provides earthquake warnings for their high-speed trains; train derailment by an earthquake could result in the loss of hundreds of lives. A proposed system for California involves a sophisticated network of seismometers and transmitters along the San Andreas fault. This system would first sense motion associated with a large earthquake and then send a warning to Los Angeles, which would then relay the warning to critical facilities, schools, and the general population (Figure 6.34). The warning time would vary from as little as 15 seconds to as long as about 1 minute, depending on the location of the epicenter of the earthquake. This could be enough time for people to shut down machinery and computers and take cover.[34] Note that this earthquake warning system is not a prediction tool; it only warns that an earthquake has already occurred.

Figure 6.34 **Earthquake warning** Idealized diagram showing an earthquake warning system. Once an earthquake is detected, a signal is sent ahead of the seismic shaking to warn people and facilities. The warning time depends on how far away an earthquake occurs. It could be long enough to shut down critical facilities and for people to take cover. *(After Holden, R., Lee, R., and Reichle, M. 1989. California Division of Mines and Geology Bulletin 101)*

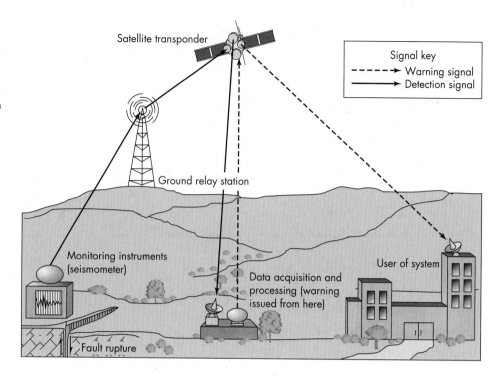

A potential problem with a warning system is the chance of false alarms. For the Japanese system, the number of false alarms is less than 5 percent. However, because the warning time is so short, some people have expressed concern as to whether much evasive action could be taken. There is also concern for liability issues resulting from false alarms, warning system failures, and damage and suffering resulting from actions taken as the result of false early warning.

Perception of Earthquake Hazard

The fact that terra firma is not so firm in places is disconcerting to people who have experienced even a moderate earthquake. The large number of people, especially children, who suffered mental distress after the San Fernando and Northridge earthquakes attests to the emotional and psychological effects of earthquakes. These events were sufficient to influence a number of families to move away from Los Angeles.

The Japanese were caught off guard by the 1995 Kobe earthquake, and the government was criticized for not mounting a quick and effective response. Emergency relief did not arrive until about 10 hours after the earthquake! They evidently believed that their buildings and highways were relatively safe compared with those that had failed in Northridge, California, one year earlier.

As mentioned earlier, a remarkable sequence of earthquakes in Turkey terminated in 1999 with two large damaging earthquakes (Figure 6.32). The first occurred on August 17 and leveled thousands of concrete buildings. Six hundred thousand people were left homeless, and approximately 38,000 people died as a result of the earthquake. Some of the extensive damage to the town of Golcuk in western Turkey is shown in Figure 6.35. Note that the very old mosque on the left is still standing, whereas many modern buildings collapsed, suggesting that earlier construction was more resistant to earthquakes. Although Turkey has a relatively high standard for new construction to withstand earthquakes, there is fear that poor construction was a factor in the collapse of the newer buildings from intense seismic shaking. It has been alleged that some of the Turkish contractors bulldozed the rubble from collapsed buildings soon after the earthquake, perhaps in an effort to remove evidence of shoddy construction. If that allegation is true,

Figure 6.35 Collapse of **buildings in Turkey** Damage to the town of Golcuk in western Turkey from the M 7.4 earthquake of August 1999. The very old mosque on the left remains standing, whereas many modern buildings collapsed. *(Enric Marti/AP/Wide World Photos)*

these contractors also tied up bulldozers that could have been used to help rescue people trapped in collapsed buildings.

The lessons learned from Northridge, Kobe, Turkey, and Pakistan were bitter ones that illuminate our modern society's vulnerability to catastrophic loss from large earthquakes. Older, unreinforced concrete buildings or buildings not designed to withstand strong ground motion are most susceptible to damage. In Kobe, reinforced concrete buildings constructed post-1990 with improved seismic building codes experienced little damage compared with those constructed in the mid-1960s or earlier. Minimizing the hazard requires new thinking about the hazard. In addition, microzonation is instrumental in the engineering response to design structures that are less vulnerable to ground motion from earthquakes.

Personal and Community Adjustments: Before, During, and After an Earthquake

A group of individuals living in an area define a community, whether it is a village, town, or city. In areas with an earthquake hazard, it is important for people to prepare for the hazard in terms of what can be done before, during, and after an earthquake.

At the community level, one important aspect of earthquake preparedness is enforcing building codes such as the Uniform Building Code of California. The objective of the earthquake section of the code is to provide a safeguard against loss of life and major structural failures through better design of buildings to withstand earthquake shaking. Another important task is the inspection of older buildings to determine if a "retrofit" is necessary to increase the strength of the building in order to better withstand earthquakes. In fact, the State of California has regulated building retrofits. A recent study concluded that many of the hospitals in southern California are in need of extensive and costly retrofitting.

Education is also an important component of earthquake preparedness at the community level. Government, state, and local agencies prepare pamphlets and videos concerning the earthquake hazard and help ensure that this information is distributed to the public. Workshops and training meetings concerning the most up-to-date information to minimize earthquake hazards are available to professionals in engineering, geology, and planning. A great deal of information at the community level is also available on the World Wide Web from a variety of sources, including the U.S. Geological Survey and scientific organizations such as the Southern California Earthquake Center, as well as other agencies interested in earthquake hazard reduction.[35]

In schools, it is important to practice what is sometimes called the "earthquake drill": everyone pretends that an earthquake is happening and students "duck, cover, and hold." After about 15 or 20 seconds, students emerge from their duck-and-cover location, take five slow, deep breaths, practice calming down, and then walk to a designated safe area.

At the personal level in homes and apartments, it is estimated that billions of dollars of damages from earthquakes could be avoided if our buildings and contents were better secured to withstand shaking from earthquakes. Before an earthquake occurs, homeowners and apartment dwellers should complete a thorough check of the building, including rooms, foundations, garages, and attics. The following are some items of a home safety check.[35]

- Be sure your chimney has been reinforced to withstand shaking from earthquakes. After the 1994 Northridge earthquake, one of the most commonly observed types of damage was collapsed chimneys.
- Be sure your home is securely fastened to the foundation and that there are panels of plywood between the wooden studs in walls. Make sure that large openings such as garage doors with a floor above are adequately braced.
- Within the house, be sure that anything that is heavy enough to hurt you if it falls on you, or is fragile and expensive, is secured. There are ways to secure tabletop objects, television sets, tall furniture, and cabinet doors.
- Large windows and sliding doors may be covered with strong polyester (Mylar®) films to make them safer and reduce the hazard from broken glass that is shattered during earthquakes.
- Ensure that your gas water heater is strapped to the building so that it will not fall over and start a fire. If necessary, change to flexible gas connections that can withstand some movement.

Probably the most important and rational thing you can do before an earthquake is to prepare a plan of exactly what you will do should a large earthquake occur. This might include the following points.[35]

- Teach everyone in your family to "duck, cover, and hold." For each room in your home, identify safe spots such as a sturdy desk or table or strong interior walls.
- Instruct everyone who might be home how to turn off the gas. However, the gas should be turned off only if a leak is detected through either hearing or smelling the leaking gas.
- After an earthquake, out-of-area phone calls are often much easier to place and receive than local calls. Therefore, establish an out-of-area person who can be contacted.
- Be sure you have necessary supplies such as food, water, first-aid kit, flashlights, and some cash to tide you over during the emergency period when it may be difficult to obtain some items.
- Canvass your neighborhood and identify elderly or disabled neighbors who may need your help should an earthquake occur, and help educate others in the neighborhood about how to prepare their own plans.

During an earthquake, the strong ground motion will greatly restrict your motion, and, as a result, your strategy is to "duck, cover, and hold." Given your knowledge of earthquakes, you may also try to recognize how strong the earthquake is likely to be and to predict what may happen during the event. For example, you know there will be several types of waves including P, S, and surface waves. The P waves will arrive first, and you may even hear them coming. However, the S and R waves, which soon follow, have bigger displacement and cause most of the damage. The length of shaking during an earthquake will vary with the magnitude. For example, during the 1994 Northridge earthquake, the shaking lasted approximately 15 seconds, but the time of shaking in a great earthquake may be much longer. For example, during the 1906 San Francisco earthquake, shaking lasted nearly 2 minutes. In addition to the "duck, cover, and hold" strategy, you need to remain calm during an earthquake and try to protect yourself from appliances, books, and other materials that may slide or fly across the room. A good strategy would be to crouch under a desk or table, roll under a bed, or position yourself in a strong doorway. During the earthquake, there may be explosive flashes from transformers and power poles, and you must obviously avoid downed power lines. At all costs, resist the natural urge to panic.[35]

After the shaking stops, take your deep breaths and organize your thinking in accordance with the plan you developed. Check on your family members and neighbors, check for gas leaks and fires. Your telephones should be used only for emergency calls. Examine your chimney, as chimneys are particularly prone to failure from earthquakes. The chimney may separate from the roof or the walls along the side of the chimney. If you are caught in a theater or stadium during a large earthquake, it is important to remain in your seat and protect your head and neck with your arms; do not attempt to leave until the shaking has stopped. Remain calm and walk out slowly, keeping a careful eye out for objects that have fallen or may fall. If you are in a shopping mall, it is important to "duck, cover, and hold," away from glass doors and display shelves of books or other objects that could fall on you. If you are outdoors and an earthquake occurs, it is prudent to move to a clear area where you can avoid falling trees, buildings, power lines, or other hazards. If you are in a mountainous area, be aware of landslide hazards, since earthquakes often generate many slides that may occur during and for some time after the earthquake. If you are in a high-rise building, you need to "duck, cover, and hold" as in any indoor location, avoiding any large windows. It is likely that the shaking may activate fire alarms and water sprinkler systems. Streets lined with tall buildings are very dangerous locations during an earthquake; glass from these buildings often shatters and falls to the street below, becoming razor-sharp shards that can cause serious damage and death to people below.[35]

Finally, after an earthquake, be prepared for aftershocks. There is a known relationship between the magnitude of the primary earthquake and the distribution of aftershocks in hours, days, months, or even years following the earthquake. If the earthquake has a magnitude of about 7, then several magnitude 6 aftershocks can be expected. Many magnitude 5 and 4 events are likly to occur. In general, the number and size of aftershocks decrease with time from the main earthquake event, and the most hazardous period is in the minutes, hours, and days following the main shock.

The good news concerning living in earthquake country in the United States is that large earthquakes are survivable. Our buildings are generally constructed to withstand earthquake shaking, and our woodframe houses seldom collapse. It is, however, important to be well informed and prepared for earthquakes. For that reason it is extremely important to develop and become familiar with your personal plan of what to do before, during, and after an earthquake.

SUMMARY

Large earthquakes rank among nature's most catastrophic and devastating events. Most earthquakes are located in tectonically active areas where lithospheric plates interact along their boundaries, but some large damaging intraplate earthquakes also occur.

A fault is a fracture or fracture system along which rocks have been displaced. Strain builds up in the rocks on either side of a fault as the sides pull in different directions. When the stress exceeds the strength of the rocks, they rupture, giving rise to seismic, or earthquake, waves that shake the ground.

Strike-slip faults exhibit horizontal displacement and are either right- or left-lateral. Dip-slip faults exhibit vertical displacement and are either normal or reverse. Some faults are buried and do not rupture the surface even when their movement causes large earthquakes. Recently a new fundamental earth process known as slow earthquakes has been discovered. Slow earthquakes may last from days to months with large areas of fault rupture, but small displacements.

A fault is usually considered active if it has moved during the past 10,000 years and potentially active if it has moved during the past 1.65 million years. Some faults exhibit tectonic creep, a slow displacement not accompanied by felt earthquakes.

The area within Earth where fault rupture begins is called the focus of the earthquake and can be from a few kilometers to almost 700 km (435 mi) deep. The area of the surface directly above the focus is called the epicenter. Seismic waves of different kinds travel away from the focus at different rates; much of the damage from earthquakes is caused by surface waves. The severity of shaking of the ground and buildings is affected by the frequency of the seismic waves and by the type of Earth material present. Buildings on unconsolidated sediments or landfill, which tend to amplify the shaking, are highly subject to earthquake damage.

The magnitude of an earthquake is a measure of the amount of energy released. The measure of the intensity of an earthquake, the Modified Mercalli Scale, is based on the severity of shaking as reported by observers and varies with proximity to the epicenter and local geologic and engineering features. Following an earthquake, shake maps based on a dense network of seismographs can quickly show areas where potentially damaging shaking occurred. This information is needed quickly to assist emergency efforts. Ground acceleration during an earthquake is important information necessary to design structures that can withstand shaking.

The hypothesized earthquake cycle for large earthquakes has four stages. A period of seismic inactivity, during which elastic strain builds up in the rocks along a fault, is followed by a period of increased seismicity as the strain locally exceeds the strength of the rocks, initiating local faulting and small earthquakes. The third stage, which does not always occur, consists of foreshocks. The fourth stage is the major earthquake, which occurs when the fault segment ruptures, producing the elastic rebound that generates seismic waves.

Human activity has caused increasing earthquake activity by loading Earth's crust through construction of large reservoirs; by disposal of liquid waste in deep disposal wells, which raises fluid pressures in rocks and facilitates movement along fractures; and by setting off underground nuclear explosions. The accidental damage caused by the first two

activities is regrettable, but what we learn from all the ways we have caused earthquakes may eventually help us to control or stop large earthquakes.

Effects of earthquakes include violent ground motion accompanied by fracturing, which may shear or collapse large buildings, bridges, dams, tunnels, and other rigid structures. Other effects include liquefaction, landslides, fires, and regional subsidence and uplift of landmasses as well as regional changes in groundwater levels. Large to great submarine earthquakes can generate a damaging catastrophic tsunami.

Prediction of earthquakes is a subject of serious research. To date, long-term and medium-term earthquake prediction based on probabilistic analysis has been much more successful than short-term prediction. Long-term prediction provides important information for land-use planning, developing building codes, and engineering design of critical facilities. Some scientists believe that we will eventually be able to make long-, medium-, and short-range predictions based on previous patterns and frequency of earthquakes as well as by monitoring the deformation of land, the release of radon gas, and existing seismic gaps. Although not currently being pursued, reports suggest that anomalous animal behavior before an earthquake may offer potential aid in earthquake prediction. A potential problem of predicting earthquakes is that their pattern of occurrence is often variable, with clustering or sequencing of events separated by longer periods of time with reduced earthquake activity.

Reduction of earthquake hazards will be a multifaceted program, including recognition of active faults and Earth materials sensitive to shaking and development of improved ways to predict, control, and adjust to earthquakes, including designing structures to better withstand shaking. Warning systems and earthquake prevention are not yet reliable alternatives, but more communities are developing emergency plans to respond to a predicted or unexpected catastrophic earthquake. Seismic zoning, including microzonation and other methods of hazard reduction, are active areas of research. At a personal level, there are steps an individual can take before, during, and after an earthquake to reduce the hazard and ease recovery.

Revisiting Fundamental Concepts

Human Population Growth

Human population growth, especially in large cities in seismically active regions, is placing more and more people and property at risk from earthquakes.

Sustainability

Minimizing the damages from earthquakes to public and private property is a component of sustainable development. The goal is to produce stable communities that are less likely to experience catastrophic losses as a result of poor earthquake preparation.

Earth as a System

Earthquakes are produced by Earth's internal tectonic systems. Landforms, including ocean basins and mountains, are the products of continental movement and resultant earthquakes. Mountains cause changes to atmospheric processes

that create deserts and regional patterns of rainfall and thus affect vegetation and erosion. This is an example of the principle of environmental unity: One change, in this case the development of mountains, causes a chain of other events.

Hazardous Earth Processes, Risk Assessment, and Perception

We cannot control processes that produce earthquakes, but how we perceive the earthquake hazard greatly influences the actions we take to minimize the risk of loss of life. If we perceive earthquakes as a real risk to our lives and those of our family and friends, we will take the necessary steps to prepare for future earthquakes.

Scientific Knowledge and Values

Scientific knowledge about earthquakes in terms of how they are produced, where and why they occur, and how to design buildings to better withstand earthquake shaking has grown dramatically in recent years. Important lessons were learned from the 1999 earthquakes in Turkey that killed about 38,000 people. Turkey has relatively high building standards, and more of their buildings should have survived the quakes. Some people believe that improper construction contributed significantly to the loss of buildings. Contractors may have destroyed evidence of inadequate construction after the earthquake occurred. Community values result in building regulations that help reduce earthquake losses. If buildings are properly constructed to withstand earthquakes, then catastrophes such as the Turkey earthquakes that caused terrible and unacceptable loss of life may be fewer in number in the future.

Key Terms

directivity (p. 181)

earthquake (p. 173)

earthquake cycle (p. 186)

epicenter (p. 162)

fault (p. 173)

focus (p. 162)

liquefaction (p. 190)

material amplification (p. 180)

Modified Mercalli Scale (p. 167)

moment magnitude (p. 162)

P waves (p. 176)

shake map (p. 167)

slow earthquakes (p. 176)

surface wave (p. 177)

S waves (p. 176)

tectonic creep (p. 175)

tsunamis (p. 192)

Review Questions

1. What is the difference between the focus and the epicenter of an earthquake?

2. How is Richter magnitude determined?

3. What is moment magnitude?

4. What factors determine the Modified Mercalli Scale?

5. What are the main differences between the Richter, Moment Magnitude, Modified Mercalli, and Instrumental scales?

6. Define a fault.

7. What are the major types of fault?

8. What is the difference between an anticline and a syncline?

9. How do we define an active fault?

10. What is tectonic creep?

11. What are the main types of seismic wave?

12. Describe the motion of P, S, and R waves. How do their physical properties account for their effects?

13. What is a shake map, how is it produced, and why are they important?

14. What is material amplification?

15. Define the earthquake cycle, and illustrate it with a simple example.

16. How has human activity caused earthquakes?

17. What are some of the major effects of earthquakes?

18. What is a tsunami and how are most produced?

19. What are some of the precursory phenomena likely to assist us in predicting earthquakes?

20. What are the major goals of earthquake hazard-reduction programs?

21. What are the main adjustments people make to seismic activity and the occurrence of earthquakes?

Critical Thinking Questions

1. Assume you are working for the Peace Corps and are in a developing country where most of the homes are built out of unreinforced blocks or bricks. There has not been a large damaging earthquake in the area for several hundred years, but earlier there were several earthquakes that killed thousands of people. How would you present the earthquake hazard to the people living where you are working? What steps might be taken to reduce the hazard?

2. You live in an area that has a significant earthquake hazard. There is ongoing debate as to whether an earthquake warning system should be developed. Some people are worried that false alarms will cause a lot of problems, and others point out that the response time may not be very long. What are your views? Do you think it is a responsibility of public officials to finance an earthquake warning system, assuming such a system is feasible? What are potential implications if a warning system is not developed and a large earthquake results in damage that could have partially been avoided with a warning system in place?

SEVEN

Volcanic Activity

Learning Objectives

There are about 1,500 active volcanoes on Earth, almost 400 of which have erupted in the twentieth century. Volcanoes occur on all seven continents, as well as in the middle of the ocean. When human beings live in the path of an active volcano, the results can be devastating. In this chapter, we focus on the following learning objectives:

- Know the major types of volcanoes, the rocks they produce, and their plate tectonic setting

- Understand the main types and effects of volcanic activity, including lava flows, pyroclastic activity, debris flows, and mudflows

- Understand the methods of studying volcanic activity, including seismic activity, topographic change, emission of gases, and geologic history, in order to better predict volcanic eruptions and minimize the hazard

Harry Glicken at Mount St. Helens Observation site in 1980 before the eruption that blew the top off the mountain. Harry is looking directly toward the bulge that was developing on the flank of the volcano. When the volcano erupted, the observation post was destroyed and the geologist at the site was killed. *(Courtesy of the Glicken Family)*

215

CASE HISTORY | Mt. Unzen, 1991

Japan has 19 active volcanoes distributed throughout most of the country's landmass. Nearly 200 years ago, Mt. Unzen in southwestern Japan erupted and killed approximately 15,000 people. After this eruption, the volcano lay dormant for 200 years. Then, on June 3, 1991, Mt. Unzen erupted violently. Authorities ordered the evacuation of thousands of people. By the end of 1993, lava had erupted, and several thousand flows of hot ash had occurred (Figure 7.1). In fact, Mt. Unzen has the dubious honor of being one of the ash flow centers of the world. This volcano has provided scientists with a natural laboratory for studying such flows.[1,2] The 1991 eruption also produced damaging mudflows. A specially designed channel was constructed to contain the mudflows, but, as seen in Figure 7.2, the flows overran the channel, burying many homes in mud.

Harry Glicken was one of the 44 people killed when Mt. Unzen erupted. Harry was a brave and dedicated young scientist who, through good fortune, had escaped death in the May 18, 1980, eruption of Mount St. Helens. Harry's death reminds us that being a volcanologist is a hazardous career path. They often risk their lives to better understand volcanoes and save the lives of people living nearby. *This chapter on volcanic processes is dedicated to Harry, who loved volcanoes and who was my friend.*

Figure 7.1 Ash flow Eruption of Mt. Unzen, Japan, June 1991. The cloud in the mountain area is moving rapidly downslope as an ash flow, and the firefighter is running for his life. *(AP/Wide World Photos)*

(a)

(b)

Figure 7.2 Mudflow (a) Mudflows from Mt. Unzen in 1991 damaged many homes in Shimabara, Japan. Flows overflowed the channels constructed to contain them. (b) The flows inundated many homes and buildings. *(Michael S. Yamashita)*

TABLE 7.1 **Selected Historic Volcanic Events**

Volcano or City	Year	Effect
Vesuvius, Italy	A.D. 79	Destroyed Pompeii and killed 16,000 people. City was buried by volcanic activity and rediscovered in 1595.
Skaptar Jokull, Iceland	1783	Killed 10,000 people (many died from famine) and most of the island's livestock. Also killed some crops as far away as Scotland.
Tambora, Indonesia	1815	Global cooling; produced "year without a summer."
Krakatoa, Indonesia	1883	Tremendous explosion; 36,000 deaths from tsunami.
Mount Pelée, Martinique	1902	Ash flow killed 30,000 people in a matter of minutes.
La Soufrière, St. Vincent	1902	Killed 2,000 people and caused the extinction of the Carib Indians.
Mount Lamington, Papua New Guinea	1951	Killed 6,000 people.
Villarica, Chile	1963–64	Forced 30,000 people to evacuate their homes.
Mount Helgafell, Heimaey Island, Iceland	1973	Forced 5,200 people to evacuate their homes.
Mount St. Helens, Washington, USA	1980	Debris avalanche, lateral blast, and mudflows killed 54 people, destroyed more than 100 homes.
Nevado del Ruiz, Colombia	1985	Eruption generated mudflows that killed at least 22,000 people.
Mount Unzen, Japan	1991	Ash flows and other activity killed 41 people and burned more than 125 homes. More than 10,000 people evacuated.
Mount Pinatubo, Philippines	1991	Tremendous explosions, ash flows, and mudflows combined with a typhoon killed more than 300 people; several thousand people evacuated.
Montserrat, Caribbean	1995	Explosive eruptions, pyroclastic flows; south side of island evacuated, including capital city of Plymouth; several hundred homes destroyed.
Mount Nyiragongo, Congo, Africa	2002	Lava flows through 14 villages and part of the city of Gomo; several hundred thousand people evacuated, about 5,000 homes destroyed, about 45 people killed.

Source: Data partially derived from C. Ollier. 1969. *Volcanoes.* Cambridge, MA: MIT Press.

7.1 Introduction to Volcanic Hazards

Fifty to sixty volcanoes erupt each year worldwide. The United States experiences eruptions two or three times a year, with most of them occurring in Alaska.[1] Eruptions often occur in sparsely populated areas of the world, causing little if any loss of life or economic damage. However, when a volcano erupts near a densely populated area, the effects can be catastrophic.[3] Several hundred million people on Earth live close to volcanoes, and as the human population grows, more and more people are living on the flanks, or sides, of active or potentially active volcanoes. In the past century almost 100,000 people have been killed by volcanic eruptions; nearly 23,000 lives were lost in the last two decades of the twentieth century alone.[1,3] Densely populated countries with many active volcanoes, such as Japan, Mexico (especially near Mexico City), the Philippines, and Indonesia, are particularly vulnerable.[2] The western United States, including Alaska, Hawaii, and the Pacific Northwest, also has many active or potentially active volcanoes; several of these volcanoes are located near large cities (Figure 7.3). Table 7.1 lists some historic volcanic events and their effects.

7.2 Volcanism and Volcanoes

Volcanic activity, or volcanism, is directly related to plate tectonics, and most active volcanoes are located near plate boundaries.[4] As spreading or sinking lithospheric plates interact with other Earth materials, **magma,** or molten rock, including a small component of dissolved gases, mostly water vapor and carbon dioxide, is produced. Magma that has emerged from a volcano onto Earth's

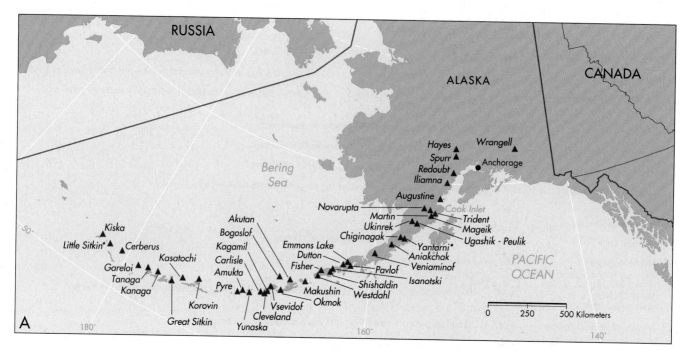

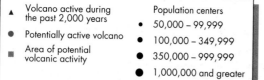

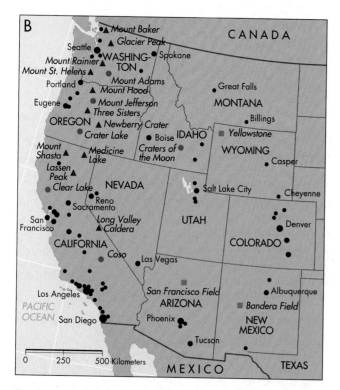

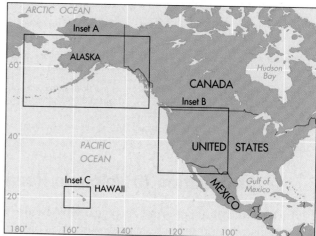

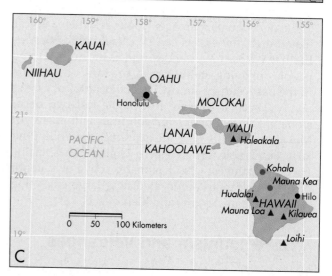

Figure 7.3 Locations of volcanoes in the United States Index maps show locations of active and potentially active volcanoes and nearby population centers (not all labeled). *(From Wright, T. L., and Pierson, T. C. 1992. U.S. Geological Survey Circular 1073)*

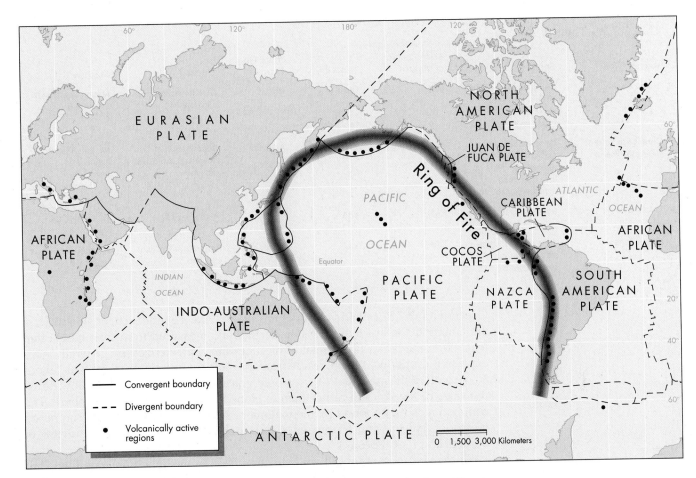

Figure 7.4 The "ring of fire" The ring of fire surrounds the Pacific plate. *(Modified from Costa, J. E., and Baker, V. R. 1981. Surficial geology. Originally published by John Wiley and Sons, New York, and republished by Tech Books, Fairfax, VA)*

surface is called **lava.** Approximately two-thirds of all active volcanoes on Earth are located in the "ring of fire" that circumscribes the Pacific Ocean, an area corresponding to subduction zones on the border of the Pacific plate (Figure 7.4).

7.3 Volcano Types

Each type of volcano has a characteristic style of activity that is partly a result of the viscosity of the magma. Viscosity describes a liquid's resistance to flow, with a high viscosity indicating a high resistance to flow. For example, honey is more viscous than water. Magma viscosity is determined by both its silica (SiO_2) content, which can vary from about 50 to 70 percent, and its temperature. The higher the silica content of a magma, the more viscous it is. Also, the hotter the magma is, the less viscous it is. The effect of temperature may make sense to you if you think about honey again. If you keep honey in the refrigerator, it becomes very difficult to pour or squeeze out of its container. However, if it is stored in a cabinet, or heated slightly before serving, it flows much more readily. Highly viscous magmas often erupt explosively, as opposed to less viscous magmas, which tend to flow.

Table 7.2 lists the types of volcanoes and their characteristics.

Shield Volcanoes

Shield volcanoes are by far the largest volcanoes. They are common in the Hawaiian Islands and are also found in Iceland and some islands in the Indian

TABLE 7.2   Types of Volcanoes

Volcano Type	Shape	Silica Content of Magma	Viscosity	Rock Type Formed	Eruption Type	Example
Shield volcano	Gentle arch, or shield shape, with shallow slopes; built up of many lava flows	Low	Low	Basalt	Lava flows, tephra ejections	Mauna Loa, Hawaii Figure 7.5
Composite volcano, or stratovolcano	Cone-shaped; steep sides; built up of alternating layers of lava flows and pyroclastic deposits	Intermediate	Intermediate	Andesite	Combination of lava flows and explosive activity	Mt. Fuji, Japan Figure 7.7
Volcanic dome	Dome shaped	High	High	Rhyolite	Highly explosive	Mt. Lassen, CA Figure 7.8
Cinder cone	Cone shaped; steep sides; often with summit crater	Low	Low	Basalt	Tephra (mostly ash) ejection	Springerville, AZ Figure 7.9

Ocean (Figure 7.5). Shield volcanoes are shaped like a gentle arch, or shield. They are among the tallest mountains on Earth when measured from their base, often located on the ocean floor. Shield volcanoes are characterized by generally non-explosive eruptions, which result from the relatively low silica content (about 50 percent) of the magma. When a shield volcano erupts, lava tends to flow down the sides of the volcano rather than explode violently into the atmosphere. The common rock type formed by the magma of shield volcanoes is *basalt,* composed mostly of feldspar and ferromagnesian minerals. (See Chapter 3 for a review of mineral and rock types.)

Shield volcanoes are built up almost entirely from numerous lava flows, but they can also produce a lot of tephra. *Tephra,* also referred to as pyroclastic debris, includes all types of volcanic debris that are explosively ejected from a volcano. Debris particles range from *ash,* less than 2 mm (0.08 in.) in diameter, to *cinders,* 4 to 32 mm (0.16 to 1.26 in.) in diameter, to *blocks* and *bombs* greater than 64 mm (2.52 in.) in diameter. Accumulation of tephra forms *pyroclastic deposits* (Greek *pyro,* "fire," and *klastos,* "broken"). Pyroclastic deposits may be consolidated to form *pyroclastic rocks.*

The slope of a shield volcano is very gentle near the top, but it increases on the flanks. This change is due to the viscosity of the flowing lava. When magma comes out of *vents,* or openings, at the top of the volcano, it is quite hot and flows easily. As it flows down the sides of the volcano it cools and becomes more viscous, so a steeper slope is needed for it to travel farther downslope.

Figure 7.5 Shield volcano
Profile of a shield volcano. Notice that the profile to the summit is very gently curved like a warrior's shield. The mountain shown is Mauna Loa, Hawaii, as viewed from the Hawaiian Volcanic Observatory. *(John S. Shelton)*

Figure 7.6 **Lava tube** 1984 Mauna Loa, Hawaii, eruption. *(Scott Lopez/Wind Cave National Park)*

In addition to flowing down the sides of a volcano, lava can move away from a vent in a number of ways. Magma often moves for many kilometers underground in *lava tubes*. These tubes are often very close to the surface, and they insulate the magma, keeping it hot and fluid. After the lava cools and crystallizes to rock, the lava tubes may be left behind as long, sinuous cavern systems (Figure 7.6). They form natural conduits for movement of groundwater and may cause engineering problems when they are encountered during construction projects.

Shield volcanoes may have a *summit caldera*, which is a steep, walled basin often 10 km (6.2 mi) or more in diameter. A summit caldera is formed by collapse of the summit of the volcano, in which a lava lake may form and from which lava may flow during an eruption. Eruptions of lava from shield volcanoes also commonly occur along linear fractures known as rift zones on the flank of a volcano. For example, rift eruptions at the Hawaiian shield volcano Kilauea, on the big island of Hawaii, during the past 24 years continue to add new land to the island.[4]

Composite Volcanoes

Composite volcanoes are known for their beautiful cone shape (Figure 7.7). Examples in the United States include Mount St. Helens and Mt. Rainier, both in Washington State. Composite volcanoes are characterized by magma with an

Figure 7.7 **Composite volcano** Mt. Fuji, Japan, is a composite volcano with beautiful steep sides. *(Jon Arnold Images/Alamy)*

intermediate silica content (about 60 percent), which is more viscous than the lower-silica magma of shield volcanoes. The common rock type formed by the magma of composite volcanoes is *andesite*, composed mostly of soda and lime-rich feldspar and ferromagnesian minerals, with small amounts of quartz. Composite volcanoes are distinguished by a mixture of explosive activity and lava flows. As a result, these volcanoes, also called *stratovolcanoes*, are composed of alternating layers of pyroclastic deposits and lava flows. Their steep flanks are due to the angle of repose, or the maximum slope angle for loose material, which for many pyroclastic deposits is approximately 30 percent to 35 percent.

Don't let the beauty of these mountains fool you. Because of their explosive activity and relatively common occurrence, composite volcanoes are responsible for most of the volcanic hazards that have caused death and destruction throughout history. As the 1980 eruption of Mount St. Helens demonstrated, they can produce gigantic horizontal blasts, similar in form to the blast of a shotgun. We should consider such volcanoes armed and dangerous.

Volcanic Domes

Volcanic domes are characterized by viscous magma with a relatively high silica content (about 70 percent). The common rock type produced by this magma is *rhyolite*, composed mostly of potassium- and sodium-rich feldspar, quartz, and minor amounts of ferromagnesian minerals. The activity of volcanic domes is mostly explosive, making these volcanoes very dangerous. Mt. Lassen in northeastern California is a good example of a volcanic dome. Mt. Lassen's last series of eruptions, from 1914 to 1917, included a tremendous lateral blast that affected a large area (Figure 7.8).

Cinder Cones

Cinder cones are relatively small volcanoes formed from tephra, mostly volcanic ash and larger particles including volcanic bombs. Bombs are formed from blobs of ejected lava that spin in the air to take on a rounded shape with tapered ends.

Figure 7.8 Eruption of Mt. Lassen June 1914. *(Courtesy of B. F. Loomis/Lassen Volcanic National Park)*

Figure 7.9 Cinder cone Cinder cone with a small crater at the top near Springerville, Arizona. The brownish black material from the center upper part of the photograph to the upper right-hand corner is a lava flow, which originates from the base of the cinder cone.
(Michael Collier)

Cinder cones grow from the accumulation of tephra near a volcanic vent. They are often found on the flanks of larger volcanoes or along normal faults and long cracks or fissures (Figure 7.9).

The Paricutín cinder cone (Figure 7.10) in the Itzicuaro Valley of central Mexico, west of Mexico City, offered a rare opportunity to observe the birth and rapid growth of a volcano at a location where none had existed before. On February 20, 1943, after several weeks of earthquakes and sounds like thunder coming from beneath the surface of Earth, an astounding event occurred. As Dionisio Pulido was preparing his cornfield for planting, he noticed that a hole in the ground he had been trying to fill for years had opened wider. As Señor Pulido watched, the surrounding ground swelled, rising over 2 m (6.5 ft), while sulfurous smoke and ash began billowing from the hole. By that night, the hole was ejecting glowing red rock fragments high into the air. By the next day, the cinder cone had grown as high as a three-story building as rocks and ash continued to be blown into the sky by the eruption. After only 5 days, the cinder cone had grown to the height of a

Figure 7.10 Rapid growth of a volcano Paricutín cinder cone, central Mexico, 1943, erupting ash and lava. A lava flow has nearly buried the village of San Juan Parangaricutiro, leaving the church steeple exposed.
(Courtesy of Tad Nichols)

30-story building. In June 1944, a fissure that had opened in the base of the now 400 m (1,312 ft) high cone erupted a thick lava flow that overran the nearby village of San Juan Parangaricutiro, leaving little but the church steeple exposed (see Figure 7.10). No one was killed by the ash or lava flow, and within a decade, Paricutín cinder cone became a dormant volcano. Nevertheless, the years of eruption from Dionisio's cornfield had significant local impacts. Crops failed; they were sometimes buried by ash faster than they could grow; and livestock became sick and died. Some villages relocated to other areas, and some villagers moved back to the area. Locating property boundaries was difficult because boundary markers were often covered by ash and lava, resulting in property disputes.[5]

7.4 Volcano Origins

We established earlier that the causes of volcanic activity are directly related to plate tectonics.[4] Understanding the tectonic origins of different types of volcanoes helps explain the chemical differences in their rock types. Figure 7.11 is an idealized diagram showing the relationship of processes at plate boundaries to the volcano types we have described. The following features are illustrated on the diagram:

1. Volcanism occurring at mid-oceanic ridges produces basaltic rocks. Basaltic rock has a relatively low silica content and wells up directly from the asthenosphere as magma. The basaltic magma mixes very little with other materials except oceanic crust, which itself is basaltic. Where these spreading ridge systems occur on land, as, for example, in Iceland, shield volcanoes are formed (Figure 7.12).

2. Shield volcanoes are formed above hot spots located below the lithospheric plates. For example, the Hawaiian volcanoes are located well within the Pacific plate rather than near a plate boundary. It is currently believed that there is a hot spot below the Pacific plate where magma is generated. Magma moves upward through the plate and produces a volcano on the bottom of the sea; this volcano may eventually become an island. The plate containing the Hawaiian Islands is moving roughly northwest over the stationary hot spot; therefore a chain of volcanoes running northwest to southeast is formed (see Chapter 2). The island of Hawaii is presently near the hot spot and is experiencing active

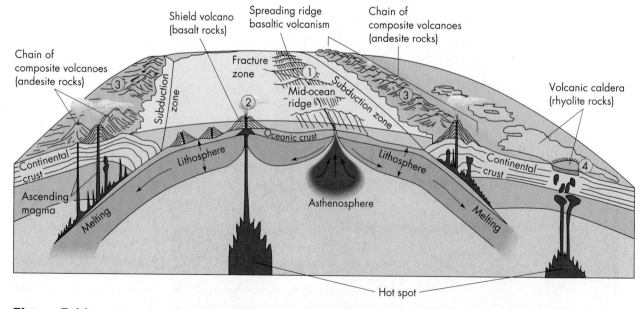

Figure 7.11 **Volcanic activity and plate tectonics** Idealized diagram showing plate tectonic processes and their relation to volcanic activity. Numbers refer to explanations in the text. *(Modified from Skinner, B. J., and Porter, S. C. 1992. The dynamic Earth, 2nd ed. New York: John Wiley)*

Figure 7.12 Fissure on the Mid-Atlantic Ridge Icelandic shield volcano (background) and large fissure (normal fault) associated with spreading of tectonic plates along the Mid-Atlantic Ridge. *(John S. Shelton)*

volcanism and growth. Islands to the northwest, such as Molokai and Oahu, have evidently moved off the hot spot, since the volcanoes on these islands are no longer active.

3. Composite volcanoes are associated with andesitic volcanic rocks and subduction zones. These are the most common volcanoes found around the Pacific Rim. For example, volcanoes in the Cascade Range of Washington, Oregon, and California are related to the Cascadia subduction zone (Figure 7.13). Andesitic volcanic rocks are produced at subduction zones, where rising magma mixes with both oceanic and continental crust. Since continental crust has a higher silica content than basaltic magma, this process produces rock with an intermediate silica content.

4. Caldera-forming eruptions may be extremely explosive and violent. These eruptions tend to be associated with rhyolitic rocks, which are produced when magma moves upward and mixes with continental crust. Rhyolitic rocks contain more silica than other volcanic rocks because of the high silica content of continental crust. Volcanic domes, although not always associated with caldera-forming eruptions, are usually found inland of subduction zones and erupt silica-rich rhyolitic magma as well.

Our brief discussion does not explain all differences in composition and occurrence of basaltic, andesitic, and rhyolitic rocks. However, it does establish the most basic relationships between plate tectonics, volcanic activity, and volcanic rocks.

7.5 Volcanic Features

Geologic features that are often associated with volcanoes or volcanic areas include craters, calderas, volcanic vents, geysers, and hot springs.

Craters, Calderas, and Vents

Depressions commonly found at the top of volcanoes are craters. Craters form by explosion or collapse of the upper portion of the volcanic cone and may be flat floored or funnel shaped. They are usually a few kilometers in diameter. *Calderas* are gigantic, often circular, depressions resulting from explosive ejection of magma and subsequent collapse of the upper portion of the volcanic cone. They may be 20 or more kilometers in diameter and contain volcanic vents as well as other volcanic features such as gas vents and hot springs. *Volcanic vents* are

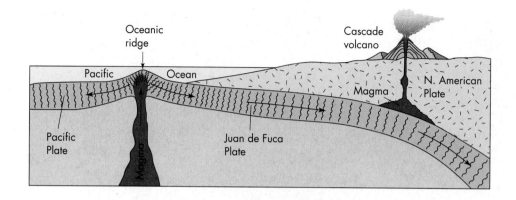

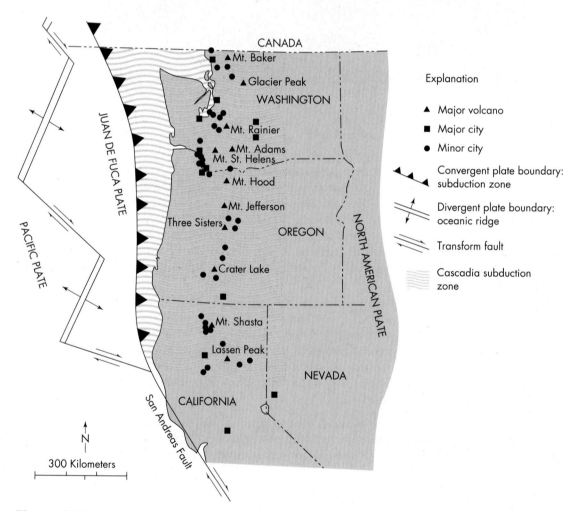

Figure 7.13 Cascade volcanoes and plate tectonics Map and plate tectonic setting of the Cascade Range, showing major volcanoes and cities in their vicinity. *(Modified after Crandell, D. R., and Waldron, H. H. 1969. Disaster preparedness. Office of Emergency Preparedness)*

openings through which lava and pyroclastic debris are erupted at the surface of Earth. Vents may be roughly circular conduits, and eruptions construct domes and cones. Other vents may be elongated fissures or rock fractures, often normal faults, which produce lava flows. Some extensive fissure eruptions have produced huge accumulations of nearly horizontal basaltic lava flows called flood basalts. The best-known flood basalt deposit in the United States is the Columbia Plateau region (Figure 7.14) in parts of Washington, Oregon, and Idaho, where basalt covers a vast area.

Figure 7.14 Flood basalt
Columbia Plateau flood basalts along
the bluff of the Columbia River in
Benton County, Washington. (*Calvin
Larsen/Photo Researchers, Inc.*)

Hot Springs and Geysers

Hot springs and geysers are hydrologic features found in some volcanic areas.
Groundwater that comes into contact with hot rock becomes heated, and in
some cases the heated water discharges at the surface as a *hot spring,* or *thermal
spring.* In rare cases, the subsurface groundwater system involves circulation and
heating patterns that produce periodic release of steam and hot water at the
surface, a phenomenon called a *geyser.* World-famous geyser basins or fields are
found in Iceland, New Zealand, and Yellowstone National Park in Wyoming
(Figure 7.15).

(a)

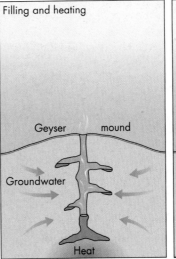

Filling and heating

Geyser   mound

Groundwater

Heat

Groundwater fills
subsurface geyser's
irregular tubes
and water is heated
until it flashes to
steam and erupts.

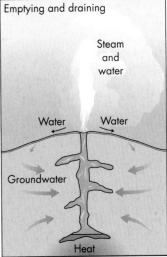

Emptying and draining

Steam
and
water

Water   Water

Groundwater

Heat

(b)

Eruption partly empties
subsurface tubes
and recycled groundwater
starts the filling process
again. Water running
off precipitates a white
silica-rich mineral called
geyserite that forms the
geyser mound.

Figure 7.15 **How a geyser works** (a) Eruption of Old Faithful Geyser, Yellowstone National Park,
Wyoming. The geyser is named for its very predictable periodic eruption. (*James Randklev/Getty Images, Inc.*)
(b) Schematic drawing of a geyser, illustrating processes that lead to eruption.

Figure 7.16 Caldera geology Map (a) and diagram (b) illustrating the volcanic hazard near Mammoth Lakes, California. The map (a) shows the location of past volcanic events, the area of uplift where magma seems to be moving up, and finally the area where earthquake swarms have occurred near Mammoth Lakes. The geological cross section (b) shows a section (northeast-southwest) through the Long Valley caldera. Shown are geologic relations inferred for the 1980 magma rise that produced uplift and swarms of earthquakes. *(From Bailey, R. A. 1983. Mammoth Lakes earthquakes and ground uplift: Precursor to possible volcanic activity? In U.S. Geological Survey Yearbook, Fiscal Year 1982.)*

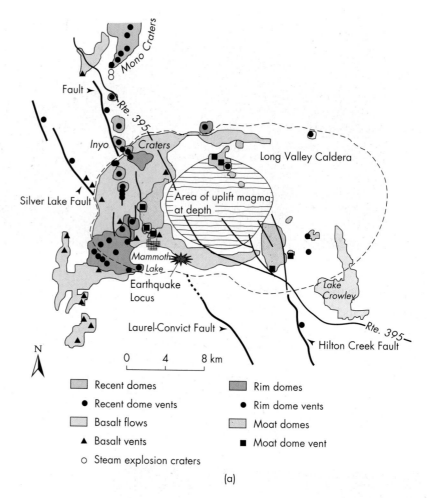

(a)

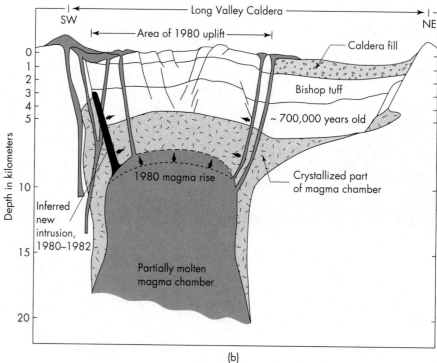

(b)

Caldera Eruptions

Calderas are produced by very rare, but extremely violent, eruptions. Although none have occurred anywhere on Earth in the last few hundred thousand years, at least 10 **caldera eruptions** have occurred in the last million years, three of them in North America. A large caldera-forming eruption may explosively extrude up to 1,000 cubic km^3 (240 mi^3) of pyroclastic debris, consisting mostly of ash. This is approximately 1,000 times the quantity ejected by the 1980 eruption of Mount St. Helens! Such an eruption could produce a caldera more than 10 km (6.2 mi) in diameter and blanket an area of several tens of thousands of square kilometers with ash. These ash deposits can be 100 m (328 ft) thick near the crater's rim and a meter or so thick 100 km (62 mi) away from the source.[6] The most recent caldera-forming eruptions in North America occurred about 600,000 years ago at Yellowstone National Park in Wyoming and 700,000 years ago in Long Valley, California. Volcanic features and geology of the Long Valley caldera are shown in Figure 7.16. Figure 7.17a shows the area covered by ash in the eruption event, which produced the Long Valley caldera near the famous Mammoth Mountain ski resort. Figure 7.17b shows the potential hazard from a future volcanic eruption at Long Valley. The most recent volcanic eruptions at Long Valley were about 400 years ago. Measurable uplift of the land accompanied by swarms of earthquakes up to M 6 in the early 1980s suggested magma was moving upward, prompting the U.S. Geological Survey to issue a potential volcanic hazard warning that was subsequently lifted. However, the future of Long Valley remains uncertain.

The main events in a caldera-producing eruption can occur quickly—in a few days to a few weeks—but intermittent, lesser-magnitude volcanic activity can

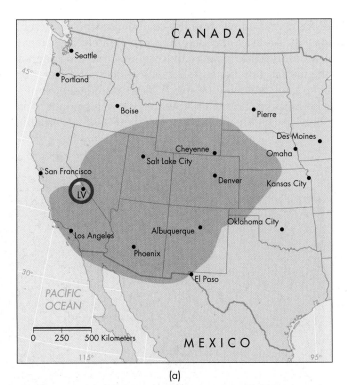

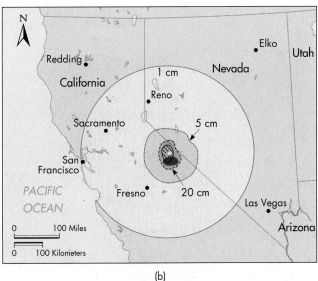

(a)   (b)

Figure 7.17 **Potential hazards** Potential hazards from a volcanic eruption at the Long Valley caldera near Mammoth Lakes, California. (a) Area covered by ash from the Long Valley caldera eruption approximately 700,000 years ago. The circle near Long Valley (LV) has a radius of 120 km (75 mi) and encloses the area that would be subject to at least 1 m (3.3 ft) of downwind ash accumulation if a similar eruption were to occur again. Also within this circle, hot pyroclastic flows (ash flows) are likely to occur and, in fact, could extend farther than shown by the circle. (b) The red area and diagonal lines show the hazard from flowage events out to a distance of approximately 20 km (12.5 mi) from recognized potential vents. Lines surrounding the hazard areas represent potential ash thicknesses of 20 cm (8 in.) at 35 km (22 mi) (dashed line), 5 cm (2 in.) at 85 km (53 mi) (dotted line), and 1 cm (0.4 in.) at 300 km (186 mi) (solid line). These estimates of potential hazards assume an explosive eruption of approximately 1 km^3 (0.24 mi^3) from the vicinity of recently active vents. *(From Miller, C. D., Mullineaux, D. R., Crandell, D. R., and Bailey, R. A. 1982. Potential hazards from future volcanic eruptions in the Long Valley–Mono Lake area, east-central California and southwest Nevada—A preliminary assessment. U.S. Geological Survey Circular 877)*

linger on for a million years. Thus, the Yellowstone event has left us hot springs and geysers, including Old Faithful, while the Long Valley event has left us a potential volcanic hazard. In fact, both sites are still capable of producing volcanic activity because magma is still present at variable depths beneath the caldera floors. Both are considered *resurgent* calderas because their floors have slowly domed upward since the explosive eruptions that formed them. The most likely future eruptions for Long Valley or Yellowstone would be much smaller than the giant caldera eruptions that occurred hundreds of thousands of years ago.

7.6 Volcanic Hazards

Volcanic hazards include the *primary effects* of volcanic activity that are the direct results of the eruption and *secondary effects*, which may be caused by the primary effects. Primary effects are lava flows; pyroclastic activity, including ash fall, ash flows, and lateral blasts; and the release of gases. Most of the gases released are steam, but corrosive or poisonous gases have also been known to be released. Secondary effects include debris flows, mudflows, landslides, or debris avalanches, floods, and fires. At the planetary level, large eruptions can cause global cooling of the atmosphere for a year or so.[6,7]

Lava Flows

Lava flows are one of the most familiar products of volcanic activity. They result when magma reaches the surface and overflows the crater or a volcanic vent along the flanks of the volcano. The three major groups of lava take their names from the volcanic rocks they form: basaltic, by far the most abundant of the three, andesitic, and rhyolitic.

Lava flows can be quite fluid and move rapidly, or they can be relatively viscous and move slowly. Basaltic lavas, composed of approximately 50 percent silica, exhibit a range of velocities. Basaltic lavas with lower viscosity and higher eruptive temperatures are the fastest moving, with a usual velocity of a meter or so per hour (around 3.2 ft per hour, but may be much faster on a steep slope). These lavas, called *pahoehoe* lavas (pronounced pa-hoy-hoy), have a smooth, *ropy* surface texture when they harden (Figure 7.18a). Cooler, more viscous basaltic

Figure 7.18a *Pahoehoe* **lava flow** With a smooth surface texture, surrounding and destroying a home at Kalapana, Hawaii, in 1990. Such flows destroyed more than 100 structures, including the National Park Service Visitor Center. These types of flows are often called *ropy lava* because of the surface texture, which looks a bit like pieces of rope lying side by side as if coiled. *(Paul Richards/UPI/CORBIS)*

Figure 7.18b *Aa* **lava flow** Blocky *aa* lava flow engulfing a building during the eruption on the island of Heimaey, Iceland. *(Solarfilma HF)*

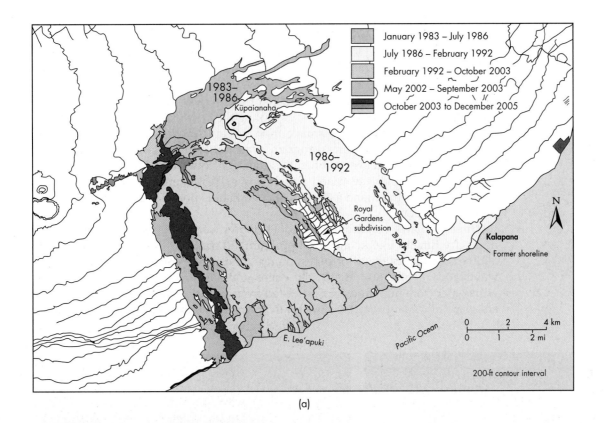

January 1983 – July 1986

July 1986 – February 1992

February 1992 – October 2003

May 2002 – September 2003

October 2003 to December 2005

1983–1986

Kūpaianaha

1986–1992

Royal Gardens subdivision

Kalapana

Former shoreline

N

E. Lee'apuki

Pacific Ocean

0   2   4 km

0   1   2 mi

200-ft contour interval

(a)

(b)

Figure 7.19 Lava flows on Island of Hawaii, 1983–2005 (a) Map of flows. (b) Lava flowing into the ocean. The white "smoke" is steam. The eruption illustrates the process that continues to build Hawaii. *(USGS Hawaiian Volcano Observatory)*

lava flows move at rates of a few meters per day and have a *blocky* texture after hardening (Figure 7.18b). This type of basaltic lava is called *aa* (pronounced ah-ah). With the exception of some flows on steep slopes, most lava flows are slow enough that people can easily move out of the way as they approach.[8]

Lava flows from rift eruptions on Kilauea, Hawaii, began in 1983 and have become the longest and largest eruptions of Kilauea in history[4] (Figure 7.19). By 2005 more than 100 structures in the village of Kalapana had been destroyed by lava flows, including the National Park Visitor Center. Lava flowed across part of the famous Kaimu Black Sand Beach and into the ocean. The village of Kalapana has virtually disappeared, and it will be many decades before much of the land is productive again. The eruptions, in concert with beach processes, have produced new black sand beaches. The sand is produced when the molten lava enters the relatively cold ocean waters and shatters into sand-sized particles.

Methods to Control Lava Flows

Several methods, such as hydraulic chilling and wall construction, have been employed to deflect lava flows away from populated or otherwise valuable areas. These methods have had mixed success. They cannot be expected to modify large flows, and their effectiveness with smaller flows requires further evaluation.

Hydraulic chilling of lava flows, or cooling the flow with water, has sometimes been successful in reducing damage from volcanic eruptions. The world's most ambitious hydraulic chilling program was initiated in January 1973 on the Icelandic island of Heimaey, when basaltic lava flows from Mt. Helgafell nearly closed the harbor of the island's main town, Vestmannaeyjar, threatening the continued use of the island as Iceland's main fishing port. The situation prompted immediate action.

Three favorable conditions existed: (1) Flows were slow moving, allowing the necessary time to initiate a control program; (2) transport by sea and local roads allowed for the transport of pipes, pumps, and heavy equipment; and (3) water was readily available. Initially, the edges and surface of the flow were cooled with water from numerous fire hoses fed from a pipe (Figure 7.20). Then bulldozers were moved up on the slowly advancing flow, making a track or road on

(a)

(b)

(c)

Figure 7.20 **People fighting lava flows** Eruptions of Mt. Helgafell on the island of Heimaey, Iceland. (a) At night from the harbor area. *(Solarfilma HF)* (b) Aerial view toward the harbor. Notice the advancing lava flow with the white steam escaping in the lower right-hand corner. The steam is the result of water being applied to the front of the flow. A water cannon is operating in the lower right-hand corner, and the stream of water is visible. *(James R. Andrews)* (c) Aerial view showing the front of the blocky lava flow encroaching into the harbor. By fortuitous circumstances the flows stopped at a point that actually has improved the harbor through better protection against storm waves. *(James R. Andrews)*

which the plastic pipe was placed. The pipe did not melt as long as water was flowing in it, and small holes in the pipe also helped cool particularly hot spots along various parts of the flow. Watering near the front of the flow had little effect the first day, but then flow began to slow down and in some cases stopped.

The action taken in Iceland undoubtedly had an important effect on the lava flows from Mt. Helgafell. It tended to restrict lava movement and thus reduced property damage in the harbor town so that after the outpouring of lava stopped in June 1973, the harbor was still usable.[9] In fact, by fortuitous circumstances, the shape of the harbor was actually improved, since the new rock provided additional protection from the sea.

Pyroclastic Activity

Pyroclastic activity describes explosive volcanism, in which tephra is physically blown from a volcanic vent into the atmosphere. Several types of pyroclastic activity can occur. In *volcanic ash eruptions*, or **ash fall,** a tremendous quantity of rock fragments, natural glass fragments, and gas is blown high into the air by explosions from the volcano. **Lateral blasts** are explosions of gas and ash from the side of a volcano that destroy part of the mountain. The ejected material travels away from the volcano at tremendous speeds. Lateral blasts can be very destructive, and sometimes the velocity of the ejected material is nearly the speed of sound (see the discussion of Mount St. Helens in Section 7.7 of this chapter). *Pyroclastic flows*, or **ash flows,** are some of the most lethal aspects of volcanic eruptions. They are avalanches of very hot pyroclastic materials—ash, rock, volcanic glass fragments, and gas—that are blown out of a vent and move very rapidly down the sides of the volcano. Pyroclastic flows are also known as hot avalanches, or *nueé ardentes*, which is French for "glowing cloud."

Ash Fall.   Volcanic ash eruptions can cover hundreds or even thousands of square kilometers with a carpet of volcanic ash. Ash eruptions create several hazards:

- Vegetation, including crops and trees, may be destroyed.
- Surface water may be contaminated by sediment, resulting in a temporary increase in acidity of the water. The increase in acidity generally lasts only a few hours after the eruption ceases.
- Structural damage to buildings may occur, caused by the increased load on roofs (Figure 7.21). A depth of only 1 cm (0.4 in.) of ash places an extra 2.5 tons of weight on a roof of an average house.
- Health hazards, such as irritation of the respiratory system and eyes, are caused by contact with the ash and associated caustic fumes.[8]

Figure 7.21 Volcanic ash on roofs Ash on buildings may increase the load on walls by several tons. Shown here are buildings that collapsed and burned from hot ash and lava during an eruption in Iceland in 1973. *(Owen Franken/Stock Boston)*

■ Engines of jetliners may "flame-out" as melted silica-rich ash forms a thin coating of volcanic glass in the engines. For example, the 1989 eruption of Redoubt Volcano in Alaska produced an ash eruption cloud that an airliner on the way to Japan encountered. Power to all four engines of the jetliner was lost. Fortunately, the engines restarted within about 5 minutes. The plane fell about 10,000 feet during the time it took to restart, but it managed to land safely in Anchorage, Alaska. Damage to the aircraft was about $80 million.[5]

Ash Flows. Ash flows, or nueé ardentes, may be as hot as hundreds of degrees Celsius and move as fast as 100 km per hour (62 mi per hour) down the sides of a volcano, incinerating everything in their path. They can be catastrophic if a populated area is in the path of the flow; fortunately, they seldom occur in populated areas. A tragic example occurred in 1902 on the West Indian island of Martinique. On the morning of May 8, a flow of hot, glowing ash, steam, and other gases roared down Mt. Pelée and through the town of St. Pierre, killing 30,000 people. A jailed prisoner was one of only two survivors, and he was severely burned and horribly scarred. Reportedly, he spent the rest of his life touring circus sideshows as the "Prisoner of St. Pierre." Flows like these have occurred on volcanoes of the Pacific Northwest and Japan in the past and can be expected in the future.

Poisonous Gases

A number of gases, including water vapor (H_2O), carbon dioxide (CO_2), carbon monoxide (CO), sulfur dioxide (SO_2), and hydrogen sulfide (H_2S), are emitted during volcanic activity. Water and carbon dioxide make up more than 90 percent of all emitted gases. Toxic concentrations of hazardous volcanic gases rarely reach populated areas. A notable tragic exception occurred in 1986 in Lake Nyos, Cameroon, West Africa. Lake Nyos is located in a 200 m deep (656 ft deep) crater on a dormant volcano; without any warning other than an audible rumbling, the seemingly dormant volcano released a dense cloud of gas, consisting primarily of carbon dioxide. Carbon dioxide is a colorless, odorless gas that is heavier than air. The gas cloud flowed 10 km (6.2 mi) down the volcano along a valley, hugging the ground and suffocating 1,700 people and 3,000 cattle (Figure 7.22a and b).[4]

Since 1986 carbon dioxide gas has continued to accumulate in the bottom of the lake, and another release could occur at any time. Although the lake area was to remain closed to all people but scientists studying the hazard, thousands of people are returning to farm the land. Scientists have installed an alarm system at the lake that will sound if carbon dioxide levels become high. They have also installed a pipe from the lake bottom to a degassing fountain on the surface of the lake (Figure 7.22c) that allows the carbon dioxide gas to escape slowly into the atmosphere. The hazard is being slowly reduced as the single degassing fountain is now releasing a little more carbon dioxide gas than is naturally seeping into the lake. At least five additional pipes with degassing fountains will be necessary to adequately reduce the hazard.

Sulfur dioxide can react in the atmosphere to produce acid rain downwind of an eruption. Toxic concentrations of some chemicals emitted as gases may be absorbed by volcanic ash that falls onto the land. Eventually, the acid rain and toxic ash are incorporated into the soil and into plants eaten by people and livestock. For example, fluorine is erupted as hydrofluoric acid; it can be absorbed by volcanic ash and be leached into water supplies.[3]

In Japan, volcanoes are monitored to detect the release of poisonous gases such as hydrogen sulfide. When releases are detected, sirens are sounded to advise people to evacuate to high ground to escape the gas.

Volcanoes can also produce a type of smog known as *vog* (volcanic material, *v*, and fog, *og*). For example, emissions of sulfur dioxide and other volcanic gases from Kilauea in Hawaii have continuously erupted since 1986. At times, these gases have traveled downwind and chemically interacted with oxygen and

(a)

(b)

(c)

Figure 7.22 **Poisonous gas from dormant volcano** (a) In 1986 Lake Nyos released immense volumes of carbon dioxide. *(T. Oban/Corbis/Sygma)* (b) The gas killed, by asphyxiation, about 1,700 people and 3,000 cattle. *(Peter Turnley/CORBIS)* (c) Gas being released with a degassing fountain in 2001. *("Nyos degassing system conceived, designed and manufactured by a French company, Data Environment and the University of Savoie, France." Courtesy of J. C. Sabroux)*

moisture in the atmosphere to produce vog conditions mixed with acid rain. At times, southeastern areas of Hawaii were blanketed with thick acidic haze, and health warnings were issued. The vog is potentially hazardous to both people and other living things, and adverse health effects are being evaluated on Hawaii.[10,11] Small acidic aerosol particulates and sulfur dioxide concentrations like those in vog can penetrate deep within human lungs and are known to induce asthma attacks and cause other respiratory problems. Atmospheric concentrations of sulfur dioxide emitted from Kilauea, at times, far exceed air quality standards (see Chapter 18). Residents and visitors have reported a variety of symptoms when exposed to vog, including breathing difficulties, headaches, sore throats, watery eyes, and flulike symptoms. In 1988, acid rain in the Kona district captured from rooftop rainwater-catchment systems produced lead-contaminated water in 40 percent of homes. The lead is leached by acid rain from roofing and plumbing materials. Some residents when tested were found to have an elevated concentration of lead in their blood, presumably from drinking the water.[12]

Debris Flows and Mudflows

The most serious secondary effects of volcanic activity are **debris flows** and **mudflows,** known collectively by their Indonesian name of *lahar*. Lahars are produced when a large volume of loose volcanic ash and other ejecta becomes saturated with water and becomes unstable, allowing it to suddenly move downslope (see Figure 7.2). The distinction between a debris flow and a mudflow depends upon the dominant size of the particles. In debris flows, more than 50 percent of the particles are coarser than sand, 2 mm (0.08 in.) in diameter. We will discuss landslides in detail in Chapter 9.

Debris Flows. Research completed at several volcanoes suggests that even relatively small eruptions of hot volcanic material may quickly melt large volumes of snow and ice. The copious amounts of melted water produce floods; the flood may erode the slope of the volcano and incorporate sediment such as volcanic ash and other material, forming debris flows. Volcanic debris flows are fast-moving mixtures of sediment, including blocks of rock and water, with the general consistency of wet concrete. Debris flows can travel many kilometers down valleys from the flanks of the volcano where they were produced.[8] For example, a pyroclastic flow that erupted from Alaska's Mt. Redoubt volcano early in 1990 rapidly melted snow and ice while moving across Drift Glacier. Voluminous amounts of water and sediment produced a debris flow that quickly moved down the valley, with a discharge comparable to that of the Mississippi River at flood stage. Fortunately, the event was in an isolated area, so no lives were lost.[3]

Mudflows. Gigantic mudflows have originated on the flanks of volcanoes in the Pacific Northwest.[7] A mudflow that is composed mostly of volcanic ash is often called a *lahar*. The paths of two ancient mudflows originating on Mt. Rainier are shown in Figure 7.23. Deposits of the Osceola mudflow are 5,000 years old. This mudflow moved more than 80 km (50 mi) from the volcano and involved more than 1.9 km^3 (0.45 mi^3) of debris, equivalent to 13 km^2 (5 mi^2) piled to a depth of more than 150 m (492 ft). Deposits of the younger, 500 year old Electron mudflow reached about 56 km (35 mi) from the volcano.

Hundreds of thousands of people now live on the area covered by these old flows, and there is no guarantee that similar flows will not occur again. Figure 7.24 shows the potential risk of mudflows, lava flows, and pyroclastic debris accumulation for Mt. Rainier. Someone in the valley facing the advancing front of such a flow would describe it as a wall of mud a few meters high, moving at about 30 to 50 km per hour (20 to 30 mi per hour). This observer would first see the flow at a distance of about 1.6 km (around 1 mi) and would need a car headed in the right direction toward high ground to escape being buried alive.[8]

The U.S. Geological Survey has recently developed an automated, solar-powered lahar-detection system for several volcanoes in the United States (Mt. Rainier), Indonesia, the Philippines, Mexico, and Japan. Detectors (acoustic-flow monitors) that sense ground vibrations from a moving lahar can warn that a flow is moving down the valley. The system replaces visual sightings or cameras that are not reliable in bad weather or at night, and require continued personal attention. Dangerous lahars can occur quickly with little warning. The important thing is to know when to evacuate. If the lahar-detection system sounds an alarm that a lahar is approaching, the warning is issued, with sufficient time to evacuate to safe (higher) ground.[12]

What may be the largest active landslides on Earth are located on Hawaii. Presently, these landslides are slow moving (approximately 10 cm, or 4 in., per year). They can be up to 100 km (62 mi) wide, 10 km (6.2 mi) thick, and 20 km (12.4 mi) long and extend from a volcanic rift zone on land to an endpoint beneath the sea. These slow-moving landslides contain blocks of rock the size of Manhattan Island. The fear is that they might, as they have in the past, become giant, fast-moving submarine debris avalanches, generating huge tsunamis that

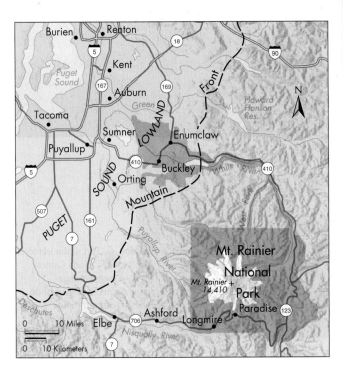

Figure 7.23 **Mudflow** Map of Mt. Rainier and vicinity showing the extent of the Osceola mudflow in the White River valley (orange) and the Electron mudflow (beige) in the Puyallup River valley. *(From Crandell, D. R., and Mullineaux, D. R., U.S. Geological Survey Bulletin 1238)*

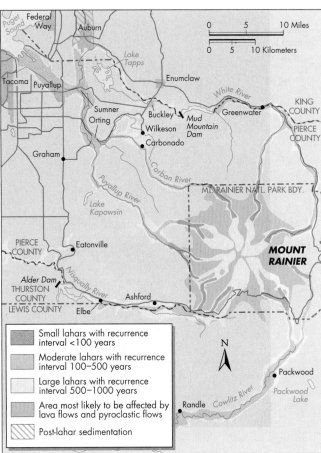

Figure 7.24 **Mudflow hazard potential** Map of Mt. Rainier and vicinity showing potential hazards from lahars, lava flows, and pyroclastic flows. *(Hoblitt and others. 1998. U.S. Geological Survey Open File Report 98-428)*

deposit marine debris hundreds of meters above sea level at nearby islands and cause catastrophic damage around the Pacific Basin. Fortunately, such high-magnitude events apparently happen only every 100,000 years or so.[4]

Other huge volcano-related landslides and debris avalanches have been documented in the Canary Islands, located in the Atlantic Ocean off the western coast of Africa. On Tenerife, the largest island, six huge landslides occurred during the last several million years (Figure 7.25). The most recent event occurred less than 150,000 years ago. Mt. Teide, with an elevation of 3.7 km (2.3 mi) is within the now-collapsed Cañada Caldera.[13,14] The landslides have produced large linear valleys, the largest of which is Orotova Valley, the site of the city of Puerto de la Cruz, which is a major coastal tourist destination and retirement center for Europeans. The youngest valley is Icod (Figure 7.25). The total volume of the slide deposits exceeds 1,000 km³ (240 mi³). The offshore seafloor north of Tenerife is covered by 5,500 km² (2,150 mi²) of landslide deposits, an area of more than twice the land surface of the island.[15]

There is concern that a future landslide from Tenerife or La Palma, another of the Canary Islands, would cause a huge tsunami known as a megatsunami. Waves with heights exceeding 100 m are possible. They would cause catastrophic damage to the eastern coast of the United States, including the city of New York. The possibility of future huge landslides is unknown, but their occurrence cannot be ruled out.

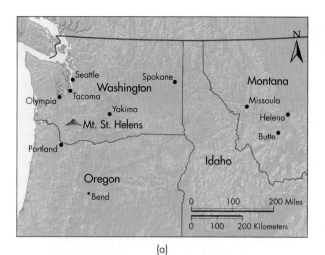

(a)

Figure 7.27 **Mount St. Helens before and after** (a) Location of Mount St. Helens. (b) Mount St. Helens before and (c) after the May 18, 1980, eruption. As a result of the eruption, much of the northern side of the volcano was blown away, and the altitude of the summit was reduced by approximately 450 m (1,476 ft). *(Photo [b] by Bruce Spainhower. Photo [c] by Harry Glicken)*

(b)

(c)

and a large cloud of ash moved over the United States, reaching as far east as New England (Figure 7.29b). The entire northern slope of the volcano, which is the upper part of the north fork of the Toutle River basin, was devastated. Forested slopes were transformed into a gray, hummocky, or hilly, landscape consisting of volcanic ash, rocks, blocks of melting glacial ice, narrow gullies, and hot steaming pits (Figure 7.30).[18]

The first of several mudflows consisted of a mixture of water, volcanic ash, rock, and organic debris, such as logs, and occurred minutes after the start of the eruption. The flows and accompanying floods raced down the valleys of the north and south forks of the Toutle River at estimated speeds of 29 to 55 km per hour (18 to 35 mi per hour), threatening the lives of people camped along the river.[18]

On the morning of May 18, 1980, two young people on a fishing trip on the Toutle River were sleeping about 36 km (22 mi) downstream from Spirit Lake. They were awakened by a loud rumbling noise from the river, which was covered by felled trees. They attempted to run to their car, but water from the rising river poured over the road, preventing their escape. A mass of mud then crashed through the forest toward the car, and the couple climbed onto the car roof to escape the mud. They were safe only momentarily, however, as the mud pushed the vehicle over the bank and into the river. Leaping off the roof, they fell into the river, which was by now a rolling mass of mud, logs, collapsed train trestles, and other debris. The water also was increasing in temperature. One of the young people got trapped between logs and disappeared several times beneath the flow but was lucky enough to emerge again. The two were carried downstream for approximately 1.5 km (0.9 mi) before another family of campers spotted and rescued them.

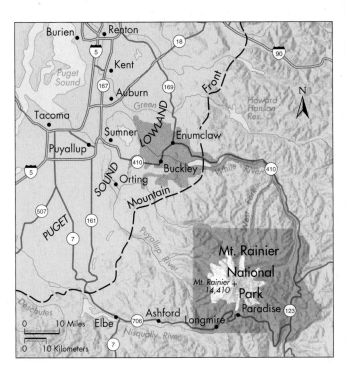

Figure 7.23 **Mudflow** Map of Mt. Rainier and vicinity showing the extent of the Osceola mudflow in the White River valley (orange) and the Electron mudflow (beige) in the Puyallup River valley. *(From Crandell, D. R., and Mullineaux, D. R., U.S. Geological Survey Bulletin 1238)*

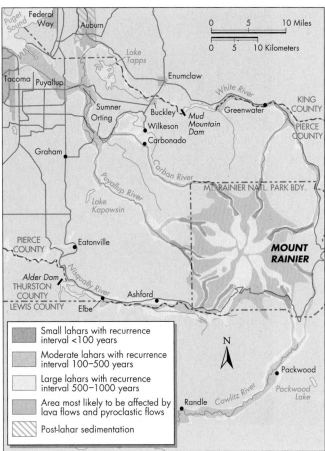

Figure 7.24 **Mudflow hazard potential** Map of Mt. Rainier and vicinity showing potential hazards from lahars, lava flows, and pyroclastic flows. *(Hoblitt and others. 1998. U.S. Geological Survey Open File Report 98-428)*

deposit marine debris hundreds of meters above sea level at nearby islands and cause catastrophic damage around the Pacific Basin. Fortunately, such high-magnitude events apparently happen only every 100,000 years or so.[4]

Other huge volcano-related landslides and debris avalanches have been documented in the Canary Islands, located in the Atlantic Ocean off the western coast of Africa. On Tenerife, the largest island, six huge landslides occurred during the last several million years (Figure 7.25). The most recent event occurred less than 150,000 years ago. Mt. Teide, with an elevation of 3.7 km (2.3 mi) is within the now-collapsed Cañada Caldera.[13,14] The landslides have produced large linear valleys, the largest of which is Orotova Valley, the site of the city of Puerto de la Cruz, which is a major coastal tourist destination and retirement center for Europeans. The youngest valley is Icod (Figure 7.25). The total volume of the slide deposits exceeds 1,000 km³ (240 mi³). The offshore seafloor north of Tenerife is covered by 5,500 km² (2,150 mi²) of landslide deposits, an area of more than twice the land surface of the island.[15]

There is concern that a future landslide from Tenerife or La Palma, another of the Canary Islands, would cause a huge tsunami known as a megatsunami. Waves with heights exceeding 100 m are possible. They would cause catastrophic damage to the eastern coast of the United States, including the city of New York. The possibility of future huge landslides is unknown, but their occurrence cannot be ruled out.

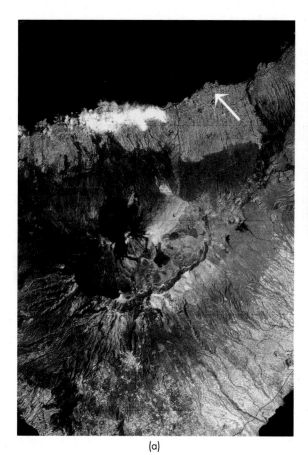

(a)

(b)

Figure 7.25 **Giant landslide on volcano** (a) Part of the island of Tenerife, Canary Islands. Labeled are the collapsed Cañada Caldera and Mt. Teide (elevation 3.7 km, or 12,100 ft). *(Espagna Instituto Geografico Nacional)* In both images, the arrow points to the landslide. (b) Landslide in the valley of Orotova. The city within the slide is Puerto de la Cruz. *(Jose Barea/Espagna Instituto Geografico Nacional)*

7.7 Two Case Histories

Mt. Pinatubo

On June 15 and 16, 1991, the second largest volcanic eruption of the twentieth century occurred at Mt. Pinatubo on Luzon Island in the Philippines (Figure 7.26). The combined effects of ash fall, debris flows, mudflows, and a typhoon resulted in the deaths of about 300 people. Most deaths were due to the collapse of buildings as heavy, wet volcanic ash—resulting from the simultaneous arrival of Typhoon Yunya—accumulated on roofs to thicknesses of 30 cm (12 in.) as far as 40 km (25 mi) from the volcano.[2] Evacuation of 250,000 people from villages and a U.S. military base within a radius of 30 km (19 mi) from the summit saved thousands of lives.[4] Teams of scientists educated local people and authorities by talking to them and showing them videos of volcanic eruptions and their potential hazards. Their efforts convinced local officials to order the evacuations and the people who were at risk to comply.[16]

The tremendous explosions at Pinatubo sent a cloud of ash 400 km (250 mi) wide to elevations of 34 km (21 mi).[4] As with similar past events of this magnitude, the aerosol cloud of ash, including sulfur dioxide, remained in the atmosphere for more than a year. The ash particles and sulfur dioxide scattered incoming sunlight and slightly cooled the global climate during the year following the eruptions.[2,16]

The 1991 success of saving lives during the Mt. Pinatubo eruptions stand in stark contrast to the 1985 eruption of Nevado del Ruiz (see Chapter 5). In Armero a volcanic hazard became a catastrophe that resulted in part from a series of human errors. It was not a huge eruption or a case of bad luck. Rather, it resulted because the science of the hazard was not effectively communicated to people and officials in the area. Nevado del Ruiz was a tragedy that could have and should have been avoided.

Figure 7.26 Evacuation
Thousands of people were evacuated, including more than a thousand from U.S. Clark Air Base, during this large ash eruption and explosion of Mt. Pinatubo in the Philippines in June 1991. *(Carlo Cortes/Reuters/ CORBIS)*

Mount St. Helens

The May 18, 1980, eruption of Mount St. Helens in the southwestern corner of Washington (Figure 7.27) exemplifies the many types of volcanic events expected from a Cascade volcano. The eruption, like many natural events, was unique and complex, making generalizations somewhat difficult. Nevertheless, we have learned a great deal from Mount St. Helens, and the story is not yet complete.

Mount St. Helens awoke in March 1980, after 120 years of dormancy, with seismic activity and small explosions as groundwater came into contact with hot rock. By May 1 a prominent bulge on the northern flank of the mountain could be clearly observed and grew at a rate of about 1.5 m per day (5 ft per day) (Figure 7.28a). At 8:32 A.M. on May 18, 1980, a magnitude 5.1 earthquake that was registered on the volcano triggered a large landslide/debris avalanche (approximately 2.3 km,[3] or 0.6 mi[3]), which involved the entire bulge area (Figure 7.28b). The avalanche shot down the north flank of the mountain, displacing water in nearby Spirit Lake; struck and overrode a ridge 8 km (5 mi) to the north; and then made an abrupt turn, moving 18 km (11 mi) down the Toutle River.

Seconds after the failure of the bulge, Mount St. Helens erupted with a lateral blast directly from the area that the bulge had occupied (Figure 7.28c). The blast moved at speeds up to 1,000 km per hour (621 mi per hour) to distances of nearly 30 km (19 mi) from its source. The blast devastated an area of about 600 km[2] (232 mi[2])[17,18] The areas of the debris avalanche, blasted-down timber, scorched timber, pyroclastic flows, and mudflows are shown on Figure 7.29a.

About one hour after the lateral blast, a large vertical cloud had risen quickly to an altitude of approximately 19 km (12 mi) (Figure 7.28d). Eruption of the vertical column continued for more than 9 hours, and large volumes of volcanic ash fell on a wide area of Washington, northern Idaho, and western and central Montana. During the 9 hours of eruption several ash flows swept down the northern slope of the volcano. The total amount of volcanic ash ejected was about 1 km[3] (0.24 mi[3]),

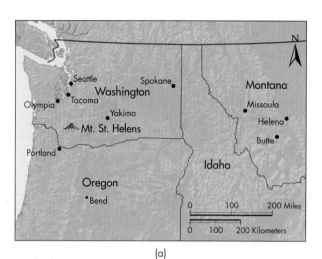

(a)

Figure 7.27 Mount St. Helens before and after (a) Location of Mount St. Helens. (b) Mount St. Helens before and (c) after the May 18, 1980, eruption. As a result of the eruption, much of the northern side of the volcano was blown away, and the altitude of the summit was reduced by approximately 450 m (1,476 ft). *(Photo [b] by Bruce Spainhower. Photo [c] by Harry Glicken)*

(b)

(c)

and a large cloud of ash moved over the United States, reaching as far east as New England (Figure 7.29b). The entire northern slope of the volcano, which is the upper part of the north fork of the Toutle River basin, was devastated. Forested slopes were transformed into a gray, hummocky, or hilly, landscape consisting of volcanic ash, rocks, blocks of melting glacial ice, narrow gullies, and hot steaming pits (Figure 7.30).[18]

The first of several mudflows consisted of a mixture of water, volcanic ash, rock, and organic debris, such as logs, and occurred minutes after the start of the eruption. The flows and accompanying floods raced down the valleys of the north and south forks of the Toutle River at estimated speeds of 29 to 55 km per hour (18 to 35 mi per hour), threatening the lives of people camped along the river.[18]

On the morning of May 18, 1980, two young people on a fishing trip on the Toutle River were sleeping about 36 km (22 mi) downstream from Spirit Lake. They were awakened by a loud rumbling noise from the river, which was covered by felled trees. They attempted to run to their car, but water from the rising river poured over the road, preventing their escape. A mass of mud then crashed through the forest toward the car, and the couple climbed onto the car roof to escape the mud. They were safe only momentarily, however, as the mud pushed the vehicle over the bank and into the river. Leaping off the roof, they fell into the river, which was by now a rolling mass of mud, logs, collapsed train trestles, and other debris. The water also was increasing in temperature. One of the young people got trapped between logs and disappeared several times beneath the flow but was lucky enough to emerge again. The two were carried downstream for approximately 1.5 km (0.9 mi) before another family of campers spotted and rescued them.

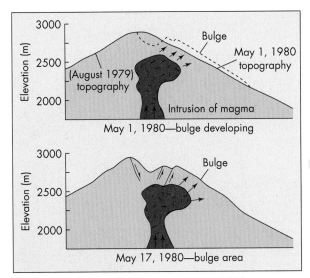

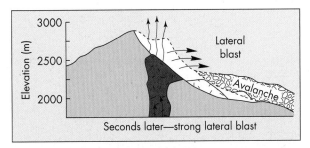

(a) Before eruption May 1 to 17, 1980

Seconds later—strong lateral blast

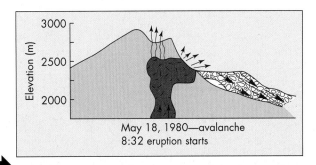

May 18, 1980—avalanche
8:32 eruption starts

(b) Eruption starts May 18, 1980

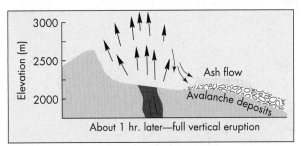

About 1 hr. later—full vertical eruption

(c) Seconds after eruption starts

Figure 7.28 Mount St. Helens erupts Diagrams and
photographs showing the sequence of events for the May 18,
1980, eruption of Mount St. Helens. *(Photographs [a], [b], and
[c] © 1980 by Keith Ronnholm, the Geophysics Program, University of
Washington, Seattle. Photograph [d] from Roger Werth/Woodfin Camp
and Associates. Drawings inspired from lecture by James Moore,
U.S. Geological Survey)*

(d) About an hour after eruption starts

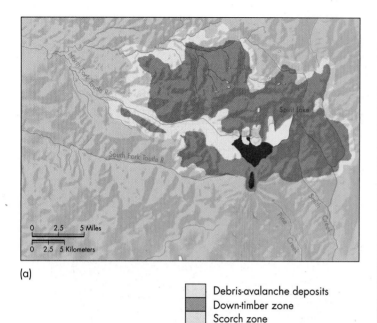

(a)

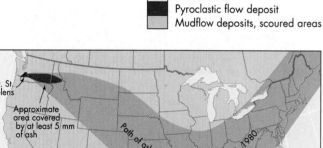

Debris-avalanche deposits
Down-timber zone
Scorch zone
Pyroclastic flow deposit
Mudflow deposits, scoured areas

(b)

Figure 7.29 **Debris avalanche and ash cloud** (a) Mount St. Helens debris-avalanche deposits, zones of timber blown down or scorched, mudflows, and pyroclastic flow deposits associated with the May 18, 1980, eruption. (b) Path of the ash cloud from the 1980 eruption. *(Data from various U.S. Geological Survey publications)*

Figure 7.30 **Barren landscape produced by eruption** The desolate, barren landscape shown here was produced by the May 18, 1980, eruption of Mount St. Helens. The debris avalanche/debris flow that moved down the Toutle River valley is shown here from the center left of the photograph to the lower right-hand corner. The entire valley is full of debris. The surface of the deposit is hummocky, characterized by a scattering of large blocks of volcanic debris. *(John S. Shelton)*

When the volcano could be viewed again after the eruption, its maximum altitude had been reduced by about 450 m (1,476 ft): The original symmetrical mountain was now a huge, steep-walled amphitheater facing northward, and Spirit Lake (lower Figure 7.27b) was filled by deposits (Figure 7.27c). The debris avalanche, lateral blast, pyroclastic flows, and mudflows devastated an area of nearly 400 km^2 (154 mi^2), killing 54 people. More than 100 homes were destroyed by the flooding, and approximately 800 million board-feet of timber were flattened by the blast (Figure 7.29a). A board-foot is a board of 1 ft^2 that is 1 in. thick. The total damage was estimated to exceed $1 billion.

After the catastrophic eruption of Mount St. Helens, an extensive program was established to monitor volcanic activity. The construction by lava flows of a dome within the crater produced by the May 18, 1980, eruption is carefully monitored. During the first 3 years following the eruption, at least 11 smaller eruptions

(a)   (b)

Figure 7.31 **The years of recovery** (a) The eruption of May 18, 1980, of Mount St. Helens produced a barren landscape. *(John S. Shelton)* (b) It is recovering, as illustrated by the flowering lupine 10 years later in July 1990. *(Gary Braasch/Woodfin Camp and Associates)*

occurred. These smaller eruptions contributed to the building of the lava dome to a height of approximately 250 m (820 ft). In each of these eruptions, lava was extruded near the top of the dome and slowly flowed toward the base. Monitoring the growth of the dome and the deformation of the crater floor, along with such other techniques as monitoring the gases emitted, was useful in predicting eruptions. Each event added a few million cubic meters to the size of the dome.[19] By 2000, 20 years after the main eruption, life had returned to the mountain, and its surrounding area was in many places green once more (Figure 7.31).[17] However, the hummocky landscape from the landslide deposits is still prominent, a reminder of the catastrophic event of 1980. Mount St. Helens has a visitors' center and is now a tourist attraction that has attracted over 1 million visitors.[20]

7.8 Forecasting Volcanic Activity

A "forecast" for a volcanic eruption is a probabilistic statement concerning the time, place, and character of an eruption before it occurs. It is analogous to forecasting the weather and is not as precise a statement as a prediction.[4] Forecasting volcanic eruptions is a major component of the goal to reduce volcanic hazards.

It is unlikely that we will be able to forecast the majority of volcanic activity accurately in the near future, but valuable information is being gathered about phenomena that occur before eruptions. One problem is that most forecasting techniques require experience with actual eruptions before the mechanism is understood. Thus, we are better able to predict eruptions in the Hawaiian Islands then elsewhere because we have had so much experience there.

The methods of forecasting volcanic eruptions include

- Monitoring of seismic activity
- Monitoring of thermal, magnetic, and hydrologic conditions
- Topographic monitoring of tilting or swelling of the volcano
- Monitoring of volcanic gas emissions
- Studying the geologic history of a particular volcano or volcanic center[3,16]

Seismic Activity

Our experience with volcanoes such as Mount St. Helens and those on the big island of Hawaii suggests that earthquakes often provide the earliest warning of an impending volcanic eruption. In the case of Mount St. Helens, earthquake activity started in mid-March before the eruption in May. Activity began suddenly, with near-continuous shallow seismicity. Unfortunately, there was no increase in earthquakes immediately before the May 18 event. In Hawaii, earthquakes have been used to monitor the movement of magma as it approaches the surface.

Several months before the 1991 Mt. Pinatubo eruptions, small steam explosions and earthquakes began.[3] Mt. Pinatubo (present elevation 1,700 m, or 5,578 ft) was an eroded ridge, and, as a result, did not have the classic shape of a volcano. Furthermore, it had not erupted in 500 years; most of the people living near it did not even know it was a volcano! Scientists began monitoring earthquake activity and studying past volcanic activity, which was determined to be explosive. Earthquakes increased in number and magnitude before the catastrophic eruption, migrating from deep beneath the volcano to shallow depths beneath the summit.[4]

Thermal, Magnetic, and Hydrologic Monitoring

Monitoring of volcanoes is based on the fact that, before an eruption, a large volume of magma moves up into some sort of holding reservoir beneath the volcano. The hot material changes the local magnetic, thermal, hydrologic, and geochemical conditions. As the surrounding rocks heat, the rise in temperature of the surficial rock may be detected by remote sensing or infrared aerial photography. Increased heat may melt snowfields or glaciers; thus, periodic remote sensing of a volcanic chain may detect new hot points that could indicate potential volcanic activity. This method was used with some success at Mount St. Helens before the main eruption on May 18, 1980.

When older volcanic rocks are heated by new magma, magnetic properties, originally imprinted when the rocks cooled and crystallized, may change. These changes can be detailed by ground or aerial monitoring of the magnetic properties of the rocks that the volcano is composed of.[3,21]

Topographic Monitoring

Monitoring topographic changes and seismic behavior of volcanoes has been useful in forecasting some volcanic eruptions. The Hawaiian volcanoes, especially Kilauea, have supplied most of the data. The summit of Kilauea tilts and swells before an eruption and subsides during the actual outbreak (Figure 7.32). Kilauea also undergoes earthquake swarms that reflect moving subsurface magma and an imminent eruption. The tilting of the summit in conjunction with the earthquake swarms was used to predict a volcanic eruption in the vicinity of the farming community of Kapoho on the flank of the volcano, 45 km (28 mi) from the summit. As a result, the inhabitants were evacuated before the event, in which lava overran and eventually destroyed most of the village.[22] Because of the characteristic swelling and earthquake activity before eruptions, scientists expect the Hawaiian volcanoes to continue to be more predictable than others. Monitoring of ground movements such as tilting, swelling, opening of cracks, or changes in the water level of lakes on or near a volcano has become a useful tool for recognizing change that might indicate a coming eruption.[3] Today satellite-based radar and a network of Global Positioning System (GPS) receivers can be used to monitor change in volcanoes, including surface deformation, without sending people into a hazardous area.[17]

Monitoring Volcanic Gas Emissions

The primary objective of monitoring volcanic gas emissions is to recognize changes in the chemical composition of the gases. Changes in both gas composition— that is, the relative amounts of gases such as steam, carbon dioxide, and sulfur dioxide—and gas emission rates are thought to be correlated with changes in subsurface volcanic processes. These factors may indicate movement of magma

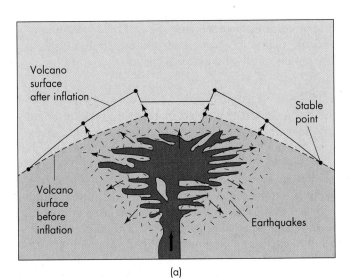

(a)

Figure 7.32 **Inflation and tilting before eruption** (a) Idealized diagram of Kilauea, illustrating inflation and surface tilting accompanied by earthquakes as magma moves up. *(U.S. Geological Survey Circular 1073, 1992)* (b) The actual data graph, showing the east–west component and the north–south component of ground tilt recorded from 1964 to 1966 on Kilauea Volcano, Hawaii. Notice the slow change in ground tilt before eruption and rapid subsidence during eruption. *(From Fiske, R. S., and Koyanagi, R. Y. 1968. U.S. Geological Survey Professional Paper 607)*

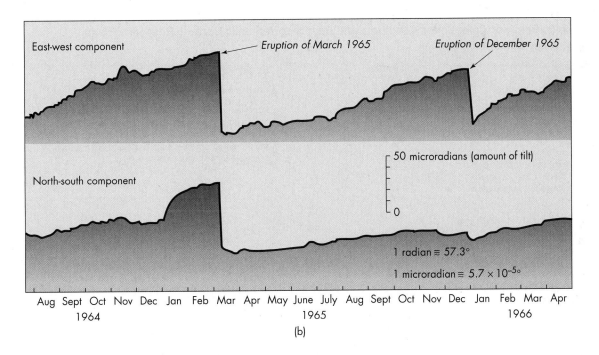

(b)

toward the surface. This technique was useful in studying eruptions at Mount St. Helens and Mt. Pinatubo. Two weeks before the explosive eruptions at Mt. Pinatubo, the emissions of sulfur dioxide increased by a factor of about 10.[3]

Geologic History

Understanding the geologic history of a volcano or volcanic system is useful in predicting the types of eruptions likely to occur in the future. The primary tool used to establish the geologic history of a volcano is geologic mapping of volcanic rocks and deposits. Attempts are made to date lava flows and pyroclastic activity to determine when they occurred. These are the primary data necessary to produce maps depicting volcanic hazards at a particular site. Geologic mapping, in conjunction with the dating of volcanic deposits at Kilauea, Hawaii, led to the discovery that more than 90 percent of the land surface of the volcano has been covered by lava in only the past 1,500 years. The town of Kalapana, destroyed by lava flows in 1990, might never have been built if this information had been known before development, because the risk might have been thought too great. The real value of geologic mapping and dating of volcanic events is that they allow development of hazard maps to assist in land-use planning and preparation for future eruptions.[3] Such maps are now available for a number of volcanoes around the world.

Volcanic Alert or Warning

At what point should the public be alerted or warned that a volcanic eruption may occur? This is an important question being addressed by volcanologists. At present there is no standard code, but one being used with various modifications has been developed by the U.S. Geological Survey. The system is color coded by condition; each color—green, yellow, orange, and red—denotes increasing concern (Table 7.3). This table is specifically for Long Valley caldera in California. Similar systems have been or are being developed for other volcanic areas, including Alaska and the Cascade Mountains of the Pacific Northwest. The color-coded system is a good start; however, the hard questions remain: When should evacuation begin? When is it safe for people to return? Evacuation is definitely necessary before condition red, but when during conditions yellow or orange should it begin?

TABLE 7.3 Geologic Behavior, Color-Coded Condition, and Response: Volcanic Hazards Response Plan; Long Valley Caldera, California

Geologic Behavior	Condition	Response
Typical behavior since 1980 includes:	*Green*	*Routine monitoring* plus
Background: As many as 10 to 20 small earthquakes with magnitudes (M) less than 3 (M < 3) per day and uplift of the resurgent dome at an average rate of about 1 inch (2.54 cm) per year	No immediate risk	*Information calls* to U.S. Geological Survey personnel, town, county, state, and federal agencies regarding locally felt earthquakes and notable changes in other monitored parameters such as ground deformation, fumarole activity, gas emissions, etc.
Weak unrest (likely to occur several times a year): For example, increased number and (or) strength of small earthquakes or a single felt magnitude M > 3 earthquake		
Moderate unrest (likely to occur about once a year): For example, a M > 4 earthquake or more than 300 earthquakes in a day		
Intense unrest (may occur about once per decade): For example, a swarm with at least one magnitude 5 earthquake and (or) evidence of magma movement at depth as indicated by an increased rate in ground deformation	*Yellow* Watch	*Intensified monitoring*: Set up emergency field headquarters at Long Valley caldera. Initial *Watch* message sent by U.S. Geological Servey to California officials, who promptly inform local authorities. (Includes above information calls.)
Eruption likely within hours or days (may occur every few hundred years). Stong evidence of magma movement at shallow depth	*Orange* Warning	*Geologic Hazard Warning* issued by U.S. Geological Survey to governors of California and Nevada and others who inform the public. (Includes *Watch* response.)
Eruption under way (may occur every few hundred years)	*Red* Alert	*Sustained on-site monitoring and communication.* Maintain intensive monitoring and continuously keep civil authorities informed on progress of eruption and likely future developments.

Notes:

Condition at a given time is keyed to successively more intense levels of geologic unrest, detected by the monitoring network.

Response for a given condition includes the responses specified for all lower levels.

Estimated Recurrence intervals for a given condition are based on the recurrence of unrest episodes in Long Valley caldera since 1980, the record of magnitude 4 or greater earthquakes in the region since the 1930s, and the geologic record of volcanic eruptions in the region over the last 50,000 years.

Expiration of Watch, Warning, and Alert: This table shows the length of time (in days) a given condition remains in effect after the level of unrest drops below the threshold that initially triggered the condition.

Condition	Expires After	Subsequent Condition*
Watch	14 days	Green (no immediate risk)
Warning (eruption likely)	14 days	Watch
Alert (eruption in progress)	1 day	Warning

*By the level of unrest at the time the previously established condition expires. In the case of the end of an episode of eruptive activity (Alert), a Warning will remain in effect for at least 14 days, depending on the level of ongoing unrest.

Source: Modified from U.S. Geological Survey, 1997.

7.9 Adjustment to and Perception of the Volcanic Hazard

Apart from the psychological adjustment to losses, the primary human adjustment to volcanic activity is evacuation. A notable exception is the previously discussed decision of the people on the island of Heimaey to successfully fight an eruption by hydraulically chilling the lava. Information concerning how people perceive the volcanic hazard is limited. People live near volcanoes for a variety of reasons, including: (1) They were born there, and, in the case of some islands such as the Canary Islands, all land on the island is volcanic; (2) the land is fertile and good for farming; (3) people are optimistic and believe an eruption is unlikely; and (4) they do not have any choice as to where they live—for example, they may be limited by economics. One study of perception evaluated volcanic activity in Hawaii and found that a person's age and length of residence near a volcanic hazard are significant factors in a person's knowledge of volcanic activity and possible adjustments.[23] One reason the evacuation of 60,000 people before the 1991 eruption of Mt. Pinatubo was successful was that the government had provided a program to educate people concerning the dangers from violent ash eruptions with debris flows. A video depicting these events was widely shown in the area before the eruption, and it helped convince local officials and residents that they faced a real and immediate threat.[4]

The science of volcanoes is becoming well known. However, good science is not sufficient. Probably the greatest payoff in terms of reducing volcanic hazards in the future will come from an increased understanding of human and societal issues that come up during an emerging potential **volcanic crisis**—that is, a situation in which the science suggests that a volcanic eruption is likely in the near future. The development of improved communication between scientists, emergency managers, educators, media, and private citizens is particularly important. The goal is to prevent a volcanic crisis from becoming a volcanic disaster or catastrophe.[17]

SUMMARY

Volcanic eruptions often occur in sparsely populated areas, but they have a high potential to produce catastrophes when they occur near populated areas. Volcanic activity is directly related to plate tectonics. Most volcanoes are located at plate boundaries, where magma is produced as spreading or sinking lithospheric plates interact with other Earth material. The "ring of fire" is a region surrounding most of the Pacific Ocean (Pacific plate) that contains about two-thirds of the world's volcanoes.

Lava is magma that has been extruded from a volcano; the activity of different types of volcano is partly determined by the differing silica content and viscosity of their lavas. Shield volcanoes occur at mid-ocean ridges, such as Iceland, and over mid-plate hot spots, such as the Hawaiian Islands. Their common rock type is basalt, and they are characterized by nonexplosive lava flows. Composite volcanoes occur at subduction zones, particularly around the Pacific Rim, for example, in the Pacific Northwest of the United States. They are composed largely of andesite rock and are characterized by explosive eruptions and lava flows. Volcanic domes occur inland of subduction zones; an example of a volcanic dome is Mt. Lassen, California. They are composed largely of rhyolite rock and are highly explosive.

Features of volcanoes include vents, craters, and calderas. Other features of volcanic areas are hot springs and geysers. Giant caldera-forming eruptions are violent, but rare, geologic events. After their explosive beginning, they often resurge and may present a volcanic hazard for a million years or longer. Recent uplift and earthquakes at the Long Valley caldera in California are reminders of the potential hazard.

Primary effects of volcanic activity include lava flows, pyroclastic hazards, and, occasionally, the emission of poisonous gases. Hydraulic chilling and the construction of walls have been used in attempts to control lava flows. These methods have had mixed success and require further evaluation. Pyroclastic hazards include volcanic ash falls, which may cover large areas with carpets of ash; ash flows, or hot avalanches, which move as fast as 100 km per hour down the side of a volcano; and lateral blasts, which can be very destructive. Secondary effects of volcanic activity include debris flows and mudflows, generated when melting snow and ice or precipitation mixes with volcanic ash. These flows can devastate an area many kilometers from the volcano. All of these effects have occurred in the recent history of the Cascade Range of the Pacific Northwest, and there is no reason to believe that they will not occur there in the future.

Sufficient monitoring of seismic activity; thermal, magnetic, and hydrologic properties; and topographic changes, combined with knowledge of the recent geologic history of volcanoes, may eventually result in reliable forecasting of volcanic activity. Forecasts of eruptions have been successful, particularly for Hawaiian volcanoes and Mt. Pinatubo in the Philippines. On a worldwide scale, however, it is unlikely that we will be able to accurately forecast most volcanic activity in the near future.

Apart from psychological adjustment to losses, the primary human adjustment to volcanic activity is evacuation.

Perception of the volcanic hazard is apparently a function of age and length of residency near the hazard. Some people have little choice but to live near a volcano. Education plays an important role in informing people about the hazards of volcanoes.

The goal of reducing the volcanic hazard is focusing on human and societal issues through communication, with the objective of preventing a volcanic crisis from becoming a volcanic disaster or catastrophe.

Revisiting Fundamental Concepts

Human Population Growth

When humans live close to active volcanoes, the consequences can be catastrophic. Volcanic soil is generally rich, and population centers often develop in broad agricultural-based valleys downslope from volcanoes. In addition, large urban centers including Seattle, Washington, and Mexico City are uncomfortably close to active volcanoes. In some cases, such as island cities like Puerto de la Cruz on Tenerife, Canary Islands, the entire island is part of a large volcanic center. There is no alternative for the island people other than to live in a potentially hazardous area. As population centers and cities near volcanoes grow and attract more people in the next 100 years, the volcanic hazard will become more serious to millions of additional people.

Earth as a System

Volcanoes are part of the internal system of Earth. Magma is generated at plate boundaries and hot spots that change position and progress only slowly over time periods of interest to people. However, when a volcano becomes active, there are important changes in the earthquakes it produces and the gases it releases as magma moves toward the surface. These changes alert us and help predict eruptions.

Hazardous Earth Processes, Risk Assessment, and Perceptions

Volcanoes are one of our most violent natural hazards. Yet people living in the shadow of a volcano that has been dormant for a few hundred years may scarcely know it is there, and the potential threat of future eruptions may be unknown or not perceived as a threat. In areas with more frequent eruptions, such as Hawaii, people generally are aware of the risk, having lived with daily news reports of ongoing eruptions that may last for decades. In other words, if eruptions are frequent, people's perception of the hazard increases.

Scientific Knowledge and Values

We have learned a great deal about how volcanoes function, the rocks they produce, and the hazards they present. However, values sometimes clash. For example, some people in the Hawaiian Islands view the volcanoes in a spiritual and religious light. Volcanic gases, magma, and lands are the breath and life of the volcanic goddess Pele, not to be used to produce geothermal energy or cover over with urban development. Removing bits of Pele's volcanic rocks from the island is believed to bring bad luck to the thief.

Key Terms

ash fall (p. 233)

ash flow (p. 233)

caldera eruption (p. 229)

cinder cone (p. 222)

composite volcano (p. 221)

debris flow/mudflow/lahar (p. 235)

lateral blast (p. 233)

lava (p. 219)

lava flow (p. 230)

magma (p. 217)

pyroclastic activity (p. 233)

shield volcano (p. 219)

volcanic crisis (p. 247)

volcanic dome (p. 222)

Review Questions

1. From a hazards perspective why it important to know the type of a volcano?

2. What is viscosity, and what determines it?

3. List the major types of volcano and the type of magma associated with each.

4. List the major types of volcano and their eruption style. Why do they erupt the way they do?

5. What is the relationship between plate tectonics and volcanoes?

6. How do lava tubes help move magma far from the erupting vents?

7. What is the relationship between the Hawaiian Islands and the hot spot below the big island of Hawaii?

8. What is the origin of a geyser?

9. Why are caldera eruptions so dangerous?

10. List the primary and secondary effects of volcanic eruptions.

11. What are some of the methods that have been attempted to control lava flows?

12. Differentiate between ash falls, lateral blasts, and ash flows.

13. What are the major gases emitted in a volcanic eruption, and which are most hazardous?

14. How are volcanic eruptions able to produce gigantic mudflows?

15. What are some of the possible methods of forcasting volcanic eruption?

Critical Thinking Questions

1. While looking through some old boxes in your grandparents' home, you find a sample of volcanic rock collected by your great-grandfather. No one knows where it was collected. You take it to school, and your geology professor tells you that it is a sample of andesite. What might you tell your grandparents about the type of volcano from which it probably came, its geologic environment, and the type of volcanic activity that likely produced it?

2. In our discussion of adjustment and perception to volcanic hazards, we established that people's perceptions and what they will do in case of an eruption are associated with both their proximity to the hazard and their knowledge of volcanic processes and necessary adjustments. With this association in mind, develop a public relations program that could alert people to a potential volcanic hazard. Keep in mind that the tragedy associated with the eruption of Nevado del Ruiz (see Chapter 5) was in part due to political and economic factors that influenced the apathetic attitude toward the hazard map prepared for that area. Some people were afraid that the hazard map would result in lower property values in some areas.

EIGHT

Rivers and Flooding

Learning Objectives

Water covers about 70 percent of Earth's surface and is critical to supporting life on the planet. However, water can also cause a significant hazard to human life and property in certain situations, such as a flood. Flooding is the most universally experienced natural hazard. Flooding in the United States is the most common and costly natural hazard we face. Floodwaters have killed more than 10,000 people in the United States since 1900, and property damage from flooding exceeds $5 billion a year. Flooding is a natural process that will remain a major hazard as long as people choose to live and work in flood-prone areas. In this chapter we focus on the following learning objectives:

- Understand basic river processes

- Understand the nature and extent of the flood hazard

- Understand the effects of urbanization on flooding in small drainage basins

- Know the major adjustments to flooding and which are environmentally preferable

- Know the potential adverse environmental effects of channelization and the benefits of channel restoration

Confluence of the Mississippi and Ohio Rivers
(Alex S. MacLean/Peter Arnold, Inc.)

CASE HISTORY | Mississippi River Flooding, 1973 and 1993

In 1973, spring flooding of the Mississippi River caused the evacuation of tens of thousands of people as thousands of square kilometers of farmland were inundated throughout the Mississippi River Valley. Fortunately, there were few deaths, but the flooding resulted in approximately $1.2 billion in property damage.[1] The 1973 floods occurred despite a tremendous investment in upstream flood-control dams on the Missouri River. Reservoirs behind these dams inundated some of the most valuable farmland in the Dakotas, and despite these structures, the flood near St. Louis was record breaking.[2] Impressive as this flood was at the time, it did not compare either in magnitude or in the suffering it caused with the flooding that occurred 20 years later.

During the summer of 1993 the Mississippi River and its tributaries experienced one of the largest floods of the century. There was more water than during the 1973 flood, and the recurrence interval exceeded 100 years. The floods lasted from late June to early August and caused 50 deaths and more than $10 billion in property damages. In all, about 55,000 km[2] (21,236 mi[2]), including numerous towns and farmlands, were inundated with water.[3,4]

The 1993 floods resulted from a major climatic anomaly that covered the upper Midwest and north-central Great Plains, precisely the area that drains into the Mississippi and lower Missouri River systems.[5] The trouble began with a wet autumn and a heavy spring snowmelt that saturated the ground in the upper Mississippi River basin. Then, early in June, a high-pressure center became stationary on the East Coast, drawing moist, unstable air into the upper Mississippi River drainage basin. This condition kept storm systems in

the Midwest from moving east. At the same time, air moving in over the Rocky Mountains initiated unusually heavy rainstorms.[5] The summer of 1993 was the wettest on record for Illinois, Iowa, and Minnesota. For example, Cedar Rapids, Iowa, received about 90 cm (35 in.) of rain from April through July—the equivalent of a normal year's rainfall in just 4 months![4] Intense precipitation falling on saturated ground led to a tremendous amount of runoff and unusually large floods during the summer. The floodwaters were high for a prolonged time, putting pressure on the flood defenses of the Mississippi River, particularly **levees,** which are earth embankments constructed parallel to the river to contain floodwaters and reduce flooding (Figure 8.1). Levees are constructed on the flat land adjacent to the river known as the floodplain.

Before construction of the levees, the Mississippi's floodplain, flat land adjacent to the river that periodically floods, was much wider and contained extensive wetlands. Since the first levees were built in 1718, approximately 60 percent of the wetland in Wisconsin, Illinois, Iowa, Missouri, and Minnesota—all hard hit by the flooding in 1993—have been lost. In some locations, such as St. Louis, Missouri, levees give way to floodwalls designed to protect the city against high-magnitude floods. Examination of Figure 8.2, a satellite image from mid-July 1993, shows that the river is narrow at St. Louis, where it is contained by the floodwalls, and broad upstream near Alton, where extensive flooding occurred. The floodwalls produce a bottleneck because water must pass through a narrow channel between the walls; the floodwaters get backed up waiting to get through. These

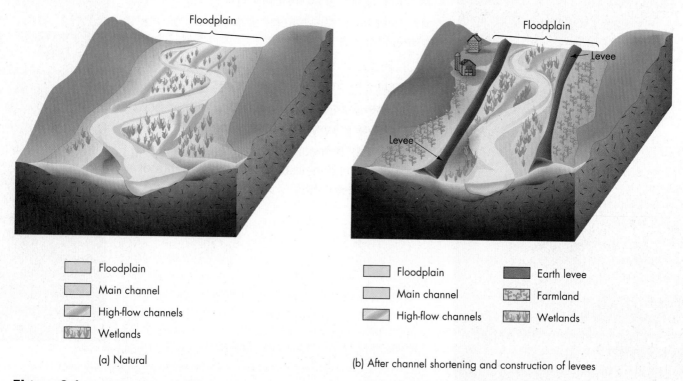

Floodplain

	Floodplain
	Main channel
	High-flow channels
	Wetlands

(a) Natural

	Floodplain			Earth levee
	Main channel			Farmland
	High-flow channels			Wetlands

(b) After channel shortening and construction of levees

Figure 8.1 **Floodplain with and without levees** Idealized diagram of (a) natural floodplain (flat land adjacent to the river produced by the river) with wetlands. (b) Floodplain after channel is shortened and levees are constructed. Land behind levees is farmed, and wetlands are generally confined between the levees.

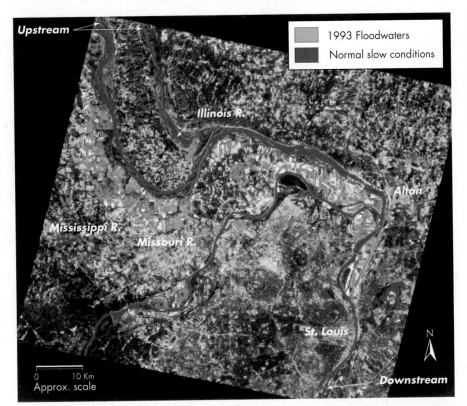

1993 Floodwaters

Normal slow conditions

Figure 8.2 Mississippi River flood of 1993 This image shows the extent of flooding from the 1993 Mississippi River floods. In the lower right-hand corner the river becomes narrow where it flows by the city of St. Louis, Missouri (orange area in lower right corner). The river is narrow here because flow is constricted by a series of floodwalls constructed to protect the city. Notice the extensive flooding upstream of St. Louis, Missouri. The city with its floodworks is a real bottleneck to the flow of water. The town of Alton, Illinois, is the first orange area upstream from St. Louis. This city has a notorious history of flooding. *(ITD-SRSC/RSI/ SPOT Image, copyright ESA/CNES/ Sygma)*

floodwaters contributed to the 1993 flooding upstream of St. Louis (Figure 8.2).

Despite the high walls constructed to prevent flooding, the rising flood peak came to within about 0.6 m (2 ft) of overtopping the floodwalls at St. Louis. Failure of levees downstream from St. Louis partially relieved the pressure, possibly saving the city from flooding. Levee failures (Figure 8.3) were very common during the flood event.[4,5] In fact, almost 80 percent of the private levees, that is, levees built by farmers and homeowners, along the Mississippi River and its tributaries failed.[4] However, most of the levees built by the federal government survived the flooding and undoubtedly saved lives and property. Unfortunately, there is no uniform building code for the levees, so some areas have levees that are higher or lower than others. Failures occurred as a result of overtopping and breaching, or rupturing, resulting in massive flooding of farmlands and towns (Figure 8.4).[4]

One of the lessons learned from the 1993 floods is that construction of levees provides a false sense of security. It is difficult to design levees to withstand extremely high-magnitude floods for a long period of time. Furthermore, the loss of wetlands allows for less floodplain space to "soak up" the floodwaters.[6] The 1993 floods caused extensive damage and loss of property; in 1995, floodwaters of the Mississippi River system inundated floodplain communities once again. Several communities along the river are rethinking strategies concerning the flood hazard and are moving to higher ground! Of course, this is exactly the adjustment that is appropriate.

Although flooding may be caused by several processes including coastal flooding from a hurricane, in this chapter we will focus on river flooding. We will discuss flooding from several perspectives, including river processes; effects of land-use changes on flooding; effects of flooding; and how flooding may be minimized.

Figure 8.3 Levee failure Failure of this levee in Illinois during the 1993 floods of the Mississippi River caused flooding in the town of Valmeyer. *(Comstock Images)*

Figure 8.4 **Damaged farmland** Damage to farmlands during the peak of the 1993 flood of the Mississippi River. *(Comstock Images)*

8.1 Rivers: Historical Use

For more than 200 years, Americans have lived and worked on floodplains, enticed to do so by the rich *alluvial* (stream-deposited) soil, abundant water supply, ease of waste disposal, and proximity to the commerce that has developed along the rivers. Of course, building houses, industry, public buildings, and farms on a floodplain invites disaster, but too many floodplain residents have refused to recognize the natural floodway of the river for what it is: part of the natural river system. The **floodplain,** the flat surface adjacent to the river channel that is periodically inundated by floodwater, is in fact produced by the process of flooding (Figures 8.1a, 8.5, and 8.6). If the floodplain and its relation to the river are not recognized, flood control and drainage of wetlands, including floodplains, become prime concerns. It is not an oversimplification to say that as the pioneers moved west they had a rather set procedure for modifying the land: First clear the land by cutting and burning the trees, then modify the natural drainage. From that history came two parallel trends: an accelerating program to control floods, matched by an even greater growth of flood damages. In this chapter, we will consider flooding as a natural aspect of river processes and examine the successes and failures of traditional methods of flood control. We will also discuss river restoration attempts that work with the natural river processes rather than against them.

8.2 Streams and Rivers

Streams and rivers are part of the water, or hydrologic, cycle, and *hydrology* is the study of this cycle. The hydrologic cycle involves the transport of water by evaporation from Earth's surface, primarily from the oceans, to the atmosphere and, via surface and subsurface runoff from the land, back again to the oceans. Some of the water that falls on the land as rain or snow infiltrates soils and rocks; some evaporates; and the rest drains, or runs off, following a course determined by the local topography. This runoff finds its way to streams, which may merge to form a larger stream or a *river*. Streams and rivers differ only in size; that is, streams are

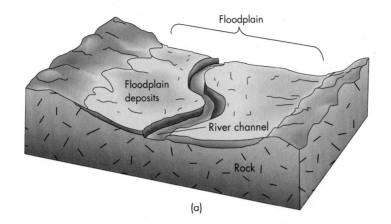

Floodplain

Floodplain deposits

River channel

Rock

(a)

Figure 8.5 Floodplain
(a) Diagram illustrating the location
of a river's floodplain. (b) Floodplain
of the Rio Grande in Colorado.
(Edward A. Keller)

(b)

(a)

(b)

Figure 8.6 **Floodplain inundation from snowmelt** Gaylor Creek, Yosemite National Park, during
spring snowmelt. (a) In the morning water is within the channel. (b) In the afternoon, during daily peak
snowmelt, flow covers the floodplain. *(Edward A. Keller)*

Figure 8.7 **Drainage basin and river profile** Idealized diagram showing (a) drainage basin, (b) longitudinal profile of the Fox River, (c) cross section of valley near headwater, and (d) cross section near base level.

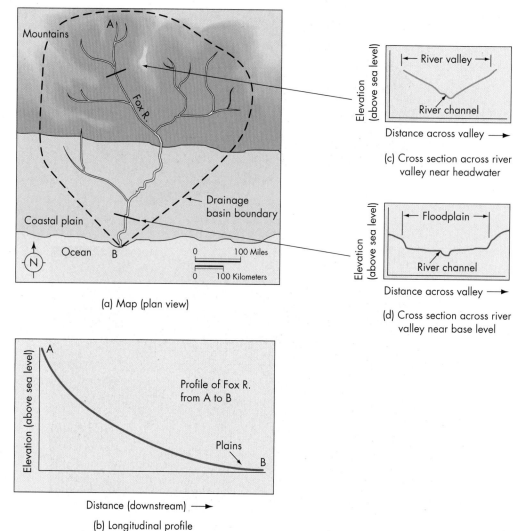

(a) Map (plan view)

(c) Cross section across river valley near headwater

(d) Cross section across river valley near base level

(b) Longitudinal profile

small rivers. However, geologists commonly use the term *stream* for any body of water that flows in a channel. The region drained by a single river or river system is called a **drainage basin,** or *watershed* (Figure 8.7a).

A river's *slope*, or gradient, is the vertical drop of the channel over some horizontal distance. In general, the slope is steepest at higher elevations in the drainage basin and levels off as the stream approaches its *base level.* The base level of a stream is the theoretical lowest level to which a river may erode. Most often, the base level is at sea level, although a river may have a temporary base level such as a lake. Rivers flow downhill to their base level, and a graph of elevation of a river against distance downstream is called the longitudinal profile (Figure 8.7b). A river usually has a steeper-sided and deeper valley at high elevations near its headwaters than closer to its base level, where a wide floodplain may be present (Figure 8.7c, d). At higher elevations, the steeper slope of the river causes deeper erosion of the valley. Increased erosion is due to the higher flow velocity of the river water produced by the steeper channel slopes.

8.3 Sediment in Rivers

The total quantity of sediment carried in a river, called its *total load*, includes the bed load, the suspended load, and the dissolved load. The *bed load* moves by the bouncing, rolling, or skipping of particles along the bottom of the channel. The bed load of most rivers, usually composed of sand and gravel, is a relatively small component, generally accounting for less than 10 percent of the total load. The *suspended load*, composed mainly of silt and clay, is carried above the streambed by the flowing water. The suspended load accounts for nearly 90 percent of the total load and makes rivers look muddy. The *dissolved load* is carried in chemical solution and is derived from chemical weathering of rocks in the drainage basin. The dissolved load may make stream water taste salty if it contains large amounts of sodium and chloride. It may also make the stream water "hard" if the dissolved load contains high concentrations of calcium and magnesium. The most common constituents of the dissolved load are bicarbonate ions (HCO_3^-), sulfate ions (SO_4^{2-}), calcium ions (Ca^{2+}), sodium ions (Na^+), and magnesium ions (Mg^+). As discussed in Chapter 3, an ion is an atom or molecule with a positive or negative charge resulting from a gain or loss of electrons. Typically, the above five ions constitute more than 90 percent of a river's dissolved load. It is the suspended and bed loads of streams that, when deposited in undesirable locations, produce the sediment pollution discussed in Chapter 13.

8.4 River Velocity, Discharge, Erosion, and Sediment Deposition

Rivers are the basic transportation system of that part of the rock cycle involving erosion and deposition of sediments. They are a primary erosion agent in the sculpting of our landscape. The velocity, or speed, of the water in a river varies along its course, affecting both erosion and deposition of sediment.

Discharge (Q) is the volume of water moving by a particular location in a river per unit time. It is reported in cubic meters per second (cms) or cubic feet per second (cfs). Discharge is calculated as

$$Q = W \times D \times V$$

where Q is discharge (cubic meters per second), W is the width of flow in meters, D is depth of flow in meters, and V is mean velocity of flow (meters per second). The equation $Q = W \times D \times V$ is known as the **continuity equation** and is one of the most important relationships in understanding the flow of water in rivers. We assume that if there are no additions or deletions of flow along a given length of river, then discharge is constant. It follows then that if the cross-sectional area of flow ($W \times D$) decreases, then the velocity of the water must increase. You can observe this change with a garden hose. Turn on the water and observe the velocity of the water as it exits the hose. Then put your thumb partly over the end of the hose, reducing the area where the water flows out of the hose, and observe the increase in the velocity. This concept explains why a narrow river channel in a canyon has a higher velocity of flow. It is also the reason that rapids are common in narrow canyons. In general, a faster-flowing river has the ability to erode its banks more than a slower-moving one. Streams that flow from mountains onto plains may form fan-shaped deposits known as *alluvial fans* (Figure 8.8). Rivers flowing into the ocean or some other body of still water may deposit sediments that form a *delta*, a triangular or irregular-shaped landmass extending into the sea or a lake (Figure 8.9). The flood hazard associated with alluvial fans and deltas is different from hazards in a river valley and floodplain environment because rivers entering alluvial fan or delta environments often split into a system of *distributary channels*. That is, the river no longer has only one main channel but has several channels that carry floodwaters to different parts of the fan or delta. Furthermore, these channels characteristically may change position rapidly during floods, producing a flood hazard that is difficult to predict.[7] For example, a large recreational vehicle (RV) park on the delta of the Ventura River in southern California flooded four times in the 1990s. The RV park was constructed across a historically active distributary channel of the Ventura River. However, before the construction of the park, the engineers mapping the potential flood hazards on the site did not recognize that the park was located on a delta. This story emphasizes the importance of studying a river's flooding history as part of flood hazard evaluation (see A Closer Look: History of a River).

Figure 8.8 **Alluvial Fan** Alluvial fan along the western foot of the Black Mountains, Death Valley. Note the road along the base of the fan. The white materials are salt deposits in Death Valley. *(Michael Collier)*

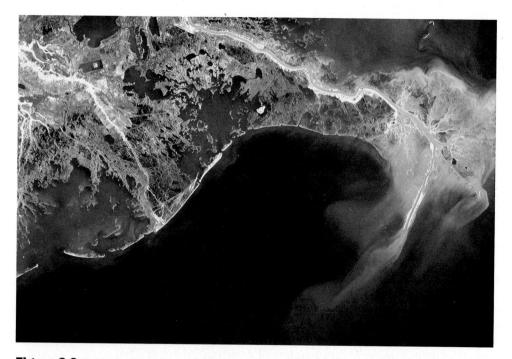

Figure 8.9 **Delta** The delta of the Mississippi River. In this false-color image, vegetation appears red and sediment-laden waters are white or light blue; deeper water with less suspended sediment is a darker blue. The system of distributary channels in the delta in the far right of the photograph looks something like a bird's foot, and, in fact, the Mississippi River delta is an example of a bird's-foot delta. The distributary channels carry sediment out into the Gulf of Mexico, and, because wave action is not strong in the gulf, the river dominates the delta system. Distance across the image is about 180 km (112 mi). Other rivers flow into a more active coastal environment. Such deltas have a relatively straight coastline, rather than bird's foot shaped, and are considered to be wave dominated. Other deltas are between the end points of river domination and wave domination, as for example the Nile, delta with its beautiful triangular shape, with convex shoreline protruding into the Mediterranean Sea. *(LANDSAT image by U.S. Geological Survey/Courtesy of John S. Shelton)*

A CLOSER LOOK | History of a River

In 1905 philosopher George Santayana said, "Those who cannot remember the past are condemned to repeat it." Scholars may debate the age-old question of whether cycles in human history repeat themselves, but the repetitive nature of natural hazards such as floods is undisputed.[8] Better understanding of the historical behavior of a river is therefore valuable in estimating its present and future flood hazards. Consider the February 1992 Ventura River flood in southern California. The flood severely damaged the Ventura Beach Recreational Vehicle (RV) Resort, which had been constructed a few years earlier on an active distributary channel of the Ventura River delta. Although the recurrence interval is approximately 22 years (Figure 8.A), earlier engineering studies suggested that the RV park would not be inundated by flood with a recurrence interval of 100 years. What went wrong?

- Planners did not recognize that the RV park was constructed on a historically active distributary channel of the Ventura River delta. In fact, early reports did not even mention a delta.

- Engineering models that predict flood inundation are inaccurate when evaluating distributary channels on river deltas where extensive channel fill and scour as well as lateral movement of the channel are likely to occur.

- Historical documents such as maps dating back to 1855 and more-recent aerial photographs that showed the channels were not evaluated. Figure 8.B shows that maps rendered from these documents suggest that the distributary channel was in fact present in 1855.[9]

Clearly, the historical behavior of the river was not evaluated as part of the flood-hazard evaluation. If it had been, the site would have been declared unacceptable for development, given that a historically active channel was present. Nevertheless, necessary permits were issued for development of the park, and, in fact, the park was rebuilt after the flood. Before 1992, the distributary channel carried discharges during 1969, 1978, and 1982. After the 1992 flood event, the channel carried floodwaters in the winters of 1993, 1995, and 1998, again flooding the RV park. During the 1992 floods, the discharge increased from less than 25 m^3 per second (883 ft^3 per second) to a peak of 1,322 m^3 per second (46,683 ft^3 per second) in only about 4 hours! This is approximately twice as much as the daily high discharge of the Colorado River through the Grand Canyon in the summer, when it is navigated by river rafters. This is an incredible discharge for a relatively small river with a drainage area of only about 585 km^2 (226 mi^2). The flood occurred during daylight, and one person was killed. If the flood had occurred at night, many more deaths would have been recorded. A warning system that has been developed for the park has, so far, been effective in providing early warning of an impending flood. The park, with or without the RVs and people, is a "sitting duck." Its vulnerability was dramatically illustrated in 1995 and 1998, when winter floods again swept through the park. Although the warning system worked and the park was successfully evacuated, the facility was again severely damaged. There is now a movement afoot to purchase the park and restore the land to a more natural delta environment: a good move!

Figure 8.A **Flooding of California's Ventura Beach RV Resort in February 1992** The RV park was built directly across a historically active distributary channel of the Ventura River delta. The recurrence interval of this flood is approximately 22 years. A similar flood occurred again in 1995. Notice that U.S. Highway 101 along the Pacific Coast is completely closed by the flood event. *(Mark J. Terrell/AP/Wide World Photos)*

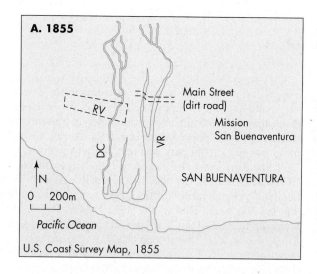

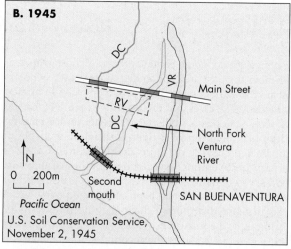

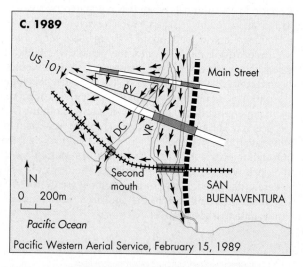

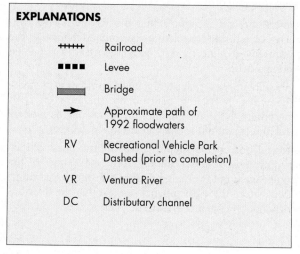

Figure 8.B **Historical maps of the Ventura River delta** The maps shows the distributary channel and the location of the RV park. *(From Keller, E. A., and Capelli, M. H. 1992. Ventura River flood of February, 1992: A lesson ignored? Water Resources Bulletin 28(5):813–31)*

The reasons erosion or deposition occurs in a specific area of river channel or on alluvial fans or deltas are complex, but they can be correlated to the physical properties of the river:

- Change in channel width, depth, or slope
- Composition of channel bed and banks (rock, gravel, sand, silt, or clay)
- Type and amount of vegetation
- Land use such as clearing forest for agriculture (discussed in Section 8.5)

For example, deposition on alluvial fans occurs in part because of changes in the shape and slope of distributary channels. They often become wider and shallower with a decreasing slope, and this change decreases the velocity of flow, favoring deposition. In general, as the velocity of flow in a river increases, the size of the bed load it can transport increases, as does the volume of suspended load consisting of silt and clay-sized particles. Specific relationships between flow velocity, discharge, and sediment transport are beyond the scope of our discussion here.

The largest particle (diameter in millimeters or centimeters) a river may transport is called its *competency*; the total load, by mass or weight, of sediment that a river carries in a given period of time is called its *capacity*.

8.5 Effects of Land-Use Changes

Streams and rivers are open systems that generally maintain a rough *dynamic equilibrium*, or steady state between the work done, that is, the sediment transported by the stream, and the load imposed, or the sediment delivered to the stream from tributaries and hill slopes. The stream tends to have a slope and cross-sectional shape that provide the velocity of flow necessary to do the work of moving the sediment load.[10] An increase or decrease in the amount of water or sediment received by the stream usually brings about changes in the channel's slope or cross-sectional shape, effectively changing the velocity of the water. The change of velocity may, in turn, increase or decrease the amount of sediment carried in the system. Therefore, land-use changes that affect the stream's volume of sediment or water volume may set into motion a series of events that results in a new dynamic equilibrium.

Consider, for example, a land-use change from forest to agricultural row crops. This change will cause increased soil erosion and an increase in the sediment load supplied to the stream because agricultural lands have higher soil erosion rates than forested lands. At first, the stream will be unable to transport the entire load and will deposit more sediment, increasing the slope of the channel. The steeper slope of the channel will increase the velocity of water and allow the stream to move more sediment. If the base level is fixed, the slope will continue to increase by deposition in the channel until the increase in velocity is sufficient to carry the new load. If the notion that deposition of sediment increases channel slope is counterintuitive to you, see Figure 8.10 for an illustration of this principle. A new dynamic equilibrium may be reached, provided the rate of sediment increase levels out and the channel slope and shape can adjust before another land-use change occurs. Suppose the reverse situation now occurs; that is, farmland is converted to forest. The sediment load to the stream from the land will decrease, and less sediment will be deposited in the stream channel. Erosion of the channel will eventually lower the slope; the lowering of the slope will, in turn, lower the velocity of the water. The predominance of erosion over deposition will continue until equilibrium between the total load imposed and work done is achieved again.

The sequence of change just described occurred in parts of the southeastern United States. On the Piedmont, between the Appalachian Mountains and the coastal plain, forestland had been cleared for farming by the 1800s. The land-use change from forest to farming accelerated soil erosion and subsequent deposition

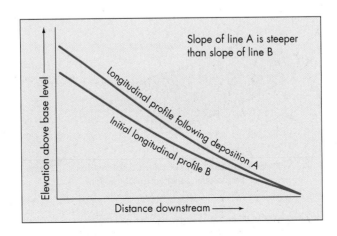

Figure 8.10 **Effect of deposition on river slope** Idealized diagram illustrating that deposition in a stream channel results in an increase in channel gradient.

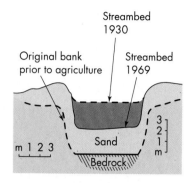

Figure 8.11 Stream bed changes from land use changes
Accelerated sedimentation and subsequent erosion resulting from land-use changes (natural forest to agriculture and back to forest) at the Mauldin Millsite on the Piedmont of middle Georgia. *(After Trimble, S. W. 1969. "Culturally accelerated sedimentation on the middle Georgia Piedmont." Master's thesis, Athens: University of Georgia. Reproduced by permission)*

of sediment in the stream (Figure 8.11). This acceleration caused the preagriculture channel to fill with sediment, as shown in Figure 8.11. After 1930, the land reverted to pine forests, and this reforestation, in conjunction with soil conservation measures, reduced the sediment load delivered to streams. Thus, formerly muddy streams choked with sediment had cleared and eroded their channels by 1969 (Figure 8.11).

Consider now the effect of constructing a dam on a stream. Considerable changes will take place both upstream and downstream of the reservoir created behind the dam. Upstream, at the head of the reservoir, the water in the stream will slow down, causing deposition of sediment. Downstream, the water coming out below the dam will have little sediment, since most of it has been trapped in the reservoir. As a result, the stream may have the capacity to transport additional sediment; if this happens, channel erosion will predominate over deposition downstream of the dam. The slope of the stream will then decrease until new equilibrium conditions are reached (Figure 8.12). (We will return to the topic of dams on rivers in Chapter 12.)

8.6 Channel Patterns and Floodplain Formation

The configuration of the channel as seen in an aerial view is called the **channel pattern.** Channel patterns can be braided or meandering, or both characteristics may be found in the same river. *Braided* channels (Figure 8.13) are characterized by numerous gravel bars and islands that divide and reunite the channel. A steep slope and coarse sediment favor transport of bed load material important in the development of gravel bars that form the "islands" that divide and subdivide the flow. The formation of the braided channel pattern, as with many other river forms, results from the interaction of flowing water and moving sediment. If the river's longitudinal profile is steep and there is an abundance of coarse bed load sediment, the channel pattern is likely to be braided. Braided channels tend to be wide and shallow compared with meandering rivers. They are often associated with steep rivers flowing through areas that are being rapidly uplifted by tectonic processes. They are also common in rivers receiving water from melting glaciers that provide a lot of coarse sediment.

Some channels contain *meanders,* which are bends that migrate back and forth across the floodplain (Figure 8.14a). Although we know what meander bends look like and what the water and sediment do in the bends, we do not know for sure why rivers meander. On the outside of a bend, sometimes referred to as the *cut bank,* the water moves faster during high flow events, causing more bank erosion; on the inside of a curve water moves more slowly and sediment is deposited, forming *point bars.* As this differential erosion and sediment deposition continues, meanders migrate laterally by erosion on the cut banks and by deposition on point

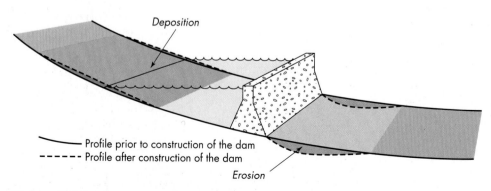

— Profile prior to construction of the dam
----- Profile after construction of the dam

Figure 8.12 Effect of a dam on erosion Upstream deposition and downstream erosion from construction of a dam and a reservoir.

(a)

(b)

Figure 8.13 **Braided channels** (a) The north Saskatchewan River, shown here, has a braided channel pattern. Notice the numerous channel bars and islands that subdivide the flow. *(John S. Shelton)* (b) Ground view of a braided channel in Granada in southern Spain with multiple channels, a steep gradient, and coarse gravel. The distance across the channel is about 7 m (21 ft). *(Edward A. Keller)*

(a)

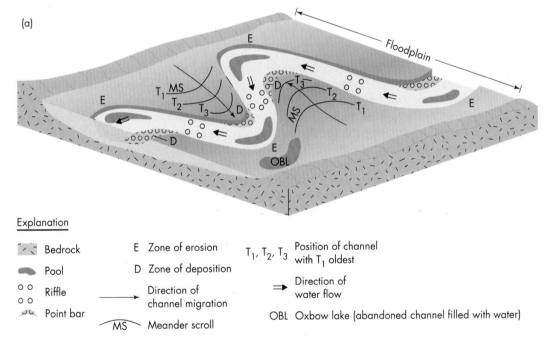

Explanation

🪨 Bedrock	E	Zone of erosion	T_1, T_2, T_3		Position of channel with T_1 oldest
🪨 Pool	D	Zone of deposition			
○○ Riffle	→	Direction of channel migration	⇒		Direction of water flow
○○ Point bar					
MS Meander scroll			OBL		Oxbow lake (abandoned channel filled with water)

(b)

Figure 8.14 **Meandering river** (a) Idealized diagram of a meandering stream and important forms and processes. Meander scrolls are low, curved ridges of sediment parallel to a meander bend. They form at the edge of a riverbank as sediment accumulates with plants. A series of scrolls marks the evolution of a meander bend. (b) Koyakuk River, Alaska, showing meander bend, point bar, and cut bank. The Oxbow Lake formed as the river eroded laterally across the floodplain and "cut off" a meander bend, leaving the meander bend as a lake. *(© Andy Deering/Omni-Photo Communications, Inc.)*

Figure 8.15 **Pool and riffle**
Well-developed pool-riffle
sequence in Sims Creek near
Blowing Rock, North Carolina.
A deep pool is apparent in the
middle distance, and shallow
riffles can be seen in the far
distance and in the foreground.
(Edward A. Keller)

bars, a process that is prominent in constructing and maintaining some flood-plains (Figure 8.14b). Overbank deposition, or deposition beyond the banks of a river, during floods causes layers of relatively fine sediments such as sand and silt to build upward; this accumulation is also important in the development of floodplains. Much of the sediment transported in rivers is periodically stored by deposition in the channel and on the adjacent floodplain. These areas, collectively called the *riverine environment*, are the natural domain of the river.

Meandering channels often contain a series of regularly spaced pools and riffles (Figure 8.15). *Pools* are deep areas produced by scour, or erosion at high flow, and characterized at low flow by relatively deep, slow movement of water. Pools are places in which you might want to take a summer swim. *Riffles* are shallow areas produced by depositional processes at high flow and characterized by relatively shallow, fast-moving water at low flow. Pools and riffles have important environmental significance: The alternation of deep, slow-moving water with shallow, fast-moving water in pools and riffles produces a variable physical and hydrologic environment and increased biological diversity.[11] For example, fish may feed in riffles and seek shelter in pools, and pools have different types of insects than are found in riffles.

Having presented some of the characteristics and processes of flow of water and sediment in rivers, we will now discuss the process of flooding in greater detail.

8.7 River Flooding

The natural process of *overbank flow* is termed **flooding** (see Figure 8.6). Most river flooding is a function of the total amount and distribution of precipitation in the drainage basin, the rate at which precipitation infiltrates the rock or soil, and the topography. Some floods, however, result from rapid melting of ice and snow in the spring or, on rare occasions, from the failure of a dam. Finally, land use can greatly affect flooding in small drainage basins.

The channel discharge at the point where water overflows the channel is called the flood discharge and is used as an indication of the magnitude of the flood (see A Closer Look: Magnitude and Frequency of Floods). The height of the water in a

A CLOSER LOOK | Magnitude and Frequency of Floods

Flooding is intimately related to the amount and intensity of precipitation and runoff. Catastrophic floods reported on television and in newspapers are often produced by infrequent, large, intense storms. Smaller floods or flows may be produced by less intense storms, which occur more frequently. All flow events that can be measured or estimated from a stream-gauging station (Figure 8.C) can be arranged in order of their magnitude of discharge, generally measured in cubic meters per second (Figure 8.D). The list of annual peak flows—that is, the largest flow each year or the annual series so arranged (see data for the Patrick River on table adjacent to Figure 8.E)—can be plotted on a discharge-frequency curve by deriving the recurrence interval (R) for each flow from the relationship

$$R = (N + 1)/M$$

where R is a recurrence interval in years, N is the number of years of record, and M is the rank of the individual flow within the recorded years (Figure 8.E).[13] For example, in Figure 8.E the highest flow for 9 years of data for the stream is approximately 280 m^3 per second (9,888 ft^3 per second) and has a rank M equal to 1.[14] The recurrence interval of this flood is

$$R = \frac{N + 1}{M} = \frac{9 + 1}{1} = 10$$

Solar power to transmit data

Gauging station instrument

Water depth sensor

Figure 8.C **Stream-gauging station** San Jose Creek, Goleta, California. *(Edward A. Keller)*

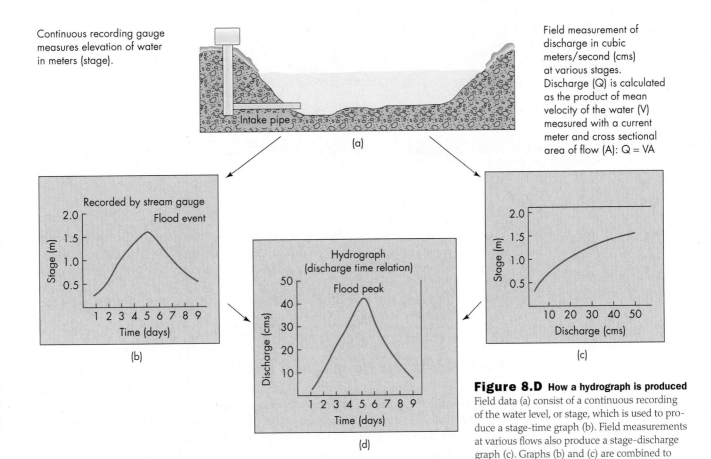

Continuous recording gauge measures elevation of water in meters (stage).

Intake pipe

(a)

Field measurement of discharge in cubic meters/second (cms) at various stages. Discharge (Q) is calculated as the product of mean velocity of the water (V) measured with a current meter and cross sectional area of flow (A): Q = VA

Recorded by stream gauge

Flood event

(b)

Hydrograph (discharge time relation)

Flood peak

(d)

(c)

Figure 8.D **How a hydrograph is produced** Field data (a) consist of a continuous recording of the water level, or stage, which is used to produce a stage-time graph (b). Field measurements at various flows also produce a stage-discharge graph (c). Graphs (b) and (c) are combined to produce the final hydrograph (d).

which means that a flood with a magnitude equal to or exceeding 280 m^3 per second can be expected about every 10 years; we call this a 10 year flood. The probability that the 10 year flood will happen in any one year is 1/10, or 0.1 (10 percent). The probability that the 100 year flood will occur in any one year is 1/100, or 0.01 (1 percent). The curve in Figure 8.E is extended by extrapolation to estimate the discharge of the 20 year flood at 450 cms. Extrapolation is risky and shouldn't extend much beyond two times the length of recorded values of discharge. Studies of many streams and rivers show that channels are formed and maintained by bankfull discharge, defined as a flow with a recurrence interval of about 1.5 years (27 m^3 per second in Figure 8.E). Bankfull is the flow that just fills the channel.

Therefore, we can expect a stream to emerge from its banks and cover part of the floodplain with water and sediment once every year or so.

As flow records are collected, we can more accurately predict floods. However, designing structures for a 10 year, 25 year, 50 year, or even 100 year flood, or, in fact, any flow, is a calculated risk since predicting such floods is based on a statistical probability. In the long term, a 25 year flood happens on the average of once every 25 years, but two 25 year floods could occur in any given year, as could two 100 year floods![15] As long as we continue to build dams, highways, bridges, homes, and other structures without considering the effects on flood-prone areas, we can expect continued loss of lives and property.

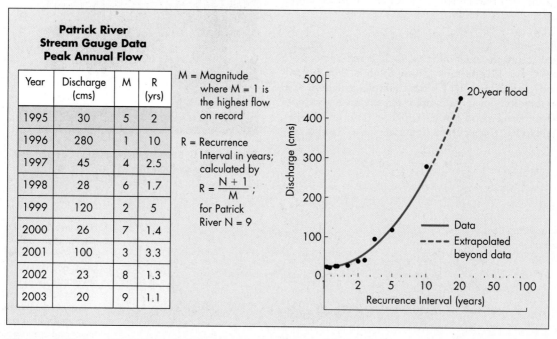

Year	Discharge (cms)	M	R (yrs)
1995	30	5	2
1996	280	1	10
1997	45	4	2.5
1998	28	6	1.7
1999	120	2	5
2000	26	7	1.4
2001	100	3	3.3
2002	23	8	1.3
2003	20	9	1.1

Patrick River Stream Gauge Data Peak Annual Flow

M = Magnitude where M = 1 is the highest flow on record

R = Recurrence Interval in years; calculated by $R = \dfrac{N+1}{M}$; for Patrick River N = 9

Figure 8.E **Example of a discharge-frequency curve** for the Patrick River on the adjacent table. The curve is extended (extrapolated) to estimate the 20 year flood at about 450 cms.

river at any given time is called the *stage*. The term *flood stage* frequently connotes that the water surface has reached a high-water condition likely to cause damage to personal property. This definition is based on human perception of the event, so the elevation that is considered flood stage depends on human use of the floodplain.[12] Therefore, the magnitude of a flood may or may not coincide with the extent of property damage.

Flash Floods and Downstream Floods

It is useful to distinguish between flash and downstream floods (Figure 8.16). **Flash floods** occur in the upper parts of drainage basins and are generally produced by intense rainfall of short duration over a relatively small area. In general, flash floods may not cause flooding in the larger streams they join downstream, although they can be quite severe locally. For example, the high-magnitude flash

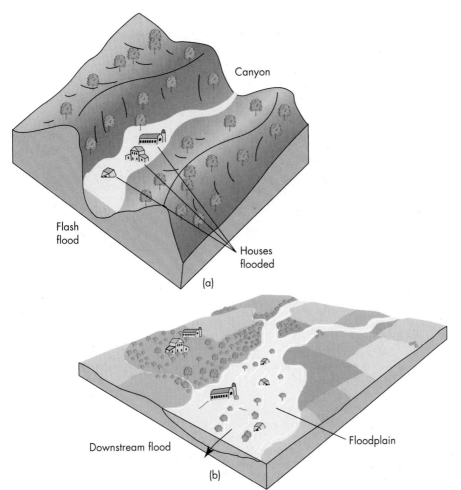

Figure 8.16 Flash floods and
downstream floods Idealized diagram comparing flash flood (a) with a downstream flood (b). Flash floods generally cover relatively small areas and are caused by intense local storms with steep topography, often in a canyon. A distinct floodplain may not be present, whereas downstream floods cover wide areas of a floodplain and are caused by regional storms or spring runoff of a floodplain. *(Modified after U.S. Department of Agriculture drawing)*

flood that occurred in July 1976 in the Front Range of Colorado was caused by violent flash floods, which are characterized by a rapid rise in floodwaters in response to precipitation. In addition, the flash floods were nourished by a complex system of thunderstorms that swept through several canyons west of Loveland, delivering up to 25 cm (9.8 in.) of rain. This brief local flood killed 139 people and caused more than $35 million in damages to highways, roads, bridges, homes, and small businesses. Most of the damage and loss of life occurred in the Big Thompson Canyon, where hundreds of residents, campers, and tourists were caught with little or no warning. Although the storm and flood were rare events in the Front Range canyons, comparable floods have occurred in the past and others can be expected in the future.[16,17,18] Interestingly, the U.S. Geological Survey reports that about half of all deaths during flash floods are related to automobiles. When people attempt to drive through shallow, fast-moving floodwater, the strong lateral force of the water may sweep automobiles off the road into deeper water, trapping people in sinking or overturned vehicles.

It is the large **downstream floods,** such as the 1993 Mississippi River flood and the 1997 Red River, North Dakota, flood that usually make television and newspaper headlines. We discussed the 1993 Mississippi River flooding in our opening case. Floodwaters of the Red River, which flows north to Canada, inundated the city of Grand Forks, North Dakota, initiating the evacuation of 50,000 people, causing a fire that burned part of the city center and more than $1 billion in damage (Figure 8.17). The Red River often floods in the spring, and it did so again

Figure 8.17 Flooded city
Flooding of the Red River at Grand
Forks, North Dakota, in 1997 caused
a fire that burned part of the city.
(Eric Hylden/Grand Forks Herald)

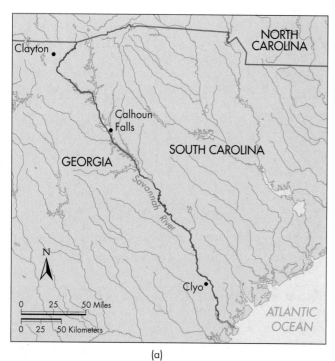

(a)

**Figure 8.18 Downstream movement
of a flood wave** Downstream movement
of a flood wave on the Savannah River,
South Carolina and Georgia. The distance
from Clayton to Clyo is 257 km (160 mi).
(a) Map of the area. (b) Volume of water
passing Clayton, Calhoun Falls, and Clyo.
(c) Volume of water per unit area at the
same points. *(After Hoyt, W. G., and
Langbein, W. B. Floods. © Copyright 1955 by
Princeton University Press, figure 8, p. 39.
Reprinted by permission of Princeton University
Press)*

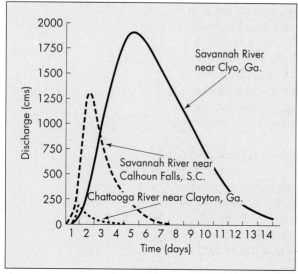

(b)

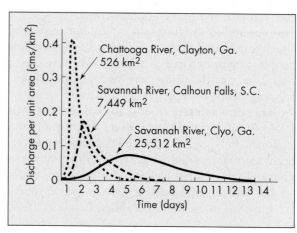

(c)

in the spring of 2001 when heavy rains melted snow and ice on frozen ground that did not allow infiltration of the rain into the forest.

Downstream floods cover a wide area and are usually produced by storms of long duration that saturate the soil and produce increased runoff. Flooding on small tributary basins is limited, but the contribution of increased runoff from thousands of tributary basins may cause a large flood downstream. A flood of this kind is characterized by the downstream movement of the floodwaters with a large rise and fall of discharge at a particular location.[19] Figure 8.18a shows an area map and 8.18b shows the 257 km (160 mi) downstream migration of a flood crest on the Chattooga–Savannah River system. It illustrates that a progressively longer time is necessary for the rise and fall of water as the flood wave proceeds downstream. In addition, it dramatically shows the tremendous increase in discharge from low-flow conditions to more than 1,700 m^3 per second (60,000 ft^3 per second) in 5 days.[20] Figure 8.18c illustrates the same flood in terms of discharge per unit area, eliminating the effect of downstream increase in discharge. This better illustrates the shape and form of the flood wave as it moves downstream.[20]

8.8 Urbanization and Flooding

Human use of land in urban environments has increased both the magnitude and frequency of floods in small drainage basins of a few square kilometers. The rate of increase is a function of the percentage of the land that is covered with roofs, pavement, and cement, referred to as *impervious cover* (Figure 8.19), and the percentage of area served by storm sewers. Storm sewers are important because they allow urban runoff from impervious surfaces to reach stream channels much more quickly than in natural settings. Therefore, impervious cover and storm sewers are collectively a measure of the degree of urbanization. The graph in Figure 8.20 shows that an urban area with 40 percent impervious surface and 40 percent of its area served by storm sewers can expect to have about three times as many floods as before urbanization. This ratio applies to floods of small and intermediate frequency. As the size of the drainage basin increases, however, high-magnitude floods with frequencies of approximately 50 years are not significantly affected by urbanization (Figure 8.21).

Floods are a function of rainfall-runoff relations, and urbanization causes a tremendous number of changes in these relationships. One study showed that urban runoff from the larger storms is nearly five times that of preurban runoff.[21]

Figure 8.19 Urbanization **increases runoff** Cities in most of the United States, such as Santa Barbara, California, shown here, have a high portion of their land covered by roofs, streets, and parking lots. These surfaces do not allow water to infiltrate the ground so surface runoff greatly increases. *(Edward A. Keller)*

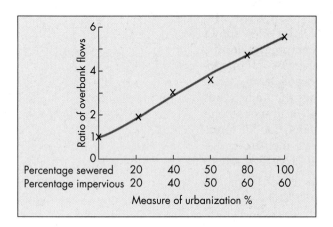

Figure 8.20 Floods before and after urbanization
Relationship between the ratio of overbank flows (after urbanization to before urbanization) and measure of urbanization. For example, a ratio of 3 to 1, or simply 3, means that after urbanization there are three floods for every one there was before urbanization; or flooding is three times as common after urbanization. This figure shows that as the degree of urbanization increases, the number of overbank flows per year also increases. *(After Leopold, L. B. 1968. U.S. Geological Survey Circular 559)*

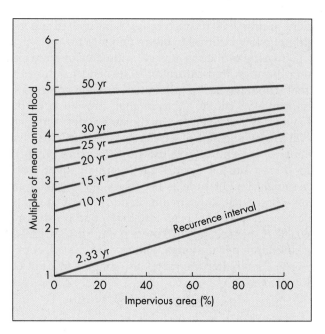

Figure 8.21 Urban flood hazard increases as impervious area increases Graph showing the variation of flood frequency with percentage of impervious area. The mean annual flood (approximately bankfull) is the average (over a period of years) of the largest flow that occurs each year. The mean annual flood in a natural river basin with no urbanization has a recurrence interval of 2.33 years. Note that the smaller floods with recurrence intervals of just a few years are much more affected by urbanization than are the larger floods. The 50-year flood is little affected by the amount of area that is rendered impervious. *(From Martens, L. A. 1968. U.S. Geological Survey Water Supply Paper 1591C)*

Estimates of discharge for different recurrence intervals at different degrees of urbanization are shown in Figure 8.22. The estimates clearly indicate the tremendous increase of runoff with increasing impervious areas and storm sewer coverage. However, it is not only the peak discharge that causes urban flooding. Long-duration storms resulting from moderate precipitation can also cause flooding if storm drains become blocked with sediment and storm debris. In this case, water begins to pond, causing flooding in low areas. It is analogous to water rising in a bathtub shower when the drain becomes partly blocked by soap.

Urbanization causes increased runoff because less water infiltrates the ground. Figure 8.23a shows a generalized hydrograph before urbanization. Of particular importance is the *lag time,* defined as the time between when most of the rainfall occurs and a flood is produced. Figure 8.23b shows two hydrographs, before and after urbanization. Note the significant reduction in lag time after urbanization. Short lag times, referred to as *flashy discharge,* are characterized by rapid rise and fall of floodwater. Since little water infiltrates the soil, the low water or dry-season flow in urban streams, normally sustained by groundwater seepage into the channel, is greatly reduced. This reduction in flow effectively concentrates any pollutants present and generally lowers the aesthetic amenities of the stream.[14]

Urbanization is not the only type of development that can increase flooding. Some flash floods have occurred because bridges built across small streams block the passage of floating debris, causing a wave of water to move downstream when the debris breaks loose (see Case History: Flash Floods in Eastern Ohio).

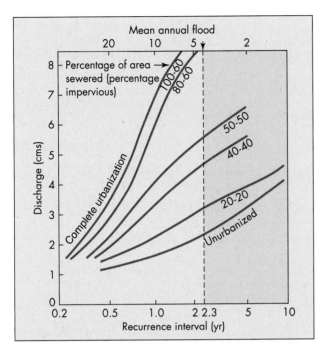

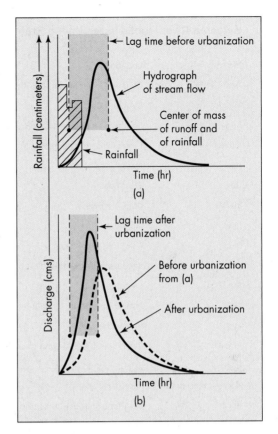

Figure 8.22 Urbanization increases flood for a particular recurrence interval Flood frequency curve for a 2.6-km² (1-mi²) basin in various states of urbanization. 100-60 means the basin is 100 percent sewered and 60 percent of surface area is impervious. The dashed line shows the increase in mean annual flood with increasing urbanization. *(After Leopold, L. B. 1968. U.S. Geological Survey Circular 559)*

Figure 8.23 Urbanization shortens lag time Generalized hydrographs. (a) Hydrograph shows the typical lag between the time when most of the rainfall occurs and the time when the stream floods. (b) Here, the hydrograph shows the decrease in lag time because of urbanization. *(After Leopold, L. B. 1968. U.S. Geological Survey Circular 559)*

8.9 The Nature and Extent of Flood Hazards

Flooding is one of the most universally experienced natural hazards. In the United States, floods were the number one type of disaster during the twentieth century, and approximately 100 lives per year (or about 10,000 in the twentieth century) are lost because of river flooding. Tragically, this figure is low compared with losses suffered by developing countries that lack monitoring facilities, warning systems, and effective disaster relief.[12,22] Table 8.1 lists some severe river floods that occurred in the United States from 1937 to 2006.

Factors That Cause Flood Damage

Factors that affect the damage caused by floods include

- Land use on the floodplain
- Magnitude, or the depth and velocity of the water and frequency of flooding
- Rate of rise and duration of flooding
- Season, for example, growing season on the floodplain
- Sediment load deposited
- Effectiveness of forecasting, warning, and emergency systems

TABLE 8.1   Selected River Floods in the United States

Year	Month	Location	No. of Lives Lost	Property Damage (Millions of Dollars)
1937	Jan.–Feb.	Ohio and lower Mississippi River basins	137	418
1938	March	Southern California	79	25
1940	Aug.	Southern Virginia and Carolinas and eastern Tennessee	40	12
1947	May–July	Lower Missouri and middle Mississippi River basins	29	235
1951	June–July	Kansas and Missouri	28	923
1955	Dec.	West Coast	61	155
1963	March	Ohio River basin	26	98
1964	June	Montana	31	54
1964	Dec.	California and Oregon	40	416
1965	June	Sanderson, Texas (flash flood)	26	3
1969	Jan.–Feb.	California	60	399
1969	Aug.	James River basin, Virginia	154	116
1971	Aug.	New Jersey	3	139
1972	June	Black Hills, South Dakota (flash flood)	242	163
1972	June	Eastern United States	113	3,000
1973	March–June	Mississippi River	0	1,200
1976	July	Big Thompson River, Colorado (flash flood)	139	35
1977	July	Johnstown, Pennsylvania	76	330
1977	Sept.	Kansas City, Missouri, and Kansas	25	80
1979	April	Mississippi and Alabama	10	500
1983	Sept.	Arizona	13	416
1986	Winter	Western states, especially California	17	270
1990	Jan.–May	Trinity River, Texas	0	1,000
1990	June	Eastern Ohio (flash flood)	21	Several
1993	June–Aug.	Mississippi River and tributaries		
1997	January	Sierra Nevada, Central Valley, California	23	Several hundred
2001	June	Houston, Texas. Buffalo Bayou (coastal river)	22	2,000
2006	June–July	Mid-Atlantic states, New York to North Carolina	16	100+

CASE HISTORY | Flash Floods in Eastern Ohio

On Friday, June 15, 1990, over 14 cm (5.5 in.) of precipitation fell within approximately 4 hours in some areas of eastern Ohio. Two tributaries of the Ohio River, Wegee and Pipe Creeks, generated flash floods near the small town of Shadyside, killing 21 people and leaving 13 people missing and presumed dead. The floods were described as 5 m (16 ft) high walls of water that rushed through the valley. In all, approximately 70 houses were destroyed and another 40 were damaged. Trailers and houses were seen washing down the creeks, bobbing like corks in the torrent.

The rush of water was apparently due to the failure of debris dams that had developed across the creeks upstream of bridges. Runoff from rainfall had washed debris into the creeks from side slopes; this debris, including tree trunks and other material, became lodged against the bridges, creating the debris dams. When the bridges could no longer contain the weight of the debris, the dams broke loose, sending surges of water downstream. This scenario has been played and replayed in many flash floods around the world. All too often, the supports for bridges are too close together to allow large debris to pass through; instead the debris accumulates on the upstream side of the bridge, damming the stream and eventually causing a flood.

Effects of Flooding

The effects of flooding may be primary, that is, directly caused by the flood, or secondary, caused by disruption and malfunction of services and systems due to the flood.[22] Primary effects include injury, loss of life, and damage caused by swift currents, debris, and sediment to farms, homes, buildings, railroads, bridges, roads, and communication systems. Erosion and deposition of sediment in the rural and urban landscape may also involve a loss of considerable soil and vegetation. Secondary effects may include short-term pollution of rivers, hunger and disease, and displacement of persons who have lost their homes. In addition, fires may be caused by shorts in electrical circuits or gas mains broken by flooding and associated erosion.[22]

8.10 Adjustments to Flood Hazards

Historically, particularly in the nineteenth century, humans have responded to flooding by attempting to prevent the problem; that is, they modified the stream by creating physical barriers such as dams or levees or by straightening, widening, and deepening the entire stream so that it would drain the land more efficiently. Every new flood control project has the effect of luring more people to the floodplain in the false hope that the flood hazard is no longer significant. We have yet to build a dam or channel capable of controlling the heaviest runoff, and when the water finally exceeds the capacity of the structure, flooding may be extensive.[22,23]

In recent years, we have begun to recognize the advantages of alternative adjustments to trying to physically prevent flooding. These include flood insurance and controlling the land use on floodplains. We will discuss each of the main adjustments with the realization that no one adjustment is best in all cases. Rather, an integrated approach to minimizing the flood hazard that incorporates the appropriate adjustments for a particular situation is preferable.

The Structural Approach

Physical Barriers.   Measures to prevent flooding include construction of physical barriers such as levees (Figure 8.24) and floodwalls, which are usually

Figure 8.24 Mississippi River levee A levee with a road on top of it protects the bank (left side of photograph) of the lower Mississippi River at this location in Louisiana.
(Comstock Images)

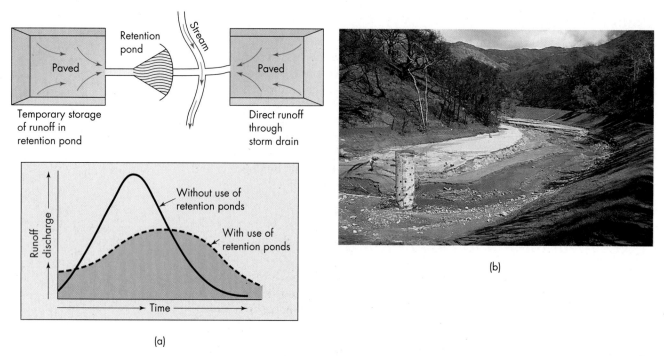

Figure 8.25 **Retention pond** (a) Comparison of runoff from a paved area through a storm drain with runoff from a paved area through a temporary storage site (retention pond). Notice that the paved area drained by way of the retention pond produces a lesser peak discharge and therefore is less likely to contribute to flooding of the stream. *(Modified after U.S. Geological Survey Professional Paper 950)* (b) Photograph of a retention pond near Santa Barbara, California. *(Edward A. Keller)*

constructed of concrete as opposed to earthen levees; reservoirs to store water for later release at safe rates; and on-site stormwater retention basins (Figure 8.25). Unfortunately, the potential benefits of physical barriers are often lost because of increased development on floodplains that are supposedly protected by these structures. For example, the winters of 1986 and 1997 brought tremendous storms and flooding to the western states, particularly California, Nevada, and Utah. In all, damages exceeded several hundred million dollars and several people died. During one of the storms and floods in 1986, a levee broke on the Yuba River in California, causing more than 20,000 people to flee their homes. An important lesson learned during this flood is that levees constructed along rivers many years ago are often in poor condition and subject to failure during floods.

The 1997 floods damaged campsites and other development in Yosemite National Park. As a result, the park revised its floodplain management policy; some camping and other facilities were abandoned, and the river is now allowed to "run free."

Some engineering structures designed to prevent flooding have actually increased the flood hazard in the long term. For example, as discussed in the case history opening this chapter, floodwalls produced a bottleneck at St. Louis that increased upstream flooding during the 1993 floods of the Mississippi River (Figure 8.2).

Recurring flooding, particularly on the Mississippi, has led to the controversial speculation that human activities have contributed to an increase in flooding. Over time equal flood discharge is producing higher flood stages. The systems of levees, floodwalls, and structures to improve river navigation for barges transporting goods down river control the smaller floods. For the largest floods, these same systems may constrain or retard flow (slow it down) and this results in higher levels of flood flow (stage).[26]

After observing causes and effects of many floods, we have learned that structural controls must go hand in hand with floodplain regulations if the hazard is to be minimized.[1,23,24,25]

Channelization.   Straightening, deepening, widening, clearing, or lining existing stream channels are all methods of **channelization.** Basically, it is an engineering technique, with the objectives of controlling floods, draining wetlands, controlling erosion, and improving navigation.[27] Of the four objectives, flood control and drainage improvement are the two most often cited in channelization projects. Thousands of kilometers of streams in the United States have been modified, and thousands of kilometers of channelization are now being planned or constructed. Federal agencies alone have carried out several thousand kilometers of channel modification. In the past, however, inadequate consideration has been given to the adverse environmental effects of channelization.

Opponents of channelizing natural streams emphasize that the practice is antithetical to the production of fish and wetland wildlife and, furthermore, the stream suffers from extensive aesthetic degradation. The argument is as follows:

- Drainage of wetlands adversely affects plants and animals by eliminating habitats necessary for the survival of certain species.
- Cutting trees along the stream eliminates shading and cover for fish and exposes the stream to the sun; the exposure results in damage to plant life and heat-sensitive aquatic organisms.
- Cutting hardwood trees on the floodplain eliminates the habitats of many animals and birds, while facilitating erosion and siltation of the stream.
- Straightening and modifying the streambed destroy both the diversity of flow patterns and the feeding and breeding areas for aquatic life while changing peak flow.
- Conversion of wetlands from a meandering stream to a straight, open ditch seriously degrades the aesthetic value of a natural area.[27] Figure 8.26 summarizes some of the differences between natural streams and those modified by channelization.

Not all channelization causes serious environmental degradation; in many cases, drainage projects are beneficial. Benefits are probably best observed in urban areas subject to flooding and in rural areas where previous land use has caused drainage problems. In addition, there are other examples in which channel modification has improved navigation or reduced flooding and has not caused environmental disruption.

Channel Restoration: Alternative to Channelization

Many streams in urban areas scarcely resemble natural channels. The process of constructing roads, utilities, and buildings with the associated sediment production is sufficient to disrupt most small streams. **Channel restoration**[28] uses various techniques: cleaning urban waste from the channel, allowing the stream to flow freely, protecting the existing channel banks by not removing existing trees or, where necessary, planting additional native trees and other vegetation. Trees provide shade for a stream, and the root systems protect the banks from erosion (Figure 8.27). The objective is to create a more natural channel by allowing the stream to meander and, when possible, provide for variable, low-water flow conditions: fast and shallow flow on riffles alternating with slow and deep flow in pools. Where lateral bank erosion must be absolutely controlled, the outsides of

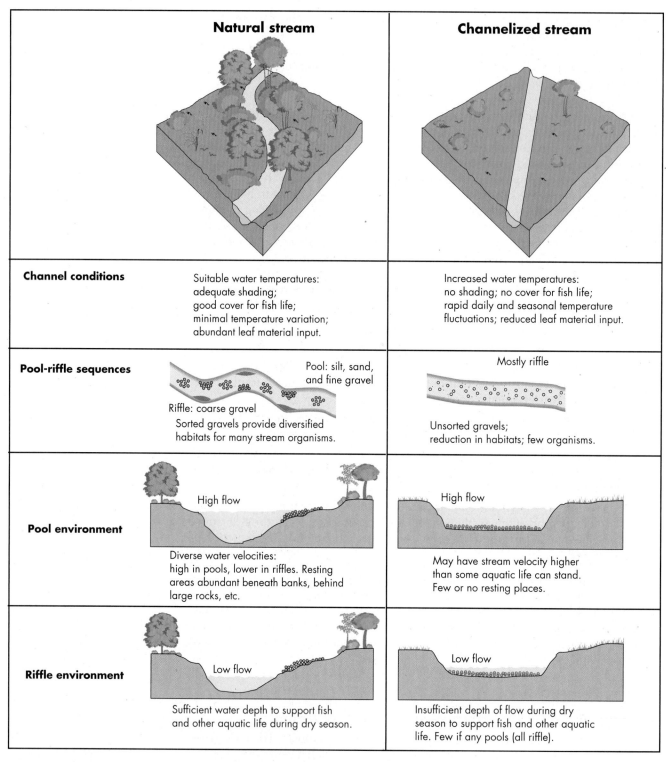

Figure 8.26 Natural versus channelized stream A natural stream compared with a channelized stream in terms of general characteristics and pool environments. *(Modified after Corning,* Virginia Wildlife, *February 1975)*

bends may be defended with large stones known as *riprap*. Design criteria for channel restoration are shown on Figure 8.28.

River restoration of the Kissimmee River in Florida may be the most ambitious restoration project ever attempted in the United States (see Chapter 4).[29] In Los Angeles, California, a group called Friends of the River has suggested that

Figure 8.27 Tree roots protecting stream bank from erosion *(Edward A. Keller)*

the Los Angeles River be restored. This will be a difficult task since most of the riverbed and banks are lined with concrete (Figure 8.29). However, a river park is planned for one section of the river where a more natural-looking channel has reappeared since channelization (Figure 8.30).

Flood Insurance

In 1968, when private companies became reluctant to continue to offer flood insurance, the federal government took over. The U.S. National Flood Insurance

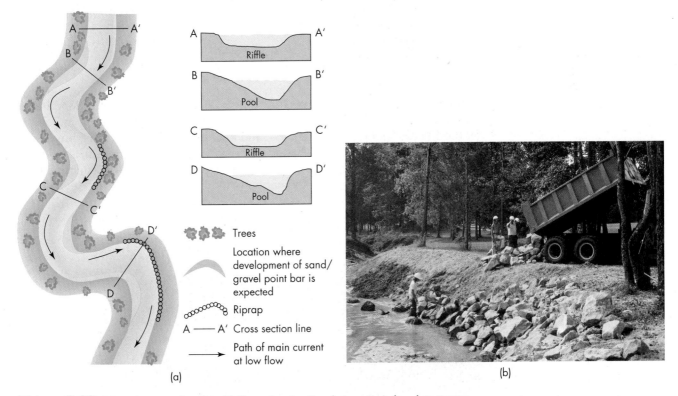

(a)   (b)

Figure 8.28 Urban stream restoration (a) Channel-restoration design criteria for urban streams, using a variable channel shape to induce scour and deposition (pools and riffles) at desired locations. *(Modified after Keller, E. A., and Hoffman, E. K. 1977. Journal of Soil and Water Conservation 32(5):237–40)* (b) Placing riprap where absolutely necessary to defend the bank, Briar Creek, Charlotte, North Carolina. Notice the planting of grass on banks with straw mulch and trees growing on banks. *(Edward A. Keller)*

Figure 8.29 Channelization versus restoration (a) Concrete channel in Los Angeles River system compared with (b) channel restoration in North Carolina. *(Edward A. Keller)*

(a)

(b)

Program makes, with restrictions including a 30 day waiting period, flood insurance available at subsidized rates. Special Flood Hazard Areas, those inundated by the 100 year flood, are designated, and new property owners must buy insurance at rates determined on the basis of the risk. The insurance program is intended to provide short-term financial aid to victims of floods as well as to establish long-term land-use regulations for the nation's floodplains. The basic risk evaluation

Figure 8.30 Los Angeles River Part of the Los Angeles River where a more natural channel has developed since channelization. *(Deirdra Walpole Photography)*

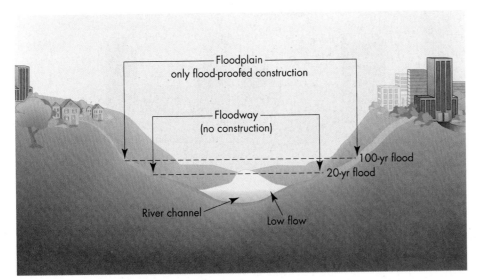

Figure 8.31 Floodplain regulation
Idealized diagram showing areas inundated by the 100 and 20 year floods used in the U.S. National Flood Insurance Program.

centers on identifying the floodplain area inundated by the 100 year flood. Only flood-proofed buildings are allowed in this area (Figure 8.31), and no construction is allowed on the portion of the floodplain inundated by the 20 year flood. For a community to join the National Flood Insurance Program, it must adopt minimum standards of land-use regulation within the 100 year floodplain, mapped by the Federal Emergency Management Agency (FEMA). Nearly all communities with a flood risk in the United States have basic flood hazard maps and have initiated some form of floodplain regulations. Several million insurance policies are presently held by property owners.[30]

By the early 1990s, it was recognized that the insurance program was in need of reform, resulting in the National Flood Insurance Reform Act of 1994. The act was passed to encourage opportunities to mitigate flood hazards, including flood-proofing, relocations, and buy-outs of properties likely to be frequently flooded.[30]

Flood-Proofing

There are several methods of flood-proofing. The most common include

- Raising the foundation of a building above the flood hazard level by using piles or columns or by extending foundation walls or earth fill[30]
- Constructing floodwalls or earth berms around buildings to seal them from floodwaters
- Using waterproofing construction such as waterproofed doors and water-proofed basement walls and windows
- Installing improved drains with pumps to keep flood waters out

There are also modifications to buildings that are designed to minimize flood damages while allowing floodwaters to enter a building. For example, ground floors along expansive riverfront properties in some communities in Germany are designed so that they are not seriously damaged by floodwaters and may be easily cleaned and made ready for reuse after a flood.[30]

Floodplain Regulation

Previously we have defined the floodplain as a landform produced by a river. When we try to regulate development on a flood-hazard area we often define the floodplain from a hydrologic point of view. Thus the 100 year floodplain is that part of a river valley that is inundated by the 100 year flood. For a particular river at a particular site that flood has a discharge (volume of flow per unit time such as cubic meters or cubic feet per second). We often determine that discharge from analyzing past flow records (see A Closer Look: Magnitude and Frequency of Floods).

Figure 8.32 Increasing flood hazard Development that encroaches on the floodplain can increase the heights of subsequent floods. *(From Water Resources Council. 1971. Regulation of flood hazard areas, vol. 1)*

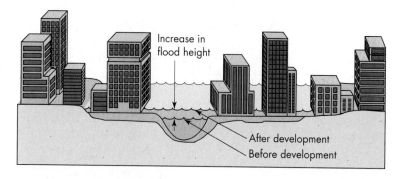

From an environmental point of view, the best approach to minimizing flood damage in urban areas is **floodplain regulation.** The big problem is that several million people in nearly 4,000 U.S. towns and cities live on floodplains with a recognized flood hazard. The objective of floodplain regulation is to obtain the most beneficial use of floodplains while minimizing flood damage and cost of flood protection.[31] Floodplain regulation is a compromise between indiscriminate use of floodplains, resulting in loss of life and tremendous property damage, and complete abandonment of floodplains, which gives up a valuable natural resource.

This is not to say that physical barriers, reservoirs, and channelization works are not desirable. In areas developed on floodplains, they will be necessary to protect lives and property. We need to recognize, however, that the floodplain belongs to the river system, and any encroachment that reduces the cross-sectional area of the floodplain increases flooding (Figure 8.32). An ideal solution would be discontinuing floodplain development that necessitates new physical barriers. In other words, the ideal is to "design with nature." Realistically, the most effective and practical solution in most cases will be a combination of physical barriers and floodplain regulations that results in fewer physical modifications of the river system. For example, reasonable floodplain zoning in conjunction with a diversion channel project or upstream reservoir may result in a smaller diversion channel or reservoir than would be necessary without floodplain regulations.

Flood-Hazard Mapping

A preliminary step to floodplain regulation is flood-hazard mapping, which is a means of providing information about the floodplain for land-use planning.[32] Flood-hazard maps may delineate past floods or floods of a particular frequency, for example, the 100 year flood (Figure 8.31). They are useful in regulating private development, purchasing land for public use as parks and recreational facilities, and creating guidelines for future land use on floodplains.

Flood-hazard evaluation may be accomplished in a general way by direct observation and measurement of physical parameters. For example, extensive flooding of the Mississippi River Valley during the summer of 1993 is clearly shown on images produced from satellite-collected data (see Figure 8.2). Floods can also be mapped from aerial photographs taken during flood events; they can be estimated from high-water lines, flood deposits, scour marks, and trapped debris on the floodplain, measured in the field after the water has receded.[33] The most common way we produce flood-hazard maps today is to use mathematical models that show the land flooded by a particular flow—often the 100 year flood.

Floodplain Zoning. Flood-hazard information is used to designate a flood-hazard area. Once the hazard area has been established, planners can establish zoning regulations and acceptable land use. Figure 8.33 shows a typical zoning map before and after establishment of floodplain regulations.

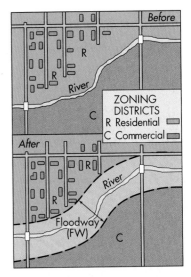

Figure 8.33 Floodplain zoning
Typical zoning map before and after the addition of flood regulations. *(From Water Resources Council. 1971. Regulation of flood hazard areas, vol. 1)*

Relocating People from Floodplains: Examples from North Carolina and North Dakota

For several years state and federal governments have been selectively purchasing homes damaged by floodwaters. The purpose is to remove homes from hazardous areas and thereby reduce future losses. In September 1999, Hurricane Floyd brought nearly 50 cm (20 in.) of rain to the North Carolina region, flooding many areas. State and federal governments decided to spend nearly $50 million to remove about 430 homes in Rocky Mount, North Carolina.

At Churchs Ferry, North Dakota there has been a wet cycle since 1992, causing nearby Devils Lake to rise approximately 8 m (26 ft). The lake has no outlet and this part of the Northern Plains is very flat. As a result, the lake has more than doubled in area and is inundating the land in the vicinity of Churchs Ferry. By late June 2000, the town was all but deserted; the population of the town has shrunk from approximately 100 to 7 people. Most of the people in the town have taken advantage of a voluntary federal buyout plan and have moved to higher ground, many to the town of Leeds, approximately 24 km (15 mi) away. The empty houses left behind will be demolished or moved to safer ground.

The lucrative buyout of $3.5 million seemed to be assured, a "slam dunk." The people who participated in the buyout program were given the appraised value of their homes plus an incentive; most considered the offer too good to turn down. There was also recognition that the town would eventually have come to an end as a result of flooding. Nevertheless, there was some bitterness among the town's population, and not everyone participated. The mayor and the fire chief of the town are among the seven people who decided to stay. The buyout program for Churchs Ferry demonstrated that the process is an emotional one; it is difficult for some people to make the decision to leave their home, even though they know it is likely to be damaged by floodwaters in the relatively near future.

Personal Adjustment: What to Do and What Not to Do

Flooding is the most commonly experienced natural hazard. Although we cannot prevent floods from happening, individuals can be better prepared. Table 8.2 summarizes what individuals can do to prepare for a flood as well as what not to do.

8.11 Perception of Flooding

At the institutional level—that is, at the government and flood-control agency level—perception and understanding of flooding are adequate for planning purposes. On the individual level, however, the situation is not as clear. People are tremendously variable in their knowledge of flooding, anticipation of future flooding, and willingness to accept adjustments caused by the hazard.

Progress at the institutional level includes mapping of flood-prone areas (thousands of maps have been prepared), of areas with a flash-flood potential downstream from dams, and areas where urbanization is likely to cause problems in the near future. In addition, the federal government has encouraged states and local communities to adopt floodplain management plans.[8] Still, planning to avoid the flood hazard by not developing on floodplains or by relocating

TABLE 8.2 What to Do and What Not to Do before and after a Flood

What to Do	**Preparing for a Flood** • Check with your local flood control agency to see if your property is at risk from flooding. • If your property is at risk, purchase flood insurance if you can and be sure that you know how to file a claim. • Buy sandbags or flood boards to block doors. • Make up a Flood Kit, including a flashlight, blankets, raingear, battery-powered radio, first-aid kit, rubber gloves, and key personal documents. Keep it upstairs if possible. • Find out where to turn off your gas and electricity. If you are not sure, ask the person who checks your meter when he or she next visits. • Talk about possible flooding with your family or housemates. Consider writing a Flood Plan, and store these notes with your Flood Kit.
What Not to Do	• Underestimate the damage a flood can do.
What to Do	**When You Learn a Flood Warning Has Been Issued** • Be prepared to evacuate. • Observe water levels and stay tuned to radio and television news and weather reports. • Move people, pets, and valuables upstairs or to higher ground. • Move your car to higher ground. It takes only 0.6 m (2 ft) of fast-flowing water to wash your car away. • Check on your neighbors. Do they need help? They may not be able to escape upstairs or may need help moving furniture. • Do as much as you can in daylight. If the electricity fails, it will be hard to do anything. • Keep warm and dry. A flood can last longer than you think, and it can get cold. Take warm clothes, blankets, a Thermos, and food supplies.
What Not to Do	• Walk in floodwater above knee level: it can easily knock you off your feet. Manholes, road works, and other hazards may be hidden beneath the water.
What to Do	**After a Flood** • Check house for damage; photograph any damage. • If insured, file a claim for damages. • Obtain professional help in removing or drying carpets and furniture as well as cleaning walls and floors. • Contact gas, electricity, and water companies. You will need to have your supplies checked before you turn them back on. • Open doors and windows to ventilate your home. • Wash water taps and run them for a few minutes before use. Your water supply may be contaminated; check with your water supplier if you are concerned.
What Not to Do	• Touch items that have been in contact with the water. Floodwater may be contaminated and could contain sewage. Disinfect and clean thoroughly everything that got wet.

Source: Modified after Environment Agency, United Kingdom. Floodline accessed 11/1/00 at www.environment_agency.gov.uk/flood/press_2.htm

present development to locations off the floodplain needs further consideration and education to be accepted by the general population. This was tragically shown by the 2006 floods in the Mid-Atlantic United States, when severe river flooding impacted the region from Virginia to New York (Figure 8.34). Over 200,000 floodplain residents were evacuated in Pennsylvania alone and damages exceeding $100 million were incurred. About 16 people lost their lives as cars were swept away by floodwaters and people drowned in flood-swollen creeks and rivers. About 70 people were rescued from rooftops. As a people we need to just "say no" to future development on floodplains. That is the most cost-effective way to reduce chronic flooding.

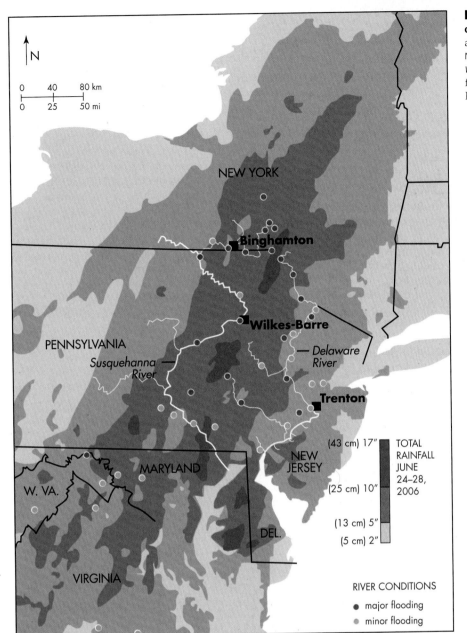

Figure 8.34 Mid-Atlantic floods
of June–July 2006 (a) Map of major
and minor flooding. *(Modified from*
New York Times *with data from National
Weather Service)* (b) Collecting mail
from a flooded home in Wilkes-Barre,
Pennsylvania. *(AP/Wide World Photos)*

N

| 0 | 40 | 80 km |
| 0 | 25 | 50 mi |

NEW YORK

Binghamton

PENNSYLVANIA

Susquehanna
River

Wilkes-Barre

Delaware
River

Trenton

NEW
JERSEY

MARYLAND

W. VA.

DEL.

VIRGINIA

(43 cm) 17"

(25 cm) 10"

(13 cm) 5"
(5 cm) 2"

TOTAL
RAINFALL
JUNE
24–28,
2006

RIVER CONDITIONS

● major flooding
○ minor flooding

(a)

(b)

SUMMARY

Streams and rivers form a basic transport system of the rock cycle and are a primary erosion agent shaping the landscape. The region drained by a stream system is called a drainage basin.

Sediments deposited by lateral migration of meanders in a stream and by periodic overflow of the stream banks form a floodplain. The magnitude and frequency of flooding are inversely related and are functions of the intensity and distribution of precipitation, the rate of infiltration of water into the soil and rock, and topography. Flash floods are produced by intense, brief rainfall over a small area. Downstream floods in major rivers are produced by storms of long duration over a large area that saturate the soil, causing increased runoff from thousands of tributary basins. Urbanization has increased flooding in small drainage basins by covering much of the ground with impermeable surfaces, such as buildings and roads, that increase the runoff of stormwater.

River flooding is the most universally experienced natural hazard. Loss of life is relatively low in developed countries that have adequate monitoring and warning systems, but property damage is much greater than in preindustrial societies because floodplains are often extensively developed. Factors that control damage caused by flooding include land use on the floodplain; the magnitude and frequency of flooding; the rate of rise and duration of the flooding; the season; the amount of sediment deposited; and the effectiveness of forecasting, warning, and emergency systems.

Environmentally, the best solution to minimizing flood damage is floodplain regulation, but it will remain necessary to use engineering structures to protect existing development in highly urbanized areas. These include physical barriers such as levees and floodwalls and structures that regulate the release of water, such as reservoirs. The realistic solution to minimizing flood damage involves a combination of floodplain regulation and engineering techniques. The inclusion of floodplain regulation is critical because engineered structures tend to encourage further development of floodplains by producing a false sense of security. The first step in floodplain regulation is mapping the flood hazards, which can be difficult and expensive. Planners can then use the maps to zone a flood-prone area for appropriate uses. In some cases, homes in flood-prone areas have been purchased and demolished by the government and people relocated to safe ground.

Channelization is the straightening, deepening, widening, cleaning, or lining of existing streams. The most commonly cited objectives of channelization are flood control and drainage improvement. Channelization has often caused environmental degradation, so new projects are closely evaluated. New approaches to channel modification using natural processes are being practiced, and in some cases channelized streams are being restored.

An adequate perception of flood hazards exists at the institutional level. On the individual level, however, more public-awareness programs are needed to help people clearly perceive the hazard of living in flood-prone areas.

Revisiting Fundamental Concepts

Human Population Growth

More and more people are living on floodplains. These flat lands adjacent to rivers, which have a high flood risk, are seen by too many people as a place to develop. To end this folly, we must be firm in establishing floodplain regulation and just say no to most floodplain development.

Sustainability

Rivers are the lifeblood of the land. They provide water resources and routes for the transport of people and goods, and they maintain important ecosystems, from wetlands to floodplains. Building a sustainable future is not possible without planning for sustainable rivers.

Earth as a System

Rivers are one of the land's major systems. They transport water and sediment while eroding the land to form most of our landscape. Over time, rivers change as a result of land use such as conversion of forest lands for urban and agricultural purposes. These changes have increased the flood hazard by: (1) filling channels in agricultural regions with sediment eroded from the land, which also depletes soils, and (2) increas-

ing runoff in urban areas, producing more and larger floods, while decreasing infiltration of water into the soil.

Hazardous Earth Processes, Risk Assessment, and Perception

Flooding is the most universally experienced hazard. It is also a hazard for which the risks are well known. A major problem is convincing people and communities that unwise land use on floodplains will lead to flood losses. The key is to educate people so that they gain a better understanding of the flood hazard and of where and why floods occur: in other words, heighten the public perception of flooding.

Scientific Knowledge and Values

The science of rivers, including their ecology and hydrology, is well advanced. However, our values often conflict with science when it comes to reducing the flood hazard. Often we choose a "technology fix" to build more flood-control dams, higher levees, more floodwalls, or to channelize rivers. These practices have damaged river ecosystems and have lured people to encroach on floodplains, leading to even greater flood losses. Floodplain management and river restoration reflect the value we place on rivers as a resource to be revered, not degraded.

Key Terms

channelization (p. 275)

channel pattern (p. 262)

channel restoration (p. 275)

continuity equation (p. 257)

discharge (p. 257)

downstream floods (p. 267)

drainage basin (p. 256)

flash floods (p. 266)

flooding (p. 264)

floodplain (p. 254)

floodplain regulation (p. 280)

levee (p. 252)

Review Questions

1. Define drainage basin.

2. What are the three components that make up the total load a stream carries?

3. What is the continuity equation?

4. What were the lessons learned from the 1992 flood of the Ventura River?

5. Differentiate between competency and capacity.

6. Differentiate between braided and meandering channels.

7. What is the riverine environment?

8. Differentiate between pools and riffles.

9. Differentiate between upstream and downstream floods.

10. What do we mean when we say a 10 year flood has occurred?

11. How does urbanization affect the flood hazard?

12. What are the major factors that control damage caused by floods?

13. What are the primary and secondary effects of flooding?

14. What do we mean by floodplain regulation?

15. Define channelization.

16. What is channel restoration?

Critical Thinking Questions

1. You are a planner working for a community that is expanding into the headwater portions of drainage basins. You are aware of the effects of urbanization on flooding and want to make recommendations to avoid some of these effects. Outline a plan of action.

2. You are working for a county flood-control agency that has been channelizing streams for many years. Although bulldozers are usually used to straighten and widen the channel, the agency has been criticized for causing extensive environmental damage. You have been asked to develop new plans for channel restoration to be implemented as a stream-maintenance program. Devise a plan of action that would convince the official in charge of the maintenance program that your ideas will improve the urban stream environment and help reduce the potential flood hazard.

3. Does the community you live in have a flood hazard. If not, why not. If there is a hazard what has/is being done to reduce or eliminate the hazard? What more could be done?

NINE

Slope Processes, Landslides, and Subsidence

Learning Objectives

Landslides, the movement of materials down slopes, constitute a serious natural hazard in many parts of the United States and the rest of the world. Landslides are often linked to other hazards such as earthquakes and volcanoes. Most landslides are small and slow, but a few are large and fast. Both may cause significant loss of life and damage to human property, particularly in urban areas. In this chapter we focus on the following learning objectives:

- Understand basic slope processes and the causes of slope failure

- Understand the role of driving and resisting forces on slopes and how these are related to slope stability

- Understand how slope angle and topography, vegetation, water, and time affect both slope processes and the incidence of landslides

- Understand how human use of the land has resulted in landslides

- Know methods of identification, prevention, warning, and correction of landslides

- Understand processes related to land subsidence

Landslide La Conchita, California, landslide in 2005 destroyed 13 homes and killed 10 people. *(AP/Wide World Photos)*

CASE HISTORY | La Conchita Landslide of 2005

The small beachside community of La Conchita, located about 80 km (50 mi) northwest of Los Angeles, California, was the site of a disaster on January 10, 2005 (see opening photograph). On that day, a fast-moving landslide, debris flow damaged or destroyed 36 homes and killed 10 people. The 2005 landslide was a reactivation of a 1995 landslide that destroyed several homes but caused no deaths. Both the 1995 and 2005 events were in turn part of an older prehistoric landslide, less than about 6,000 years old, on the steep slope directly above La Conchita. As we shall see this 6,000 year old slide is part of an even older, larger prehistoric slide on the mountain above.

The winter of 2004–2005 was particularly wet, with high-intensity rainfall occurring at times. Neither residents nor local officials recognized that another landslide was imminent. The 2005 debris flow was different from that of 1995 in that the flow was faster, at 45 km/hr (30 mph), and moved further into the community. A number of people were trapped in their homes while others ran for their lives.

Directly behind the community is a 200 m (600 ft) high slope that produces continuing landslide hazards for humans living below it. La Conchita should never have been constructed at the foot of that slope. It has been known that landslides have occurred in the area for about 100 years. More telling, the community is built on about 15 m (50 ft) of old landslide deposits. Landslides above La Conchita and to the east and west have been occurring for thousands of years, long before people made the decision to build beach homes at the base of the steep slope.

A study of La Conchita and the surrounding area suggests that the debris flows and slides of 1995, 2005, and older events are a part of a much larger prehistoric landslide less than about 40,000 years old that had not been recognized at the time of the more recent events (Figure 9.1). There is no evidence that the much larger, older prehistoric slide is moving as a mass, but parts are active, especially at the western margin of the slide.[1]

It is not *if*, but *when* future landslides will occur on the slopes above La Conchita. No part of the La Conchita community exempt from a future landslide.[2]

A potential solution to reduce the hazard to people and houses is to transform the land use of La Conchita from an area where people live to a coastal park. Not all people living there will be happy to hear such a suggestion, but society could help relocate people with fair compensation for valuable coastal property. The result would be a transformation of a hazardous site into a resource for future generations. At the least, we need to be diligent in the future concerning land-use planning and avoid unwise development on the large prehistoric slide above La Conchita. A similar, but smaller, prehistoric slide was reactivated in Malibu, California, in the mid-1980s, causing claimed damages in excess of $200 million.

What has been learned so far at La Conchita is that both prehistoric and active landslides can be recognized and their future activity evaluated. We now need to be proactive and take steps to reduce risk from future landslides.

9.1 Introduction to Landslides

Landslides and related phenomena cause substantial damage (Figure 9.2) and loss of life. In the United States, between 25 and 50 people are killed each year by landslides. This number increases to between 100 and 150 if collapses of trenches and other excavations are included. The total annual cost of damages is about $3.5 billion.[3]

Landslides and other types of ground failure are natural phenomena that would occur with or without human activity. However, human land use has led to an increase in these events in some situations and a decrease in others. For example, landslides may occur on previously stable hillsides that have been modified for housing development; on the other hand, landslides on naturally sensitive slopes are sometimes averted through the use of stabilizing structures or techniques.

Mass wasting is a comprehensive term for any type of downslope movement of Earth materials. In its more restricted sense, mass wasting refers to a downslope movement of rock or soil as a more or less coherent mass. In this chapter, we consider landslides in the restricted sense. We will also discuss the related phenomena of earthflows and mudflows, rockfalls, and snow or debris avalanches. For the sake of convenience, we sometimes refer to all of these as **landslides.** We will also discuss **subsidence,** a type of ground failure characterized by nearly vertical deformation, or the downward sinking of Earth materials. Subsidence often produces circular surface pits but may produce linear or irregular patterns of failure.

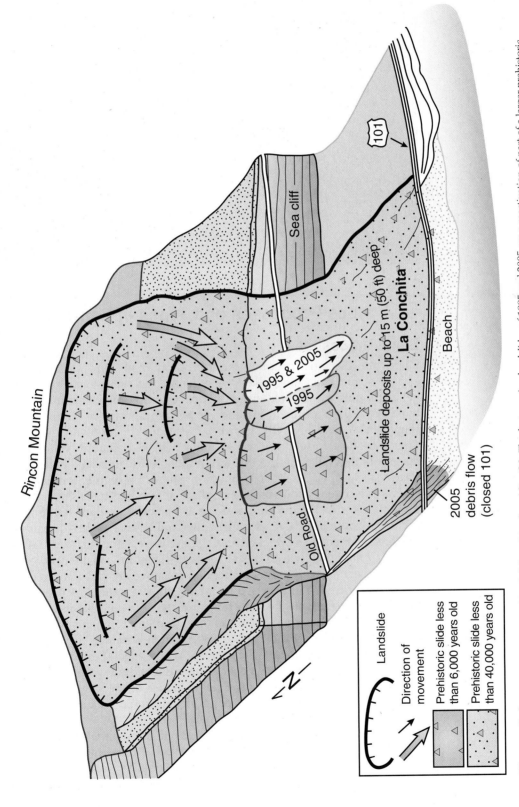

Figure 9.1 Idealized diagram of landslides at La Conchita, CA The fast-moving landslides of 1995 and 2005 are a reactivation of part of a larger prehistoric slide that is less than 6,000 years old, which is part of a much larger, older (few tens of thousands of years) landslide called the Rincon Mountain landslide.

Figure 9.2 Homes destroyed
Aerial photograph of the lower part of the Portuguese Bend landslide in southern California (1950s). Note the kink in the pier near the end of the landslide. Eventually, most of the homes as well as the swim club and pier shown here were destroyed by the slow-moving landslide.
(John S. Shelton)

9.2 Slope Processes and Types of Landslides

Slope Processes

Slopes are the most common landforms on Earth, and although most slopes appear stable and static, they are dynamic, evolving systems. Slopes are not generally uniform in their shape but are composed of segments that are straight or curved. Two different-looking slopes are shown in Figure 9.3. The first (Figure 9.3a) has a high cliff, or *free face*, a straight, nearly vertical slope segment. This is a slope formed on hard granite rock in Yosemite National Park. Rock fragments that fall from the free-face may accumulate at the base of the slope to form a *talus* slope. The free-face and talus slopes are segments of the slope. Notice the surface cover of soil, missing on the free face and talus slopes of Figure 9.3a. The hillslopes in Figure 9.3b are formed on rocks that are not as strong as the granite in Yosemite. The slopes are gentler and lack a free face. In this photo, there are three segments: an upper convex slope, a lower concave slope, and, separating the two, a straight slope. Thus, we see that slopes are usually composed of different slope segments. The five segment types above are sufficient to describe most slopes you will encounter in nature. Which slope segments are present on a particular slope depends on the rock type and the climate. Free-face development is more common on strong hard rocks or in arid environments where there is little vegetation. Convex and concave slopes are more common on softer rocks or with a more humid wet climate where thick soil and vegetation are present. But there are many exceptions to these general rules, depending upon local conditions. For example, the gentle, convex, red-colored slopes in the lower part of Figure 9.3b have formed on weak, easily eroded metamorphic rock (schist) in a semiarid climate on Santa Cruz Island, California.

Material on most slopes is constantly moving down the slope at rates that vary from an imperceptible creep of soil and rock to thundering avalanches and rockfalls that move at tremendous velocities. These slope processes are one significant reason that valleys are much wider than the streams they contain.

Types of Landslides

Earth materials may fail and move or deform in several ways (Figure 9.4). Rotational slumps involve sliding along a curved slip plain producing slump blocks

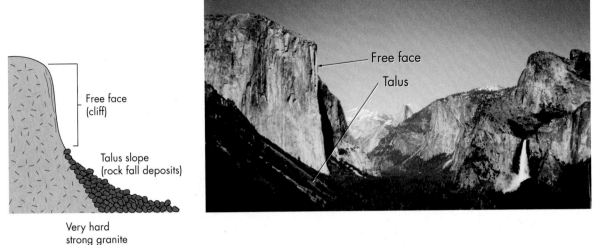

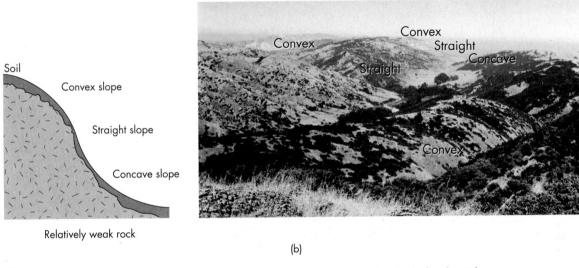

Figure 9.3 **Slope segments** (a) Slope on hard granite in Yosemite National Park with free face (several thousand feet high) and talus slope. (b) Slopes on Santa Cruz Island, California, on relatively weak schist (red) and volcanic rocks (white), with convex, straight, and concave slope segments. *(Edward A. Keller)*

(Figure 9.4a, b). Translational sliding is downslope movement of Earth materials along a planar slip plane such as a bedding plane or fracture (Figure 9.4c, d). Rock fall is the free fall of Earth materials from a free face of a cliff (Figure 9.4e). Flows are the downslope movement of unconsolidated (usually saturated) materials in which particles move about and mix within the mass. Very slow flowage of rock or soil is called creep (Figure 9.4f); rapid flowage may be an earthflow, mudflow, or debris flow. An earthflow (Figures 9.4g and 9.5) often originates on a slope where soil partially liquefies and runs out. The source area becomes a bowl-shaped depression and a depositional area spreads out at the toe of the slope, giving the event an "hourglass shape." A debris flow or mudflow (Figure 9.4h) is a mixture of rock, soil, and organic matter that mixes with air and water to flow rapidly down slope. The upper part of the flow is often confined to a channel or valley but may spread out when it is no longer confined. A debris flow has less than 50 percent fines (sand, silt, and clay), whereas a mudflow has more than 50 percent fines by volume. A debris avalanche (Figure 9.4i) is a very rapid to extremely rapid debris flow. Large debris avalanches can cause catastrophic damage and loss of life. Lateral spreads (Figure 9.4j) are a type of landslide that often occurs on nearly flat slopes or very gentle slopes. The movement starts with liquefaction of silts, clays, or fine sands during earthquake shaking or other disturbance. The actual

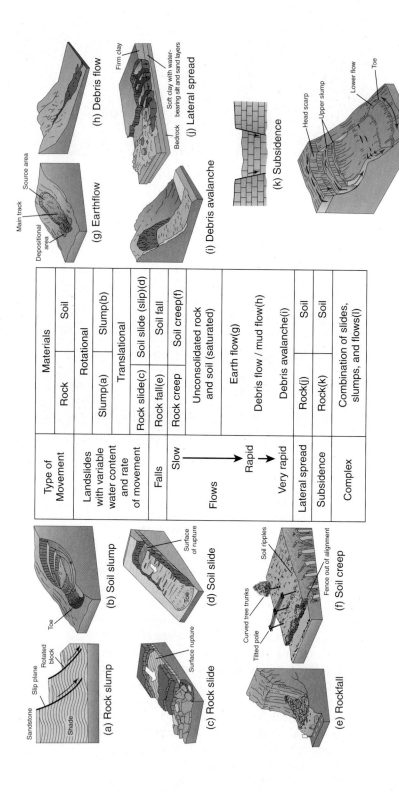

Figure 9.4 Types of landslides Classification of landslides based on type of movement, materials, water content, and rate of movement.
(Modified after U.S. Geological Survey 2004. Fact Sheet 2004-3072)

Figure 9.5 **Earthflow** on soft shale near Santa Barbara, California. *(Edward A. Keller)*

movement is lateral extension. If stronger coherent rock or soil is at the surface and is over a soil layer that liquefies, the stronger material may fracture, translate, rotate, or disintegrate and flow. Lateral spreads often start suddenly and then become larger in a slower, progressive manner.[3] Subsidence may occur on slopes or on flat ground and involves the sinking of a mass of Earth material below the level of the surrounding surface (Figure 9.4k).

Landslides are commonly complex combinations of sliding and flowage. As an example, Figures 9.4l and 9.6 show failures consisting of an upper *slump* that is transformed to a flow in the lower part of the slide. Such complex landslides may form when water-saturated Earth materials flow from the lower part of the slope, undermining the upper part and causing slumping of blocks of Earth materials.

Important variables in classifying downslope movements are the type of movement (slide, fall, flow, slump, or complex movement), slope material type, amount of water present, and rate of movement. In general, the movement is considered rapid if it can be discerned with the naked eye; otherwise, it is classified as slow (Figure 9.4). Actual rates vary from a slow creep of a few millimeters or centimeters per year to very rapid, at 1.5 m (5 ft) per day, to extremely rapid, at 30 m (98 ft) or more per second.[4]

Figure 9.6 **Complex landslide** at La Conchita, California (1995). This slide, which had an upper slump block and a lower flow, destroyed the three-story home in its path. *(Edward A. Keller)*

9.3 **Slope Stability**

Forces on Slopes

To determine the causes of landslides, we must examine slope stability, which can be expressed in terms of the forces that act on slopes. The stability of a slope expresses the relationship between *driving forces*, which move Earth materials down a slope, and *resisting forces*, which oppose such movement. The most common driving force is the downslope component of the weight of the slope material, including anything superimposed on the slope such as vegetation, fill material, or buildings. The most common resisting force is the strength, or the resistance to failure by sliding or flowing, of the slope material acting along potential slip planes. Potential *slip planes* are geologic surfaces of weakness in the slope material; for example, foliation planes in a slope composed of schist, bedding planes in sedimentary rocks, and fractures in all rock types are potential slip planes.

Slope stability is evaluated by computing a **safety factor (SF),** defined as the ratio of the resisting forces to the driving forces. If the safety factor is greater than 1, the resisting forces exceed the driving forces and the slope is considered stable. If the safety factor is less than 1, the driving forces exceed the resisting forces and a slope failure can be expected. Driving and resisting forces are not static: As local conditions change, these forces may change, increasing or decreasing the safety factor.

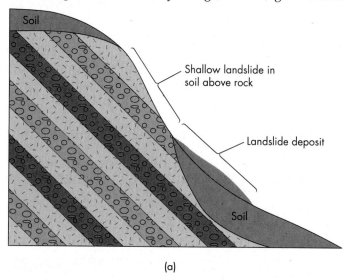

Soil

Shallow landslide in soil above rock

Landslide deposit

Soil

(a)

Soil slips

(b)

Figure 9.7 **Multiple soil slips**
(a) Diagram of a shallow soil slip.
(b) Shallow soil slips on steep slopes in southern California. The vegetation is low brush known as chaparral.
(Edward A. Keller)

Driving and resisting forces on slopes are determined by the interrelationships of the following variables:

- Type of Earth materials
- Slope angle and topography
- Climate
- Vegetation
- Water
- Time

The Role of Earth Material Type

The material composing a slope affects both the type and the frequency of down-slope movement. Slides have two basic patterns of movement, rotational and translational. In **rotational slides,** or slumps, the sliding occurs along a curved slip surface (Figure 9.4a, b). Because the movement follows a curve, slump blocks, the blocks of Earth material that are undergoing slump, tend to produce topographic benches, which can be rotated and tilted in the upslope direction like those in Figure 9.4b. Slumps are most common on soil slopes, but they also occur on some rock slopes, most often in weak rock such as shale. **Translational slides** are planar; that is, they occur along inclined slip planes within a slope (Figure 9.4c, d) (see A Closer Look: Translation Slides Along Bedding Planes). Common translation slip planes in rock slopes include fractures in all rock types, bedding planes, weak clay layers, and foliation planes in metamorphic rocks. *Soil slips*, another type of translational slide, can also occur in some areas. Soil slips are very shallow slides in soil over rock that occur parallel to the slope (Figure 9.7). For soil slips,

A CLOSER LOOK | Translation Slides Along Bedding Planes

Bedding planes are potential slip planes for landslides when they are inclined downslope and "daylight"—that is, are exposed on the surface of a slope. A slope of a seacliff with daylighting bedding planes in shale is diagramatically shown on Figure 9.A. Several months later in 2003 the slope failed, perhaps as a result of water added to the top of the slope where grass was planted and watered. The landslide deposits cover part of the sandy beach (Figure 9.B) and a catastrophe was narrowly avoided as a beach party was happening a short distance away.

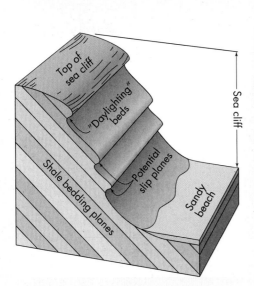

Figure 9.A Daylighting beds Bedding planes that intersect the surface of the land on a slope are said to "daylight." Such beds are potential slip planes.

Figure 9.B Translation slide This slide occurred in late 2003. Failure was along a "daylight" bedding plane. Slide deposits cover part of the beach. *(Edward A. Keller)*

Vegetation removed by slide

Retaining wall—to keep landslide deposits off the road

(a)

Head of flow

Track of flow

Flow deposits

(b)

Figure 9.8 Shallow slides (a) Shallow soil slip, North Carolina. (b) Shallow debris flow, Klamath River, California. Note the long narrow track and debris on the bank of the river. The logging road near the bend of the failure may have helped destabilize the slope. *(Edward A. Keller)*

the slip plane is usually above the bedrock but below the soil within a slope material known as colluvium, a mixture of weathered rock and other material (Figure 9.8).

Material type is a factor in falls as well as slides. If a resistant rock forms a very steep slope, weathering or erosion along fractures may cause a rockfall. Such failures on nearly vertical high slopes in hard granite present a continuous, chronic hazard in Yosemite National Park (Figure 9.9).

The type of materials composing a slope may greatly influence the type of slope failure that occurs. For example, on shale slopes or slopes on weak volcanic pyroclastic materials, failure commonly occurs as *creep*, the very slow downslope

Figure 9.9 Rockfall, Yosemite
National Park, California This
rockfall occurred at 6:52 pm, on
July 10, 1996 at Happy Isles along the
Merced River. The rockfall fell from
near Glacer Point 665 m (2,180 ft) to
the valley floor, reaching a speed or
110 m/s (250 mph). One person was
killed and over 1,000 trees were blown
down by the blast of air. The volume of
rock was about 30,000 m³. If the fall
had occurred earlier when numerous
people were at the Happy Isles Visitors
Center, many deaths could have
resulted. *(Edwin L. Harp/U.S. Geological
Survey/U.S. Department of the Interior)*

movement of soil or rock; *earthflows*, or mudflows, the downslope flow of saturated Earth materials; slumps; or soil slips. Slopes formed in resistant rock such as well-cemented sandstone, limestone, or granite do not experience the same problems. Therefore, before developing on shale or other weak rock slopes, people must give careful consideration to the potential landslide hazard.

The Role of Slope and Topography

The hillslope angle, which is a measure of how steep a hillslope is, is usually called the *slope*. Slope greatly affects the relative magnitude of driving forces on slopes. As the slope of a hillside or potential slip plane within a slope increases, say from 15 degrees to 45 degrees or steeper, the driving force also increases; therefore, landslides are more frequent on steep slopes. A study of landslides that occurred during two rainy seasons in California's San Francisco Bay area established that 75 percent to 85 percent of landslide activity is closely associated with urban areas on steep slopes.[5] Within the United States, the coastal mountains of California and Oregon, the Rocky Mountains, and the Appalachian Mountains have the greatest frequency of landslides. All of the types of downslope movement shown on Figure 9.4 occur on steep slopes in those locations.

(a)

(b)

Figure 9.10 **Shallow soil slips can kill** (a) Shallow soil slips on steep southern California, vegetated slopes. *(Edward A. Keller)* (b) A home in southern California destroyed by a shallow soil slip, debris flow that originated as a soil slip. This 1969 event claimed two lives. *(Courtesy of John Shadle, Los Angeles Department of Building and Safety)*

House destroyed

Steep slopes are often associated with rockfalls and *debris avalanches*, which are the very rapid downslope movement of soil, rock, and organic debris. In southern California, shallow soil slips are common on steep saturated slopes. Once they move downslope, these soil slips are often transformed into earthflows or debris flows, which can be extremely hazardous (Figure 9.10). Earthflows can occur on moderate slopes, and creep can be observed on very gentle slopes.

Debris flows are the downslope flow of relatively coarse material; more than 50 percent of particles in a debris flow are coarser than sand. Debris flows can move very slowly or rapidly, depending on conditions. Debris flows, debris avalanches, and mudflows vary in size: they can be relatively small to moderate events, confined to a single valley of slope with a few hundred to hundreds of thousands of cubic meters of debris. However, they can also be huge events involving an entire flank of a mountain, measured in cubic kilometers of material (see volcanic mudflows and debris flows, discussed in Chapter 7).

The Role of Climate

Climate can be defined as characteristic weather at a particular place or region over seasons, years, or decades. Climate is more than the average air temperature and amount of precipitation. Climate includes seasonal patterns of weather such as winter rains along the West Coast of the United States, summer thunderstorm activity in the southwestern United States, and hurricane activity of the southeastern United States. The subject of climate and how it changes is discussed in detail in Chapter 19.

The role of climate is important in our discussion of landslides because climate influences the amount and timing of water, in the form or rain and snow, that may infiltrate or erode a hillslope as well as the type and abundance of vegetation that grows on a hillslope. Hillslopes in arid and semiarid climates tend to have sparse vegetation and soils with a lot of bare rock exposed. Slope elements such as the free-face and talus slopes tend to be more common because small differences in resistance of rock to weathering and erosion are emphasized. Common landslide activity in arid and semiarid regions includes rockfall, debris flows, and shallow soil slips.

The Role of Vegetation

In the more subhumid to humid regions of the world, vegetation is abundant, thick soil cover develops, and the slopes have many more convex and concave slope segments. Landslide activity includes deep complex landslides, earthflows, and soil creep. The role of vegetation in landslides and related phenomena is complex. Vegetation in an area is a function of several factors, including climate, soil type, topography, and fire history, each of which also influences what happens on slopes. Vegetation is a significant factor in slope stability for three reasons:

- Vegetation provides a cover that cushions the impact of rain falling on slopes, facilitating infiltration of water into the slope while retarding grain-by-grain erosion on the surface.
- Vegetation has root systems that tend to provide an apparent cohesion (like iron bars in concrete) to the slope materials, which increases resistance to landsliding.
- Vegetation adds weight to the slope.

In some cases, the presence of vegetation increases the probability of a landslide, especially for shallow soil slips on steep slopes. In southern California coastal areas, one type of soil slip occurs on steep-cut slopes covered with low vegetation called ice plant (Figure 9.11). During especially wet winter months, the shallow-rooted ice plants take up water, adding considerable weight to steep slopes—each leaf stores water and looks like a small canteen—thereby increasing the driving forces. The plants also cause an increase in the infiltration of water into the slope, which decreases the resisting forces. When failure occurs, the plants and several centimeters of roots and soil slide to the base of the slope.

Soil slips on natural steep slopes are a serious problem in southern California. Chaparral, the dense shrubs or brush shown in Figure 9.7, facilitates an increase in water infiltrating into the slope, lowering the safety factor.[6]

The Role of Water

Water is almost always directly or indirectly involved with landslides, so its role is particularly important.[7] When studying a landslide, we look first to examine what the water on or in the slope is doing. There are three basic ways in which water

(a)

Ice plants
slipped down slope

(b)

Figure 9.11 **Ice plants on slopes are often unstable** Shallow soil slips on steep slopes covered with shallow-rooted ice plants near Santa Barbara, California: (a) an embankment on a road; (b) a home site. The plastic sheet is an attempt to reduce infiltration of rain water. *(Edward A. Keller)*

(b)

(a)

Figure 9.12 **Water eroding the toe of a slope causes instability**
(a) Stream-bank erosion caused this failure, which damaged a road,
San Gabriel Mountains, California. *(Edward A. Keller)* (b) Beachfront home
being threatened by a landslide, Cove Beach, Oregon. *(Gary Braasch/Getty
Images Inc.)*

on or in the slope can affect stability: (1) landslides such as shallow soil slips can
develop during rainstorms when slopes become saturated; (2) landslides such as
slumps or translational slides can develop months or even years after infiltration
of water deep into the slope; and (3) water can erode the base or toe of a slope,
thereby decreasing slope stability.

Water's ability to erode affects the stability of slopes. Stream or wave erosion on
a slope may remove material and create a steeper slope, thus reducing the safety
factor (Figure 9.12). This problem is particularly critical if the base of the slope
is an old, inactive landslide that is likely to move again if stability is reduced
(Figure 9.13). Therefore, it is important to recognize old landslides along potential

Curve in coastline identifies slide

Head of slide

House destroyed

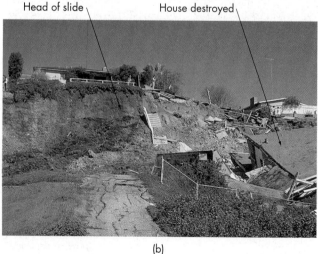

(a)

(b)

Figure 9.13 **Reactivation of a slide** (a) Aerial view of a landslide along the Santa Barbara coastal
area. The arrow points to the location of the slide. (b) Closeup of the slide, which destroyed two homes.
The slide is a reactivation of an older failure. *(Courtesy of Don Weaver)*

road cuts and other excavations before construction in order to isolate and correct potential problems.

Another way that water can cause landslides is by contributing to spontaneous liquefaction of clay-rich sediment, or *quick clay*. When disturbed, some clays lose their shear strength, behave as a liquid, and flow. The shaking of clay below Anchorage, Alaska, during the 1964 earthquake produced this effect and was extremely destructive. In Quebec, Canada, several large slides associated with quick clays have destroyed numerous homes and killed about 70 people. The slides occurred on river valley slopes when initially solid material was converted into a liquid mud as the sliding movement began.[8] These slides are especially interesting because the liquefaction of clays occurs without earthquake shaking. The slides are often initiated by river erosion at the toe of the slope and, although they start in a small area, may develop into large events. Since they often involve the reactivation of an older slide, future problems may be avoided by restricting development in these areas.

The Role of Time

The forces on slopes often change with time. For example, both driving and resisting forces may change seasonally as the moisture content or water table position alters. Much of the chemical weathering of rocks, which slowly reduces their strength, is caused by the chemical action of water in contact with soil and rock near Earth's surface. Water (H_2O) is often acidic because it reacts with carbon dioxide (CO_2) in the atmosphere and soil to produce weak carbonic acid (H_2CO_3) This chemical weathering is especially significant in areas with limestone, which is susceptible to weathering and decomposition by carbonic acid. Changes due to weathering are greater in especially wet years, as reflected by the increased frequency of landslides during or after wet years. In other slopes, there may be a continuous reduction in resisting forces over time, perhaps due either to weathering, which reduces the cohesion in slope materials, or to a regular increase in pore water pressure in the slope from natural or artificial conditions. A slope that is becoming less stable with time may have an increasing rate of creep until failure occurs. The case history of the Vaiont Dam illustrates this concept (see Case History: Vaiont Dam).

CASE HISTORY | Vaiont Dam

The world's worst dam disaster occurred on October 9, 1963, when approximately 2,600 lives were lost at the Vaiont Dam in Italy (Figure 9.C). The disaster involved the world's highest thin-arch dam, yet, strangely, no damage was sustained by the main shell of the dam or the abutments.[9] The tragedy was caused by a huge landslide in which more than 238 million cubic meters (0.06 mi^3) of rock and other debris moved at speeds of about 95 km per hour (59 mi per hour) down the north face of the mountain above the reservoir. Slide material completely filled the reservoir for 1.8 km (1.1 mi.) along the axis of the valley to heights of nearly 152 m (500 ft) above the reservoir level. The rapid movement created a tremendous updraft of air and propelled rocks and water up the north side of the valley, higher than 250 m (820 ft) above the reservoir level. The slide and its accompanying blasts of air, water, and rock produced strong earthquakes recorded many kilometers away. It blew the roof off one man's house well over 250 m (820 ft) above the reservoir and pelted the man with rocks and debris. The filling of the reservoir produced waves of water more than 90 m (295 ft) high that swept over the abutments of the dam.

The waves were still more than 70 m (230 ft) high more than 1.5 km (0.9 mi) downstream; in fact, everything for many kilometers downstream was completely destroyed. The entire event—slide and flood—was over in less than 7 minutes.

The landslide followed a 3 year period of monitoring the rate of creep on the slope, which varied from less than 1 to as many as 30 cm (12 in.) per week, until September 1963, when it increased to 25 cm (10 in.) per day. Finally, on the day before the slide, it was about 100 cm (39 in.) per day. Although engineers expected a landslide, they did not realize until the day before the slide that a large area was moving as a uniform, unstable mass. Interestingly, animals grazing on the slope had sensed danger and moved away on October 1, over a week before the landslide.

The slide was caused by a combination of factors. First, adverse geologic conditions—including weak rocks and limestone with open fractures, sinkholes, and weak clay layers that were inclined toward the reservoir—produced unstable blocks (Figure 9.D). Second, very steep topography created a strong driving force. Third, water pressure was increased in

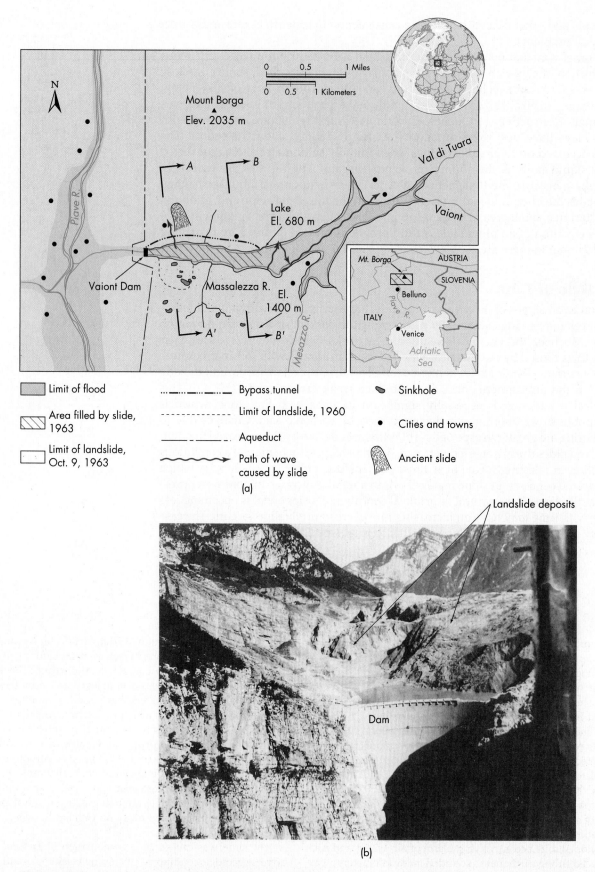

Limit of flood

Area filled by slide, 1963

Limit of landslide, Oct. 9, 1963

·—··—··—··— Bypass tunnel

------------- Limit of landslide, 1960

——·——·——·— Aqueduct

〜〜〜〜〜→ Path of wave caused by slide

Sinkhole

Cities and towns

Ancient slide

(a)

Landslide deposits

Dam

(b)

Figure 9.C **Sketch map of the Vaiont Reservoir** (a) showing the 1963 landslide that displaced water that overtopped the dam and caused severe flooding and destruction over large areas downstream. A-A' and B-B' are the cross sections shown in Figure 8.D. *(After Kiersch, G. A. Civil Engineering 34:32–39)* (b) Photograph of the Vaiont Dam after the landslide. Notice that the concrete dam is still intact but the reservoir above the dam is completely filled (or nearly so) with landslide deposits. *(ANSA)*

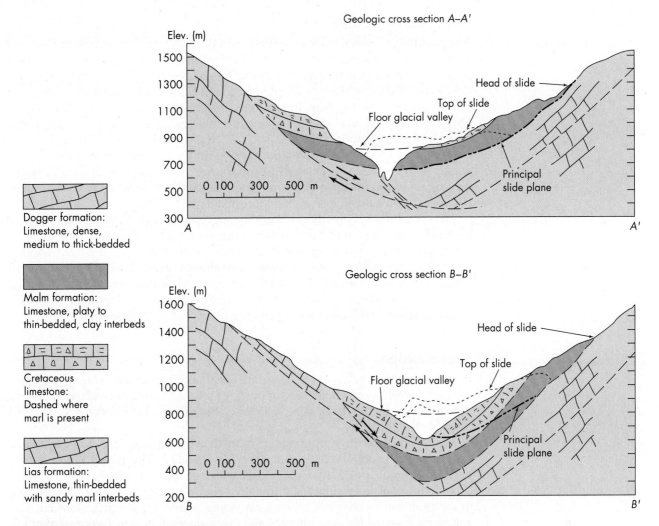

Geologic cross section A–A'

Dogger formation: Limestone, dense, medium to thick-bedded

Malm formation: Limestone, platy to thin-bedded, clay interbeds

Cretaceous limestone: Dashed where marl is present

Lias formation: Limestone, thin-bedded with sandy marl interbeds

Geologic cross section B–B'

Figure 9.D **Generalized geologic cross sections** through the slide area of the Vaiont River Valley. The locations of the sections are shown in Figure 8.C. *(After Kiersch, G. A. Civil Engineering 34:32–39)*

the valley rocks because of the water in the reservoir. The rate of creep before the slide increased as the groundwater level rose in response to higher reservoir levels. Fourth, heavy rains from late September until the day of the disaster further increased the weight of the slope materials, raised the water pressure in the rocks, and produced runoff that continued to fill the reservoir even after engineers tried to lower the reservoir level.

Officials concluded that the disaster was caused by an increase in the driving forces accompanied by a great decrease in the resisting forces, as rising groundwater in the slope increased along zones of weakness in the rock.[9]

9.4 Human Use and Landslides

The effect of human use on the magnitude and frequency of landslides varies from nearly insignificant to very significant. In cases in which our activities have little to do with the magnitude and frequency of landslides, we need to learn all we can about where, when, and why they occur to avoid developing in hazardous areas and to minimize damage. In cases in which human use has increased the number and severity of landslides, we need to learn how to recognize, control, and minimize their occurrence wherever possible.

Many landslides have been caused by interactions of adverse geologic conditions, excess moisture, and artificial changes in the landscape and slope material. The Vaiont Dam and Reservoir slide of 1963 in Italy is a classic example (see Case History: Vaiont Dam). Other examples include landslides associated with timber harvesting as well as numerous slides in urban areas.

Timber Harvesting

The possible cause-and-effect relationship between timber harvesting and erosion in northern California, Oregon, and Washington is a controversial topic. There is evidence to support the hypothesis that landslides, especially shallow soil slips, debris avalanches, and more deeply seated earthflows, are responsible for much of the erosion in these areas. In fact, one study in the western Cascade Range of Oregon concluded that shallow slides are the dominant erosion process in the area. Timber-harvesting activities, such as clear-cutting and road building over approximately a 20 year observation period on geologically stable land did not greatly increase landslide-related erosion. During that same time period, however, logging on weak, unstable slopes did increase landslide erosion by several times compared with landslide erosion on forested land.[10]

The construction of roads in areas to be logged is an especially serious problem because roads may interrupt surface drainage, alter subsurface movement of water, and adversely change the distribution of mass on a slope by cut-and-fill, or grading, operations.[10] As we learn more about erosional processes in forested areas, we are developing improved management procedures to minimize the adverse effects of timber harvesting. Nevertheless, we are not yet out of the woods with respect to landslide erosion problems associated with timber harvesting.

Urbanization

Human activities are most likely to cause landslides in urban areas where there are high densities of people and supporting structures such as roads, homes, and industries. Examples from Rio de Janeiro, Brazil, and Los Angeles, California, illustrate the situation.

Rio de Janeiro, with a population of more than 6 million people, may have more slope-stability problems than any other city its size.[11] The city is noted for the beautiful granite peaks that spectacularly frame the urban area (Figure 9.14). Combinations of steep slopes and fractured rock covered with thin soil contribute to the problem. In the past, many such slopes were logged for lumber and fuel and to clear space for agriculture. Landslides associated with heavy rainfall followed the logging activity. More recently, lack of room on flat ground has led to increased urban development on slopes. Vegetation cover has been removed, and roads leading to development sites at progressively higher areas are being built.

Slide areas

Figure 9.14 Landslides are common in the Rio de Janeiro area Panoramic view of Rio de Janeiro, Brazil, showing the steep (sugarloaf) hills. The combination of steep slopes, fractured rock, shallow soils, and intense precipitation contributes to the landslide problem, as do human activities such as urbanization, logging, and agriculture. Virtually all of the bare rock slopes were at one time vegetated, and that vegetation has been removed by landsliding and other erosional processes. *(Getty Images Inc.)*

Excavations have cut the base of many slopes and severed the soil mantle at critical points. In addition, placing slope fill material below excavation areas has increased the load on slopes already unstable before the fill. Because this area periodically experiences tremendous rainstorms, it is apparent that Rio de Janeiro has a serious problem.

In February 1988, an intense rainstorm dumped more than 12 cm (4.7 in.) of rain on Rio de Janeiro in 4 hours. The storm caused flooding and mudslides that killed about 90 people, leaving about 3,000 people homeless. Restoration costs exceeded $100 million. Many of the landslides were initiated on steep slopes where housing was precarious and control of stormwater runoff nonexistent. It was in these hill-hugging shantytown areas that most of the deaths from mudslides occurred. In addition, 25 patients and staff of a nursing home located in a more affluent mountainside area were killed when the home was knocked down by the landslide. If future disasters are to be avoided, Rio de Janeiro is in dire need of measures to control storm runoff and increase slope stability.

Los Angeles in particular and southern California in general have experienced a remarkable frequency of landslides associated with hillside development. Landslides in southern California result from complex physical conditions, in part because of the great local variation in topography, rock and soil types, climate, and vegetation. Interactions between the natural environment and human activity are complex and notoriously unpredictable. For this reason, the area has the sometimes dubious honor of showing the ever-increasing value of the study of urban geology."[12] Los Angeles has led the nation in developing building codes concerning grading for development.

In southern California, the grading process, in which benches, referred to as pads, are cut into slopes for home sites, has been responsible for many landslides. It took natural processes many thousands, if not millions, of years to produce valleys, ridges, and hills. In this century, we have developed the machines to grade them. F. B. Leighton writes: "With modern engineering and grading practices and appropriate financial incentive, no hillside appears too rugged for future development."[12] No Earth material can withstand the serious assault of modern technology. Thus, human activity is a geological agent capable of carving the landscape as do glaciers and rivers, but at a tremendously faster pace. Almost overnight, we can convert steep hills into a series of flat lots and roads, and such conversions have led to numerous artificially induced landslides. As shown in Figure 9.15, oversteepened slopes in conjunction with increased water from sprinkled lawns or septic systems, as well as the additional weight of fill material and a house, make formerly stable slopes unstable. As a rule, any project that steepens or saturates a slope, increases its height, or places an extra load on it may cause a landslide.[12]

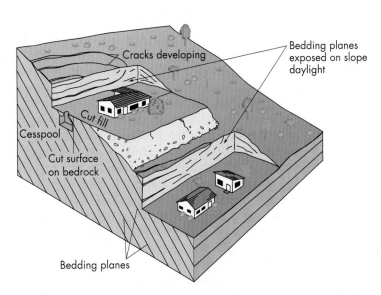

Figure 9.15 **Urbanization and landslide potential** Development of artificial translational landslides. Stable slopes may be made unstable by a variety of alterations including removing support from the bedding plane surfaces, adding water to the slope, steepening the slope, and adding fill on the slope. The cracks shown in the upper part of the diagram are an early sign that a landslide is likely to occur soon. *(Reprinted, with permission, from Leighton F. B. 1966. "Landslides and urban development." In* Engineering geology in southern California. *Los Angeles: Los Angeles Section of the Association of Engineering Geology)*

Landslides on both private and public land in Hamilton County, Ohio, have been a serious problem. The slides occur in glacial deposits, composed mostly of clay, lakebed sediments, and unstratified material called till as well as colluvium and soil formed on shale. The average cost of damage exceeds $5 million per year. Major landslides in Cincinnati, Ohio, have damaged highways and several private structures.[2]

Modification of sensitive slopes associated with urbanization in Allegheny County, Pennsylvania, is estimated to be responsible for 90 percent of the landslides in the area. An average of $2 million in damages results each year from these landslides. Most of the landslides are slow moving, but one rockfall in an adjacent county crushed a bus and killed 22 passengers. Most of the landslides in Allegheny County are caused by construction activity that loads the top of a slope, cuts into a sensitive location such as the toe of a slope, or alters water conditions on or beneath the surface of a slope.[13]

9.5 Minimizing the Landslide Hazard

Minimizing the landslide hazard requires identifying areas in which landslides are likely to occur, designing slopes or engineering structures to prevent landslides, warning people in danger areas of impending slides, and controlling slides after they have started moving. The most preferable and least expensive option to minimize the landslide hazard is to avoid development on sites where landslides are occurring or are likely to occur.

Identifying Potential Landslides

Identifying areas with a high potential for landslides is the first step in developing a plan to avoid landslide hazards. Slide tendency can be recognized by examining both geologic conditions in the field and aerial photographs to identify previous slides. This information can then be used to evaluate the risk and produce slope stability maps.

Once a landslide hazard is identified, it must be evaluated. A landslide inventory, which may be a reconnaissance map showing areas that have apparently experienced slope failure, is prepared. This inventory may be done by aerial photographic interpretation followed by an on-site check. At a more detailed level the landslide inventory may be a map that shows definite landslide deposits in terms of their relative activity. Figure 9.16a shows an example of such a map for part of Santa Clara County, California. Information concerning past landslide activity may then be combined with land-use considerations to develop a slope stability or landslide hazard map with recommended land uses, as shown in Figure 16b. The latter map is of most use to planners, whereas the former supplies useful information for the engineering geologist. These maps do not take the place of detailed fieldwork to evaluate a specific site but serve only as a general guideline for land-use planning and more detailed geologic evaluation. Determining the landslide risk and making landslide risk maps is more complicated, since it involves the probability of occurrence and assessment of potential losses.[14]

Grading codes to minimize the landslide hazard have been in effect in the Los Angeles area since 1963. These codes were instituted in the aftermath of the landsliding in the 1950s and 1960s that resulted in great losses of life and property. Since the grading codes have been in effect and detailed engineering geology studies have been required, the percentage of hillside homes damaged by landslides and floods has been greatly reduced. Although initial building costs are greater because of the strict codes, they are more than balanced by the reduction of losses in subsequent wet years. Landslide disasters during extremely wet years will continue to plague us; however, the application of geologic and engineering information before hillside development can help minimize the hazard.

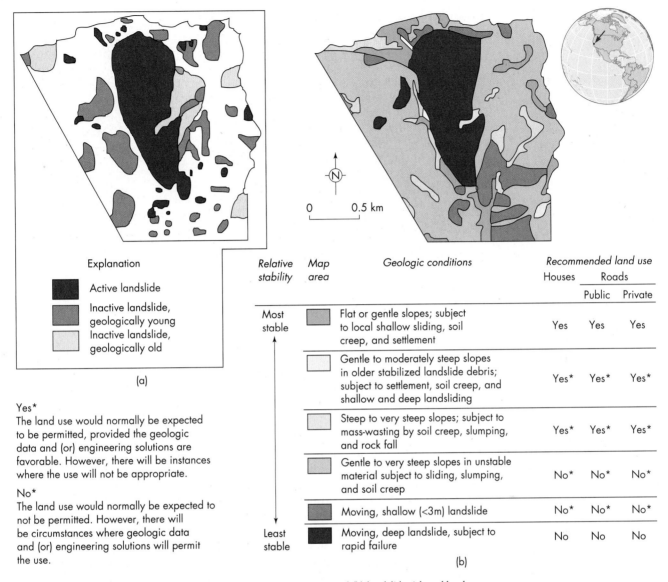

Figure 9.16 **Landslide hazard map** (a) Landslide inventory map and (b) landslide risk and land-use map for part of Santa Clara County, California. *(After U.S. Geological Survey. 1982. Goals and tasks of the landslide part of a ground-failure hazards reduction program. U.S. Geological Survey Circular 880)*

Preventing Landslides

Preventing large, natural landslides is difficult, but common sense and good engineering practices can help to minimize the hazard. For example, loading the top of slopes, cutting into sensitive slopes, placing fills on slopes, or changing water conditions on slopes should be avoided or done with caution.[13] Common engineering techniques for landslide prevention include provisions for surface and subsurface drainage, removal of unstable slope materials, construction of retaining walls or other supporting structures, or some combination of these.[4]

Drainage Control. Surface and subsurface drainage control are usually effective in stabilizing a slope. The objective is to divert water to keep it from running across or infiltrating into the slope. Surface water may be diverted around the slope by a series of surface drains. This practice is common for roadcuts (Figure 9.17a). The amount of water infiltrating a slope may also be controlled by covering the slope with an impermeable layer such as soil-cement, asphalt, or even plastic (Figure 9.17b). Groundwater may be inhibited from entering a slope by constructing subsurface drains. A drainpipe with holes along its length is

Figure 9.17 **Two ways to increase slope stability** (a) Drains on a roadcut to remove surface water from the cut before it infiltrates the slope. (b) Covering a slope with a soil-cement in Greece to reduce infiltration of water and provide strength. *(Edward A. Keller)*

surrounded with permeable gravel or crushed rock and is positioned underground so as to intercept and divert groundwater away from a potentially unstable slope.[4]

Grading. Although grading of slopes for development has increased the landslide hazard in many areas, carefully planned grading can be used to increase slope stability. In a single cut-and-fill operation, material from the upper part of a slope is removed and placed near the base. The overall gradient is thus reduced, and material is removed from an area where it contributes to the driving forces and placed at the toe of the slope, where it increases the resisting forces. However, this method is not practical on very steep, high slopes. As an alternative, the slope may be cut into a series of benches or steps. The benches are designed with surface drains to divert runoff. The benches reduce the overall slope of the land and are good collection sites for falling rock and small slides (Figure 9.18).[4]

Slope Supports. Retaining walls constructed from concrete, stone-filled wire baskets, or piles (long concrete, steel, or wooden beams driven into the ground)

Figure 9.18 **Benching** (upper right quadrant) a slope along the Pacific Ocean to reduce the overall steepness of the slope and provide for better drainage. *(Edward A. Keller)*

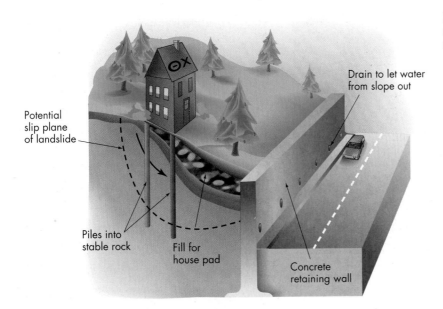

Figure 9.19 **How to support a slope** Some types of slope support: retaining wall, piles, and drains.

Drain to let water from slope out

Potential slip plane of landslide

Piles into stable rock

Fill for house pad

Concrete retaining wall

are designed to provide support at the base of a slope (Figure 9.19). They should be anchored well below the base of the slope, backfilled with permeable gravel or crushed rock (Figure 9.20), and provided with drain holes to reduce the chances of water pressure building up in the slope (Figure 9.20). The evolution of a retaining wall is shown on Figure 9.21. A shallow landslide along a road causes a problem (Figure 9.21a). The wall is shown during construction in 1999 in Figure 9.21b. The finished wall in 2001 now stabilizes the slope (Figure 9.21c).

Preventing landslides can be expensive, but the rewards can be well worth the effort. It has been estimated that the benefit-to-cost ratio for landslide prevention ranges from approximately 10 to 2,000. That is, for every dollar spent on landslide prevention, the savings will vary from $10 to $2,000.[15] The cost of *not* preventing a slide is illustrated by the massive landslide in Utah known as the Thistle slide. In April 1983, this slide moved across a canyon, creating a natural dam about 60 m (197 ft) high and flooding the community of Thistle, the Denver–Rio Grande Railroad and its switchyard, and a major U.S. highway (Figure 9.22).[15] The landslide and resultant flooding caused approximately $200 million in damages.

Figure 9.20 **Retaining wall** (concrete cribbing) with backfill used to help stabilize a roadcut. *(Edward A. Keller)*

Figure 9.21 **Steps in making a retaining wall** (a) Shallow slide in the early 1990s. (b) Retaining wall being constructed in 1999 to correct the problem. (c) Finished wall in 2001. *(Edward A. Keller)*

The Thistle slide involved a reactivation of an older slide, which had been known for many years to be occasionally active in response to high precipitation. Therefore, it could have been recognized that the extremely high amounts of precipitation in 1983 would cause a problem. In fact, a review of the landslide history suggests that the Thistle landslide was recognizable, predictable, and preventable! Analysis of the pertinent data suggests that emplacement of subsurface drains and control of surface runoff would have lowered the water table in the slide mass enough to have prevented failure. The cost of preventing the landslide was estimated to be between $300,000 and $500,000, a small amount compared with the damages caused by the slide.[15] Because the benefit-to-cost ratio in landslide prevention is so favorable, it seems prudent to evaluate active and potentially active landslides in areas where considerable damage may be expected and possibly prevented.

Warning of Impending Landslides

Landslide warning systems do not prevent landslides, but they can provide time to evacuate people and their possessions and to stop trains or reroute traffic. Surveillance provides the simplest type of warning. Hazardous areas can be visually inspected for apparent changes, and small rockfalls on roads and other areas can be noted for quick removal. Human monitoring of the hazard has the advantages of reliability and flexibility but becomes disadvantageous during adverse weather and in hazardous locations.[16] Other warning methods include electrical systems, tilt meters, and geophones that pick up vibrations from moving rocks. Shallow

Figure 9.22 Landslide blocks a canyon Thistle landslide, Utah. This landslide, which occurred in 1983, involved the reactivation of an older slide. The landslide blocked the canyon, creating a natural dam, flooding the community of Thistle, the Denver–Rio Grande Railroad, and a major U.S. highway. *(Michael Collier)*

Slide blocks canyon

wells can be monitored to signal when slopes contain a dangerous amount of water. In some regions, monitoring rainfall is useful for detecting when a threshold precipitation has been exceeded and shallow soil slips become more probable.

Correcting Landslides

After a slide has begun, the best way to stop it is to attack the process that started the slide. In most cases, the cause of the slide is an increase in water pressure, and in such cases an effective drainage program must be initiated. This may include surface drains at the head of the slide to keep additional surface water from infiltrating and subsurface drainpipes or wells to remove water and lower the water pressure. Draining tends to increase the resisting force of the slope material, thereby stabilizing the slope.[7]

9.6 Snow Avalanches

A **snow avalanche** is the rapid downslope movement of snow and ice, sometimes with the addition of rock, soil, and trees. Thousands of avalanches occur every year in the western United States. As more people venture into avalanche-prone areas and more development occurs in these areas, the loss of life and property due to avalanches increases. The most damaging avalanches occur when a large

(a)

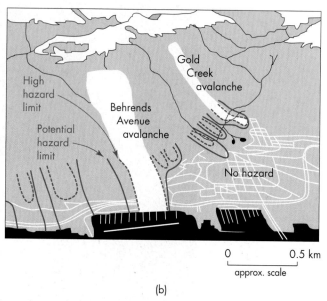

(b)

Figure 9.23 **Avalanche hazard** (a) Avalanche chute or track in the Swiss Alps. *(Edward A. Keller)* (b) Map of part of Juneau, Alaska, showing the avalanche hazard. *(After Cupp, D. 1982.* National Geographic *162:290–305)*

slab of snow and ice, weighing millions of tons, fails owing to the overloading of a slope with fresh snow or to development of zones of weakness within the snow-pack. These slabs move rapidly downslope at velocities of up to 100 km per hour (62 mi per hour). Avalanches tend to move down tracks, called *chutes*, that have previously produced avalanches (Figure 9.23). As a result, maps delineating the hazard may be developed. Avoiding hazardous areas is obviously the preferred and least expensive adjustment to avalanches. Other adjustments include clearing excess snow with carefully placed explosives, constructing buildings and structures to divert or retard avalanches, or planting trees in avalanche-prone areas to better anchor the snow on slopes.

9.7 Subsidence

Interactions between geologic conditions and human activity have been factors in numerous incidents of subsidence, the very slow to rapid sinking or settling of Earth materials (Figure 9.4e). Most subsidence is caused either by the withdrawal of fluids from subsurface reservoirs or by the collapse of surface and near-surface soil and rocks over subterranean voids.

Withdrawal of Fluids

The withdrawal of fluids—such as oil with associated gas and water, ground-water, and mixtures of steam and water for geothermal power—have all caused

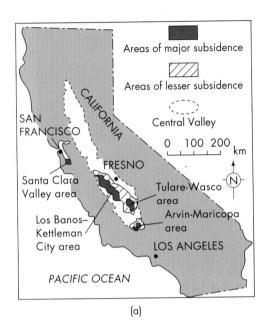

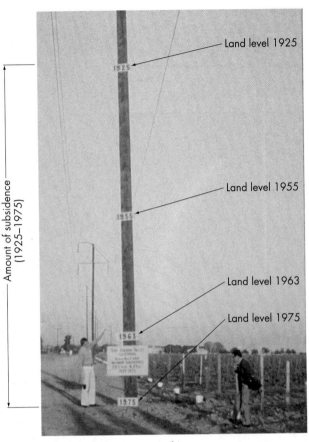

Figure 9.24 Land subsidence from groundwater extraction
(a) Principal areas of land subsidence in California resulting from
groundwater withdrawal. *(After Bull, W. B. 1973. Geological Society of
America Bulletin 84. Reprinted by permission)* (b) Photograph illustrating
the amount of subsidence in the San Joaquin Valley, California. The
marks on the telephone pole are the positions of the ground surface
in recent decades. The photo shows nearly 8 m (26 ft) of subsidence.
(Courtesy of Ray Kenny)

subsidence.[17] In all cases, the general principles are the same: Fluids in Earth
materials below Earth's surface have a high fluid pressure that tends to support
the material above. This is why a large rock at the bottom of a swimming pool
seems lighter: Buoyancy produced by the liquid tends to lift the rock. If support or
buoyancy is removed from Earth materials by pumping out the fluid, the support
is reduced, and surface subsidence may result.

Thousands of square kilometers of the central valley of California have subsided as a result of overpumping groundwater in the area (Figure 9.24a). More
than 5,000 km² (1,930 mi²) in the Los Banos–Kettleman City area alone have
subsided more than 0.3 m (1 ft), and within this area, one stretch of valley 113 km
(70 mi) long has subsided an average of more than 3 m (10 ft), with a maximum of
about 9 m (30 ft) (Figure 9.24b). As the water was mined, the pore pressure was
reduced and the grains were compacted;[18,19] the effect at the surface was subsidence (Figure 9.25). Similar examples of subsidence caused by overpumping are
documented near Phoenix, Arizona; Las Vegas, Nevada; Houston–Galveston,
Texas; and Mexico City, Mexico. The subsidence can cause extremely long, deep
surface fissures (open cracks) to form in sediments.[19]

Sinkholes

Subsidence is also caused by removal of subterranean Earth materials by natural
processes. Voids, large open spaces such as caves, often form by chemical weathering within soluble rocks such as limestone and dolomite, and the resulting lack
of support for overlying rock may cause it to collapse. The result is the formation
of a **sinkhole,** a circular area of subsidence caused by the collapse of a near-
surface subterranean void or room in a cavern.

Sinkholes have caused considerable damage to highways, homes, sewage facilities, and other structures. Natural or artificial fluctuations in the water table are

TEN

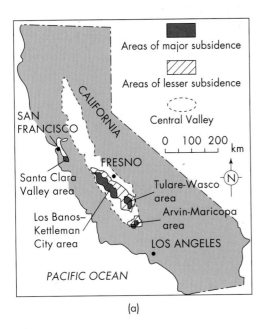

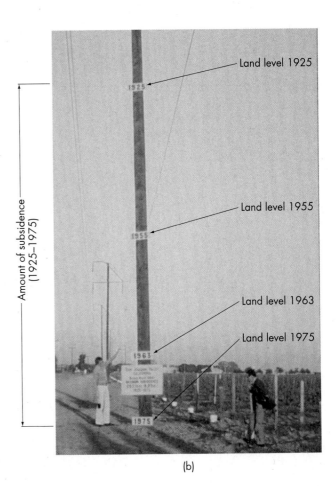

(a)

Figure 9.24 Land subsidence from groundwater extraction
(a) Principal areas of land subsidence in California resulting from
groundwater withdrawal. *(After Bull, W. B. 1973.* Geological Society of
America Bulletin 84. *Reprinted by permission)* (b) Photograph illustrating
the amount of subsidence in the San Joaquin Valley, California. The
marks on the telephone pole are the positions of the ground surface
in recent decades. The photo shows nearly 8 m (26 ft) of subsidence.
(Courtesy of Ray Kenny)

(b)

subsidence.[17] In all cases, the general principles are the same: Fluids in Earth
materials below Earth's surface have a high fluid pressure that tends to support
the material above. This is why a large rock at the bottom of a swimming pool
seems lighter: Buoyancy produced by the liquid tends to lift the rock. If support or
buoyancy is removed from Earth materials by pumping out the fluid, the support
is reduced, and surface subsidence may result.

Thousands of square kilometers of the central valley of California have subsided as a result of overpumping groundwater in the area (Figure 9.24a). More
than 5,000 km² (1,930 mi²) in the Los Banos–Kettleman City area alone have
subsided more than 0.3 m (1 ft), and within this area, one stretch of valley 113 km
(70 mi) long has subsided an average of more than 3 m (10 ft), with a maximum of
about 9 m (30 ft) (Figure 9.24b). As the water was mined, the pore pressure was
reduced and the grains were compacted;[18,19] the effect at the surface was subsidence (Figure 9.25). Similar examples of subsidence caused by overpumping are
documented near Phoenix, Arizona; Las Vegas, Nevada; Houston–Galveston,
Texas; and Mexico City, Mexico. The subsidence can cause extremely long, deep
surface fissures (open cracks) to form in sediments.[19]

Sinkholes

Subsidence is also caused by removal of subterranean Earth materials by natural
processes. Voids, large open spaces such as caves, often form by chemical weathering within soluble rocks such as limestone and dolomite, and the resulting lack
of support for overlying rock may cause it to collapse. The result is the formation
of a **sinkhole,** a circular area of subsidence caused by the collapse of a near-
surface subterranean void or room in a cavern.

Sinkholes have caused considerable damage to highways, homes, sewage facilities, and other structures. Natural or artificial fluctuations in the water table are

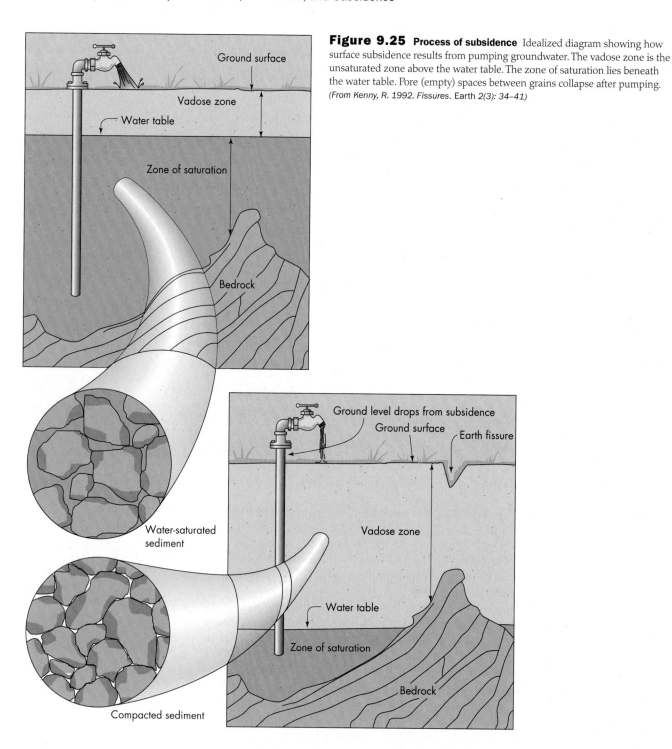

Figure 9.25 Process of subsidence Idealized diagram showing how surface subsidence results from pumping groundwater. The vadose zone is the unsaturated zone above the water table. The zone of saturation lies beneath the water table. Pore (empty) spaces between grains collapse after pumping. *(From Kenny, R. 1992. Fissures. Earth 2(3): 34–41)*

probably the trigger mechanism. High water table conditions enlarge the cavern closer to the surface of Earth by dissolving material, and the buoyancy of the water helps support the overburden. Lowering of the water table eliminates some of the buoyant support and facilitates collapse. On May 8, 1981, this process was dramatically illustrated in Winter Park, Florida, when a large sinkhole began developing. The sink grew rapidly for 3 days, swallowing part of a community swimming pool, parts of two businesses, several automobiles, and a house (Figure 9.26). Damage caused by the sinkhole exceeded $2 million. Sinkholes form nearly every year in central Florida when the groundwater level is lowest. The Winter Park sinkhole formed during a drought, when groundwater levels were at a record low. Although the exact positions of sinkholes cannot be predicted, their occurrence is greater during droughts; in fact, several smaller sinkholes

Figure 9.26 Sinkhole swallows part of a town The Winter Park, Florida, sinkhole that grew rapidly for 3 days, swallowing part of a community swimming pool as well as two businesses, a house, and several automobiles. *(Leif Skoogfors/Woodfin Camp and Associates)*

developed at about the same time as the Winter Park event.[20] On June 23, 1986, a large subsidence pit developed at the site of an unrecognized, filled sinkhole in Lehigh Valley near Allentown, in eastern Pennsylvania. Within a period of only a few minutes, the collapse left a pit approximately 30 m (100 ft) in diameter and 14 m (46 ft) deep. Fortunately, the damage was confined to a street, parking lots, sidewalks, sewer lines, water lines, and utilities. Seventeen residences adjacent to the sinkhole narrowly escaped damage or loss; subsequent stabilization and repair costs were nearly half a million dollars. Figure 9.27 shows the generalized geology of Lehigh Valley. The northern part of the valley is underlain by shale, whereas the southern portion is underlain by limestone. The valley is bounded by resistant sandstone rocks to the north and resistant Precambrian igneous rocks to the south (Figure 9.27).[21]

Photographs from the 1940s to 1969 provide evidence of the sinkhole's history. In the 1940s, the sinkhole was delineated by a pond approximately 65 m (213 ft) in diameter. By 1958 the pond had dried up, the sinkhole was covered by vegetation, and the surrounding area was planted in crops. Ground photographs in 1960 suggest that people were using the sinkhole as a site to dump tree stumps, blocks of

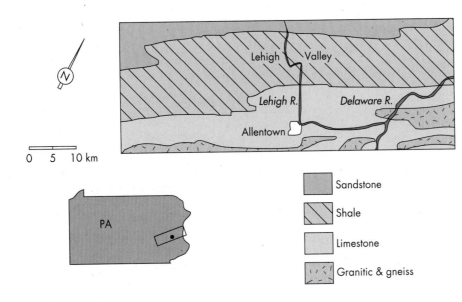

Sandstone

Shale

Limestone

Granitic & gneiss

Figure 9.27 Geology of a valley with sinkhole hazard Generalized geologic map of the Lehigh Valley in eastern Pennsylvania where a large collapse sinkhole formed suddenly. *(Modified from Dougherty, P. H., and Perlow, M., Jr. 1987.* Environmental Geology and Water Science *12(2):89–98)*

asphalt, and other trash. By 1969 there was no surface expression of the sinkhole; it was completely filled, and corn was planted over it.

Even though the sinkhole was completely filled with trash and other debris, it continued to receive runoff water that was later increased in volume by urbanization. Sources of water included storm runoff from adjacent apartments and townhouses, streets, and parking lots. It is also suspected that an old, leaking water line contributed to runoff into the sinkhole area. In addition, urbanization placed increased demand on local groundwater resources, resulting in the lowering of the water table. Geologists believe that hydrologic conditions contributed to the sudden failure. The increased urban runoff facilitated the loosening or removal of the plug—that is, the soil, clay, and trash that filled the sinkhole— while the lowering of the groundwater reduced the overlying support, as was the case with the Winter Park sinkhole. Sinkholes are discussed further in Chapter 12 with groundwater processes.

Salt Deposits

Serious subsidence events have been associated with salt mining. Salt is often mined by solution methods: Water is injected through wells into salt deposits, the salt dissolves, and water supersaturated with salt is pumped out. Because the removal of salt leaves a cavity in the rock and weakens support for the overlying rock, it may lead to large-scale subsidence.

On November 21, 1980, a bizarre example of subsidence associated with a salt mine occurred in southern Louisiana. Lake Peigneur, a shallow lake with an average depth of 1 m (3.3 ft), drained after the collapse of the salt mine below it. The collapse occurred after an oil-drilling operation apparently punched a hole into an abandoned mine shaft of the Jefferson Island Salt Dome, a still-active multimillion-dollar salt mine located about 430 m (1,410 ft) below the surface. As water entering the mine enlarged the hole, pillars of salt were scoured and dissolved, causing the roof of the mine to collapse and producing a large subsidence pit (Figure 9.28).

The lake drained so fast that 10 barges, a tugboat, and an oil-drilling barge disappeared in a whirlpool of water into the mine. Fortunately, the 50 miners and 7 people on the oil rig escaped. The subsidence also claimed more than 0.25 km^2 (0.1 mi^2) of Jefferson Island, including historic botanical gardens, greenhouses, and

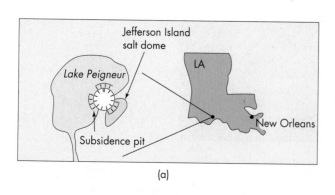

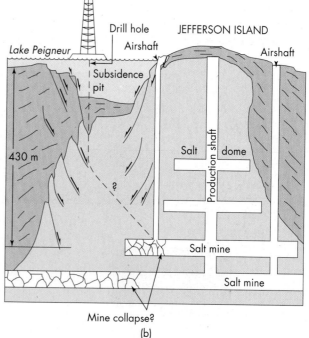

Figure 9.28 **A bizarre subsidence event** (a) Location of Lake Peigneur. (b) Idealized diagram showing the Jefferson Island salt dome collapse that caused a large subsidence pit to form in the bottom of Lake Peigneur, Louisiana.

a $500,000 private home. The remaining gardens were disrupted by large fractures that dropped the land down to the new edge of the lake. These fractures are formed as the land sinks and are commonly found on the margins of large subsidence pits.

Lake Peigneur immediately began refilling with water from a canal connecting it to the Gulf of Mexico, and nine of the barges popped to the surface 2 days later. There was fear at first that even larger subsidence would take place as pillars of salt holding up the roof of the salt dome dissolved. However, the hole was apparently sealed by debris in the form of soil and lake sediment that was pulled into the mine. Approximately 15 million cubic meters (530 million cubic feet) of water entered the salt dome, and the mine was a total loss. The previously shallow lake now has a large, deep hole in the bottom, which undoubtedly will change its aquatic ecology. In a 1983 out-of-court settlement, the salt-mining company was reportedly compensated $30 million by the oil company involved. The owners of the botanical garden and private home apparently were compensated $13 million by the oil company, drilling company, and mining company.

The flooding of the mine raises important questions concerning the structural integrity of salt mines. The federal Strategic Petroleum Reserve Program is planning to store 75 million barrels of crude oil in an old salt mine of the Weeks Island salt dome about 19 km (12 mi) from Jefferson Island. Although the role of the draining lake in the collapse is very significant, few salt domes have lakes above them. The Jefferson Island subsidence was thus a rare event.

Coal Mining

Subsurface mining of coal in the western and eastern United States has produced serious subsidence problems. The subsidence is most common where underground mining is close to the surface of the land or where the rocks left as pillars after mining are weak or intensely fractured. Usually, only 50 percent of the coal is removed, leaving the remainder as pillars that support the roof, formed from the rocks overlying the mine. Over time, the pillars weather, weaken, and collapse, producing the surface subsidence.[22] In the United States, more than 8,000 km² (3,090 mi²) of land has subsided owing to underground coal mining, and subsidence continues today, long after mining terminated. In 1995, a coal mine that was last operated in the 1930s collapsed beneath a 600 m (1,970 ft) length of Interstate Highway 70 in Ohio; repairs took 3 months.[23] Subsidence most often affects farmland and rangelands but has also damaged buildings and other structures in towns and cities, including Scranton, Wilkes-Barre, and Pittsburgh, Pennsylvania (Figure 9.29).[22,23]

Figure 9.29 **Subsidence over coal mines** An underground fire in an abandoned coal mine in Pennsylvania has melted the snow near this vent. Note the smoke escaping to the atmosphere. *(National Institute for Occupational Safety & Health)*

9.8 Perception of the Landslide Hazard

A common reaction of homeowners concerning landslides is, "It could happen on other hillsides, but never this one.[12] Just as flood hazard mapping does not prevent development in flood-prone areas, landslide hazard maps will probably not prevent all people from moving into hazardous areas. Prospective hillside occupants who are initially unaware of the hazards may not all be swayed by technical information. The infrequency of large slides tends to reduce awareness of the hazard where evidence of past events is not readily visible. Unfortunately, it often takes catastrophic events that claim numerous expensive homes to bring the problem to people's attention. In the meantime, people in many parts of the Rocky Mountains, Appalachian Mountains, California, and other areas continue to build homes in areas subject to future landslides.

What You Can Do to Minimize Your Landslide Hazard

Consider the following advice if purchasing property on a slope:

- Landslides often involve complex geology, and a geologic evaluation by a professional geologist of any property on a slope is recommended.
- Avoid homes at the mouth of a canyon, even a small one, where debris flows or mudflows may originate from upstream slopes and travel down the canyon.
- Consult local agencies, such as city or county engineering departments, that may be aware of landslides in your area.
- Watch out for "little landslides" in the corner of the property—they usually get larger with time.
- If purchasing a home, look for cracks in house walls and for retaining walls that lean or are cracked. Be wary of doors or windows that stick or floors that are uneven. Check foundations for cracks or tilting. If cracks in the walls or the foundation can be followed to the land outside the house, be especially concerned that a landslide may be present.
- Be wary of leaks in a swimming pool, trees tilted downslope, utility wires that are taut or sagging.
- Be wary if small springs are present, as landslides tend to leak water. Look for especially "green areas," where more subsurface water is present.
- Walk the property and surrounding property, if possible, looking for linear or curved cracks (even small ones) that might indicate incipient instability of the land.
- Active landslides often have hummocks or steplike ground features.
- Fixing a potential landslide problem is often cost effective, but it can still be expensive and much of the fix is below ground, where you will never see the improvement. It is better to not purchase land with a potential landslide hazard!
- The presence of one or more of the above features is not proof of the presence of a landslide. For example, cracks in walls and foundations can be caused by soils that shrink and swell. However, further investigation is warranted if the above features are present.

SUMMARY

Landslides and related phenomena cause substantial damages and loss of life. Although they are natural events, their occurrence can be increased or decreased by human activity.

The most common landforms are slopes—dynamic, evolving systems in which surface material is constantly moving downslope, or mass wasting, at rates varying from imperceptible creep to thundering avalanches. Slope failure may involve flowage, slumping, sliding, or falling of Earth materials; landslides are often complex combinations of sliding and flowage.

The forces that produce landslides are determined by the interactions of several variables: the type of Earth material on the slope, topography and slope angle, climate, vegetation, water, and time. The cause of most landslides can be determined by examining the relations between forces that tend to make Earth materials slide, the driving forces, and forces that tend to oppose movement, the resisting forces. The most common driving force is the weight of the slope materials, and the most common resisting force is the strength of the slope materials. The safety factor of a slope is the ratio of resisting forces to driving forces; a ratio greater than 1 means that the slope is stable; a ratio less than 1 indicates potential slope failure. The type of rock or soil on a slope influences both the type and the frequency of landslides.

Water has an especially significant role in producing landslides. Water in streams, lakes, or oceans erodes the base of slopes, increasing the driving forces. Excess water increases the weight of the slope materials while raising the water pressure. Increased water pressure, in turn, decreases the resisting forces in the slope materials. The effects of human use on the magnitude and frequency of landslides vary from insignificant to very significant. When landslides occur independently of human activity, we need to learn enough about them to avoid development in hazardous areas or to provide protective measures. In other cases, when human use has increased the number and severity of landslides, we need to learn how to minimize these occurrences. In some cases,

dams and reservoirs have increased migration of groundwater into slopes, resulting in slope failure. Logging operations on weak, unstable slopes have increased landslide erosion. Grading of slopes for development has created or increased erosion problems in many urbanized areas of the world.

Minimizing landslide hazards requires identification, prevention, and correction procedures. Monitoring and mapping techniques facilitate identification of hazardous sites. Identification of potential landslides has been used to establish grading codes, and landslide damage in these areas has been decreased. Prevention of large natural slides is very difficult, but careful engineering practices can do much to minimize the hazard when it cannot be avoided. Engineering techniques for landslide prevention include drainage control, proper grading, and construction of supports such as retaining walls. Correction of landslides must attack the processes that started the slide; this usually means initiating a drainage program that lowers water pressure in the slope.

Snow avalanches present a serious hazard on snowcovered, steep slopes. Loss of human life due to avalanches is increasing as more people venture into mountain areas for winter recreation.

Withdrawal of fluids such as oil and water and subsurface mining of salt, coal, and other minerals have caused widespread subsidence. In the case of fluid withdrawal, the cause of subsidence is a reduction of water pressures that tend to support overlying Earth materials. In the case of solid-material removal, subsidence may result from loss of support for the overlying material. The latter situation may occur naturally when voids are formed in soluble rock such as limestone and the collapse of overlying Earth material produces sinkholes.

Most people perceive the landslide hazard as minimal, unless they have prior experience. Furthermore, hillside residents, like floodplain occupants, are not easily swayed by technical information. Nevertheless, the wise person will have a geologist inspect property on a slope before purchasing.

Revisiting Fundamental Concepts

Human Population Growth

By 2050 the world's population will have increased by about 50 percent. Three billion more people to feed and house will place increased stress on the environment. More people will be living on steep slopes where landslides are likely to occur. Therefore, we must learn more about how to minimize the landslide hazard to reduce potential losses of human life and property.

Sustainability

For future generations to inherit a quality environment, we must work to minimize the magnitude and frequency of

human-induced landslides. When we remove trees from hillslopes or intensively urbanize hillslopes, landslides increase in both size and number. Landsliding is an erosion process, removing vegetation and soils along with valuable land resources that future generations will need.

Earth as a System

Hillslopes are complex systems with physical, chemical, hydrologic, biological, and human components. When we change one part of the system through land-use change, we cause other changes that may affect the stability of the slope. This is an example of the principle of environmental unity.

Slopes are very sensitive to changes in hydrology, and when trees are removed the amount of water that infiltrates the soil or runs off the land changes. Trees use a lot of soil water through the process of transpiration, which is loss of water through leaves and other plant tissue. When trees or other vegetation are removed this process stops, and the amount of water in the soil may increase. Wet slopes are more prone to landsliding than are dry slopes. If trees are logged, the roots may die and decay. When they do, the soil loses some strength because root systems bind the soil together. The combined effects of wetter soils and reduced soil strength as tree roots decay result in more shallow landslides. This concept explains the significant increase in shallow landslides that occurs several years after logging.

Hazardous Earth Processes, Risk Assessment, and Perception

People are basically optimistic about where they wish to live, and their perception of landslide risk reflects this optimism. Education about potential risks and construction of hazard maps are necessary to help inform people about the potential hazards of hillside development.

Scientific Knowledge and Values

Physical, chemical, and biological processes operating on hillslopes are fairly well understood. We also know the risks associated with hillside development and the methods to reduce the landslide hazard. As more people in the future are forced by economics or lured by the view to live in hazardous areas on steep hillslopes, we need to examine our values. How can we provide safe housing for growing urban areas, especially in developing countries? Too often "shantytowns" develop on undesirable steep lands that have been cleared of forest. These areas, when hit by intense rainstorms or shaken by earthquakes, commonly produce slides that may kill hundreds to thousands of people. This scenario does not have to be replayed again and again if we place value on human life and apply principles of sound land-use planning for people moving to urban centers.

Key Terms

landslide (p. 288)

mass wasting (p. 288)

rotational landslide (p. 295)

safety factor (SF) (p. 294)

sinkholes (p. 313)

snow avalanche (p. 311)

subsidence (p. 288)

translational landslide (p. 295)

Review Questions

1. What is a landslide?

2. What are the main ways that materials on a slope may fail?

3. What is the safety factor, and how is it defined?

4. Differentiate between rotational slides (slumps) and translational slides and shallow slips.

5. How does the slope angle affect the incidence of landslides?

6. What are the three ways that vegetation is important in slope stability?

7. How may spontaneous liquefaction occur?

8. Why does time play an important role in landslides?

9. What is the main lesson learned from the Vaiont Dam disaster?

10. How might the process of urbanization increase or decrease the stability of slopes?

11. What are the main steps we can take to prevent landslides?

12. What is the process that causes subsidence due to withdrawal of fluid such as groundwater or petroleum?

13. What was the role of groundwater in the formation of the Winter Park sinkhole?

Critical Thinking Questions

1. Your consulting company is hired by a national park in your region to estimate the future risk from landsliding. Develop a plan of attack that outlines what must be done to achieve this objective.

2. Why do you think that many people are not easily swayed by technical information concerning hazards such as landslides? Assume you have been hired by a community to make the citizens more aware of the landslide hazard in their very steep topographic area. Outline a plan of action and defend it.

TEN

Coastal Processes

Learning Objectives

In this chapter we focus on one of the most dynamic environments on Earth—the coast, where the sea meets the land. The beauty of the coastal zone, with its salty smells and the sight and sound of wind and waves striking the land, has inspired poets and artists for thousands of years. Beaches composed of sand or pebbles and rocky coastlines continue to attract tourists like few other areas. Yet most of us have little understanding of how ocean waves form and change the coastlines of the world. A major purpose of this chapter is to remove the mystery of how coastal areas are formed and maintained while retaining the wonder. We seek to understand the hazards resulting from wind, waves, and storms, and how we can learn to live in the ever-changing coastal environment while sustaining its beauty. We will focus on the following learning objectives:

- Know the basic terminology of waves, and processes of waves
- Be able to define the basic components of a beach
- Understand the process of littoral transport of sediment
- Know what rip currents are and why they are a serious hazard to swimmers
- Know the major processes related to coastal erosion
- Understand the various engineering approaches to shoreline protection
- Understand how human activities affect coastal erosion
- Understand why we are at a crossroads with respect to adjustments to coastal erosion
- Understand what tropical cyclones are and the hazards they produce

The Outer Banks View to the south of Cape Hatteras and the Outer Banks of North Carolina. The Outer Banks extend far out in the Atlantic. They are a ribbon of sand tied to a continent, ever shifting and changing in response to wind and waves, pounded by storms and warmed by the sun. A magical place! *(NASA/The Image Bank/Getty Images Inc.)*

CASE HISTORY | The Cape Hatteras Lighthouse Controversy

In North Carolina, a dramatic collision of opinions concerning beach erosion is now being played out. Historically, North Carolina has adhered to the philosophy that beach erosion is a natural process with which its residents can live. When erosion began to threaten the historic Cape Hatteras Lighthouse (Figure 10.1a), this philosophy was tested. The lighthouse is located near Buxton, on North Carolina's barrier islands known as the Outer Banks. When the lighthouse was originally constructed in the late nineteenth century, it was approximately 0.5 km (0.3 mi) from the sea. By the early 1990s it was only 100 m (328 ft) away from the sea and in danger of being destroyed by a major storm. Officials in the area had to weigh the following options:

- Artificially control coastal erosion at the site, and reverse state policy of yielding to erosion. The U.S. Army Corps of Engineers originally proposed to protect the lighthouse by constructing a $5.6 million seawall around the base.[1]

- Do nothing and eventually lose the lighthouse and, thus, an important bit of American history.

- Move the lighthouse inland. Many local people opposed this plan, fearing the lighthouse would collapse if moved.[1]

Following much discussion, argument, and controversy over what to do, the decision was made by the National Park Service to move the lighthouse inland approximately 500 m (1,640 ft) from the eastern shore of Hatteras Island at a cost of about $12 million. The decision to move the lighthouse was based on several factors. It was consistent with the philosophy of flexible coastal zone planning to avoid hazardous zones rather than attempting to control natural processes. It was also consistent with North Carolina's policy to preserve historic objects, such as the lighthouse, for the enjoyment of future generations. The lighthouse was successfully moved during the summer of 1999 (Figure 10.1b). Given the present rate of coastal erosion, the lighthouse should be safe in its new location until the middle or end of the twenty-first century. Hurricane Dennis struck the island in 1999 after the lighthouse was moved, and the historic structure was not significantly damaged.

Another lighthouse battle is looming on the East Coast. Hundreds of homes and other buildings along the coast on Long Island, New York, are now close to the actively eroding shoreline. Included is the famous Montauk Lighthouse, which was constructed in 1746 on a bluff 100 m (300 ft) from the ocean and is now about 25 m (75 ft) from the ocean. One group argues for a hard structure consisting of a multimillion dollar, 280 m (840 ft) long rock revetment (rock seawall) to protect the base of the bluff from erosion. Another group is arguing to move the lighthouse inland. Moving the lighthouse would be a challenge and the rock revetment might change the coastline, causing problems in adjacent areas. The rock solution might also set a precedent for hard structural control of beach erosion on Long Island and other East Coast areas.

(a)

(b)

Figure 10.1 **Lighthouse is moved** (a) Cape Hatteras Lighthouse before being moved. *(Don Smetzer/Getty Images Inc.)* (b) Cape Hatteras Lighthouse being moved in the summer of 1999. *(Hart Matthews/Reuters/Getty Images Inc.)*

10.1 Introduction to Coastal Hazards

Coastal areas are dynamic environments that vary in their topography, climate, and vegetation. Continental and oceanic processes converge along coasts to produce landscapes that are characteristically capable of rapid change. The East Coast of the United States is a passive margin coastline far from a convergent plate boundary (see Chapter 2). The coastline is characterized by a wide continental shelf with barrier islands (see opening photograph) with wide sandy beaches. Rocky coastlines are mostly restricted to the New England coast where the Appalachian Mountains merge with the Atlantic Ocean. The west coast is close to the convergent boundary between the North American and Pacific plates (active margin coast). Mountain building has produced a coastline with sea cliffs and rocky coastlines. Long sandy beaches are present, but not as abundant as along the East Coast.

The impact of hazardous coastal processes is considerable, because many populated areas are located near the coast. In the United States, it is expected that most of the population will eventually be concentrated along the nation's 150,000 km (93,000 mi) of shoreline, including the Great Lakes. Today, the nation's largest cities lie in the coastal zone, and approximately 75 percent of the population lives in coastal states.[2] Coastal problems will thus increase because so many more people will live in coastal areas where the hazards occur. Once again, our activities continue to conflict with natural processes! Hazards along the coasts may become compounded by the fact that global warming and the accompanying global rise in sea level are increasing the coastal erosion problem. (Climate change and sea level rise are discussed in Chapter 19.)

The most serious coastal hazards are the following:

- Rip currents generated in the surf zone
- Coastal erosion, which continues to produce considerable property damage that requires human adjustment
- Tsunamis, or seismic sea waves (discussed in Chapter 5), which are particularly hazardous to coastal areas of the Pacific Ocean
- Tropical cyclones, called hurricanes in the Atlantic and typhoons in the Pacific, which claim many lives and cause enormous amounts of property damage every year

10.2 Coastal Processes

Waves

Waves that batter the coast are generated by offshore storms, sometimes thousands of kilometers from the shoreline where they will expend their energy. Wind blowing over the water produces friction along the air-water boundary. Since the air is moving much faster than the water, the moving air transfers some of its energy to the water, resulting in waves. The waves, in turn, eventually expend their energy at the shoreline. The size of the waves produced depends on the following:

- The velocity, or speed, of the wind. The greater the wind velocity, the larger the waves.
- The duration of the wind. Storms of longer duration have more time to impart energy to the water, producing larger waves.
- The distance that the wind blows across the surface, or *fetch.* The longer the fetch, the larger the waves.

Within the area of the storm, the ocean waves have a variety of sizes and shapes, but as they move away from their place of origin, they become sorted out into groups of similar waves. These groups of waves may travel for long distances across the ocean and arrive at distant shores with very little energy loss.

The basic shape, or wave form, of waves moving across deep water is shown in Figure 10.2a. The important parameters are *wave height,* which is the difference in height between the wave's trough and its peak, and *wave length,* the distance between successive peaks. The *wave period* (P) is the time in seconds for successive waves to pass a reference point. If you were floating with a life preserver in deep

Figure 10.2 **Waves and beaches** (a) Deep-water wave form (water depth is greater than 0.5 L, where L is wave length). The ratio of wave height to wave length is defined as wave steepness. If wave height exceeds about 10 percent (0.1) of the wave length, the wave becomes unstable and will break. Our drawing exaggerates wave height for illustrative purposes. The steepness of the waves in the drawing is about 1/3 or 0.33, which would be very unstable and not exist long in nature. (b) Motion of water particles associated with wave movement in deep water. (c) Motion of water particles in shallow water at a depth less than 0.25 L. Water at the beach moves up and back in the swash zone, the very shallow water on the beach face.

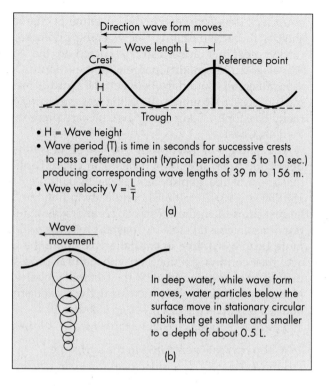

- H = Wave height
- Wave period (T) is time in seconds for successive crests to pass a reference point (typical periods are 5 to 10 sec.) producing corresponding wave lengths of 39 m to 156 m.
- Wave velocity $V = \dfrac{L}{T}$

(a)

In deep water, while wave form moves, water particles below the surface move in stationary circular orbits that get smaller and smaller to a depth of about 0.5 L.

(b)

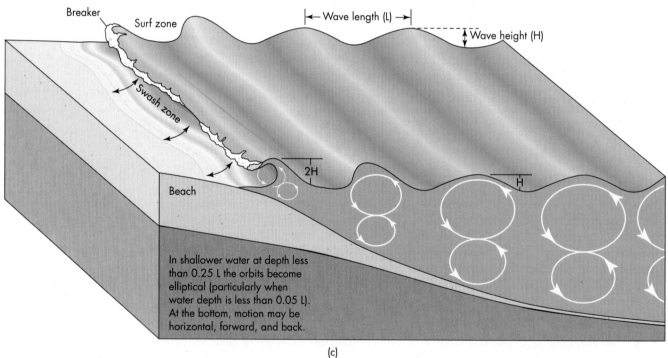

In shallower water at depth less than 0.25 L the orbits become elliptical (particularly when water depth is less than 0.05 L). At the bottom, motion may be horizontal, forward, and back.

(c)

water and could record your motion as waves moved through your area, you would find that you bob up, down, forward, and back in a circular orbit, returning to about the same place. If you were below the surface with a breathing apparatus, you would still move in circles, but the circle would be smaller. That is, you would move up, down, forward, and back in a circular orbit that would remain in the same place while the waves traveled through. This concept is shown in Figure 10.2b. When waves enter shallow water at a depth of less than about one-half their wavelength (L), they "feel bottom." The circular orbits change to become ellipses; the motion at the bottom may be a very narrow ellipse, or essentially horizontal, that is, forward and back (Figure 10.2c). You may have experienced this phenomenon if you have stood or have swum in relatively shallow water on a beach and felt the water repeatedly push you toward the shore and then back out toward the sea.

The wave groups generated by storms far out at sea are called *swell*. As the swell enters shallower and shallower water, transformations take place that eventually lead to the waves' breaking on the shore. For deep-water conditions, there are equations to predict wave height, period, and velocity based on the fetch, wind velocity, and length of time that the wind blows over the water. This information has important environmental consequences: by predicting the velocity and height of the waves, we can estimate when waves with a particular erosive capability generated by a distant storm will strike the shoreline.

We have said that waves expend their energy when they reach the coastline. But just how much energy are we talking about? The amount is surprisingly large. For example, the energy expended on a 400 km (250 mi) length of open coastline by waves with a height of about 1 m (3.3 ft) over a given period of time is approximately equivalent to the energy produced by one average-sized nuclear power plant over the same time period.[3] Wave energy is approximately proportional to the square of the wave height. Thus, if wave height increases to 2 m (6.6 ft), the wave energy increases by a factor of 2^2, or 4. If wave height increases to 5 m (16 ft), which is typical for large storms, then the energy expended, or wave power, increases 5^2, or 25 times over that of waves with a height of 1 m (3.3 ft).

When waves enter the coastal zone and shallow water, they impinge on the bottom and become steeper. Wave steepness is the ratio of wave height to wave length. Waves are unstable when the wave height is greater than about 10 percent (0.1) of the wave length. As waves move into shallow water, the wave period remains constant, but wave length and velocity decrease and wave height increases. The waves change shape from the rounded crests and troughs in deep water to peaked crests with relatively flat troughs in shallow water close to shore. Perhaps the most dramatic feature of waves entering shallow water is their rapid increase in height. The height of waves in shallow water, where they break, may be as much as twice their deep-water height (Figure 10.2c). Waves near the shoreline, just outside the surf zone, reach a wave steepness that is unstable. The instability causes the waves to break and expend their energy on the shoreline.[4]

Although wave heights offshore are relatively constant, the local wave height may increase or decrease when the *wave front* (see Figure 10.3a) reaches the nearshore environment. This change can be attributed to irregularities in the offshore topography and the shape of the coastline. Figure 10.3a is an idealized diagram showing a rocky point, or headland, between two relatively straight reaches of coastline. The offshore topography is similar to that of the coastline. As a wave front approaches the coastline, the shape of the front changes and becomes more parallel to the coastline. This change occurs because, as the waves enter shallow water, they slow down first where the water is shallowest, that is, off the rocky point. The result is a bending, or *refraction*, of the wave front. In Figure 10.3a, the lines drawn perpendicular to the wave fronts, with arrows pointing toward the shoreline, are known as *wave normals*. Notice that, owing to the bending of the wave fronts by refraction, there is a *convergence* of the wave normals at the headland, or rocky point, and a *divergence* of the wave normals at the beaches, or embayments. Where wave normals converge, wave height increases; as a result, wave energy expenditure at the shoreline also increases. Figure 10.3b shows a photograph of large waves striking a rocky headland.

The long-term effect of greater energy expenditure on protruding areas is that wave erosion tends to straighten the shoreline. The total energy from waves reaching a coastline during a particular time interval may be fairly constant, but there may be considerable local variability of energy expenditure when the waves break on the shoreline. In addition, breaking waves may peak up quickly and plunge or surge, or they may gently spill, depending on local conditions such as

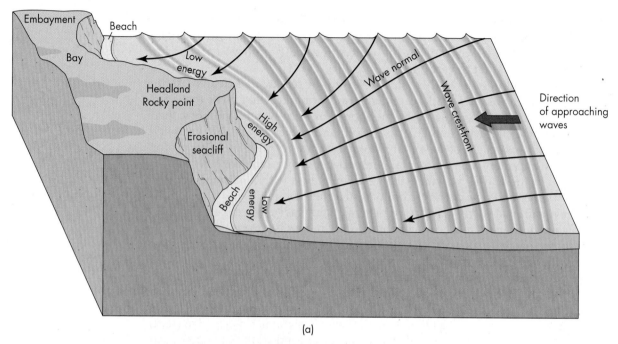

(a)

(b)

Figure 10.3 Convergence and divergence of wave energy (a) Idealized diagram of the process of wave refraction and concentration of wave energy at rocky points, or headlands. The refraction, or bending of the wave fronts causes the convergence of wave normals on the rocky point and divergence at the bay. (b) Photograph of large waves striking a rocky headland. *(Douglas Faulkner/ Photo Researchers, Inc.)*

the steepness of the shoreline (Figure 10.4) and the height and length of waves arriving at the shoreline from a distant storm. *Plunging breakers* tend to be highly erosive at the shoreline, whereas *spilling breakers* are more gentle and may facilitate the deposition of sand on beaches. The large plunging breakers that occur during storms cause much of the coastal erosion we observe.

Beach Form and Beach Processes

A **beach** is a landform consisting of loose material, such as sand or gravel, that has accumulated by wave action at the shoreline. Beaches may be composed of a variety of loose material in the shore zone, the composition of which depends on the environment. For example, many Pacific island beaches include broken bits of shell and coral; Hawaii's black sand beaches are composed of volcanic rock; and grains of quartz and feldspar are found on the beaches of southern California. Figure 10.5 shows the basic terminology of an idealized nearshore environment. The landward extension of the beach terminates at a natural topographic and morphologic change such as a seacliff or a line of sand dunes. The *berms* are flat backshore areas on beaches formed by deposition of sediment as waves rush up and expend the last of their energy. Berms are where you will find people sunbathing. The *beach face* is the sloping portion of the beach below the berm, and the part of the beach face that is exposed by the uprush and backwash of waves is

Figure 10.4 Types of breakers Idealized diagram and photographs showing (a) plunging breakers on a steep beach and (b) spilling breakers on a gently sloping beach. *([a] Peter Cade/Getty Images Inc.; [b] Penny Tweedie/Getty Images Inc.)*

called the *swash zone*. The *surf zone* is that portion of the seashore environment where turbulent translational waves move toward the shore after the incoming waves break; the *breaker zone* is the area where the incoming waves become unstable, peak, and break. The *longshore trough* and *longshore bar* are an elongated depression and adjacent ridge of sand produced by wave action. A particular beach, especially if it is wide and gently sloping, may have a series of longshore bars, longshore troughs, and breaker zones.[4]

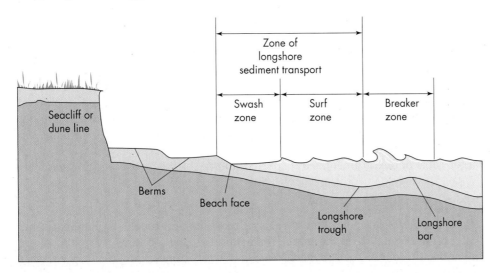

Figure 10.5 Beach terms Basic terminology for landforms and wave action in the beach and nearshore environment.

Figure 10.6 **Transport of sediment along a coast** Block diagram showing the processes of beach drift and longshore drift, which collectively move sand along the coast in a process known as longshore sediment transport. Sediments transported in the swash zone and surf zone follow paths shown by the arrows.

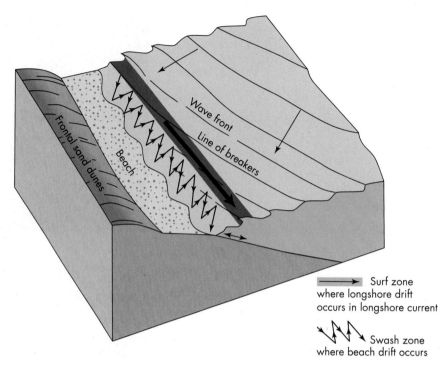

Transport of Sand

The sand on beaches is not static; wave action keeps the sand moving along the beach in the surf and swash zones. A longshore current is produced by incoming waves striking the coast at an angle (Figure 10.6). Because the waves strike the coast at an angle, a component of wave energy is directed along the shore. If waves arrive at a beach perfectly parallel to the beach, then no longshore current is generated. The longshore current is a stream of water flowing parallel to the shore in the surf zone. This current can be surprisingly strong. If you are swimming on a beach and wading in and out of the surf zone, you may notice that the longer you go in and out through the surf zone the further away you are from where you started and left your beach towel and umbrella. As you move in and out through the surf and swash zone, the current will move you along the coast, and the sand is doing exactly the same thing.

The process that transports sand along the beach, called **longshore sediment transport,** has two components: (1) Sand is transported along the coast with the longshore current in the surf zone; and (2) the up-and-back movement of beach sand in the swash zone causes the sand to move along the beach in a zigzag path (Figure 10.6). Most of the sand is transported in the surf zone by the longshore current.

The direction of transport of sand along beaches in the United States is generally from the north to the south for beaches on both the East and West Coasts of the country. Although most of the transport is to the south, it can be variable and depends upon the wave action and in which direction they strike the shore. The amount of sand transported along a beach, whether we are talking about Long Island, New York, or Los Angeles, California, is surprisingly large, at several hundred thousand cubic meters of sediment per year. Having said this, the amount of sand transported on a given day or period of days is extremely variable. On many days little sand is being transported and on others the amount is much larger. Most of the sediment is transported during storms by the larger waves.

Rip Currents

When a series of large waves arrives at a coastline and breaks on the beach, the water tends to pile up on the shore. The water does not return as it came in, along the entire shoreline, but is concentrated in narrow zones known as **rip currents**

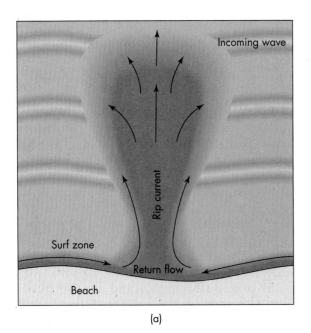

(a)

(b)

Figure 10.7 Rip current (a) Bird's-eye view of the surf zone, showing a rip current, which is the return flow of water that forms as a result of incoming waves. (b) Rip current at Santa Cruz Island, CA. The floating kelp shows the rip current. *(Edward A. Keller)*

(Figure 10.7). Beachgoers and lifeguards call them riptides or undertow. They certainly are not tides, and they do not pull people under the water, but they can pull people off shore. In the United States, up to 200 people are killed and 20,000 people are rescued from rip currents each year. Therefore, rip currents constitute a serious coastal hazard to swimmers, killing more people in the United States on an annual basis than do hurricanes or earthquakes; the number of deaths caused by rip currents is equivalent to the number caused by river flooding. People drown in rip currents because they do not know how to swim or because they panic and fight the current by trying to swim directly back to shore. Winning a fight with a rip current is nearly impossible because the current can exceed 6 km per hour (4 mi per hour), a speed that even strong swimmers cannot maintain for long. A swimmer trying to fight a rip current soon becomes exhausted and may not have the energy to keep swimming. Fortunately, rip currents are usually relatively narrow, a few meters to a few tens of meters wide, and they dissipate outside of the surf zone, within tens to hundreds of meters offshore. To safely escape a rip current, a swimmer must first recognize the current and then swim parallel to the shore until he or she is outside the current. Only then should the swimmer attempt to swim back to shore. The key to survival is not to panic. When you swim in the ocean, watch the waves for a few minutes before entering the water and note the "surf beat," the regularly arriving sets of small and larger waves. Rip currents can form quickly after the arrival of a set of large waves. They can be recognized as a relatively quiet area in the surf zone where fewer incoming waves break. You may see the current as a mass of water moving out through the surf zone. The water in the current may also be darker, carrying suspended sediment. Remember, if you do get caught in a rip current, don't panic; swim parallel to the shore until you are outside the current, then head back to the beach. If there are lifeguards in the area, yell for assistance!

10.3 Coastal Erosion

As a result of the global rise in sea level and inappropriate development in the coastal zone, coastal erosion is becoming recognized as a serious national and worldwide problem. Coastal erosion is generally a more continuous, predictable process than other natural hazards such as earthquakes, tropical cyclones, or floods, and large sums of money are spent in attempts to control it. If extensive development of coastal areas for vacation and recreational living continues, coastal erosion will certainly become a more serious problem.

Beach Budget

An easy way to visualize beach erosion at a particular beach is to take a **beach-budget** approach. An analogy to the budget is your bank account. You deposit money at regular or not so regular times. Some money is in storage and that's your account balance, and you periodically withdraw funds, which is output. Similarly, we can analyze a beach in terms of input, storage, and output of sand or larger sediment that may be found on the beach. Input of sediment to a beach is by coastal processes that move the sediment along the shoreline (Figure 10.6) or produce sand from erosion of a sea cliff or sand dunes on the upper part of a beach. The sediment that is in storage on a beach is what you see when you visit the site. Output of sediment is that material that moves away from the site by coastal processes similar to those that brought the sediment to the beach. If input exceeds output, then the beach will grow as more sediment is stored and the beach widens. If input and output are relatively equal to one another, the beach will remain in a rough equilibrium at about the same width. If output of sediment exceeds input, then the beach will erode and there will be fewer grains of sediment on the beach. Thus we see the budget represents a balance of sand on the beach over a period of years. Short-term changes in sediment supply due to the attack of storm waves will cause seasonal or storm-related changes to the supply of sediment on a beach. Long-term changes in the beach budget caused by climate change or human impact cause long-term growth or erosion of a beach.

Erosion Factors

The sand on many beaches is supplied to the coastal areas by rivers that transport it from areas upstream, where it has been produced by weathering of quartz- and feldspar-rich rocks. We have interfered with this material flow of sand from inland areas to the beach by building dams that trap the sand. As a result, some beaches have become deprived of sediment and have eroded.

Damming is not the only reason for erosion. For example, beach erosion along the East Coast is a result of tropical cyclones (hurricanes) and severe storms, known as Northeasters or Nor'easters;[3] a rise in sea level; and human interference with natural shore processes.[5] Sea level is rising around the world at the rate of about 2 to 3 mm (0.08 to 0.12 in.) per year, independent of any tectonic movement. Evidence suggests that the rate of rise has increased since the 1940s. The increase is due to the melting of the polar ice caps and thermal expansion of the upper ocean waters, triggered by global warming that is in part related to increased atmospheric carbon dioxide produced by burning fossil fuels. Sea levels could rise by 700 mm (28 in.) over the next century, ensuring that coastal erosion will become an even greater problem than it is today.

Seacliff Erosion

When a **seacliff** (a steep bluff or cliff) is present along a coastline, additional erosion problems may occur because the seacliff is exposed to both wave action and land erosion processes such as running water and landslides. These processes may work together to erode the cliff at a greater rate than either process could alone. The problem is further compounded when people interfere with the seacliff environment through inappropriate development.

Figure 10.8 shows a typical southern California seacliff environment at low tide. The rocks of the cliff are steeply inclined and folded shale. A thin veneer of sand and coarser material, such as pebbles and boulders, near the base of the cliff covers the wave-cut platform, which is a nearly flat bench cut into the bedrock by wave action. A mantle of sand approximately 1 m (3.3 ft) thick covers the beach during the summer, when long, gentle, spilling breakers construct a wide berm

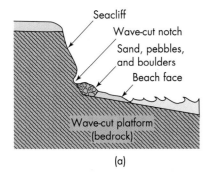

Figure 10.8 Seacliff and beach
(a) Generalized cross section and
(b) photograph of seacliff, beach, and
wave-cut platform, Santa Barbara,
California. *(Courtesy of Donald Weaver)*

while protecting the seacliff from wave erosion. During the winter, plunging breakers, which have a high potential to erode beaches, remove the mantle of sand and expose the base of the seacliff. Thus, it is not surprising that most erosion of the seacliffs in southern California takes place during the winter.

A variety of human activities can induce seacliff erosion. Urbanization, for example, results in increased runoff. If the runoff is not controlled, carefully collected, and diverted away from the seacliff, serious erosion can result. Drainpipes that dump urban runoff from streets and homes directly onto the seacliff increase erosion. Drainpipes that route runoff to the base of the seacliff on the face of the beach result in less erosion (Figure 10.9a). Watering lawns and gardens on top of a seacliff may add a good deal of water to the slope. This water tends to migrate downward through the seacliff toward the base. When water emerges as small

Landslide deposits

Figure 10.9 Seacliff erosion (a) The pipe in this photograph, from the early 1990s, carries surface runoff from the top of the seacliff down to the beach. Notice that the house at the top of the seacliff has a layer of cement partway down the cliff. The purpose of the cement is to retard erosion. (b) The same location in 2001. Note the landslide that has occurred, removing some of the cement and rock of the seacliff. *(Edward A. Keller)*

Figure 10.10 **Students living on the edge** Apartment buildings on the edge of the seacliff in the university community of Isla Vista, California. The sign states that the outdoor deck is now open. Unfortunately, the deck is not particularly safe because it is overhanging the cliff by at least 1 m (3.3 ft). Notice the exposed cement pillars in the seacliff, which were intended to help support the houses. These decks and apartment buildings are in imminent danger of collapsing into the sea. *(Edward A. Keller)*

seeps or springs from a seacliff, it effectively reduces the stability of the seacliff, facilitating erosion, including landslides (Figure 10.9b).[6]

Structures such as walls, buildings, swimming pools, and patios near the edge of a seacliff may also decrease stability by adding weight to the slope, increasing both small and large landslides (Figure 10.10). Strict regulation of development in many areas of the coastal zone now forbids most risky construction, but we continue to live with some of our past mistakes.

The rate of seacliff erosion is variable, and few measurements are available. Near Santa Barbara, California, the rate of seacliff erosion averages 15 to 30 cm (6 to 12 in.) per year. These erosion rates are moderate compared with those in other parts of the world. Along the Norfolk coast of England, for example, erosion rates in some areas are about 2 m (6.6 ft) per year. The rate of erosion is dependent on the resistance of the rocks and the height of the seacliff.[6] The rate of coastal erosion can be determined by a new remote sensing technique (see A Closer Look: Measuring Coastal Change).

Seacliff erosion is a natural process that cannot be completely controlled unless large amounts of time and money are invested, and even then there is no guarantee that erosion will cease. Therefore, it seems we must learn to live with some erosion. It can be minimized by applying sound conservation practices, such as controlling the water on and in the cliff and not placing homes, walls, large trees, or other structures that contribute to driving forces close to the top edge of a cliff.

A CLOSER LOOK | Measuring Coastal Change

Technology has recently provided a remote sensing method of measuring and monitoring changes in the coastal environment. Light Detection and Ranging (LIDAR) is an aircraft-mounted laser system that can record several thousand elevation measurements per second, with vertical resolution of better than 15 cm (6 in.). Once a baseline set of elevations is recorded, subsequent flights can detect changes in the coastal zone such as change in shape of the beach coastal dunes or seacliff. Figure 10.A shows a digital elevation model determined by LIDAR for the Hilton Head, South Carolina, coastal area in the vicinity of a hotel which is very close to the active beach. Figure 10.B shows the seacliff environment near Pacifica, California, where 12 homes were condemned as unsafe following winter storms in early 1998. The amount of seacliff erosion from LIDAR from October 1997 to April 1998 is shown on Figure 10.C. Total erosion was about 10 m (30 ft) landward and a meter or two on the beach.

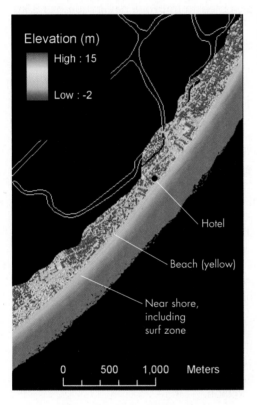

Figure 10.A **Coastal Topography** Digital elevation map of part of Hilton Head, South Carolina. Data from Light Detection and Ranging (LIDAR), an aircraft mounted laser remote sensing technique. The hotel is very close to the beach source—modified from NOAA. *(http://www.csc.noaa.gov/products/sccoasts/html/tutlid.htm)*

Figure 10.B **Severe coastal erosion** near Pacifica, California, that occurred early in 1998. *(Courtesy of Monty Hampton, USGS http://coastal.er.usgs.gov/lidar/AGU_fall98/)*

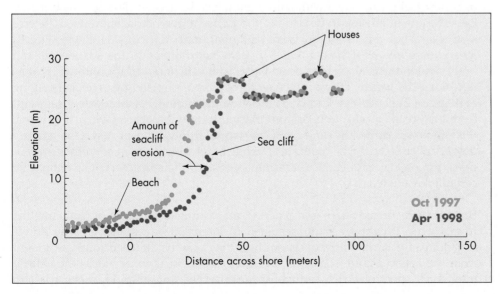

Figure 10.C **About 10 meters (30 ft) of seacliff erosion** near Pacifica, California (LIDAR data). Several houses were condemned and seven were torn down. *(Modified after USGS 1998 http://coastal.er.usgs.gov/lidar/AGU_fall98)*

10.4 **Coastal Hazards and Engineering Structures**

Efforts to stabilize a beach can be generalized into three approaches.

- Hard stabilization: Engineering structures to protect a shoreline from waves
- Soft stabilization: Adding sand to a beach (beach nourishment)
- Managed retreat: Living with beach erosion with perhaps a mixture of hard and soft stabilization

Hard Stabilization

Engineering structures in the coastal environment, including seawalls, groins, breakwaters, and jetties, are primarily designed to improve navigation or retard erosion. However, because they tend to interfere with the littoral transport of sediment along the beach, these structures all too often cause undesirable deposition and erosion in their vicinity.

Seawalls. **Seawalls** are structures constructed parallel to the coastline to help retard erosion. They may be constructed of concrete, large stones, wood, or other materials. Seawalls constructed at the base of a seacliff may not be particularly effective; considerable erosion of the seacliff results from mass-wasting processes on the cliff itself (see Figure 10.9b) as well as wave erosion at the base. Seawall use has been criticized because seawalls are often vertical structures that reflect incoming waves, or bounce them back from the shore. The reflection of waves enhances beach erosion and over several decades produces a narrower beach with less sand. Unless carefully designed to complement existing land uses, seawalls generally cause environmental and aesthetic degradation. Seawalls, in addition to causing a narrowing of a beach, may result in a reduction in biodiversity of the beach ecosystem (see Chapter 4). The design and construction of seawalls must be carefully tailored to specific sites. Some geologists believe that seawalls cause more problems than they solve and should be used rarely, if ever.[7]

Groins. **Groins** are linear structures placed perpendicular to the shore, usually in groups called *groin fields* (Figure 10.11). Each groin is designed to trap a portion of the sand that moves in the littoral transport system. A small accumulation of sand will develop updrift of each groin, thus building an irregular but wider beach. The wider beach protects the shoreline from erosion.

However, there is an inherent problem with groins: Although deposition occurs updrift of a groin, erosion tends to occur in the downdrift direction. Thus, a groin or a groin field results in a wider, more protected beach in a desired area but may cause a zone of erosion to develop in the adjacent downcoast shoreline. The erosion results because as a groin traps sediment on its updrift side, the downdrift area is deprived of sediment. Once a groin has trapped all the sediment it can hold, sand in the groin is transported around its offshore end to continue its journey along the beach. Therefore, erosion may be minimized by artificially filling each groin. This process, known as **beach nourishment,** requires extracting sand from the ocean or other sources and placing it onto the beach. When nourished, the groins will draw less sand from the natural littoral transport system, and the downdrift erosion will be reduced.[4] Despite beach nourishment and other precautions, groins may still cause undesirable erosion; therefore, their use should be carefully evaluated.

Breakwaters. Breakwaters and jetties protect limited stretches of the shoreline from waves. **Breakwaters** are designed to intercept waves and provide a protected area, or harbor, for boat moorings; they may be attached to, or separated from, the beach (Figure 10.12a, b). In either case, a breakwater blocks the natural littoral transport of beach sediment, causing the configuration of the coast to

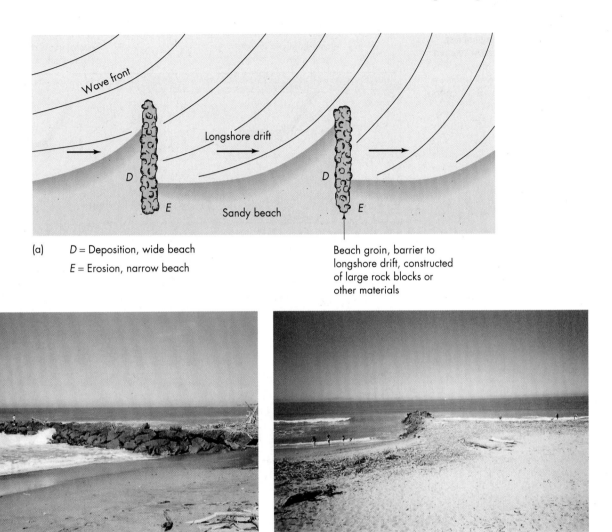

(a)   *D* = Deposition, wide beach

E = Erosion, narrow beach

Beach groin, barrier to longshore drift, constructed of large rock blocks or other materials

(b)

(c)

Figure 10.11 **Beach groins** (a) Diagram of two beach groins. (b) Closeup of downdrift zone of erosion. (c) Updrift deposition (right) and downdrift erosion (left). Deposited sediment builds a wide beach in the updrift direction; in the downdrift direction, the sparse amount of sediment for transport can cause erosion. *(Edward A. Keller)*

change locally as new areas of deposition and erosion develop. In addition to possibly causing serious erosion problems in the downdrift direction, breakwaters act as sand traps that accumulate sand in the updrift direction. Eventually the trapped sand may fill or block the entrance to the harbor as a deposit called a *sand spit* or *bar* develops. As a result, a dredging program, with an *artificial bypass*, is often necessary to keep the harbor open and clear of sediment. The sand that is removed by dredging, unless it is polluted, is transported in a pipe with water and released on the beach downdrift of the breakwater to rejoin the natural littoral transport system, thus reducing the erosion problem. It is the transport of the sand from the site where it is dredged to the downdrift location that is referred to as an artificial bypass.

Jetties. **Jetties** are often constructed in pairs at the mouth of a river or inlet to a lagoon, estuary, or bay (Figure 10.12c). They are designed to stabilize the channel, prevent or minimize deposition of sediment in the channel, and generally protect it from large waves.[4] Jetties tend to block the littoral transport of beach sediment, thus causing the updrift beach adjacent to the jetty to widen while downdrift beaches erode. The deposition at jetties may eventually fill the channel, making it useless,

Figure 10.12 Engineering structures constructed in the surf zone cause change Diagrams illustrating the effects of breakwaters and jetties on local patterns of deposition and erosion.

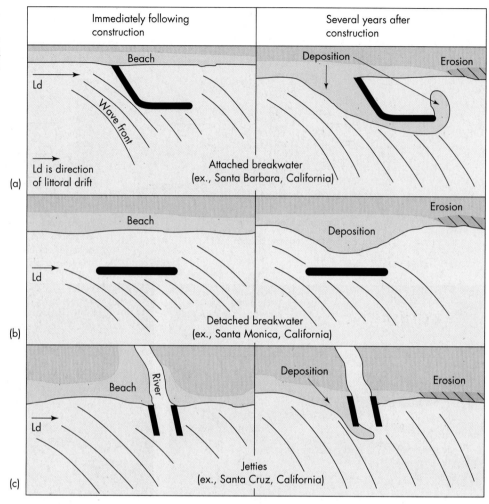

while downcoast erosion damages coastal development. Dredging the sediment minimizes but does not eliminate all undesirable deposition and erosion.

Unfortunately, it is impossible to build a breakwater or jetty that will not interfere with longshore movement of beach sediment. These structures must therefore be carefully planned and must incorporate protective measures that eliminate or at least minimize adverse effects. These protective measures may include installation of a dredging and artificial sediment-bypass system, a beach nourishment program, seawalls, riprap (large rocks), or some combination of these.[4]

Soft Stabilization

Beach Nourishment: An Alternative to Engineering Structures. In the discussion above, we introduced the topic of beach nourishment as an adjunct to engineering structures in the coastal zone. Beach nourishment can also be an alternative to engineering structures. In its purest form, beach nourishment consists of artificially placing sand on beaches in the hope of constructing a positive *beach budget*. When you budget your money you probably hope for a positive budget, allowing you some extra cash. Similarly, a positive beach budget means that when all the sand that enters and leaves the beach is accounted for, there is enough sand left to maintain the beach itself. Beach nourishment is sometimes referred to as "soft" stabilization to control beach erosion, as contrasted with "hard" stabilization, such as constructing groins or seawalls. Ideally, the presence of the nourished beach protects coastal property from the attack of waves.[4] The procedure has distinct advantages: It is aesthetically preferable to many engineering

(a)   (b)

Figure 10.13 Beach nourishment Miami Beach (a) before and (b) after beach nourishment.
(Courtesy of U.S. Army Corps of Engineers)

structures, and it provides a recreation beach as well as some protection from shoreline erosion.

In the mid-1970s the city of Miami Beach, Florida, and the U.S. Army Corps of Engineers began an ambitious beach nourishment program to reverse a serious beach erosion problem that had plagued the area since the 1950s. The program was also intended to provide protection from storms. The natural beach had nearly disappeared by the 1950s, and only small pockets of sand could be found associated with various shoreline protection structures, including seawalls and groins. As the beach disappeared, coastal resort areas, including high-rise hotels, became vulnerable to storm erosion.[8] The nourishment program was designed to produce a positive beach budget, which would widen the beach and provide additional protection from storm damage. The project cost approximately $62 million over 10 years and involved nourishment of about 160,000 m³ (209,300 yd³) of sand per year to replenish erosion losses. By 1980, about 18 million m³ (23.5 million yd³) of sand had been dredged and pumped from an offshore site onto the beach, producing a 200 m wide (656 ft wide) beach.[8] Figure 10.13 shows Miami Beach before and after the nourishment. The change is dramatic.

A cross-sectional design of the project shows the wide berm and frontal dune system, which is a line of sand dunes just above the high-tide line that functions as a buffer to wave erosion and storm surge (Figure 10.14). The Miami project was expanded in the mid- to late 1980s to include dune restoration, which involved establishing native vegetation on the dune. Special wooden walkways allow public access through the dunes while other areas of the dunes are protected. The successful Miami Beach nourishment project has functioned for more than 20 years, surviving major hurricanes in 1979 and 1992,[4] and is certainly preferable to the fragmented erosion-control methods that preceded it.

More than 600 km (373 mi) of coastline in the United States have received some sort of beach nourishment. Not all of this nourishment has had the positive effects reported for Miami Beach. For example, in 1982 Ocean City, New Jersey, nourished a stretch of beach at a cost of just over $5 million. After a series of storms that struck the beach, the sand was eroded in just two and a half months. The beach sands at Miami may be eroded at a much greater cost. Beach nourishment remains

Figure 10.14 Miami Beach shape after nourishment Cross section of the Miami Beach nourishment project. The dune and beach berm system provide protection against storm attack. *(Courtesy of U.S. Army Corps of Engineers)*

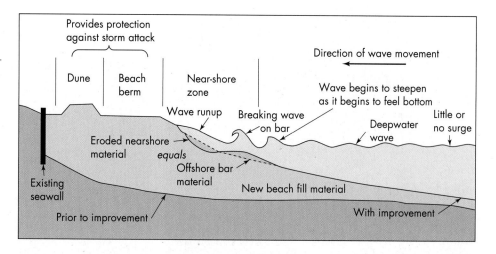

controversial, and some consider beach nourishment nothing more than "sacrificial sand" that will eventually be washed away by coastal erosion.[7] Miami Beach is an exception with respect to its apparent success and our discussion oversimplifies positive aspects of nourishment. The annual cost to nourish Miami Beach is about $3 million. Over 20 million tourist visit the beach each year. Foreign tourism alone brings in about $2 billion per year, over 650 times the cost of the nourishment.

Important issues of finding sand compatible with the site, how the nourishment will be paid for, and possible disruption of beach ecosystems must be carefully evaluated prior to any nourishment project. Nevertheless, beach nourishment has become a preferred method of restoring or even creating recreational beaches and protecting the shoreline from coastal erosion around the world. Additional case histories are needed to document the success or failure of the projects. Also needed is more public education to inform people of what can be expected from beach nourishment.[4]

10.5 Human Activity and Coastal Erosion: Some Examples

Human interference with natural shore processes has caused considerable coastal erosion. Most problems arise in areas that are highly populated and developed. As we have discussed, artificially constructed barriers often retard the movement of sand, causing beaches to grow in some areas and erode in others, resulting in damage to valuable beachfront property.

The Atlantic Coast

The Atlantic Coast from northern Florida to New York is characterized by *barrier islands*, long narrow islands separated from the mainland by a body of water (Figure 10.15). Many barrier islands have been altered to a lesser or greater extent by human use and interest.

The barrier island coast of Maryland illustrates some effects of human activity on coastal processes. Demand for the 50 km (30 mi) of Atlantic oceanfront beach in Maryland is very high, and the limited resource is used seasonally by residents of Washington, D.C., and Baltimore, Maryland (Figure 10.16). Since the early 1970s, Ocean City on Fenwick Island has promoted high-rise condominium and hotel development on its waterfront. The natural frontal dune system of this narrow island has been removed in many locations, resulting in a serious beach erosion problem. Even more ominous is the almost certain possibility that a future hurricane will cause serious damage to Fenwick Island. The Ocean City inlet found to the south of Ocean City formed during a hurricane in 1933. There is no guarantee, despite attempts to stabilize the inlet by coastal engineering, that a new inlet will not form, destroying, by erosion, part of Ocean City in the future.[9]

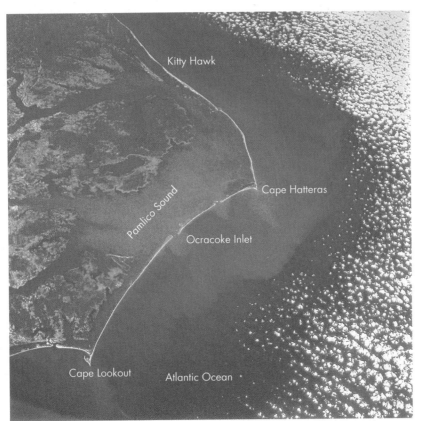

Figure 10.15 The Outer Banks of North Carolina appear in this image from the *Apollo 9* as a thin white ribbon of sand. The Barrier Islands are separated from the mainland by the Pamlico Sound. The brown color in the water is the result of sediment suspended in the water moving within the coastal system. Notice the fan-shaped plume of sediment just seaward of Ocracoke Inlet. The distance from Cape Lookout to Cape Hatteras is approximately 100 km (62 mi). *(National Aeronautics and Space Administration)*

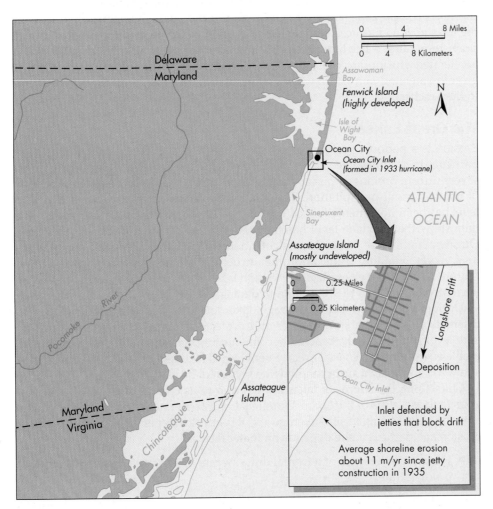

Figure 10.16 Urban development with jetty construction increases beach erosion The barrier island coast of Maryland. Fenwick Island is experiencing rapid urban development, and there is concern for potential hurricane damage. What will happen if a new inlet forms during a hurricane at the site of Ocean City? The inset shows details of the Ocean City inlet and effects of jetty construction.

Assateague Island is located to the south, across the Ocean City inlet. It encompasses two-thirds of the Maryland coastline. In contrast to the highly urbanized Fenwick Island, Assateague Island is in a much more natural state. The island is used for passive recreation such as sunbathing, swimming, walking, and wildlife observation. However, the two islands are in the same littoral cell, meaning they share the same sand supply. At least, that was the case until 1935, when the jetties were constructed to stabilize the Ocean City inlet. Since construction of the jetties, coastal erosion in the northern few kilometers of Assateague Island has averaged about 11 m (36 ft) per year, which is nearly 20 times the long-term rate of shoreline retreat for the Maryland coastline. During the same time, beaches immediately north of the inlet became considerably wider, requiring the lengthening of a recreational pier.[10]

Observed changes on Maryland's Atlantic coast are clearly related to the pattern of longshore drift of sand and human interference. Longshore drift is to the south at an average annual volume of about 150,000 m³ (196,500 yd³). Construction of the Ocean City inlet jetties interfered with the natural southward flow of sand and diverted it offshore rather than allowing it to continue southward to nourish the beaches on Assateague Island. Starved of sand, the northern portions of the island have experienced serious shoreline erosion during the past 50 years. This example of beach erosion associated with engineering structures that block longshore transport of sediment has been cited as the most severe that can be found in the United States.[10] Thus, the fundamental principle involving environmental unity holds: Earth is a system, a set of components that function together as a whole; a change in one part of the system affects all other parts.

The Gulf Coast

Coastal erosion is also a serious problem along the Gulf of Mexico. One study in the Texas coastal zone suggests that in the last 100 years, human modification of the coastal zone has accelerated coastal erosion by 30 percent to 40 percent over prehistoric rates.[11] The human modifications that appear most responsible for the accelerated erosion are coastal engineering structures, subsidence as a result of groundwater and petroleum withdrawal, and damming of rivers that supply sand to the beaches.

The Great Lakes

Erosion is a periodic problem along the coasts of the Great Lakes and has been particularly troublesome along the Lake Michigan shoreline. Damage is most severe during prolonged periods of high lake levels that follow extended periods of above-normal precipitation. The relationship between precipitation and lake level has been documented since 1860 by the U.S. Army Corps of Engineers. The data show that the lake level has fluctuated about 2 m (6.6 ft) during this time. During a high-water stage, there is considerable coastal erosion, and many buildings, roads, retaining walls, and other structures are destroyed by wave erosion (Figure 10.17).[12] For example, in 1985 high lake levels due to fall storms caused an estimated $15 million to $20 million in damage.

During periods of below-average lake level, wide beaches develop that dissipate energy from storm waves and protect the shore. However, with rising lake-level conditions, the beaches become narrow, and storm waves exert considerable energy against coastal areas. Even a small lake-level rise on a gently sloping shore will inundate a surprisingly wide section of beach.[12]

Cliffs along the shores of lakes are referred to as coastal bluffs and are analogous to the seacliffs of the ocean shoreline. Long-term rates of coastal bluff erosion at many Lake Michigan sites average about 0.4 m (1.3 ft) per year.[13] The severity of erosion at a particular site depends on many factors:

- Presence or absence of a frontal dune system. Dune-protected bluffs erode at a slower rate.

Figure 10.17 Lakes have coastal erosion problems Coastal erosion along the shoreline of Lake Michigan has destroyed this home. *(Steve Leonard/Getty Images, Inc.)*

- Orientation of the coastline. Sites exposed to high-energy storm winds and waves erode faster.
- Groundwater seepage. Seepage along the base of a coastal bluff causes slope instability, increasing the erosion rate.
- Existence of protective structures. Structures may be locally beneficial but often accelerate coastal erosion in adjacent areas.[12,13]

In recent years, beach nourishment has been attempted for some Great Lakes beaches. In some cases, sands that are coarser than the natural sands that had eroded have been added in the hope that the coarser, heavier sand will reduce the erosion potential.

10.6 Tropical Cyclones

Tropical cyclones are known as *typhoons* in most of the Pacific Ocean and **hurricanes** in the Atlantic. Tropical cyclones have taken hundreds of thousands of lives in a single storm. In November 1970, a tropical cyclone struck the northern Bay of Bengal in Bangladesh, producing a 6 m (20 ft) rise in the sea. Flooding caused by this sea-level rise killed approximately 300,000 people, caused $63 million in crop losses, and destroyed 65 percent of the total fishing capacity of the coastal region.[14] Another devastating cyclone hit Bangladesh in the spring of 1991, killing more than 100,000 people while causing more than $1 billion in damage. Hurricane Mitch, known as the deadliest Atlantic hurricane since 1780, was responsible for more than 11,000 deaths in Honduras and Nicaragua in 1998, and Hurricane Katrina in 2005 caused nearly 2,000 deaths with damages of about $100 billion to New Orleans and the Gulf Coast (see Chapter 5).

Hurricane Form and Process

The origin for the word *hurricane* possibly comes from a Caribbean Indian word for "big wind" or "evil spirit." They certainly fit the bill for the big wind! To be called a hurricane, the storm must have sustained winds of at least 119 km/hr (74 mph). Hurricanes are a variation of the tropical cyclone, which is the general term for a huge, complex series of thunderstorms that rotate around an area of low pressure, forming over warm, tropical ocean water.

Hurricanes begin as tropical disturbances, which are large areas of unsettled weather with a diameter as large as 600 km (370 mi). Within this area, there exists an

Figure 10.18 **Hurricane form and process** Energy from warm ocean water is transformed into the storm, which may have a diameter of 600 km (370 mi). Main components are rotating bands of thunderstorms; rising warm air around an eye of clear sky with subsiding air; and the eye wall as the boundary between the rotating rain bands and eye.

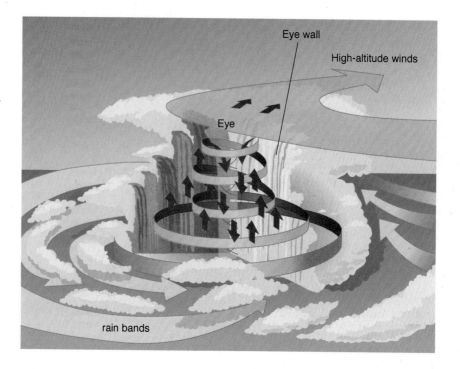

organized mass of thunderstorms with a general low pressure in which there is initial rotation caused by the movement of the storm and rotation of Earth. A tropical depression may grow in size and strength as warm, moist air is drawn into the depression and begins to rotate counterclockwise in the Northern Hemisphere and clockwise in the Southern Hemisphere. The process that increases the intensity of the storm is caused by warm water that evaporates from the sea and is drawn into the storm. As this process occurs, the energy of the storm increases. As warm seawater evaporates, it is transformed from liquid water in the sea to water vapor (gas) in the air mass of the storm. When this happens, potential energy in the form of *latent heat* enters the storm. Latent heat of water is the amount of heat required to change liquid water to water vapor. Latent heat is one of the major sources of the power (time rate of energy expenditure) of a hurricane. As the moist air rises, condensation (rain) occurs as the latent heat is released, warming the air and making it lighter. As the lighter air rises, more energy from warm water is drawn in and the storm may increase in size, strength, and intensity. If wind speeds in the storm reach 63 km (39 mi), then the depression is called a tropical storm and receives a name.

Hurricanes are very high energy storms. Their size can be huge by human standards. It is not unusual to observe on satellite images a hurricane moving toward the United States that has an area nearly the size of the Gulf of Mexico, stretching from Florida to Texas and beyond.

A generalized diagram of a hurricane and the processes that occur within it are shown in Figure 10.18. The hurricane has bands of spinning storms and thunder cells with an eye where air is subsiding in the center around what is called the eye wall. As the moist air rotates within the storm, intense rain bands are produced. When a hurricane moves across an island, it is deprived of some of its energy and often weakens. When a hurricane makes landfall on a continent and moves inland, it weakens and eventually dies, but the storm and rainfall can cause serious river flooding well inland. The intense rainfall on slopes may well generate numerous landslides, producing an additional hazard.

Hurricanes are classified based upon their size and intensity. Table 10.1 is the Saffir-Simpson Hurricane Scale. It is used to estimate potential wind damage and flooding. There are five categories, with Category One being the smallest and Category Five the largest. Category Five hurricanes are capable of producing catastrophic damage, but even a Category One storm is a very serious and dangerous event.

TABLE 10.1   The Saffir-Simpson Hurricane Scale

The Saffir-Simpson Hurricane Scale is a 1 to 5 rating based on the hurricane's present intensity. This is used to give an estimate of the potential property damage and flooding expected along the coast from a hurricane landfall. Wind speed is the determining factor in the scale, as storm surge values are highly dependent on the slope of the continental shelf in the landfall region, where the hurricane is expected to come on land.

Category One Hurricane	Winds 119 to 153 km/hr (74–95 mph). Storm surge generally 1.2 to 1.5 m (4 to 5 ft) above normal. No real damage to building structures. Damage primarily to unanchored mobile homes, shrubbery, and trees. Some damage to poorly constructed signs. Also, some coastal road flooding and minor pier damage. Hurricanes Allison of 1995 and Danny of 1997 were Category One hurricanes at peak intensity.
Category Two Hurricane	Winds 154 to 177 km/hr (96 to 110 mph). Storm surge generally 1.8 to 2.4 m (6 to 8 ft) above normal. Some roofing material, door, and window damage of buildings. Considerable damage to shrubbery and trees, with some trees blown down. Considerable damage to mobile homes, poorly constructed signs, and piers. Coastal and low-lying escape routes flood 2 to 4 hours before arrival of the hurricane center. Small craft in unprotected anchorages break moorings. Hurricane Bonnie of 1998 was a Category Two Hurricane when it hit the North Carolina coast; Hurricane George of 1998 was a Category Two Hurricane when it hit the Florida Keys and the Mississippi Gulf Coast.
Category Three Hurricane	Winds 178 to 209 km/hr (111 to 130 mph). Storm surge generally 2.7 to 3.7 m (9 to 12 ft) above normal. Some structural damage to small residences and utility buildings, with a minor amount of wall failures. Damage to shrubbery and trees, with foliage blown off trees and large trees blown down. Mobile homes and poorly constructed signs are destroyed. Low-lying escape routes are cut off by rising water 3 to 5 hours before arrival of the hurricane center. Flooding near the coast destroys smaller structures, and larger structures are damaged by battering by floating debris. Terrain continuously lower than 1.5 m (5 ft) above mean sea level may be flooded inland 13 km (8 mi) or more. Evacuation of low-lying residences within several blocks of the shoreline may be required. Hurricane Roxanne of 1995 was a Category Three hurricane at landfall on the Yucatan Peninsula of Mexico; Hurricane Fran of 1996 was a Category Three hurricane at landfall in North Carolina.
Category Four Hurricane	Winds 210 to 249 km/hr (131 to 155 mph). Storm surge generally 4 to 5.5 m (13 to 18 ft) above normal. More extensive wall failures, with some complete roof structure failures on small residences. Shrubs, trees, and all signs are blown down. Complete destruction of mobile homes. Extensive damage to doors and windows. Low-lying escape routes may be cut off by rising water 3 to 5 hours before arrival of the hurricane center. Major damage to lower floors of structures near the shore. Terrain lower than 3.1 m (10 ft) above sea level may be flooded, requiring massive evacuation of residential areas as far inland as 10 km (6 mi). Hurricane Luis of 1995 was a Category Four hurricane while moving over the Leeward Islands; Hurricanes Felix and Opal of 1995 also reached Category Four at peak intensity.
Category Five Hurricane	Winds greater than 249 km/hr (155 mph). Storm surge generally greater than 5.5 m (18 ft) above normal. Complete roof failure on many residences and industrial buildings. Some complete building failures, with small utility buildings blown over or away. All shrubs, trees, and signs blown down. Complete destruction of mobile homes. Severe and extensive window and door damage. Low-lying escape routes are cut off by rising water 3 to 5 hours before arrival of the hurricane center. Major damage to lower floors of all structures located less than 4.6 m (15 ft) above sea level and within 458 m (500 yd) of the shoreline. Massive evacuation of residential areas on low ground within 8 to 16 km (5 to 10 mi) of the shoreline may be required. Hurricane Mitch of 1998 was a Category Five hurricane at peak intensity over the western Caribbean. Hurricane Gilbert of 1988 was a Category Five hurricane at peak intensity and is the strongest Atlantic tropical cyclone of record.

Source: Modified after Spindler, T., and Beven, J. 1999. Saffir-Simpson Hurricane Center, National Oceanic and Atmospheric Administration. www.nhc.noaa.gov.aboutsshs.html. Accessed 5/23/01.

We know when a hurricane is coming. Images from satellites and high-altitude aerial photographs show the locations of tropical depressions and tropical storms that we may monitor to see if they will become hurricanes. In order to verify satellite data, special airplanes are flown through storms, recording wind speed, air temperature, and air pressure. Once a hurricane has formed, its continual movement (called its path) is monitored. The paths of four hurricanes in 2005 are shown on Figure 10.19.

As hurricanes get close to landfall, moving over shallower water, they generally slow down. However, if they encounter warmer water, they may actually increase in intensity.

The most dangerous aspect of hurricanes is not the wind itself, although violent winds can be lethal. Often what causes the most flooding and loss of life is the **storm surge,** which arrives with the storm (Figure 10.20). The storm surge is a local rise in sea level that results when hurricane winds push water toward the coast. Storm surges may arrive as a quick rise in sea level or series of waves that may increase water depths near the coast by several meters to more than 10 m (30 ft). Storm surge is generally highest on the part of the rotating storm that pushes landward. For example, along the Gulf Coast, the storm surge and wind speed are higher to the east than west (the right front quadrant of the storm) as the hurricane rotates in a counterclockwise direction. During Hurricane Katrina (see opening Case History in Chapter 5), the storm surge east of New Orleans was significantly higher than in the central or western part of the storm. Should the storm arrive co-incidentally with a high tide, then a storm surge of even greater height occurs. Most of the deaths from hurricanes are caused by storm surges as people are drowned or

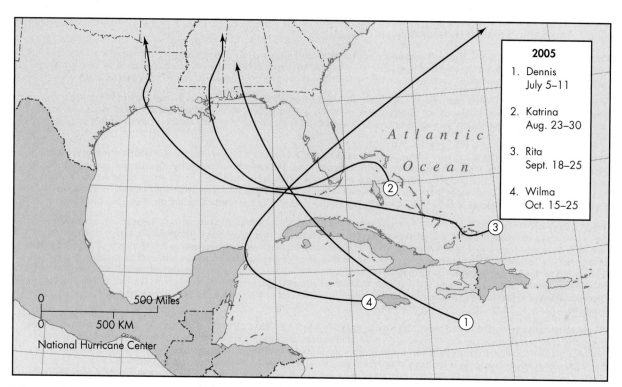

Figure 10.19 Paths of four hurricanes in 2005 *(National Hurricane Center)*

struck by solid objects within the surge. When the surge moves ashore, it is not a tall advancing line of water but is more like a continual increase in rise of sea level as a hurricane approaches and makes landfall. Having said this, the actual storm surge in places may well appear as a series of wavelike forms that inundate the land.

Most hurricanes and typhoons form in a belt between 8 degrees north and 15 degrees south of the equator, where warm surface-water temperatures exceed about 27°C (80°F). During an average year, approximately five hurricanes will develop that might threaten the Atlantic and Gulf Coasts. One of three likely storm tracks may develop (Figure 10.21):

1. A storm heads toward the east coast of Florida, sometimes passing over islands such as Puerto Rico, and then, before striking the land, it moves out into the Atlantic to the northeast.

Figure 10.20 Storm surge and high waves produced by Hurricane Gloria on September 27, 1985, in New Jersey. *(Ryan Williams/International Stock Photography Ltd.)*

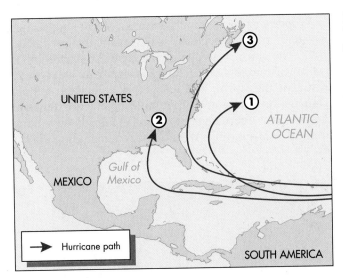

Figure 10.21 Hurricane paths
Three common paths of hurricanes in the Atlantic.

2. A storm travels over Cuba and into the Gulf of Mexico to strike the Gulf Coast.

3. A storm skirts along the East Coast and may strike land from central Florida to New York. Storms that may develop into hurricanes are closely monitored from both satellites and specially designed aircraft that fly through the storms.

Hazards presented by hurricanes include high winds that may rip shingles and rafters from roofs, blow over large trees and utility lines, and generally wreak havoc on structures built by people. The processes that kill most people and often cause the most damage, however, are (1) flooding resulting from intense precipitation and landward transport of wind-driven waves of ocean waters and, as previously mentioned, (2) storm surges.[15]

The probability that a hurricane will strike a particular 80 km (50 mi) coastal segment of the Atlantic or Gulf Coast in a given year is shown in Figure 10.22. Notice that the probabilities are particularly high in southern Florida and on the Louisiana coastline.

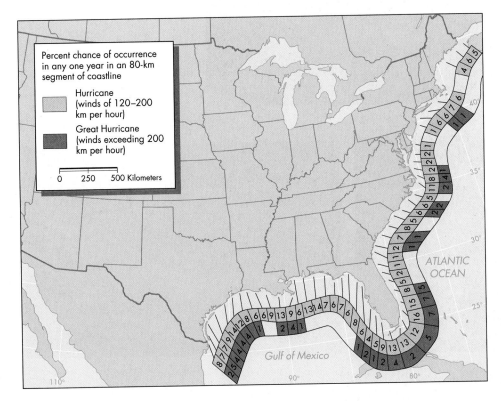

Figure 10.22 Hurricane hazard map Probability that a hurricane will strike a particular 80 km (50 mi) south Atlantic coastal segment in a given year. *(From Council on Environmental Quality, 1981. Environmental trends)*

Property damage from hurricanes can be staggering (recall Hurricane Katrina). As another example, consider Hurricane Andrew, which struck Florida in August 1992. The storm was one of the costliest storms in U.S. history, with estimated damages in excess of $25 billion. Despite evacuation, 23 lives were lost as a direct result of the storm, and 250,000 people were temporarily rendered homeless. Damages to homes and other buildings were extensive: About 25,000 homes were destroyed as entire neighborhoods in Florida were flattened (Figure 10.23).[16] More than 100,000 buildings were damaged, including the National Hurricane Center in

(a)

Figure 10.23 **Hurricane across Florida and the Gulf of Mexico**
(a) Hurricane Andrew, August 25, 1992, as shown on a multispectral image. The storm has left south Florida, where it did extensive damage, and is moving toward Louisiana. *(Courtesy of Hasler, Pierce, Palaniappan, Manyin/NASA Goddard Laboratory for Atmospheres)*
(b) Lighthouse at Cape Florida in Biscayne Bay, Miami, Florida, before the arrival of the hurricane. *(Wingstock/Comstock Images)*
(c) The same coastline after the hurricane. *(Cameron Davidson/ Comstock Images)*

(b)

(c)

Florida, where a radar antenna housed in a protective dome was torn from the roof.[16] Two hurricanes in the summer of 1996 struck the North Carolina coast. The second storm in August caused about $3 billion in property damage and killed more than 20 people. Although the population has increased along the Atlantic and Gulf Coasts, the loss of lives from hurricanes has decreased significantly because of more effective detection and warning. The amount of property damage, however, has greatly increased.

There is concern for large cities such as Miami and New Orleans (especially after experiencing Katrina in 2005). Unsatisfactory evacuation routes, building codes, and disaster preparedness may contribute to another catastrophic hurricane along the Atlantic or Gulf Coast.

10.7 Perception of and Adjustment to Coastal Hazards

Perception of Coastal Erosion

An individual's past experience, proximity to the coastline, and probability of suffering property damage all play primary roles in the perception of coastal erosion as a natural hazard. One study of coastal erosion of seacliffs near Bolinas, California, 24 km (15 mi) north of the entrance to the San Francisco Bay, established that people living close to the coast in an area likely to experience damage in the near future are generally very well informed and see the erosion as a direct and serious threat.[17] People living a few hundred meters from a possible hazard, although aware of the hazard, know little about its frequency of occurrence, severity, and predictability. Still farther inland, people are aware that coastal erosion exists but have little perception of it as a hazard.

Adjustment to Coastal Hazards

Tropical Cyclones. People adjust to the tropical cyclone hazard either by doing nothing and bearing the loss or by taking some kind of action to modify potential loss. For example, homes in hurricane-prone areas may be constructed to allow the storm surge to pass under the house (Figure 10.24). Community adjustments include attempts to modify potential loss by strengthening the environment with protective structures and land stabilization and by adapting better land-use zoning, evacuation procedures, and warning procedures.[18] Some general guidelines of what to do before, during, and after a hurricane are listed in Table 10.2.

Figure 10.24 Hurricane resistant house Home in the Florida Keys, constructed with strong blocks to withstand hurricane force wind and space below to allow the flow of a storm surge beneath the building.
(Edward A. Keller)

TABLE 10.2 What to Do Before, During, and After a Hurricane

Before a Hurricane	• Probably the most important thing you can do before a hurricane is to plan your evacuation route. The plan should include information concerning the safest evacuation route as well as locations of nearby shelters. Be prepared to drive inland to a safe place 13 to 31 km (20 to 50 mi).
	• Prepare a disaster kit, including flashlight with extra batteries, portable battery-operated radio, first-aid kit, emergency food and water, can opener, necessary medicines, cash and credit cards, and change of clothes, including sturdy shoes.
	• Make arrangements to care for your pets. Pets may not be allowed in emergency shelters.
	• Have a family plan on how to respond after a hurricane, and be sure that members of the family know how to turn off household gas, electricity, and water.
	• Teach young children when and how to call for emergency assistance from police or fire departments and which radio stations to listen to for emergency information.
	• Protect your windows with permanent shutters or be prepared to use plywood panels cut to fit each window. It is important to put up window protection long before the storm arrives.
	• Homeowner insurance policies do not cover damage from flooding that accompanies a hurricane, so obtain flood insurance.
	• Develop a family emergency communication plan. Family members may be separated during a disaster. Your plan should include how you are going to get back together.
	• Ask an out-of-state relative or friend to serve as your family contact. After a hurricane, it is often easier to call long distance than within your area. Be sure everyone in your family knows the name, address, and phone number of the family contact.
	• Hurricanes are generally spotted far out at sea days before they strike land. A hurricane watch is issued when there is a threat that hurricane conditions will occur within 24 to 36 hours. A hurricane warning is issued when hurricane conditions are expected in 24 hours or less.
During a Hurricane Watch	• Listen to your television or radio for hurricane progress reports.
	• Check emergency supplies.
	• Be sure there is fuel in your car.
	• Bring inside outdoor objects such as lawn furniture, toys, and garden tools. Objects that cannot be brought inside should be anchored.
	• Secure your home and other buildings by closing and boarding up windows.
	• Remove outside antennas or satellite dishes.
	• Because power outages are likely during a storm, turn your refrigerator and freezer to the coldest possible settings and open them only when absolutely necessary.
	• Store drinking water in clean bathtubs, bottles, and cooking utensils.
	• Review with your family your evacuation plan.
	• If you have a boat, be sure it is secure or moved to a designated safe place.
During a Hurricane Warning	• Listen constantly to television or radio for official instructions.
	• If you live in a mobile home, check your tiedowns and evacuate immediately.
	• Store your valuables and personal papers in waterproof containers in the highest level of your home.
	• Avoid elevators.
	• Stay inside and away from windows, skylights, and glass doors.
	• Keep a supply of flashlights and extra batteries on hand. Avoid using open flames such as candles or kerosene lamps.
	• If electrical power is lost, turn off major appliances to reduce power surges when electricity is restored.
	• If officials indicate that evacuation is necessary, leave as soon as possible, avoiding flooded roads and washed-out bridges. Secure your home by unplugging appliances and turning off electricity and the main water valve. Let someone outside the storm area know where you are going. If time permits, move furniture to protect it from flooding (if possible, move it to a higher floor). Load preassembled emergency supplies and warm protective clothing, including sleeping bags and blankets. Finally, lock up your home and leave!
After a Hurricane	• Stay tuned to local radio and television for information.
	• Assist injured or trapped persons. Do not move seriously injured persons unless they are in immediate danger. Call for help.
	• Return to your home only after authorities have advised you that it is safe to do so.
	• When you return home, be aware of the possibility of dangling power lines, and enter your home with caution. Be aware of snakes, insects, and animals that may have been driven to higher ground by floodwaters. Open your windows and doors to ventilate and dry your home. Check your refrigerator for foods that may have spoiled. Finally, take pictures of the damage to both the house and its contents for insurance purposes.

Source: Modified after Federal Emergency Management Agency Fact Sheet: Hurricanes. www.fema.gov/library/hurricaf.htm. Accessed 5/10/2001.

Coastal Erosion. Adjustments to coastal erosion generally fall into one of three categories:

- Beach nourishment that tends to imitate natural processes, the "soft solution"
- Shoreline stabilization through structures such as groins and seawalls, the "hard" solution
- Land-use change that attempts to avoid the problem by not building in hazardous areas or by relocating threatened buildings

A preliminary process in any approach to managing coastal erosion is estimating the rates of erosion. Esimates of future erosion rates are based on historical shoreline change or statistical analysis of the oceanographic environment, such as the waves, wind, and sediment supply that affect coastal erosion. Recommendations are then made concerning setbacks considered to be minimum standards for state or local coastal erosion management programs. A setback is the distance from the shoreline to where development, such as homes, is allowed. A small number of states (including Florida, New Jersey, New York, and North Carolina) use a setback distance for buildings based on the rate of erosion (see A Closer Look: E-Lines and E-Zones).[19] The concept of setback has real merit in coastal erosion management and is at the heart of land-use planning to minimize damage from coastal erosion.

We are at a crossroads today with respect to adjustment to coastal erosion. One road leads to ever-increasing coastal defenses in an attempt to control the processes of erosion. The second road involves learning to live with coastal erosion through flexible environmental planning and wise land use in the coastal zone.[20,21] The first road follows history in our attempt to control coastal erosion through the construction of engineering structures such as seawalls. In the second road, structures in the coastal zone, with such exceptions as critical facilities in certain areas, are considered temporary and expendable. Development in the coastal zone must be in the best interests of the general public rather than for a few who profit from developing the oceanfront. This philosophy is at odds with the viewpoint of developers, who consider the coastal zone "too valuable not to develop." In fact, development in the coastal zone is not the problem; rather, the problem lies in the inappropriate development of hazardous areas and areas better suited for uses other than building. In other words, beaches belong to all people to enjoy, not only to those fortunate enough to purchase beachfront property. The state of Hawaii has taken this idea to heart: There, all beaches are public property, and local property owners cannot deny access to others.

Accepting the philosophy that, with minor exceptions, coastal zone development is temporary and expendable and consideration should first be to the general public requires an appreciation of the following five principles:[20]

1. Coastal erosion is a natural process rather than a natural hazard; erosion problems occur when people build structures in the coastal zone. The coastal zone is an area where natural processes associated with waves and moving sediment occur. Because such an environment will have a certain amount of natural erosion, the best land uses are those compatible with change. These include recreational activities such as swimming and fishing.

2. Any shoreline construction causes change. The beach environment is dynamic. Any interference with natural processes produces a variety of secondary and tertiary changes, many of which may have adverse consequences. Adverse consequences are particularly likely when engineering structures, such as groins and seawalls, that affect the storage and flow of sediment along a coastal area are used.

3. Stabilization of the coastal zone through engineering structures protects the property of relatively few people at a large expense to the general public.

A CLOSER LOOK | E-Lines and E-Zones

Recently, a special committee of the National Research Council (NRC) at the request of the Federal Emergency Management Agency (FEMA) developed coastal zone management recommendations,[19] some of which follow.

- Future erosion rates should be estimated on the basis of historical shoreline change or statistical analysis of the oceanographic environment (that is, the waves, wind, and sediment supply that affect coastal erosion).

- E-lines and E-zones based on erosion rates should be mapped (Figure 10.D). The E stands for erosion; for example, the E-10 line is the location of expected erosion in 10 years. The E-10 zone is considered to be an imminent hazard where no new habitable structures should be allowed. The setback distance depends on the erosion rate. For example, if the rate is 1 m (3.3 ft) per year, the E-10 setback is 10 m (33 ft).

- Movable structures are allowed in the intermediate and long-term hazard zones (E-10 to E-60) (see Figure 10.D).

- Permanent large structures are allowed at setbacks greater than the E-60 line.

- New structures built seaward of the E-60 line, with the exception of those on high bluffs or seacliffs, should be constructed on pilings to withstand erosion associated with a high-magnitude storm with a recurrence interval of 100 years.

NRC recommendations concerning setbacks are considered to be minimum standards for state or local coastal erosion management programs. A small number of states (including Florida, New Jersey, New York, and North Carolina) use a setback based on the rate of erosion; however, most states do not require this type of setback. Nevertheless, the concept of E-lines and E-zones based on erosion-designated setbacks and allowable construction has real merit in coastal zone management.

Figure 10.D Erosion hazard zones Idealized diagram illustrating the concept of the E-lines and E-zones based on the rate of coastal erosion from a reference point such as the seacliff or dune line. The width of the zone depends on the rate of erosion and defines a setback distance. Of course, even with setbacks based on 60 years of expected erosion (E-60 line), eventually, 60 years down the road, structures will be much closer to the shoreline and will become vulnerable to erosion. It is a form of planned obsolescence. *(From National Research Council. 1990. Managing coastal erosion. Washington, DC: National Academy Press)*

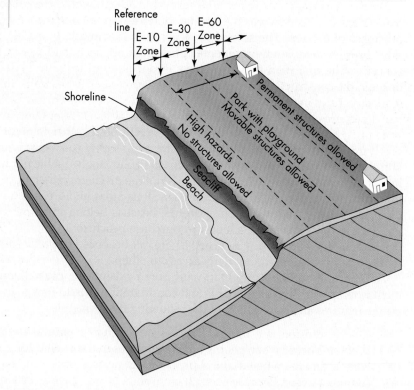

Engineering structures along the shoreline are often meant to protect developed property, not the beach itself. It has been argued that the interests of people who own shoreline property are not compatible with the public interest and that it is unwise to expend large amounts of public funds to protect the property of a few.

4. Engineering structures designed to protect a beach may eventually destroy it. Engineering structures often modify the coastal environment to such an

extent that it may scarcely resemble a beach. For example, construction of large seawalls causes reflection of waves and turbulence that eventually narrow the beach.

5. Once constructed, shoreline engineering structures produce a trend in coastal development that is difficult, if not impossible, to reverse. Engineering structures often lead to additional repairs and larger structures, with spiraling costs. In some areas, the cost of the structures eventually exceeds the value of the beach property itself. For these and other reasons, several states have recently imposed severe limitations on future engineering construction intended to stabilize the coastline. As sea levels continue to rise and coastal erosion becomes more widespread, the nonstructural alternatives to the problem should continue to receive favorable attention because of both financial necessity and the recognition that the amenities of the coastal zone should be kept intact for future generations to enjoy.

If you consider purchasing land in the coastal zone, remember these guidelines:[21] (1) Allow for a good setback from the beach or seacliff; (2) be high enough above sea level to avoid flooding; (3) construct buildings to withstand adverse weather, especially high winds; and (4) if hurricanes are a possibility, be sure there are adequate evacuation routes. Remember, it is always risky to buy property where land meets water.

SUMMARY

The coastal environment is one of the most dynamic areas on Earth, and rapid change is a regular occurrence. Migration of people to coastal areas is a continuing trend, and approximately 75 percent of the population in the United States now live in coastal states.

Ocean waves are generated by storms at sea and expend their energy on the shoreline. Irregularities in the shoreline account for local differences in wave erosion; these irregularities are largely responsible for determining the shape of the coast. Beaches are most commonly formed by accumulations of sand or gravel that are deposited at the coast by rivers and shaped by wave action. Actually, beaches can form from any loose material, such as broken shells or coral or volcanic rock, located in the shore zone. Waves striking a beach at an angle result in longshore transport of the beach sediments.

Rip currents are a serious hazard to swimmers, killing up to 200 people a year in the United States. They can be recognized and avoided, and it is possible to escape from them if you do not panic.

Although coastal erosion causes a relatively small amount of property damage compared with other natural hazards such as river flooding, earthquakes, and tropical cyclones, it is a serious problem along all the coasts of the United States, including the shorelines of the Great Lakes. Factors contributing to coastal erosion include river damming, high-magnitude storms, and the worldwide rise in sea level.

Human interference with natural coastal processes such as the building of seawalls, groins, breakwaters, and jetties is occasionally successful, but in many cases it has caused considerable coastal erosion. Sand tends to accumulate on the updrift side of the structure and erode on the downdrift side.

Most problems occur in areas with high population density, but sparsely populated areas along the Outer Banks in North Carolina are also experiencing trouble with coastal erosion. Beach nourishment has had limited success in restoring or widening beaches, but it remains to be seen whether it will be effective in the long term.

The most catastrophic coastal hazard is the tropical cyclone. Also called typhoons and hurricanes, tropical cyclones are violent storms that bring high winds, storm surges, and river flooding. They continue to take thousands of lives and to cause billions of dollars in property damage.

Perception of the coastal erosion hazard depends mainly on the individual's experience with and proximity to the hazard. Community and individual adjustments to tropical cyclones generally attempt to modify the environment by building protective structures designed to lessen potential damage or to encourage change in people's behavior by better land-use zoning, evacuation, and warning.

Adjustment to coastal erosion in developed areas is often the "technological fix": building seawalls, groins, and other structures or (more recently) using beach nourishment. These approaches to stabilizing beaches have had mixed success and may cause additional problems in adjacent areas. Engineering structures are very expensive, require maintenance, and, once in place, are difficult to remove. The cost of engineering structures may eventually exceed the value of the properties they protect; such structures may even destroy the beaches they were intended to save. Managing coastal erosion will benefit from careful land-use planning that emphasizes establishment of designated setbacks and allowable construction determined by predicted rates of coastal erosion.

Revisiting Fundamental Concepts

Human Population Growth

Many populated areas are located near the coast, and human population in the coastal zone is expected to continue to increase. As a result, potential impacts of coastal hazards will increase.

Sustainability

Coastal areas contain some of the most scenic and valuable property on Earth. In order to maintain a quality environment in the coastal zone, we must develop plans to sustain our coast for future generations. Sustaining the coast will involve learning to live with and adjust to coastal hazards through land-use planning that maintains the integrity of the coastal environment.

Earth as a System

Coastal environments are complex systems where rapid change is often the norm. Learning to live with change in the coastal environment is a necessity.

Hazardous Earth Processes, Risk Assessment, and Perception

The most hazardous processes in the coastal environment are coastal erosion and erosion and flooding associated with tropical cyclones. The impacts of these hazards are increasing as a result of increased development in the coastal zone and global climate change, which is causing a rise in sea level. In general, people living in the coastal zone are aware of potential hazards and there have been many studies in coastal areas to evaluate risk and appropriate adjustments so that loss of life and property may be minimized.

Scientific Knowledge and Values

It is clear that people value the coastal environment. Scientific knowledge concerning coastal processes is a mature field of study, and in general we know where hazards are most likely to occur and what their potential impacts are. Solutions to reducing coastal hazards vary from building hard engineering structures to reduce damage, to softer approaches that allow us to live in the coastal zone and adjust to hazards. The solution we choose for a particular site depends in part upon how we value the coastal zone. For example, if a row of beach homes is being threatened by erosion, the choices may be building a seawall, which would eventually cause the loss of the beach, or moving the homes inland out of harm's way.

Key Terms

beach (p. 328)

beach-budget (p. 332)

beach nourishment (p. 336)

breakwater (p. 336)

groins (p. 336)

hurricane (p. 343)

jetties (p. 337)

longshore sediment transport (p. 330)

rip current (p. 330)

seacliff (p. 332)

seawall (p. 336)

storm surge (p. 345)

tropical cyclone (typhoon) (p. 343)

Review Questions

1. How does wave refraction at a rocky point result in concentration of wave energy at the point?

2. What is the difference between plunging and spilling breakers?

3. What are the processes of longshore transport of sand?

4. What are some of the human activities that can increase seacliff erosion?

5. What are some of the important differences in coastal processes and beach erosion between the East and West Coasts of the United States?

6. What are the major alternatives to stabilize a coast? Which is preferred in a particular situation? Why?

7. What are seawalls and groins, and why are they constructed? What is their effect on erosion and coastal processes?

8. What are the processes important in the formation of a hurricane?

9. What is the process of beach nourishment, and what is its objective?

10. What are the major factors causing erosion problems for the Great Lakes?

11. What is storm surge, and how is it produced?

12. What are the three major adjustments to coastal erosion?

13. What are the five general principles that should be accepted if we choose to live with rather than control coastal erosion?

Critical Thinking Questions

1. Do you think that human activity has increased the coastal erosion problem? Outline a research program that could test your hypothesis.

2. Do you agree or disagree with the statements that all structures in the coastal zone (with the exceptions of critical facilities) should be considered temporary and expendable and that any development in the coastal zone must be in the best interest of the general public rather than the few who developed the oceanfront? Explain your position.

3. A beach park is experiencing coastal erosion. Some want to protect the park with a restaurant, parking lots, outhouse and lawn at any cost—not loose a blade of grass. That would require a hard solution such as a seawall. Others want to maintain the sand beach and use a more flexible approach to the erosion. They argue for beach nourishment and planned retreat. Both groups have clearly stated their values. What are the pros and cons for each position? Can the two views be considered simultaneously?

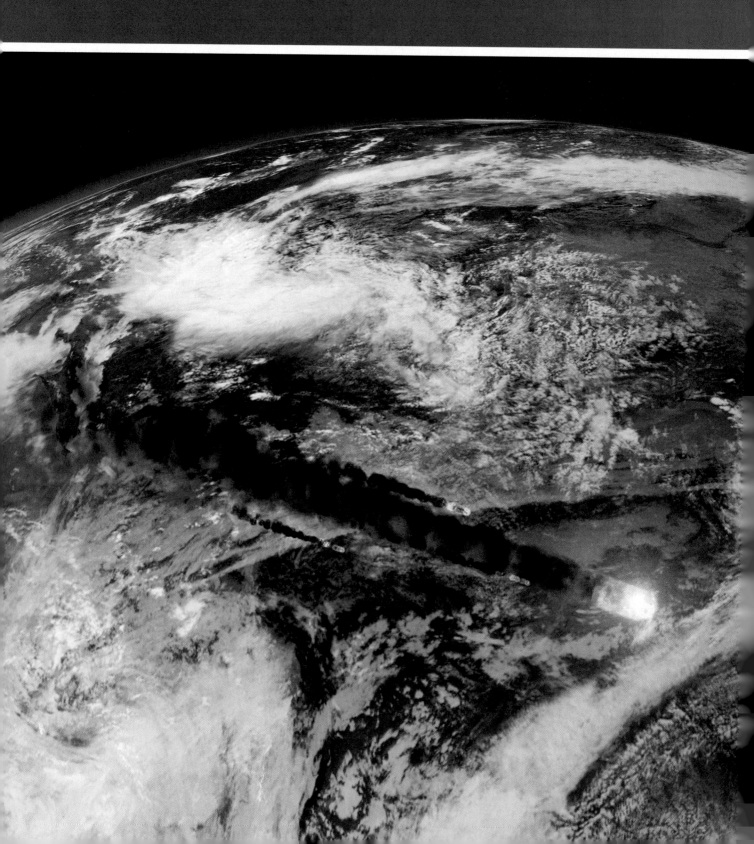

ELEVEN

Impact of Extraterrestrial Objects

Written with the assistance of Frank Spera

Learning Objectives

Bombardment of Earth by objects from space has been occurring since the birth of our planet. Such impacts have been linked to the extinction of many species, including the dinosaurs. The risk of impact from asteroids, comets, and meteoroids continues today. Learning objectives for this chapter include

- Know the difference between asteroids, meteoroids, and comets
- Understand the physical processes associated with aerial burst and impact craters
- Understand the possible causes of mass extinction
- Know the evidence for the impact hypothesis producing the late Cretaceous mass extinction
- Know the likely physical, chemical, and biological consequences of impact from a large asteroid or comet
- Understand the risk of impact or aerial burst of extraterrestrial objects and how that risk might be minimized

Siberian impact Tunguska event (artist image) just prior to the atmospheric 10-megaton (equivalent TNT) blast and destruction of a 30-meter diameter asteroid several kilometers above Siberia. The blast flattened and burned over 2,000 square kilometers (770 square miles) of forest, but left no crater. *(© Joe Tucciarone)*

CASE HISTORY | The Tunguska Event

On June 30, 1908, shortly after 7 A.M., witnesses in Siberia reported observing a blue-white fireball with a glowing tail descending from the sky. The fireball exploded above the Tunguska River Valley in a heavily forested, sparsely populated area. Later, calculations would show that the explosion had the force of 10 megatons of TNT, equivalent to 10 hydrogen bombs. Though there were few witnesses close to the event, the sounds from the explosion were heard hundreds of kilometers away, and the blast wave was recorded at meteorological recording sites throughout Europe. The tremendous air blast caused more than 2,000 square kilometers (770 square miles) of forest to be flattened and burned (Figure 11.1).

A herdsman in the vicinity of the blast was one of the few people who witnessed the devastation on the ground. His hut was completely flattened by the blast and its roof was blown away. Other witnesses a few tens of kilometers from the explosion reported that they were physically blown into the air and knocked unconscious; they awoke to find a transformed landscape of smoke and burning trees that had been blasted to the ground.

At the time of the explosion, Russia was in the midst of political upheaval. As a result, there was no quick response to or investigation of the Tunguska event. Finally, in 1924, geologists who were working in the region interviewed surviving witnesses and determined that the blast from the explosion was probably heard over an area of at least a million square kilometers (386,000 square miles) and that the fireball had been witnessed by hundreds of people. Russian scientists went into the area in 1927, expecting to find an impact crater produced by the asteroid that had apparently struck the area. Surprisingly, they found no crater, leading them to conclude that the devastation had been caused by an aerial explosion, probably at an elevation of about 7 km (4.3 mi). Later calculations estimate the size of the asteroid responsible for the explosion to be about 25 to 50 meters (82 to 164 feet) in diameter. It was most likely composed of relatively friable (easy to crumble) stony material.[1,2]

The people of Earth were lucky that the Tunguska event occurred in a sparsely populated, forested region. If the blast had occurred over a city such as London, Paris, or Tokyo, the lives of millions of people would have been lost. Tunguska-type events are thought to occur on the order of every 1,000 years.[3] A widely reported near miss of a potentially dangerous asteroid occurred in March 2004. The asteroid was 30 m in diameter and passed between Earth and the moon at a distance of about 43,000 km (25,000 mi).*

Figure 11.1 Tunguska forest, Siberia 1908 An aerial blast downed trees over an area of about 2,000 km². (© AP/Wide World Photos)

*For a great interactive Web site for near-Earth objects see http://neo.jpl.nasa.gov

11.1 Earth's Place in Space

We introduced a quick history of the origin of the universe and Earth in Chapter 1, and you may wish to review that discussion. The story is a magnificent one with much of the mystery now removed, but the wonder remains. A story of how Earth, our home, evolved from a barren, nonliving planet into one filled with and modified by life (Figure 11.2). It is a story of change, with periodic drama caused by continued impact of material from outer space.[4,5]

Asteroids, Meteoroids, and Comets

There are literally trillions of particles in our solar system, ranging in size from dust that is a fraction of a millimeter in diameter to larger bodies such as **asteroids** that range in diameter from about 10 meters (32 feet) to 1,000 kilometers (621 miles). For the most part, asteroids are found in the *asteroid belt*, which is a region between Mars and Jupiter (Figure 11.3). The asteroids, which are composed of either rock material, metallic material, or rocky-metal mixtures, would pose no threat to Earth if they remained in the asteroid belt. Unfortunately, they move around and collide with one another, and a number of them are now in orbits that intersect Earth's orbit. When asteroids break into smaller particles they are known as **meteoroids** (Table 11.1), which range in size from dust particles to objects a few meters in diameter. When a meteoroid enters Earth's atmosphere, it is known as a **meteor.** As they streak through the atmosphere, meteors—sometimes called shooting stars—produce light that results from frictional heating. Sometimes meteors occur in large numbers, producing the well-known *meteor showers*.

The orbits of **comets,** which range in size from a few meters to a few hundred kilometers in diameter, also sometimes intersect Earth's orbit. Comets are thought to be composed of a rocky core surrounded by ice and covered in carbon-rich dust. They are believed to have originated far out in the solar system beyond Neptune in the Kuiper belt and Oort Cloud (Figure 11.3), which extends as far as 50,000 times the distance from the Earth to the Sun.[2] Figure 11.4 shows the comet Hale-Bopp, visible from Earth in 1997. The characteristic tail of the comet is composed of dust and gas that escape as the comet is warmed by the Sun and moves through space.

Figure 11.2 Earth rise from the moon Earth is the blue planet compared to a lifeless moon. *(© World Perspectives/Stone/ Getty Images)*

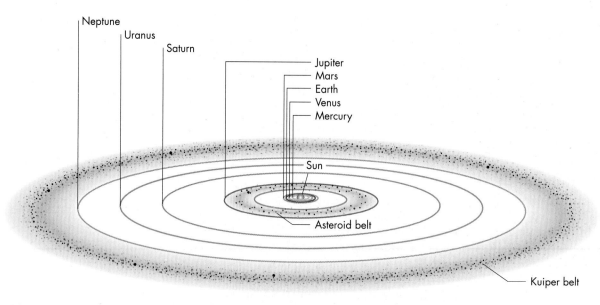

Figure 11.3 **Solar system** Diagram of our planetary system showing the asteroid belt and Kuiper belt. The Oort Cloud is too far out to show. Orbits of planets are not to scale. *(Modified after Marshak, S. 2001. Earth. W. W. Norton & Co. New York)*

Early in the history of the evolution of Earth, bombardment by asteroids and comets contributed the building blocks of our planet, which was built from the collision of innumerable bodies. Asteroids and comets contain water and impacts through geologic time delivered the water that later formed our oceans by volcanic degassing.

11.2 Aerial Bursts and Impacts

When entering Earth's atmosphere, asteroids, comets, and meteoroids travel at velocities which range from about 12 to 72 km per second (26,000 to 156,000 miles per hour).[1] A speeding bullet from a high powered rifle would initially be as much as 1.5 km per second. Asteroids and meteoroids are variable in composition (Table 11.1). Some contain carbonaceous material while others are composed of

TABLE 11.1 Meteorites and Related Objects

Type	Diameter	Composition	Comments
Asteroid	10 m–1,000 km	Metallic or rocky	Strong and hard if metallic or solid rock. Some hard types may impact Earth. If they are friable or weak they likely will break up in atmosphere of Earth at elevations of several km to hundreds of kilometers. Most originate in the asteroid belt between Mars and Jupiter.
Comet	Few meters to few hundred kilometers	Frozen water and/or carbon dioxide plus small rock fragments and dust; "dirty snowball"; core is ice surrounded by rock particles.	Weak, porous, will often explode in atmosphere of earth at elevations of several kilometers to several hundreds of kilometers. Most originate in the outer solar system such as the Oort Cloud 50,000 AU[1] from the Sun or from the Kuiper belt of comets. The comet tail is produced as ices sublimate (vaporize) and gases and dust particles are shed from the object.
Meteoroid	Less than 10 m to more than dust size	Stony, metallic, or carbonaceous (contains carbon)	Most originate from collisions of asteroids or comets. May be strong or very weak.
Meteor	Centimeter to dust size	Stony, metallic, carbonaceous, or icy	Are destroyed in Earth's atmosphere. "Shooting star" light produced by frictional heating in the atmosphere.
Meteorite	Variably larger than dust to asteroid size	Stony or metallic	Actually impact Earth's surface. Most abundant type of stony meteorite is called chondrite.[2]

[1] 1 AU is the distance from Earth to the Sun, about 93 million miles (150 million km).

[2] There are many types of chondrites. They contain chondrules, which are small (less than 1 mm) spheroidal inclusions that are glassy or crystalline. It is the chondrite meteorites (asteroids) from which planets are constructed.

Source: Data from Rubin, A. F. 2002. *Disturbing the solar system.* Princeton, NJ: Princeton University Press.

Figure 11.4 Comet Hale-Bopp 1997. (© Aaron Horowitz/ Corbis)

native metals such as iron and nickel. Others are stony, consisting of silicate minerals such as olivine and pyroxene—common minerals in igneous rocks. Stony meteoroids and asteroids are said to be *differentiated*, meaning that they have undergone igneous and sometimes metamorphic processes as part of their geologic histories. As previously mentioned, meteoroids and asteroids come from the asteroid belt which is between the orbits of Mars and Jupiter, whereas comets come from the Kuiper belt or Oort Cloud. Regardless of where they come from, when they intersect Earth's orbit and enter our atmosphere, meteoroids, asteroids, and comets undergo remarkable changes as they heat up due to friction during descent, producing bright light. If the object actually strikes Earth, then we speak of it as a **meteorite.** Many meteorites have been collected from around the world, particularly from Antarctica. Figure 11.5 shows a hypothetical diagram of a meteoroid entering the atmosphere at about 85 kilometers (53 miles) above Earth's surface. The meteoroid becomes a meteor and emits light. It will either result in an airburst in the atmosphere at an altitude between 12 and 50 km (7.5 and 31 miles) or it will collide with Earth to become a meteorite. The Tunguska event that opened this chapter was a giant airburst, but there is a significant number of meteorite craters on the surface of Earth, and over a hundred have been identified.[1]

Impact Craters

The most direct and obvious evidence for impacts on the surface of Earth comes from studying the craters they produce. The 50,000-year-old Barringer Crater in Arizona is perhaps the most famous meteorite crater in the United States. Known as "Meteor Crater," it is an extremely well-preserved, bowl-shaped depression with a pronounced, upraised rim (Figure 11.6a). The rim of the crater is overlain by an ejecta blanket (or layer of debris blown out of the crater upon impact) which today can be identified as hummocky terrain surrounding the crater. Figure 11.6b shows the present crater is not nearly as deep as the initial impact crater. This is because of material falling back into the crater and because of the partial collapse of the

Figure 11.5 Meteoride
Idealized diagram showing a meteoroid entering Earth's atmosphere. It may become a meteor, cause an airburst, or impact the surface of Earth. *(Modified after R. Baldini: http://www-th.bo.infn.it/tunguska/impact/fig1_2.jpg)*

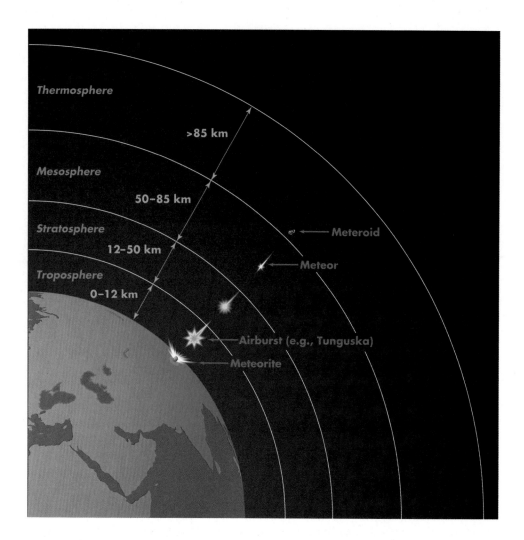

crater. The rocks that the asteroid impacted were shattered and deformed, forming what is known as brecciated rock, composed of angular broken pieces of rocks produced during the impact. Barringer Crater is approximately 1.2 km (0.75 miles) in diameter with a depth of about 180 meters (590 feet). The rim of the crater rises about 260 meters (850 feet) above the surrounding topography. When Americans first discovered Barringer Crater in the late nineteenth century, there was a lot of debate concerning its origin. Ironically, G. K. Gilbert, the famous geologist who postulated that the majority of the moon's craters were formed by impacts, did not believe that Barringer Crater was formed by impact. Only through careful study and evaluation was it finally concluded that Barringer Crater in fact resulted from the impact of a small asteroid, probably about 25 to 100 meters (80 to 330 feet) in diameter.[6]

The process of impact differentiates impact craters from craters that result from other processes, such as volcanic activity. This is because the processes related to impacts involve extremely high velocity, energy, pressure, and temperature, which normally are not experienced or produced by other geologic processes. Most of the energy of the impact is in the form of kinetic energy, or energy of movement. This energy is transferred to Earth's surface through a shock wave which propagates into Earth. The shock wave compresses, heats, melts, and excavates Earth materials. It is this transfer of kinetic energy that produces the crater.[6] The shock can metamorphose some of the rocks in the impact area while others are melted or vaporized and mixed with the materials of the impacting object itself. Most of the metamorphism consists of high-pressure modifications of minerals such as quartz. Such modifications are characteristic of meteorite impact and

(a)

(b)

Figure 11.6 Crater in Arizona
(a) Barringer Crater, Arizona (about 50,000 years old). The crater is about 1.2 km across and 180 m deep. *(© Charles O'Rear/Corbis)* (b) Diagram showing some of the features of the crater. *(Modified after Grieve, R., and Cintala, M. 1999. Planetary impacts. In Weissman, P. R., McFaddden, L., and Johnson, T. V. (eds.) Encyclopedia of the solar system. San Diego, CA: Academic Press)*

are extremely helpful in confirming the origin of a meteorite crater. A typical small crater, a few kilometers in diameter, is idealized in Figure 11.7a. The same process of vaporization, melting, ejection of material, formation of ejecta rims, and infilling of the crater also occurs for large, more complex craters, but the shape of a larger crater may be quite different (Figure 11.7b). Complex craters over a period of seconds to several minutes following impact may grow to sizes of tens of kilometers to over 100 kilometers (62 miles) in diameter. A pattern of more complete rim collapse and central crater uplift occurs following the impact. In general, impact craters on Earth that are larger than about 6 km (3.7 miles) are complex, whereas smaller craters tend to be more simple.

Geologically, ancient impact craters are difficult to identify because they have often been either eroded or filled with sedimentary rocks that are younger than the impact. For example, subsurface imaging and drilling below the present Chesapeake Bay have identified a crater about 85 km (53 miles) in diameter, now buried by about 1 km (0.62 miles) of sediment (Figure 11.8). The crater was produced by impact of a comet or asteroid about 3 to 5 km (1.9 to 3 miles) in diameter about 35 million years ago.[7] Compaction and faulting above the buried crater may be in part responsible for the location of the bay.

A good example of an eroded impact crater is found near Quebec, Canada. A ring-shaped lake about 70 km (45 miles) across is eroded along impact-brecciated (broken) rocks, marking most of the crater, which is about 100 km (62 miles) in diameter (Figure 11.9).[8]

Figure 11.7a Simple crater
Idealized diagram of the evolution of a
simple crater over a time interval of
seconds to a minute or so. *(Modified
after Grieve, R., and Cintala, M. 1999.
Planetary impacts. In Weissman, P. R.,
McFaddden, L., and Johnson, T. V. (eds.)
Encyclopedia of the solar system.
San Diego, CA: Academic Press)*

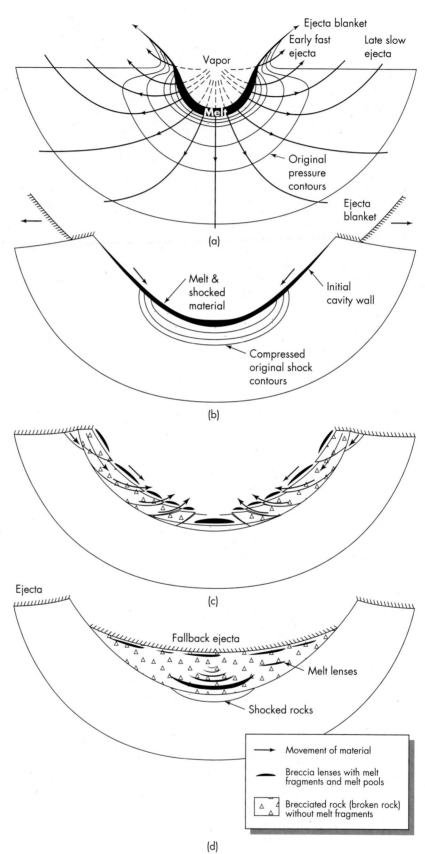

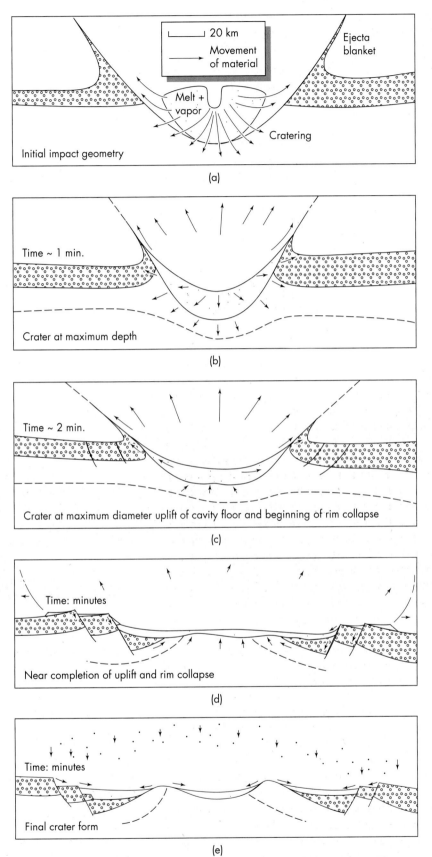

(a)

(b)

(c)

(d)

(e)

Figure 11.7b Complex crater
Idealized diagram of the evolution of a
complex crater over several minutes.
*(Modified after Grieve, R., and Cintala, M.
1999. Planetary impacts. In Weissman,
P. R., McFaddden, L., and Johnson, T. V.
(eds.) Encyclopedia of the solar system.
San Diego, CA: Academic Press)*

Figure 11.8 Chesapeake Bay
Map and idealized diagram of the
Chesapeake Bay impact structure that
formed about 35 million years ago.
*(Williams, S., Barnes, P., and Prager,
E. J. 2000. U.S. Geological Survey Circular
1199)*

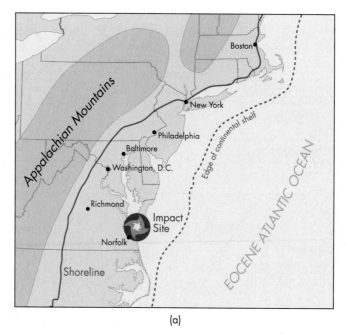

(a)

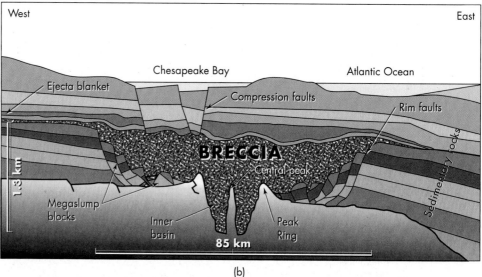

(b)

Most detailed remote sensing studies of craters come from evaluation of these
features on the moon and Mars. Impact craters are much more common on the
moon than on Earth because as smaller objects enter Earth's atmosphere, they
tend to burn up and disintegrate before actually striking the surface. Studying
impact features on the moon and Mars has greatly increased our understanding of
the process of impact and the associated surface features.

Early in 1994, a comet known as Shoemaker-Levy 9 (discovered in 1993) provided visual evidence of the most tremendous impacts ever witnessed. The comet
was unusual in that it was composed of several discrete comet fragments with
several bright tails. It was reported as a misshapen comet. It was determined that
the comet was a member of the Jupiter family of comets (comets that circle between Jupiter and the Sun), about 50 of which are known. The orbit of the comet
was tied to that of Jupiter, and after years of orbiting the planet it separated into
21 fragments known as a "string of pearls." As the fragments of the comet entered
Jupiter's atmosphere at speeds of about 60 km/sec (37 mi/sec), they exploded,
releasing energy on the order of 10,000 megatons for the smaller fragments and
as much as 100,000 megatons for the larger fragments (Figure 11.10). The total

Figure 11.9 **Canada impact structure** Satellite image and diagram of the Manicougan impact structure in Quebec, Canada. The ring-shaped lake is about 70 km (45 mi) across. *(NASA)*

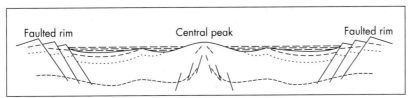

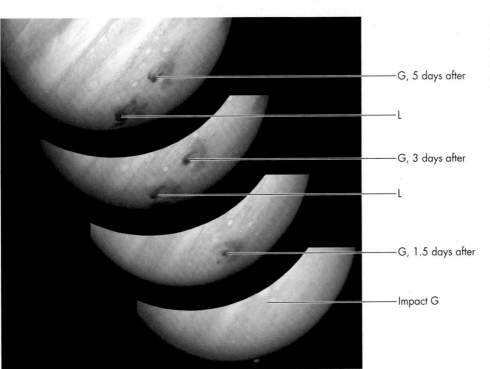

Figure 11.10 **Impact on Jupiter** Shoemaker-Levy 9G impact on Jupiter, "string of pearls" comet. *(R. Evans, J. Trauger, H. Hammel and the HST Comet Science Team/NASA)*

G, 5 days after

L

G, 3 days after

L

G, 1.5 days after

Impact G

amount of energy released was estimated to be more than would be released by Earth's entire store of nuclear weapons if detonated at one time. Hot, compressed gaseous emissions expanded violently upward from the lower part of Jupiter's atmosphere at speeds as high as 10 km/sec (6.2 miles/sec). These plumes of gas from the larger impacts reached elevations of more than 3,000 km (1,864 miles), as tremendously large rings that exceed the diameter of Earth developed around the sites of the impact. It was truly a remarkable show for astronomers and a sobering event for those who consider that impacts such as this might one day occur on Earth.[2]

Now that we have viewed the impact of the fragmented comet on Jupiter and studied Barringer Crater in Arizona as well as several others on Earth, the idea of the possibility of catastrophic impacts on Earth is finally being accepted. When the Tunguska event that opened this chapter occurred, several bizarre ideas were suggested to explain it, including nuclear explosions and even the explosion of an alien spaceship! The idea that impacts by asteroids and comets might cause catastrophes on Earth, and even mass extinction of life, was greatly resisted by scientists until very recently.

11.3 **Mass Extinctions**

A mass extinction is characterized by the sudden loss of large numbers of plants and animals relative to the number of new species being added.[9] Because the geologic time scale was originally based on fossils and the appearance and disappearance of different species, mass extinctions correlate to boundaries in the geologic time scale as shown in Table 1.1 and Figure 1A. Many hypotheses have been suggested for mass extinction events including (1) relatively rapid climate change; (2) plate tectonics, a relatively slow process that moves the position of continents and thus environments to different locations; (3) extremely large volcanic events that erupted huge volumes of volcanic ash into the atmosphere and changed the climate; and (4) impact of a large extraterrestrial object.

During the past 550 million years of Earth history, there were six major mass extinction events. The four earliest extinction events are being studied intensively to test the hypothesis that impacts of large asteroids or comets were involved. As dating of extinction events improves we may to be able to determine how fast the extinction occurred. A fast mass extinction favors impact by an extraterrestrial object as a cause of extinction.

The earliest mass extinction occurred approximately 443 million years ago, near the end of the Ordovician period. Although not much is known about the cause of this extinction, it may have been related to global cooling. Approximately 100 families and their associated species became extinct during this period. The next mass extinction occurred about 90 million years later near the end of the Devonian. That extinction was responsible for the death of about 70 percent of all (marine) invertebrate species and was also probably related to climate change and global cooling. The third mass extinction occurred near the end of the Permian— about 245 million years ago—when 95 percent of all marine species died. Although there is now an argument for impact at the Permian-Triassic boundary, it is also believed that this mass extinction possibly spanned a period of about 7 million years. It is thought that global cooling, followed by rapid global warming with large variations in climate, may have been responsible. There was also significant volcanic activity during this time, and the tremendous amount of volcanic ash and gases in the atmosphere probably contributed to the cooling. At the end of the Cretaceous period (K-T boundary), another tremendous mass extinction event occurred. This event was sudden, and there is abundant evidence to suggest that it was caused by the impact of a giant asteroid. Another mass extinction occurred near the end of the Eocene about 35 million years ago. There is limited evidence that impact from an asteroid or comet may have occurred at that

time, but many scientists link the extinction to cooling and glaciation, which occurred approximately 40 to 30 million years ago. Finally, near the end of the Pleistocene and into the present, there is an ongoing mass extinction of mammals, reptiles, amphibians, birds, fish, and plants. It is possible that overhunting by Stone Age man may have been a partial cause of this event. However, the loss of so many large mammals in North America, including some that were not hunted, may point to other causes more catastrophic in nature. It is possible we may find evidence for an impact in the late Pleistocene that played a role in the extinction event. More recently, in the past two hundred years, loss of habitat as a result of land-use changes, widespread deforestation, and application of chemicals has contributed to the ongoing extinction.[9]

While some of these extinction events are believed to have been related to climate changes, the case for impact-related mass extinction 65 million years ago at the end of K-T boundary is well documented. The K-T extinction brought an end to the dinosaurs, which had been at the top of the food chain for 100 million years or more. Their demise allowed small mammals to expand and evolve into many species, including humans, that are present on nearly all land areas of Earth today. A similar evolution also occurred in the oceans of the world. We will now address the K-T extinction and the impact hypothesis in more detail.

Late Cretaceous: K-T Boundary Mass Extinction

One of the great geologic detective stories of the past 50 years is the investigation of the K-T mass extinction. We now believe that 65 million years ago, a comet or asteroid with a diameter of about 10 km (6.2 miles) impacted Earth in the vicinity of what is now the Yucatan Peninsula. That event changed Earth history forever. Although after the event much of the physical landscape of Earth remained unchanged, the planet's inhabitants were changed forever. The dinosaurs disappeared, as did many species of plants and animals in both the oceans and on land. Approximately 70 percent of all genera on the planet and their associated species died off. Somehow, animals such as turtles, alligators, and crocodiles survived as well as some birds, plants, and smaller mammals. It is not known why some life forms survived the mass extinction and others did not. What we do know is the demise of the dinosaurs on land and reptiles and dinosaurs in the oceans set the stage for the evolution of mammals that eventually produced primates and humans. What would the world look like today if the K-T extinction had never happened? There's a good chance that humans never would have evolved!

Now let's back up and look at the history and development of the hypothesis that the K-T mass extinction event was indeed caused by the impact of a large comet or asteroid. The story is full of intrigue, suspense, rivalries, and cooperation—typical of many of the great scientific discoveries.[10] A number of scientists with backgrounds in geology, physics, chemistry, biology, geophysics, and astronomy worked together to develop and test the hypothesis that the K-T mass extinction was triggered by an impact. Walter Alvarez asked the question that started it all: "What is the nature of the boundary between rocks of the Cretaceous and Tertiary periods (65 million years ago)?"*

Alvarez was interested in Earth history and was particularly interested in reading that history as recorded by rocks. Early in his studies of the K-T boundary, Alvarez teamed up with his physicist father, Luis, and nuclear chemists Frank Asaro and Helen Michel. They decided to measure the concentration of a platinum metal called iridium in the thin clay layer that represented the K-T boundary in Italy. Walter Alvarez and colleagues initially went to the site in Italy to study the magnetic history of Earth. What they found at the K-T boundary was a very thin

---

*Walter Alvarez wrote a book entitled *T. Rex and the Crater of Doom* which was published in 1997 (Alvarez, 1997). While our discussion will present some of the highlights of this story, interested readers are invited to see Alvarez's book for the complete story.

Figure 11.11 **Evidence for impact** The K-T boundary in Italy recognized as a thin clay layer in these rocks with greatly increased concentration of the metal iridium, consistent with the impact of an asteroid. *(Walter Alvarez)*

layer of clay (Figure 11.11). It looked as if the extinction of many species was very abrupt at the clay layer. Fossils found in rocks below the clay were not in rocks above the clay. They then asked the question: "How much time was involved in the deposition of the clay layer? Was it a few years, a few thousand years, or millions of years?"

They decided to try to determine the timing of the clay deposition by measuring the amount of iridium in the clay. Iridium is found in very small concentrations in meteorites; assuming that the rate of accumulation of small meteoritic dust on Earth is constant, it would be possible to determine the age of the clay from its oldest to its youngest part by looking at the amount of iridium. What they found was entirely unexpected. The team anticipated measuring approximately 0.1 parts per billion of iridium in the clay layer, which they thought would represent slow accumulation through time. If the clay layer was deposited rapidly, then the amount of iridium would be even less. What they actually found was about three parts per billion—30 times more than expected. Even three parts per billion is a very small amount, but it was much more than could be explained by their previous hypothesis of slow deposition over time. They reevaluated the data. This time they included samples that were removed for treatment before measurement and got a final value of about nine parts per billion, which is nearly 100 times greater than expected. This discovery led them to a new hypothesis—that the iridium might be the result of an asteroid impact.

The team's iridium discovery, along with the hypothesis that there might be an extraterrestrial cause for the extinction at the K-T boundary, was published in 1980.[11] In that paper, they also reported elevated concentrations of iridium in deep sea sediments in Denmark and New Zealand, all of which are at the K-T boundary. With the discovery that the iridium anomaly occurred at several places around the world, the team became more confident of their impact hypothesis; however, they had no crater. This ushered in a period of research from other scientists directed at finding potential craters that formed 65 million years ago. There was some concern that the crater might have been completely filled with sediment and no longer recognizable, or that the crater was at the bottom of the ocean and may have been destroyed by plate tectonics. Fortunately, the site of the crater was identified in 1991.[12]

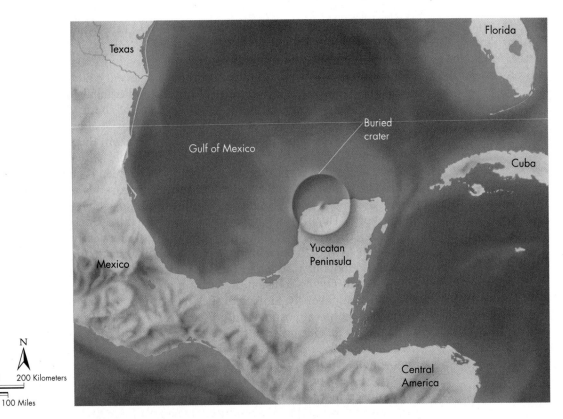

Figure 11.12 Crater in Mexico Map showing location of the buried impact crater from the asteroid that caused the K-T mass extinction. (© *D. Van Ravenaswaay/Science Photo Library/Photo Researchers, Inc.*)

Geologists in Mexico studying the structural geology of the Yucatan Peninsula discovered what they determined to be a buried impact crater with a diameter of approximately 180 km (110 mi). The crater is nearly circular, and there is a clear boundary between unfractured rocks within the crater and fractured rocks outside of the crater. About half of the crater lies in the Gulf of Mexico and half on the northern end of the Yucatan Peninsula (Figure 11.12). On land, the researchers found a semicircular pattern of sinkholes, known to the Maya people as *cenotes*, that correspond directly to the edge of the proposed impact crater. The *cenotes* range from about 50 to 500 meters (164 to 1,640 feet) in diameter and were presumably formed by chemical weathering of the limestone by water flowing through the fractured rock on the outside of the crater boundary. It was reasoned that the fracturing that created the ring of *cenotes* must be related to the circular structure since there is no other known structure that could explain the curved pattern of the features. The crater, which is now filled with other rocks, is believed to have been as deep as 30 to 40 km (18 to 25 mi) at the time of impact. However, subsequent slumping and sliding of materials from the sides soon filled in much of the crater, and sedimentation over the last 65 million years completely buried the structure.

Drill cores were taken and samples of glassy melt rock were retrieved from below another massive layer that was interpreted to be impact breccia within the proposed crater. The force of the impact excavated the crater, fractured the rock on the outside, and produced the breccia, which is a rock composed of angular pieces of rocks that have been cemented together. The glassy melt rock suggests that sufficient heat was present to actually melt rocks as a result of the impact.[13] Another study of the crater found glass mixed with and overlain by the breccia as well as evidence of shock metamorphism commonly associated with impact features.[14] The results of these studies of the Yucatan Peninsula crater (known as the Chicxulub crater) are accepted, and it is believed that the impact from the asteroid that struck the area 65 million years ago did in fact cause the K-T mass extinction.

After identifying the site of the crater and the evidence to support its existence, questions naturally arose regarding how such an event could cause a global mass extinction.[10] The asteroid that struck the Yucatan Peninsula 65 million years ago was huge; it is estimated to have had a diameter of about 10 km (6 miles). As a comparison, jet airplanes travel at cruising altitudes of about 30,000 ft, which is about 10 km above Earth's surface. The summit of Mt. Everest is not as high as 10 km, but it's close. Consider, too, that the asteroid struck the atmosphere of Earth at a speed of about 30 km/sec (19 mi/sec), which is about 150 times faster than a jet airliner travels. The amount of energy released is estimated to be about 100 million megatons—roughly 10,000 times as great as the entire nuclear arsenal of the world. The likely sequence of events of the impact and aftermath is shown in Figure 11.13. The hole blasted in Earth's crust was nearly 200 km (125 miles) across and 40 km (25 mi) deep. At an altitude of 10 km (6 mi) in the upper atmosphere and moving at about 30 km/sec (19 mi/sec), it would have taken less than half a second for the collision on Earth to take place, producing the large crater almost instantaneously. Moving at such high speed, the asteroid would have penetrated the entire atmosphere in only a second or two. When contact occurred, shock waves quickly crushed the rocks beneath, filling in all the cracks, partially melting the rocks, producing the breccia, and blasting bits and pieces high into the atmosphere. All this probably took about 2 seconds as the shock wave and heat vaporized rocks on the outer fringes of the impact. A tremendous amount of debris built up in a huge blanket around the crater. A gigantic cloud of vaporized rocks and gases would have produced an equally huge fireball that rose, producing a huge mushroom cloud. The explosion itself and the rising materials would probably have been sufficient to accelerate and eject material far beyond the surface of Earth. Particles of rock were blasted into ballistic trajectories before they fell back to the ground. The fireball would have produced sufficient heat to set fires around the globe. Vaporization of the limestone bedrock, which contained some sulfur, produced sulfuric acid in the atmosphere. Additional acids were added as a result of burning nitrogen in the atmosphere. Thus, following the impact, a long period of acid rainfall likely occurred. The dust in the atmosphere circled Earth, and for months there was essentially no sunlight reaching the lower atmosphere. The lack of sunlight stopped photosynthesis on land and in the ocean, and the acid rain was toxic to many living things, particularly terrestrial and shallow marine plants and animals. As a result, the food chain virtually stopped functioning because the base of the chain was greatly damaged. The impact occurred partly in the marine environment and significantly disturbed the seafloor, generating tsunamis that could have reached heights of 1 km (3,000 ft). These waves would have raced across the Gulf of Mexico and inundated parts of North America.[10]

In summary, the impact of the asteroid caused a global catastrophic killing which we refer to as a mass extinction. While there is some evidence that some species of dinosaurs were dying off before the impact, this event certainly seems to be responsible for wiping out the remaining dinosaurs. Also, so many other species of animals and plants on land and in the oceans died that there is little doubt that a massive impact was the likely cause of the extinction. Should such an event occur again, the loss of life would be as significant. It might well mean the extinction of humans and many of the larger mammals and birds on the planet, leading the way to yet another pattern of evolution.

What we have learned from the K-T extinction is sobering. On the other hand, we know that impacts from objects as large as 10 km (6 miles) in diameter are very rare, occurring perhaps every 40 to 100 million years or so. However, large impacts are not the only hazard from comets, asteroids, and meteoroids. Smaller impacts are much more probable, and they can wreak havoc on a region, causing great loss of life and damage.

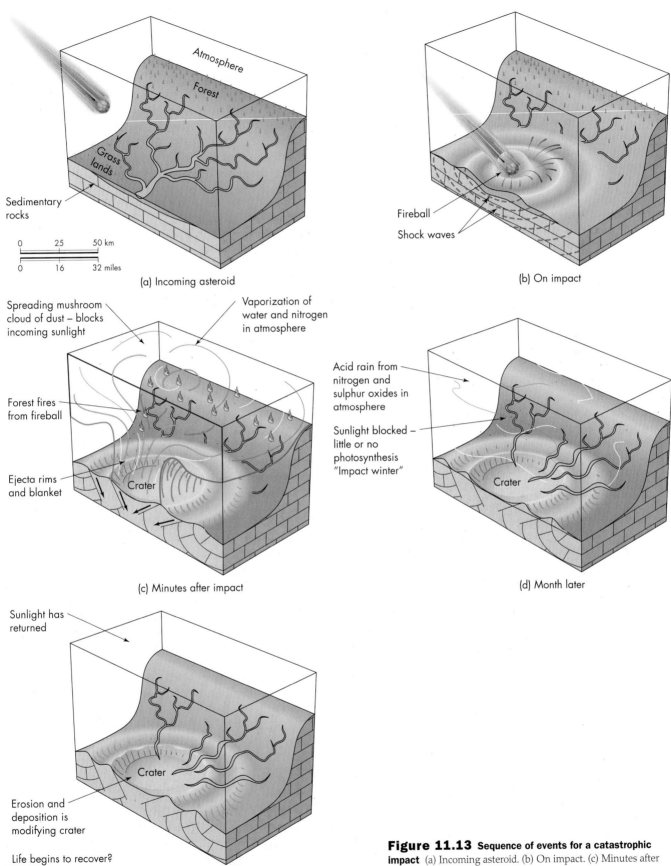

(a) Incoming asteroid

Atmosphere

Forest

Grass lands

Sedimentary rocks

0   25   50 km

0   16   32 miles

(b) On impact

Fireball

Shock waves

(c) Minutes after impact

Spreading mushroom cloud of dust – blocks incoming sunlight

Vaporization of water and nitrogen in atmosphere

Forest fires from fireball

Ejecta rims and blanket

Crater

(d) Month later

Acid rain from nitrogen and sulphur oxides in atmosphere

Sunlight blocked – little or no photosynthesis "Impact winter"

Crater

(e) After dust and acid rains out (several months)

Sunlight has returned

Crater

Erosion and deposition is modifying crater

Life begins to recover?

Figure 11.13 Sequence of events for a catastrophic impact (a) Incoming asteroid. (b) On impact. (c) Minutes after impact. (d) Months later.

11.4 Minimizing the Impact Hazard

Risk Related to Impacts

The risk of an event is related to both the probability of an event occurring and the consequences should it occur. The consequences of aerial bursts or direct impact from extraterrestrial objects several kilometers in diameter would be catastrophic. Although 7 out of 10 such events would likely occur in the oceans, results would be felt worldwide because of the enormous size of the object. Certainly there would be significant differences depending on the site of impact, but the overall consequences would constitute a global catastrophe with high potential for mass extinction. Such events probably occur on Earth with return periods of tens to hundreds of millions of years (Figure 11.14). Whether smaller objects on the order of a few tens of meters produce an aerial blast or cause an impact crater, they would produce a regional catastrophe if the event occurred on land near a populated area. The size of the area devastated would be on the order of several thousand square kilometers and could cause millions of deaths if the event occurred over or in an urbanized region. We term these smaller but regionally significant events "Tunguska-type" events.

A recent study that evaluated aerial blasts from asteroids with diameters of about 50 to 100 meters suggests that asteroids that are capable of causing catastrophic damages to a region occur on average about every 1,000 years (Figure 11.14). Using the Tunguska event as a characteristic event with a hit every thousand years somewhere on Earth, an urban area is likely to be destroyed every few tens of thousands of years. The dataset for this analysis comes from the distribution of 300 small asteroids that exploded in the atmosphere. Extrapolation of the data allows estimates to be made for larger, more damaging events.[3]

There is tremendous statistical variability in trying to predict the likelihood and type of future impacts, and when such estimates are made the death tolls over a typical century may range from zero to as high as several hundred thousand.

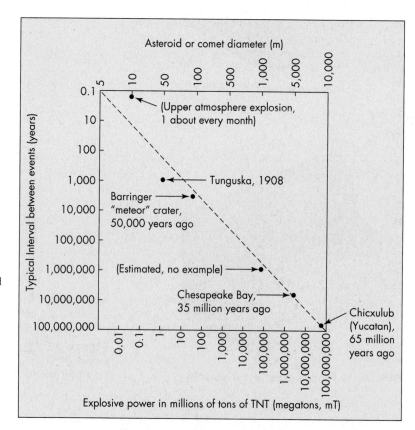

Figure 11.14 Energy from impact Estimate of relationship between energy of Earth impact and interval between events for various sizes of asteroids or comets. Also listed is the age of the events, which is not the same as the interval between events. *(Weissman, P. R., McFadden, L., and Johnson, T. V. (eds.) 1999. Encyclopedia of the solar system, San Diego, CA: Academic Press; and Brown, P, Spalding, R. E., ReVelle, D. O., Tagliaferri, E., and Worden, S. P. 2002. The flux of near-Earth objects colliding with the Earth. Nature 420:294–296)*

Computer-run simulations suggest that over a given century there may be approximately 450 deaths per year due to impacts. A truly catastrophic event could kill millions of people. When this is averaged over thousands of years, it results in a relatively high average annual death toll. For example, if a large urban area with 10 million people is devastated, and such events are thought to occur every 30,000 years, then the average annual death toll for such an event is over 300 people per year. Put this way, the risk from impacts is relatively high. For example, the probability that you will be killed by an impact-driven catastrophe at the global level is approximately 0.01 percent to 0.1 percent. By comparison, the probability that you will be killed in a car accident is approximately 0.008 percent and by drowning is 0.001 percent. According to the probabilities, it appears that the risk of dying from a large impact or aerial blast from a comet or asteroid is considerably greater than other risks we normally face in life. However, we emphasize again that the risks related to impacts are spread out over thousands of years. Although the average death toll in any one year may appear high, it is just that—an *average*. Remember that such events (and related deaths) actually occur very infrequently. Certainly there is a risk, but the time period between events is so long that we shouldn't lose any sleep worrying about being involved in a global catastrophic event caused by an extraterrestrial object.

Minimizing the Impact Hazard

Now that we have some inkling of the probability of an impact of an asteroid or comet, what can we do to minimize the hazard? First and foremost is to identify near-Earth objects that might threaten Earth (see A Closer Look: Near-Earth Objects). Identification and categorization of comets and asteroids that cross Earth's orbit are already in progress and could be scaled up to include objects of different size classes, including those with diameters less than 50 meters (164 feet), those 50 meters to several hundred meters, and those with a diameter of several kilometers. A program known as Spacewatch, which has been operating since 1981, is attempting to take inventory of the region surrounding Earth with expansion to the entire solar system. Based on the inventory to date, scientists believe there are around 135,000 objects with a diameter of 100 meters (328 feet) or less that are Earth-crossing asteroids. Another program known as the NEAT (Near-Earth Asteroid Tracking) system began in 1996. The objective of this program, which is supported by the National Aeronautics and Space Administration (NASA), is to study the size distribution and dynamic processes associated with near-Earth objects and specifically to identify those objects with a diameter of about 1 km (3,280 feet). Both observation programs utilize cameras and telescopes. Images are analyzed to identify fast-moving objects.[8] The programs and systems to identify near-Earth objects are expected to intensify in the future, and more objects will be cataloged. This is a first step toward evaluating the potential hazard from near-Earth objects. However, evaluation will take a long time because many of the objects have orbits that may not bring them close to Earth for decades, and the average amount of time between potentially catastrophic impacts is at least thousands of years for the smallest objects. The good news is that most of the objects identified as being potentially hazardous to Earth will likely not collide with our planet until several thousand years after they have been discovered. Therefore, we will have an extended period of time to learn about a particular extraterrestrial object and to attempt to develop appropriate technology to minimize the hazard.[1]

By one estimate, there are about 20 million extraterrestrial bodies in near-Earth orbits that have the potential for a significant impact.[1] Only about 4 percent of these bodies are likely to penetrate Earth's atmosphere and excavate a crater. Over 50 percent are structurally weak and are prone to explode at altitudes of about 30 km (18.6 miles) above Earth's surface. While these are spectacular explosions, they are not a significant hazard at the surface of Earth. The remaining objects,

A CLOSER LOOK | Near-Earth Objects

Near-Earth objects (NEOs) are asteroids that either reside and orbit between Earth and the Sun or have orbits that intersect with Earth's orbit. There are literally millions of meteoroids, but it is the larger objects such as asteroids and comets that vary in size from a few tens of meters to a few kilometers in diameter that are of most significance to our discussion. If a near-Earth object with a diameter of a few tens of meters were to impact the atmosphere or surface of Earth, we would experience a Tunguska-type event. A near-Earth object of several kilometers in diameter would cause a global catastrophe. As mentioned earlier and shown in Figure 11.3, the origin of most asteroids is the asteroid belt, located between Mars and Jupiter. If the asteroids remained there, they would not become NEOs; however, their orbits may become disturbed because of collisions or near misses with other objects. This may cause the orbit of one or more bodies to become more elliptical and cross into the space between Earth and the Sun, or even cross the orbit of Earth. It is estimated that the number of Earth-crossing asteroids with diameters larger than 100 meters (328 feet) is about 135,000. Larger Earth-crossing asteroids are more scarce, and it is estimated that there are about 1,500 with a diameter greater than 1 kilometer (3,280 feet) and about 20 with a diameter greater than 5 kilometers (3.1 miles).[8]

Comets are generally a few kilometers in diameter and can be described as giant snowballs in that they consist of mixtures of rock and ice (with more ice than rock by volume). Rock particles and dust cover the outside. Most comets are thought to originate far out in the solar system at a distance of about 50,000 AU (1 AU is the distance from the Earth to the Sun, or about 150 million kilometers/93 million miles) in a spherical cloud known as the Oort Cloud. Comets are best known for the beautiful light they create in the night sky, which results because of ice evaporating from the surface, releasing particles and vapor. The expanding gas and dust produces a spherical cloud around the comet that trails behind it, sometimes for great distances. This release of particles and gas produces streams of meteors that also light up the sky as showers if they enter Earth's atmosphere. One of the most famous comets is Halley's Comet, which is perhaps the best studied of all comets because of a 1986 spacecraft mission to observe it. That expedition found that the comet is a fluffy, porous body with very little strength. In fact, the entire nucleus of Halley's Comet is comprised of only about 20 percent water, and therefore must consist of 80 percent empty space which is a network of cracks and voids and loosely cemented materials.[1] Halley's Comet visits the space above Earth approximately once every 76 years, giving every generation a chance to view it (Figure 11.A).

Near-Earth objects, both asteroids and comets, apparently have a relatively short lifetime. As a result, there is a continuous production of these objects because asteroids are continuously kicked out of the asteroid belt, some becoming NEOs. Likewise, comets from the outer solar system have their orbits perturbed by planets or other objects to eventually become NEOs as well.

Figure 11.A Halley's Comet (© AP/Wide World Photos)

which constitute about 40 percent of the total population of objects, are moderately strong chondritic asteroids (Table 11.1). These are relatively slow-moving and can penetrate the atmosphere and produce a serious threat at the surface of Earth. Such bodies produce the Tunguska-type events. Identifying all of these objects will be extremely difficult; there are believed to be about 10 million of them. With a diameter of only 25 meters (82 feet), they are relatively small objects, making them difficult to identify and track.[1] It is impossible for us to identify and catalog 10 million small asteroids or comets. We are much better prepared to identify and track objects of a few hundred meters to a kilometer or so in diameter.

Once it has been determined that a large near-Earth object is on a collision path with our planet, options available to avoid or minimize the hazard from an aerial burst or crater-forming event are somewhat limited. In the event of a large comet or asteroid colliding with Earth, there will be no place in which to escape on the planet. Living things, including people, within the blast area will be killed immediately, while those further away are likely to be killed in the ensuing months from the cold and the destruction of the food chain. Even if we could identify and intercept the object, blowing it apart into smaller pieces would likely cause more damage than would one impact from the larger body because each of the smaller pieces would rain on Earth. A more thoughtful approach would be to try to gently divert the object so that it misses Earth. Let's assume we identify a 400-meter (1,300-foot) asteroid that we believe will strike Earth approximately 100 years in the future. In all likelihood, the body has been crossing Earth's orbit for millions of years without an impact. If it were possible to nudge it and change its orbit, it would miss rather than strike Earth. This is not an unlikely scenario. There is a 99 percent probability that we would identify the object at least 100 years before impact. We have the potential technology to change the orbit of a threatening asteroid with small nuclear explosions that are close enough to the asteroid to nudge it, but far enough away to avoid breaking it up. To accomplish this mission would require cooperation among the world's militaries and space agencies. The cost of such an expedition would likely exceed $1 billion. However, this seems a small price to pay considering the potential damages from a Tunguska-type event if it were to occur in an urban area.

Another option for smaller events might be evacuation. If we could predict precisely where the event would occur months in advance, evacuation is *theoretically* possible. However, evacuating an area of several thousand square kilometers would be a tremendous, if not impossible, undertaking.[1]

In summary, we continue to catalog extraterrestrial objects that intersect Earth's orbit. We are beginning to think about options to minimize the hazard. Given the potential long-term warning before an object would actually strike the surface of Earth, it is very possible that we will be able to devise methods to intercept and minimize the hazard by nudging the object into a different orbit so that it misses Earth.

SUMMARY

Asteroids, meteoroids, and comets are extraterrestrial objects that may intercept Earth's orbit. Small objects may burn up in the atmosphere and be visible as meteors at night. Depending on their size, velocity, and composition, large objects from a few meters to 1,000 kilometers may disintegrate in the upper or lower atmosphere in an aerial burst or impact the surface of Earth. Large objects can cause local to global catastrophic damage, including mass extinction of life. The best documented impact occurred 65 million years ago (K-T boundary), likely producing the mass extinction of species, including the dinosaurs, recorded in the rock record for that time.

The risk from extraterrestrial aerial burst or direct impact is a function of the probability of an event happening and the consequences should it occur. Relatively small events such as the 1908 Tunguska explosion will occur somewhere on Earth about every 1,000 years. Seventy percent will occur in or over the oceans. A significantly larger event would be capable of causing catastrophic damage to an urban area. Such events can be expected to occur every few tens of thousands of years. Programs such as Spacewatch and NEAT (Near-Earth Asteroid Tracking) will, with high certainty, identify near-Earth objects of diameters greater than a few hundred meters at least 100 years prior to possible impact. This should allow sufficient time to intercept and divert the object, using nuclear explosions. There are about 10 million smaller objects (potential Tunguska-type objects) that could produce catastrophic damage to urban areas. Identifying all these objects is extremely difficult. Thus we are particularly vulnerable to these smaller objects.

Revisiting Fundamental Concepts

Human Population Growth

As our population increases the potential loss from impact of an extraterrestrial object increases.

Sustainability

Sustainability as we know it can be destroyed by impact of a large asteroid.

Earth as a System

Earth is a part of the larger solar system. Extraterrestrial processes continue to change our planet, from adding water from comets to changing the landscape and life from impacts of asteroids and comets.

Hazardous Earth Processes, Risk Assessment, and Perception

Impact of a large asteroid or comet is the ultimate Earth hazard capable of causing a mass extinction including humans. The long-term risk is appreciable but our perception of the hazard is not great as the time between catastrophic events is many millions of years.

Scientific Knowledge and Values

Our understanding of our solar system and processes that could impact Earth are better understood than ever before. We place value on this understanding by monitoring the sky to identify potential hazards and forming plans should a threat be identified.

Key Terms

asteroids (p. 359)

comet (p. 359)

meteor (p. 359)

meteorite (p. 361)

meteoroid (p. 359)

Review Questions

1. What is the difference between an asteroid, meteor, comet, and meteroid?

2. What are the characteristics of an impact crater?

3. Differentiate between a simple and a complex crater.

4. What evidence supports the hypothesis of an asteroid impact at the K-T boundary?

5. How is the risk of impact determined?

6. Why was the Tungeska event a wake-up call?

Critical Thinking Question

1. Describe the likely results if a Tunguska-type event were to occur over or in central North America. If the event were predicted with 100 years warning, what could be done to mitigate the effects if changing the object's orbit were not possible? Outline a plan to minimize death and destruction.

PART THREE

Resources and Pollution

eople on Earth are absolutely dependent on natural resources associated with the geologic environment, including water (Chapters 12 and 13), minerals (Chapter 14), energy (Chapter 15), soils (Chapter 16), waste management (Chapter 17), and air (Chapter 18). Our main objective in Part 3 is to present basic information concerning our natural resources and to identify potential environmental problems and solutions associated with the use of resources. Two fundamental certainties come to mind: (1) Earth is the only suitable habitat that we have; and (2) our resources are limited. The fact that our resources are limited is true not only for our mineral resources and fossil fuels, which are recognized as nonrenewable, but also for water, trees, and some forms of energy, which are considered renewable. What is renewable or nonrenewable depends on the time frame as well as on our pattern of resource allocation and utilization. Resources we commonly think of as completely

renewable, such as water, are really not renewable if the resource is spoiled, used unwisely, or not available when it is needed at a particular location. The nonrenewable/renewable aspect of water resources is particularly true for groundwater, which may take tens to hundreds of years to be replaced by natural processes.

It is important to develop mineral resource utilization and management strategies that will sustain our resources so that future generations will have their fair share. Thus, sustainability is the environmental solution, and it is not a one-size-fits-all situation. That is, the requirement for sustainable water resources is quite different from the requirement for sustaining energy resources, minerals, or our soils. Sustaining nonrenewable resources such as metals and oil is not possible because supplies are finite. The best we can do is to conserve and recycle the nonrenewable resources so they can be used as long as possible and so future generations will have their fair share of resources.

TWELVE

Central Park

Manhattan Island

Hudson River

New Jersey

Wester

Brooklyn

Staten Island

Coney Island

Fire Isl

Water Resources

Long Island

Queens

Learning Objectives

Water is one of our most basic resources. Ensuring that we maintain an adequate, safe, sustainable supply of water is one of our most important environmental objectives. In this chapter, we will consider the topics of water supply, use, and management. We will focus on the following learning objectives:

- Understand the water cycle and the basic concepts associated with our water supply

- Understand the main types of water use

- Understand basic surface water and groundwater processes

- Be able to discuss some of the key principles associated with water management

- Know what wetlands are and understand their environmental significance

- Know why we are facing a global water shortage linked to food supply

High-altitude image of Long Island, New York Image showing Brooklyn and part of Queens. Red color is healthy vegetation. The red rectangle on Manhattan Island is Central Park. *(Science Source/Photo Researchers, Inc.)*

CASE HISTORY | Long Island, New York

The western end of Long Island contains Brooklyn and Queens, home to millions of people (Figure 12.1). Groundwater pollution has been a serious problem on the western end of the island since at least the beginning of the twentieth century, and people living there depend on water imported from upstate New York. Nassau and Suffolk counties on the eastern end of Long Island do not import water. The water resources are carefully monitored in those counties, yet they are experiencing water pollution problems.[1]

Groundwater in Nassau County is a tremendous resource, but, despite the huge quantity of underground water, intensive pumping in recent years has caused water levels to decline as much as 15 m (50 ft) in some areas. Streams that used to flow all year as a result of groundwater seeping into their channels now experience less flow or have dried up and flow only in response to precipitation.[2] The decline in water level has allowed the infiltration of salty groundwater, a problem known as *salt-water intrusion*. However, the most serious water pollution problem in Nassau County is associated with urbanization. There are many sources of pollution, including urban runoff, household sewage from septic tanks, salt used to de-ice highways, and industrial and other solid waste. These pollutants have a tendency to be released into surface waters and then migrate downward to groundwater resources. It has been noted that the greatest concentrations of pollutants are located beneath densely populated urban areas. Water levels in these areas have dramatically declined, and nitrates from sources such as septic tanks and fertilizers are routinely introduced into the subsurface environment.[1,2] Finally, landfills containing urban waste in the near-surface environment have been of particular concern because the waste often contains many pollutants. Groundwater pollution is inevitable, and, as a result, most landfills on Long Island have been closed.

The ongoing lesson learned from Long Island's groundwater problems is that the need for water resources goes hand in hand with the threat of water pollution. Urban areas are often in need of groundwater resources; at the same time, urban processes are likely to pollute these resources. Water pollution is discussed in detail in Chapter 13.

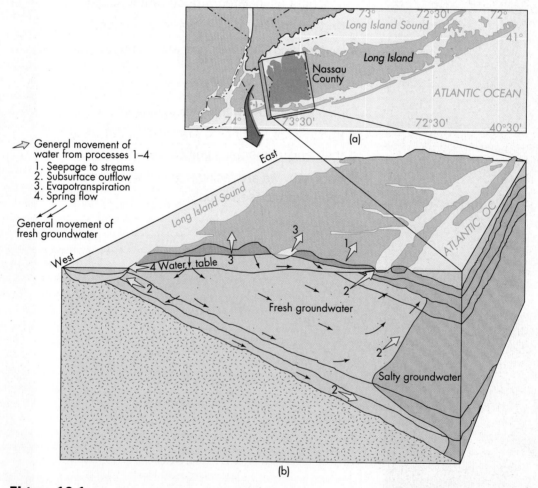

Figure 12.1 Groundwater movement at Long Island, NY The general movement of surface and groundwater for Nassau County, Long Island, New York. *(From Alley, W. M., Reilly, T. E., and Franke, O. L. 1999. U.S. Geological Survey Circular 1186)*

12.1 **Water: A Brief Global Perspective**

The global **water cycle,** or *hydrologic cycle,* involves the movement or transfer of water from one of Earth's storage compartments, such as the ocean, lakes, and the atmosphere, to another. In its simplest form, the water cycle can be viewed as water moving from the oceans to the atmosphere, falling from the atmosphere as rain and snow, and then returning to the oceans as surface runoff and subsurface flow or to the atmosphere by evaporation. The cyclic nature of this global movement of water is illustrated in Figure 12.2. The major processes are evaporation, precipitation, transpiration (loss of water by plants to the atmosphere), surface runoff, and subsurface groundwater flow. These are quantitatively shown in Figure 12.3. The annual volume of water transferred from the ocean to the land is balanced by the same volume returning by river and groundwater flow to the ocean, and there is a balance between total evaporation and precipitation.[2] The water that returns to the ocean is changed because it carries with it gravel, sand, silt, and clay that has eroded from the land. The return flow also carries many chemicals. Most of the chemicals are natural; however, many human-made and human-induced compounds are trapped in the return flow, such as organic waste and nutrients, as well as thousands of chemicals used in our agricultural, industrial, and urban processes.

Water is a heterogeneous resource that can be found in liquid, solid, or gaseous form on or near Earth's surface. Water's residence time may vary from a few days to many thousands of years, depending on its specific location (Table 12.1). Furthermore, more than 99 percent of Earth's water is unavailable or unsuitable for beneficial human use either because of its salinity, as with seawater, or its form and location, as is the case with water stored in ice caps and glaciers. Thus, all people compete for less than 1 percent of Earth's water supply!

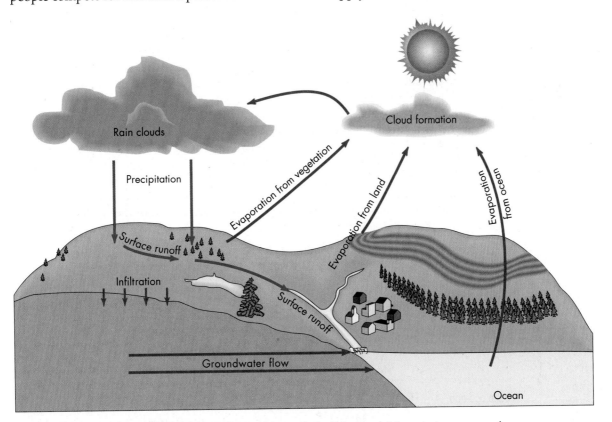

Figure 12.2 Hydrologic cycle Idealized diagram showing the hydrologic cycle's important processes and transfer of water. In many areas winter precipitation is in the form of snow, and spring and summer snowmelt contribute to surface runoff and infiltration of water to enter the groundwater system. *(Modified after Council on Environmental Quality and Department of State. 1980. The Global 2000 Report to the President. Vol. 2)*

Figure 12.3 **Global transfer of water** Movement of water in the global water cycle. Units are thousands of cubic kilometers (km³) per year. *(Data from Gleick, P. H. 1993. An introduction to global fresh water issues. In Water in Crisis, ed. P. H. Gleick, pp. 3–12. New York: Oxford University Press)*

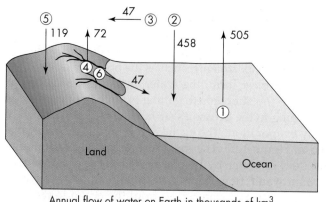

Example of global water balances:

- Water transferred to land from atmosphere ③ = Water returned to ocean ⑥

- Total evaporation ①+④ = Total precipitation ②+⑤

Annual flow of water on Earth in thousands of km³

① Evaporation from oceans to the atmosphere

② Precipitation to oceans

③ Transfer of water from atmosphere to land

④ Evaporation to atmosphere from land

⑤ Precipitation to land

⑥ Runoff of surface water and groundwater from land to oceans

12.2 Surface Water

Surface Runoff and Sediment Yield

Surface runoff has important effects on both the transport of materials and erosion. Surface water can dislodge soil and rock particles on impact (Figure 12.4). Water can move these materials either in a dissolved state or as suspended particles. The number and size of the suspended particles moved by surface waters depend on the volume and depth of the water as well as the velocity of flow. The faster a stream or river flows, the larger the particles it can move, resulting in more material being transported. Therefore, the factors that affect runoff also affect sediment erosion, transport, and deposition.

The flow of water on land is divided into watersheds. A *watershed,* or **drainage basin,** is an area of land that contributes water to a particular stream or river. A drainage basin is a basic unit of the landscape. Figure 12.5 shows two side-by-side drainage basins (A and B) with similar geology. The boundary between them is called a *drainage divide.*

Large drainage basins can be subdivided into smaller ones. For example, the Mississippi River drainage basin drains about 40 percent of the United States but

TABLE 12.1 The World's Water Supply (Selected Examples)

Location	Surface Area (km²)	Water Volume (km³)	Percentage of Total Water	Estimated Average Residence Time
Oceans	361,000,000	1,230,000,000	97.2	Thousands of years
Atmosphere	510,000,000	12,700	0.001	9 days
Rivers and streams	—	1,200	0.0001	2 weeks
Groundwater; shallow to depth of 0.8 km	130,000,000	4,000,000	0.31	Hundreds to many thousands of years
Lakes (freshwater)	855,000	123,000	0.009	Tens of years
Ice caps and glaciers	28,200,000	28,600,000	2.15	Up to tens of thousands of years and longer

Data from U.S. Geological Survey

(a)   (b)

Figure 12.4 **Soil erosion** (a) A raindrop falling in a cornfield causes soil particles to be lifted into the air, initiating the erosion process. *(Runk-Schoenberger/Grant Heilman Photography, Inc.)* (b) Surface runoff often causes the formation of small gullies such as those shown here. *(Courtesy of U.S. Department of Agriculture)*

contains many subbasins such as the Ohio and Missouri basins. Drainage basins such as the Ohio may be further divided into smaller basins. Two drops of rain separated by only centimeters on a continental divide, which is a large-scale drainage divide from each side of which river systems flow in opposite directions, may end up a few weeks later in different oceans, thousands of kilometers apart. We may also think of a drainage basin as the land area that contributes its runoff to a specific *drainage net*, which is the set of channels that makes up a drainage basin. The drainage net in basin A in Figure 12.5 includes all the channels from the small headwater streams at the top of the drainage basin to the single large channel at the bottom of the drainage basin. Thus, the term *drainage basin* refers to an area of land, whereas the term *drainage net* refers to the actual river and stream channels in the drainage basin.

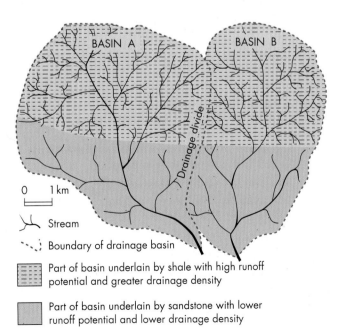

0   1 km

⟍ Stream

⋅⋅⋅ Boundary of drainage basin

Part of basin underlain by shale with high runoff potential and greater drainage density

Part of basin underlain by sandstone with lower runoff potential and lower drainage density

Figure 12.5 **Two drainage basins** Water falling on one side of the central boundary will drain into basin A; on the other side, water will drain into basin B. The two drainage basins have similar geology consisting of shale and sandstone.

Factors Affecting Runoff and Sediment Yield

The amount of surface-water runoff and sediment carried by the runoff varies significantly among drainage basins and rivers. The variation results from geologic, topographic, climatic, vegetation, and land-use characteristics of a particular drainage basin and changes in these factors over time. Even the most casual observer can see the difference in the amount of sediment carried by the same river in flood state and at low flow, since floodwaters are usually muddier.

Geologic Factors. The major geologic factors affecting surface-water runoff and sedimentation include rock and soil type, mineralogy, degree of weathering, and structural characteristics of the soil and rock. Fine-grained, dense clay soils on shale and exposed rock types with few fractures generally allow little water to move downward and become part of the subsurface flow. The runoff from precipitation falling on such materials is comparatively rapid, and there are usually many streams. Conversely, sandy soils on sandstone and well-fractured rocks absorb a larger amount of precipitation and have less surface runoff and fewer streams. These principles are illustrated in Figure 12.5. The upper parts of basins A and B are underlain by shale, and the lower parts are underlain by sandstone. Because the shale has a greater potential to produce runoff than the more porous sandstone, the *drainage density*—that is, the length of stream channel per unit area—is much greater in the shale areas than in the sandstone areas. The drainage density for any area of land is determined by measuring the total length of stream channel from all streams in the area and dividing this length by the area.

Topographic Factors. *Relief* refers to the difference in elevation between the highest and lowest points of any landform of interest. The greater the relief of a drainage basin, the more likely the streams in the basin are to have a steep gradient and a high percentage of steep, sloping land adjacent to the channel. Relief and slope are important because they affect the velocity of water in a stream, the rate at which water infiltrates the soil or rock, and the rate of overland flow. These characteristics then affect the rate at which surface and subsurface runoff enters a stream.

Climatic Factors. Climatic factors affecting runoff and sediment transport include the type of precipitation that occurs, the intensity of the precipitation, the duration of precipitation with respect to the total annual climatic variation, and the types of storms. In general, production of large volumes of water and sediment is associated with infrequent high-magnitude storms that occur on steep, unstable topography underlain by soil and rocks with a high erosion potential.

Vegetation Factors. Vegetation can influence runoff and sediment yield. Vegetation is capable of affecting stream flow in several ways:

- Vegetation may decrease runoff by increasing the amount of rainfall intercepted and removed by evaporation.
- Decrease or loss of vegetation due to climatic change, wildfire, or land use, such as grazing by sheep, will increase runoff or production of sediment or both (Figure 12.6).
- Streamside vegetation reduces stream-bank erosion because its roots bind and hold soil particles in place.
- In forested watersheds, large woody debris, such as stems and pieces of wood, may profoundly affect the stream-channel form and processes. In steep mountain watersheds, many of the pool environments that are important for fish habitats may be produced by large woody debris (Figure 12.7).

(a)   (b)

Figure 12.6 **Wildfire and overgrazing increase soil erosion** (a) Vegetation loss by wildfire in a forest in the Yukon Territory, Canada, in 1995. *(Stephen J. Krasemann/Photo Researchers, Inc.)* (b) Grazing of sheep (here in South Africa) can increase runoff, resulting in soil erosion. *(Nigel J. Dennis/Photo Researchers, Inc.)*

Land-Use Factors. Agriculture and urban development are land uses that have an effect on runoff and sediment yield. Agriculture generally increases runoff and sediment yield as land is plowed for crops (see Chapter 16). Urbanization, with construction of streets, parking lots, and buildings for homes and industry may also greatly increase runoff because of the large amount of impervious pavement covering the land (see Chapters 8 and 16).

12.3 Groundwater

The major source of groundwater is precipitation that infiltrates the surface of the land and into soil and rock. The two major zones of groundwater are the **vadose zone** and the **zone of saturation** (Figure 12.8). The vadose zone includes the Earth

Figure 12.7 **Large woody debris in streams** The large redwood stem in Prairie Creek, California, shown here is responsible for the development of a scour pool important in providing fish habitat for the stream ecosystem. Such large woody debris may reside in the stream channel for centuries. *(Charles A. Lauzy/ Getty Images Inc.)*

Figure 12.8 Groundwater zones
Idealized diagram showing the two main zones of groundwater: vadose zone and zone of saturation. Also shown are the water table and directions of groundwater movement.

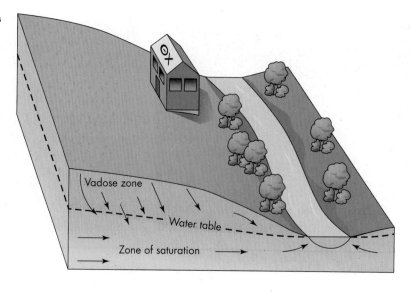

→ Direction of movement of groundwater

Sandy soil not saturated, vadose zone

Sandy soil saturated, zone of saturation

- - - Water table, boundary between vadose zone and zone of saturation

material above the **water table,** which is the boundary between the two zones, including soil, alluvium, and rock. The vadose zone is seldom saturated. Until recently, the vadose zone was called the *unsaturated zone,* but we now know that some saturated areas may exist there at times as water moves through after precipitation events. The vadose zone has special significance because potential pollutants infiltrating at the surface must percolate through the vadose zone before they enter the saturated zone below the water table. Thus, during environmental subsurface monitoring, the vadose zone is an early warning area for potential pollution to groundwater resources.

Water that reaches the zone of saturation begins with infiltration from the surface. Factors that influence the rate of infiltration include

- **Topography:** With steeper topography, more water runs off, which reduces infiltration.
- **Soil and rock type:** Soils and rocks with lots of open space due to fractures or pore spaces between grains have higher infiltration rates.
- **Amount and intensity of precipitation:** Low-intensity precipitation or snowmelt favors infiltration. High-intensity precipitation favors runoff.
- **Vegetation:** Leaves and stems intercept precipitation. The water then falls more gently to the ground and infiltration increases.
- **Land use:** Urban lands with pavement or roofs reduce infiltration. Agricultural practices generally increase surface runoff and soil erosion, decreasing infiltration. Harvesting of timber, particularly clear-cut logging, reduces vegetation cover, increasing soil erosion and runoff while decreasing infiltration.

Water that percolates through the vadose zone will enter the zone of saturation, where true groundwater flow occurs. As the name implies, the Earth material in

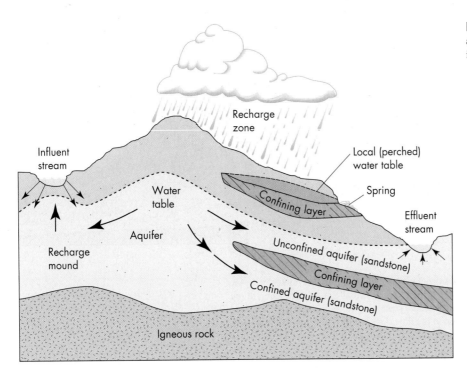

Figure 12.9 An unconfined **aquifer,** a perched water table, and influent and effluent streams.

the zone of saturation has all of the spaces between grains in soil and rock filled with water—this is the definition of *saturated*. The upper surface of the zone of saturation is the water table.

Aquifers

Earth material capable of supplying groundwater at a useful rate from a well is called an **aquifer** (Figure 12.9). Gravel, sand, soils, and fractured sandstone, as well as granite and metamorphic rocks with high porosity from open fractures, are good aquifers if groundwater is present. A *confining layer* (such as a clay or shale layer) restricts or blocks the movement of groundwater.

An aquifer is called an *unconfined aquifer* if there is no confining layer restricting the upper surface of the zone of saturation at the water table (Figure 12.9). If a confining layer is present the aquifer is called a confined aquifer; both confined and unconfined aquifers can be found within the same area. A *perched aquifer* is a local zone of saturation above a regional water table.

The water beneath a confined aquifer may be under pressure, forming *artesian* conditions. Artesian conditions are analogous in their effect to a water tower that produces water pressure for homes (Figure 12.10a). Groundwater in an artesian system (Figure 12.10b) moves downward and laterally. It is confined and under pressure. The water may rise upward through rock fractures to form an artesian spring. The resultant rise may also create an *artesian well* (Figure 12.10b).

In a general sense *groundwater recharge* is any process that adds water to the aquifer. Groundwater recharge can result from natural processes, most commonly precipitation (Figure 12.9) infiltrating the ground. There is also human-induced infiltration, as, for example, groundwater recharge from leakage and infiltration of water from a broken water line, leakage from a canal, or irrigation of crops. *Groundwater discharge* is any process that removes groundwater from an aquifer. Natural discharge from a spring is an example of groundwater discharge. A *spring* forms when water flowing in an aquifer intersects the Earth's surface. Spring discharge can form the beginning of a stream or river (Figure 12.11). Groundwater discharge also occurs when water is pumped from a well.

When water is pumped from a well, a *cone of depression* forms in the water table or artesian pressure surface (Figure 12.12). A large cone of depression can alter the

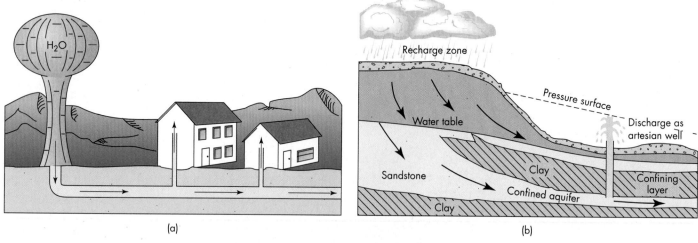

(a)   (b)

Figure 12.10 **Development of an artesian well system** (a) Water rises in homes because of pressure created by the water level in the tower. If there is only a small amount of friction in the pipes, there will be little drop in pressure. (b) The pressure surface, or water table, in natural systems declines away from the source because of friction in the flow system, but water may still rise above the surface of the ground if impervious particles such as clay create a confining layer and cap the groundwater.

Figure 12.11 **Spring** Discharge of groundwater from Fern Spring at the southern end of Yosemite Valley, California. This spring emerges at the base of a hillslope, as do many springs. The small stream emerging from the spring pool in a short cascade or falls is about 2 m (6.6 ft) wide. *(Edward A. Keller)*

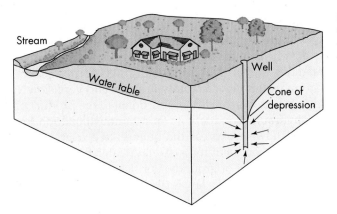

Figure 12.12 Pumping ground-water Cone of depression in the water table, resulting from pumping water from a well.

direction of groundwater movement within an area. Overpumping an aquifer causes the water table to drop deeper within Earth. This drop necessitates lowering the pump settings or drilling deeper wells. These adjustments are often costly, and they may not work, depending on the hydrologic conditions. For instance, continued deepening to correct for overpumping of wells that tap igneous and metamorphic rocks is limited. Water from these wells is pumped from open fracture systems that tend to close or diminish in number and size with increasing depth. Also, the quality of groundwater may be degraded if it is extracted from deeper water containing more dissolved minerals.

Groundwater Movement

The rate and direction of groundwater movement in an aquifer in part depend on both the gradient of the water table, or *hydraulic gradient,* and the type of material found in the aquifer. In general, the hydraulic gradient for an unconfined aquifer is approximately the slope of the water table. The ability of a particular material to allow water to move through it is called its *hydraulic conductivity,* which is expressed in units of cubic meters (m^3) of water per day through a cross section of 1 square meter (m^2) that is ($m^3/day/m^2$), which reduces to meters per day. Hydraulic conductivity is in part dependent upon the size of the open spaces between grains in an aquifer and how well they are connected. The percentage of empty space, called void space, in sediment or rock is called *porosity.* It varies from 1 percent for granite with few fractures to 50 percent for clay. Values of porosity and hydraulic conductivity for selected Earth materials are listed on Table 12.2. Notice that some of the most porous materials, such as clay, have low hydraulic

TABLE 12.2   Porosity and Hydraulic Conductivity of Selected Earth Materials

	Material	Porosity (%)	Hydraulic Conductivity[1] (m/day)
Unconsolidated	Clay	50	0.041
	Sand	35	32.8
	Gravel	25	205.0
	Gravel and sand	20	82.0
Rock	Sandstone	15	28.7
	Dense limestone or shale	5	0.041
	Granite	1	0.0041

[1]In older works, may be called coefficients of permeability.

Modified after Linsley, Kohler and Paulhus, 1958. *Hydrology for Engineers.* New York: McGraw-Hill. Copyright © 1958 by McGraw-Hill Book Company. Used by permission of McGraw-Hill Book Company.

conductivity. Its conductivity is low because the spaces between clay particles are very small and hold water tenaciously. Sand and gravel have high porosity, with relatively large openings between grains and a high hydraulic conductivity. These relations explain why sand and gravel form aquifers, and clay forms aquitards. Groundwater moves rapidly in sands and very slowly in clays. The rate of flow of groundwater is directly proportional to the product of the hydraulic gradient and the hydraulic conductivity in a famous equation known as **Darcy's law.** Using Darcy's law to express the quantitative relationship of hydraulic gradient and hydraulic conductivity to groundwater flow allows us to solve many problems such as the rate of groundwater flow and how fast groundwater is being used or replenished. Darcy's law is discussed further in Appendix E, where an example of groundwater movement is presented.

Groundwater Supply

Nearly half the population of the United States uses groundwater as a primary source of drinking water. Therefore, protecting groundwater resources is an environmental problem of particular public concern. Fortunately, the total amount of groundwater available in the United States is enormous. Within the contiguous United States the amount of groundwater within 0.8 km (0.5 mi) of the land surface is estimated to be between 125,000 and 224,000 km^3 (30,000 to 54,000 mi^3). Put in perspective, the lower estimate is about equal to the total flow of the Mississippi River during the last 200 years. Unfortunately, the cost of water pumping and exploration reduces the total quantity of groundwater that is available.[3]

In many parts of the country, groundwater withdrawal from wells exceeds natural inflow or recharge. In such cases water is being "mined" and can be considered a nonrenewable resource. Groundwater overdraft is a serious problem in the Texas-Oklahoma-High Plains areas, California, Arizona, Nevada, New Mexico, and isolated areas of Louisiana, Mississippi, Arkansas, and the South Atlantic-Gulf region (Figure 12.13a). In the Texas-Oklahoma-High Plains area alone, the overdraft is approximately equal to the natural flow of the Colorado River.[4] The Ogallala aquifer, composed of water-bearing sands and gravel, underlies this region from South Dakota into Texas (Figure 12.13b). Although the aquifer holds a tremendous amount of groundwater, it is being used in some areas at a rate that is up to 20 times that of natural recharge by precipitation infiltration. The water level in many parts of the aquifer has declined in recent years, and eventually a significant portion of land now being irrigated may return to dry farming if the resource is depleted.

Some towns and cities in the High Plains also have water supply problems. Along the Platt River in northern Kansas groundwater levels are still high, and there is enough water for now (Figure 12.13). Further south in southwest Kansas and the panhandle in western Texas where water levels have declined the most, supplies may last only another decade or so. The situation for urban water in Ulysses, Kansas (population 6,000), and Lubbock, Texas (population 200,000), is getting to be a problem. South of Ulysses, Lower Cimerron Springs, which was a famous water hole along a dry part of the Santa Fe Trail, dried up decades ago due to pumping groundwater, a symptom of what was to come. Both Ulysses and Lubbock are now facing water shortages and will need to spend millions of dollars to find alternative sources.

To date, only a few percent of the United States total groundwater resource has been depleted, but water levels are declining in parts of Kansas, Oklahoma, New Mexico, and Texas. As the water table becomes lower, yields from wells decrease and energy costs to pump the water increase. More than half a century of irrigation in the High Plains area and Ogallala aquifer has created the water problems seen in these areas.

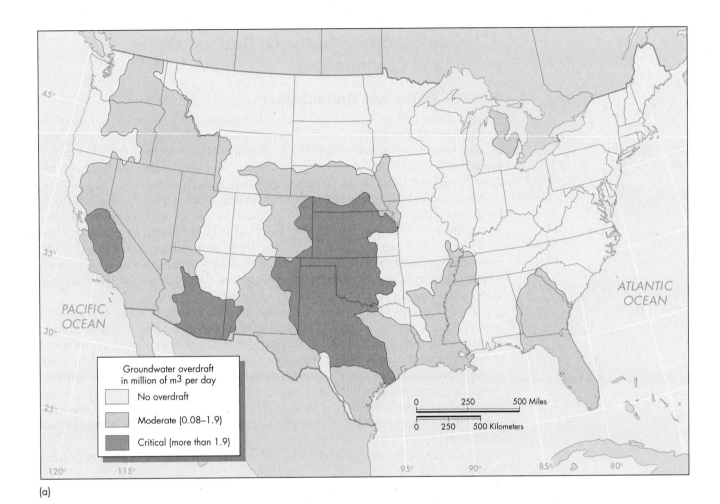

(a)

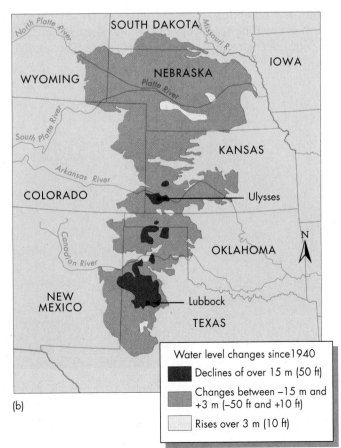

(b)

Water level changes since 1940
  ■ Declines of over 15 m (50 ft)
  ■ Changes between −15 m and +3 m (−50 ft and +10 ft)
  □ Rises over 3 m (10 ft)

Figure 12.13 **Mining groundwater** (a) Groundwater overdraft for the contiguous United States. (b) A detail of water level changes in the Ogallala aquifer, Texas-Oklahoma-High Plains area. Ulysses, Kansas and Lubbock, Texas are two communities located where declines in groundwater are most severe. Both are facing urban water supply problems. *(U.S. Geological Survey)*

12.4 Interactions between Surface Water and Groundwater

Surface Water and Groundwater

Surface water and groundwater are so interrelated that we need to consider the two as part of the same resource.[5] Some linkages between surface water and groundwater are shown on Figure 12.14. Surface water infiltrates from the lake to groundwater, river flow receives groundwater seepage, and groundwater at the coast comes to the surface in wetlands and seeps into the oceans.

Nearly all natural surface-water environments such as rivers, lakes, and wetlands, as well as human-constructed water environments such as reservoirs, have strong linkages with groundwater. Withdrawal of groundwater by pumping from wells can reduce stream flow, lower lake level, and reduce water in wetlands. Conversely, withdrawal of surface water can deplete groundwater resources. As a result, groundwater management requires that the linkages between surface water and groundwater be known and understood.[5]

Figure 12.9 shows some of the interactions between surface water and groundwater. In particular, two types of streams may be defined. *Effluent streams* tend to be perennial, that is, to flow all year. During the dry season, groundwater seeps into the channel, maintaining stream flow (Figure 12.9, right). *Influent streams* are often above the water table everywhere along their channel, and flow only in direct response to precipitation. Water from influent streams moves down through the vadose zone to the water table, forming a recharge mound (Figure 12.9, left). Influent streams may be intermittent or ephemeral in that they flow only part of the year.

From an environmental standpoint, influent streams are particularly important because water pollution in the stream may move downward through the streambed and eventually pollute the groundwater below. Dry river beds are particularly likely to experience this type of problem. For example, the Mojave River in the desert of southern California is dry most of the time in the vicinity of Barstow. The river bottom was the site of a railroad yard where trains and other equipment were cleaned with solvents. These solvents infiltrated down through

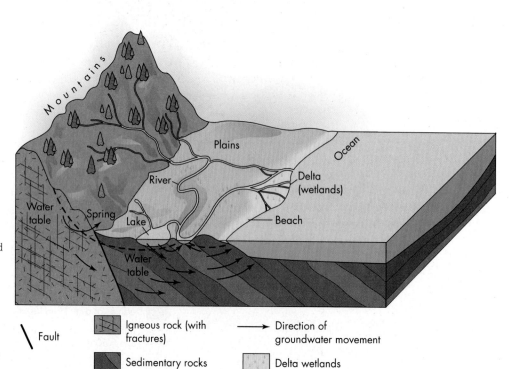

Figure 12.14 Surface-groundwater interactions Idealized diagram showing some of the ways surface water and groundwater interact in the landscape from the mountains to the sea. For example, groundwater moves up along the fault to discharge as a spring; seeps into lakes and rivers on the plains, delta and beaches, and offshore into the ocean.

Figure 12.15 Collapsed
sinkhole Golly Hole, Alabama.
(Geological Survey of Alabama)

the vadose zone to contaminate groundwater used by several communities for drinking and other municipal purposes.

Karst. Some of the more interesting interrelationships between surface water and groundwater occur in areas underlain by soluble rocks such as limestone. Often the limestone is dense, thin bedded, and well jointed and has an abundance of fractures. A particular type of topography can result from this type of limestone that is the result of the diversion of surface waters to subterranean routes. Known as **karst topography,** these regions are very common in the United States, where approximately 25 percent of the surface land area is underlain by limestone. The land surface in karst areas is often dotted with pits formed by chemical weathering, known as **sinkholes,** that vary in size from one to several hundred meters in diameter. Sinkholes can result from one of two processes: (1) solutional weathering at the surface of the limestone, with water diverted to subterranean routes below the sinkhole, or (2) pits produced by collapse of surface material into part of an underground cavern system, forming spectacular collapse sinkholes (Figure 12.15). The dissolution of limestone is a chemical weathering process, and limestone is particularly susceptible to chemical weathering from acids commonly found in the natural environment (see Chapter 3, A Closer Look: Weathering). As solutional pits enlarge and water moves downward through limestone, a series of caverns may be produced. The pockmarked surface of a karst area resulting from sinkhole development often forms a *karst plain*. The Mitchell Plain in southern Indiana, shown in Figure 12.16, is an example of a karst plain. Because limestone

Sinkholes

Figure 12.16 The Mitchell
karst plateau in southern Indiana,
with numerous sinkholes.
*(Samuel S. Frushour, Indiana
Geological Survey)*

Figure 12.17 Cave formations in Carlsbad Caverns, New Mexico, include stalactites, which hang from the ceiling; stalagmites, which grow up from the ground; and flowstone, which forms as water slowly flows down the walls or across an incline. The term "stalagmite" has a letter "g" in it to remind us it forms from the ground up. *(Bruce Roberts/Photo Researchers, Inc.)*

Flowstone
Stalactite
Column
Stalagmite

is so abundant, all states in the contiguous United States have some karst features; Carlsbad Caverns in southeastern New Mexico is a spectacular example. Major belts of karst topography in the United States are: (1) a region extending through the states of Tennessee, Virginia, Maryland, and Pennsylvania; (2) south-central Indiana and west-central Kentucky; (3) the Salem-Springfield plateaus of Missouri; (4) central Texas; and (5) central Florida. One of the most famous cave systems in the United States is Mammoth Cave in Kentucky. Substantial research has been dedicated to explaining the origin of caverns and their development through time. It is clear that groundwater moving through rocks is the primary mechanism that forms caves. Cavern systems tend to form at or near the present water table where there is a continuous replenishment of water that is not saturated with the weathering products of the limestone. However, because many cavern systems contain a number of levels, it is believed that each level may represent a different time of cave formation and thus may be related to a fluctuating water table. Fundamentally, caves are enlarged as groundwater moves through limestone along bedding plains or fractures, eventually forming a cavern. Later, if the water table moves to a lower level, water seeping into the cavern starts to deposit calcium carbonate on the sides, floor, and ceiling, forming the beautiful cave formations such as *flowstone, stalagmites*, and *stalactites* (Figure 12.17).

From an environmental perspective, karst topography causes many problems, including the following:

- Water pollution occurs where sinkholes have been used for waste disposal. The bottom of many sinkholes is near the water table; this proximity directly injects water pollutants into the groundwater system.

- Cavern systems are prone to collapse, producing sinkholes that may form in areas that damage buildings on the ground surface, roads, and other facilities (see Chapter 9).

- In many areas underlain by limestone, such as the Edwards Plateau in Texas, groundwater is being mined. As a result of the mining, important karst springs where water emerges from caverns are being changed; the springs are experiencing less discharge or may dry up completely. These changes have important ecological consequences because many of these springs harbor forms of life unique to the spring environment. Their loss causes a reduction in biodiversity and contributes to the increasing number of endangered species (see Case History: The Edwards Aquifer, Texas—Water Resource in Conflict).

CASE HISTORY | **The Edwards Aquifer, Texas—Water Resource in Conflict**

Mark Twain wrote, "Whisky is for drinking and water is for fighting over." In central Texas, the Edwards aquifer, a karst flow system developed in limestone, has created an intense conflict over water. The conflict has escalated into near-open warfare from a legal, political, and economic viewpoint.[7]

The geology of the Edwards aquifer is interesting. Water that recharges the aquifer is primarily from stream flow over the aquifer, except where it is confined by layers of clay beds and shales. The combination of geologic structures, including folds and a zone of normal faulting, along with the karst flow system results in a complex aquifer with points of natural discharge at major karst springs (Figure 12.A). These springs exemplify the conflicting interest in water resources. For example, the Barton Springs Swimming Pool at Austin, Texas, is supplied by water that discharges from submerged openings in fractured limestone, visible on the right bank in Figure 12.B. In addition to providing important areas for recreation, Barton Springs is the only environment for the endangered Barton Springs salamander. The city of Austin dramatically illustrates conflicts between increasing urbanization and water resources. How can Texans continue to maintain a quality environment in the face of increasing water demands from cities such as Austin and San Antonio? It is clear that future decisions concerning the use of the Edwards aquifer and its abundant water resources will have significant social, economic, and environmental ramifications. In order to make these decisions, decision makers will have to conduct an accurate hydrogeologic evaluation of the entire aquifer system and how its waters may be allocated to the various users without degrading environmental systems.

The Edwards aquifer is one of the most prolific in North America. It provides water for more than 2 million people, including all the water used by the city of San Antonio and any smaller municipalities, industries, and agriculture. Water yields from a single well in the aquifer can be incredibly high. A well drilled in 1991 is one of the world's largest. It flows without pumping with a natural yield of 25,000 gallons per minute; enough to fill a good-sized back yard swimming pool.[7]

The Edwards aquifer is recharged primarily through influent streams that flow over the recharge zone where water sinks into the limestone (Figure 12.A). The recharge is dependent on the duration and volume of stream flows. The rate of pumping from the aquifer has increased steadily over the years. Natural recharge for the most part has exceeded pumpage and spring flow, two of the major sources of discharge. However, owing to increased water demand for growing urban areas and for irrigation, the water needs of the region now exceed the historical availability of water. For example, during a long drought from 1947 to 1956, two of the largest springs experienced serious reduction in discharge. Comal Springs ceased to flow for 4 months in 1956, and the discharge from San Marcos Springs dropped to a level that threatened important ecosystems dependent upon water in the springs.[7] Since the drought, several species have been listed under the Endangered Species Act of 1973. The San Marcos salamander of San Marcos Springs, the fountain darter fish, and Texas wild rice have all been categorized as endangered. As a

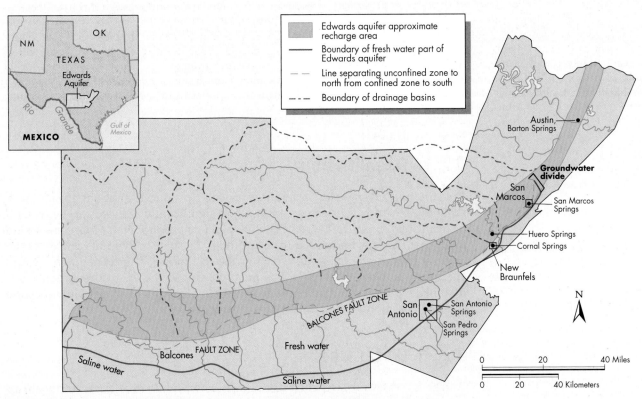

Figure 12.A **Edwards aquifer, Texas,** showing the location of the recharge area and major springs. *(Modified and simplified after Loaiciga, H. A., Maidment, D. R., and Valdes, J. B. 1999. Climate-change impacts in a regional karst aquifer, Texas, USA. Journal of Hydrology 227:173–94)*

Figure 12.B Barton (karst) spring near Austin, Texas. *(Marshall Frech/TEC)*

result of concern for these endangered species, a 1991 lawsuit was initiated by the Sierra Club against the U.S. Fish and Wildlife Service and other agencies to ensure adequate flow of the springs in order to maintain the ecosystems. The litigation established minimum stream flows required for preservation of the spring environment. Unfortunately, historical data clearly demonstrate that spring flows cannot be maintained during serious droughts, so the conflict will continue. In addition, increasing population growth has intensified water demands, and there are no potential alternative water resources that can provide abundant, high-quality, inexpensive water when compared with the Edwards aquifer. Potential resolutions to the water conflicts of the Edwards aquifer are likely to be expensive, not readily available, and certainly not agreeable to all parties.

Further complicating matters, downstream from the recharge zone, surface-water resources remain important to cities such as Corpus Christi. However, flows are required to maintain the ecological integrity and health of Gulf Coast estuaries in Texas. This flow is absolutely necessary to maintain habitat in a healthy estuarine ecosystem.[8]

Water shortages are not the only problems threatening the aquifer. For example, in the Austin area, studies of Barton Springs suggest that both groundwater and surface waters in some places contain a higher than normal concentration of several pollutants, including sediment, hydrocarbons, pesticides, nitrates, heavy metals, and bacteria. Examination of the distribution and levels of these pollutants relative to local land use suggest that there is a correspondence between contamination and those parts of the aquifer where urban growth and development have been the greatest.[7]

Potential solutions to future water shortages in the Edwards aquifer region include increasing water supplies, decreasing water demands, and managing existing water resources more efficiently through measures such as water conservation. The last solution seems to be the most likely approach. Presently, over one-half of the water use of the aquifer is for urban supply, and agricultural and rural supply uses the rest.[7] This use distribution is atypical; agricultural water demands generally far exceed urban demands. Water conservation efforts, including water reuse and augmentation of spring flow during droughts, could help alleviate water shortages in the Edwards aquifer region.

The Edwards aquifer is an important case history for water resources management. At the center of the case is an important natural water resource in which geologic, hydrologic, biological, political, legal, social, and economic factors are all playing a role. Clearly, the aquifer will continue to be a source of conflict as the various interests sort out the future of the precious water resources in this part of Texas. Water managers have predicted that by the year 2050 the total water withdraws from the aquifer may increase by about 50 percent. However, water withdrawals necessary to maintain the ecosystem and health of the environment are thought to be about 10 percent less than is currently being withdrawn. Of course, there are natural ebbs and flows as a result of variable stream flow; discharge from wells may be greater in some years than others. However, during droughts all users of the water will be stressed. Therefore, groundwater management plans for the Edwards aquifer are being developed with the assumption that deficits in the water supply will steadily grow over time.

12.5 Desalination

Desalination of seawater is a source of water produced at about 15,000 plants around the world. The salt content must be significantly reduced for the water to be drinkable, or *potable*. Large desalination plants produce water that costs several times that paid for traditional water supplies. Because the various processes that actually remove the salt require energy, the cost of the water is tied to increasing energy costs. New technology has greatly reduced the energy requirements to desalinate water. As a result the use of desalinated water will certainly increase, especially as more traditional sources of water are stressed.

12.6 Water Use

To discuss water use, we must distinguish between offstream uses and instream uses. **Offstream use** is water that is removed or diverted from its source. Examples include water for irrigation, thermoelectric power generation, industrial

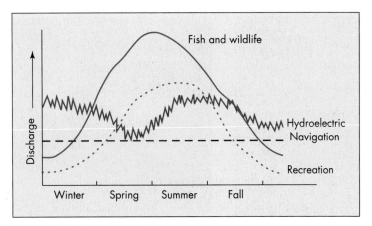

Figure 12.18 Various uses of water in a river are in conflict

Diagram comparing instream water uses and the varying discharges for each use. Discharge is the amount of water passing by a particular location and is measured in cubic meters per second (cms). The zigzag pattern for hydroelectric use results from rapid (often daily) change in water released from a dam to produce electric power. Larger seasonal changes reflect winter and summer power demands for heating and air conditioning. The "flat line" hydrograph is preferred for boat or barge traffic.

processes, and public supply. **Consumptive use** is a type of offstream use in which water does not return to the stream or groundwater resource immediately after use. This is the water that evaporates, is incorporated into crops or products, or is consumed by animals and humans.[4,6] In **instream use,** the water that is used is not withdrawn from its source. Examples include use of river water for navigation of vessels, hydroelectric power generation, fish and wildlife habitats, and recreation.

Multiple instream uses of rivers and streams usually create controversy because each use requires different conditions to prevent damage or detrimental effects. Fish and wildlife require seasonal fluctuations in water levels and flow rates for maximum biological productivity, and these levels and rates differ from the requirements for hydroelectric power generation, which requires large daily fluctuations in discharges to match power needs. Similarly, both of these may conflict with requirements for shipping and boating. The discharge necessary to move the sediment load in a river may require yet another pattern of flow. Figure 12.18 illustrates the seasonal patterns of discharge for some of these uses.

Movement of Water to People

In our modern civilization, water is often moved vast distances from areas with abundant rainfall to areas of high usage. In California, demands are made on northern rivers for reservoir systems to supply the cities in the southern part of the state. Two-thirds of California's runoff occurs north of San Francisco, where there is a surplus of water, while two-thirds of the water use occurs south of San Francisco, where there is a deficit. In recent years, canals constructed by both the California Water Project and the Central Valley Project have moved tremendous amounts of water from the northern to the southern part of the state (Figure 12.19). The diversion of waters has adversely affected ecosystems, especially fisheries in some northern California rivers.

Los Angeles is not unique; many large cities in the world must seek water from areas increasingly farther away. For example, New York City has imported water from nearby areas for more than 100 years. Water use and supply in New York City represent a repeating pattern. Originally, local groundwater, streams, and the Hudson River were used. However, water needs exceeded local supply; in 1842 the first large dam was built more than 48 km (30 mi) north of the city. As the city expanded rapidly from Manhattan to Long Island, water needs again increased. The sandy aquifers of Long Island were at first a source of drinking water, but this water was removed faster than rainfall replenished it. Local cesspools contaminated the groundwater, and salty ocean water intruded (see the chapter-opening Case History: Long Island, New York). A larger dam was built at Croton in 1900, but further expansion of the population brought repetition of the same pattern: initial use of groundwater, pollution, salinification, and exhaustion of

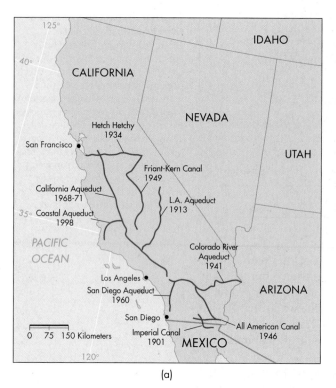

(b)

Figure 12.19 **Moving water to people** (a) California aqueducts. (b) California aqueduct in San Joaquin Valley. *(Allan Pitcairn/Grant Heilman Photography Inc.)*

this resource, followed by building new, larger dams in forested areas farther upstate. The trend of cities to import water continues. For example, San Diego, California, is bargaining with agricultural water districts to the east in the Coachella Valley to purchase water. Additional water sources are needed to fuel the city's development and establish a water supply independent of Los Angeles, its large neighbor to the north, from which it currently receives water. Development in urban centers lacking an adequate local water supply now or in the future is becoming a problem for many cities including Atlanta, New York, Miami, Chicago, Austin, Denver, Los Angeles, and San Diego. Perhaps development and populations of people in such areas should be limited, and more people should move closer to the water supply. This solution may be preferable from an environmental perspective because less water would need to be transported long distances. More water could be used locally near its site of origin. Developing areas near water sources would allow more flexibility in water use, and less water would be lost through evaporation and seepage from aqueducts, leaving more water for ecosystems.

Trends in Water Use

Water use on a global basis is about 70 percent for agriculture, 20 percent for industry, and 10 percent for urban and rural homes. Trends in water withdrawals in the United States provide insight that is both interesting and necessary for managing our water resources. Figure 12.20a shows trends in freshwater withdrawals from 1950 through 1995 (the most recent government data). These data suggest

- Surface water withdrawals far exceed groundwater withdrawals.
- Water withdrawals increased until 1980 and have since decreased and leveled off. The population of the United States was about 151 million in 1950 and continued to increase, reaching 267 million in 1995. Thus, during the

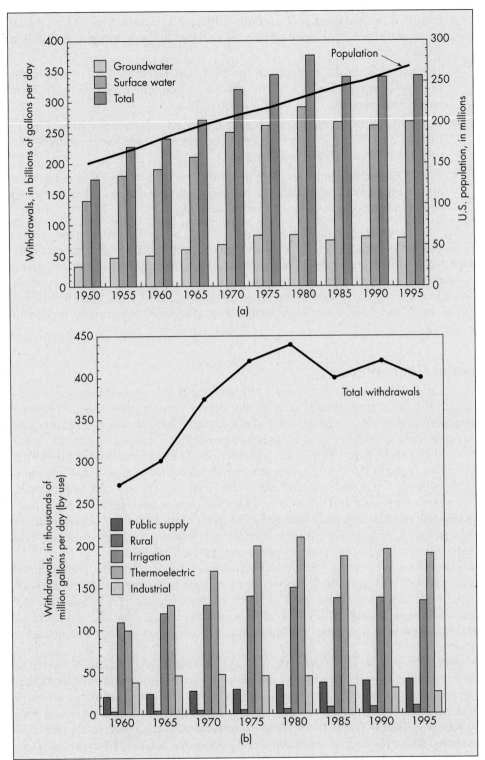

Figure 12.20 Trends in U.S. **water withdrawals** (a) Freshwater withdrawals by water-source category (1950–1995) and (b) total (fresh and saline) withdrawals by water-use category (1960–1995). *(Solley, W. B., Pierce, R. P., and Perlman, H. A. 1998.* Estimated use of water in the United States in 1995. *U.S. Geological Survey Circular 1200)*

period when water withdrawals decreased and leveled off (1980–1995), population was increasing. This trend suggests better water management and conservation during the 15-year period.[6]

Figure 12.20b shows trends in freshwater and saline water withdrawals by water-use category from 1960 through 1995. These data show that

■ Irrigation needs and the thermoelectric industry are the primary consumers of water.

- Use of water by the public in both urban and rural sectors has increased through the period, a trend presumably related to the increase in population of the country.

- The use of water by agriculture for irrigation leveled off in 1980 and has slightly decreased since then. This change presumably is related to efforts in water conservation.

- Water used for thermoelectric power increased dramatically from 1960 to 1980, as numerous power plants began operating; usage has since decreased because of more efficient use of water.

- Since 1980 industry has used significantly less fresh water. This decrease is due, in part, to new technologies that require less water as well as improved plant efficiencies and increased water recycling.

There are encouraging indications that the public is generally more aware of our water resources and the need to conserve them. As a result, water demands have been reduced in many states. Another encouraging sign is that use of reclaimed wastewater is now much more common, about 1 billion gallons per day, which is about five times what it was 50 years ago.[6] More significantly, the trend is continuing as more innovative ways are found to reuse water, particularly reclaimed wastewater.

Water Conservation

What can be done to use water more efficiently and reduce withdrawal and consumption? Since irrigation is one of the largest consumptive uses, improved agricultural irrigation could reduce water withdrawals by between 20 percent and 30 percent. Poor irrigation practices waste a tremendous amount of water. In some cases only a small percentage of the irrigation water used actually goes to the target plant. Techniques to improve water conservation for agricalatural irrigation include using lined and covered canals that reduce seepage and evaporation; computer monitoring and scheduling of water releases from canals; a more integrated use of surface waters and groundwaters; night irrigation, which reduces the amount of evaporation; improved irrigation systems such as sprinklers and drip irrigation; and better land preparation for water application.

Domestic use of water (urban and rural) accounts for only 10 percent of the total national withdrawals. However, domestic use is concentrated, and it poses major local problems. Withdrawal of water for domestic use may be substantially reduced at a relatively small cost with more efficient bathroom and sink fixtures, watering lawns and gardens at night, and drip irrigation systems for domestic plants.

Water removal for thermoelectricity could be reduced as much as 25 percent to 30 percent by using cooling towers designed to use less or no water. Manufacturing and industry could curb water withdrawals by increasing in-plant treatment and recycling of water or by developing new equipment and processes that require less water. The field of water conservation is changing so rapidly, it is expected that a number of innovations will reduce the total withdrawals of water despite increased consumption.[4]

12.7 Water Management in the Future

Managing water resources is a complex issue that will become more difficult as the demand for water increases (see A Closer Look: Management of the Colorado River). Although water supply problems are most serious in the southwestern United States and other arid and semiarid parts of the world, large cities in more humid regions such as New York and Atlanta also face problems. However, there

A CLOSER LOOK | Management of the Colorado River

The Colorado River is the primary river in the southwestern United States, and its water is an important resource. Managing the Colorado River has two aspects: (1) managing the water resources for people and (2) managing the river in the Grand Canyon to maintain the river environment. We will discuss each of these in turn.

Managing the Water

No discussion of water resources and water management would be complete without a mention of the Colorado River basin and the controversy that surrounds the use of its water. People have been using the water of the Colorado River for about 800 years. Early Native Americans in the basin had a highly civilized culture with a sophisticated water-distribution system. Many of their early canals were later cleared of debris and used by settlers in the 1860s.[10] Given this early history, it is somewhat surprising to learn that the Colorado was not completely explored until 1869, when John Wesley Powell, who later became director of the U.S. Geological Survey, navigated wooden boats through the Grand Canyon.

Although the waters of the Colorado River basin are distributed by canals and aqueducts to many millions of urban residents and to agricultural areas such as the Imperial Valley in California, the basin itself, with an area of approximately 632,000 km^2 (244,017 mi^2), is only sparsely populated. Yuma, Arizona, with approximately 42,000 people, is the largest city on the river, and within the basin only the cities of Las Vegas, Phoenix, and Tucson have more than 50,000 inhabitants.

Nevertheless, only about 20 percent of the total population of the basin is rural. Vast areas of the basin have extremely low densities of people, and in some areas measuring several thousand square kilometers there are no permanent residents.[10] Rod Nash, writing about the wilderness values of the river, states that at the confluence of the Green and Colorado Rivers, it is 80 km to the nearest video game and you are in the heart of a national park.[11]

The headwaters of the Colorado River are in the Wind River Mountains of Wyoming, and in its 2,300 km (1,430 mi) journey to the sea the river flows through or abuts seven states—Wyoming, Colorado, Utah, New Mexico, Nevada, Arizona, and California—and Mexico (Figure 12.C). Although the drainage basin is very large, encompassing much of the southwestern United States, the annual flow is only about 3 percent of that of the Mississippi River and less than a tenth of that of the Columbia. Therefore, for its size, the Colorado River has only a modest flow; yet it has become one of the most regulated, controversial, and disputed bodies of water in the world. Conflicts that have gone on for decades extend far beyond the Colorado River basin itself to involve large urban centers and developing agricultural areas of California, Colorado, New Mexico, and Arizona. The need for water in these semiarid areas has resulted in overuse of limited supplies and deterioration of water quality. Interstate agreements, court settlements, and international pacts have periodically eased or intensified tensions among people who use the waters along the river. The legacy of laws and court decisions, along with changing water-use patterns, continues

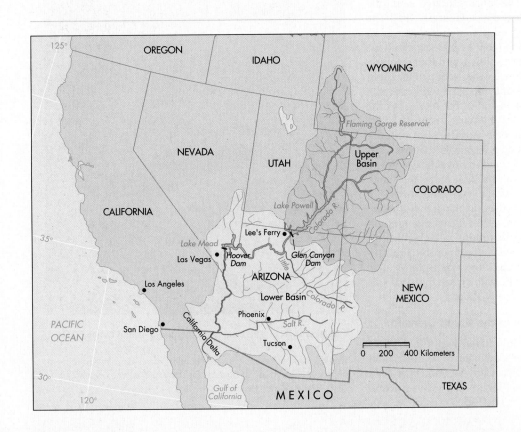

Figure 12.C The Colorado River basin, showing major reservoirs and the division of the watershed for management purposes. The delta, once a large wetland area, at the head of the Gulf of Mexico has been severely degraded as a result of diversion of water for various uses.

to influence the lives and livelihood of millions of people in both Mexico and the United States.[12]

Waters of the Colorado River have been appropriated among the various users, including the seven states and the Republic of Mexico. This appropriation has occurred through many years of negotiation, international treaty, interstate agreements, contracts, federal legislation, and court decisions. As a whole, this body of regulation is known as the Law of the River. Two of the more important early documents in this law were the Colorado River Compact of 1922, which divided water rights in terms of an upper and a lower basin, and the treaty with Mexico in 1944, which promised an annual delivery of 1.85 km³ (1.5 million acre-feet [1 acre-foot is the volume of water covering 1 acre to a depth of 1 ft], or 325,829 gallons) of Colorado River water to Mexico. More recent was a 1963 U.S. Supreme Court decision involving Arizona and California. Arizona refused to sign the 1922 compact and had a long conflict with California concerning the appropriation of water. The court decided that southern California must relinquish approximately 0.74 km³ (600,000 acre-feet) of Colorado River water. Finally, in 1974 the Colorado River Basin Salinity Control Act was approved by Congress. The act authorized procedures to control adverse salinity of the Colorado River water, including construction of desalinization plants to improve water quality.

Issues of water management in the Colorado remain complex because all the water is spoken for as demand for more water increases. The history of water management in the Colorado illustrates several major questions likely to be faced in other parts of the Southwest in coming years:

- How can we appropriate limited water resources?

- How can we maintain water quality?

As an example of a complex water management issue consider the All American Canal. The canal is located along the U.S. side of the border with Mexico (see Figure 12.19a). The canal is lined only with earth and as a result water from the Colorado River that flows through the canal seeps into the ground. That water is a source of argument and is developing into a political water war. A 36-km (23-mi) section of the canal that delivers water to San Diego may be lined with concrete. The objective is to stop the loss of water that seeps into the ground from the canal. However, this will adversely affect Mexican farmers. Billions of gallons of water seep into the ground and replenish groundwater resources that Mexican farmers have used to irrigate their fields for over 60 years. The treaty of 1944 with Mexico doesn't prohibit lining the canal, but Mexicans argue that it is unethical to take water they have used for decades. The Mexican farmers say that if they lose the water, farms will fail, leading to increased illegal immigration into the United States. Anything that reduces the opportunities in Mexico may lead to increased immigration, but the ethical argument is stronger. Water used for decades in Mexico, even though the water was not anticipated when the canal was constructed, sets a precedent we need to honor.

Managing the Colorado River in the Grand Canyon

The Grand Canyon of the Colorado River (Figure 12.D) provides a good example of a river's adjustment to the impact of a large dam. In 1963 the Glen Canyon Dam was built upstream from the Grand Canyon (Figure 12.C). Construction

of the dam drastically altered the pattern of flow and channel process downstream; from a hydrologic viewpoint, the Colorado River had been tamed. Before the Glen Canyon Dam was built, the river reached a maximum flow in May or June during the spring snowmelt; then flow receded during the remainder of the year, except for occasional flash floods caused by upstream rainstorms. During periods of high discharge, the river had a tremendous capacity to transport sediment (mostly sand and silt) and vigorously scoured the channel. The high floods also moved large boulders off the rapids, which formed because of shallowing of the river where it flows over alluvial fan or debris flow deposits delivered from tributary canyons to the main channel. As the summer low flow approached, the stream was able to carry less sediment, so deposition along the channel formed large bars and terraces, known as *beaches* to people who rafted the river.

After the dam was built, the mean annual flood (the average of the highest flow each year) was reduced by about 66 percent, and the 10 year flood was reduced by about 75 percent. On the other hand, the dam did control the flow of water to such an extent that the median discharge actually increased about 66 percent. The flow is highly unstable,

Figure 12.D **The Colorado River in the Grand Canyon** The sandbar "beach" in the lower left corner is being used by river rafters whose numbers have increased and affect the canyon. The number of people allowed to raft through the canyon is now restricted. *(Larry Minden/Minden Pictures)*

however, because of fluctuating needs to generate power, and the level of the river may vary by as much as 5 m (16 ft) per day. The dam and reservoir trap sediment from moving downstream, greatly reducing sediment load of the river immediately downstream from the dam. As a result, some of the large sandbars, which are valuable habitat, are starved of sand and are eroding. A lesser reduction in sediment load occurred farther downstream because tributary channels continued to add sediment to the channel.[13]

The change in hydrology of the Colorado River in the Grand Canyon has greatly changed the river's morphology. The rapids may be becoming more dangerous because large floods that had previously moved some of the large boulders farther downstream no longer occur.

Changes in the river flow, mainly deleting the high flows, have also resulted in vegetational shifts. Before the dam was built, three nearly parallel belts of vegetation were present on the slopes above the river. Adjacent to the river and on sandbars grew ephemeral plants, which were scoured out by yearly spring floods. Above the high-water line were clumps of thorned trees (mesquite and catclaw acacia) mixed with cactus and Apache plume. Higher yet could be found a belt of widely spaced brittle brush and barrel cactus.[14] Closing the dam in 1963 tamed the spring floods for 20 years, and plants not formerly found in the canyon, including tamarisk (salt cedar) and indigenous willow, became established in a new belt along the riverbanks.

In June 1983, a record snowmelt in the Rocky Mountains forced the release of about three times the amount of water normally released and about the same as an average spring flood before the dam. The resulting flood scoured the riverbed and riverbanks, releasing stored sediment that replenished the sediment on sandbars and scoured out or broke off some of the tamarisk and willow stands. The effect of the large release of water was beneficial to the river environment and emphasizes the importance of the larger floods in maintaining the system in a more natural state. Later, as an experiment, between March 26 and April 2 of 1996, a "test flood" was released from the dam in order to redistribute the sand supply. The floods resulted in the formation of 55 new beaches and added sand to 75 percent of the existing beaches. It also helped rejuvenate marshes and backwaters, which are important habitats to native fish and some endangered species. Water released from the dam is cold and a trophy trout fishery has been established. Native fish that prefer warmer water including large (up to 6 feet long) squaw fish are threatened with extinction due to habitat transformation and introduction of the trout. The experimental flood was hailed a success,[15] although a significant part of the new sand deposits was subsequently eroded away.[16]

The 1996 test flood remobilized sand, scouring it from the channel bottom and banks of the Colorado River below Glen Canyon Dam and depositing it on sandbars. However, little new sand was added to the river system from tributaries to the Colorado River, because they were not in flood flow during the test flood. The sand was mined from the river below the dam; as such, it is a limited, nonrenewable source that cannot supply sand to sandbars on a sustainable basis. A new, creative idea has recently been suggested.[16] The plan is to use the sand delivered to the Grand Canyon by the Little Colorado River, a relatively large river with a drainage area of

$67,340 \text{ km}^2$ ($26,000 \text{ mi}^2$) (Figure 12.C), which joins the Colorado River in the canyon downstream from Lee's Ferry. In 1993, a flood on the Little Colorado River delivered a large volume of sand to the Colorado River in the Grand Canyon, and prominent beaches were produced. Unfortunately, a year later the beaches had been almost entirely eroded away by the flow of the Colorado River. The problem was that the beaches were not deposited high enough above the bed of the Colorado and so were vulnerable to erosion from normal postdam flows. The idea suggested in the new study is to time the releases of flood flows from Glen Canyon Dam with sand-rich spring floods of the Little Colorado River. The resulting combined flood of the two rivers would be larger, and the new sand from the Little Colorado (Figure 12.E) would be deposited higher above the channel bed and would be less likely to be removed by lower flows of the Colorado. Evaluation of the hydrology of the Little Colorado River suggests that the opportunity to replenish sand on the beaches occurs, on average, once in 8 years. The proposed plan would help restore or re-create river flow and sediment transport conditions that formed and maintained the natural ecosystems of the canyon to be more as they were before the construction of Glen Canyon Dam.[16]

Another impact of the Glen Canyon Dam is the increase in the number of people rafting through the Grand Canyon. Although rafting is now limited to 15,000 people annually, the long-range impact on canyon resources is bound to be appreciable. Before 1950, fewer than 100 explorers and river runners had made the trip through the canyon.

We must concede that the Colorado River of the 1970s and 1980s is a changed river. Despite the 1983 and 1996 floods that pushed back some of the changes, river restoration efforts cannot be expected to return it to what it was before construction of the dam.[13,14,16] On the other hand, better management of the flows and sediment transport will improve and better maintain river ecosystems.

Finally, as a result of upstream water use the delta of the Colorado River, once a mjor wetland, has been degraded. Some years no flow reaches the sea.

Figure 12.E The Little Colorado River channel has abundant sand to help maintain sandbar "beaches" in the Colorado River. *(Edward. A. Keller)*

are options available, including locating alternative supplies, better protecting and managing existing supplies, or controlling population growth.

Cities in need of water are beginning to treat water as a commodity such as oil or gas that can be bought and sold on the open market. If cities are willing to pay for water and are allowed to avoid current water regulation, then allocation and pricing, as they are now known, will change. If the cost rises enough, "new water" from a variety of sources may become available. For example, irrigation districts in agricultural areas may contract with cities to supply water to urban areas. This arrangement could be done without negatively affecting crops by using conservation measures to minimize present water loss through evaporation and seepage from unlined canals. Currently, most irrigation districts do not have the capital to finance expensive conservation methods, but cities in need of water could finance such projects. Clearly, water will become much more expensive in the future, and, if the price is right, many innovative programs are possible.

Luna Leopold, a leader in the study of rivers and water resources, has suggested a new philosophy of **water management.** This new philosophy is based on geologic, geographic, and climatic factors as well as on the traditional economic, social, and political factors. The management of water resources cannot be successful as long as it is naively perceived primarily from an economic and political standpoint. The term *water use* is more appropriate because we seldom really manage water.[9] The essence of Leopold's water management philosophy is that surface water and groundwater are both subject to natural flux. In wet years, surface water is plentiful, and the near-surface groundwater resources are replenished. During these years, we hope that our flood-control structures, bridges, and storm drains will withstand the excess water. These structures are designed to withstand a particular flow; exceeding that flow may cause damage or flooding. Leopold concludes that we are much better prepared to handle floods than water deficiencies. Specific strategies to minimize hardships must be in place to combat water deficiencies during the dry years. For example, subsurface waters in various locations in the western United States are either too deep to be economically extracted or have marginal water quality. Furthermore, these waters may be isolated from the present hydrologic cycle and therefore may not be subject to natural recharge. However, water from these sources could be available if plans were in place for drilling wells and connecting them to existing water lines. Treatment of wastewater for reuse could also be an option in an emergency. Reuse of water on a regular basis might be too expensive or objectionable for other reasons, but advance planning to reuse more treated water during emergencies might be wise.[9]

In wet years, groundwater is naturally replenished. Surface water should be used when it is available, while reserving groundwater for dry years. In other words, groundwater could be pumped out at a rate exceeding the replenishment rate in dry years and be replenished by natural discharge and artificial recharge during wet years. This water management plan recognizes that excesses and deficiencies in water are natural and can be planned for.

12.8 Water and Ecosystems

The major ecosystems of the world have evolved in response to physical conditions that include climate, nutrient input, soils, and hydrology. Changes in these factors affect ecosystems; in particular, human-induced changes may have far-reaching effects. Throughout the world, with few exceptions, people are degrading natural ecosystems. Hydrologic conditions, particularly surface water processes and quality, are becoming factors that limit the continued existence of some ecosystems, particularly wetlands[17] (see A Closer Look: Wetlands).

A CLOSER LOOK | Wetlands

The term **wetlands**[18] refers to a variety of landscape features such as

- Swamps: wetland that is frequently or continuously inundated by water
- Marshes: wetland that is frequently or continuously inundated by water
- Bogs: wetland that accumulates peat deposits
- Prairie potholes: small marshlike ponds
- Vernal pools: shallow depressions that occasionally hold water

Some of these features are shown in Figure 12.F. The common feature and operational definition of wetlands is that they are either inundated by water or the land is saturated to a depth of a few centimeters for at least a few days most years. Hydrology, or wetness, and the types of vegetation and soil determine the presence or absence of wetlands. Of these three factors, hydrology is often the most difficult to define because some wetlands are wet for only a very short period each year. However, the presence of water, even for short periods on a regular basis, does give rise to characteristic wetland soils and specially adapted vegetation. Recognition of wetland soils and vegetation greatly assists in identifying the wetland itself in many cases.[19,20]

Wetlands and their associated ecosystems have many important environmental features:

(a)

(b)

(c)

Figure 12.F Several types of wetlands (a) Chesapeake Bay salt marsh. *(Comstock Images)* (b) Freshwater cypress swamp in North Carolina. *(Carr Clifton/Minden Pictures)* (c) Prairie potholes in the Dakotas. *(Jim Brandenburg/Minden Pictures)*

- Coastal wetlands such as salt marshes provide a buffer for inland areas from coastal erosion associated with storms and high waves.

- Wetlands are one of nature's natural filters. Plants in wetlands may effectively trap sediment, nutrients, and pollutants.

- Freshwater wetlands are a natural sponge. During floods they store water, helping to reduce downstream flooding. The stored water is slowly released after the flood, nourishing low flows of river systems.

- Wetlands are often highly productive lands where many nutrients and chemicals are naturally cycled while providing habitat for a wide variety of plants and animals.

- Freshwater wetlands are often areas of groundwater recharge to aquifers. Some of them, a spring-fed marsh, for example, are points of groundwater discharge.

Although most coastal marshes are now protected in the United States, freshwater wetlands are still threatened in many areas. It is estimated that 1 percent of the nation's total wetlands is lost every 2 years. Freshwater wetlands account for nearly all of this loss. In just the past 200 years about half of the total wetlands in the United States, including about 90 percent of the freshwater wetlands, have disappeared as a result of being drained for agricultural purposes or urban development.

The damage and degradation of wetlands have led to a growing effort to restore them. Unfortunately, restoration is not usually an easy task, since wetlands are a result of complex hydrologic conditions that may be difficult to restore if the water has been depleted or is being used for other purposes. Researchers are carefully documenting the hydrology of wetlands as well as the movement of sediment and nutrients. As more information is gathered concerning how wetlands work, restoration is likely to be more successful.

Water resources development, such as construction of dams, reservoirs, canals, or aqueducts, have a variety of environmental impacts at a variety of scales. Dams and accompanying reservoirs are often designed to be multifunctional, providing water for agriculture, a mechanism for flood control, and sites for recreational activities. Reconciling these various uses at a given reservoir site is often difficult. For example, water demands for agriculture are generally high during the summer, resulting in lowering of the water level in the reservoir and production of extensive mudflats. People interested in recreation find the low water level and mudflats to be aesthetically displeasing; moreover, the effects of the high water demand and drawdown of the reservoir water may interfere with wildlife by damaging or limiting a fish's spawning opportunities (see the discussion of dams and reservoirs with water power in Chapter 15). Dams and reservoirs also tend to instill a false sense of security in persons living downstream of the facilities, who believe they are protected from floods. However, the three major environmental effects of dams are

- Land, with its cultural and biological resources, is lost in the area flooded by the reservoir.

- The reservoir traps sediment that is transported from upstream by the rivers and streams that enter the reservoir. The trapped sediment reduces the water-storage capacity of the reservoir. More significantly, the reservoir holds sediment that would normally be transported downstream, nourishing the river and coastal environment. When rivers and coasts are deprived of sediment, river erosion occurs downstream of dams and coastal erosion occurs on beaches.

- The downstream hydrology and sediment transport system change the entire river environment, affecting the organisms that live there (see A Closer Look: Management of the Colorado River).

12.9 Emerging Global Water Shortages

Only recently have we realized that isolated shortages of water are an indication of a global pattern of a resource being depleted.[21] Around the Earth, groundwater is being depleted, mined from aquifers mostly used for agriculture; large lakes are drying up, as for example the Aral Sea mentioned in Chapter 1; and some large rivers including the Colorado in the U.S. and the Yellow in China run dry in some years and don't reach the sea, while others such as the Nile in Africa have seen their discharge to the sea reduced by over 90 percent. Water demand in the past 50 years has tripled as grain (wheat) production to feed people has tripled. During the same 50 years human population more than doubled. In the next 50 years human population is expected to increase another 2 to 3 billion and there is concern that without very careful management of water resources, there won't be enough water to grow the food for the expected 8 to 9 billion people in 2050. Thus we see that the emerging water shortage is linked to a potential food shortage. The problem is that the tremendous increase in grain production has been dependent on irrigation water mostly from groundwater resources. These same resources are being depleted nearly everywhere grain is grown—the United States, China, India, Pakistan, and Mexico for examples.[21] As water shortages occur—food shortages will follow. The solution is clear: control human population growth and conserve water. This chapter has outlined a number of ways to conserve our water resources through using less and reclaiming more. To avoid a food crisis more needs to be done. The good news is that a solution is possible, but it will take a lot of proactive work.

SUMMARY

The global water cycle involves the movement, storage, and transfer of water from one part of the cycle to another. The movement of water on land—that is, surface runoff and subsurface flow—is the part of the cycle of most direct concern to people.

Drainage basins, or watersheds, are a basic unit of the landscape. Surface-water runoff varies greatly from one drainage basin to another and is influenced by geographic, climatic, and biological factors.

The major source of groundwater is precipitation that infiltrates the recharge zone on the land surface and moves down through the vadose zone, which is seldom saturated, to the zone of saturation. An aquifer is a zone of Earth material capable of supplying water at a useful rate from a well. Interactions and linkages between surface water and groundwater are important environmentally because pollution in surface water may eventually contaminate the groundwater. Karst areas are particularly vulnerable to pollution.

Water supply is limited, even in areas of high precipitation and runoff, by our inability to store all runoff and by the large annual variation in stream flows. In many areas, ground-

water is being mined, resulting in withdrawals that exceed natural replenishment. In some areas mining of groundwater has permanently changed the character of the land.

Water uses are categorized as offstream, including consumptive, and instream. Multiple instream uses for hydroelectric power, recreation, and fish and wildlife habitat, often have conflicting requirements; partitioning water resources to meet the various uses is a controversial subject.

Water resource management needs a new philosophy that considers geologic, geographic, and climatic factors and utilizes creative alternatives.

Water is an integral part of ecosystems, and its increasing use by people is a major contributor to the degradation of ecosystems. Loss or damage of wetlands is an area of particular environmental concern in the United States because a significant portion of these ecosystems has already been lost, including 90 percent of the freshwater wetlands.

We are facing a global water shortage linked to food supply. Proactive water management is needed to produce an adequate food supply in the next 50 years as human population grows by 2 to 3 billion new people.

Revisiting Fundamental Concepts

Human Population Growth

There is plenty of water on Earth for our present population. However, it is not uncommon to have water shortages because people tend to reside in areas where water resources are limited. Population centers do not coincide with places where there is the most fresh water available. As a result, as human population continues to grow, water shortages will become more apparent, particularly in the more arid parts of

the world. It is expected that during the twenty-first century very serious water shortages will result in some regions, and these are likely to lead to conflicts between nations.

Sustainability

One way to avoid water shortages in the future is to ensure that we sustain our present sources. Sustainability means that we will leave to future generations water resources that have not been degraded by human processes. We also wish to ensure that the ecosystems of the world have enough water to remain healthy in the future. Therefore, development of sustainable water management is of prime concern to persons responsible for managing our water resources. Aspects of the sustainable water management program will include more efficient use of present water resources, water conservation, and protection of drainage basins so that they will continue to yield high-quality surface water and groundwater.

Earth as a System

Our water resources at the global, regional, and local levels are maintained by complex hydrologic systems that involve interactions between surface water and groundwater, both of which are subject to change over a variety of time scales, from several years to several hundred years or longer. As a result, in order to effectively manage water resources, we must understand various parts of the hydrologic system, including climate, geology, and land use. Human change that causes diversion of water from rivers and streams will likely degrade the ecosystems that the water nourished. As a result, to effectively use our water resources, we must pay close attention to how biological systems interact with the hydrologic system.

Hazardous Earth Processes, Risk Assessment, and Perception

There are significant risks and hazards associated with water shortages. For example, much of the food we consume is produced by irrigated lands, and our water resources are generally managed to provide sufficient water over a period of years. Use of water resources affects several natural hazards, the foremost of which is flooding, which is the most universally experienced natural hazard.

Scientific Knowledge and Values

The science of the water cycle and how it interacts with biological systems is a mature science. We have painstakingly learned how to utilize water for the benefit of people. However, we also recognize that other living things need water, too. To support ecosystems that require water resources, such as wetlands, we need to examine how we value the world around us. When we build a dam or overuse an aquifer, we change the hydrologic conditions and the life that depends upon water resources. Thus, we face a value judgment as to whether we wish to develop more water resources or better protect ecosystems that are dependent upon the water, such as the unique organisms in the large springs that discharge from the Edwards aquifer.

Key Terms

aquifer (p. 389)

consumptive use (p. 399)

Darcy's law (p. 392)

desalination (p. 398)

drainage basin (watershed) (p. 384)

instream use (p. 399)

karst (p. 395)

offstream use (p. 398)

sinkhole (p. 395)

vadose zone (p. 387)

water cycle (p. 383)

water management (p. 406)

water table (p. 388)

wetlands (p. 407)

zone of saturation (p. 387)

Review Questions

1. What is a watershed or drainage basin?

2. Define the vadose zone, zone of saturation, and water table.

3. What is an aquifer?

4. What major factors control the movement of groundwater?

5. What is Darcy's law?

6. What are some of the important interactions between surface water and groundwater?

7. Distinguish between instream and offstream water uses.

8. What is consumptive use?

9. Who are the biggest users of fresh water?

10. What are some of the ways we may conserve water?

11. Define Luna Leopold's philosophy concerning water management.

12. Define wetlands.

13. List some of the important environmental features associated with wetlands.

14. Why might we be facing a global food shortage based on water use?

Critical Thinking Questions

1. What sort of wetlands are found in your region? Outline a plan to inventory the wetlands and make an assessment of how much of the resource has been lost or damaged. Is wetlands restoration possible in your region, and what would you need to do to make it successful?

2. Evaluate the looming groundwater crisis over the Edwards aquifer in Texas. Consider social, economic, and environmental aspects.

3. Do you think that we are living in a "food-bubble" that may burst in the next few decades? That is, will links between food and water lead to a drop in food production with increasing human tragedy?

THIRTEEN

Water Pollution

Learning Objectives

In Chapter 12, we learned that the utilization of our water resources often goes hand in hand with the threat of water pollution. For example, we learned that the urban land use of Long Island, New York, has resulted in groundwater pollution. However, by far the most serious water resource problem is the lack of a pollution-free water supply for billions of people in many regions of the world. In this chapter, we will focus on the following learning objectives:

- Define water pollution and discuss some of the common water pollutants

- Understand the methods for treating groundwater pollution

- Understand the important processes related to wastewater treatment and renovation

Pig farms in North Carolina flooded in 1999 during Hurricane Floyd. *(Courtesy of Rick Dove)*

413

CASE HISTORY | North Carolina's Bay of Pigs

Hurricane Floyd drove through the heart of North Carolina's pig country in September 1999. The hurricane killed nearly 50 people, flooded thousands of homes, forced 48,000 people into shelters, and destroyed 5.7 million hectares (2.3 million acres) of crops. The storm produced up to 50 cm (20 in.) of rain in the eastern part of the state, flooding an area barely beginning to recover from Hurricane Dennis, which had struck just 2 weeks earlier.[1]

A water pollution catastrophe occurred as a result of the floodwaters from Hurricane Floyd. However, it was a preventable catastrophe that was anticipated by environmentalists. Although North Carolina has a long history of hog production, the population of pigs swelled from about 2 million in 1990 to over 10 million by 1997, making the area the second-largest pig-farming region in the nation (Figure 13.1). In the early 1990s, the state allowed large commercial pig farms to expand. Large automated confining farms that housed hundreds or thousands of pigs were built. North Carolina did not impose restrictions on their locations, and many of these "factory farms" were constructed on floodplains or reclaimed wetlands. Each pig produces approximately 2 tons of waste a year, so the North Carolina pigs were putting out approximately 20 million tons of waste a year. Waste consisting of manure and urine was flushed out of the farms, into

4-m (13-ft) deep, unlined lagoons the size of football fields. The only waste treatment in these lagoons was natural bacterial decay. The farmers argued that after the waste was degraded in the lagoons, the liquid could be sprayed onto crops as fertilizer. In reality, however, few crops were grown in the area; in addition, heavy rains sometimes caused the ponds to overflow and run into the rivers. No precautions were taken for the flood hazard and, as a result, large parts of eastern North Carolina became cesspools with lagoons. An early warning came in June 1995; the citizens of Onslow County, on North Carolina's coastal plain, were rudely awakened by a very unpleasant sight. A pig-waste holding lagoon had collapsed, sending approximately 95 million liters (25 million gallons) of concentrated pig feces and urine across a road and over fields into the New River. During the next 24 hours or so the noxious mass of waste traveled approximately 35 km (22 mi) down the river, slowing up near the city of Jacksonville. Some of the waste found its way into the New River estuary, where its adverse effects on marine life lasted for approximately 3 months. Researchers investigating this spill in the New River found numerous carcasses of fish littering the bank and hanging in stream-side brush. The water had turned murky brown and produced a nauseating stench.[2]

The favorable regulatory climate and availability of inexpensive waste-disposal systems were responsible for the phenomenal growth in the North Carolina pig population in the 1990s. Most of the factory-style animal operations are located in the lower Cape Fear and New River watersheds (Figure 13.1). The concentration of pig farms in this area has transformed the region into a natural laboratory for examining the impact of industrial-scale animal production and its waste products on river and estuarian systems. When Hurricane Floyd struck in 1999, more than 38 pig lagoons washed out and perhaps as much as 95,000 m^3 (250 million gallons) of pig waste was dumped into creeks, rivers, and wetlands (Figure 13.2). After the hurricane, approximately 250 pig operations were flooded out or had overflowing lagoons. Many of the flooded lagoons were upriver from towns that flooded, and the waste moved with the floodwater through schools, churches, homes, and businesses. People along the

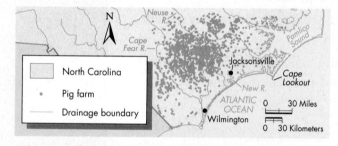

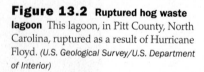

Figure 13.1 **Several thousand factory-style hog farms** Map of part of the pig farming region in North Carolina. *(Modified after Mallin, M. A. 2000. Impacts of industrial animal production on rivers and estuaries. American Scientist 88(1): 26–27)*

Figure 13.2 **Ruptured hog waste lagoon** This lagoon, in Pitt County, North Carolina, ruptured as a result of Hurricane Floyd. *(U.S. Geological Survey/U.S. Department of Interior)*

Figure 13.3 **Hogs killed by flooding** Floodwaters from Hurricane Floyd inundated many pig farms and pig carcasses were buried in shallow pits. *(Chris Seward/News & Observer Publishing Co., Raleigh, North Carolina)*

agricultural facilities lost in 1995. They also asked for exemption from the Clean Water Act for a period of 6 months so that waste from lagoons might be discharged directly into streams. Fortunately these requests were not granted.[1]

Even before the catastrophe caused by Hurricane Floyd, the pig farm situation in North Carolina had grown into a scandal reported by newspapers, local television shows, and the national news show *60 Minutes*. Finally, by 1997 a state law was enacted that prohibited building new waste lagoons and sewage plants on floodplains. In the spring of 1999, the governor of North Carolina proposed a 10-year plan that would introduce new waste-treatment technology and phase out the state's 4,000 animal waste lagoons. Unfortunately, Hurricane Floyd occurred before these changes could be enacted. Pig farms used the hurricane as a scapegoat, blaming it for the environmental catastrophe. However, it was clearly a human-induced catastrophe. States such as Iowa have large pig farms but have much more stringent environmental controls. Given North Carolina's frequent hurricanes, restricting factory farms from floodplains would seem to be a "no brainer." However, this is only the first step to control the waste problem posed by the pig farms in North Carolina. In order to prevent a similar catastrophe in the future, pig farmers will have to replace the use of waste lagoons with alternative waste-management practices.[1,2] Several years ago citizens formed the "Hog Roundtable," a coalition of health, civic, and environmental groups to control hog farming. Results included a mandate to phase out hog lagoons; expanded requirements for a buffer zone between hog farms and wells supplying water for people; requirements for larger buffers between fields being sprayed with pig waste and surface waters; and halting construction of a proposed slaughterhouse that would allow even more hog farms to be constructed.

Hurricane Floyd proved that large-scale agriculture is vulnerable to environmental catastrophes. Economic growth and livestock production must be carefully planned to anticipate problems; waste-management facilities must be designed that will not harm the environment. Of course, North Carolina is not the only place where environmental damage from pig production has occurred. Fish kills have occurred in Missouri and Iowa as well. Clearly, these examples demonstrate the need for the design of environmentally safe industrial agricultural practices.

rivers reported that the stench was overwhelming. The State Department of Agriculture estimated that approximately 30,000 hogs, 2 million chickens, and 735,000 turkeys had died. Most of the poultry carcasses were piled up and left to rot. Mobile incinerators that were moved to the area to burn floating hog carcasses were overwhelmed, leaving the farmers to bury the animals in shallow pits (Figure 13.3). The pits were supposed to be dry and at least 1 m (3.3 ft) deep; unfortunately, there was not always time to find dry ground, and for the most part the pits were dug and filled with no oversight. Many of the burials took place on floodplains, and it can be expected that, as the pig carcasses rot, bacteria will leak into groundwater and surface waters for some appreciable time.[1]

After the 1995 spills, environmentalists raised warnings concerning the location of the pig farms. They pointed out that hurricanes and flooding were common on North Carolina's coastal plain. However, the pig farm corporations are very rich and politically powerful; they fight long and hard to resist any change in their operations. Incredibly, they asked for approximately $1 billion in grants to repair or replace the

13.1 An Overview of Water Pollution

Water pollution refers to degradation of water quality as measured by biological, chemical, or physical criteria. This degradation is judged according to the intended use of the water, its departure from the norm, and public health or ecological impacts. From a public health or ecological point of view, a **pollutant** is any substance that, in excess, is known to be harmful to desirable living organisms. In fact, the primary water pollution problem facing billions of people today, especially in the developing world, is the lack of clean drinking water that is free of disease-causing organisms or substances. The concentration at which a material becomes harmful to living things is the subject of toxicology, discussed in Chapter 20.

The greatest water pollution problem in the world today is lack of disease-free drinking water for about 20 percent of the world's population. Another 20 percent

TABLE 13.1 Selected Chemicals That Agricultural, Industrial, and Municipal Processes Produce, Use, or Release to Impact Water Quality. Rates Are Millions (10^6) Tons per Year

• *Nutrients in rivers of the world*	
Inorganic nitrogen	16
Phosphorus	3.4
• *Heavy metals in water systems*	
Mercury, Lead, Zinc, Copper, Cadmium, Nickel, Chromium	0.1–1
• *Global production of other chemicals affecting water qualities*	
Fertilizer	140
Synthetic Organic Chemicals	300
Oil Spills	0.4

Source: Data from Schwarzenbach, R. P., and six others. 2006. The challenge of micropollutants in aquatic systems. *Science* 313: 1072–77.

have poor sanitation conditions that favor waterborne diseases that kill about 2 million people a year. Most of the deaths are of children under the age of 5. Chemical pollution is also an emerging problem on a global scale that occurs almost everywhere where people live. Agricultural, industrial, and municipal processes release chemical pollutants whose long-term effects on the environment and human health are largely unknown.[3] Table 13.1 lists annual releases (flux) of common chemicals that cause water pollution. Some of the pathways for water pollutants in the environment are shown in Figure 13.4.

13.2 Selected Water Pollutants

Many materials can pollute surface water or groundwater. Our discussion will focus on oxygen-demanding waste; pathogenic, or disease-causing, microorganisms; nutrients; oil; synthetic organic chemicals; heavy metals; radioactive materials; and sediment. We will also discuss a nonmaterial form of pollution: thermal pollution.

Oxygen-Demanding Waste

Dead plant and animal matter, called organic matter, in streams decays over time; that is, it is consumed by bacteria, which require oxygen. These are called *aerobic* bacteria, meaning they require oxygen to live. If there is enough bacterial activity, the oxygen in the water can be reduced to levels so low that fish and other organisms die. The amount of oxygen used for bacterial decomposition is the **biochemical oxygen demand (BOD),** a commonly used measure in water-quality management. A high BOD indicates a high level of decaying organic matter in the water.

Dead organic matter in streams and rivers comes from natural sources, such as fallen leaves, as well as from agriculture and urban sewage. Approximately 33 percent of all BOD results from agricultural activities, but urban areas, particularly those with sewer systems that combine sewage and storm-water runoff, may add considerable BOD to streams during floods. Sewers entering treatment plants can be overloaded and overflow into streams, producing pollution.

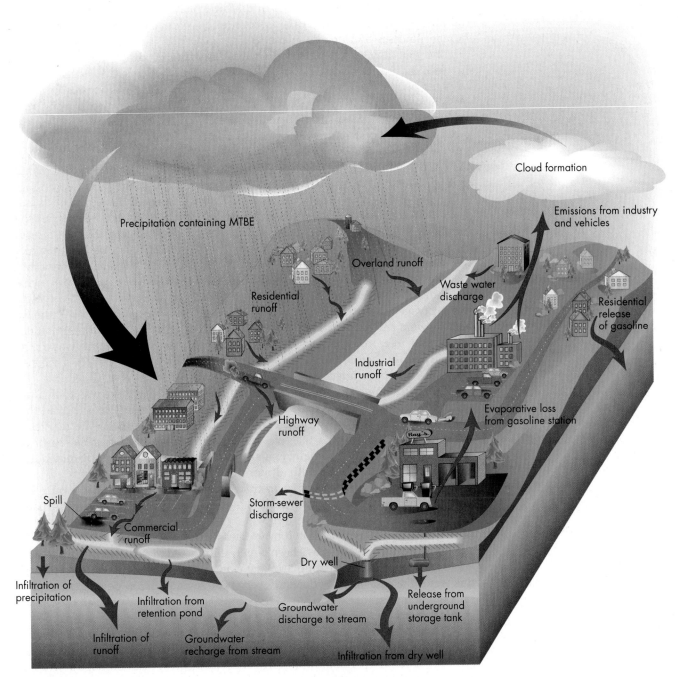

Figure 13.4 **Pathways for chemical pollutants** within the hydrologic cycle of the environment. *(Modified from Delzer, G. C., Zogorski, J. S., Lopes, T. J., and Basshart, R. S. 1996. Occurrence of gasoline oxygenate MTBE and BTEX compounds in urban storm water in the United States, 1991–95. U.S. Geological Survey Water-Resources Investigation Report 96–4145)*

Pathogenic Organisms

Pathogenic microbes or microorganisms, which are those that can be seen only with a microscope, are important biological pollutants. Cholera, typhoid infections, hepatitis, and dysentery are all waterborne diseases caused by pathogenic microorganisms. It is often difficult to monitor the pathogens directly; instead, the level of human **fecal coliform bacteria** is used as a common measure of biological pollution and as a standard measure of microbial pollution. Fecal coliform bacteria are usually harmless, part of the normal constituents of human intestines and found in all human waste.

However, not all forms of fecal coliform bacteria are harmless. *Escherichia coli* (also known as *E. coli* 0157), a strain of *E. coli* bacteria, was responsible for human illnesses and deaths in the 1990s. *E. coli* 0157 produces strong toxins in humans that may lead to bloody diarrhea, dehydration, kidney failure and death. In 1993, outbreaks of disease, apparently caused by *E. coli* 0157, occurred as a result of people's eating contaminated meat at a popular fast-food restaurant. In 1998 *E. coli* apparently contaminated the water in a Georgia water park and a Wyoming town's water supply, causing illness and one death.

One of the worst outbreaks of *E. coli* bacterial infection in Canadian history unfolded in May 2000 in Walkerton, Ontario. It is believed that the likely cause of the contamination in Walkerton was the result of *E. coli* bacteria in cow manure that washed into the public water supply during heavy rains and flooding that occurred on May 12, 2000. The local Public Utility Commission was aware as early as May 18 that water from wells serving the town was contaminated, but they did not report this contamination immediately to health authorities. As a result, people were not advised to boil water until it was too late to avoid the outbreak of disease. By May 26, 5 people had died, over 20 were in the intensive care unit of the local hospital, and approximately 700 were ill with severe symptoms, including cramps, vomiting, and diarrhea. The old and very young are most vulnerable to the ravages of the disease, which can damage the kidneys, and two of the first victims were a 2 year old baby and an 82 year old woman. Government officials finally took over management of the water supply, and bottled water was distributed. Tragically, before the outbreak was over at least seven people had died and over 1,000 had been infected.

Authorities launched an investigation, focusing on why there was such a long delay between identifying the potential problem and warning people. Had there not been such a long delay, illnesses might have been avoided. The real lesson learned from Walkerton is that we remain vulnerable to contamination of our water supply. We must remain vigilant in testing our waters and immediately report problems to public health authorities if any problems arise.

In the fall of 2006 *E. coli* 0157 was traced to farms in northern California. Contaminated spinach was shipped to 23 states. About 150 people became sick and one person died.

In the past, epidemics of waterborne diseases have killed thousands of people in U.S. cities. Such epidemics have been largely eliminated by separating sewage water and drinking water and treating drinking water before consumption. Unfortunately, this is not the case worldwide, and every year several billion people, particularly in poor countries, are exposed to waterborne diseases. For example, an epidemic of cholera occurred in South America in the 1990s. Although developing nations are more vulnerable, the risk of waterborne diseases is a potential threat in all countries.

The threat of an outbreak of a waterborne disease is exacerbated by disasters such as earthquakes, floods, and hurricanes; these events can damage sewer lines or cause them to overflow, resulting in contamination of water supplies. For example, after the 1994 Northridge earthquake, people in the San Fernando Valley of the Los Angeles Basin were advised to purify municipal water by boiling because of the threat of bacterial contamination.

Nutrients

Nutrients released by human activity may lead to water pollution. Two important nutrients that can cause problems are phosphorus and nitrogen, both of which are released from a variety of materials including fertilizers, detergents, and the products of sewage-treatment plants. The concentration of phosphorus and nitrogen in streams is related to land use. Forested land has the lowest concentrations of phosphorus and nitrogen, while the highest concentrations are found in agricultural areas such as fertilized farm fields and feedlots. Urban areas can also add phosphorus and nitrogen to local waters, particularly where wastewater-treatment plants discharge treated waters into rivers, lakes, or the ocean. These plants are effective in reducing organic pollutants and pathogens, but without advanced treatment, nutrients pass through the system.

High human-caused concentrations of nitrogen and phosphorus in water often result in the process known as **cultural eutrophication.** *Eutrophication* (from the Greek for "well fed"), a natural process, is characterized by a rapid increase in the abundance of plant life, particularly algae. Blooms of algae form thick mats that sometimes nearly cover the surface of the water in freshwater ponds and lakes. The algae block sunlight to plants below, and they eventually die. In addition, the algae consume oxygen as they decompose, thereby lowering the oxygen content of the water, and fish and aquatic animals may die as well.

In the marine environment, nutrients in nearshore waters may cause blooms of seaweed, referred to as marine algae, to flourish. The marine algae become a nuisance when they are torn loose and accumulate on beaches. Algae may also damage or kill coral in tropical areas. For example, the island of Maui in the Hawaiian Islands has a cultural eutrophication problem resulting from nutrients entering the nearshore environment from waste-disposal practices and agricultural runoff. Beaches in some areas become foul with algae that washes up on the shore, where it rots and creates a stench, providing a home for irritating insects that eventually drive away tourists (Figure 13.5).

A serious and ongoing cultural eutrophication problem is occurring in the Gulf of Mexico offshore of Louisiana. A so-called dead zone (Figure 13.6) develops in the summer, over a large area about the size of New Jersey. Water in the zone has low concentrations of oxygen, killing shellfish and crabs, and blooms of algae

(a)

Figure 13.5 **Algae contaminated beaches in Hawaii**
(a) Ocean-front condominium on the island Maui, Hawaii. The brown line along the edge of the beach is an accumulation of marine algae (locally called seaweed). (b) On the beach itself, the algae pile up, sometimes to a depth of about 0.5 m (1.7 ft), and people using the beach avoid the areas of algae piles. (c) Condominium complexes often have small wastewater-treatment plants, such as the one shown here, inside the plant-covered fence, that provide primary and secondary treatment. After this treatment, the water is injected underground at a relatively shallow depth. The treatment does not remove nutrients such as phosphorus and nitrogen that apparently encourage the accelerated growth of marine algae in the nearshore environment. *(Edward A. Keller)*

(b)

(c)

Figure 13.6 **Dead zone in Gulf of Mexico** Area in the Gulf of Mexico in July 2001 with bottom water with less than 2 mg/L dissolved oxygen. *(Modified after Rabalais, Turner, and Wiseman, 2001. Hypoxia studies)*

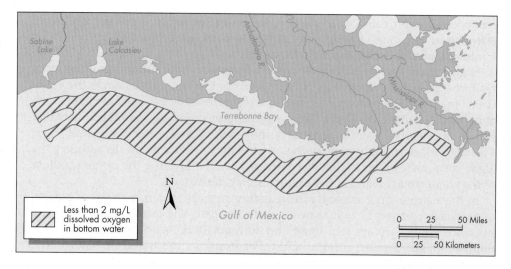

occur. The cause of the cultural eutrophication is believed to be the Mississippi River. The Mississippi drains about 40 percent of the lower 48 states, and much of the land use in the drainage basin is agricultural. The nutrient believed to cause the problem is nitrogen, which is used in great amounts to fertilize fields. The problem will not be easy to solve as long as agriculture continues to use tremendous amounts of fertilizer. Part of the solution will be modification of agricultural practices to use less nitrogen by using it more efficiently so that less of the nutrient runs off the land into the river.[4]

Oil

Oil discharged into surface water (rivers, lakes, and the ocean) has caused major pollution problems. The largest oil discharges have usually involved oil-tanker accidents at sea. For example, just after midnight on March 24, 1989, the oil tanker *Exxon Valdez* ran aground on Bligh Reef, 40 km (25 mi) south of Valdez, Alaska, in Prince William Sound. Crude oil poured out of the ruptured tanks of the vessel at a rate of approximately 20,000 barrels per hour (Figure 13.7). The *Exxon Valdez* was loaded with 1.2 million barrels of crude oil, and, of this, more than 250,000 barrels

(a)

(b)

Figure 13.7 **Oil spill from the *Exxon Valdez* in Alaska, 1989** (a) Aerial view of oil being offloaded from the leaking tanker *Exxon Valdez* on the left to the smaller *Exxon Baton Rouge* on the right. Floating oil is clearly visible on the water. *(Michelle Barns/Getty Images Inc.)* (b) Attempting to clean oil from the coastal environment by scrubbing and spraying with hot water. *(I. L. Atlan/Corbis/Sygma)*

(11 million gallons) gushed from the hold of the 300-m (984-ft) tanker. The oil remaining in the *Exxon Valdez* was loaded into another tanker.[5]

The oil spilled into what *was* considered one of the most pristine and ecologically rich marine environments of the world, and the accident is now known as the worst oil spill in the history of the United States. Short-term impacts were very significant; commercial fisheries, sport fisheries, and tourism were disrupted. In addition, many seabirds and mammals were lost. Lessons learned from the *Exxon Valdez* spill have resulted in better management strategies for both the shipment of crude oil and emergency plans to minimize environmental degradation.

A large oil spill in 2006 was caused by the war in Lebanon when a coastal power plant was bombed and over 100,000 barrels of fuel oil entered the Mediterranean Sea. Over half of Lebanon's tourist beaches were polluted, including a popular public beach visited by people from the capital city Beirut.

Toxic Substances

Many substances that enter surface water and groundwater are toxic to organisms. We will discuss three general categories of toxic substances—synthetic organic chemicals, heavy metals, and radioactive waste.

Synthetic Organic Chemicals. Organic compounds are compounds of carbon that are produced naturally by living organisms or synthetically by industrial processes. Up to 100,000 new chemicals are now being used or have been used in the past. It is difficult to generalize concerning the environmental and health effects of synthetic organic compounds because there are so many of them and they have so many uses and produce so many different effects.

Synthetic organic compounds have many uses in industrial processes, for pest control, pharmaceuticals, and food additives. Some of these compounds are called persistent organic pollutants, also known as POPs. Many of these chemicals were produced decades ago before their harm to the environment was known and a number have now been banned or restricted. Table 13.2 lists some of the common persistent organic pollutants and their uses. POPs have several general properties useful in defining them.[6] First, they have a carbon-based structure and often contain reactive chlorine. Second, most are produced by human processes and thus are synthetic chemicals. Third, they persist in the environment, do not break down easily, are polluting and toxic, and tend to accumulate in living tissue. Fourth, they occur in a number of forms that allow them to be easily transported by water and wind with sediment for long distances.

TABLE 13.2   Selected Persistent Organic Pollutants (POPs)

Chemical	Example of Use
Aldrin[a]	Insecticide
Atrazine	Herbicide
DDT[a]	Insecticide
Dieldrin[a]	Insecticide
Endrin[b]	Insecticide
PCBs[a]	Liquid insulators in electric transformers
Dioxins	By-product of herbicide production

[a]Banned in the United States and many other countries.
[b]Restricted or banned in many countries.

Source: Data in part from McGinn, Anne Platt, "Phasing Out Persistent Organic Pollutants," in Lester R. Brown et al., *State of the world 2000* (New York: Norton, 2000).

A significant example of water polluter is the chemical MTBE (methyl tertbutyl ether). The Clean Air Act Amendments that were passed in 1990 required cities with air pollution problems to use what are known as "oxygen additives" in gasoline. MTBE is added to gasoline with the objective of increasing the oxygen level of the gasoline and decreasing emissions of carbon monoxide from gasoline-burning cars. It is used because MTBE is more economical than other additives, including alcohol. MTBE is very soluble in water and is a commonly detected volatile organic compound (VOC) in urban groundwater. It is hypothesized that the MTBE detected in shallow groundwater originates from three sources: urban storm-water runoff, leaking underground gasoline tanks, and leakage occurring at service stations when car tanks are being filled.

It is ironic that a gasoline additive intended to improve air quality is now contaminating the groundwater that is used as a source of drinking water for approximately 15 million people in California. In 1997, MTBE-polluted groundwater in Santa Monica, California, forced the city to stop pumping groundwater, eliminating approximately 50 percent of the total drinking water supply for the city. Concentrations of MTBE in Santa Monica's groundwater ranged from about 8 to 600 micrograms (μg) per liter. The Environmental Protection Agency has stated that concentrations of 20 to 40 μg of MTBE per liter of water are sufficient to cause objectionable taste and odor. MTBE in that concentration smells like turpentine or fresh paint and is nauseating to some people. Studies are under way concerning the toxicity of MTBE, and some researchers fear it is a carcinogenic chemical. As a result of the contamination, some states, such as California, have terminated the use of MTBE. Figure 13.4 illustrates some of the pathways of MTBE as well as other volatile organic compounds in the hydrologic cycle of an urban area.[7]

Heavy Metals. Heavy metals such as lead, mercury, zinc, cadmium, and arsenic are dangerous pollutants that are often deposited with natural sediment in the bottoms of stream channels. If these metals are deposited on floodplains, they may become incorporated into plants, including food crops, and animals. Once the metal has dissolved in water used for agricultural or domestic use, heavy-metal poisoning can result.

As an example, consider mercury contamination of aquatic ecosystems. It has been known for decades that mercury is a significant pollutant of aquatic ecosystems, including ponds, lakes, rivers, and the ocean.[7]

Perhaps the best-known case history of mercury toxicity comes from Minamata, Japan. Minamata is a coastal town on the island of Kyushu and was the site of a serious illness that was first recognized in the middle of the twentieth century. It was first called the disease of the dancing cats because the illness was first observed in cats that seemingly went mad and ran in circles foaming at the mouth. It was also noticed that birds flew into buildings or fell to the ground. People were subsequently affected and most were families of fishermen. Some of the first symptoms were fatigue, irritability, numbness in arms and legs, and headaches, as well as difficulty in swallowing. Some of the more severe symptoms included blurred vision, loss of hearing, and loss of muscular coordination. Some people complained of a metallic taste in their mouths and suffered from diarrhea. By the time the disease ran its course, over 40 people died and over 100 were severely disabled. The people affected by the disease lived in a relatively small area and their diet mostly came from fish harvested from Minamata Bay.

The disease was eventually traced to a vinyl chloride factory on Minamata Bay that used mercury in its production processes. Inorganic mercury was released as waste into the bay and it was believed that the mercury would not get into the food chain. However, the inorganic mercury was converted by bacterial activity in Minamata Bay to methylmercury (discussed in more detail below). Methylmercury readily passes through cell membranes and is transported throughout the body by red blood cells. It can enter and damage brain cells. The harmful effects of methylmercury depend on a number of factors that include the amount of exposure and intake, the duration of the exposure, and the species affected. The effects of the mercury are often delayed from several weeks to months in people from the time of ingestion. Furthermore, if the intake of mercury ceases, some of the symptoms may disappear, but others are difficult to reverse.[8]

The disease of the dancing cats eventually became known as Minamata disease and nearly 800 people were officially recognized as having the disease, but as many as several thousand may have been involved. The mercury pollution in the bay ceased in 1968. As recently as the 1990s, some of the people afflicted by the disease were still being compensated for damages.

There are several natural sources of mercury, including input from volcanoes and erosion of natural mercury deposits. In most cases, however, we are most concerned with the input of mercury into the environment through processes such as burning coal, incinerating waste, and processing metals. Although the rates of mercury input into the environment by humans are poorly understood, it is believed that human activities have doubled or tripled the amount of mercury in the atmosphere, and it is increasing at about 1.5 percent per year.[9] Deposition from the atmosphere through rainfall is the primary source of mercury in most aquatic ecosystems. Once ionic mercury (Hg^{2+}) is in surface water it enters into complex cycles, during which a process known as methylation may occur. Bacterial activity changes the inorganic mercury to methylmercury (CH_3Hg^+). This process is important from an environmental viewpoint because methylmercury is much more toxic than is ionic mercury. Furthermore, living things require longer periods of time to eliminate methylmercury from their systems than they do to eliminate inorganic mercury. As the methylmercury works its way through food chains, a process known as *biomagnification* occurs in which concentrations of methylmercury increase in higher levels of the food chain. Thus, big fish in a pond contain higher concentrations of mercury than do the smaller fish and aquatic insects that the large fish feed upon. Some aspects of the mercury cycle in aquatic ecosystems are illustrated in Figure 13.8. The input side of the mercury cycle shows the deposition of inorganic mercury through the formation of methylmercury. On the output side of the cycle, mercury entering fish may be taken up by the organism that eats the fish (Figure 13.9). Sediment may also release mercury by a variety of processes, including resuspension in the water; this can eventually result in the mercury's entering the food chain or being released back into the atmosphere through volatilization, the process of converting a liquid or solid to a vapor.

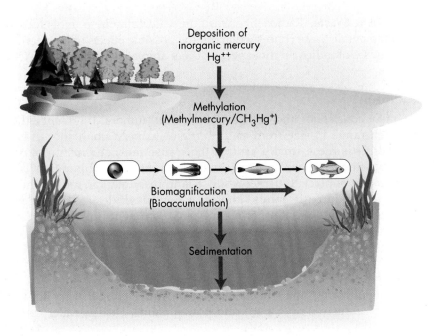

Figure 13.8 **Mercury in the environment** Input and changes of mercury in aquatic ecosystems. *(Modified from U.S. Geological Survey. 1995. Mercury contamination of aquatic ecosystems. U.S. Geological Survey FS-216-95)*

Figure 13.9 Fish may contain toxic metals People cooking and eating fish, here in the Fiji Islands, are taking in chemicals, including metals that the fish have in their tissue. Mercury is a potential problem with fish such as tuna and swordfish. *(Peter Arnold Inc.)*

Arsenic is an example of a highly toxic natural metal that is found in soil, rock, and water. There are many industrial and commercial uses of arsenic compounds including the processing of glass, pesticides, and wood preservatives. Arsenic may enter our water supplies through a number of processes including natural rain, snowmelt, or groundwater flow. It may also be released with industrial wastewater and agricultural processes. Finally, it may be released through the production of pesticides, the burning of fossil fuels, and as a by-product of mining.

Arsenic has been known as a deadly poison since ancient times and more recently it has been recognized that elevated levels of arsenic in drinking water may cause a variety of health problems that affect organs such as the bladder, lung, and kidney. It may also cause disease to the central nervous system. Finally, arsenic is known to be a carcinogen (capable of causing or promoting cancer).

The occurrence of arsenic in drinking water is now recognized as a global problem. It certainly is not found in all water supplies, but it is found in many around the world. For example, arsenic in groundwater in Bangladesh has affected many millions of some of the poorest people on Earth. Ongoing research has the objective to identify those locations where arsenic pollution occurs and to develop appropriate technology or methods to avoid or reduce the hazard of exposure to arsenic.[10]

Radioactive Waste. Radioactive waste in water may be a dangerous pollutant. Environmentalists are concerned about the possible effects of long-term exposure to low doses of radioactivity to people, other animals, and plants (see Chapter 15).

Sediment

Sediment consists of unconsolidated rock and mineral fragments, the smallest of which range in size from sand particles to very small silt- and clay-sized particles. It is these small particles that cause most sediment pollution problems. Sediment is our greatest water pollutant by volume; it is clearly a resource out of place. It depletes soil, a land resource; can reduce the quality of the water resource it enters; and may deposit undesired materials on productive croplands or on other useful land. Sediment pollution is discussed in detail in Chapter 16.

Thermal Pollution

Thermal pollution is the artificial heating of waters, primarily by hot-water emission from industrial operations and power plants. Heated water causes several problems. First, heated water contains less oxygen than cooler water; even water only several degrees warmer than the surrounding water holds less oxygen. Second, warmer water favors different species than cooler water and may increase the growth rates of undesirable organisms, including certain water plants and fish. In some cases, however, the warm water may attract and allow better survival of certain desirable fish species, particularly during the winter.

13.3 Surface-Water Pollution and Treatment

Pollution of surface waters occurs when an excess of an undesirable or harmful substance flows into a body of water and exceeds the natural ability of that ecosystem to utilize, remove, or convert the pollutant to a harmless form. Water pollutants are emitted from localized sources, known as point sources, or by diffuse sources, known as nonpoint sources.

Point Sources of Surface-Water Pollution

Point sources are discrete and confined, such as pipes that empty into streams or rivers from industrial or municipal sites (Figure 13.10). In general, point-source pollutants from industries are controlled through on-site treatment or disposal and are regulated by permit. Municipal point sources are also regulated by permit.

Nonpoint Sources of Surface-Water Pollution

Nonpoint sources are diffuse and intermittent; they are influenced by such factors as land use, climate, hydrology, topography, native vegetation, and geology. Common urban nonpoint sources include runoff from streets or fields. Rural sources of nonpoint pollution are generally associated with agriculture, forestry, or mining (see A Closer Look: Acid Mine Drainage).

Pollution from nonpoint sources, or polluted runoff, is difficult to control and contains all sorts of pollutants, from heavy metals to chemicals and sediment

Figure 13.10 Point source Pipe discharging partially treated waste from the Climax Molybdenum Mine in Colorado. Point sources are restricted and controlled by law in the U.S. *(Jim Richardson/Richardson Photography)*

A CLOSER LOOK | Acid Mine Drainage

Acid mine drainage refers to acidic water with elevated concentrations of dissolved metals that drains from coal or metal mines. Specifically, acid mine drainage is water with a high concentration of sulfuric acid (H_2SO_4) that drains from some mining areas to pollute surface-water resources. Acid mine drainage is produced by complex geochemical and microbial reactions when sulfide minerals associated with coal or a metal, such as zinc, lead, or copper, come into contact with oxygen-rich water near the surface. For example, pyrite (FeS_2) is a common sulfide often associated with coal; when pyrite oxidizes in the presence of water and microbes, sulfuric acid is formed. The sources for the water may be surface water that infiltrates into mines or shallow groundwater that moves through mines. Similarly, surface and shallow groundwaters that come into contact with mining waste, called tailings, may also react with sulfide minerals found there to form acid-rich waters.

Once waters with a high concentration of sulfuric acid and dissolved metals migrate away from a mining area, they can pollute surface and groundwater resources. If the acid-rich water runs into a natural stream or lake, significant ecological damage may result. The acid water is extremely toxic to plants and animals in aquatic ecosystems; acidic waters may also mobilize other potentially harmful chemicals. Acid mine drainage is a significant water pollution problem in many parts of the United States, including parts of Wyoming, Illinois, Indiana, Kentucky, Tennessee, Missouri, Kansas, Oklahoma, West Virginia, Maryland, Pennsylvania, Ohio, and Colorado. The total impact of acid mine drainage is significant; thousands of kilometers of streams have been polluted (Figure 13.A).

The Tar Creek area of Oklahoma was at one time designated by the Environmental Protection Agency as the nation's worst example of acid mine drainage. The creeks in the area were severely polluted by acid-rich water from abandoned mines in the Tri-State Mining District of Kansas, Oklahoma, and Missouri. Sulfide deposits containing both lead and zinc were first mined there in the late nineteenth century, and

mining ended in some areas in the 1960s. During operation of the mines, subsurface areas were kept dry by pumping out groundwater that was constantly seeping in. After mining stopped, the groundwater tables naturally rose again. Subsequently, some of the mines flooded and overflowed, polluting nearby streams. The Tri-State mining district remains an area of concern, and a lot of work has been done to reduce the impact of mines. The likely solution to reducing acid mine drainage in Tar Creek and other areas in the region is to use passive methods of treatment that utilize naturally occuring chemical and/or biological reactions in controlled environments to treat acid mine drainage. The simplest controlled environment is an open limestone channel. The acid-rich water reacts with crushed limestone and the acid is neutralized.

Figure 13.A **Acid mine drainage** Water seeping from this Colorado mine is an example of acid mine drainage. The water is also contaminated by heavy metals, including iron compounds producing the orange color. *(Tim Haske/Profiles West/Index Stock Photography, Inc.)*

(Figure 13.11). Even when you wash your car in your driveway and the detergent and oil on the surface run down a storm drain that enters a stream, you are contributing to polluted runoff. Polluted runoff is also produced when rainwater washes insecticides from the plants in your garden, then runs off into a stream or infiltrates the surface to contaminate groundwater. Similarly, rain and runoff from factories and storage yards are a source of nonpoint pollution.[11]

Reduction of Surface-Water Pollution: The Cuyahoga River Success Story

In the United States, a concerted effort is underway to reduce water pollution and thereby improve water quality. The assumption is that people have a basic right to safe water for drinking, swimming, and use in agriculture and industry. At one time water quality near major urban centers was considerably worse than it is today. Consider the story of the Cuyahoga River. The continental glacial ice sheets that formed the Great Lakes last receded from the United States over 10,000 years ago. In what is now northern Ohio a river formed, flowing into what was later

(a)

(b)

Figure 13.11 **Examples of water pollution** (a) Severe water pollution producing a health hazard. This ditch carries sewage and toxic waste to the Rio Grande in Mexico. *(Jim Richardson/Richardson Photography)* (b) Sediment being removed by heavy equipment (shown in the background) after the 1995 flood in Goleta, California. The sediment came from a nearby stream that overflowed its banks and deposited it in the lot of a new car dealership. *(Rafael Moldanado/Santa Barbara News-Press)*

named Lake Erie. Native Americans who arrived over 2,000 years ago to the Ohio Valley named the river Cuyahoga, which in their language meant a meandering river (that is, having a sinuous form with many bends). The river was of great use to the early Native Americans who lived along the river and depended on it for a natural route for transporting goods that they consumed and traded with other people. In the seventeenth and eighteenth centuries, fur traders arrived from Europe to trade with the Native Americans and the war of 1812 displaced Native Americans from the Cuyahoga Valley. A few decades later industrialization arrived in the river valley, and canals that linked the river to Lake Erie were constructed that in their time carried a tremendous volume of freight. With the growth of industrialization of the river valley and in particular the city of Cleveland, Ohio, pollution of the river became more and more common. At that time laws to prohibit the dumping of waste into the river did not exist, and with the arrival of the petroleum industry in Cleveland the river became even more polluted. As a result of the pollution, fish and other living things in the river died, and the river became one of the most polluted in the United States. In 1969 the Cuyahoga River made environmental history when sparks from a train ignited oil-soaked water in the river, setting the very surface of the river on fire! The burning of an American river became a rallying point for growing environmental consciousness.

The Cuyahoga River today is cleaner and is no longer flammable.[12] Parts of the river have been transformed into a green belt that has changed the river from an open sewer to a valuable public resource as well as a focal point for economic and environmental renewal. However, in downtown Cleveland and Akron the river remains an industrial stream and parts are still polluted. Large portions of the upper part of the river have been designated as a State of Ohio Scenic River and the Cuyahoga Valley National Recreation Area is sited in the valley between the cities of Cleveland and Akron. Finally, portions of the middle and lower river valley were designated as National Heritage Sites. What will be the final outcome for the Cuyahoga River and its scenic valleys and urban areas? It seems that the river system is much revered by the people of Ohio and the river will hopefully continue to be a showcase of positive river restoration, demonstrating how even very polluted rivers may be restored.

13.4 **Groundwater Pollution and Treatment**

Approximately one-half of all people in the United States depend on groundwater as their source of drinking water. We are therefore concerned about the introduction into aquifers of chemical elements, compounds, and microorganisms that do not occur naturally within the water. The hazard presented by a particular groundwater pollutant depends on several factors, including the volume of pollutant discharged, the concentration or toxicity of the pollutant in the environment, and the degree of exposure of people or other organisms.

Most of us have long believed that our groundwater is pure and safe to drink, so many of us find it alarming to learn that it may be easily polluted by any one of several sources (Table 13.3). In addition, even the very toxic pollutants may be difficult to recognize.

Currently, the problem of groundwater pollution in the United States is becoming more apparent as water testing becomes more common. For example, Atlantic City, New Jersey, and Miami, Florida, are two eastern cities threatened by polluted groundwater that is slowly migrating toward their wells. It is estimated that 75 percent of the 175,000 known waste-disposal sites in the country may be producing plumes, or bodies of contaminated groundwater. Many of the chemicals found in our groundwater are toxic or suspected carcinogens. It therefore appears that we have been conducting a national large-scale experiment concerning the effects of chronic low-level exposure of people to potentially harmful chemicals! Unfortunately, the final results of the experiment will not be known for many years.[13] Preliminary results suggest we had better act now before a hidden time bomb of health problems explodes.

National Water-Quality Assessment Program

The past quarter century has seen significant financial investments and great improvements in manufacturing, processing, and wastewater-treatment facilities with the objective of reducing the amount of contaminants emitted into our water resources. These programs have significantly improved water quality across the United States; however, considerable apprehension continues concerning the effects of water pollutants such as nutrients, toxins, and pathogens on human health and the health of aquatic ecosystems. In response to this need, the U.S. Geological Survey in 1991 began a program to assess water quality throughout the country. The program integrates both surface-water and groundwater systems that monitor and study aquatic ecosystems. The goals of the program are (1) to carefully describe current water-quality conditions for many of the freshwater streams and aquifers in the United States, (2) to monitor and describe water-quality changes over time, and (3) to increase understanding concerning the human and natural factors that affect the nation's water quality.

TABLE 13.3 **Common Sources of Groundwater Pollution and Contamination**

Leaks from storage tanks and pipes

Leaks from waste-disposal sites such as landfills

Seepage from septic systems and cesspools

Accidental spills and seepage (train or truck accidents, for example)

Seepage from agricultural activities such as feedlots

Intrusion of saltwater into coastal aquifers

Leaching and seepage from mine spoil piles and tailings

Seepage from spray irrigation

Improper operation of injection wells

Seepage of acid water from mines

Seepage of irrigation return flow

Infiltration of urban, industrial, and agricultural runoff

Figure 13.12 River basin studied to monitor and describe water quality Delaware River basin with physiographic provinces. *(After U.S. Geological Survey, 1999. National water-quality assessment program, Delaware River basin. U.S. Geological Survey FS-056-99)*

Physiographic provinces

Appalachian Plateaus

Valley and Ridge

Piedmont

Coastal Plain

Physiographic-province boundary

▲ Streamflow-gaging station at Montague, NJ

N

0   15   30 Miles

0   15   30 Kilometers

One of the larger water systems under study is the Delaware River basin, which includes parts of Pennsylvania, New Jersey, New York, and Delaware (Figure 13.12). The drainage basin contains parts of several distinct physiographic provinces, which are regions that share similar topography, rock types, and geologic history. Provinces vary from the relatively flat Coastal Plain and broader uplands of the Piedmont, where 80 percent of the people living in the basin reside, to the more rugged topography of the Valley and Ridge province and the Appalachian Plateaus in the northern portions of the drainage basin. The Delaware River basin includes forested lands (60 percent), agricultural lands (24 percent), urban and residential areas (9 percent), and land taken up by surface water bodies and other miscellaneous land types (7 percent).[14]

Some of the major water-quality issues being addressed in the Delaware River basin study[14] include

- Effects of the natural river system on the distribution, fate, and effects of contaminants in water, sediment, and living things
- Relationships between the flow of water in the river and concentrations of nutrients, contaminants, and pathogens
- Presence of contaminants, including pathogens and pesticides, in drinking water supplies and in water used for recreational activities
- Development of water-management plans and strategies for the protection of areas of the river basin that have high water quality
- Effects of septic systems, used to treat sewage from individual homes, on water quality and river ecology
- Effects of groundwater withdrawals on water quality
- Effects of discharge from coal mines on water quality and river ecosystems

In addition, the Delaware River basin is being studied with the objective of determining what data are useful for predicting the impacts of climate change.[15] The plan is to monitor the river for changes in response to global warming related to storage of water for New York City, maintaining stream flow requirements for a variety of water uses in the basin, and controlling the migration of saltwater into the Delaware estuary as sea level rises.[15]

Saltwater Intrusion

Aquifer pollution does not result solely from the disposal of wastes on the land surface or in the ground. Overpumping or mining groundwater allows inferior waters to migrate from both adjacent aquifers and the sea, also causing contamination problems. Hence, human use of public or private water supplies can accidentally result in aquifer pollution. Intrusion of saltwater into freshwater supplies has caused problems in coastal areas of New York, Florida, and California, among other areas.

Figure 13.13 illustrates the general principle of saltwater intrusion. The groundwater table is generally inclined toward the ocean, while a wedge of saltwater is

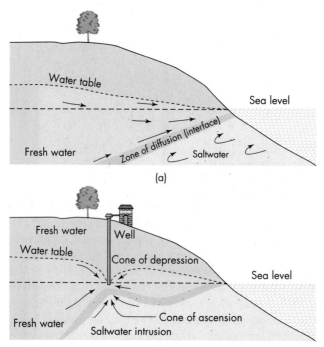

Figure 13.13 Saltwater intrusion (a) The groundwater system near the coast under natural conditions. (b) A well with both a cone of depression and a cone of ascension. If pumping is intensive, the cone of ascension may be drawn upward, delivering saltwater to the well.

TABLE 13.4 · **Methods of Treating Groundwater and Vadose-Zone Water**

Extraction Wells	Vapor Extraction	Bioremediation	Permeable Treatment Bed
Pumping out contaminated water and treatment by filtration, oxidation, air stripping (volatilization of contaminant in an air column), or biological processes	Use of vapor-extraction well and then treatment	Injection of nutrients and oxygen to encourage growth of organisms that degrade the contaminant in the groundwater	Use of contact treatment as contaminated water plume moves through a treatment bed in the path of groundwater movement; neutralization of the contaminant by chemical, physical, or biological processes

inclined toward the land. Thus, with no confining layers, salty groundwater near the coast may be encountered below the land surface. Because fresh water is slightly less dense than saltwater, fresh water will float on top of saltwater; the density difference explains why a layer of fresh water can sometimes be found in the ocean offshore from a river's delta. When wells are drilled, a *cone of depression* develops in the freshwater table (Figure 13.13b), which may allow intrusion of saltwater as the interface between fresh water and saltwater rises and forms a *cone of ascension* in response to the loss of fresh water.

Groundwater Treatment

The difficulty of detecting groundwater pollution, the long-term residency of groundwater, the degradation of the polluted aquifer, and the difficulty and expense of aquifer recovery establish a strong argument: No wastes or possible pollutants should be allowed to enter any part of the groundwater system. However, preventing the entry of pollutants requires careful protection and management of recharge areas and may in fact be impossible. Accidents, such as a broken pipeline, will happen. In order to reduce groundwater pollution, we should learn more about the natural processes that treat wastes, so that when soil and rocks cannot naturally treat, store, or recycle wastes, we can develop processes to make the pollutants treatable, storable, or recyclable. Table 13.4 briefly outlines some of the methods for treating groundwater. The specific treatment selected depends on variables such as type of contaminant, method of transport, and characteristics of the local environment, such as depth to the water table and geologic characteristics.

13.5 Water-Quality Standards

A question people commonly ask is, "How safe is our water supply?" Americans believe that they have high-quality drinking water, some of the best in the world. For the most part, we do have high-quality drinking water, but in recent years we have gained an ability to detect specific contaminants in parts per billion (ppb) or, in some cases, even parts per trillion (ppt) of water. The question that then arises is, "How dangerous might some of these chemicals be?" You may think that such small amounts of contaminants cannot possibly be dangerous, but as the U.S. Environmental Protection Agency (EPA) reminds us, a single microscopic virus can cause a disease. Physicians are able to delineate what diseases are caused by particular viruses, but we are less sure about the effects of long-term exposure to very small amounts of chemicals.

In response to this concern, the U.S. Congress has mandated that the EPA establish minimum national drinking water standards for a variety of chemicals and other materials. In 1986, Congress expanded the Safe Drinking Water Act of 1974 to include 83 contaminants. Among other regulations, the new legislation banned the use of lead in the installation or repair of water systems used for drinking water. Health effects associated with lead toxicity are very well known. At high concentrations, lead causes damage to the nervous system and the kidneys and is particularly toxic to infants and pregnant women.[16]

The EPA has set standards for a number of contaminants that might be found in our drinking water. However, only two substances for which these standards have been set are thought to pose an immediate health threat when standards are exceeded. These are[16]

- Coliform bacteria, possibly indicating that the water is contaminated by harmful disease-causing organisms
- Nitrate, because contamination above the standard is an immediate threat to young children. In some children younger than a year old, high levels of nitrate may react with their blood to produce an anemic condition known as "blue baby."

Table 13.5 is an abbreviated list illustrating some of the contaminants included in the EPA's list of National Primary Drinking Water Standards and related health problems. The purposes of the standards and regulations concerning drinking water are[16]

- To ensure that our water supply is treated to remove harmful contaminants
- To regularly test and monitor the quality of our water supply
- To provide information to citizens so that they are better informed concerning the quality and testing of their water supply

TABLE 13.5   National Drinking Water Standards: Some Examples

Contaminant	Maximum Contaminant Level (MCL)(mg/L)	Comments/Problems
Inorganics		
Arsenic	0.05	Highly toxic
Cadmium	0.01	Kidney
Lead	0.015[1]	Highly toxic
Mercury	0.002	Kidney, nervous system
Selenium	0.01	Nervous system
Asbestos	7 MFL[2]	Benign tumors
Fluoride	4	Skeletal damage
Organic chemicals		
Pesticides		
Endrin	0.0002	Nervous system, kidney
Lindane	0.004	Nervous system, kidney, liver
Methoxychlor	0.1	Nervous system, kidney, liver
Herbicides		
2,4D	0.07	Liver, kidney, nervous system
Silvex	0.05	Nervous system, liver, kidney
Volatile organic chemicals		
Benzene	0.005	Cancer
Carbon tetrachloride	0.005	Possible cancer
Trichloroethylene	0.005	Probable cancer
Vinyl chloride	0.002	Cancer risk
Microbiological organisms		
Fecal coliform bacteria	1 cell/100 mL	Indicator—disease-causing organisms

[1]The action level for lead related to treatment of water to reduce lead to the safe level. There is no MCL for lead.
[2]Million fibers per liter with fiber length >10 micrometers.

Source: U.S. Environmental Protection Agency

13.6 Wastewater Treatment

Water that is used for municipal and industrial purposes is often degraded by a variety of contaminants including oxygen-demanding materials, bacteria, nutrients, salts, suspended solids, and other chemicals. In the United States, our laws dictate that these contaminated waters must be treated before they are released back into the environment. Wastewater treatment in the United States is big business, costing several tens of billions of dollars per year. In rural areas, the conventional method of treatment uses septic-tank disposal systems. In larger communities, wastewaters are generally collected and centralized in water-treatment plants that collect the wastewater from a sewer system.

In many parts of the country, water resources are being stressed. As a result, innovative systems are being developed to reclaim wastewaters so that they can be used for such purposes as irrigating fields, parks, or golf courses, rather than being discharged into the nearest body of water. New technologies are also being developed to convert wastewaters into a resource that can be used. Those developing the new technologies say that sewage-treatment sites should not have to be hidden from people's eyes and noses. Rather, we should come to expect sewage to be reclaimed at a small cost while producing flowers and shrubs in a more gardenlike setting.[17]

Septic-Tank Sewage Disposal

In the United States, the population continues to move from rural to urban, or urbanizing, areas. Although municipal sewers and sewage-treatment facilities are the most effective method of sewage disposal, construction of an adequate sewage system has often not kept pace with growth. As a result, the individual *septic-tank disposal system* continues to be an important method of sewage disposal. There are more than 22 million systems in operation, and about half a million new systems are added each year. As a result, septic systems are used by about 30 percent of the people in the United States.[18] Not all land, however, is suitable for installation of a septic-tank disposal system, so evaluation of individual sites is necessary and often required by law before a permit can be issued.

The basic parts of a **septic tank** disposal system are shown in Figure 13.14. The sewer line from the house or small business leads to an underground septic tank in the yard. Solid organic matter settles to the bottom of the tank, where it is digested and liquefied by bacterial action. This is part of the treatment that reduces solid organic material to a more liquid state. The liquid wastewater discharges into either a drain field, also called an absorption field, which is a system of shallowly buried perforated piping, or a large-diameter, deep, gravel-filled "dry well" through which the wastewater seeps into the surrounding soil. As the wastewater moves through the soil, it is further treated and purified by natural processes of filtering and oxidation.

Geologic factors that affect the suitability of a septic-tank disposal system in a particular location include type of soil, depth to the water table, depth to bedrock, and topography. These variables are generally included in the soil descriptions found in the soil survey of an area. Soil surveys are published by the Soil Conservation Service and are extremely valuable when evaluating potential land use, such as suitability for a septic system. However, the reliability of a soils map for predicting the limitations of soils is confined to an area no larger than a few thousand square meters, and soil types can change within a few meters, so it is often necessary to have an on-site evaluation by a soil scientist or soils engineer. To calculate the size of the drain field needed, one must know the rate at which water infiltrates into the soil, which is best determined by testing the soil.

Sewage drain fields may fail for several reasons. The most common cause is poor soil drainage, which allows the wastewater to rise to the surface in wet

Figure 13.14 **Septic tank**
Sewage disposal system for a
home. (a) Plan (map) view.
(b) Cross section.

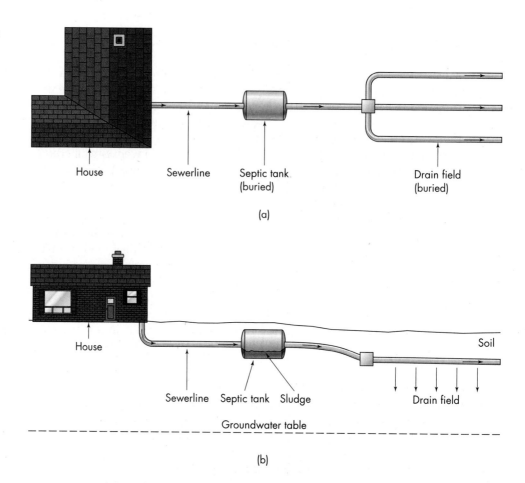

House   Sewerline   Septic tank (buried)   Drain field (buried)

(a)

House   Sewerline   Septic tank   Sludge   Soil   Drain field

Groundwater table

(b)

weather. Poor drainage can be expected to be present in areas with clay-rich soils or rock such as shale with low hydraulic conductivity (Chapter 12), in areas with a water table that is near the surface, or in areas of frequent flooding.

When septic systems fail, waste materials often surface above the drain field, producing a potential health hazard. This sort of failure is easy to see. Unfortunately, what is happening beneath the ground is not so easy to see, and if extensive leaching of waste occurs at the water table, then groundwater resources may be polluted. Septic systems that serve small commercial and industrial activities are of particular concern. These tend to cause severe problems of groundwater pollution because of the potentially hazardous nature of the waste being disposed of by these sytems. Possible contaminants include nutrients such as nitrates; heavy metals such as zinc, copper, and lead; and synthetic organic chemicals such as benzine, carbon tetrachloride, and vinyl chloride. In recent years, the EPA has identified a number of commercial and industrial septic systems that caused substantial water pollution that necessitated cleanup.[18]

Wastewater-Treatment Plants

The main purposes of wastewater treatment for municipal sewage from homes and industry are to break down and reduce the amount of organic solids and BOD and to kill bacteria in the wastewater. In addition, new techniques are being developed to remove nutrients and harmful dissolved inorganic materials that may be present.

Existing wastewater treatment generally has two or three stages (Figure 13.15):

- *Primary treatment.* This stage includes screening, which removes the grit composed of sand, stones, and other large particles; and sedimentation, in which much of the remaining particulate matter (mostly organic) settles out to form a mudlike sediment called sludge. The sludge is piped to the digester, and the partially clarified wastewater goes on to the secondary stage

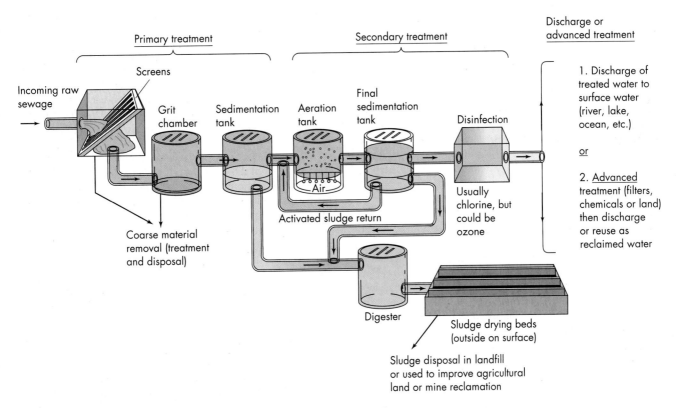

Figure 13.15 **Sewage treatment** Idealized diagram showing activated sludge sewage treatment with (or without) advanced treatment.

of treatment. Primary treatment removes 30 percent to 40 percent of the pollutants from the wastewater.[19]

- *Secondary treatment.* Although there are several methods, the most common secondary treatment is known as activated sludge. Wastewater from primary treatment enters the aeration tank, where air is pumped in and aerobic, or oxygen-requiring, bacteria break down much of the organic matter remaining in the waste. The wastewater is then pumped to the final sedimentation tank, where sludge settles out and is pumped to the digester. Some of the sludge, rich in bacteria, is recycled back to the aeration tank as "activated sludge" to act as a starter. Thus, the bacteria are used over and over again. The digester provides an oxygen-poor environment in which anaerobic bacteria, those that do not require oxygen, can break down organic matter in the sludge from both the primary and secondary sedimentation tanks. This anaerobic digestion produces methane gas, a by-product that can be used as a fuel to help heat or cool the plant or to run equipment. After secondary treatment, about 90 percent of the organic material (BOD) in the waste has been removed. However, this treatment does not remove nitrogen, phosphorus, heavy metals or some human-made chemicals, such as solvents and pesticides.[19] The final part of secondary treatment is disinfection of the wastewater. This is usually done with chlorine, but sometimes ozone is used. The treated wastewater is usually discharged to surface waters (rivers, lakes, or the ocean, where water quality problems can still occur; see A Closer Look: Boston Harbor—Cleaning up a National Treasure), but in some places it is discharged to disposal wells, as for example, in Maui, Hawaii.

- *Advanced treatment.* Used to remove nutrients, heavy metals, or specific chemicals, this additional stage may be required if higher-quality treated wastewater is needed for particular uses, such as for wildlife habitat or irrigation of golf courses, parks, or crops. The treated wastewater for such uses

A CLOSER LOOK | Boston Harbor—Cleaning up a National Treasure

The city of Boston is steeped in early American history. The names of Samuel Adams and Paul Revere immediately come to mind when in the late 1700s the colonies were struggling to obtain freedom from Britain. In 1773 Samuel Adams led a group of patriots aboard three British ships, dumping their cargo of tea into Boston Harbor. The issue that the patriots were emphasizing was what they believed to be an unfair tax on tea, and the event be came became known as the "Boston Tea Party." The tea dumped into the harbor by patriots in the Boston Tea Party did not pollute the harbor, but the growing city and dumping of all sorts of waste eventually did. For about 300 years Boston harbor has been a disposal site for dumping of sewage, treated wastewater, and water contaminated from sewer overflows during storms into the bay. Late in the twentieth century, court orders demanded that measures be taken to cleanup the bay.

After studying Boston Harbor and further offshore in Massachusetts Bay, it was decided to relocate the areas of discharge of waste (called outfalls) further offshore from Boston Harbor (Figure 13.B). Pollution of the harbor resulted because the waste that was being placed there moved into a small, shallow part of the Massachusetts Bay. Although there is vigorous tidal action between the harbor and the bay, the flushing time is about one week. The input of wastewater from the sewage outfalls was sufficient to cause water pollution. Study of Massachusetts Bay suggested that the outfalls, if placed further offshore where water is deeper and currents are stronger, would lower the pollution levels in Boston Harbor.[20]

Moving the wastewater outfall offshore was definitely a step in the right direction, but the long-term solution to pollutants entering the marine ecosystem will require additional measures. Pollutants in the water, even when placed further offshore with greater circulation and greater water depth, will eventually accumulate and cause environmental damage. As a result, any long-term solution must include source reduction of pollutants. To this end the Boston Regional Sewage Treatment Plan called for a new treatment plant designed to significantly reduce the levels of pollutants that are discharged into the bay. This acknowledges that dilution by itself cannot solve the urban waste management problem. Today the new $3.8 billion Deer Island Sewage Treatment Plant collects and treats sewage from 43 greater Boston communities. Moving the sewage outfall offshore when combined with source reduction of pollutants is a positive example of what can be done to better manage our waste and reduce environmental problems.[20]

Figure 13.B Boston **Harbor and Massachusetts Bay** showing old sewage outfalls (red squares) and new outfall (green rectangle) 15 km offshore. *(Modified after U.S. Geological Survey FS-185-97, 1997)*

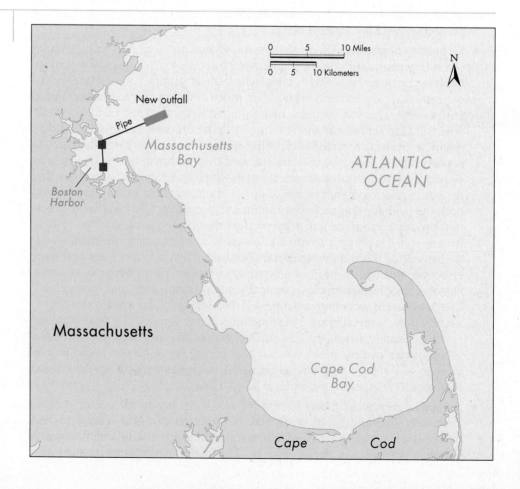

is often referred to as **reclaimed water.** Methods of advanced treatment include use of chemicals, sand filters, or carbon filters. After advanced treatment, up to 95 percent of the pollutants in the wastewater has been removed.

A troublesome aspect of wastewater treatment is the handling and disposal of sludge. Sludge generated from industrial wastewater may contain heavy metals as well as other toxic materials and is a hazardous waste. Many communities now require industry to pretreat their sewage to remove heavy metals before it is sent to a municipal treatment plant.

The amount of sludge produced in the treatment process is conservatively estimated at about 54 to 112 g per person per day, and sludge disposal accounts for 25 percent to 50 percent of the capital and operating cost of a treatment plant.

Sludge handling and disposal have four main objectives:[21]

- To convert the organic matter to a relatively stable form
- To reduce the volume of sludge by removing liquid
- To destroy or control harmful organisms
- To produce by-products, the use or sale of which reduces the cost of processing and disposal of the sludge

Final disposal of sludge is accomplished by incineration, burying it in a landfill, using it for soil reclamation, or dumping it in the ocean. From an environmental standpoint, the best use of sludge is to improve soil texture and fertility in areas disturbed by activity such as strip mining and poor soil conservation. Although it is unlikely that all of the tremendous quantities of sludge generated by large metropolitan areas can ever be used for beneficial purposes, many industries, institutions, and agricultural activities can take advantage of municipal and animal wastes by converting them into resources.

Wetlands as Wastewater-Treatment Sites

Natural and human-constructed wetlands are potentially good places to treat or partially treat wastewater or other poor-quality water. Wetland treatment is attractive to communities that have difficulty purchasing expensive traditional wastewater-treatment plants or desire an alternative to the traditional plants. The states of Louisiana and Arizona, with very different climates—warm-humid and hot-dry, respectively—have both had success using wetlands to treat wastewater.

Advanced treatment of wastewater in Louisiana, by applying nitrogen- and phosphorus-rich wastewater to coastal wetlands, is improving water quality as the wetland plants use these nutrients in their life cycle. In Louisiana, use of coastal wetlands to remove nutrients has resulted in considerable economic savings over use of conventional advanced wastewater treatment.[22]

Near Phoenix, Arizona, human-constructed wetlands are used to treat poor-quality agricultural wastewater with high nitrogen concentration. The artificial wetland is within a residential community and consists of ponds with wetland plants and bacteria that reduce the nitrogen to acceptable levels before discharging the water to a nearby river channel, where it seeps into the ground to become a groundwater resource.

Wastewater Renovation

The process of recycling liquid waste, called the **wastewater renovation and conservation cycle,** is shown schematically in Figure 13.16. The primary processes in the cycle (Figure 13.16) are: the return of treated wastewater to crops by a sprinkler or other irrigation system; renovation by natural purification of wastewater as it slowly seeps through soil to eventually recharge the groundwater resource with clean water; and reuse, or *conservation*, of the water by pumping it out of the

Figure 13.16 The wastewater renovation and conservation cycle
(From Pasizek, R. R., and Myers, E. A. 1968. Proceedings of the Fourth American Water Resources Conference)

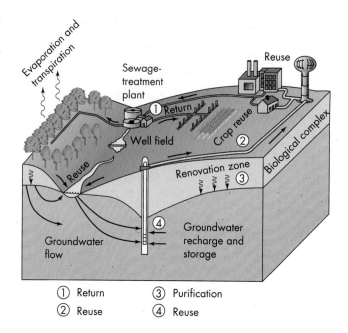

① Return ③ Purification
② Reuse ④ Reuse

ground for municipal, industrial, institutional, or agricultural purposes.[23] Of course, not all aspects of the cycle are equally applicable to a particular wastewater problem.

Wastewater renovation from cattle feedlots differs considerably from renovation of water from industrial or municipal sites. However, the general principle of renovation is valid, and the processes are similar in theory.

The return and renovation processes are crucial to wastewater recycling, and soil and rock type, topography, climate, and vegetation play significant roles. In the planning of return and renovation processes, it is particularly important that the soil be able to safely assimilate waste, that the selected vegetation can use the nutrients in the wastewater, and that the volume of wastewater that can safely be applied to the area is known.[23]

In Clayton County, Georgia, just south of Atlanta, a large water renovation and conservation cycle project was recently initiated. The project handles wastewater that is applied to a pine forest. Trees will be harvested on a 20 year rotation. The forest is part of the watershed that supplies water to the area; therefore, wastewater is recycled to become part of the drinking-water supply.[24]

13.7 Federal Legislation

In the United States, the mid-1990s was a time of debate and controversy regarding water pollution. Republicans, in control of the House of Representatives and the Senate since 1994, attempted to rewrite major environmental laws, including the Clean Water Act of 1972 (amended in 1977). Their purpose was to provide industry greater flexibility in choosing how to comply with environmental regulations concerning water pollution. Industry is in favor of proposed new regulations that, in their estimation, would be more cost effective without causing an increase in environmental degradation. Environmentalists viewed the attempts to rewrite the Clean Water Act as a step backward in the nation's 20 year fight to clean up our water resources. Apparently, the Republican majority misinterpreted the public's position on this issue. Survey after survey established that there is strong support for a clean environment in the United States today. People are willing to pay to have clean air and clean water for this and future generations. There has been a strong backlash and criticism of steps taken to weaken environmental laws, particularly as they pertain to important resources such as water. Thus the debate went on. The Water Quality Act of 1987 established national policy to control nonpoint sources of water pollution, but it did not go far enough in protecting water resources.[25]

In July 2000, President Clinton, in defiance of Congress, imposed new water pollution controls to protect thousands of streams and lakes from nonpoint sources of agricultural, industrial, and urban pollution. The rules are administered by the Environmental Protection Agency, which works with states and local communities to develop detailed programs with the objective of reducing pollution to those streams, rivers, lakes, and estuaries that do not now meet the minimum standards of water quality. The rules imposed demonstrate that nonpoint sources of water pollution are recognized as a serious problem that is difficult to regulate. The plan will take at least 15 years to implement completely. It has been opposed for years by Congress as well as by some agricultural groups, the utility industry, and even the U.S. Chamber of Commerce. In 2001, the newly elected President Bush temporarily suspended some regulations, including control of arsenic in water, over the objections of many who were concerned that some people were drinking contaminated water. The primary objections of those who oppose controls are that (1) the requirements would be costly (billions of dollars) and (2) local and state governments are better suited to implement water pollution rules. Nevertheless, the regulations are now in place and are another step in water pollution control measures started many years ago in the United States.

13.8 What Can Be Done to Reduce Effects of Water Pollution?

This chapter has documented a long list of water pollutants from disease-causing organisms to chemical pollutants. We now turn to what is needed to better address water pollution problems. Several suggestions follow:[3]

- Develop and refine better ways to evaluate water pollution problems and their impact on aquatic life and the health of people.
- Implement new and innovative, cost-effective water treatment technologies.
- Develop products and processes that minimize production of water pollutants and their release into the environment.

SUMMARY

Water pollution is the degradation of water quality as measured by physical, chemical, or biological criteria. These criteria take into consideration the intended use for the water, departure from the norm, effects on public health, and ecological impacts. The most serious water pollution problem today is the lack of noncontaminated drinking water for billions of people.

Surface-water pollutants have point or nonpoint sources. The major water pollutants are oxygen-demanding waste, measured by biochemical oxygen demand (BOD); pathogens, measured by the fecal coliform bacteria count; nutrients that lead to eutrophication, in which overgrowth of algae deprives water of oxygen and sunlight; oil; toxic substances, including synthetic organic and inorganic compounds, heavy metals, and radioactive materials; heat; and sediment. Acid mine drainage refers to water with a high concentration of sulfuric acid that drains from some coal or metal mining areas, causing surface water and groundwater pollution.

Development of water-quality standards in the United States has been mandated by federal legislation and involves setting of maximum contaminant levels (MCLs) for contaminants that might be found in our drinking water. The major purposes of the standards are to ensure that our water supply is treated to remove harmful contaminants and that water quality is regularly tested and monitored.

Wastewater-treatment facilities include septic-tank sewage disposal systems and wastewater-treatment plants. Septic-tank systems, used by homes and small commercial and industrial activities, are very common in the United States today. Failure of these systems may cause significant pollution to groundwater resources. Wastewater treatment plants collect and process water from municipal sewage systems. The use of reclaimed water is growing fast in the United States today, particularly in areas where water shortages are most likely to occur.

Revisiting Fundamental Concepts

Human Population Growth

As human population has grown, particularly in large urban regions, ground- and surface waters are often polluted. Pollution reduces the water resources available and makes it necessary in some urban areas to import water, sometimes from far away. As human population continues to increase during the next decades, we will need to find creative ways to avoid continual pollution of water resources that we depend on. Groundwater is particularly vulnerable and difficult to treat once polluted.

Sustainability

At the very heart of sustainability is the desire to pass on a quality environment to future generations. Certainly, a quality environment includes the water we drink and use to grow our crops. It is apparent that our present management practices to maintain high water quality are not sustainable. That is, if we continue present practices, more and more of our water resources will become contaminated and polluted, becoming unfit for human use. Therefore, it is of utmost importance to find ways to use water resources responsibly in a sustainable way. Recent regulations and laws to restrict water pollutants are steps in the right direction.

Earth as a System

Our water resources are part of complex hydrologic, physical, chemical, and biological systems that are prone to change and disturbance. When we introduce pollutants into our surface water or groundwater resources, we cause further changes to ecosystems that support people and other living things. We have learned that we cannot do just one thing. For example, when we introduce toxic materials, such as MTBE, into the groundwater, we cause far-reaching changes to the environment. These changes often cause further changes.

Hazardous Earth Processes, Risk Assessment, and Perception

The largest and most significant water pollution hazard is the lack of noncontaminated water for billions of people on Earth. Epidemics caused by poor water quality and pathogenic organisms have been nearly eliminated in developed countries, but they are still a serious problem around the world where people do not have the financial ability to treat water before it is consumed. We have the perception in the United States that our water supply is clean and free of disease-causing agents. From time to time, though, we learn differently, and, although outbreaks of disease have been isolated, there is concern for our future supply of clean drinking water.

Scientific Knowledge and Values

The role of particular toxins and pollutants in our water resources is a science still in its infancy. In particular, we have trouble making decisions concerning what the toxic level of a particular chemical is and what the effects are likely to be in the future. How we decide to deal with living with this uncertainty is in part a function of our value system. We may choose to be conservative and restrict the use of potentially toxic chemicals, or we can assume that our present practices will not cause problems to future generations. We have learned in the past that making the assumption that problems will not result from a particular practice has been costly. For example, a decision was recently made not to control arsenic in groundwater used by people in parts of California and other areas. This decision was made despite the fact that arsenic is known to be very toxic. Part of the reason for choosing not to restrict the level of arsenic in drinking water was that treatment of the water to remove the arsenic would be overly costly to society. The decision not to remove the arsenic constitutes an experiment on how the arsenic in the environment will affect people and ecosystems.

Key Terms

acid mine drainage (p. 426)

advanced treatment (p. 435)

biochemical oxygen
demand (BOD) (p. 416)

cultural eutrophication (p. 419)

fecal coliform bacteria (p. 417)

nonpoint sources (p. 425)

point sources (p. 425)

pollutant (p. 415)

pollution (p. 425)

primary treatment (p. 434)

reclaimed water (p. 437)

secondary treatment (p. 435)

septic tank (p. 433)

wastewater renovation and
conservation cycle (p. 437)

water pollution (p. 415)

Review Questions

1. Define water pollution.

2. Define biochemical oxygen demand.

3. What is the role of fecal coliform bacteria in determining water pollution?

4. Define cultural eutrophication.

5. Differentiate between point source and nonpoint sources of water pollution.

6. What is acid mine drainage?

7. What is saltwater intrusion?

8. Differentiate among primary, secondary, and advanced wastewater treatment.

9. What are some of the ways that a septic-tank disposal system may fail?

10. Define the wastewater renovation and conservation cycle.

Critical Thinking Questions

1. For your community, develop an inventory of point and nonpoint sources of water pollution. Carefully consider how each of these might be eliminated or minimized as part of a pollution abatement strategy.

2. Visit a wastewater-treatment plant. What processes are used at the plant, and could the concept of resource recovery or wastewater renovation and conservation be used? What would be the advantages and disadvantages of using a biological system such as plants as part of the wastewater-treatment procedures?

3. How safe do you think your water supply is? Upon what are you basing your answer? What do you need to know to give an informed answer?

FOURTEEN

Mineral Resources

Learning Objectives

Our modern society completely depends on the availability of mineral resources. As world population increases, we face an ever-increasing resource crisis. It is feared that Earth may have reached its capacity to absorb environmental degradation related to mineral extraction, processing, and use. In this chapter, we will focus on the following learning objectives:

- Understand the relationship between human population and resource utilization

- Understand why minerals are so important to modern society

- Understand the difference between a resource and a reserve and why that difference is important

- Know some of the factors that control the availability of mineral resources

- Understand the environmental impact of mineral development

- Know the potential benefits that biotechnology can offer to environmental cleanup associated with mineral extraction and production

- Understand the economic and environmental role of recycling mineral resources

- Understand the relationship between sustainability and mineral use

Fossil Trace Golf Course was, until recently, a mining site for clay. *(Courtesy Fossil Trace Golf Course)*

CASE HISTORY | Mine Near Golden, Colorado, Is Transformed into a Golf Course

An award-winning golf course is now located on land that was for about 100 years an open-pit mine (quarry) in limestone rock near the city of Golden, Colorado. The mine produced clay from layers between the limestone beds to make bricks. The bricks were used as a building material for buildings in the Denver area, including the Colorado governor's mansion. The mining site included unsightly pits with vertical limestone walls as well as a landfill for waste disposal. The area has spectacular views of the foothills and the Rocky Mountains. Today the limestone cliffs left by the mining with their exposed plant and dinosaur fossils have been transformed into golf greens and fairways. The name of the golf course is Fossil Trace, reflecting its geologic

heritage. Trails lead to the best locations to see fossils—an added incentive to visit the area. Constructed wetlands and three lakes store runoff of floodwater, helping protect Golden from flash floods. The reclamation project started with the desire to have a public golf course. The mine reclamation demonstrates that previous mines can be reclaimed and transformed into valuable property.

Fossil Trace Golf Club is a unique instance of mine reclamation. However, each potential site for restoration offers opportunities based on local physical, hydrological, and biological conditions. This chapter discusses the origin of mineral deposits, environmental consequences of mineral development, and sustainable mineral use.

14.1 Minerals and Human Use

Our society depends on the availability of mineral resources.[1] Consider the mineral products found in a typical American home (Table 14.1). Specifically, consider your breakfast this morning. You probably drank from a glass made primarily of quartz-sand, ate food from dishes made from clay, flavored your food with salt mined from Earth, ate fruit grown with the aid of fertilizers such as potassium carbonate (potash) and phosphorus, and used utensils made from stainless steel, which comes from processing iron ore and other minerals. If you read a magazine or newspaper while eating, the paper was probably made using clay fillers. If the phone rang and you answered, you were using more than 40 minerals in the telephone. When you went to school or work, you may have turned on a computer or other equipment made largely of minerals.[2,3]

Minerals are extremely important to people; all other things being equal, one's standard of living increases with the increased availability of minerals in useful

TABLE 14.1   A Few of the Mineral Products in a Typical American Home

Building materials	Sand, gravel, stone, brick (clay), cement, steel, aluminum, asphalt, glass
Plumbing and wiring materials	Iron and steel, copper, brass, lead, cement, asbestos, glass, tile, plastic
Insulating materials	Rock, wool, fiberglass, gypsum (plaster and wallboard)
Paint and wallpaper	Mineral pigments (such as iron, zinc, and titanium) and fillers (such as talc and asbestos)
Plastic floor tiles, other plastics	Mineral fillers and pigments, petroleum products
Appliances	Iron, copper, and many rare metals
Computers, phones, videos	Petroleum products, and many minerals
Furniture	Synthetic fibers made from minerals (principally coal and petroleum products); steel springs; wood finished with mineral varnish
Clothing	Natural fibers grown with mineral fertilizers; synthetic fibers made from minerals (principally coal and petroleum products)
Food	Grown with mineral fertilizers; processed and packaged by machines made of metals
Drugs and cosmetics	Mineral chemicals
Other items	Windows, screens, light bulbs, porcelain fixtures, china, utensils, jewelry: all made from mineral products

Source: U.S. Geological Survey Professional Paper 940, 1975.

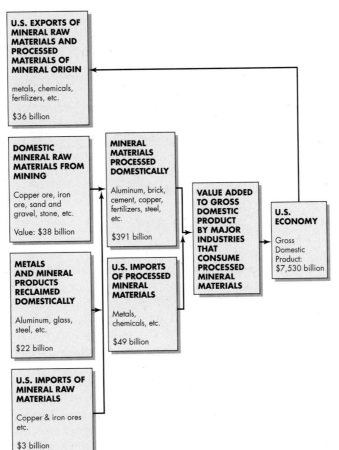

Figure 14.1 The role of nonfuel minerals in the U.S. economy (for 1996). Since 1996 the U.S. import of processed mineral products from countries such as China and India has increased dramatically. *(U.S. Geological Survey 1997. Accessed on 1/15/07 @minerals.usgs.gov.)*

forms. Furthermore, the availability of mineral resources is one measure of the wealth of a society. Societies that have been successful in the location, extraction, or importation and use of minerals have grown and prospered. Without mineral resources, modern technological civilization as we know it would not be possible.

The important role of nonfuel minerals in the U.S. economy is shown in Figure 14.1. The data in this diagram suggest that[4]

- Processed materials from minerals have an annual value of several hundred billion dollars, which is about 5 percent of the U.S. gross domestic product.

- The value of reclaimed metals and mineral products is about one-half of the value of domestic mineral raw materials, a significant contribution.

Minerals can be considered our nonrenewable heritage from the geologic past. Although new deposits are forming from present Earth processes, these processes are too slow to be of use to us today. Mineral deposits tend to be hidden in small areas. Deposits must therefore be discovered; unfortunately, most of the easy-to-find deposits have already been exploited. If our civilization, with its science and technology, were to vanish, future civilizations would have a much harder time discovering minerals than our ancestors did. Unlike biological resources, which are renewable, minerals are consumed and cannot be managed with the assumption that Earth processes will replace the source in a useful period of time. Recycling and conservation will help extend our mineral resources, but eventually the supply will be exhausted.

Resources and Reserves

Mineral resources can be defined broadly as elements, compounds, minerals, or rocks concentrated in a form that can be extracted to obtain a usable commodity. From a practical viewpoint, this definition is unsatisfactory because a resource will not normally be extracted unless extraction can be accomplished at a profit.

Figure 14.2 **Resources and reserves** Classification of mineral resources used by the U.S. Geological Survey and the U.S. Bureau of Mines. *(After U.S. Geological Survey Circular 831, 1980)*

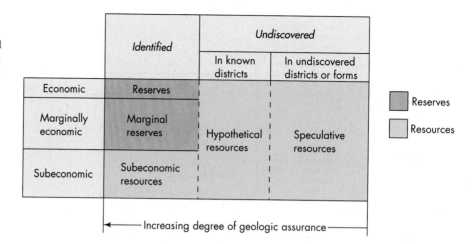

A more pragmatic definition is that a mineral **resource** is a concentration of a naturally occurring material (solid, liquid, or gas) in or on the crust of Earth in a form that can now or *potentially* be extracted at a profit. A **reserve** is that portion of a resource that is identified and is *currently* available to be legally extracted at a profit. The distinction between resources and reserves, therefore, is based on current geologic, economic, and legal factors (Figure 14.2).[5]

Not all resource categories are reserves (Figure 14.2). As an analogy to help clarify this point, think about your personal finances. Your reserves are your liquid assets, such as money in your pocket or in the bank, whereas your resources include the total income you can expect to earn during your lifetime. This distinction is often critical to you because resources are "frozen" assets or next year's income and cannot be used to pay this month's bills.[5]

Silver's value as a reserve can illustrate some important points about resources and reserves. Earth's crust contains almost 2 trillion metric tons of silver. This is Earth's crustal resource of silver—an amount much larger than the annual world use, approximately 10,000 metric tons. If this silver existed as pure metal concentrated into one large mine, it would represent a supply sufficient for several hundred million years at current levels of use. Most of this silver, however, exists in extremely low concentrations—too low to be extracted economically with current technology. The known reserve of silver, reflecting the amount we could obtain immediately with known techniques, is about 200,000 metric tons, or a 20 year supply at current use levels.

Availability and Use of Mineral Resources

The availability of a mineral in a certain form, in a certain concentration, and in a certain total amount at that concentration is determined by Earth's history; it is a geologic issue that we will consider in Section 14.2. Types of mineral resources and the limits of their availability are technological and social questions that we will consider here.

Types of Mineral Resources. Some mineral resources are necessary for life. One example is salt, or halite (NaCl). Primitive peoples traveled long distances to obtain salt when it was not available locally. Other mineral resources, such as diamonds, are desired for their beauty, and many more are necessary for producing and maintaining a certain level of technology.

Earth's mineral resources can be divided into several broad categories based on how we use them:

- Elements for metal production and technology, which can be classified according to their abundance. The abundant metals include iron, aluminum, chromium, manganese, titanium, and magnesium. Scarce metals include copper, lead, zinc, tin, gold, silver, platinum, uranium, mercury, and molybdenum.

- Building materials such as *aggregate* sand, gravel, and crushed stone for concrete; clay for tile; and volcanic ash for cinder blocks to construct walls of buildings.

- Minerals for the chemical industry, such as the many minerals used in the production of petrochemicals, which are materials produced from natural gas or crude oil, such as plastics.

- Minerals for agriculture, such as fertilizers.

When we think of mineral resources, we usually think of the metals used in structural materials, but, in fact, with the exception of iron, the predominant mineral resources are not metals. Consider the annual world consumption of a few selected elements. Sodium and iron are used at a rate of approximately 0.1 billion to 1 billion tons per year. Nitrogen, sulfur, potassium, and calcium are used at a rate of approximately 10 million to 100 million tons per year. These four elements are used primarily as soil conditioners or fertilizers. Zinc, copper, aluminum, and lead have annual world consumption rates of about 3 million to 10 million tons, whereas gold and silver have annual consumption rates of 10,000 tons or less. Of the metallic minerals, iron makes up 95 percent of all the metals consumed; nickel, chromium, cobalt, and manganese are used mainly in alloys of iron, such as stainless steel, which are mixtures of iron and other elements. Therefore, we can conclude that nonmetallic mineral resources, with the exception of iron, are consumed at much greater rates than elements used for their metallic properties.

Responses to Limited Availability. The fundamental problem associated with the availability of mineral resources is not actual exhaustion or extinction; rather, it is the cost of maintaining an adequate reserve, or stock, within an economy through mining and recycling. At some point, the costs of mining exceed the worth of the material. When the availability of a particular mineral becomes a limitation, several solutions are possible:

- Find more sources.
- Find a substitute.
- Recycle what has already been obtained.
- Use less and make more efficient use of what we have.
- Do without.

We can use a particular mineral resource in several ways: rapid consumption, consumption with conservation, or consumption and conservation with recycling. The option selected depends in part on economic, political, and social criteria. Figure 14.3 shows the hypothetical depletion curves corresponding to these three options. Historically, resources have been consumed rapidly, with the exception of

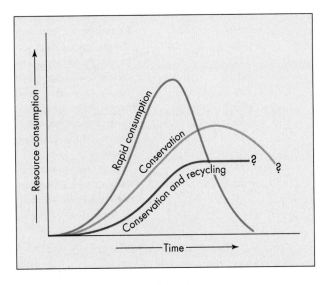

Figure 14.3 **Depletion curves** Diagram of three hypothetical depletion curves for the use of mineral resources.

TABLE 14.2   U.S. Import Reliance (by Percentage) for Selected Nonfuel Mineral Resources, 2001

Mineral	U.S. Reliance on Import From (%)	Major Sources (1996–1999)[1]
Arsenic trioxide	100	China, Chile, Mexico
Asbestos	100	Canada
Bauxite and alumina	100	Australia, Guinea, Jamaica, Brazil
Columbium (niobium)	100	Brazil, Canada, Germany, Russia
Fluorspar	100	China, South Africa, Mexico
Graphite (natural)	100	China, Mexico, Canada
Manganese	100	South Africa, Gabon, Australia, France
Mica, sheet (natural)	100	India, Belgium, Germany, China
Quartz crystal	100	Brazil, Germany, Madagascar
Strontium	100	Mexico, Germany
Thallium	100	Belgium, Canada, Germany, United Kingdom
Thorium	100	France
Yttrium	100	China, Hong Kong, France, United Kingdom
Gemstones	99	Israel, India, Belgium
Bismuth	95	Belgium, Mexico, United Kingdom, China
Antimony	94	China, Mexico, South Africa, Bolivia
Tin	86	China, Brazil, Peru, Bolivia
Platinum	83	South Africa, United Kingdom, Russia, Germany
Stone (dimension)	80	Italy, Canada, Spain, India
Tantalum	80	Australia, China, Thailand, Japan
Chromium	78	South Africa, Kazakhstan, Russia, Zimbabwe
Titanium concentrates	76	South Africa, Australia, Canada, India
Cobalt	74	Norway, Finland, Zambia, Canada
Rare earths	72	China, France, Japan, United Kingdom
Barite	71	China, India, Mexico, Morocco
Potash	70	Canada, Russia, Belarus
Iodine	69	Chile, Japan, Russia
Tungsten	68	China, Russia, Bolivia
Titanium (sponge)	62	Russia, Japan, Kazakhstan, China
Zinc	60	Canada, Mexico, Peru
Nickel	58	Canada, Norway, Russia, Australia
Peat	52	Canada
Silver	52	Canada, Mexico, Peru
Silicon	48	Norway, South Africa, Russia, Canada
Diamond (dust, grit, and powder)	47	Ireland, China, Russia
Magnesium compounds	45	China, Canada, Austria, Australia
Magnesium metal	40	Canada, Russia, China, Israel
Copper	37	Canada, Chile, Mexico
Beryllium	35	Russia, Canada, Kazakhstan, Germany
Aluminum	33	Canada, Russia, Venezuela, Mexico
Pumice	33	Greece, Turkey, Ecuador, Italy
Lead	24	Canada, Mexico, Peru, Australia
Gypsum	22	Canada, Mexico, Spain
Sulfur	22	Canada, Mexico, Venezuela
Nitrogen (fixed), Ammonia	21	Trinidad and Tobago, Canada, Mexico, Venezuela
Cement	20	Canada, China, Spain, Venezuela
Iron ore	19	Canada, Brazil, Venezuela, Australia
Iron and steel	17	European Union, Canada, Japan, Mexico
Mica, scrap and flake (natural)	17	Canada, India, Finland, Japan
Perlite	15	Greece
Salt	15	Canada, Chile, Mexico, The Bahamas
Talc	12	China, Canada, France, Japan
Cadmium	6	Canada, Belgium, Australia
Phosphate rock	1	Morocco

[1]In descending order of import share.

Source: U.S. Geological Survey. *Minerals Information, 2001.* Mineral Commodity Summaries 2001.

precious metals. However, as more resources become limited, increased conservation and recycling are expected. Certainly the trend toward recycling is well established for metals such as copper, lead, and aluminum.

As the world population and the desire for a higher standard of living increase, the demand for mineral resources expands at a faster and faster rate. The more developed countries in the world contain only 16 percent of Earth's population, yet they consume a disproportionate share of mineral resources. For example, the vast majority of the aluminum, copper, and nickel extracted is used by the United States, Japan, China, and Western Europe.[6] If the world per capita consumption rate of these metals were to rise to the U.S. level, production would have to increase to several times the present rate. Because such an increase in production is very unlikely, affluent countries will have to find substitutes for some minerals or use a smaller proportion of the world annual production. Fortunately, these alternatives are being implemented. For example, in the United States, per capita nonfuel mineral consumption decreased significantly between about 1980 and 2000.

Domestic supplies of many mineral resources in the United States and other affluent nations are insufficient for current use and must be supplemented by imports from other nations. Table 14.2 shows the U.S. reliance on importation of selected nonfuel minerals and the major foreign sources for the needed minerals. Industrial countries are particularly concerned about the possibility that the supply of a much desired or needed mineral may become interrupted by political, economic, or military instability of the supplying nation. The United States, along with many other countries, is currently dependent on a steady supply of imports to meet the mineral demand of industries. Of course, the fact that a mineral is imported into a country does not mean that it does not exist within the country in quantities that could be mined. Rather, it suggests that there are economic, political, or environmental reasons that make it easier, more practical, or more desirable to import the material.

14.2 Geology of Mineral Resources

The geology of mineral resources is intimately related to the rock cycle (see Chapter 3). Nearly all aspects and processes of the cycle as well as biogeochemical processes are involved to a lesser or greater extent in producing local concentrations of useful materials.

Local Concentrations of Metals

The term **ore** is sometimes used for useful metallic minerals that can be mined at a profit; locations where ore is found have anomalously high concentrations of these minerals. The concentration of metal necessary for a particular mineral to be classified as an ore varies with technology, economics, and politics. Before smelting, extraction of metal by heat, was invented, the only metal ores available were those in which the metals appeared in their pure form; for example, gold was originally obtained as a pure, or native, metal. Now gold mines extend deep beneath the surface, and the recovery process reduces tons of rock to ounces of gold. Although the rock contains only a minute amount of gold, we consider it a gold ore, because we can extract the gold profitably.

The **concentration factor** of a metal is a measure of its concentration necessary for profitable mining. It is how many times the average crustal concentration of a metal must be increased to be mined at a profit. For example, aluminum has an average concentration of about 8 percent in Earth's crust, but it needs to be found at concentrations of about 30 percent to be mined economically, giving it a concentration factor of about 4. Mercury, on the other hand, has an average concentration of only a tiny fraction of 1 percent (0.00001%) and must have a concentration factor of about 10,000 to be mined economically. Nevertheless, mercury ores (rocks with at least 0.1 percent mercury) are common in certain regions, where they and other metallic ores are deposited by tectonic processes (see A Closer Look: Plate Tectonics and Minerals). Finally, gold is naturally

A CLOSER LOOK | Plate Tectonics and Minerals

In a broad-brush approach to the geology of mineral resources, tectonic plate boundaries are related to the origin of ore deposits such as iron, gold, copper, and mercury. The ores are formed by processes operating at the plate boundaries (see Chapter 2).

The origin of metallic ore deposits at divergent plate boundaries is related to the migration of ocean water. Cold, dense ocean water moves down through numerous fractures in the basaltic rocks at oceanic ridges and is heated by contact with heat from nearby molten rock (Figure 14.A, part a). The warm water is lighter and more chemically active than the cold water. The warm water moves through the fractured

rocks and leaches out metals, which are carried in solution and precipitated as metallic sulfides at black smokers, so called because of their dark color (Figure 14.A, parts b and c).[7] Many hot-water vents with sulfide deposits have been discovered along oceanic ridges, and undoubtedly numerous others will be located. These deposits on the ocean floor are known as massive sulfide deposits because they are composed of unusually high concentrations of iron and copper sulfide minerals. To date, all such known deposits on the seabed at divergent plate boundaries are too small in volume to be mined at a profit. However, on land, massive sulfide deposits, formed by tectonic processes that squeezed up

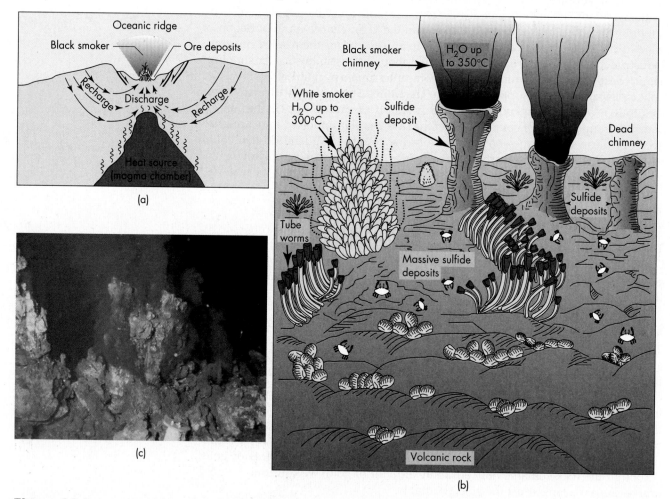

Figure 14.A **Ore deposits on divergent plate boundaries** (a) Oceanic ridge hydrothermal system. (b) Detail of black smoker where massive sulfide deposits form. (c) Photograph of black smoker. *(R. Haymon)*

concentrated in the crust at about one two-hundredth the concentration of mercury. However, because it is so valuable (over $600/oz in 2006) a rock (ore) with only 0.001 percent gold can be mined at a profit. Thus the concentration factor for gold is about 5,000, about half that of mercury. The percentage of a metal in a rock necessary for profitable mining, and thus its concentration factor, is subject to change as the demand for the metal changes. For example, if both the

oceanic crust and even mantle rocks, were mined several thousand years ago on the island of Cyprus. Interestingly, the word *copper* is derived from the Greek word *cyprus*.[8]

The origin of metallic ore deposits at convergent plate boundaries (subduction zones; see Chapter 2) is hypothesized as being the result of the partial melting of seawater-saturated rocks of the oceanic lithosphere in a subduction zone. The high heat and pressure that cause the rocks to melt also facilitate the release and movement of metals from the partially molten rock. The metals become concentrated and ascend as metal-rich fluids in the magma. The metal-rich fluids are eventually released from the magma, and the metals are deposited in a host rock.[7]

Perhaps the best example of metallic deposits at subduction or volcanic zones is the global occurrence of known mercury deposits (Figure 14.B). The belts of productive mercury deposits are associated with volcanic systems, located near convergent plate boundaries. It has been suggested that the mercury, originally found in oceanic sediments of the crust, is distilled out of the downward-plunging plate and deposited at a higher level above the subduction zone. Convergent plate junctions characterized by volcanism and tectonic activities are therefore likely places to find mercury. A similar argument can be made for other ore deposits; however, there is danger in oversimplification since many deposits are not directly associated with plate boundaries.

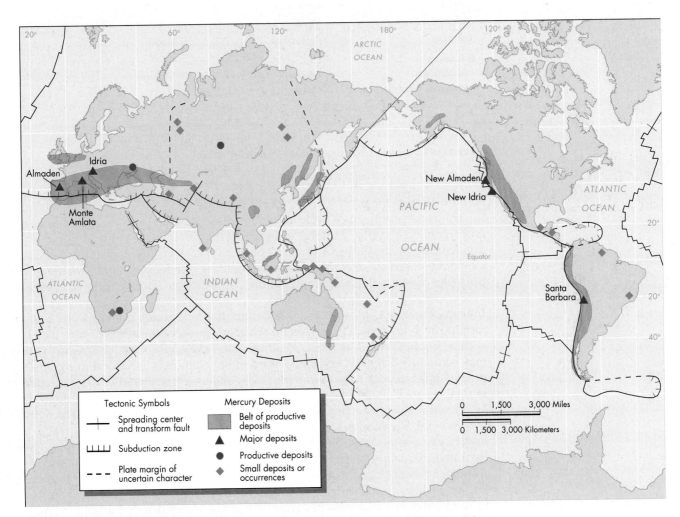

Figure 14.B **Mercury deposits** Relationship between mercury deposits and recently active subduction zones. *(From Brobst, D. A., and Pratt, W. P., eds. 1973. U.S. Geological Survey Professional Paper 820)*

demand for and the price of a particular metal increase, then rocks with a lower percentage of the metal may be mined.

Genesis of Some Common Mineral Resources

The geology of economically useful deposits of mineral and rock materials is as diversified and complex as the processes responsible for their formation or

TABLE 14.3 Types of Mineral Resources

Type	Example	Location
Igneous		
Disseminated	Diamonds	South Africa
Crystal settling[1]	Chromite	Stillwater, Montana
Late magmatic	Magnetite	Adirondack Mountains, New York
Pegmatite[2]	Beryl and lithium	Black Hills, South Dakota
Hydrothermal	Copper	Butte, Montana
Metamorphic		
Contact metamorphism	Lead and silver	Leadville, Colorado
Regional metamorphism	Asbestos	Quebec, Canada
Sedimentary		
Evaporite (lake or ocean)	Potassium	Carlsbad, New Mexico
Placer (stream)	Gold	Sierra Nevada foothills, California
Glacial	Sand and gravel	Northern Indiana
Deep ocean	Manganese oxide nodules	Central and southern Pacific Ocean
Biological	Phosphorus	Florida
Weathering		
Residual soil	Bauxite	Arkansas
Secondary enrichment	Copper	Utah

[1]Heavier crystals sink to bottom of magma.
[2]Very coarse-grained igneous rock.

Source: Modified from Foster, R. J. 1983. *General geology*, 4th ed. Columbus, OH: Charles E. Merrill

accumulation in the natural environment. Most deposits, however, can be related to various parts of the rock cycle under the influence of the tectonic, geochemical, and hydrologic cycles. Mineral resources with commercial value can be subdivided into several categories based on the type of process that formed them, including igneous, metamorphic, sedimentary, biological, and weathering processes. Table 14.3 lists examples of ore deposits from each of these categories.

Igneous Processes. Most of the world's ore deposits result from igneous rock–forming and enrichment processes that concentrate economically desirable metals such as copper, nickel, or gold. In some cases, the entire mass of igneous rock may contain *disseminated* crystals. For example, diamonds are found in a coarse-grained igneous rock called kimberlite, where they are scattered or disseminated within the rock (Figure 14.4). Perhaps the most common type of ore deposits associated with igneous processes are *hydrothermal deposits*. Hydrothermal activity involves hot, chemically active fluid associated with magma that gives rise to a variety of minerals including gold, silver, copper, mercury, lead, and zinc. The hydrothermal solutions that form ore deposits are mineral-rich fluids that migrate through a host rock and recrystallize the ore minerals as veins or small intrusions known as dikes (see Chapter 3). It is speculated that circulating groundwater that is heated and enriched with minerals after contact with deeply buried magma is also responsible for some of these deposits (Figure 14.5).[6]

Metamorphic Processes. Cooling magma interacts with the surrounding rock it has intruded and often causes elevated heat and pressure and the production of chemically active fluids; this process causes changes (metamorphism) in the surrounding rock. Metamorphism that occurs on a local scale of a few meters to a few hundreds of meters from contact with magma is *contact metamorphism;* metamorphism that occurs on a regional scale of thousands of square kilometers is *regional metamorphism*. Both types can produce a variety of mineral deposits, including both metallic deposits and nonmetallic deposits, such as asbestos and talc.

Kimberlite pipe

Lake sediments

Intrusive
kimberlite
breccia

Massive
kimberlite

Diamond mine

Waste dumps

Open pit

Mine
shaft

Broken ore

Ore is drawn
out here

Ore bin

Crusher

Figure 14.4 Diamond pipe
(a) Idealized diagram showing a typical
South African diamond pipe and mine.
Diamonds are scattered throughout
the cylindrical body of igneous rock,
kimberlite. *(From Kesler, S. E. 1994.
Mineral resources, economics and the environment. New York: Macmillan)* (b) Aerial
view of Diamond Mine, Kimberly,
South Africa. This is one of the largest
hand-dug excavations in the world.
*(Helen Thompson/Animals Animals/Earth
Scenes)*

(a)

(b)

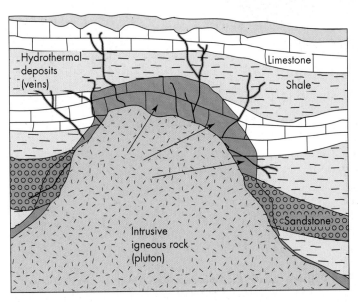

Hydrothermal
deposits
(veins)

Limestone

Shale

Intrusive
igneous rock
(pluton)

Sandstone

Contact metamorphic zone
where mineral deposits may be
present. Notice the zone is
wider in the limestone
than in the sandstone or shale.
This results because limestone
is chemically more active
under contact metamorphism.

Figure 14.5 Igneous and metamorphic ore deposits Formation of hydrothermal (hot-water) and
contact metamorphic ore deposits. Hydrothermal ore deposits generally appear as veins that form when
chemicals in the hydrothermal solutions crystallize as minerals. Contact metamorphic ore deposits form as
the result of elevated temperature in chemically active rocks in contact with hot magma from an intrusive
body known as a pluton.

453

Sedimentary Processes. Sediment deposits resulting from physical or chemical processes produce a variety of important mineral resources in the United States. Sand and gravel deposits produced by stream processes constitute a multibillion-dollar industry. Most sand and gravel is obtained from river deposits and water-worked *glacial deposits.* Another important sedimentary resource is **evaporite deposits,** which result when shallow marine basins or lakes dry up. As evaporation progresses, the dissolved materials in the ocean or lake precipitate, become solid, forming a wide variety of compounds and minerals that have important commercial value. These include halite, or common salt (NaCl); gypsum ($CaSO_4 \cdot 2\,H_2O$) used for industrial purposes; and potassium minerals, used for a variety of industrial and agricultural activities (Figure 14.6). Finally, stream processes may concentrate a variety of heavy metals, including gold, weathered from rocks. These are known as *placer deposits* (Figure 14.7).

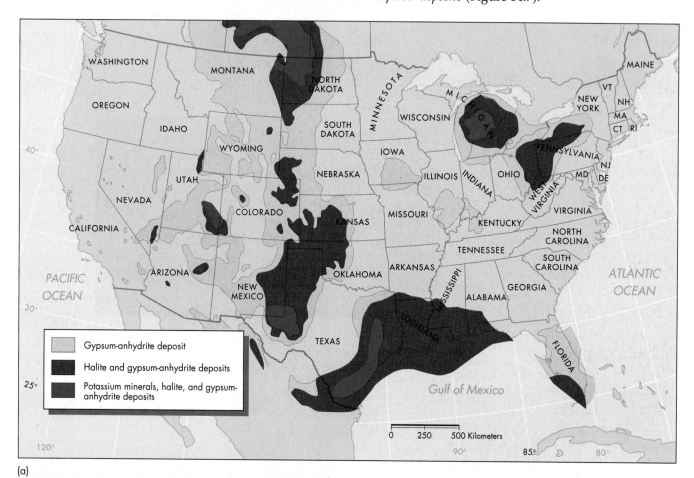

Gypsum-anhydrite deposit

Halite and gypsum-anhydrite deposits

Potassium minerals, halite, and gypsum-anhydrite deposits

(a)

(b)

Figure 14.6 **Evaporite deposits** (a) Marine evaporite deposits in the United States. *(After Brobst, D. A., and Pratt, W. P., eds. 1973. U.S. Geological Survey Professional Paper 820)* (b) Salt deposits, Death Valley, California. Salt is one of several common evaporite minerals that are deposited when a water body such as a lake or shallow sea dries up by evaporation. Other types of sedimentary deposits include sand and gravel, the quarrying of which annually is a multibillion-dollar industry in the United States. *(Willard Clay/Getty Images Inc.)*

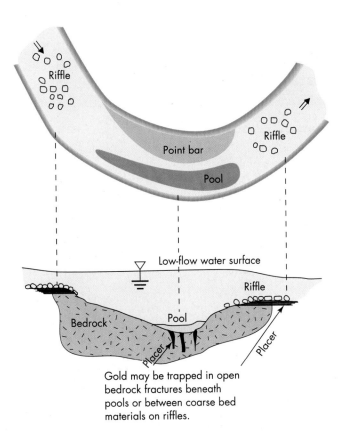

Figure 14.7 **Placer gold**
Diagram of a stream channel and bottom profile showing areas where placer deposits of gold are likely to occur. Mining for placer gold was called a "poor man's" method because all the miner needs is a shovel and a pan.

Gold may be trapped in open bedrock fractures beneath pools or between coarse bed materials on riffles.

Biological Processes. Organisms are able to form many types of useful minerals, including calcium and magnesium carbonate for shells and calcium phosphate in the bones of fish and other organisms. Accumulations of phosphate-rich fish bones and teeth form some of the richest phosphate deposits in the world (Figure 14.8). Fish and other marine organisms extract the phosphate from seawater, and the mineral deposits result from sedimentary accumulation of phosphate-rich fish remains that

Figure 14.8 **Large phosphate mine, Florida** The phosphate minerals in the deposit were extracted from seawater by fish and other marine organisms and deposited as part of their bones and shells with other marine sediments. Piles of mining waste with standing water dominate the landscape shown here. Some land has been reclaimed for use as pasture in the upper part of the photograph. *(William Felger/Grant Heilman Photography, Inc.)*

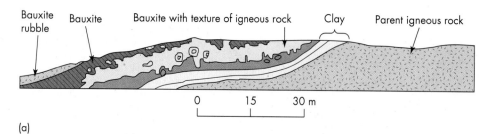

(a)

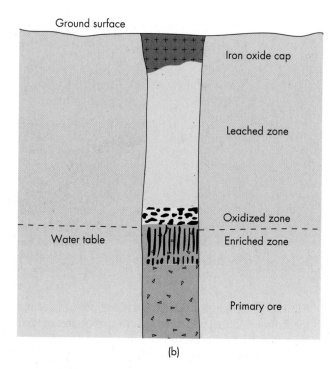

(b)

Figure 14.9 Ore deposits enriched by weathering (a) Cross section of the Pruden bauxite (aluminum) mine in Arkansas. The bauxite was formed by intensive weathering of the aluminum-rich igneous rocks. *(After Mackenzie, G., Jr., et al. 1958. U.S. Geological Survey Professional Paper 299)* (b) Typical zones that form during secondary enrichment processes. Sulfide ore minerals in the primary ore vein are oxidized, altered, and then leached from the oxidized zone and redeposited in the enriched zone. The iron oxide cap is generally a reddish color and may be helpful in locating ore deposits that have been enriched. *(After Foster, R. J. 1983. General geology, 4th ed. Columbus, OH: Charles E. Merrill)*

are deposited with other sediments at the bottom of the ocean. Such sediments eventually formed the sedimentary rocks from which phosphate deposits are mined today.

Weathering Processes. Physical, chemical, and biochemical weathering processes may concentrate some materials to the point at which they can be extracted at a profit. These processes can produce residual ore deposits in weathered material. **Residual deposits** result from intense weathering of rocks and soils that, along with erosional processes, leaves behind the less soluble material with economic value. Intensive weathering of some rocks in a tropical climate forms a particular type of soil known as laterite, often derived from aluminum and iron-rich igneous rocks. The weathering process concentrates oxides of aluminum and iron. The residual aluminum oxide forms an ore of aluminum known as bauxite (Figure 14.9a). Weathering can also cause secondary enrichment processes that increase the concentration of a metal (such as copper) in an *enriched zone*, as shown on Figure 14.9b.

14.3 Environmental Impact of Mineral Development

Many scientists and other observers fear that, as population increases place more demands on mineral resources, the world will be facing a resource crisis. Furthermore, this crisis will come at a time when Earth may be close to the limit of its ability to absorb mineral-related pollution of air, water, land, and biological resources.[6] With this possibility in mind, we will discuss environmental impacts related to nonenergy mineral development. Development of minerals as energy sources is discussed in Chapter 15.

The environmental impact of mineral exploitation depends on factors such as mining procedures, local hydrologic conditions, climate, rock types, size of operation, topography, and many more interrelated factors. Furthermore, the impact varies with the stage of development of the resource. The exploration and testing stage involves considerably less environmental impact than the extraction and processing stages. In the United States, the time period from discovery to mine production may be several years or longer, owing in part to environmental concerns. In countries that have less environmental control, the time from discovery to production may be much shorter. However, forgoing environmental control can lead to significant damage to the air, water, land, and ecosystems.

Impact of Mineral Exploration and Testing

Exploration and testing activities for mineral deposits vary from collecting and analyzing data gathered by remote sensing to fieldwork involving surface mapping, drilling, and gathering of geochemical and geophysical data. In general, exploration has a minimal impact on the environment provided care is taken in sensitive areas such as some arid lands, marshlands, and permafrost areas (areas underlain by permanently frozen ground). Some arid lands are covered by a thin layer of pebbles over several centimeters of fine silt. The layer of pebbles, called desert pavement, protects the finer material from wind erosion. When the pavement is disturbed by road building or other activity, the fine silts may be eroded, impairing physical, chemical, and biological properties of the soil in the immediate environment and scarring the land for many years. In other areas, such as marshlands and the northern tundra, wet, organically rich soils render the land sensitive to even light traffic.

Impact of Mineral Extraction and Processing

Mining and processing mineral resources are likely to have a considerable adverse impact on land, water, air, and biological resources. In addition to their direct environmental effects, these activities can initiate adverse social impacts on the environment by increasing the demand for housing and services in mining areas. These effects are part of the price we pay for the benefits of mineral consumption. It is unrealistic to expect that we can mine our resources without affecting some aspect of the local environment, but we must keep environmental degradation to a minimum. Minimizing environmental degradation can be very difficult because while the demand for minerals continues to increase, deposits of highly concentrated minerals are decreasing. Therefore, to provide more minerals, we are developing larger operations to mine ever-poorer deposits. In the year 2000, the cumulative land use for mining on Earth was approximately 0.2 percent of the land area, or about 300,000 km^3 (115,830 mi^2).

Currently, in the United States, less than 0.3 percent, or 29,000 square km^2 (11,200 mi^2), of the total land area is dedicated to surface mines and quarries. In comparison, wilderness, wildlife, and national park lands cover approximately 500,000 km^2 (193,050 mi^2) of land. However, environmental degradation tends to extend beyond the excavation and surface plant areas of both surface and subsurface mines. Large mining operations change the topography by removing material in some areas and dumping waste in others. At best, these actions produce severe aesthetic degradation; often they produce significant environmental degradation as well. The impact of a single mining operation is a local phenomenon, but numerous local occurrences eventually constitute a larger problem.

Waste from Mines. In the United States, approximately 60 percent of the land dedicated to mining is used for mineral extraction. The remaining 40 percent is used for waste disposal. Most of the waste is overburden, the rock removed to get to the ore. This is an enormous waste-disposal problem, representing 40 percent of all of the solid waste generated in the country.[6] During the past 100 years or so,

Figure 14.10 Bingham Canyon Copper Mine near Salt Lake City, Utah. Notice the large volume of mine waste, the large piles of light-colored material in the left and lower parts of the photograph. *(Michael Collier)*

an estimated 50 billion tons of mining waste has accumulated in the United States, and the annual production of mining waste is 1 billion to 2 billion tons.[9]

Types of Mining and Their Impact. A major practical issue in mining is determining whether surface or subsurface mines should be developed in a particular area. Surface mining is more economical but has more direct environmental effects. The trend in recent years has moved away from subsurface mining and toward large, open-pit surface mines such as the Bingham Canyon copper mine in Utah (Figure 14.10). This mine is one of the world's largest human-made excavations, covering nearly 8 km^2 (3 mi^2) to a maximum depth of nearly 800 m (2,625 ft).

Sometimes leaching is used as a mining technique. Leaching is the process of dissolving materials by percolating liquid through a deposit. For example, some gold deposits contain such finely disseminated gold that extraction by conventional methods is not profitable. For some of these deposits, a process known as heap leaching is used. A dilute cyanide solution, which is applied by sprinklers over a heap of crushed gold ore, dissolves the gold as it seeps through the ore. The gold-bearing solutions are collected in a plastic-lined pond and treated to recover the gold. Because cyanide is extremely toxic, the mining process must be carefully controlled and monitored. Should an accident occur, the process has the potential to create a serious groundwater pollution problem. Research is ongoing to develop in-place cyanide leaching to eliminate the need for removing ore from the ground. However, controlling and monitoring the leaching solution will still present a difficult problem.[10]

Water Pollution. Water resources are vulnerable to degradation from mining. Surface drainage is often altered at mine sites, and runoff from precipitation may infiltrate waste material, leaching out trace elements and minerals. Trace elements leached from mining wastes and concentrated in water, soil, or plants may be toxic, causing diseases in people and other animals who drink the water, eat the plants, or use the soil. These potentially harmful trace elements include cadmium, cobalt, copper, lead, molybdenum, and zinc. The white streaks in Figure 14.11 are mineral deposits apparently leached from the tailings of a zinc mine in Colorado. Similar-looking deposits can cover rocks in rivers for many kilometers downstream from some mining areas. Specially constructed ponds to collect polluted runoff from mines can help, but they cannot be expected to eliminate all problems.

Groundwater may also be polluted by mining operations when waste comes into contact with slow-moving subsurface waters. Surface-water infiltration or

Figure 14.11 **Runoff from mine tailings** A zinc mine in Colorado. The white streaks are mineral deposits apparently leached from tailings. Many sites such as this one are the result of past mining practices in the U.S. that are not allowed today. *(Edward A. Keller)*

groundwater movement through mining waste piles causes leaching of sulfide minerals that may pollute groundwater. The polluted groundwater may eventually seep into streams and pollute surface water. Groundwater problems are particularly troublesome because reclamation of polluted groundwater is very difficult and expensive.

Even abandoned mines can cause serious problems. For example, subsurface mining for lead and zinc in the Tri-State area of Kansas, Missouri, and Oklahoma started in the late nineteenth century. Although it ceased in some areas in the 1960s, it has since then caused serious water pollution problems. The mines, extending to depths of 100 m (330 ft) below the water table, were pumped dry when the mines were in production. Since mining stopped, however, some mines have flooded and overflowed into nearby creeks. The water is very acidic; the sulfide minerals in the mine react with oxygen and groundwater to form sulfuric acid, a problem known as acid mine drainage (see A Closer Look: Acid Mine Drainage, in Chapter 13). The problem was so severe in the Tar Creek area of Oklahoma that in 1982 the Environmental Protection Agency designated it as the nation's foremost hazardous waste site. Another example of acid mine drainage is in Butte, Montana, where the Berkeley Pit, a copper strip mine more than 200 m (656 ft) deep, is filling with acidic, toxic water and forming a lake (Figure 14.12). Pumps that kept the pit dry were turned off when the mine was closed in 1982. Some people think the lake is sure to be a tourist attraction, but to others it is a

Figure 14.12 **Closed mine filling with toxic water** The Berkeley Pit near Butte, Montana, a polluted lake in a closed copper mine. *(Calvin Larsen/Photo Researchers, Inc.)*

growing threat to the environment. Wildfowl in the area have to be scared away; birds that land in the lake drink the water and die. There is increasing concern that the lake may some day leak and pollute groundwater in the Butte area.[11] Plans were developed in the 1990s to cleanup the site. Included is a facility to treat contaminated water, that started up in 2003.

Acid mine drainage in some instances may be linked with tourism and skiing. The link results because ski resorts, such as those in mining districts of Colorado use artificial snow makers both early and late in the season when natural snowfall is less likely. Making snow requires water, and when water is withdrawn from rivers contaminated by acid mine drainage the pollution may spread to the land when the snow melts. Ski resorts may face a no-win situation. If they withdraw acid-rich waters from streams to make snow in order to keep the ski runs open, then there is a risk of contamination of the land. The choice, to not produce artificial snow, may limit the use of resorts and result in financial loss. Ski resorts that make snow artificially from unpolluted water withdrawn from streams may also cause problems. Withdrawing the clean water from streams to make snow at resorts may reduce potential downstream dilution of pollutants by clean water.

As a final example of the effects of water pollution from mining areas, consider the event in Spain in the spring of 1998. The event started close to the Aznalcóllar village and mines near Seville. Acidic mining wastewater containing a toxic mixture of cadmium, zinc, lead, arsenic, cyanide, and other heavy metals was suddenly released when a 50 m (165 ft) wide section of an earth dam containing the waste failed. About 7 million cubic meters of toxic wastewater were released downstream into the Guadiamar River, which flows into one of the most ecologically rich wetlands in Europe. Fortunately, the spill occurred where groundwater is naturally discharging to the surface of the river from the Donana aquifer rather than recharging into it. As a result, groundwater pollution may not be as extensive as it could have been if the toxic wastewater had moved into the aquifer.

The affected area potentially includes the Donana National Park, Europe's largest nature reserve and a major tourist attraction spanning 75,000 hectares (209 mi^2) of land. The park is home to rare birds, the Iberian lynx (a wild cat), turtles, and other species. Engineers moved quickly to construct barriers to contain the toxic spill and to collect and treat toxic mud at a site near the dam to prevent it from entering the national park.

The toxic spill reportedly damaged fruit trees and other crops as toxic floodwaters and mud inundated the land (E. P. Martinez, personal communication, 1998). The release of the toxic chemicals produced a chain of toxicity. Dead and dying crabs, fish, and other animals, including several horses that probably drank toxic river water or were in contact with toxic mud, attracted birds and other scavengers, further spreading the toxicity through the ecosystem. To minimize the spread of toxicity by scavengers, workers used explosives and gun blasts to scare birds away from toxic areas, and hundreds of workers and volunteers collected a great number of dead eels, fish, crabs, and frogs in the days following the event.

The event occurred suddenly and it was not entirely unexpected; for a number of years, there were early warning signs, including questions about the handling of toxic materials so close to the national park. It was also reported that the foreign company running the mine had experienced similar environmental problems in South America (J. Chacon, personal communication, 1998). This event certainly raised the environmental consciousness of Spain and other parts of the world, including the mining regions of the western United States, regarding the potentially serious environmental dangers related to the management of toxic mining waste.

Air Pollution. Both extraction and processing operations have adverse effects on air quality. Smelting has released enormous quantities of pollutants into the atmosphere, including sulfur dioxide, a major constituent of acid rain and snow. Acid precipitation is discussed in Chapter 13. Dust from mineral mines may affect air resources despite the fact that care is often taken to reduce dust by sprinkling water on roads and other dust-producing areas.

As with water pollution, mines can contribute to air pollution problems after production has stopped. For example, toxic gases from abandoned mines in a coal-mining area in Russia have infiltrated homes by seeping into basements. There are many places in the world where mines will eventually close as they become unprofitable. Thus, planning to avoid future air and water pollution as a result of past mining activities is an important goal for people living in regions where mining has been or is still a widespread land use.

Impact on the Biological Environment. Physical changes in the land, soil, water, and air associated with mining directly and indirectly affect the biological environment. Direct impacts include the death of plants, animals, or people caused by mining activity or contact with toxic soil or water from mines (see A Closer Look: Mining and Itai-Itai Disease). Indirect impacts include changes in nutrient cycling; in the total mass of all living matter, called biomass; in species diversity; and in ecosystem stability. These indirect impacts are due to alterations in groundwater or surface-water availability or quality. The periodic or accidental discharge of pollutants through the failure of barriers, ponds, or water diversions or through a

A CLOSER LOOK | Mining and Itai-Itai Disease

A serious chronic disease known as Itai-Itai has claimed many lives in Japan's Zintsu River basin. This extremely painful disease (*itai-itai* means "ouch, ouch") attacks bones, causing them to become so thin and brittle that they break easily. The disease broke out near the end of World War II, when the Japanese industrial complex was damaged and good industrial-waste disposal practices were largely ignored. Mining operations for zinc, lead, and cadmium dumped mining waste into the rivers, and farmers used the contaminated water downstream for domestic and agricultural purposes. The cause of the disease was unknown for years, but in 1960 bones and tissues of victims were examined and found to contain large concentrations of zinc, lead, and cadmium.[12]

Measurement of heavy-metal concentrations in the Zintsu River basin showed that the water samples generally contained less than 1 part per million (ppm) cadmium and 50 ppm zinc. These metals were selectively concentrated in

the river sediment and concentrated even more highly in plants. This increase in concentration from water to sediment to plants is an example of biomagnification. One set of data for five samples shows an average of 6 ppm cadmium in polluted soils. In plant roots, the average was 1,250 ppm, and in the harvested rice it was 125 ppm. Subsequent experiments showed that rats fed a diet containing 100 parts per million cadmium lost about 3 percent of their total bone tissue, and rats fed a diet containing 30 ppm cadmium, 300 ppm zinc, 150 ppm lead, and 150 ppm copper lost an equivalent of about 33 percent of their total bone tissue.[13]

Although measurements of heavy-metal concentrations in the water, soil, and plants of the Zintsu River basin produce somewhat variable results, the general tendency is clear. Scientists are fairly certain that heavy metals, especially cadmium, in concentrations of a few parts per million in the soil and rice produce Itai-Itai disease.[13]

breach of barriers during floods, earthquakes, or volcanic eruptions also damages local ecological systems.

Social Impact. The social impact of large-scale mining results from a rapid influx of workers into areas unprepared for growth. Stress is placed on local services, including water supplies, sewage and solid-waste disposal systems, schools, and rental housing. Land use quickly shifts from open range, forest, or agriculture to urban patterns. More people also increase the stress on nearby recreation and wilderness areas, some of which may be in a fragile ecological balance. Construction activity and urbanization affect local streams through sediment pollution, reduced water quality, and increased runoff. Air quality is reduced as a result of more vehicles, dust from construction, and generation of power to run machinery and equipment.

Adverse social effects may result when miners are displaced by mine closures or automation, because towns surrounding large mines come to depend on the income of employed miners. In the old American West, mine closures produced the well-known ghost towns. Today, the price of coal and other minerals directly affects the lives of many small towns, especially in the Appalachian Mountain region of the United States, where coal mine closures have taken a heavy toll. These mine closings result in part from both lower prices for coal and rising mining costs.

One of the reasons for the rising cost of mining is increased environmental regulation of the mining industry. The abuse of both the miners and the land mined led to the establishment of unions and mining policies. Regulations have helped make mining safer and have facilitated land reclamation. Some mining companies, however, believe the regulations are not flexible enough, and there is some truth to their arguments. For example, if the original hills have been leveled, some areas could be reclaimed for use as farmland after mining. Environmental regulation, however, may require that the land be restored to its original hilly state, even though hills make inferior farmland.

Minimizing the Impact of Mineral Development. Technologically developed countries are making a good deal of progress in reversing the environmental damage done by mineral mining in the past and in minimizing the effects of new extraction and processing operations. It is the responsibility of developed countries to transfer this knowledge to the developing countries where much of the mining is taking place. Environmental laws regulate emissions and waste disposal and mandate restoration measures following mining operations. In addition, innovative technologies, particularly biotechnology, are providing more environmentally friendly ways of mining.

Environmental Regulation. Most of the serious environmental degradation associated with mining in the more developed countries is a relic of past practices that are now forbidden or restricted by environmental laws (remember Chapter 1's case history of Ducktown, Tennessee). Another example of severe environmental degradation occurred in the Sudbury, Ontario, area. One hundred years of smelting nickel ore in the area produced an area of barren land of approximately 100 km^2 (40 mi^2). Another 350 km^2 (135 mi^2) of land was extensively damaged by air pollutants from the smelters (Figure 14.13). The deposition of mercury, arsenic, and cadmium, among other metals, caused devastating effects on the land, water, and biological resources in the vicinity. Emissions from the smelters also contained tremendous quantities of sulfur dioxide, providing a major source of acid precipitation.[6]

Today, in the United States, smelters must adhere to air-quality emission standards of the Clean Air Act. As a result, smelters in the United States recover almost all of the sulfur dioxide from their emissions. Canada has enacted similar clean-air legislation. Smelters in the Sudbury area have reduced their emissions of sulfur dioxide by about 50 percent, and additional reductions are slated for the future. As a result of the decrease in the pollutants from the smelters, there has been some natural recovery. This has been augmented by planting trees and adding lime to lakes to help neutralize acids. These restoration measures along with natural

Figure 14.13 Impact of
smelters Barren land near Lake
St. Charles, Sudbury, Ontario, one
of the largest sources of acid rain in
North America. Vegetation is killed
by acid rain and deposition of toxic
heavy metals resulting from emissions from smelters. Tall stacks
from the smelters are just visible
on the horizon. Considerable
vegetation has been restored since
this photograph. *(Bill Brooks/
Masterfile Corporation)*

recovery have resulted in revegetation of approximately 40 percent of the barren
ground that surrounded the smelters.[6]

After mining activities have ceased, land reclamation, consisting of preparing
the land for future uses, is necessary if the mining has had detrimental effects and
if the land is to be used for other purposes. Reclamation of land used for mining
is required by law today, and approximately 50 percent of the land utilized by
the mining industry in the United States has been reclaimed. Methods of mine
reclamation will be discussed in Chapter 15 when we consider the impact of coal
mining on the environment.

Biotechnology. Several biological processes used for metal extraction and processing are likely to have important economic and environmental consequences.
Biotechnology, using processes such as biooxidation, bioleaching, biosorption,
and genetic engineering of microbes, has enormous potential for both extracting
metals and minimizing environmental degradation (see A Closer Look: Homestake Mine, South Dakota). Biotechnology is still in its infancy, and its potential
uses are just beginning to be realized by the mining and metals industries.[14,15]

A CLOSER LOOK | Homestake Mine, South Dakota

The Homestake Gold Mine in South Dakota which closed in
2001 after 125 years of mining, provides an important example
of the recent use of biotechnology to clean up an environment
degraded by mining activity. The objective of the Homestake
study is to test the use of bacterial biooxidation to convert contaminants in water to substances that are environmentally safe.[15]

The mining operation at Homestake discharges water from
the gold mine to a nearby trout stream, and the untreated
wastewater contains cyanide in concentrations harmful to
the trout. The treatment process developed at the Homestake
mine uses bacteria that have a natural capacity to oxidize the
cyanide to harmless nitrates.[14] The bacteria are collected from
mine-tailing ponds and cultured to allow biological activity at
higher cyanide concentrations. They are then colonized on
special rotating surfaces through which the contaminated
water flows before being discharged to the stream. The
bacteria also extract precious metals from the wastewater that
can be recovered by further processing.[14] The system at
Homestake has reduced the level of cyanide in the wastewater
from about 10 parts per million to less than 0.2 parts per
million, which is below the level required by water-quality
standards for discharge into trout streams. Because the process
of reducing the cyanide produces excess ammonia in the water,
a secondary bacteria treatment was designed that converts the
ammonia to nitrate compounds, so that the discharged water
now meets stream-water quality criteria.[15]

One promising biotechnology is bioassisted leaching, or bioleaching, which uses microorganisms to recover metals. In this technique, bacteria oxidize crushed gold ore in a tank, releasing finely disseminated gold that can then be treated by cyanide leaching. Although the cyanide is recycled during mining, it is very toxic and can contaminate groundwater resources if it is accidentally released into the environment. A commercial plant in Nevada was constructed to produce 50,000 troy ounces (a traditional unit of measurement for gold) of gold per year by bioassisted leaching. From both an economic and an environmental viewpoint, this method is an attractive alternative to the cyanide-leaching process for gold extraction.[14]

Biotechnology developed and tested by the U.S. Bureau of Mines is being used to treat acid mine drainage. Constructed or engineered wetlands at several hundred sites have utilized acid-tolerant plants to remove metals and neutralize acid by biological activity (Figure 14.14). Both oxidizing and sulfate-reducing

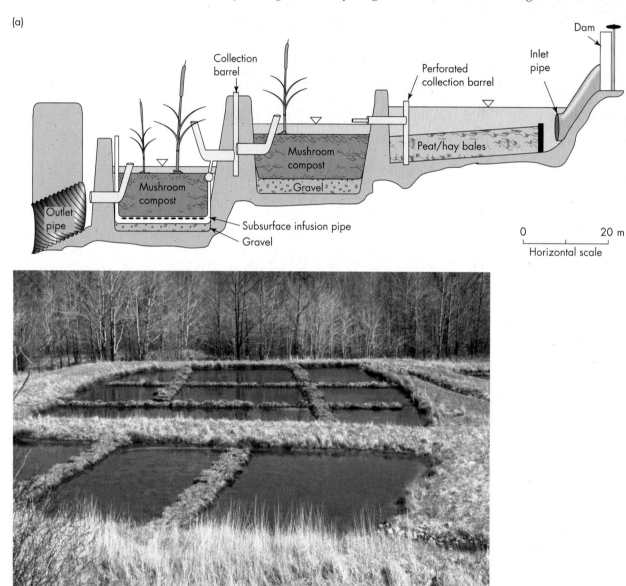

Figure 14.14 Biotechnology and mining waste (a) Idealized diagram of wetlands constructed to use biotechnology for environmental cleanup of wastewater from mines. The plan calls for several shallow ponds lined with compost, topsoil, or crushed limestone. The plants are cattails. Bacteria are living in the compost at the bottom of the ponds. (b) Photograph of artificially constructed wetlands. *(From U.S. Bureau of Mines, Pittsburgh Research Center)*

bacteria play an important role in the wetlands. Research is ongoing to develop an improved wetland design that requires little maintenance.[14]

In summary, minimization of environmental effects associated with mineral development may take several paths:

- *Environmental regulations that address problems such as sediment, air, and water pollution resulting from all aspects of the mineral cycle.* Additional regulations may address reclamation of land used for mining.

- *On-site and off-site treatment of waste.* Minimizing on-site and off-site mining problems by controlling sediment, water, and air pollution through good engineering and conservation practices is a significant goal.

- *Practicing the three Rs of waste management.* That is, reduce the amount of waste produced; reuse materials in the waste stream as much as possible; and maximize recycling opportunities.

14.4 Recycling Mineral Resources

A diagram of the mineral resources cycle reveals that many components of the cycle are connected to waste disposal (Figure 14.15). In fact, the primary environmental impacts of mineral resource utilization are related to its waste products. Wastes produce pollutants that may be toxic to humans, are dangerous to natural ecosystems and the biosphere, are aesthetically undesirable, and may degrade air, water, and soil. Waste and other minerals that are not recycled also deplete nonrenewable mineral resources with no offsetting benefits for human society. Recycling of resources is one way to reduce these wastes.

Waste from some parts of the mineral cycle may be referred to as ore, because it contains materials that might be recycled and used again to provide energy or useful products.[16,17] The notion of reusing waste materials is not new, and such metals as iron, aluminum, copper, and lead have been recycled for many years. For example, in 2004 the total value of recycled metals in the United States was about 24 billion. Recycling and reuse of iron and steel amounted to approximately 90 percent by weight and 50 percent of the total value of recycled metals. Iron and steel are recycled in such large volumes for three reasons:[18] First, the market for iron and steel is huge, and as a result there is a large scrap collection and processing industry; second, an enormous economic burden would result if recycling was not done; and third, significant environmental impacts related to disposal of over 50 million tons of iron and steel would result if we did not recycle.

Other metals that are recycled in large quantities, in terms of the total metal used, include lead (84 percent), aluminum (36 percent), and copper (29 percent).[18]

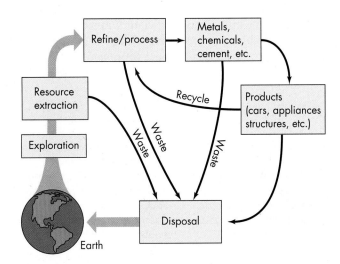

Figure 14.15 Waste is generated in many parts of a mineral cycle Simplified flowchart of the mineral resources cycle.

Recycling aluminum reduces our need to import raw aluminum ore and saves approximately 95 percent of the energy required to produce new aluminum from bauxite. It is estimated that each ton of recycled steel saves 1,136 kg (2,500 lb) of iron ore, 455 kg (1,000 lb) of coal, and 18 kg (40 lb) of limestone. In addition, only one-third as much energy is required to produce steel from recycled scrap as from native ore. Finally, the metal from almost all of the millions of automobiles that are discarded annually in the United States is recycled.[19]

14.5 Minerals and Sustainability

Sustainable development and mineral use do not appear to be compatible. This results because sustainability is a long-term concept that includes finding ways to provide future generations a fair share of Earth's resources, and over time nonrenewable mineral resources are consumed. However, it may be possible to find solutions to nonrenewable mineral resource use that meets the spirit if not the letter of sustainability. This is possible because often it is not the mineral we need so much as the uses for that mineral. For example, we mine copper to use it in wires to transmit electricity or electronic pulses, and it's the properties of copper, not the metal itself, that we desire. For telephone wires we can use glass fiber cables, eliminating the need for copper. Similarly, digital cameras have eliminated the need for film development that uses silver. The photography industry is not interested in silver, but rather the specific function of silver in photography. Therefore, it is possible to compensate for a nonrenewable mineral by finding new, innovative ways to do things. We are also learning that we may use raw mineral materials more efficiently. For example, the Eiffel Tower could be constructed today with a fourth of the amount of steel used when it was built in the late 1800s.[20]

Finding substitutes or ways to use nonrenewable resources more efficiently usually takes several decades of research and development. The time available for finding a solution to the depletion of a nonrenewable mineral is the R to C ratio (R ÷ C), where R is the known reserve and C is the rate of consumption. However, the size of a reserve is not a constant quantity through time because it changes with technology and economics. Therefore, the R to C ratio is not the time a reserve will last at the present rate of consumption. For example, during the past 50 years R to C ratios for metals such as zinc and copper have fluctuated at about 30 years, and during that time, consumption of the metals increased by about three times. This was possible because we discovered new deposits of the metals. The ratio does provide a view of how scarce a particular mineral resource may be. Those metals with relatively small ratios are viewed as being in short supply, and it is those resources for which we need to find substitutes.[20]

In summary, to reconcile sustainable development with the use of nonrenewable mineral resources we need to find ways to more wisely use resources. This includes developing more efficient ways of: exploring for and mining resources; using available resources; recycling; and applying human ingenuity to find substitutes for minerals in short supply.

SUMMARY

Availability of mineral resources is one measure of the wealth of a society. Modern technological civilization as we know it would not be possible without exploitation of these resources. However, we cannot maintain exponential population growth on a finite resource base. Because of slow-moving geologic processes, mineral deposits must be considered nonrenewable resources.

A mineral resource is a concentration of a naturally occurring Earth material in a form that makes extraction currently or potentially feasible. A mineral reserve is that portion of a mineral resource that is currently available, legally and economically, for extraction. It is important to remember that not all resources are reserves. Unless discovered and captured, resources cannot be used to solve present shortages.

Mineral resources may be classified according to their use as metals, building materials, minerals for the chemical industry, and minerals for agriculture. Nonmetallic minerals are consumed at much greater rates than any metal except iron. When a mineral becomes scarce, the choices are to find more sources, recycle what has already been obtained, find a substitute, use less, or do without. The United States and other affluent nations have insufficient domestic supplies of many mineral resources for current use and must supplement them by imports from other countries. As these countries industrialize and develop, the imports may be more difficult to obtain, and affluent countries may have to find substitutes for some minerals or use a smaller portion of the world's annual production. These adjustments are already occurring in the United States, where there has been a reduction in per capita nonfuel mineral consumption since the 1980s.

The geology of mineral resources is complex and intimately related to various aspects of the geologic cycle. Mineral resources are generally extracted from ores, the name given to naturally occurring anomalously high concentrations of Earth materials that can be profitably extracted. To be classified as an ore, a mineral-bearing deposit must have a specific concentration factor, which is the ratio of the mineral's concentration in ore to its average concentration in Earth's crust. The concentration factor of a given mineral reflects both geologic and economic circumstances.

The environmental impact of mineral exploitation depends upon many factors, including mining procedures, local hydrologic conditions, climate, rock types, size of operation, topography, and many more interrelated factors. In addition, the impact varies with the stage of development of the resource. In general, mineral exploration and testing do little damage, except in particularly fragile areas. On the other hand, mineral mining may have major adverse effects on the land, water, air, and biological resources and may initiate social impacts on the environment due to increasing demand for housing and services in mining areas.

Because the demand for mineral resources is increasing, we must strive to implement good engineering and conservation practices that will minimize both on-site and off-site problems caused by mineral development. The recent application of biotechnology to metal extraction and pollution reduction has real promise. Environmental degradation associated with mining and mineral processing in more-developed countries has been significantly reduced in recent years owing to development of pollution abatement strategies and legislation to mandate improved pollution control measures and land reclamation. Such technologies and regulations often are not present in less-developed countries that are striving to develop their mineral resources. It is the responsibility of the more-developed countries to transfer technology so that environmental degradation related to mining activities is minimized at the local, regional, and global levels.

Recycling of mineral resources is one way to delay or partly alleviate a crisis caused by the convergence of a rapidly rising population and a limited resource base.

Sustainable development and use of nonrenewable mineral resources need not be as incompatible as first might be expected. We need to find ways to use our resources more wisely by finding substitutes, recycling, and conservation.

Revisiting Fundamental Concepts

Human Population Growth

As human population has increased, so has our use of mineral resources. Developed countries have less than 20 percent of the world's population, yet use most of the mineral resources. As developing countries strive to raise their standard of living, the amount of minerals they use will increase. We in the developed world will need to continue to find ways to use a smaller proportion of the world's mineral resources.

Sustainability

Mineral resources are part of our nonrenewable heritage from the geologic past. Because they are nonrenewable, they are not sustainable in the long run. They are finite and will eventually be depleted. The objective of recycling and conservation is to extend the resource base as long as possible.

Earth as a System

Through time, global geologic systems such as the rock, water, and biogeochemical cycles have produced the mineral resources of the world. Society has produced technologic, economic, and political systems to exploit and use mineral resources. It is these systems that must continuously change if we intend to use our mineral resources wisely. The objective is to find ways to extend the supply of mineral resources.

Hazardous Earth Processes, Risk Assessment, and Perception

Mineral extraction and use present many hazards to people. Land above underground mines may subside; miners are killed and injured in mining operations; and toxins may be released into the environment by the mining and processing of minerals. Because the risks of mining are well known, mining activities are well regulated in the United States. Less well known are the long-term consequences of the release of toxic materials into the air, water, and soil. Major problems such as acid mine drainage are generally well studied, but the environmental impact of the entire mineral industry is less well known.

Scientific Knowledge and Values

There is an inherent conflict between our need for minerals and our desire to preserve a healthy environment. We value minerals and their use to society; without minerals our standard of living would be much different. However, we also value clean air, water, and soil and know that the mining and processing of minerals may cause pollution of these resources. The question is, how much are we willing to pay to minimize environmental degradation that naturally occurs from our use of minerals?

Key Terms

biotechnology (p. 463)
concentration factor (p. 449)
evaporite deposits (p. 454)

ore (p. 449)
reserve (p. 446)

residual deposits (p. 456)
resource (p. 446)

Review Questions

1. What is the difference between a resource and a reserve?

2. Define the term ore.

3. Define concentration factor.

4. Why do metallic ore deposits form at divergent plate boundaries?

5. List some of the major environmental impacts associated with mining.

6. What is the role of biotechnology in the mining industry?

7. Why do we recycle so much iron?

Critical Thinking Questions

1. Today, technological changes are coming from all directions. The increasing use of electronic mail (e-mail) threatens to lead us to a paperless society. We now have e-money, a technology that allows us to pay our bills via personal computers. Journals and newspapers we subscribe to come over the Internet, and we can print out only what we really want to keep. What is the impact of this technology on resource utilization? To answer this question, start by constructing a list of the resources from mineral sources necessary to support a paper-driven society. Then develop a plan to reduce the impact on natural resources of the transformation to a more paperless society.

2. Since we know that mineral resources are finite, there are two ways to look at our present and future use of these resources. One view is that we are headed toward a mineral crisis as the number of people on Earth increases. The other is that, as the number of people increases, the possibility for innovations increases, and we will therefore find ways to adjust our use of minerals to the increasing population. How could these two hypotheses be tested?

3. Biotechnology and genetic engineering are potential tools for cleaning up the environment. We discussed some examples of biotechnology in this chapter: Bacteria can be cultured to neutralize acids in effluent from mines, and artificial wetlands can be constructed in which biological processes purify water polluted by mineral processing and mining. What do you think of biotechnology and genetic engineering? How might the technology be transferred to the mineral industry in the United States and other countries?

FIFTEEN

Energy Resources

Learning Objectives

The highly technical society in the United States consumes a disproportionate share of the total energy produced in the world. With only 5 percent of the world's population, the United States consumes approximately 25 percent of the world's total energy. How can we break our dependency on energy sources, such as coal and oil, that harm the environment without sacrificing our standard of living? In this chapter we explore possible answers to this important question. The focus will be on the following learning objectives:

- Understand the concept of peak oil and how it might impact our economic and social environment

- Know the general patterns of energy consumption in the United States

- Know the types of major fossil fuels and the environmental impact associated with their development

- Understand nuclear energy and its associated important environmental issues

- Understand geothermal energy, how it is produced, and its future as an energy source

- Know the main types of alternative and renewable energy and their environmental significance

- Know the important issues related to energy policy, particularly the difference between the hard path and the soft path and the concept of sustainable energy development

Gasoline price shocks An abrupt rise in gas prices comes as a shock to many consumers, including people at this station in southern California in 2006. Prices continue to yo-yo but generally rise in the early twenty-first century. *(Edward Keller)*

CASE HISTORY | Peak Oil: When Will It Occur and What Is Its Importance?

The United States is prosperous and people are living longer as a result of abundant low-cost energy in the form of crude oil. While the benefits of oil are undeniable, so also are environmental problems, ranging from air and water pollution to global warming. That is changing, and we are about to learn what life with is like with less, more expensive oil. The question is not if the peak in production of oil will come, but when, and what the economic and political consequences to society will be.[1] **Peak oil** is the term for the time when one-half of Earth's oil has been extracted and used.

The global history of oil in terms of rate of discovery and consumption (Figure 15.1) is an important factor in peak oil. Nearly five times more oil was discovered in 1960 than was consumed. However, by 1980 the amount discovered was equivalent to the amount consumed; and in 2000 the consumption of oil was three times the amount discovered. This trend is not sustainable. The unfortunate truth is that oil is being consumed rapidly relative to new resources being found. In 2006, oil companies reported the largest discovery of oil ever in the United States. At 15 billion barrels, when extracted and consumed by users of the world market, it will provide about 6 months of oil. If all 15 billion barrels is saved for the United States, it would be only about a 2 year supply.

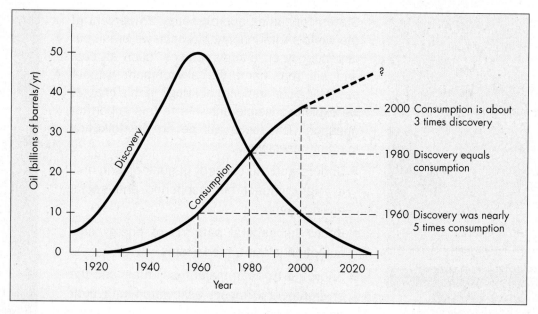

Figure 15.1 **History of U.S. oil discovery and consumption** Discovery peaked by about 1960 and consumption exceeded discovery after 1980. *(Modified after Aleklett, K. 2006, Oil: A bumpy road ahead. World Watch 19(1): 10–12)*

15.1 Worry over Energy Sources Is Nothing New: Energy Shocks Past and Present

Two thousand years ago the homes of affluent Roman citizens had central heating that consumed vast amounts of wood—perhaps as much as 125 kg (275 lb) every hour. Wood was the fuel of the day, much as oil and gas are in the United States today. The exhaustion of local supplies of wood produced shortages and a shock to the Roman society. To combat the shortages, the Romans had to import wood from distances as far away as 1,600 km (995 mi). As distant supplies of wood became scarce, the Romans faced a second energy shock. They turned to solar energy as an alternative, with much success. Ancient Romans became so efficient in the use of solar energy that they established laws to protect a person's right to his or her share of the solar energy. In some cases, it was illegal to construct a

The concept of peak oil production is shown in Figure 15.2. We aren't confident when the peak production will be, because it depends in part on technology, conservation, and use of alternative fuels such as ethanol (biofuel). For this discussion, let's assume it to be about 50 billion bbl per year, and that the peak arrives some time between 2020 and 2050. The growth rate for oil in 2004 was 3.4 percent. Increasing from the present production rate of about 30 billion bbl to 50 in a few decades seems reasonable. When peak production occurs and demand increases, a gap between production and demand will become evident. Cost will increase, and if we haven't prepared for the peak, then economic and political disruption to society is inevitable. We still have time to prepare for the eventual peak and use remaining fossil fuels to help transition to other energy sources. In the most optimistic scenario, the transition from oil to alternative fuels will not occur until we have cost-competitive choices in place.[2]

When the peak in oil production occurs, we will not run out of oil, but it will become much more expensive. When it happens, the peak will be different from any problem we have faced in the past. Human population will increase by several billion more in coming decades as countries with growing economies including China and India increase their oil consumption. China may double its import of oil in the next 5 years! As supplies tighten, economic, social, and political ramifications of competition for oil will be enormous. Planning to conserve oil and transition to alternative energy sources will be critical in the coming decades. We cannot afford to leave behind the age of oil until alternatives are in place. The remainder of this chapter will discuss sources of energy, environmental consequences of developing and using energy, and sustainable energy development.

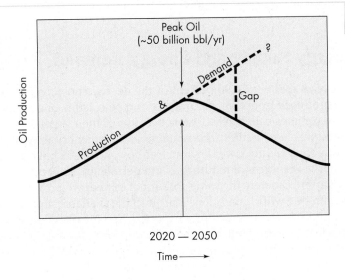

Figure 15.2 Idealized diagram illustrating concept of "peak oil" when one-half of world's oil will have been produced. When demand is greater than production, a gap (shortage) develops. This may happen between 2020 and 2050.

building that shaded a building next door because the shade would rob the neighbors of their share of the solar resource.[3]

During the summer of 2006, U.S. citizens were shocked by the rapid price increase of gasoline in only a few months. The sharp price increase followed years of low prices (particularly in the 1990s) due to an oil and gasoline glut. The reasons for the price increase are complex and based upon several factors: Oil-producing countries reduced production rates, there is political unrest and war in the Middle East and the U.S. government instituted regulations that require oil producers to reduce air pollution by reformulating gasoline with additives that produce cleaner burning fuel. These factors, as well as suspected price gouging by oil companies, resulted in shortages. As dramatic as the price increases were, they were not the first; there had been several such oil shocks in the space of less than 30 years.

The price increase in gasoline was preceded by the "California energy crisis," with its rolling blackouts, in 2001. This crisis was a wake-up call to the entire country. The crisis resulted in part because, as California prospered and grew in population, demand for energy increased. California imports a large percentage of its electrical energy and failed to bring sufficient new sources of energy online. Utility companies that had been deregulated a few years before were forced to purchase emergency power from suppliers that charged as much as 900 percent more than normal. One of the state's largest power companies went bankrupt and the governor of California initiated an investigation of illegal market control and price gouging by energy suppliers. More power plants are being constructed, but the short-range future is uncertain.

The Romans' use of solar energy illustrates that energy problems are not new. As one source of energy becomes scarce, and energy shocks occur, we search for alternatives. Today we fear that our fossil-fuel resources will be exhausted, and, like the Romans, we will need alternatives; perhaps it will be solar energy again. In this chapter we will explore present energy use, environmental effects of energy consumption, and what the future might hold in terms of establishing sustainable energy development.

August 14, 2003, brought an energy shock to New York City and the surrounding region as a massive eight-state blackout occurred. Over 50 million people were affected. Some were trapped in elevators and electric trains underground. Power was restored within 24 hours to most places. The power failure started at a U.S. power plant and quickly cascaded to widespread failure. The blackout emphasized our dependence on an aging centralized power distribution system and need for improved energy management.

15.2 Energy Supply and Energy Demand

Figure 15.3a reveals that nearly 90 percent of the energy consumed in the United States today is produced from coal, natural gas, and petroleum, or oil. These energy resources are sometimes called **fossil fuels** because of their organic origin. These fuels are nonrenewable resources. The remaining 10 percent of energy consumed originates mostly from hydropower and nuclear power. We still have huge reserves of coal, but major new sources of natural gas and petroleum are becoming scarce. In fact, we import approximately the same amount of oil that we produce, making us vulnerable to changing world conditions that affect the available supply of crude oil. Few new large hydropower plants can be expected, and the planning and construction of new nuclear power plants have become uncertain for a variety of reasons. More recently, alternative energy sources such as solar power for homes, farms, and offices are becoming economically more feasible and thus more common.

Changes in energy consumption in the United States are shown in Figure 15.3. From 1950 through 1974 there was a sharp increase in energy consumption; the rate of increase declined after the shortages in the mid-1970s to mid-1980s. The rate of energy consumption increased again during the last decade of the twentieth century owing to abundant cheap oil. Total U.S. energy consumption has increased by about 25 exajoules (EJ) since 1985 (see A Closer Look: Energy Units,

A CLOSER LOOK | Energy Units

The basic unit for energy expenditure is the joule (J). One joule is defined as a force of 1 newton applied over a distance of 1 m (3.3 ft). A newton (N) is defined as the force necessary to produce an acceleration of 1 m per second per second to a mass of 1 kg (2.2 lb). From 1950 through 1974, total U.S. energy consumption increased from 30 to approximately 75 exajoules (EJ) (1 EJ is equal to 10^{18} J and approximately equal to 1 quad, or 10^{15} Btu). The units for power are time rate of energy or joules per second (J/sec); 1 J/sec is a watt (W). A large power plant produces about 1,000 megawatts (MW), which is 1 billion watts (1 gigawatt, or 1 GW). Electrical energy is commonly sold by kilowatt-hours (kWh). This unit is 1,000 W applied over 1 hour (3,600 sec). Thus, 1 kWh is 3,600,000 J, or 3.6 MJ.

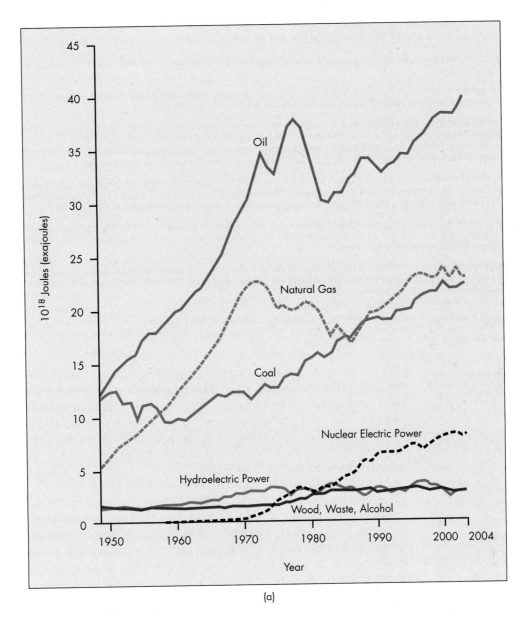

(a)

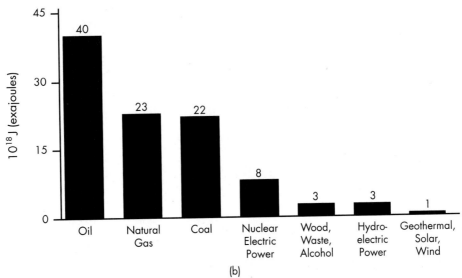

(b)

Figure 15.3 **U.S. energy** (a) U.S. energy consumption 1950–2004 by source. (b) U.S. energy consumption for 2004 by source. (Modified after Energy Information Admin. Annual Energy Review, 2004)

TABLE 15.1 Some Provisions of the Energy Policy Act of 2005[a]

1. **Promotes conventional energy sources:** Recommends and supports using more coal and natural gas with objective to reduce our reliance on energy from foreign countries.

2. **Promotes nuclear power:** Recommends that the United States start building nuclear power plants again by 2010. Recognizes that nuclear power plants can generate large amounts of electricity without emitting air pollution or greenhouse gases.

3. **Encourages alternative energy:** Authorizes support or subsidies for wind energy and other alternative energy sources such as geothermal, hydrogen, and biofuels (ethanol and biodiesel). The legislation also recognizes wave power and tidal power as renewable energy technology for the first time in U.S. energy policy. The act contains provisions to help make geothermal energy more competitive with fossil fuels in generating electricity and increases the amount of biofuel (ethanol) that must be mixed with gasoline sold in the United States.

4. **Promotes conservation measures:** Sets higher efficiency standards for federal buildings and for household products. Directs federal attention to recommend fuel-efficiency standards for cars and trucks and SUVs. Homeowners can claim new tax credits to install energy-efficient windows and appliances. The legislation also provides a tax credit on purchasing a fuel-efficient hybrid or clean-diesel vehicle.

5. **Promotes research:** The act authorizes research into finding innovative ways to improve coal plants and to help construct cleaner coal plants; developing zero-emission coal burning power plants; determining how to tap into the vast amounts of oil trapped in oil shale and tar sands; and developing hydrogen-powered, pollution-free automobiles.

6. **Provides for energy infrastructure:** Incentives are provided for oil refineries to expand their capacity. The act helps ensure that electricity is received over dependable modern infrastructure and makes electric reliability standards mandatory. The legislation gives federal officials the authority to select sites for new power lines, more independent of local pressure.

[a]The act has been criticized for making most of the substantial incentives and subsidies to fossil fuels, especially coal and nuclear energy, at the expense of energy conservation and alternative energy. Regardless of perceived favoritisms, politics, or lack of depth in understanding of our energy crisis, the Energy Policy Act of 2005 initiates a new round of debate on the future of U.S. energy policy.

for an explanation of exajoules). Energy conservation policies, such as requiring new automobiles to be more fuel efficient and buildings to be better insulated, have been at least partially successful. Nevertheless, at the end of the twentieth century, we remained hooked on fossil fuels. In 2005, the president of the United States signed the Energy Policy Act of 2005 that emphasized oil, coal, and natural gas—the fossil fuels (see Table 15.1).

Projections of energy supply and demand are difficult at best because the technical, economic, political, and social assumptions that underlie these projections are constantly changing. It is clear, however, that we must continue to research, develop, and evaluate potential energy sources and conservation practices to ensure sufficient energy to maintain our industrial society along with a quality environment. In the United States, space heating is the main use of energy for applications below 300°C (572°F) (Figure 15.4). With these ideas in mind, we will explore some selected geologic and environmental aspects of well-known energy resources, including coal, oil, and nuclear sources, as well as potentially important sources such as oil shale, tar sands, and geothermal resources. We will also discuss renewable energy sources, such as hydropower, wind, and solar power.

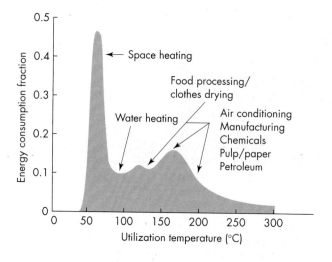

Figure 15.4 Energy use below 300°C in the United States
(From Los Alamos Scientific Laboratory. L.A.S.L. 78-24, 1978)

15.3 **Fossil Fuels**

The origin of fossil fuels—coal, oil, and gas—is intimately related to the geologic cycle. These fuels are essentially solar energy stored in the form of organic material that has been transformed by physical and biochemical processes after burial. The United States is dependent on fossil fuels for nearly all the energy we consume. This situation is changing, but slowly.

The environmental disruption associated with the exploration and development of fossil fuels must be weighed against the benefits gained from the energy, but it is not an either-or proposition. Good conservation practices combined with pollution control and reclamation can help minimize the environmental disruption associated with fossil fuels.

Coal

Coal is one of the major fossil fuels. Burning coal accounts for about 20 percent of the total energy consumption in the United States. As we shall discuss, the environmental costs of coal consumption are significant.

Geology of Coal. Like the other fossil fuels, coal is composed of organic materials that have escaped oxidation in the carbon cycle. Essentially, coal is the altered residue of plants that flourished in ancient freshwater or brackish swamps, typically found in estuaries, coastal lagoons, and low-lying coastal plains or deltas.

Coal-forming processes begin in swamps rich in plant life, where water-saturated soils exclude much of the oxygen normally present in soil (Figure 15.5). The plants partially decompose in this oxygen-deficient environment and slowly accumulate to form a thick layer of material called *peat*. The swamps and accumulations of peat may then be inundated by a prolonged, slow rise in sea level, caused either by an actual rise of sea level or by sinking of the land; sediments such as sand, silt, clay, and carbonate-rich material are then deposited on the peat. As more and more sediment is deposited, water and organic gases, or volatiles, are squeezed out, and the percentage of carbon increases in the compressed peat. As this process continues, the peat is eventually transformed to coal. Because there are often several layers of coal in the same area, scientists believe that the sea level may have alternately risen and fallen, allowing development and then drowning of coal swamps.

Classification and Distribution of Coal. Coal is commonly classified according to type and sulfur content. The type is generally based on both the percentage of carbon in the coal and its heat value on combustion. The percentage of carbon in coal increases among the four types found in Earth, from lignite to subbituminous to bituminous to anthracite (Figure 15.6). Figure 15.6 also shows that heat content is maximum in bituminous coal, which has relatively few volatiles such as oxygen, hydrogen, and nitrogen, and low moisture content. Heat content is minimum in lignite, which has a high moisture content. The distribution of the common types of coals in the contiguous United States is shown in Figure 15.7a. The distribution of world coal reserves, which amount to about 1,000 billion metric tons, is shown in Figure 15.7b. The United States has about 25 percent of these reserves. The annual world consumption of about 5 billion tons suggests that, at the present rate of consumption, known reserves are sufficient for over 200 years.[4] It is likely, however, that in countries undergoing intensive industrialization, such as China, reserves may be depleted sooner.[5]

The sulfur content of coal may generally be classified as low (zero to 1 percent), medium (1.1 to 3 percent), or high (greater than 3 percent). With all other factors equal, the use of low-sulfur coal as a fuel for power plants releases less sulfur

(a) Coal swamp forms.

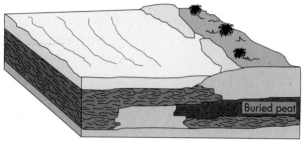

(b) Rise in sea level buries swamp in sediment.

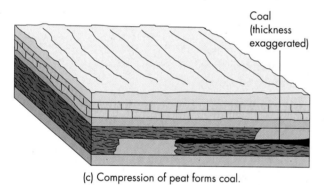

(c) Compression of peat forms coal.

Figure 15.5 **How coal forms** The processes that convert buried plant debris, or peat, into coal. Considerable lengths of geologic time must elapse before the transformation is complete.

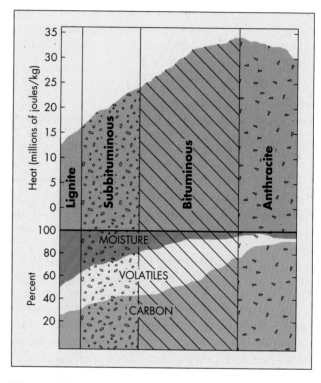

Figure 15.6 **Types of coal** Generalized classification of types of coal based on their relative percent content of moisture, volatiles, and carbon. The heat values of the different types of coal are also shown. *(After Brobst, D. A., and Pratt, W. P., eds. 1973. U.S. Geological Survey Professional Paper 820)*

dioxide (SO_2) and causes the least air pollution. Most coal in the United States is the low-sulfur variety; however, by far the most common low-sulfur coal is a relatively low-grade, subbituminous variety found west of the Mississippi River (Table 15.2).[6] To avoid air pollution, thermal electric power plants on the highly populated East Coast must continue to either treat local coal to lower its sulfur content before burning or capture the sulfur after burning. Treatment increases the cost of thermal electric power, but it may be more economical than transporting low-sulfur coal from west of the Mississippi River.

Impact of Coal Mining. Most of the coal mining in the United States is done by open-pit mining, or *strip mining*. Strip mining began in the late nineteenth century and has steadily increased, whereas production from underground mines has stabilized. As a result, most coal today is obtained from surface mines west of the Mississippi River. In recent years strip mining in the Appalachians using a technique known as mountain top removal has become more common. Coal is mined from tops of mountains and the waste rock from mining is placed in valleys, where coal sludge produced from processing the coal is stored behind coal-waste

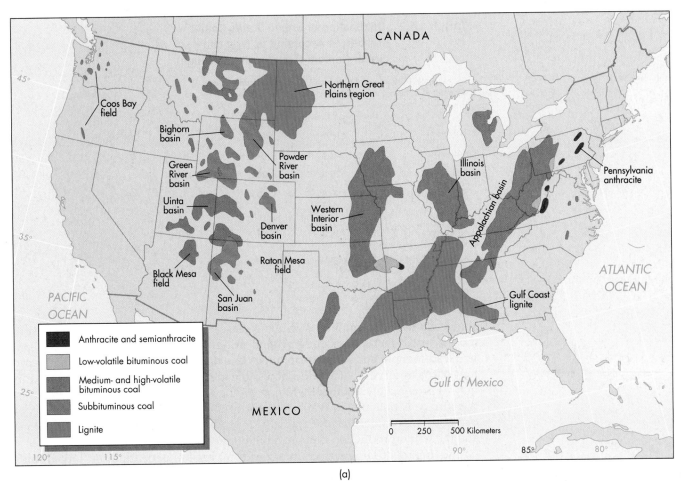

(a)

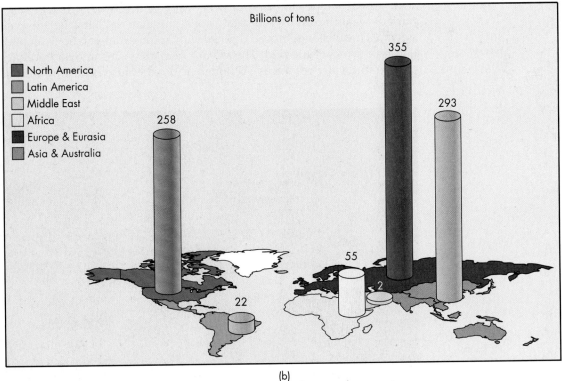

(b)

Figure 15.7 **World coal reserves** (a) Coal areas of the contiguous United States. *(Carbini, S., and Schweinfurth, S. P. 1986. U.S. Geological Survey Circular 979)* (b) World coal reserves (billions of metric tons) in 2002. The United States has about 25 percent of the total reserves. Unlike oil, coal reserves are more evenly distributed around the world. *(British Petroleum Company. 2003. BP statistical review of world energy)*

TABLE 15.2   Distribution of United States Coal Resources According to Type and Sulfur Content (Low, Medium, or High).

For example, 97 percent of anthracite is low sulfur content, and 43 percent of bituminous coal has a high sulfur content.

Type	Sulfur Content[1]		
	Low	Medium	High
Anthracite	97.1	2.9	—
Bituminous coal	29.8	26.8	43.4
Subbituminous coal	99.6	0.4	—
Lignite	90.7	9.3	—
All ranks	65.0	15.0	20.0

[1]Low = 0–1% sulfur; Medium = 1–3% sulfur; High = >3% sulfur.

Source: U.S. Bureau of Mines Circular 8312, 1966.

sludge dams. Strip mining is in many cases technologically and economically more advantageous than underground mining. The increased demand for coal will lead to more and larger strip mines to extract the estimated 40 billion metric tons of coal reserves that are now accessible to surface mining techniques. In addition, approximately another 90 billion metric tons of coal within 50 m (165 ft) of the surface is potentially available for stripping if need demands. To put this in perspective, consider that the annual U.S. production is about 1 billion metric tons and the world production is about 5 billion metric tons per year.[4]

The impact of large strip mines varies by region, depending on topography, climate, and, most important, reclamation practices. In humid areas with abundant rainfall, mine drainage of acid water is a serious problem (Figure 15.8). Surface water infiltrates the spoil banks (the material left after the coal or other minerals have been removed), reacts with sulfide minerals such as pyrite (FeS_2), and produces sulfuric acid. The sulfuric acid then runs into and pollutes streams and groundwater resources. Although acid water also drains from underground

Figure 15.8 Acid mine drainage
Acid-rich water draining from an exposed coal seam is polluting this stream in Ohio. *(Kent & Donna Dannen/Photo Researchers, Inc.)*

Figure 15.9 Mining coal
Large backhoe at the Trapper Mine, Colorado, removing the coal and loading it into large trucks for delivery to a power plant just off the mining site. *(Edward A. Keller)*

mines, roadcuts, and areas where coal and pyrite are abundant, the problem is magnified when large areas of disturbed material remain exposed to surface waters. Acid drainage can be minimized through proper use of water diversion practices that collect surface runoff and groundwater before they enter the mined area and divert them around the potentially polluting materials. This practice reduces erosion, pollution, and water-treatment cost.[7]

In arid and semiarid regions, water problems associated with mining are not as pronounced as in wetter regions. However, land may be more sensitive to mining activities such as exploration and road building. In some arid areas, the land is so sensitive that even tire tracks across the land survive for years. Soils are often thin, water is scarce, and reclamation work is difficult.

All methods of strip mining have the potential to pollute, but reclamation practices can minimize the environmental damage of strip mining (Figure 15.9). Reclamation begins with segregation of the soil and *overburden*, the rock above the coal. After mining, the topsoil that was removed is replaced. This method is widely used, and it is a successful way to minimize water pollution when combined with regrading and revegetation.[7]

Federal guidelines govern strip mining of coal in the United States. They require that mined land be restored to support its premining use. Restoration includes disposing of wastes, contouring the land, and replanting vegetation (Figure 15.10). The hope is that after reclamation the mined land will appear and function as it did before extraction of the coal. However, this is a difficult task and, in general, is not completely successful. The regulations also prohibit strip mining on prime agricultural land and give farmers and ranchers the opportunity to restrict or veto mining on their land, even if they do not own the mineral rights.

Underground mining of coal and other resources has also produced hazards and caused considerable environmental degradation. For example, (1) acid water draining from mines has polluted streams; (2) subsidence, collapse of the surface of the land into pits over underground mines, has damaged buildings, roads, and other structures; and (3) coal fires that start in mines can burn and smoke for years.

Future Use and Environmental Impacts of Coal. Limited resources of oil and natural gas and concerns about the energy supply in California and throughout the nation are increasing the demand for coal, and the coal industry is planning for increased mining activity. The crunch on oil and gas supplies is still years

Figure 15.10 **Reclaimed land** at the Trapper Mine, Colorado. The site in the foreground has just had the soil replaced after mining, whereas the vegetated sites have been entirely reclaimed. *(Edward A. Keller)*

away, but when it occurs, it will put pressure on the coal industry to open more and larger mines in both the eastern and the western coal beds of the United States. This solution to oil and gas shortages could have significant environmental impacts for several reasons:[8]

- More and more land will be strip mined and will thus require careful restoration.

- Unlike oil and gas, burned coal leaves ash (5 percent to 20 percent of the original amount of the coal) that must be collected and disposed of in landfills. Some ash can be used as fill in land development or for other purposes, but about 85 percent is presently useless.

- Handling of tremendous quantities of coal through all stages—mining, processing, disposal of mining waste, shipping, burning, and disposing of ash—has potentially adverse environmental effects, including water pollution (see A Closer Look: Coal Sludge in the Appalachian Mountains); air pollution; release of trace elements likely to cause serious health problems into the water, soil, and air; and aesthetic degradation at mines, power plants, and other facilities associated with the coal industry.

- Several billion metric tons of coal are burned worldwide, releasing huge amounts of carbon dioxide (CO_2) into the atmosphere. China, the United States, and Russia account for the majority of the carbon dioxide released. Carbon dioxide is a "greenhouse gas" thought to be significant in producing global warming. (Global warming is discussed in Chapter 19.)

Environmental problems associated with coal, although significant, are not necessarily insurmountable, and careful planning could minimize them. As world trade increases, new mines with cleaner coal will become more available. For example, coal mines being developed in Borneo are producing a low-grade but very low-sulfur coal that has been termed "solid natural gas" because it is so pure a source of energy. These clean coal deposits in Borneo are thick and have only a few meters of overburden.[10] Regardless of sources and quality, there may be few alternatives to mining tremendous quantities of coal. In the future, vast amounts of coal will be needed to feed thermoelectric power plants and to provide oil and gas by gasification and liquefaction of coal. An important objective is to find ways to use coal that minimize environmental disruption.

Hydrocarbons: Oil and Gas

Oil, called crude oil, and natural gas are hydrocarbons, made up of carbon, hydrogen, and oxygen. **Natural gas** is mostly methane (CH_4). Methane usually

A CLOSER LOOK | Coal Sludge in the Appalachian Mountains

Mining and processing coal in the Appalachian Mountains produce a lot of waste. Waste material known as coal sludge or slurry generated by processing coal is a thick sludge composed of water with particles of coal, rock, and clay. Coal sludge is often stored in impoundments behind a dam. There are hundreds of coal sludge impoundments in West Virginia, Pennsylvania, and Kentucky. There have been nine failures of coal sludge impoundments in the past 30 years. A coal-waste dam on Buffalo Creek, West Virginia, containing coal sludge failed in 1972. The resulting flood killed 118 people while destroying about 500 homes. The Buffalo Creek event resulted in suggestions, recommendations, and regulations to minimize the chances of future events. Nevertheless, other events have occurred, including that of October 11, 2000, when about 1 million cubic meters (250 million gallons) of thick black coal

sludge was released into the Big Sandy River system in southeastern Kentucky. The sludge was stored in an impoundment located over an abandoned coal mine. Collapse from the floor of the impoundment into the mine allowed the sludge to flow through the mine and into the environment over people's yards and roads and into a tributary to the Big Sandy River. Over 100 km (65 mi) of river were severely contaminated by the sludge. Life (including several hundred thousand fish) in the river was destroyed. The spill was one of the worst-ever environmental disasters in the southern United States. The same impoundment had a similar but smaller sludge spill in 1994 and was inspected as recently as three weeks before the October 2001 spill. Recommendations following the 1994 spill may not have been followed and there was not sufficient hard rock above the abandoned mine to support the heavy sludge (Figure 15.A).[9]

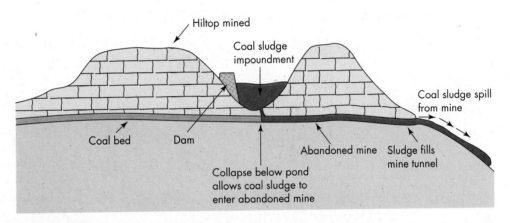

Figure 15.A Contamination by a coal sludge spill Idealized diagram of how a spill from a storage site with coal slurry might happen. In this case collapse over an abandoned coal mine allows the slurry to spill.

makes up more than 80 percent of the energy gases present at a location. Other natural-energy gases include ethane, propane, butane, and hydrogen.[11] Like coal, the hydrocarbons are, for the most part, fossil fuels in that they form from organic material that escaped complete decomposition after burial. Oil and gas are found in concentrated deposits, which have been heavily mined from wells. They are also found, in less readily available form, in oil shale and tar sands.

Geology of Oil and Gas Deposits. Next to water, oil is the most abundant fluid in Earth's crust, yet the processes that form it are only partly understood. Most Earth scientists accept that oil and gas are derived from organic materials that were buried with marine or lake sediments. Favorable environments where organic debris might escape oxidation include nearshore areas characterized by rapid deposition that quickly buries organic material or deeper-water areas characterized by a deficiency in oxygen at the bottom that limits aerobic decomposition. In addition, the locations of oil and gas formation are generally classified as subsiding, depositional basins in which older sediment is continuously buried by younger sediment; this sequence progressively subjects the older, more deeply buried material to higher temperatures and pressures.[12]

Oil and gas originate primarily in fine-grained, organic-rich sediment referred to as *source rock*. It is buried to depths of 1 to 3 km (0.6 to 2 mi) and subjected to heat and pressure that physically compress it. The elevated temperature and pressure, along with other processes, start the chemical transformation of organic debris into thermogenic oil and gas (Figure 15.11). Natural gas that forms close to the surface of Earth by biological processes is called *biogenic gas*. An example is the

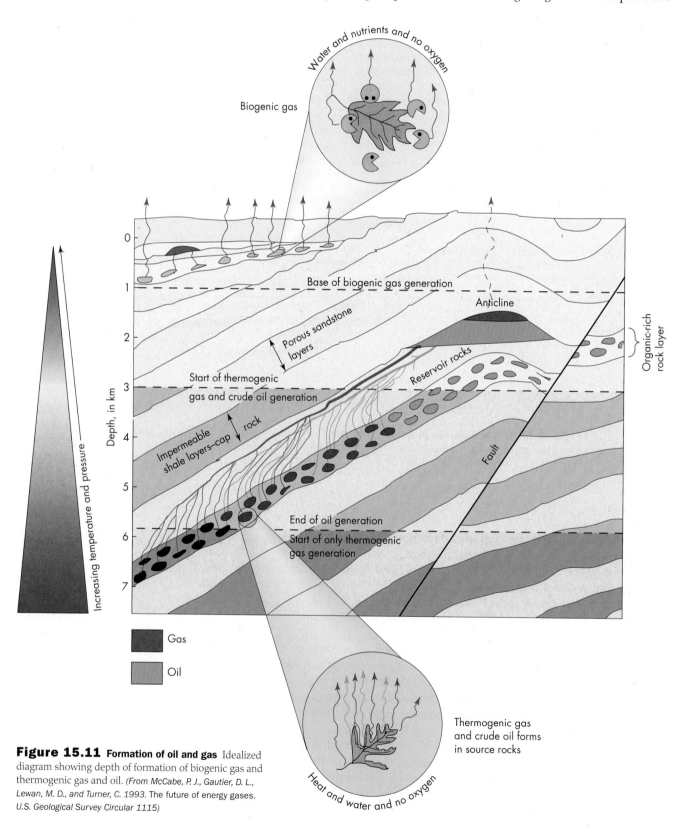

Figure 15.11 **Formation of oil and gas** Idealized diagram showing depth of formation of biogenic gas and thermogenic gas and oil. *(From McCabe, P. J., Gautier, D. L., Lewan, M. D., and Turner, C. 1993. The future of energy gases. U.S. Geological Survey Circular 1115)*

methane produced in landfills as waste decays. After their formation, the thermo-genic hydrocarbons begin an upward migration to a lower-pressure environment with increased porosity. This primary migration through the source rock merges into secondary migration as the hydrocarbons move more freely into and through coarse-grained, more permeable rocks. These porous, permeable rocks, such as sandstone or fractured limestone, are called *reservoir rocks*. If the path is clear to the surface, the oil and gas migrate and escape there, and this may explain why most oil and gas are found in geologically young rocks (less than 100 million years old): Hydrocarbons in older rocks have had a longer time in which to reach the surface and leak out.[12] The entire process from initial deposition to deep burial often takes millions of years.[11] When oil and gas are impeded in their upward migration by a relatively impervious barrier, they accumulate in the reservoir rocks. If the barrier, or *cap rock*, has a favorable geometry, such as a dome or an anticline, the oil and gas will be trapped in their upward movement at the crest of the dome or anticline below the cap rock (Figure 15.11).[12] Figure 15.12 shows an anticlinal trap and two

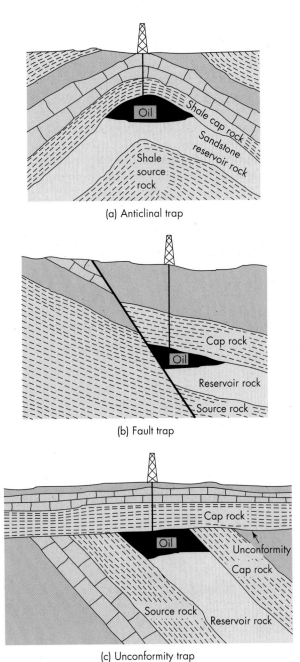

(a) Anticlinal trap

(b) Fault trap

(c) Unconformity trap

Figure 15.12 Types of oil traps (a) Anticlinal, (b) fault, (c) unconformity.

other possible traps caused by faulting or by an unconformity, which is a buried erosion surface. These are not the only possible types of traps; any rock that has a relatively high porosity and permeability that is connected to a source rock containing hydrocarbons may become a reservoir. A reservoir will form if upward migration of the oil and gas is impeded by a cap rock that is positioned to entrap the hydrocarbons at a central high point.[10]

Oil Production. Production wells in an oil field recover oil through primary or enhanced recovery methods. *Primary recovery* uses natural reservoir pressure to move the oil to the well, where it can then be pumped to the surface. Normally, this pumping delivers no more than 25 percent of the total petroleum in the reservoir. Increasing the recovery rate to 50 percent to 60 percent or more requires *enhanced recovery* methods. The enhancement manipulates reservoir pressure by injection of natural gas, water, steam, or chemicals, or some combination of those, into the reservoir, pushing petroleum to wells where it can be lifted to the surface by means of the familiar "horse head" bobbing pumps, submersible pumps, or other lift methods.

Oil production always brings to the surface a variable amount of salty water, or brine, along with the oil. After the oil and water have been separated, the brine must be disposed of, because it is toxic to the surface environment. Disposal can be accomplished either by injection as part of secondary recovery, evaporation in lined open pits, or by deep-well disposal outside the field. Disposal sites in oil fields were some of the first hazardous-waste disposal facilities, and some eventually became sites for the disposal of hazardous industrial wastes.

Distribution and Quantity of Oil and Gas. The distribution of oil and gas in space and geologic time is rather complex, but in general, three principles apply:

- First, commercial oil is produced almost exclusively from sedimentary rocks deposited during the last 500 million years of Earth's history.
- Second, although there are many oil fields in the world, most of the total production plus reserves occurs in a few percent of the producing fields.
- Third, most of the world's giant oil fields are located near plate junctions that are known to have been active in the last 70 million years.

It is difficult to assess the petroleum reserves of the United States, much less those of the entire world. However, Figure 15.13 shows a recent estimate of the

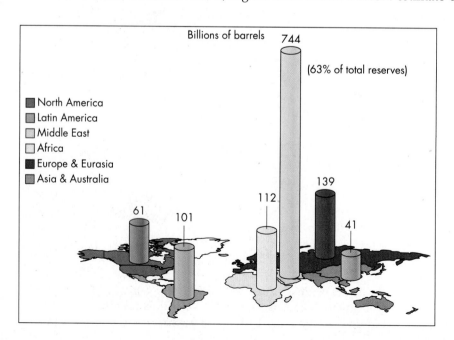

Figure 15.13 Oil reserves
Proven world oil reserves (billions of barrels) in 2004. The Middle East dominates with about two-thirds of total reserves. *(British Petroleum Company, 2005. BP statistical review of world energy)*

Billions of barrels 744
(63% of total reserves)

- North America
- Latin America
- Middle East
- Africa
- Europe & Eurasia
- Asia & Australia

61   101   112   139   41

world's reserves of crude oil. Most (63%) of the oil reserves are in the Middle East, and 70 percent of the natural gas reserves are in the former Soviet Union and the Middle East. The numbers do not suggest that we are going to run out of crude oil soon. However, there is uncertainty concerning the estimated size of the oil reserves and potential resources, and the peak in oil production will likely occur between 2020 and 2050 (see opening case history).

Natural gas is a relatively clean fuel compared with coal and oil, and it has the potential to replace coal and oil during the transition from fossil fuels to alternative energy sources. The worldwide amount of natural gas is very large, enough to last about one century at recent rates of consumption. Exploration for natural gas in the United States is ongoing, and considerable new gas is being found and developed. Two recent developments, coal-bed methane and methane hydrates, are discussed as examples.

Coal-Bed Methane. An "energy rush" is taking place in the coal fields of the western United States. The coal contains a large amount of methane stored on surfaces of the organic matter in the coal. Methane reserves in the Wyoming region including the Wasatch and Power River basins are sufficient to provide all the natural gas used by the United States for five years. Over 10,000 wells to recover the methane have already been drilled and there may eventually be 10 times that many. **Coal-bed methane** wells are shallow and much more economical to drill than the typical oil well. This relatively new source of methane is promising and will reduce the amount of energy we must import. However, there are environmental concerns associated with intensive development of coal-bed methane:[13]

- disposal of salty water that is produced with the methane
- mining of groundwater (reduction of water resources) asssociated with methane extraction
- migration of methane (a flammable, explosive gas) away from well sites
- reduction of crop production if salty water is used to irrigate crops
- pollution of stream water and loss of spring flow as groundwater is extracted and disposal of mine water enters streams and rivers
- erosion and runoff of land disturbed for roads, well sites, processing, and transportation of methane

In summary, coal-bed methane is an important energy source. However, its extraction, processing, and transportation to users must be closely evaluated to determine how best to reduce or eliminate potential adverse environmental effects. Then steps to protect the environment should be implemented into laws and policy, as we attempt to do for other types of energy exploitation and use. Extraction of any fossil fuel is inherently a process with potential to harm the environment, and coal-bed methane is not an exception.

Methane Hydrate. A potential source of natural gas is known as methane hydrate. Methane is a likely source of energy to replace oil during the transition from the fossil fuels to alternative, renewable energy sources. From an environmental perspective, burning natural gas to produce electricity and to transport people in vehicles is preferable to burning other fossil fuels. Burning gas, especially methane, is preferable because gas produces fewer air pollutants, such as sulfur, carbon dioxide, and particulates, than do other fossil fuels.

It has been known for years that methane hydrate deposits exist at depths of about 1,000 m (3,300 ft) beneath the sea. *Methane hydrate* is a white, icelike material composed of molecules of methane gas surrounded by "cages" of frozen water. The frozen water cages are formed as a result of microbial digestion of organic matter in the sediments. Methane hydrate deposits are also found on land; the first land deposits discovered were in permafrost areas of Siberia and North America, where they are called marsh gas.

Methane hydrate deposits are found throughout both the Pacific and Atlantic Oceans. Methane hydrates in the marine environment are a potential resource, with approximately twice as much carbon as all the other known natural gas, oil, and coal deposits on Earth.[14] Methane hydrate deposits have the potential to supply the energy needs of the world in the future if they can be captured. However, this is a big if. Methane hydrates look particularly attractive to countries such as Japan that rely exclusively on foreign oil and coal for fossil fuel needs.

Mining methane hydrates will be a difficult task. Hydrates tend to be found along the lower parts of the continental slopes, where water depths are often greater than 1 km (0.62 mi), deeper than most drilling rigs can safely operate. The deposits themselves extend below the ocean floor sediments for another few hundred meters. Development of a method to produce the gas and transport it to land will be complicated. Methane hydrates are found in areas that are both highly pressurized and extremely cold because of the depth. They are not stable in the ocean at lower pressure and higher temperature. At a water depth of about 500 m (1,640 ft) the methane hydrates decompose rapidly, allowing methane gas to travel in a flow of bubbles up to the surface, where it can enter the atmosphere. This feature poses an environmental threat because methane is a strong greenhouse gas with the potential to contribute to global warming (see Chapter 19). Some scientists believe that decomposition of methane hydrates in the past has contributed to global warming; moreover, gas continues to be released today. In the summer of 1998, researchers from Russia discovered unstable methane hydrates off the coast of Norway. They believe the deposits caused one of the largest identifiable releases of methane. During the release, scientists documented large (one-half kilometer high) plumes of methane gas being emitted into the atmosphere from the deposits on the seafloor.

Some scientists believe that methane from hydrates rapidly released from a single large source can create potential climate changes over a short period of time. It is believed that such releases of methane could have been responsible for the rapid increase in temperature that occurred within a few decades during the last Ice Age, approximately 15,000 years ago. There is evidence that there had been large emissions of methane from the seafloor. Fields of large depressions resembling bomb craters pock-mark the bottom of the sea in the vicinity of methane hydrate deposits. The size of some of the craters, about 30 m (98 ft) deep and 700 m (2,296 ft) in diameter, suggests that they were produced by an explosive eruption of methane.

We will probably not attempt to exploit methane hydrates until the price of oil is so prohibitively high that we have no choice but to look for alternatives. This may be soon! Several potential schemes for exploiting the deposits are being considered, and over the next few years progress is likely.[14]

Impact of Exploration and Development. The environmental impact of oil and gas exploration and development varies from negligible—such as remote-sensing techniques used in exploration—to significant—such as the unavoidable impact of projects like the Trans-Alaska Pipeline. The impact of exploration for oil and gas can include building roads, exploratory drilling, and building a supply line to remote areas. Except in sensitive areas, including some semiarid-to-arid environments and some permafrost areas, these activities generally cause few adverse effects to the landscape and resources.

Development of oil and gas fields involves drilling wells on land or beneath the sea; disposing of wastewater brought to the surface with the petroleum; transporting the oil by tankers, pipelines, or other methods to refineries; and converting the crude oil into useful products. At every step, there is a well-documented potential for environmental disruption, including problems associated with wastewater disposal, accidental oil spills, leaking pipes in oil fields, shipwrecks of tankers, and air

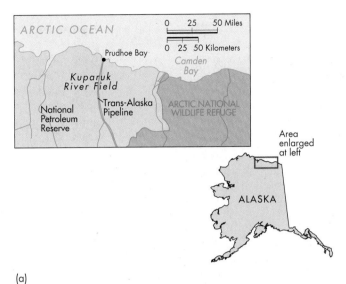

(a)

Figure 15.14 Arctic National
Wildlife Refuge (a) Map with location
of the Arctic National Wildlife Refuge.
(b) Mountains, plains, and wildlife
of the area. *(Ted Kerasote/Photo
Researchers, Inc.)*

(b)

pollution at refineries. For instance, serious oil spills along the coastlines of Europe and North America have spoiled beaches, estuaries, and harbors; have killed marine life and birds; have polluted groundwater and surface water; and have caused economic problems for coastal areas that depend on tourist trade.

As a result of current energy shortages, there is pressure to develop new oil fields in offshore areas and wilderness areas such as the Arctic National Wildlife Refuge (ANWR) in Alaska, where several billion barrels of oil are located (Figure 15.14).

The argument in favor of exploration and development of oil in ANWR is threefold. First, the United States needs more oil, and the oil in ANWR would reduce the amount of oil we import. Second, new oil development will bring jobs and economic growth to Alaska. Third, new techniques for exploration and drilling oil wells have caused less adverse environmental damage than previous development (Figure 15.15a).

These techniques include

■ Using directional drilling from central locations where many wells form a single site spread out underground like the spokes of a bicycle wheel (Figure 15.15a). This technique reduces the land area disturbed by drilling to sites of about 40 ha (100 acres).

■ Elevating pipelines above ground to allow animal migration (Figure 15.15b).

■ Avoiding construction of permanent roads. Roads are made of ice in the winter and melt in the summer (Figure 15.15a).

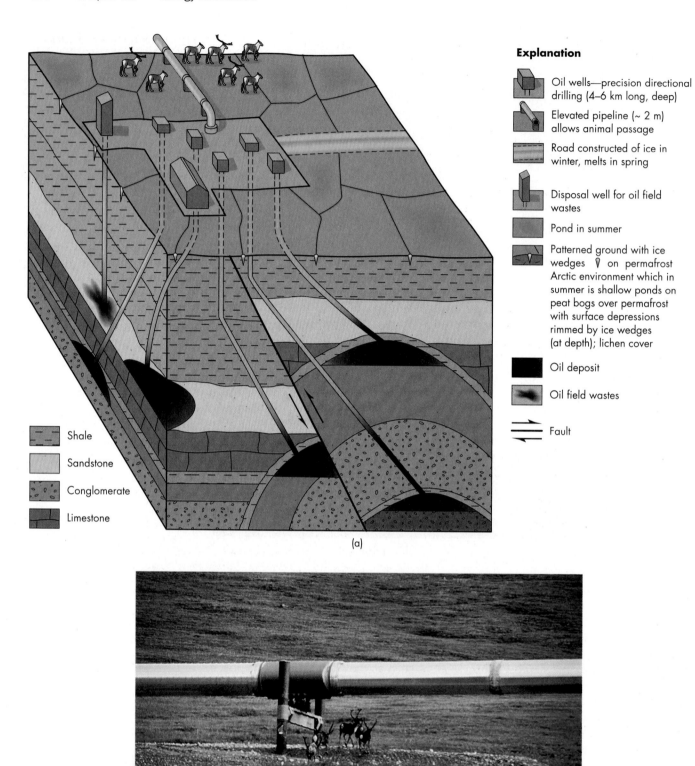

Explanation

Oil wells—precision directional drilling (4–6 km long, deep)

Elevated pipeline (~ 2 m) allows animal passage

Road constructed of ice in winter, melts in spring

Disposal well for oil field wastes

Pond in summer

Patterned ground with ice wedges ⊽ on permafrost Arctic environment which in summer is shallow ponds on peat bogs over permafrost with surface depressions rimmed by ice wedges (at depth); lichen cover

Oil deposit

Oil field wastes

Fault

Shale

Sandstone

Conglomerate

Limestone

(a)

(b)

Figure 15.15 New oil drilling and transport technology (a) Some of the new technology that could reduce the impact of developing oil at the Arctic National Wildlife Refuge. (b) Caribou crossing beneath an elevated pipeline. *(U.S. Fish and Wildlife Service/Getty Images Inc.)*

Opponents of the exploration and development of oil in ANWR state

- Even the most favorable technology will adversely affect ANWR.
- Some wilderness areas should remain wilderness, and drilling will permanently change ANWR.
- Development of oil fields is inherently damaging because it requires an extensive industrial complex of machines, vehicles, pipelines, and people.
- The oil beneath ANWR will not be a significant contribution to the U.S. oil supply. Although it would be spread out over decades, providing a few percent of our total oil consumption during that time, in total the ANWR oil itself would provide only about a 6-month to 2 year supply of oil.

The decision concerning development of oil at ANWR reflects the relationship between science and values. Science says we can develop the oil with less environmental damage than in the past. But even "less than in the past" may be too much. Our values will be determined by balancing our economic need for oil against our desire to preserve a pristine wilderness area.

Impact of Use. A familiar and serious impact associated with oil is air pollution produced in urban areas when fossil fuels are burned to produce energy for electricity, heat, and automobiles. Also, as does burning coal, burning oil and gas releases carbon dioxide, the major human-produced greenhouse gas (see Chapter 19).

Oil Shales and Tar Sands. Recovering petroleum from surface or near-surface oil shale and tar sands involves the use of well-established techniques. What are needed are economic methods for developing oil shale resources that will minimize environmental disruption.

Oil shale is a fine-grained sedimentary rock containing kerogen, a solid or semisolid hydrocarbon mineraloid substance, formed from organic matter. When heated, the kerogen in oil shale yields significant amounts of hydrocarbons that are otherwise insoluble in ordinary petroleum solvents.[15] This process is known as destructive distillation. The best known oil shales in the United States are found in the 50-million-year-old Green River Formation that underlies parts of Colorado, Utah, and Wyoming. The Green River Formation consists of oil shale interbedded with variable amounts of sandstone, siltstone, claystone, and compacted volcanic ash. The oil shale resource is huge. About one hundred billion barrels of oil could potentially be recovered. Mining of oil shale using known technology is expensive. Nevertheless as oil prices rise, oil-shale mining is being reconsidered.

Tar sands are rocks that are impregnated with tar oil, asphalt, or other petroleum materials. The recovery of any of these petroleum materials by traditional methods is not commercially possible. The term *tar sand* is somewhat confusing because it includes several rock types, such as shale and limestone, as well as unconsolidated and consolidated sandstone. However, these rocks all contain a variety of semiliquid, semisolid, and solid petroleum products. Some of these products ooze from the rock; others are difficult to remove even with boiling water.[16]

Although similar to oil pumped from wells, the oil from tar sands is much more viscous and therefore more difficult to recover. One possible conclusion concerning the geology of tar sands is that they form in essentially the same way that the more fluid oil forms, but because much more of the volatiles and accompanying liquids in the reservoir rocks have escaped, the more viscous components remain as tar sands.

Large accumulations of tar sands have been identified. For example, the Athabasca Tar Sands of Alberta, Canada, cover an area of approximately 78,000 km^2 (30,116 mi^2) and contain an estimated reserve of 35 billion barrels of oil that might be recovered.[10,17] These are now yielding about 1 million barrels of synthetic crude oil per day from two large strip mines.[10] Smaller tar-sand resources are known in Utah (1.8 billion barrels), Venezuela (900 million barrels), and California (100 million barrels).[10,17]

15.4 Future of Oil

Recent estimates of proven oil reserves in the world suggest that, at present production rates, oil and natural gas will last a few decades.[1,18] However, the important question to ask is not how long oil will last at present and future production rates, but at what point will we reach peak production? This question is important because once we have reached peak production, less oil will be available and there will be shortages and price shocks. The world oil production peak is likely to occur within the lifetime of many people living today,[19] much sooner than generally expected by most people. As a result, there may be little time left to adjust to potential changes in lifestyle and economies in a postpetroleum era. Walter Youngquist, an energy expert, argues that we are fortunate to be living in a prosperous period of human history made possible by our inheritance of a 500 million year period of oil-forming processes.[19] We will never entirely run out of crude oil, but people of the world depend upon oil for nearly 40 percent of their energy, and significant shortages will cause major problems.[20]

The following factors provide evidence that we are heading toward a potential crude oil crisis:

- We are approaching the time when approximately one-half of the total crude oil from traditional oil fields on Earth will have been consumed.[19] Revisit the case history on peak oil that opened this chapter. Recent estimates suggest that the world may have about 20 percent more oil awaiting discovery than predicted a few years ago.,[21] Furthermore, more oil exists in known fields than was earlier thought. However, the United States has already consumed approximately half of its oil resources, necessitating additional oil imports in the future.[18]

- Proven world reserves total approximately 1 trillion barrels.[1,20] It is estimated that about 3 trillion barrels of crude oil may ultimately be recovered from remaining oil resources.

- World consumption today is about 30 billion barrels per year (82 million barrels per day), and for every 3 barrels of oil consumed, only 1 barrel is discovered.[20]

- The predicted decline in oil production is based on the estimated amount of oil that may ultimately be recoverable as well as on projections of new discoveries and rates of future consumption. The peak in world crude oil production (50 billion barrels per year) is predicted to occur from about 2020 to 2050, depending upon whether you are an optimist or a pessimist.[18,20] Production of 50 billion barrels per year is nearly a 50 percent increase over production today. Whether you believe the increased production will be a problem depends on your view of predicted shortages from past history of oil consumption. We have survived several predicted shortages in the past. However, most oil experts believe that peak oil is only a few decades away.[1,2]

- It is predicted that significant production of oil in the United States will not extend beyond the year 2090. The world production of oil will be nearly exhausted by 2100.[20]

What is an appropriate response to the above statements? We need to begin with a major educational program that informs both people and governments of the potential oil depletion in the twenty-first century. We are currently operating

under ignorance or denial in the face of a potentially serious situation. Planning and education are important to avoid future military confrontations like the Gulf War and food shortages due to the reduced availability of oil to produce fertilizers for agriculture. Before significant oil shortages occur, energy sources such as large-scale gasification and liquefaction of our tremendous coal reserves should be developed. Other energy sources include oil and gas from oil shale and atomic energy. Certainly we should rely more on alternative energy sources including solar energy, wind power, and hydrogen. These changes will strongly affect our present petroleum-based society. However, there appears to be no insurmountable problem if we implement meaningful short- and long-range plans to simultaneously phase out oil, transition to natural gas, and phase in alternative energy sources.

With a better understanding of the nature and abundance of the fossil fuels, we will turn to the topic of acid rain, a major environmental consequence of burning fossil fuels.

15.5 Fossil Fuel and Acid Rain

Acid rain is thought to be a regional to global environmental problem related to the burning of fossil fuels. Although acid rain is caused primarily by coal burning, gasoline combustion also contributes to the problem. *Acid rain* refers to both wet and dry acid deposition. *Wet* acid deposition occurs when pollutants, acid precursors such as sulfur dioxide (SO_2) and nitrogen oxides (NO_x), react with water vapor in the atmosphere, producing acids. *Dry* acid deposition occurs when the particles containing acid precursors fall to Earth and then react with water to produce acids. Therefore, although *acid rain* is more commonly used, *acid deposition* is a more precise term. Acid rain is defined as precipitation in which the pH is below 5.6. The pH is a numerical value of the relative concentration of hydrogen ions (H^+) in a solution that is used to describe the solution's acidity. A solution with pH of 7 is a neutral solution, one that is not acidic. Natural rainfall is slightly acidic and has a pH of approximately 5.6, caused by water in the atmosphere combining with carbon dioxide to produce a weak carbonic acid ($H_2O + CO_2 \rightarrow H_2CO_3$). The pH scale is a negative logarithmic scale; for example, a pH value of 3 is 10 times more acidic than a pH value of 4 and 100 times more acidic than a pH value of 5 (Figure 15.16). Rainfall in Wheeling, West Virginia, was once measured at a pH value of 1.5, nearly as acidic as stomach acid, and pH values as low as 3 have been recorded in other locations.

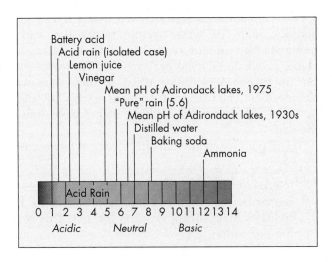

Figure 15.16 The pH scale
Values for a variety of materials are shown. *(Modified after U.S. Environmental Protection Agency, 1980)*

Figure 15.17 How acid rain forms Paths and processes associated with acid rain. *(Modified after Albritton, D. L., as presented in Miller, J. M.)*

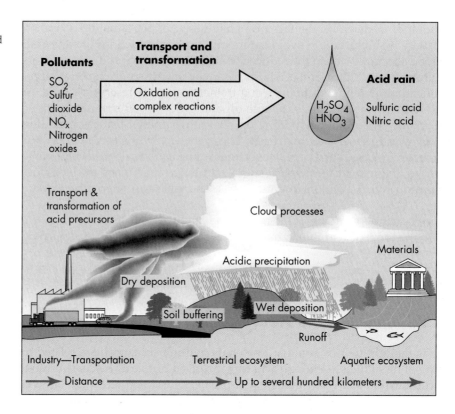

Today, fossil fuel burning in the United Sates annually releases about 20 million tons each of nitrogen oxide and sulfur dioxide into the atmosphere. After emission, these oxides are transformed to sulfate (SO_4) or nitrate (NO_3) particles that can combine with water vapor to eventually form sulfuric (H_2SO_4) and nitric (HNO_3) acids. These acids may travel long distances with prevailing winds and be deposited as acid rain (Figure 15.17). The acid rain problem we are most familiar with results from sulfur dioxide, which is primarily emitted from burning coal in power plants that produce electricity in the eastern United States.

Environmental Effects of Acid Rain

Geology, climate patterns, type of vegetation, and composition of soil all affect potential impacts of acid rain. Figure 15.18 shows areas in the United States that are sensitive to acid rain, and identification of these areas is based on some of the above-mentioned factors. Particularly sensitive areas are those in which the bedrock, soils, or water cannot buffer the acid input. For instance, areas dominated by granitic rocks have little buffering action. *Buffering* refers to the ability of a material to neutralize acids. These materials, known as *buffers*, include calcium carbonate or calcite ($CaCO_3$), found in many types of soils and rock such as limestone. The calcium carbonate reacts with and removes the hydrogen ions in the acidic water by forming bicarbonate ions (HCO_3^-) and neutralizing the acid.

The major environmental effects of acid rain include

- Damage to vegetation, especially forest resources such as evergreen trees in Germany and red spruce trees in Vermont. The soil's fertility may be reduced, either because nutrients are leached out by the acid or because the acid releases toxic substances into the soil. This is the main effect acid rain has on vegetation.

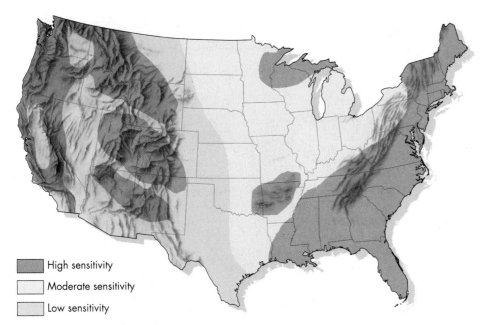

Figure 15.18 Areas in the United States sensitive to acid rain

(From U.S. Environmental Protection Agency, 1980)

High sensitivity

Moderate sensitivity

Low sensitivity

- Damage to lake ecosystems. Acid rain may damage lake ecosystems by (1) disrupting the life cycles of fish, frogs, and crayfish and (2) interfering with the natural cycling of nutrients and other chemical elements necessary for life. The acid rain tends to keep the nutrients in solution so that they leave the lake rather than being cycled in the system. As a result, aquatic plants may not grow, and animals that feed on these plants have little to eat. This degradation is passed up the food chain to the fish and other larger animals. Examples of adverse effects of acid rain on lake systems have been described in Canada and Scandinavia.

- Damage to human structures. Acid rain damages building materials, plastics, cement, masonry, galvanized steel, and several types of rocks, especially limestone, sandstone, and marble (Figure 15.19). In cities around the world, irreplaceable statues and buildings have been significantly damaged, resulting in losses that reach billions of dollars per year.

A Solution to the Acid Rain Problem

Lake acidification can be offset by periodic addition of a buffer material such as calcium carbonate. Although this remedy has been used in several areas, including New York State, Sweden, and Canada, adding buffer material to lakes is a short-term, expensive solution to lake acidification. The only practical long-term solution to the acid rain problem is to reduce the emissions of the chemicals that cause the problem. From an environmental viewpoint, the best way to reduce emissions is to practice strong energy conservation, which would result in lower emissions. The second best way is to treat the coal before, during, and after burning to intercept the sulfur dioxide before it is released into the environment. Reduction of nitrogen oxide is more difficult because it is primarily caused by gasoline burning in automobiles. Nevertheless, control strategies do exist to reduce emissions of SO_2 and NO_x (see Chapter 18). It is encouraging that the reduction of pollutants causing acid rain is a national and international goal. Unfortunately, the acid rain problem will not soon go away even with lower emissions of acid precursors. Acid deposition has been accumulating in soils for decades, and its effects will linger for decades to come.[22]

Figure 15.19 **Acid rain damages stone** Air pollution and acid rain are damaging buildings and statues in many urban regions. Shown here is the Acropolis in Athens, Greece. Statues here have been damaged to such an extent that originals have been placed inside buildings in specially constructed glass containers. *(Peter Christopher/ Masterfile Corporation)*

15.6 Nuclear Energy

Energy from Fission

Production of **nuclear energy** relies mostly on fission. *Nuclear fission* is the splitting of atomic nuclei by neutron bombardment. Fission of the nucleus of a uranium-235 atom, for example, releases three neutrons, fission fragments composed of the nuclei of radioactive elements that are lighter than uranium, and energy in the form of heat (see A Closer Look: Radioactivity). The released neutrons strike other U-235 atoms, releasing more neutrons, fission products, and heat. The process continues in a *chain reaction*, and as more and more uranium is split, it releases ever more neutrons (Figure 15.20). An uncontrolled chain reaction—the kind used in nuclear weapons—leads quickly to an explosion. However, sustained or stable nuclear reactions in reactors are used to provide heat for the generation of electricity.

The first controlled nuclear fission was demonstrated in 1942, leading the way for the use of uranium in explosives and as a heat source that provides steam for electricity generation. Fission of 1 kg (2.2 lb) of uranium releases approximately the same amount of energy as the burning of 16 metric tons of coal.

Three isotopes of uranium are found in a naturally occurring uranium sample: U-238, which accounts for approximately 99.3 percent of natural uranium; U-235, which makes up just under 0.7 percent; and U-234, which makes up about 0.005 percent. Uranium-235 is the only naturally occurring fissionable material and is therefore essential to the production of nuclear energy. Naturally occurring uranium is processed to increase the amount of U-235 from 0.7 percent to about 3 percent before

CHAIN REACTION

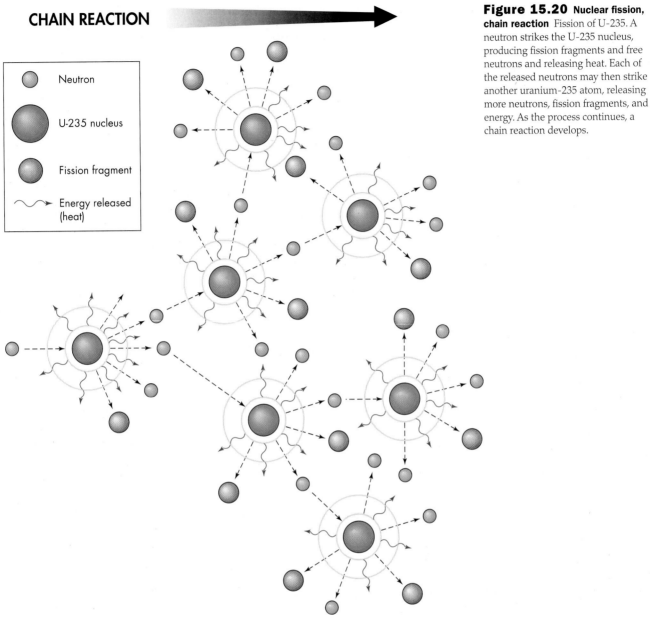

Figure 15.20 Nuclear fission, chain reaction Fission of U-235. A neutron strikes the U-235 nucleus, producing fission fragments and free neutrons and releasing heat. Each of the released neutrons may then strike another uranium-235 atom, releasing more neutrons, fission fragments, and energy. As the process continues, a chain reaction develops.

Legend: Neutron; U-235 nucleus; Fission fragment; Energy released (heat)

it is used in a reactor. The processed fuel is called enriched uranium. Uranium-238 is not naturally fissionable, but it is "fertile material" because upon bombardment by neutrons it is converted to plutonium-239, which is fissionable.[23]

Geology and Distribution of Uranium

The natural concentration of uranium in Earth's crust averages approximately 2 parts per million (ppm). Uranium originates in magma and is concentrated to about 4 ppm in granitic rock, where it is found in a variety of minerals. Some uranium is also found in late-stage igneous rocks such as pegmatites. Uranium must have a concentration factor of 400 to 2,500 times the natural concentration to be mined at a profit.[23]

Fortunately, uranium forms a large number of minerals, many of which have been found and mined. Three types of deposits have produced most of the uranium in the last few years: sandstone impregnated with uranium minerals, veins of uranium-bearing materials localized in rock fractures, and as 2.2 billion-year-old placer deposits, now coarse-grained sedimentary rock.[24]

Reactor Design and Operation

Most reactors today consume more fissionable material than they produce and are known as *burner reactors*. The reactor itself is part of the nuclear steam supply system, which produces the steam to run the turbine generators that produce electricity.[25] (Figure 15.21).

The main components of the reactor shown in Figure 15.22 are the core, control rods, coolant, and reactor vessel. The core of the reactor, where the chain reaction takes place, is contained in a heavy stainless steel reactor vessel. For increased safety and security, the entire reactor is contained in a reinforced concrete building called a containment structure. Fuel pins, which consist of enriched uranium pellets in hollow tubes less than 1 cm (0.4 in.) in diameter, are packed together (40,000 or more in a reactor) into fuel subassemblies in the core. A stable fission chain reaction is maintained by controlling fuel concentration and the number of released neutrons that are available to cause fission. A minimum fuel concentration is necessary to keep the chain reaction self-sustaining, or *critical*; control of the number of neutrons is necessary to regulate the reaction rate. The number of neutrons is

A CLOSER LOOK | Radioactivity

All atoms of the same element have the same atomic number; that is, they have the same number of protons in the nucleus. Isotopes are atoms of an element that have different numbers of neutrons and therefore different atomic masses, defined by the number of protons plus the number of neutrons in the nucleus (see Chapter 3). For example, two isotopes of uranium are $^{235}U_{92}$ and $^{238}U_{92}$. The atoms of both of these isotopes have an atomic number of 92, but their atomic mass numbers are 235 and 238, respectively. These isotopes may be written as uranium-235 and uranium-238, or U-235 and U-238, respectively.

Atomic mass

$$^{235}U_{92}$$

Atomic number

Some isotopes, called radioisotopes, are radioactive and spontaneously undergo nuclear decay. Nuclear decay occurs when a radioisotope undergoes a nuclear change while emitting one or more forms of radioactive radiation. The three major kinds of radiation emitted during nuclear decay are called alpha particles (α), beta particles (β), and gamma radiation (γ). Each radioisotope has its own characteristic emissions; some isotopes emit only one type of radiation; others emit a mixture of radiation types.

Alpha particles consist of two protons and two neutrons, making them much more massive than other types of radioactive emission. Because alpha decay, emission of an alpha particle, changes the number of protons and neutrons in the atom's nucleus, the isotope is changed into an isotope of a different element. For example, a radon-222 atom, which has 86 protons, emits an alpha particle and is thereby transformed into a polonium-218 atom, which has 84 protons. Because of their great mass, alpha particles are the slowest-moving of the radioactive emissions and have the lowest energy. They travel the shortest distances—approximately 5 to 8 cm (2 to 3 in.) in

air—and penetrate solid matter less deeply than do beta or gamma emissions.

Beta particles are energetic electrons that have a small mass compared with alpha particles. Beta decay occurs when one of the neutrons in the nucleus of the isotope spontaneously changes.[26] Note that the electron emitted is a product of the transformation; remember electrons are found surrounding the nucleus, not in it (see Chapter 3).

In gamma radiation, gamma rays are emitted from the isotope, but the number of protons and the number of neutrons in the nucleus are unchanged. Gamma rays are similar to medical X-rays, but they are usually more energetic and more penetrating. Gamma rays have the highest energy of all radioactive emissions; they travel faster and farther and penetrate more deeply than do alpha or beta particles.

An important characteristic of a radioisotope is its half-life, which is the time required for one-half of a given amount of the isotope to decay to another form. Every radioisotope has a unique characteristic half-life. Radon-222, for example, has a relatively short half-life of 3.8 days. Carbon-14, a radioactive isotope of carbon, has a half-life of 5,570 years; U-235 has a half-life of 700 million years; and U-238 has a half-life of 4.5 billion years.

Some radioisotopes, particularly those of very heavy elements, undergo a series of radioactive decay steps, until they finally become a stable, or nonradioactive, isotope. Figure 15.B shows the decay chain from U-238 through radium-226, radon-222, and polonium-218 to the stable isotope lead-206. The two most important facts about each transformation are the type of radiation emitted and the half-life of the isotope that is transformed (Figure 15.B). The decay from one radioisotope to another is often stated in terms of parent and daughter atoms. For example, the parent radon-222, a gas with a half-life of 3.8 days, decays by alpha emission to its daughter, polonium-218, a solid with a half-life of 187 seconds.

Radiation emitted			Radioactive elements	Half–life		
Alpha	Beta	Gamma		Minutes	Days	Years
●		●	← Uranium–238 ↓			4.5 billion
	●	●	← Thorium–234 ↓		24.1	
	●	●	← Protactinium–234 ↓	1.2		
●		●	← Uranium–234 ↓			247,000
●		●	← Thorium–230 ↓			80,000
●		●	← Radium–226 ↓			1,622
●			← Radon–222 ↓		3.8	
●	●		← Polonium–218 ↓	3.0		
	●	●	← Lead–214 ↓	26.8		
	●	●	← Bismuth–214 ↓	19.7		
●			← Polonium–214 ↓	0.00016 (sec.)		
	●	●	← Lead–210 ↓			22
	●		← Bismuth–210 ↓		5.0	
●		●	← Polonium–210 ↓			138.3
None			← Lead–206			Stable

Figure 15.B Decay chain and half-lives from radioactive U-238 to stable lead-206. (From Schroyer, F., ed. 1985. Radioactive waste, *2nd ed. American Institute of Professional Geologists*)

controlled by the control rods, which contain materials that capture neutrons, preventing them from bombarding other nuclei. When the rods are pulled out of the core, the chain reaction speeds up; when they are inserted into the core, the reaction slows down.

Pumps circulate a coolant, usually water, through the reactor, extracting the heat produced by fission in the reactor. The rate of generation of heat must match the rate at which heat is carried away by the coolant. Matching the rates is usually not difficult, and reactors run smoothly and stably. However, the major nuclear accidents that have occurred happened when something went wrong and heat in the reactor core built up. A *meltdown* refers to a nuclear accident in which the nuclear fuel becomes so hot that it forms a molten mass. The containment of the reactor fails, and radioactivity contaminates the environment.[27]

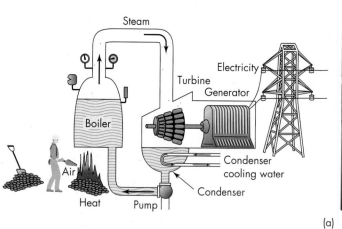

(a)

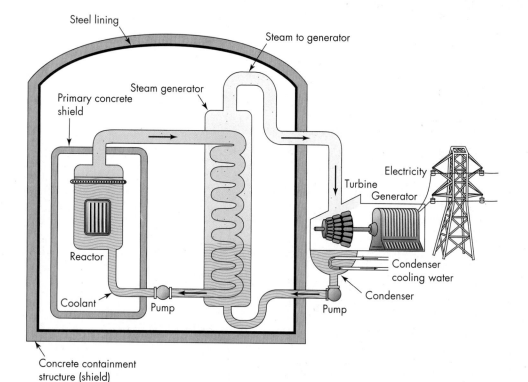

(b)

Figure 15.21 Comparing fossil fuel and nuclear power plants
(a) Fossil fuel power plant and (b) nuclear power plant with a boiling-water reactor. Notice that the nuclear reactor has exactly the same function as the boiler in the fossil fuel power plant. *(Reprinted, by permission, from* Nuclear power and the environment, American Nuclear Society, 1973) Photographs: (a) This fossil fuel power plant, Skytell Bridge in Tampa Bay, Florida, burns coal. Components include the storage of coal in the lower right-hand corner, the power plant itself in the center, cooling water leaving the power plant on the left, and the series of electric power lines leading away from the power plant. *(Wingstock/ Comstock Images)* (b) Diablo Canyon Nuclear Power Plant near San Luis Obispo, California. Reactors are in the dome-shaped buildings, and cooling water is escaping to the ocean. The siting of this power plant has been and remains very controversial because of its proximity to faults capable of producing earthquakes that might damage the facility. *(Comstock Images)*

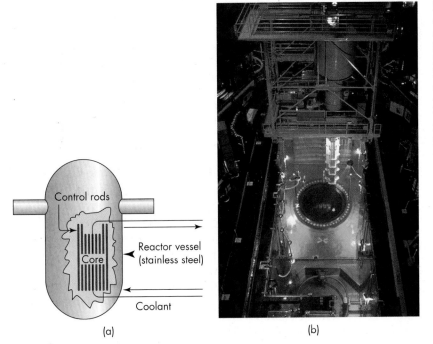

Figure 15.22 Nuclear reactor
(a) Diagram of the main components of a nuclear reactor. (b) Part of the interior of a nuclear reactor is shown here. The site is the Flamanville Nuclear Power Station, France. The core of the reactor is in the blue water pool in the center. The crane above is used to carry fuel rods to change the fuel before the reactor is operated. The power plant is a pressurized water reactor. Nuclear power accounts for approximately 75 percent of France's electricity. *(Catherine Pouedras/Science Photo Library/Photo Researchers, Inc.)*

Other parts of the nuclear power system are the primary coolant loops and pumps, which circulate the coolant through the reactor, and heat exchangers or steam generators, which use the fission-heated coolant to make steam (Figure 15.21b).

An emerging trend in the nuclear industry is to design less complex, smaller reactors that are safer. Large nuclear power plants, which produce about 1,000 MW of electricity, require a set of pumps and backup equipment to ensure that adequate cooling is available to the reactor. Smaller reactors can be designed with cooling systems that work under the influence of gravity and are not as vulnerable to failure from loss of electrical power or pump failure. The term for such a system is *passive stability*, and these reactors are said to be *passively safe*.[28] Another approach is to use helium gas to cool reactors. These reactors have specially designed fuel capsules capable of withstanding temperatures of 1,800°C (about 3,300°F). Fuel assemblies are designed to be incapable of holding enough fuel to reach this temperature, preventing the possibility of a core meltdown.

A gas-cooled reactor known as "pebble-bed reactor" may be available in the next few years. The reactor uses fuel elements called pebbles that are about the size of a pool ball. Pebbles are fed into the reactor using an analogy to a gum ball machine; one ball is realeased as another takes its place. This is a safety feature as the reactor only has the amount of fuel at any one time necessary for optimal energy production.[29]

Risks Associated with Fission Reactors

Nuclear energy and the possible adverse effects associated with it have been subjects of vigorous debate. The debate is healthy, because we should continue to examine the consequences of nuclear power generation very carefully.

Nuclear fission uses and produces radioactive isotopes. Various amounts of radiation are released into the environment at every step of the nuclear cycle: mining and processing of uranium, controlled fission in reactors, reprocessing of nuclear fuel, and final disposal of the radioactive wastes. Serious hazards are associated with transporting and disposing of nuclear material as well as with supplying other nations with reactors. Furthermore, since the plutonium produced by nuclear reactors can be used to make nuclear weapons, terrorist activity and the possibility of irresponsible actions by governments add a risk that is present in no other form of energy production.

An uncontrolled chain reaction—a nuclear explosion—cannot occur in a nuclear reactor because the fissionable material is not used in the concentrated form necessary for an explosion. However, unwanted chemical reactions in a reactor can produce explosions that release radioactive substances into the environment. Although the chance of a disastrous accident is estimated to be very low, it increases with every reactor put into operation. Major accidents have already occurred, including Three Mile Island near Harrisburg, Pennsylvania, in 1979 and the disastrous accident at Chernobyl in the former Soviet Union, now Ukraine, in 1986.

Three Mile Island. One of the most serious events in the history of U.S. nuclear power occurred on March 28, 1979, at the Three Mile Island nuclear power plant near Harrisburg, Pennsylvania. The malfunction of a valve and human errors at the nuclear plant resulted in a partial core meltdown, with the release of radioisotopes into the environment. Although intense radiation was released into the interior of the containment structure, it functioned as designed, and only a relatively small amount of radiation was released into the environment.

The Three Mile Island incident clearly demonstrated that there are potential problems with nuclear power. Historically, nuclear power had been relatively safe, and the state of Pennsylvania was unprepared for the accident. One of the serious impacts of the incident was fear, but, surprisingly, no staff member from the Department of Health was allowed to sit in on briefing sessions.

Because the long-term chronic effects of exposure to low levels of radiation are not well understood, the effects of Three Mile Island exposure, although apparently small, are not easy to estimate. Furthermore, the Three Mile Island accident illustrates that our society needs to improve the way in which it handles crises arising from the release of pollutants created by our modern technologies.[30]

Chernobyl. Lack of preparedness was dramatically illustrated by events that started on the morning of Monday, April 28, 1986. Workers at a nuclear facility in Sweden, frantically searching for the source of elevated levels of radiation near their power plant, concluded that it was not their facility that was leaking radiation but that the radioactivity was coming from the Soviet Union by way of prevailing winds. Confronted, the Soviets admitted that there had been an accident at their nuclear power plant at Chernobyl on April 26. This was the first notice that the world's worst accident in the history of nuclear power had occurred.

It is believed that the system that supplies cooling waters for the Chernobyl reactor failed as a result of human error. This caused the temperature of the reactor core to rise to over 3,000°C (about 5,400°F), melting the uranium fuel. Explosions blew off the top of the building over the reactor, and the graphite surrounding the fuel rods ignited. The fires produced a cloud of radioactive particles that rose high into the atmosphere. There were 237 confirmed cases of acute radiation sickness, and 31 people died of radiation sickness.[31] In the days following the accident, about 3 billion people in the Northern Hemisphere received varying amounts of radiation from Chernobyl. With the exception of the 30 km (19 mi) zone surrounding Chernobyl, the human exposure was relatively small. In Europe, where exposure was highest, it was less than the amount of natural radiation received over the course of 1 year.[31]

Approximately 115,000 people in that 30 km (19 mi) zone were evacuated, and as many as 24,000 people were estimated to have received an average radiation dosage of several hundred times higher than a natural annual exposure. This group of people is being studied carefully. Studies have found that the annual number of childhood thyroid cancer cases has risen steadily in the three countries of Belarus, Ukraine, and the Russian Federation (those most affected by Chernobyl) since the accident. In 1994, there were 132 new thyroid cancer cases, and, since the accident, a total of 653 thyroid cancer cases have been diagnosed in children and adolescents.

These cancer cases are thought to be linked to the accident, although other factors, including environmental pollution, may also play a role. It is predicted that a few percent of the 1 million children exposed to the radiation eventually will contract thyroid cancer.[32] Outside the 30 km (19 mi) zone the increased risk of contracting cancer is small. However, one estimate is that over the next 50 years Chernobyl will have been responsible for approximately 16,000 deaths worldwide.[28]

In the area surrounding Chernobyl, radioactive materials continue to contaminate soils, vegetation, surface water, and groundwater, presenting a hazard to plants and animals. The evacuation zone may be uninhabitable for a very long time unless a way is found to remove the radioactivity.[31] Estimates of the total cost of the Chernobyl accident vary widely but will probably exceed $200 billion.

Although the Soviets were accused of not paying sufficient attention to reactor safety and of using outdated equipment, people are still wondering if such an accident could happen again elsewhere. Because there are several hundred reactors producing power in the world today, the answer is yes. The Chernobyl accident follows a history of about 10 accidents that released radioactive particles between 1967 and 2001. Therefore, although Chernobyl is the most serious nuclear accident to date, it certainly was not the first and is not likely to be the last. As a result of the Chernobyl accident, risk analysis in nuclear power is now a real-life experience rather than a computer simulation.

Chernobyl was finally shut down on December 15, 2000, when reactor no. 3 was switched off, 14 years after the accident in reactor no. 4. Reactor no. 2 was closed in 1991 after a fire, and no. 1 was shut down in 1996. The closure occurred as a result of international pressure, with the West agreeing to pay for cleanup work and to construct two new nuclear power plants at other sites in the Ukraine.

The area around the accident may not be safe for hundreds of years. The city of Prypyat, 5 km (3 mi) from Chernobyl, is a "ghost city." At the time of the accident, the population of Prypyat was 48,000. Today, the town is abandoned, with blocks of vacant apartment buildings and rusting vehicles. The roads are cracking, and trees and other vegetation are transforming the once-urban land back to green fields. Prypyat was evacuated within 2 days of the accident. This was just 3 days before May Day, which was the celebration of Soviet power. Today, cases of thyroid cancer are still increasing. The final impact of the world's most serious nuclear accident is yet to completely unfold.[33]

The Future of Energy from Fission

Nuclear power produces about 20 percent of our electricity in the United States today. About 103 reactors are now in operation, more than 80 percent of which are in the eastern United States. Many more reactors are needed to realize uranium's potential energy contribution.

The United States has recently suggested, as national policy, that nuclear power should be expanded in the future. Evidently, more Americans agree with this suggestion than in previous years; however, there remains significant opposition to the idea. At the same time, some European countries are moving in the opposite direction. For example, in Germany, where about one-third of the country's electrical energy is produced from nuclear power, the decision has been made to shut down all their nuclear power plants by 2025 as the plants become obsolete. Environmentalists argue that the time line for the shutdown is too long. Others, who favor nuclear power, hope that a future government will overturn the decision. They fear that Germany will be forced to rely more heavily on fossil fuels that pollute the air and increase emission of carbon dioxide, potentially contributing to global warming (see Chapter 19).

Nuclear fission may indeed be one of the answers to our energy problems. Its use is being seriously evaluated for future increase because it is an alternative to fossil fuels that does not release carbon dioxide into the atmosphere and does not release pollutants that cause acid rain. However, with the use of nuclear power comes the responsibility of ensuring that it will be used for people, not against them, and that future generations will inherit a quality environment free from worry about hazardous nuclear waste. There is also concern that as nuclear power plants age and are eventually decommissioned there will be significant new costs. The actual cost of decommissioning, or taking apart, a reactor, disposing of waste, and cleaning up the site is several hundred million dollars. Finally, with the use of nuclear energy from fission comes the problem of waste management. This is the subject of our next discussion.

Radioactive Waste Management

Radioactive waste management, the safe disposal of radioactive waste, is a significant environmental issue facing the United States and the rest of the world. Radioactive wastes are by-products that must be expected as electricity is produced from nuclear reactors or weapons are produced from plutonium. Radioactive waste may be grouped into three general categories: low-level waste, transuranic waste, and high-level waste.

Low-Level Radioactive Wastes. *Low-level radioactive wastes* contain only small amounts of radioactive substances. Low-level wastes include a wide variety of items, such as residues or solutions from chemical processing; solid or liquid plant waste, sludges, and acids; and slightly contaminated equipment, tools, plastic, glass, wood, fabric, and other materials.[34] Before disposal, liquid low-level radioactive waste is solidified or packaged with material capable of absorbing at least twice the volume of liquid present.[35]

Radioactive decay of low-level waste does not generate a great deal of heat, and a general rule is that the material must be isolated from the environment for about 500 years to ensure that the level of radioactivity does not produce a hazard. In the United States, the philosophy for management of low-level waste has been "dilute and disperse." Experience suggests that low-level radioactive waste can be buried safely in carefully controlled and monitored near-surface burial areas in which the hydrologic and geologic conditions severely limit the migration of radioactivity.[35] Such waste has been buried at 15 main sites in states including Washington, Nevada, New Mexico, Missouri, Illinois, Ohio, Tennessee, Kentucky, South Carolina, and New York.

Despite precautions, several of the burial sites for low-level radioactive waste, including those in Tennessee and Washington, have not provided adequate protection of the environment. This failure has been due at least in part to a poor understanding of the local hydrologic and geologic environment. For example, a study of the Oak Ridge National Laboratory in Tennessee suggests that in places the water table is less than 7 m (23 ft) below the ground surface. The investigation identified migration of radioactive materials from one of the burial sites and concluded that containment of the waste in that area is difficult because of the short residence time of water in the vadose zone. In other words, liquid radioactive materials released from the disposal sites do not take long to infiltrate the vadose zone and percolate down to the groundwater.[35] However, near Beatty, Nevada, the depth to the water table is approximately 100 m (300 ft). The low-level radioactive waste-disposal facility there has apparently been successful in its containment of radioactive waste. Its success is based partly on the long length of time for any pollutants generated to enter the groundwater environment.[35]

Transuranic Waste. *Transuranic waste* is nuclear waste composed of human-made radioactive elements heavier than uranium. Most transuranic waste is industrial trash, such as clothing, rags, tools, and equipment, that has been contaminated. Although the waste is low level in terms of intensity of radioactivity, plutonium has a long half-life (the time required for the radioactivity to be reduced to one-half

(a)

Figure 15.23 New Mexico
nuclear waste disposal facility
Idealized diagram of WIPP (waste
isolation pilot plant) disposal site
in New Mexico for transuranic
nuclear waste. *(U.S. Department
of Energy)*

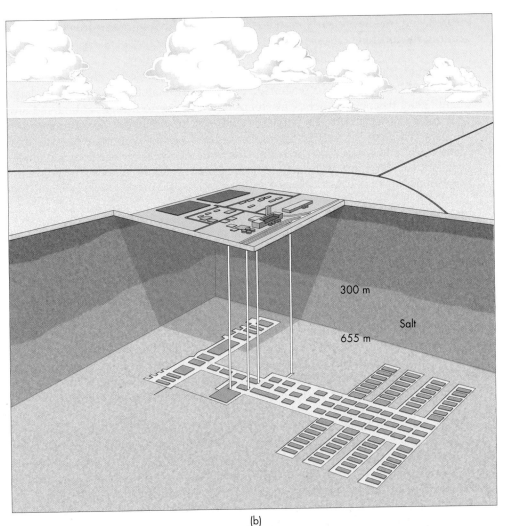

(b)

its original level) and requires isolation from the environment for about 250,000 years. Most transuranic waste is from the production of nuclear weapons and from cleanup of former nuclear weapons facilities. As of 1999, these wastes are being transported to a disposal site near Carlsbad, New Mexico. The waste is isolated at a depth of 655 m (2,150 ft) in saltbeds that are several hundred meters thick (Figure 15.23). Rock salt at the New Mexico site has several advantages:[36,37]

- The salt is about 225 million years old, and the area is geologically stable, with very little earthquake activity.

- The salt is easy to mine and has no flowing groundwater. Rooms excavated in the salt that are about 10 m (33 ft) wide and 4 m (13 ft) high are used for disposal.
- The rock salt flows slowly into mined openings. As a result, the slow-flowing salt will naturally close the waste-filled spaces in the storage facility in 100 to 200 years, sealing the waste.

The New Mexico disposal site is important because it is the first geologic disposal site for nuclear waste in the United States. It is a pilot project that is being evaluated very carefully. Safety is the main concern. Procedures to transport the waste, as safely as possible, to the disposal site and place it underground in the disposal facility have been established. The waste will be hazardous for many thousands of years; therefore, there are uncertainties concerning future languages and cultures at the site. Clear warnings above and below ground have been created, and the site is clearly marked to help ensure that human intrusion will not occur in the future.[37]

High-Level Radioactive Wastes. *High-level radioactive wastes* are produced as fuel assemblages in nuclear reactors become contaminated or clogged with large quantities of fission products. This spent fuel must periodically be removed and reprocessed or disposed of. Fuel assemblies will probably not be reprocessed in the near future in the United States, since reprocessing is more expensive than mining and processing new uranium; therefore, the present waste-management problems involve removal, transport, storage, and eventual disposal of spent fuel assemblies.[38]

The Scope of the High-Level Disposal Problem. Hazardous radioactive materials produced from nuclear reactors include fission products such as krypton-85, strontium-90, and cesium-137. Each of these radioactive elements has a different half-life. One of the biggest challenges faced when managing radioactive waste disposal is the various half-lives associated with fission products. In general, at least 10 half-lives, and preferably more, are required before a material is no longer considered a health hazard. Therefore, a mixture of the fission products mentioned above would require hundreds of years of confinement from the biosphere. Reactors also produce a small amount of plutonium-239 (half-life of 24,000 years), which is a human-produced isotope that does not occur naturally. Because plutonium and its fission products must be isolated from the biological environment for a quarter of a million years or more, their permanent disposal is a geologic problem.

Disposal of High-Level Waste in the Geologic Environment. There is fair agreement that the geologic environment can provide the most certain safe containment of high-level radioactive waste. The Nuclear Waste Policy Act of 1982 initiated a comprehensive federal and state, high-level nuclear waste–disposal program. The Department of Energy was responsible for investigating several potential sites, and the act originally called for the president to recommend a site by 1987. In December 1987, Congress amended the act to specify that only the Yucca Mountain site in southern Nevada would be evaluated to determine if high-level radioactive waste could be disposed of there. Some scientists and others believe that the site was chosen not so much for its geology, although the rock type at the site does have several favorable qualities for disposal, but because it is an existing nuclear reservation and therefore might draw minimal social and political opposition.[39,40]

The rock at the Yucca Mountain site is densely compacted tuff (a rock composed of compacted volcanic ash). Fortunately, the site is located in an extremely dry region. Precipitation is about 15 cm (6 in.) per year, and most of this runs off or evaporates. Hydrologists have estimated that less than 5 percent of the precipitation infiltrates the surface and eventually reaches the water table, which is several hundred meters below the surface. The depth to the potential repository is about

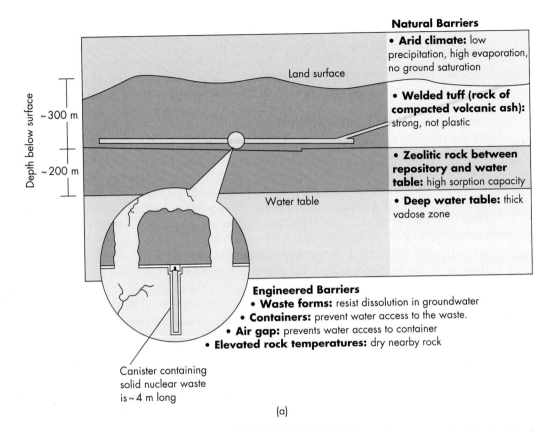

Natural Barriers

- **Arid climate:** low precipitation, high evaporation, no ground saturation

- **Welded tuff (rock of compacted volcanic ash):** strong, not plastic

- **Zeolitic rock between repository and water table:** high sorption capacity

- **Deep water table:** thick vadose zone

Land surface

~ 300 m

~ 200 m

Depth below surface

Water table

Engineered Barriers
- **Waste forms:** resist dissolution in groundwater
- **Containers:** prevent water access to the waste.
- **Air gap:** prevents water access to container
- **Elevated rock temperatures:** dry nearby rock

Canister containing solid nuclear waste is ~ 4 m long

(a)

Figure 15.24 Yucca mountain site (a) Idealized diagram of proposed Yucca Mountain repository. Natural and engineered barriers are listed. The waste would be retrievable for 50 years. *(From Lawrence Livermore Laboratory. 1988. LLL-TB-92)* (b) Aerial view of the Yucca Mountain site, which is being investigated for the disposal of high-level nuclear waste. The site is on the boundary of the Nevada Test Site, where the United States has conducted underground nuclear tests. *(Mark Marten/U.S. Department of Energy/Photo Researchers, Inc.)*

(b)

300 m (984 ft) below the mountain's surface, and, as a result, the repository could be constructed well above the water table in the vadose zone.[41]

The Department of Energy and the U.S. Geological Survey have completed an extensive scientific evaluation of the Yucca Mountain site. The study has helped determine how well the geologic and hydrologic setting can isolate high-level nuclear waste from the environment. The site is attractive because there are several natural barriers present, including (Figure 15.24)

- An arid climate, which greatly restricts downward movement of water
- A strong rock composed of welded tuff that has a high capacity to hold radioactive material
- A thick vadose zone, indicating a deep water table

Engineering methods will provide additional barriers to prevent the waste from escaping the storage canisters.

If the final studies indicate that the Yucca Mountain site can safely isolate radioactive waste, then the Department of Energy will apply to the U.S. Nuclear Regulatory Commission for a license to construct a disposal facility. If opposition from the state of Nevada is negotiated and environmental and safety factors are met, the repository might be ready for acceptance of high-level radioactive waste by the year 2010.

Long-Term Safety. A major problem with the disposal of high-level radioactive waste remains: How credible are long-range geologic predictions—that is, predictions of conditions thousands to millions of years in the future?[42] There is no easy answer to this question because geologic processes vary over both time and space. Climates change over long periods of time, as do areas of erosion, deposition, and groundwater activity. For example, large earthquakes occurring hundreds or even thousands of kilometers from a site may permanently change groundwater levels. The known seismic record for the western United States dates back only 100 years; estimates of future earthquake activity are tenuous at best. Ultimately, geologists can evaluate the relative stability of the geologic past, but they cannot guarantee future stability. Therefore, decision makers and not geologists need to evaluate the uncertainty of prediction in light of pressing political, economic, and social concerns.[42] These problems do not mean that the geologic environment is not suitable for safe containment of high-level radioactive waste, but care must be taken to ensure that the best possible decisions are made on this critical and controversial issue.

Energy from Fusion

Contrasted with nuclear fission, *nuclear fusion* combines the nuclei of lighter elements to produce heavier ones. The process of nuclear fusion releases energy and is the source of energy in our Sun and other stars. Harnessing nuclear fusion is a research objective with some success, but it is not yet certain whether commercial fusion power plants can be constructed that are economically competitive with other energy sources. From an environmental view, fusion appears attractive because little radioactive waste is produced, and mining and transportation impacts are small compared with those for fossil fuels and nuclear fission. The fuel for fusion is hydrogen, the supply of which is nearly unlimited and, as a result, fusion has the potential of being a nearly unlimited source of energy for the future. However, we do not yet have the technology to harness the hydrogen for our energy needs.[4]

15.7 Geothermal Energy

The use of **geothermal energy**—natural heat from Earth's interior—is an exciting application of geologic knowledge and engineering technology. The idea of harnessing Earth's internal heat is not new: Geothermal power was developed in Italy using dry steam in 1904 and is now used to generate electricity at numerous sites around the world, including a few in the western United States and Hawaii. At many other sites, geothermal energy that is not hot enough to produce electrical power is used to heat buildings or for industrial purposes. Existing geothermal facilities use only a small portion of the total energy that might eventually be tapped from Earth's reservoir of internal heat; the geothermal resource is vast. If only 1 percent of the geothermal energy in the upper 10 km (6.2 mi) of Earth's crust could be captured, it would amount to 500 times the total global oil and gas resource.[43]

Geology of Geothermal Energy

Natural heat production within Earth is only partially understood. We do know that some areas have a higher flow of heat from below than others and that, for the most part, these locations are associated with tectonic processes. Divergent and convergent plate boundaries are areas where this natural heat flow from Earth is anomalously high. The coincidence of geothermal power plant locations and areas of known active volcanism is no accident.

Temperature increases with depth below Earth's surface; it is measured in degrees per kilometer and is referred to as the *geothermal gradient*. In the United States, the geothermal gradient varies from about 12°C to 47°C (53°F to 116°F) per kilometer (0.6 mi) (Figure 15.25). In general, the steeper the gradient, the greater the heat flow to the surface. A steep geothermal gradient indicates that hot rock is closer to the surface than usual. A moderate gradient of 30°C to 45°C (86°F to 113°F) per kilometer is found over vast areas in the western United States Basin and Range, especially in the Battle Mountain region (Figure 15.25a). This area, with some exceptions, is clearly a good prospect for geothermal exploration. Figure 15.25b shows heat flow for the western United States, along with locations of existing geothermal power plants. Total power production is about 3,000 MW, more than 90 percent of which is in California. About one-third of total production is in the Geysers facility 145 km (90 mi) north of San Francisco, California, where electrical energy has been produced from steam for many years (G on Figure 15.25b, and see Figure 15.26). A typical commercial geothermal well will produce between 5 and 8 MW of electrical power.[43] (Remember 1 watt = 1 joule per second. See A Closer Look: Energy Units.)

Several geothermal systems may be defined on the basis of geologic criteria. We will discuss two systems here: hydrothermal convection systems and groundwater systems.[44,45,46] Each system has a different origin and different potential as an energy source.

Hydrothermal Convection Systems. Hydrothermal convection systems are characterized by a geothermal basin in which a variable amount of hot water circulates. There are two basic types: vapor-dominated systems and hot-water systems. Vapor-dominated hydrothermal convection systems are geothermal reservoirs in which both water and steam are present (Figure 15.27). Near the surface, the pressure is lower than at deeper levels and water changes quickly to superheated steam, which can be tapped and piped directly into turbines to produce electricity. These systems characteristically have a slow recharge of groundwater, meaning that the hot rocks boil off more water than can be replaced in the same amount of time by natural recharge or by injection of water from a condenser after power generation.[47] Vapor-dominated systems are not very common. Only three have been identified in the United States: the Geysers, California; Mt. Lassen National Park, California; and Yellowstone National Park, Wyoming.

In the United States, hot-water hydrothermal convection systems are about 20 times more common than vapor-dominated systems. Hot-water systems, with temperatures greater than 150°C (300°F), have a zone of circulating hot water that, when tapped, moves up to a zone of reduced pressure, yielding a mixture of steam and water at the surface. The water must be removed from the steam before the steam can be used to drive the turbine.[45] As shown in Figure 15.28a, the water can be injected back into the reservoir to be reheated.

Groundwater Systems. The idea of using groundwater at normal shallow underground temperatures is a relatively new one. At a depth of about 100 m (328 ft), groundwater typically has a temperature of about 13°C (55°F). This is cold if you want to use it for a bath, but warm compared with winter air temperature in the eastern United States. Compared with summer temperatures, it is cool. Heat pumps, devices that can raise or lower the temperature of air or water, can use

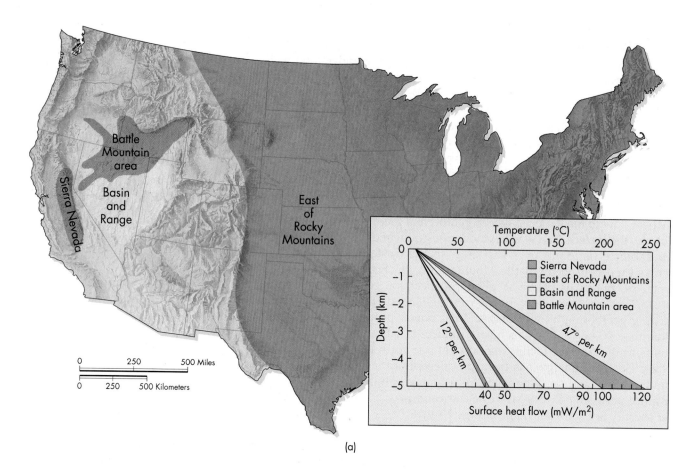

(a)

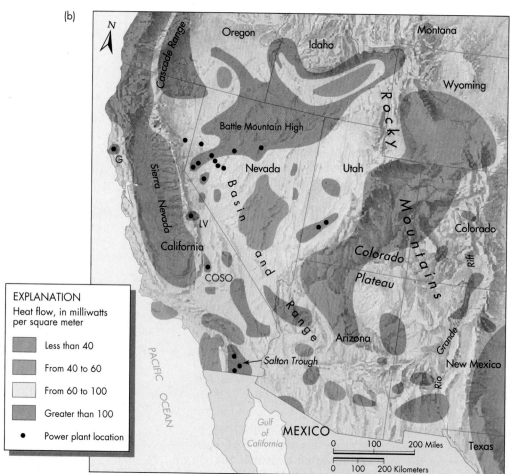

Figure 15.25 **Heat flow in the United States**
(a) Geothermal gradients (12° to 47°C per kilometer C, or 53° to 116°F per 0.6 mi) and generalized heat flow in the United States. (b) A more detailed map of heat flow for the western United States and the locations of geothermal power plants. One milliwatt per square meter is equivalent to 10,000 watts per square kilometer. *(From Duffield, W. A., Sass, J. H., and Sorey, M. L. 1994. Tapping Earth's natural heat. U.S. Geological Survey Circular 1125)*

Figure 15.26 Geothermal power plant Aerial view of the Geysers Power Plant north of San Francisco, California. The facility is the world's largest geothermal electricity development. *(Courtesy of Pacific Gas and Electricity)*

these temperature differences to heat buildings in the winter and cool them in the summer by transferring heat between groundwater and the air in a building. Although initially expensive because of well drilling, geothermal systems using constant-temperature groundwater are in service in numerous Midwestern and eastern U.S. locations. The technology for these systems is well known, and the equipment is easily obtained. As energy costs increase, such systems will become even more attractive.

Environmental Impact of Geothermal Energy Development

The adverse environmental impact of intensive geothermal energy development is less severe than that of other energy sources, but it is nonetheless considerable. Environmental problems associated with geothermal energy include on-site noise, gas emissions, and scars on the land. Fortunately, development of geothermal

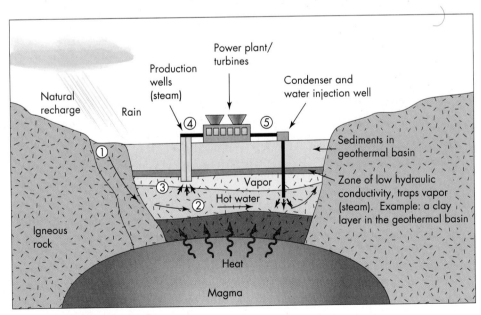

→ Direction of water flow

1. Natural recharge of water from rain
2. Hot water produced by Earth processes
3. Steam to production well
4. Steam to turbines to produce electricity
5. Water is injected back into ground

Figure 15.27 Vapor-dominated geothermal system and powerplat Idealized diagram of a vapor-dominated geothermal system. Wells produce steam that runs turbines to make electricity.

Figure 15.28 A hot-water geothermal system (a) At the power plant, the steam is separated from the water and used to generate electrical power. The water is injected back into the geothermal system through a disposal well. *(Courtesy of Pacific Gas and Electricity [PG&E])* (b) Mammoth Lakes, California, hot-water geothermal power plant. The plant uses 12 production wells and 6 injection wells to producing 40 MW, enough to provide power for about 40,000 homes.

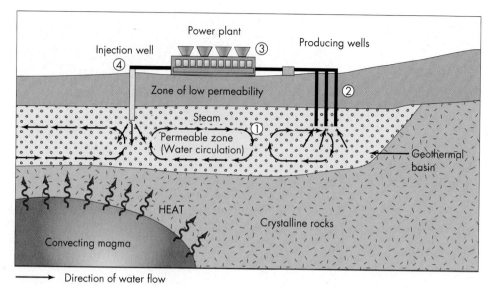

Power plant

Injection well ④   Producing wells ③

Zone of low permeability   ②

Steam   ①

Permeable zone
(Water circulation)

Geothermal basin

HEAT   Crystalline rocks

Convecting magma

→ Direction of water flow

1. Water circulating in geothermal basin
2. Wells pump out water and steam
3. Turbines in plant produce electricity
4. Water is injected back into basin

(a)

(b)

energy does not require the extensive transportation of raw materials or refining that is typical of the development of fossil fuels. Geothermal plants generate less than 1 percent of the nitrogen oxides and only 5 percent of the carbon dioxide created by coal-burning power plants producing comparable amounts of power.[43] Finally, geothermal energy does not produce either the atmospheric particulate pollutants associated with burning fossil fuels or radioactive waste.

Geothermal energy production does have associated hazards. With the exception of vapor-dominated systems, geothermal development produces considerable thermal pollution from hot wastewaters. The wastewaters can be saline, mineralized, or highly corrosive to pipes, pumps, and other equipment. The plan is to dispose of these waters by reinjecting them into the geothermal reservoir;

however, that kind of disposal also has problems. Injecting fluids may activate fracture systems in the rocks and cause earthquakes. In addition, the original withdrawal of fluids may compact the reservoir, causing surface subsidence. Subsidence could also occur as the heat in the system is extracted and the cooling rocks contract.[45] Finally, geothermal energy development may adversely affect nearby geyser activity by reducing or changing the heat source driving the geysers. As a result, federal legislation has been passed to protect the geysers, including the famous geyser Old Faithful, and hot springs of Yellowstone National Park by prohibiting geothermal development in national parks (see Figure 7.15a). What is not known is what constitutes adequate protection. How large a buffer zone is necessary to ensure that geothermal development outside the park boundary does not damage Yellowstone's geysers and hot springs? Research in the park is helping us understand this question and derive answers.[43]

Future of Geothermal Energy

Geothermal energy is a viable site-specific energy source. Over a 30-year period, the estimated yield from this vast resource far exceeds that of hundreds of modern nuclear power plants. If we disregard cost, the resource is both identified and recoverable at this time; in addition, it is expected that many more systems with presently recoverable energy are yet to be discovered.[44] Furthermore, geothermal energy in the form of hot water using lower-temperature groundwater is a future source of energy to help heat homes and other buildings in many areas.

At present, geothermal energy supplies only a small fraction of 1 percent of the electrical energy produced in the United States. With the exception of the unusual vapor-dominated systems such as the Geysers in California, it is presently more expensive to produce electricity from geothermal reservoirs than from fossil fuels. Commercial development of geothermal energy will not proceed rapidly until the economics are equalized. Even if they are, the total output from geothermal sources is not likely to exceed a few percent—10 percent at most—of electrical output in the near future. This outlook is true even for California, where geothermal energy has been produced and where expanding facilities are likely.[45] Nevertheless, the growth in power produced from geothermal sources has increased dramatically in the past 35 years. About 60 new geothermal energy projects are now being developed in the U.S.

15.8 Renewable Energy Sources

Fossil fuels are the main energy sources used today; they supply approximately 90 percent of the global energy consumed by people. All other sources are designated **alternative energy** and are subdivided into two groups: *renewable energy* and *nonrenewable energy*. Nonrenewable alternative energy sources include nuclear energy and geothermal energy, which were discussed previously. The renewable sources are solar power, water or hydropower, hydrogen, wind power, and energy derived from biomass.

Use of alternative sources, particularly solar and wind power, is growing at tremendous rates. Alternative energy sources have now proved that they can compete with fossil fuel burning that pollutes our urban air, causes acid rain, and warms our climate to an unacceptable level. Alternative renewable energy sources such as solar and wind power do not cause rapid or other climate change. Solar and wind power do not alter weather to produce killer storms or droughts; nor do they raise sea levels around the world, increasing coastal erosion and threatening low-lying areas, including islands. Alternative, renewable energy sources offer our best chance to break our addiction to fossil fuels and develop a sustainable energy policy that will not harm Earth.[48]

Figure 15.29 **Types of renewable solar energy** Selected examples with growth and potential.

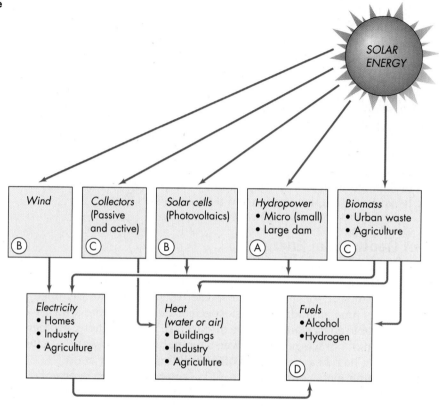

(A) Produces most electricity from renewable solar energy

(B) Rapidly growing, strong potential; wind and solar are growing at 30% per year!

(C) Used today; important energy source

(D) Potentially a very important fuel to transition from fossil fuels

A primary directive to ensure the success of alternative energy is to match renewable energy sources to sites where the natural resources for that source are of the highest quality. For instance, site solar energy power plants in the southwestern United States, where sunlight is most intense; site wind farms in the Great Plains, Texas, the Northeast, and California, where the strength of wind is strong and steady; and use biomass at existing coal-burning power plants in locations where forest and agriculture fuel resources are abundant.[49]

Renewable energy sources are usually discussed as a group because they are derivatives of the Sun's energy. That is, **solar energy**, broadly defined, comprises many of the renewable energy sources, as shown in Figure 15.29. They are renewable because they are regenerated by the Sun within a time period that is useful to people. Renewable energy sources have the advantage of being inexhaustible and are often associated with minimal environmental degradation. With the exception of burning biomass or its derivative, urban waste, solar energy does not use fuel burning and therefore does not pose a threat of increasing atmospheric carbon dioxide and modifying the climate. Another important aspect of renewable sources is that the lead time necessary to implement the technology is often short relative to the development of new sources or the construction of power plants that utilize fossil or nuclear fuels.

The total amount of solar energy reaching Earth's surface is huge. On a global scale, 10 weeks of solar energy is roughly equivalent to the energy stored in all known reserves of coal, oil, and natural gas on Earth. On the average, 13 percent of the Sun's original energy entering the atmosphere arrives at the land surface of

Earth. The actual amount is variable, depending on the time of the year and the cloud cover.[50]

Solar Energy

Solar energy is used directly through passive solar systems or active solar systems. *Passive* solar energy systems involve architectural design that enhances absorption of solar energy and takes advantage of the natural changes in solar energy that occur throughout the year without requiring mechanical power. A simple technique is to design overhangs on buildings that block high-angle summer sunlight but allow low-angle winter sunlight to penetrate and warm rooms. Another method is to build a wall that absorbs solar energy and emits it into a room, thus warming it. Numerous homes and other buildings in the southwestern United States as well as other parts of the country now use passive solar systems for at least part of their energy needs.[49] *Active* solar energy systems require mechanical power, usually pumps and other apparatuses, to circulate air, water, or other fluids from solar collectors to a heat sink, where the heat is stored until used.

Solar Collectors. Solar collectors are usually flat panels consisting of a glass plate over a black background where water is circulated through tubes. Solar radiation enters the glass and is absorbed by the black background, heating the water in the circulating tubes to 38° to 93°C (100° to 200°F).[51] The number of systems using these collectors in the United States continues to grow.

Photovoltaics. The technology that converts sunlight directly into electricity using a solid semiconductor material is known as *photovoltaics*. Photovoltaics, at a growth rate of 35 percent per year, are the most rapidly growing source of energy in the world today. The systems use photovoltaic or solar cells made of silicon or other materials and solid-state electronic components with few or no moving parts. The cells are constructed in standardized modules, which can be combined to produce systems of various sizes. As a result, power output can be matched to the intended use. Electricity is produced when sunlight strikes the cell, causing electrons to flow out of the cell through electrical wires.

Photovoltaics are emerging as a significant energy source in developing countries that, with relatively poor economies, do not have the financial ability to build large central power plants that burn fossil fuels. We now recognize that solar technology is simple and relatively inexpensive; it is also capable of meeting energy uses for people in most places in the world. One solar company in the United States is equipping villages in several countries with photovoltaic systems that power lights and televisions at an installed cost of a few hundred dollars per household.[52] Solar roofing tiles allow the roof of a building to become its own power plant.[53] Panels of solar cells can also be placed on a building's roof, walls, or window glass.[49]

Although there are specific instances in which the cost of using photovoltaics is comparable to grid-connected power, photovoltaics are still more expensive than conventional energy sources. However, the gap is narrowing.[54, 55]

Solar Energy and the Environment. The use of solar energy has a relatively low impact on the environment. The major disadvantage is that solar energy is relatively dispersed; a large land area is required to generate a large amount of energy. This problem is negligible when solar collectors can be combined with existing structures, as for example the addition of solar hot-water heaters on roofs of existing buildings. The impact of solar energy systems can be minimized by locating centralized systems in areas not used for other purposes and by making use of dispersed solar energy collectors on existing structures wherever possible.

Hydrogen

Hydrogen is the fuel burned by our Sun. It is the lightest, most abundant element in the universe. Hydrogen gas may be the fuel of the future and the key to clean energy. Hydrogen is a high-quality fuel that can be easily used in any of the ways in which we normally use fossil fuels, such as to power automobile and truck engines or to heat water or buildings. When used in **fuel cells,** similar to batteries, it can produce electricity. Hydrogen, like natural gas, can be transported in pipelines and stored in tanks, and it can be produced using solar and other renewable energy sources to split water into oxygen and hydrogen. Hydrogen is a clean fuel; the combustion product of burning hydrogen is water, so it does not contribute to global warming, air pollution, or acid rain. It is expected that experimentation with hydrogen will continue and that the fuel produced may be substantially reduced in price in the future. Hawaii, which imports oil for 88 percent of its energy needs, has abundant geothermal, solar, and wind sources that could be used to produce hydrogen. Hawaii eventually hopes to be a net exporter of energy in the form of hydrogen. Iceland in 1999 announced it intends to be the first hydrogen society, exporting hydrogen to Europe by 2050.[56,57,58]

Water Power

Water power is an ancient source of energy. Water has been used successfully as a source of power at least since the time of the Roman Empire. Waterwheels were turning in Western Europe in the seventeenth century, harnessing the energy of moving water and converting it to mechanical energy. During the eighteenth and nineteenth centuries, large waterwheels provided the energy to power grain mills, sawmills, and other machinery in the United States.

Hydroelectric Power. Today, hydroelectric power plants provide about 10 percent of the total electricity produced in the United States. Although the total amount of electrical power produced by running water will increase somewhat in the coming years, the percentage may be reduced as the production of other energy sources such as nuclear, direct solar, and geothermal increases more quickly.

Most of the acceptable sites for large dams to produce hydropower are already being utilized. However, small-scale hydropower systems may be more common in the future. These are systems designed for individual homes, farms, or small industries. They will typically have power outputs of less than 100 kW and are termed microhydropower systems (Figure 15.30).[59] Microhydropower is one of the world's oldest and most common energy sources. Numerous sites in many areas have potential for producing small-scale electrical power; mountainous areas are particularly promising because potential energy from stream water is readily available. Microhydropower development is by its nature very site specific, depending on local regulations, the area's economic situation, and hydrologic limitations.

Hydropower can be used to generate either electrical power or mechanical power to run machinery. Hydroelectric plants may help cut the high cost of importing energy and help small operations become more independent of local utility providers.[59]

Tidal Power. Another form of water power might be derived from ocean tides in a few places where there is favorable topography, such as the Bay of Fundy region of the northeastern United States and Canada. The tides in the Bay of Fundy have a maximum rise of about 15 m (50 ft). A minimum rise of about 8 m (26 ft) is necessary to even consider developing **tidal power.**

Tidal power is harnessed by building a dam across the entrance to a bay, creating a basin on the landward side of the dam. The dam creates a difference in water level between the ocean and the basin. Then, as the water in the basin rises or falls, it can be used to turn hydraulic turbines that produce electricity.[60] A tidal power station on the river Rance near Saint Malo, France, produces more than 200,000 kW of electricity from 24 power units across the dam.

Figure 15.30 Microhydropower
Rush Creek microhydropower station at Silver Lake in the Sierra Nevada of California. *(Edward A. Keller)*

Water Power and Environment. Water power is clean power. It requires no fuel burning, does not pollute the atmosphere, produces no radioactive or other waste, and is efficient. There is an environmental price to pay, however. Water falling over high dams may trap nitrogen gas, which is the major gas in air. The nitrogen then enters the blood of fish, expands, and kills them. This process is analogous to what happens to scuba divers when they rise too fast and get "the bends." Nitrogen has killed many migrating game fish in the Pacific Northwest. Furthermore, dams trap sediment that would otherwise reach the sea and replenish the sand on beaches. Building dams to harness water's power can also cause the displacement of people, loss of land to the reservoir, loss of wildlife, and adverse changes to the river ecology and hydrology downstream. In addition, many people do not want to turn wild rivers into a series of lakes by building intermittent dams. In fact, in the United States, several dams have been removed and others are being considered for removal as a result of the adverse environmental impacts their presence is causing.

Despite the inherent problems associated with dams, the world's largest dam has been constructed in China. The Three Gorges Dam on the Yangtze River displaced about 2 million people from their homes, drowning cities, farm fields, archeology sites, and highly scenic river gorges (Figure 15.31). The dam contributed to the extinction of the Yangtze river dolphin. The habitat for many other plants and animals was fragmented as mountain tops became islands in the giant reservoir, resulting in loss of species diversity. The dam is about 185 m (607 ft) high and more than 1.6 km (1 mi) wide, and when filled produce a reservoir nearly 600 km (373 mi) long. There is concern that the reservoir will become polluted by raw sewage and industrial pollutants currently disposed of in the river, turning the long narrow reservoir into an open sewer. It might also degrade or eliminate deep-water shipping harbors at the upstream end of the reservoir, where sediments will most likely be deposited. On a more positive note, 26 giant turbines will produce about 18,000 MW of electricity, the equivalent of 18 large nuclear or coal-burning power plants. However, opponents of the dam point out that a series of dams on tributaries to the Yangtze could have produced electrical power while not causing environmental damage to the main river.[61]

While the future growth of large-scale water power is limited because of objections to dam building and the fact that many good sites for dams are already

Figure 15.31 Three Gorges Dam on China's Yangtze River (a) Qutang Gorge and (b) construction of temporary locks near the dam site on Xiling Gorge. *(Bob Sacha/Bob Sacha Photography)*

(a)

(b)

utilized, there seems to be increased interest in microhydropower, or small dams for supplying electricity or mechanical energy. However, the environmental impact of numerous microhydropower installations in an area may be considerable. The sites change the natural stream flow, affecting the stream biota and productivity. Small dams and reservoirs also tend to fill more quickly with sediment than larger installations, so their useful life is shorter.

Because microhydropower development can adversely affect the stream environment, careful consideration must be given to its development over a wide region. A few such sites may cause little environmental degradation, but if the number becomes excessive, the impact over a wider region may be appreciable. This is a consideration that must be given to many forms of technology that involve small sites. The impact of a single site on a broad region may be nearly negligible, but as the number of sites increases, the total impact may become significant.

Wind Power

Wind power, like solar power, has evolved over a long period of time, beginning with early Chinese and Persian civilizations. Wind has propelled ships and has driven windmills to grind grain or pump water. More recently, wind has been used to generate electricity. Winds are produced when differential heating of Earth's surface creates air masses with differing heat contents and densities. The potential for energy from the wind is tremendous; yet there are problems with its use because wind tends to be highly variable in time, place, and intensity.[62]

Wind prospecting has become an important endeavor. On a national scale, the areas with the greatest potential for wind energy are the Pacific Northwest coastal area, the coastal region of the northeastern United States, and a belt extending from northern Texas through the Rocky Mountain states and the Dakotas. There are other good sites, however, such as mountainous areas in North Carolina and northern Coachella Valley in southern California.

At a particular site, the direction, velocity, and duration of wind may be quite variable, depending on the local topography and on the regional to local magnitude of temperature differences in the atmosphere.[62] For example, wind velocity often increases over hilltops, or wind may be funneled through a mountain pass. The increase in wind velocity over a mountain is due to a vertical convergence of the wind, whereas in a pass the increase is partly due to a horizontal convergence (Figure 15.32a–c). Because the shape of a mountain or a pass is often related to the local or regional geology, prospecting for wind energy is a geologic as well as a geographic and meteorological problem.

Significant improvements in the size of windmills and the amount of power they produce were made in recent years when many European countries and the United States became interested in large-scale wind-driven generators. In the United States, thousands of windmills are located on wind farms (Figure 15.32d). Large state-of-the-art wind turbines are much larger, each producing 3 to 5 MW, enough power for several thousand homes. Advantages of wind power are that wind is a widespread, abundant, inexhaustable resource; wind power has become an inexpensive srouce of energy (cost competitive with fossil fuels and less expensive than nuclear); and wind power is a clean source of electricity that doesn't cause air pollution or release carbon dioxide that changes climate. For these

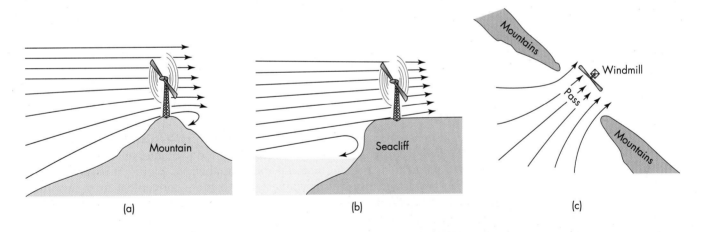

(a)   (b)   (c)

Figure 15.32 Areas with increased wind resource How wind may be converged, and velocity increased by topography, vertically (a,b) or horizontally (c). Tall windmills are necessary on hilltops or on the top of a seacliff to avoid near-surface turbulence. (d) Older wind farm in California. Newer windmills are larger. *(Glen Allison/Getty Images Inc.)*

(d)

reasons, wind power is the fastest growing energy source in the world. However, wind power does have some adverse effects:

- Windmills may kill birds, especially large birds of prey, such as hawks and falcons that fly into the blades while focused on prey.
- Large windmill farms require land for roads, windmill pads, and other equipment.
- Windmills may degrade an area's scenic resources.

The growth of wind power in the last decade has been astounding—approximately 30 percent per year, compared with the 1 percent to 2 percent per year growth of oil. It is believed that in just three states—Texas, South Dakota, and North Dakota—sufficient wind energy resources exist to satisfy the electricity needs of the entire country. The world's largest wind farm, located in Washington and Oregon on ridges above the Columbia River, will produce 300 MW of electricity at a unit cost of energy comparable to natural gas. However, many wind power sources remain untapped. China burns tremendous amounts of coal at a tragic environmental cost, including the exposure of millions of people in cities to mixtures of deadly gases and particulates. Similarly, rural China's exposure to the smoke from burning coal in homes has increased the threat of lung cancer by a factor of 9 or more. These environmental costs are endured despite the fact that China could probably double its current capacity by generating electricity with wind alone![53,63]

Wind power is being taken seriously. Sufficient wind power is generated in the United States to supply electricity to several million suburban homes. Although wind now provides less than 1 percent of the world's demand for electricity, its growth rate suggests that it will be a major power supplier in the relatively near future. One scenario suggests that wind power could supply 10 percent of the world's electricity in the next few decades. In the long run wind power could provide more energy than hydropower, which currently supplies approximately 20 percent of the electricity in the world.[63]

The total installed wind power in the world today is 48,000 MW. This is equivalent to 48 large fossil fuel or nuclear power plants.

Biofuels

Biofuel is a new name for the oldest fuel used by humans. Our Pleistocene ancestors burned wood to keep warm and to cook food. Biofuel energy sources are organic matter such as plant material and animal waste. Biofuel is organic matter that can be burned directly or converted to a more convenient form and then burned. For example, we can burn wood in a stove or convert it to charcoal and then burn it. Biomass has provided a major source of energy for human beings throughout most of the history of civilization. When North American forests were cleared for agriculture a technique known as girdling was used. Girdling involved cutting through the bark all the way around the base of a tree. After girdling had killed the tree, the settlers would then burn the forests to clear the land for farming.

Firewood is the best known and most widely used biofuel, but there are many types of biomass fuel. In India and other countries, cattle dung is burned for cooking. Peat, a form of compressed dead vegetation, provides heating and cooking fuel in northern countries such as Scotland, where it is abundant.

Today, more than 1 billion people in the world use wood as their primary source of energy for heat and cooking. Energy from biomass can be generated via several routes: direct burning of biomass either to produce electricity or to heat water and air; heating of biomass to form a gaseous fuel by gasification; or distillation or processing of biomass to produce biofuels such as ethanol, methanol, or methane.[64]

The primary sources of biofuels in North America are forest products, agricultural products, and combustible urban waste. Manure from livestock or other organic waste can be digested by microorganisms to form methane and then burned to produce electricity or used in fuel cells. Using biogenic methane from manure is highly preferable to allowing its release into the atmosphere, where it contributes to global warming.[65]

Today there are a number of facilities in the United States that process urban waste to generate electricity or to be used as a fuel. Presently, only about 1 percent of the nation's municipal solid wastes are being recovered for energy. However, if all the plants were operating at full capacity and if additional plants under construction were completed and operating, about 10 percent of the country's waste, or 24 million metric tons per year, could be used to extract energy. The United States has been slower than other countries to utilize urban waste as an energy source. For example, in Western Europe a number of countries now utilize one-third to one-half of their municipal waste for energy production. As the supply of cheap, available fossil fuels ends, additional energy recovery systems utilizing urban waste will emerge. However, burning urban waste is a controversial process with potentially adverse environmental impacts that include emission of air pollutants and production of ash that needs to be disposed of. In summary, biofuels in their various forms appear to have a future as an energy source. However, questions remain about the amount of energy they can provide and their rate of depletion. Any use of biofuel must be part of a general plan for all uses of the land's products.[64,65]

15.9 Conservation, Efficiency, and Cogeneration

Earlier in this chapter we established that we must become accustomed to living with uncertainty concerning the availability, cost, and environmental effects of energy use. Furthermore, we can expect that serious social, economic, and political shocks will continue to occur, disrupting the flow of energy to various parts of the world.

Supply and demand for energy are difficult to predict because the technical, economic, political, and social assumptions underlying projections are constantly changing. Large annual variations in energy consumption must also be considered: energy consumption peaks during the winter, with a secondary peak occurring in the summer. Future changes in population or intensive conservation measures may change this pattern: Better building design and more reliance on solar energy can also contribute to changing the existing energy usage pattern.

There has been a strong movement to change patterns of energy consumption through measures such as conservation, increased efficiency, and cogeneration. **Conservation** of energy refers to a moderation of our energy demand. Pragmatically, this means adjusting our energy uses to minimize the expenditure of energy necessary to accomplish a given task. **Efficiency** entails designing and using equipment that yields more power from a given amount of energy, while wasting less energy.[66] Finally, **cogeneration** refers to a number of processes that capture and use some of the waste heat produced by power generation and industrial operations rather than simply releasing it into the atmosphere or to water, where it may cause thermal pollution.

The three concepts of *conservation, increased efficiency*, and *cogeneration* are interrelated to a great extent. For example, when electricity is produced at large coalburning power stations, sizable amounts of heat may be emitted into the atmosphere. Typically, three units of fuel burn to produce one unit of electricity, an energy loss of about 67 percent. The use of a "unit of fuel" is arbitrary. It could, for example, be a barrel of oil or a ton of coal. Cogeneration, which involves recycling of that waste heat, can increase the efficiency of a typical power plant from 33 percent to as high as 75 percent. Put another way, cogeneration reduces energy loss from 67 percent to as little as 25 percent.[66,67]

15.10 **Energy Policy for the Future**

Hard Path Versus Soft Path

Energy policy today is at a crossroads. One road leads to development of so-called hard technologies, which involve finding ever-greater amounts of fossil fuels and building larger centralized power plants. Following this **hard path** means continuing "business as usual." This is the more comfortable approach; it requires no new thinking or realignment of political, economic, or social conditions. It also involves little anticipation of the inevitable depletion of the fossil fuel resources on which the hard path is built.

Proponents of the hard path argue that environmental problems have occurred in some countries because people have had to utilize local resources, such as wood, for energy rather than, say, for land conservation and erosion control. Hard path supporters believe that the way to solve these problems is to provide people with cheap energy that utilizes more intensive industrialization and technology. Furthermore, the United States and other countries with sizable resources of coal or petroleum should exploit these resources to prevent environmental degradation of their own countries. Proponents of this view maintain that allowing the energy industry the freedom to develop available resources ensures a steady supply of energy and less total environmental damage than if the government regulates the energy industry. They point to the present increase in the burning of firewood across the United States as an early indicator of the effects of strong governmental controls on energy supplies. The eventual depletion of forest resources, they maintain, will have a detrimental effect on the environment, as it has in so many other countries. The hard path continues to dominate energy planning in the United States.

The other road is designated as the **soft path.**[68] One of the champions of this choice has been Amory Lovins, who argues that the soft path involves energy alternatives that are renewable, flexible, decentralized, and environmentally more benign than those of the hard path. A promising technology for the soft path is the development and use of fuel cells that produce electricity from chemical reactions. Cells, as discussed earlier, use hydrogen as a fuel and, like batteries, may be arranged in a series to power vehicles or to provide electricity for homes or other buildings. The only wastes produced are oxygen and water. Fuel cells can be combined with photovoltaics; in fact, electricity from solar cells may be used to split water into oxygen and hydrogen, providing fuel for fuel cells (Figure 15.33).[57]

In the United States today, we annually consume approximately 100 EJ of energy. Projections suggest that U.S. energy consumption in the year 2030 may be as high as 120 EJ or as low as 60. Why is there such a big discrepancy? If we stay on the hard path, the high value is probably appropriate. The soft path, which advocates intensive energy conservation and increased efficiency, predicts that annual consumption of energy could be cut in half. Actual energy consumption in the year 2030 will probably not be as low as 60 EJ; we hope it will not exceed 100 EJ by much. Given the expected population increase, achieving this level will require a substantial commitment to energy conservation and increased energy efficiency.

Sustainable Energy Policy

Energy planning for the future is complicated; we know that burning fossil fuels is degrading our global environment. The soft path is called the environmental path because it would help reduce environmental degradation by reducing emissions of carbon dioxide and air pollutants. To be fair, some advocates of the hard path propose increased energy conservation, increased efficiency, and cogeneration to reduce consumption and environmental problems associated with burning fossil fuels. Although there is sufficient coal to last hundreds of years, proponents of the

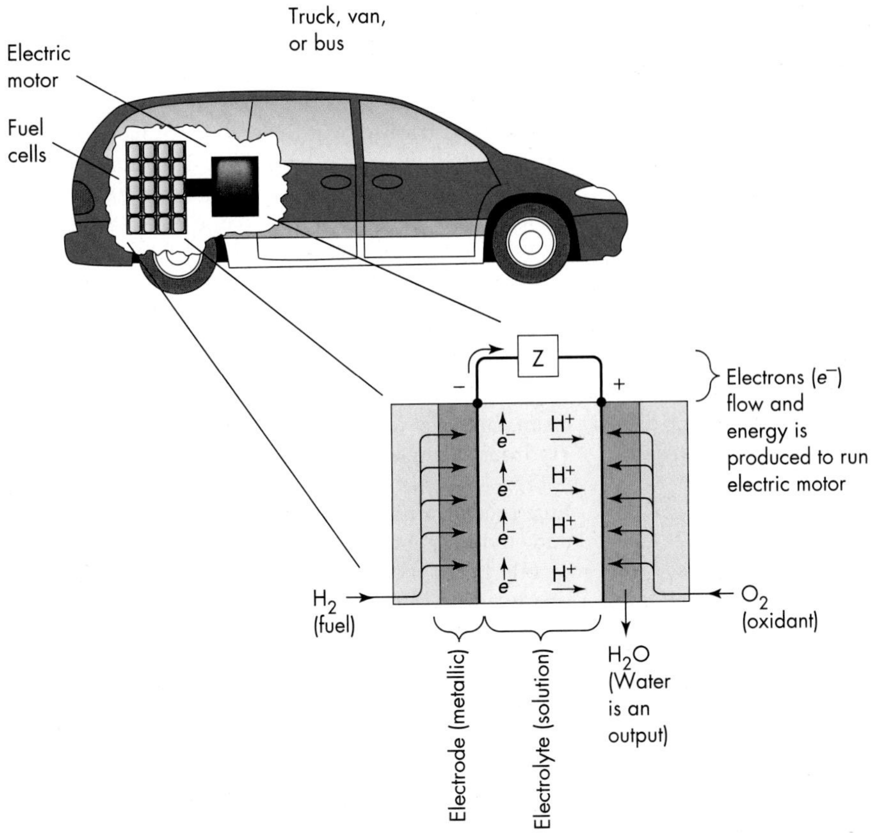

Figure 15.33 Fuel cell

Idealized diagram showing how a fuel cell works.

Truck, van, or bus

Electric motor

Fuel cells

Z

$-$ $+$

e^- H^+
e^- H^+
e^- H^+
e^- H^+

Electrons (e^-) flow and energy is produced to run electric motor

H_2 (fuel)

O_2 (oxidant)

H_2O (Water is an output)

Electrode (metallic)

Electrolyte (solution)

soft path would prefer to use this source as a transitional rather than a long-term energy source. In their view, development of a **sustainable energy policy** means finding useful sources of energy that can be maintained and do not pollute the atmosphere, cause climatic perturbations such as global warming, or present an unacceptable risk.

A transition from the hard to the soft path would presumably involve continued utilization of fossil fuels. Electrical power will continue to be essential for some purposes. The energy path we take must be one capable of supplying the energy we require for human activities without endangering the planet.[69] This is the heart of the concept of sustainable energy policy.

SUMMARY

The ever-increasing world population's appetite for energy is staggering. It is time to seriously question the need and desirability of an increasing demand for electrical and other forms of energy in industrialized societies. Quality of life is not necessarily directly related to greater consumption of energy.

The fossil fuels—coal, oil, and gas—are essentially stored solar energy in the form of organic material that has escaped destruction by oxidation. Although ongoing geologic processes formed these fuels, they are too slow to be of use to us; fossil fuels are therefore considered to be nonrenewable resources. The environmental disruption associated with exploration and development of these resources must be

weighed against the benefits gained from the energy. This development is not an either-or proposition; good conservation practices combined with pollution control and reclamation can help minimize the environmental disruption associated with fossil fuels.

There are still vast supplies of coal in the world, 25 percent of which are located in the United States. The grade, or carbon content, of coal determines its value as a fuel, while its sulfur content determines how much it pollutes the atmosphere with sulfur oxides.

Oil and gas are found in large deposits called fields; most fields are located near tectonic plate boundaries. Oil fields have been extensively mined by oil wells that pump oil and gas to the surface. We are approaching the time known as peak oil when we will have used about half of Earth's total oil. Following the peak, production will decrease and there may be a gap between demand and production. Barring discovery of many major new fields, shortages of oil and gas will occur in the future. The impending shortages have resulted in pressure to develop new oil fields, some in sensitive wilderness areas such as the Arctic National Wildlife Refuge. Oil can be recovered from Earth materials called oil shales and tar sands, but mining techniques that are economically and environmentally sound have not been fully developed. The potential for environmental disruption exists at every stage of oil development and use; disruptions include oil spills from tankers and air pollution from burning of petroleum products in automobiles and power plants. Fossil fuel burning also produces sulfur dioxide emissions that are partially responsible for acid rain.

Nuclear fission produces vast amounts of heat that can be used to generate electricity in a nuclear power plant. It also produces radioactive wastes that must be safely disposed of. Fission will remain an important source of energy but will be sustainable only if breeder reactors are developed. The possibility of environmental and health hazards as well as the increasing cost of constructing large nuclear power plants remain as factors to be considered. Finally, we continue to struggle with the problem of radioactive waste disposal—a scientific, economic, social, and political issue involving risk management.

Use of geothermal energy will become more widespread in the western United States, where natural heat flow from Earth is relatively high. Although the electrical energy produced from the internal heat of Earth will probably not exceed 10 percent of the total electrical power generated, it can still be significant. Geothermal energy has an environmental price. Surface subsidence may be caused by the withdrawal of fluids and heat; in addition, earthquakes may be caused by the injection of hot wastewater back into the ground.

Renewable sources of energy depend on solar energy and can take a variety of forms, including direct solar, water, wind, and energy from biomass, including recycled biomass from urban waste. These energy sources are generally used to produce electrical power. They will not be depleted and are thus dependable in the long term. They have varying attributes but generally cause little environmental disruption, although the burning of biomass does pollute the atmosphere. However, most of these sources are local and intermittent, and some are still expensive to produce. Continued growth and development of solar energy and wind power, as well as development of new technology and innovations such as fuel cells, will become more important in the future. Solar and wind power, because they are abundant, inexhaustible and clean, are the world's fastest growing sources of energy. Hydropower will undoubtedly continue to be an important source of electricity in the future, but it is not expected to grow much in the United States because of lack of potential sites and environmental considerations.

Energy for the future will continue to be uncertain. It does seem certain, however, that we will continue to look more seriously at conservation, energy efficiency, and cogeneration. The most likely targets for energy efficiency, conservation, and cogeneration are in the area of space heating for homes and for various manufacturing processes and automobiles. These areas collectively account for approximately 60 percent of the total energy used in the United States today.

We may be at a crossroads concerning energy policy. The choice is between the "hard path," characterized by centralized, high-technology energy sources, and the "soft path," characterized by decentralized, flexible, renewable energy sources. Perhaps the best path will be a mixture of the old and the new, ensuring a rational, smooth shift from our dependence on fossil fuels. Our goal, whatever path we take, should be to develop a sustainable energy plan that supplies the energy we need but does not harm the environment.

Revisiting Fundamental Concepts

Human Population Growth

Human population during the twentieth century increased from less than 2 billion people to about 6.6 billion. During the past several decades, world consumption of energy has doubled, while the number of people on Earth has also doubled. The tremendous increase in people and consumption of energy, most of which is supplied by the fossil fuels, has resulted in significant environmental degradation on a global scale. However, the finite fossil fuels, particularly oil, will become scarce in coming decades, so major shifts in energy sources are expected.

Sustainability

The key to minimizing environmental degradation associated with consumption of energy is to develop sustainable energy policy. Such a policy involves development of energy sources that are renewable, do not cause air pollution, do not cause climatic change such as global warming, and do not present an unacceptable risk to society.

Earth as a System

Development of energy sources as alternatives to burning fossil fuels requires knowledge concerning Earth systems and how they change from day to day, seasonally, and over periods of years. For example, both solar and wind energy, the use of which is growing dramatically, require direct use of solar energy in collection devices such as photovoltaic cells or using wind caused by differential heating of Earth by the Sun. Major challenges with solar and wind energy are how to maximize production and how to store the energy effectively. Solving these problems requires knowledge concerning solar input, which is variable, and the variables that control the nature and extent of wind resources.

Hazardous Earth Processes, Risk Assessment, and Perception

Hazards, risk assessment, and perception of risk are at the very heart of the debate over the future of nuclear energy. The risk is the product of the probability of an accident's occurring times the consequences. Because the consequences of a nuclear accident, such as Chernobyl or Three Mile Island, may be very large, the risk is perceived as being large even though the probability of an accident may be very low at a particular nuclear power plant. A major objective of those in favor of future expansion of nuclear power is the development of safer nuclear power plants. In the United States, nuclear power is being debated again, and more people than in previous years are in favor of expanding nuclear power. In some European countries, the risk is deemed too large, and recent decisions have resulted in the likelihood that nuclear power will be phased out entirely in some countries.

Scientific Knowledge and Values

We now possess the knowledge needed to develop sustainable energy. This scientific knowledge is sufficient to continue increasing the use of alternative energy sources with the objective of phasing out the use of fossil fuels, which are well known to cause serious environmental degradation. In the United States, new energy policy is being developed. The emphasis so far has been on the energy sources we have depended upon in the past such as oil, natural gas, coal, and nuclear energy. Insufficient attention has been given to alternative sources and energy conservation. Nevertheless, the people of the United States have repeatedly indicated, by responding to questionnaires, that they place high value on having a clean environment and, in particular, having clean air to breathe. Thus, our values are being tested by the perceived need of some of our political leaders who wish to supply abundant energy regardless of its environmental impacts.

Key Terms

acid rain (p. 493)

alternative energy (p. 513)

biofuel (p. 520)

coal (p. 477)

coal-bed methane (p. 487)

cogeneration (p. 521)

conservation (p. 521)

efficiency (p. 521)

fossil fuels (p. 474)

fuel cell (p. 516)

geothermal energy (p. 508)

hard path (p. 522)

natural gas (p. 482)

nuclear energy (p. 496)

oil (p. 482)

peak oil (p. 472)

soft path (p. 522)

solar energy (p. 514)

sustainable energy policy (p. 523)

tidal power (p. 516)

water power (p. 516)

wind power (p. 518)

Review Questions

1. What is peak oil and why is it important?

2. What determines the type of a coal, and why is type important?

3. What are methane hydrates, and why are they likely to be important?

4. What are the major environmental impacts of exploration for and development of oil and gas?

5. Define nuclear fission.

6. What are the major environmental concerns associated with nuclear energy?

7. What is geothermal energy?

8. How can groundwater at normal temperatures be used as an energy source?

9. What are the major renewable energy sources?

10. Are there any environmental concerns associated with hydroelectric power? If so, what are they?

11. What are the sources and potential environmental impacts associated with biomass fuel?

12. What is cogeneration?

13. What are the major differences between the hard path and soft path?

14. What do we mean by sustainable energy policy?

Critical Thinking Questions

1. When we first started using fossil fuels, particularly oil, we did not know, nor were we particularly concerned about potential environmental impacts of developing and burning oil. Suppose at that time we had completed an environmental impact report concerning the use of oil and had been able to predict consequences such as air pollution, toxicity, and acid rain. Do you think we would have developed the use of oil as fast as we did and become so dependent on it? Justify your answer.

2. Do you think that peak oil will be a defining moment in human history? Why or why not?

3. Sustainable energy development means developing an energy policy and energy sources that will provide the energy society needs without harming the environment. Do you think this is possible? Outline a plan of action to move the United States toward sustainable energy development.

SIXTEEN

Soils and Environment

Learning Objectives

Soils are an important part of our environment. Virtually all aspects of the terrestrial environment interact with soils at some level. For example, we depend upon fertile soil to grow our food, and soil properties determine, in part, the suitability of land for uses such as construction and waste disposal. As a result, the protection of soil resources is an important objective. With this in mind, we will focus on the following learning objectives for this chapter:

- Understand soil terminology and the processes responsible for the development of soils

- Understand soil fertility and the interactions of water in soil processes

- Become familiar with soil classification

- Understand the primary engineering properties of soils

- Understand relationships between land use and soils

- Know what sediment pollution is and how it can be minimized

- Understand how soils affect land-use planning

- Understand how we can sustain soil resources

Town abandoned when soil pollution is discovered The town of Times Beach, Missouri, in the 1980s became a ghost town: evacuated, abandoned, fenced off, and bulldozed following discovery of dioxin contamination. (Tom McHugh/Photo Researchers, Inc.)

CASE HISTORY | Times Beach, Missouri

Times Beach, Missouri, population 2,400, was a river town located just west of St. Louis. In 1983 the town was evacuated and purchased for $36 million by the government; Times Beach became a ghost town when it was discovered that oil sprayed on the town's road to control dust contained dioxin (Figure 16.1). Dioxin is a colorless crystal composed of oxygen, hydrogen, carbon, and chlorine. It is known to be extremely toxic to mammals and is suspected of being a carcinogen in humans. There are approximately 75 types of dioxin, which is produced as a by-product during the production of organic chemicals such as herbicides.[1] Dioxin became a household word during and after the Vietnam War; it is a component of Agent Orange, the name given to herbicides that were used to defoliate large areas in the war zones. Military personnel as well as civilians were exposed to Agent Orange and the dioxin within it. Lawsuit settlements pertaining to dioxin's role in causing diseases in people exposed to it compensated 250,000 Vietnam veterans and their families.

At Times Beach, Missouri along the Meramec River, 32 km (20 mi) southwest of St. Lewis, tests determined that the entire area had been contaminated with dioxin that had seeped into the soil from the oil applied to the roads. The decision was made in 1985 to evacuate the town and bulldoze the buildings. Following clean up and planting trees, the area today is a state park and bird refuge.

The effects of human exposure to dioxin is a controversial subject. Apparently, there is a cancer risk to workers who handle chemicals that contain dioxin, but it is not thought to be a widespread and significant cancer threat to people who are exposed to very low levels of the chemical.[2] Some scientists, including the person who ordered the evacuation, have since stated that the evacuation at Times Beach may have been an overreaction by the government to the perceived hazard of dioxin. Research concerning the potential hazard presented from exposure to dioxin continues, and the controversy concerning its potential harmful effects to people and ecosystems is still being debated.

(a)

(b)

Figure 16.1 **Dioxin pollutes soil in Missouri** (a) Times Beach, Missouri, showing a deserted building after dioxin was found to be contaminating the area. *(Corbis/Sygma)* (b) Examination of soils contaminated by dioxin at Times Beach, Missouri. The town was evacuated because of the dioxin scare. *(O. Franken/Corbis/Sygma)*

16.1 Introduction to Soils

Soil may be defined in several ways. Soil scientists define soil as solid Earth material that has been altered by physical, chemical, and biological processes such that it can support rooted plant life. Engineers define soil as solid Earth material that can be removed without blasting. Both of these definitions are important in environmental geology. Geologists must be aware of the different definitions; they must also be aware of the research concerning both soil-producing processes and the role of soils in environmental problems. Engineers have developed the field of soil mechanics to quantify engineering properties of soils such as soil strength and moisture content.

The study of soil properties, particularly with reference to land-use limitations, is becoming an important aspect of environmental work in the following ways:

- In land-use planning, the suitability of land for a particular use, or *land capability*, is often determined in part by the soils present. Soil properties are especially important for uses such as urbanization, timber management, and agriculture.

- Soils are critical when we consider waste-disposal problems. Interactions between waste, water, soil, and rock often determine the suitability of a particular site to receive waste.

- The study of soils helps land-use planners evaluate natural hazards, including floods, landslides, and earthquakes. Floodplain soils differ from upland soils, and consideration of soil properties helps delineate natural floodplains. Evaluating the relative ages of soils on landslide deposits may provide an estimate of the frequency of slides; this information assists in planning to minimize their impact. Soil studies have also been a powerful tool in establishing the age of Earth materials deformed by faulting; this information leads to better estimation of earthquake recurrence intervals.

- Soils often carry a climatic signal, something that indicates what the past climate was like. For example, calcium carbonate accumulates in desert soils, and organic matter such as bits of plants accumulates in some tropical soils. These constituents of soils assist in understanding regional and global climate change. Soil studies also provide important data used to understand how biological and geologic processes were linked in the development of both soils and ecosystems during the past few million years.[3]

16.2 Soil Profiles

Soil development is a complex process. The rock and hydrologic cycles interact to produce weathered rock materials that are basic ingredients of soils. *Weathering* is the physical and chemical breakdown of rocks and the first step in soil development (see Chapter 3). Weathered rock is further modified into soil by the activity of soil organisms. This process forms residual or transported soil. Weathered material that remains essentially in place is modified to form a **residual soil** on bedrock. The red soils of the Piedmont in the southeastern United States, formed on igneous and metamorphic rocks, are an example of residual soil. Eroded rock particles that are transported by water, wind, or glaciers and then modified in their new deposition location form **transported soil.** The fertile soils formed on glacial deposits in the Midwestern United States are transported soils.

A soil can be considered an open system that interacts with other components of the geologic cycle. The characteristics of a particular soil are a function of climate, topography, parent material (the rock from which the soil is formed), maturity (age of the soil), and biological processes.

Soil Horizons

Vertical and horizontal movements of the materials in a soil system create a distinct layering, parallel to the surface, collectively called a **soil profile.** The layers are called zones or **soil horizons.** Our discussion of soil profiles mentions only the horizons most commonly found in soils. Additional information is available from detailed soils texts.[4,5,6]

Figure 16.2a shows the common master, or prominent, soil horizons. The *O horizon* and *A horizon* contain highly concentrated organic material such as

(a)

Soil

Rock

O. Horizon is composed mostly of organic materials including decomposed or decomposing leaves, twigs, etc. The color of the horizon is often dark brown or black.

A. Horizon is composed of both mineral and organic materials. The color is often light black to brown. Leaching, defined as the process of dissolving, washing, or draining Earth materials by percolation of groundwater or other liquids, occurs in the A horizon and moves clay and other material such as iron and calcium to the B horizon.

E. Horizon is composed of light-colored materials resulting from leaching of clay, calcium, magnesium, and iron to lower horizons. The A and E horizons together constitute the zone of leaching.

B. Horizon is enriched in clay, iron oxides, silica, carbonate, or other material leached from overlying horizons. Horizon is known as the zone of accumulation.

C. Horizon is composed of partially altered (weathered) parent material; rock is shown here but the material could also be alluvial in nature, such as river gravels in other environments. The horizon may be stained red with iron oxides.

R. Unweathered (unaltered) parent material.

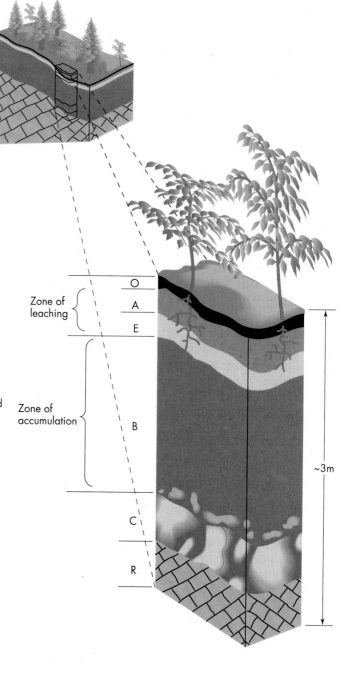

Zone of leaching

O
A
E

Zone of accumulation

B

C

R

~3m

(b)

Figure 16.2 **Soil profiles** (a) Idealized diagram showing a soil profile with soil horizons. (b) Soil profile showing a black A horizon, a light-red B horizon, a white K horizon rich in calcium carbonate, and a lighter C horizon. *(Edward A. Keller)*

decomposing plants. The differences between these two layers reflect the amount of organic material present in each. In general, the O horizon consists almost entirely of plant litter and other organic material, whereas the underlying A horizon contains a good deal of both organic and mineral material. Below the O or A horizon, some soils have an *E horizon*, or zone of leaching, a light-colored layer that is leached of iron-bearing components. This horizon is light in color because it contains less organic material than the O and A horizons and little inorganic coloring material such as iron oxides.

The *B horizon*, or zone of accumulation, underlies the O, A, or E horizons and consists of a variety of materials translocated downward from overlying horizons. Several types of B horizon have been recognized. Probably the most important type of B horizon is the *argillic B* horizon. An argillic B horizon is enriched in clay minerals that have been translocated downward by soil-forming processes. Environmental geologists are also interested in the *Bk* horizon, which is characterized by accumulation of calcium carbonate. The carbonate coats individual soil particles and may fill some pore spaces within the soil. It does not, however, dominate the structure of the horizon. If a soil horizon is impregnated with calcium carbonate to the extent that its morphology is dominated by the carbonate, it is designated a *K horizon* (Figure 16.2b). Carbonate completely fills the pore spaces in K horizons, often forming layers parallel to the surface. The term *caliche* is often used for an irregular accumulation or layers of calcium carbonate in Earth material near the surface.

The *C horizon* consists of parent material partially altered by weathering processes. It lies directly over the *R horizon*, which is unaltered parent material composed of consolidated bedrock that underlies the soil. Some of the fractures and other pore spaces in the bedrock may contain clay that has been translocated downward.[4]

The term *hardpan* is often used to refer to a hard compacted soil horizon, usually part of the B horizon. Hardpan is often composed of compacted clay or cemented with calcium carbonate, iron oxide, or silica. Hardpan horizons may be nearly impermeable and thus restrict the downward movement of soil water.

16.3 Soil Properties

Color

One of the first things we notice about a soil is its color, or the colors of its horizons. The O and A horizons tend to be dark because of their abundant organic material. The E horizon, if present, may be almost white because of the leaching of iron and aluminum oxides. The B horizon shows the most dramatic differences in color, varying from yellow-brown to light red-brown to dark red, depending on the presence of clay minerals and iron oxides. Bk horizons may be light-colored because of their carbonates, but they are sometimes reddish as a result of iron oxide accumulation. If a true K horizon has developed, it may be almost white because of the abundance of calcium carbonate. Although soil color can be an important diagnostic tool for analyzing a soil profile, one must be cautious about calling a red layer a B horizon. The original parent material, if rich in iron, may produce a very red soil even when there has been relatively little soil profile development.

Soil color may be an important indicator of how well a soil drains. Well-drained soils are well aerated, creating excellent oxidizing conditions; for example, in a well-aerated soil, iron produces soil with a red color. Poorly drained soils are wet, and these soils tend to have a yellow color. This distinction is important because

Figure 16.3 Soil texture classes The classes are defined according to the percentage of clay-, silt-, and sand-sized particles in the soil sample. The area defined by the point connected by dashed lines represents a soil composed of 40 percent sand, 40 percent silt, and 20 percent clay, which is classified as loam. *(From U.S. Department of Agriculture)*

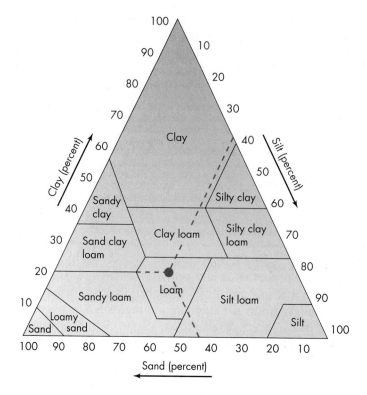

poorly drained soils are associated with environmental problems, such as lower slope stability and an inability to be utilized as a disposal medium for household sewage systems.

Texture

The *texture* of a soil depends upon the relative proportions of sand-, silt-, and clay-sized particles (Figure 16.3). Clay particles have a diameter of less than 0.004 mm (0.0002 in.); silt particles have diameters ranging from 0.004 to 0.063 mm (0.0002 to 0.003 in.); and sand particles are 0.063 to 2.0 mm (0.003 to 0.08 in.) in diameter. Earth materials with particles larger than 2.0 mm (0.08 in.) in diameter are called gravel, cobbles, or boulders, depending on the particle size.

In the field, soil texture is commonly identified by estimation and then refined in the laboratory by separating and determining the proportions of the sand, silt, and clay. A useful field technique for estimating the size of sand-sized or smaller soil particles is as follows: It is sand if you can see individual grains, silt if you can see the grains with a 10× hand lens, and clay if you cannot see grains with such a hand lens. Another method is to feel the soil. Sand is gritty; it crunches between the teeth. Silt feels like baking flour, and clay is cohesive. When mixed with water, smeared on the back of the hand, and allowed to dry, clay cannot easily be dusted off, whereas silt or sand can.

Structure

Soil particles often cling together in aggregates called *peds* that are classified according to shape into several types. Figure 16.4 shows some of the common structures of peds found in soils. The type of structure present is related to soil-forming processes, but some of these processes are poorly understood.[4] For example, granular structure is fairly common in A horizons, whereas blocky and prismatic structures are most likely to be found in B horizons. Soil structure is an important diagnostic tool for evaluating the development and approximate age of soil profiles. In general, as the profile develops with time, structure becomes more complex and may go from granular to blocky to prismatic as the clay content in the argillic B horizons increases.

Types of peds	Typical size range	Horizon usually found in	Comments
Granular 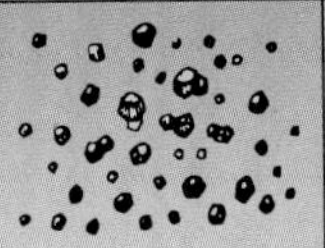	1–10 mm	A	Can also be found in B and C horizons
Blocky 	5–50 mm	B_t	Are usually designated as angular or subangular
Prismatic 	10–100 mm	B_t	If columns have rounded tops, structure is called columnar
Platy 	1–10 mm	E	May also occur in some B horizons

Figure 16.4 Soil structure
Types and description of soil structure (peds).

16.4 Soil Fertility

A soil may be considered a complex ecosystem. A single cubic meter of soil may contain millions of living things, including small rodents, insects, worms, algae, fungi, and bacteria. These organisms are important for mixing and aerating the soil particles. They also help release or convert nutrients in soils into forms that are useful for plants.[7] **Soil fertility** refers to the capacity of the soils to supply nutrients, such as nitrogen, phosphorus, and potassium, needed for plant growth when other factors are favorable.[8]

Soils that developed on some floodplains and glacial deposits contain sufficient nutrients and organic material to be naturally fertile. Other soils, developed on highly leached bedrock or on loose deposits with little organic material, may be nutrient poor and have low fertility. Soils are often manipulated to increase plant yield by applying either fertilizers to supply nutrients or materials that improve the soil's texture and moisture retention. Soil fertility can be reduced by soil erosion or leaching that removes nutrients, by interruption of natural processes that supply nutrients, such as flooding, or by continued use of pesticides that alter or damage soil organisms.

The development and maintenance of many terrestrial ecosystems depend upon soil as a basic component of the system. Interactions between physical and biochemical processes operate over variable scales of time (hundreds to thousands of years) and space (from less than 1 km^2 to thousands of square kilometers) to produce soils that help support complex ecosystems. Soils go through stages of building, sustaining, and degrading.[3] Building is a relatively fast process that starts with a new substrate, or layer, such as sediment from a volcanic eruption or flood. Chemical weathering of minerals releases chemicals that can be used by a variety of organisms, including plants, that build soils. Sustaining soils involves longer time lines—thousands to several millions of years during which additional minerals in the soils continue to weather, forming clays. Soil degradation occurs as minerals necessary to support an ecosystem are depleted by a variety of near-surface physical, hydrologic, and biological processes. A physical process, soil erosion can remove upper soil horizons and their nutrients. In some cases, soil degradation processes may be reduced. For example, nutrients contained in dust from the deserts of Africa may reach islands of the Pacific, helping maintain soil fertility after weathering and water moving through the soil have removed some of the original mineral nutrients in soils.

16.5 Water in Soil

If you analyze a block of soil, you will find it is composed of bits of solid mineral and organic matter with pore spaces between them. The pore spaces are filled with gases, mostly air, or liquids, mostly water. If all the pore spaces in a block of soil are completely filled with water, the soil is said to be in a saturated condition; otherwise, it is said to be unsaturated. Soils in swampy areas may be saturated year-round, whereas soils in arid regions may be only occasionally saturated.[9]

The amount and movement of water through soils are important research topics. Both are linked to water pollution problems, such as the movement of gasoline from leaking underground tanks or the migration of liquid pollutants from waste-disposal sites.

16.6 Soil Classification

Terminology and classification of soils pose unique problems in environmental geology. We are often interested in both soil processes and human use of soil. A classification system, or *taxonomy*, that includes engineering as well as physical and chemical properties would be most appropriate, but none exists. We must therefore be familiar with two separate systems of soil classification: soil taxonomy, used by soil scientists, and the engineering classification, which groups soils by material types and engineering properties.

Soil Taxonomy

Soil scientists have developed a comprehensive and systematic classification of soils known as *soil taxonomy*, which emphasizes the physical and chemical properties of the soil profile. Soil taxonomy is useful for agricultural and related land-use purposes.

Engineering Classification of Soils

The *unified soil classification system*, widely used in engineering practice, is shown in Table 16.1. Because all natural soils are mixtures of coarse particles, including gravel and sand, fine particles, including silt and clay, and organic material, the major divisions of this system are coarse-grained soils, fine-grained soils, and organic soils. Each group is based on the predominant particle size or the abundance of organic material in the soil. Organic soils have a high organic content and are identified by their black or gray color and sometimes by an odor of hydrogen sulfide, which smells like rotten eggs.

16.7 Engineering Properties of Soils

The water table acts as a transition zone within Earth. Pores in rocks below the water table are saturated; those above are not. Soil above the water table has three distinct parts, or phases: solid material, liquid, and gas, such as air or carbon dioxide. A soil's usefulness is greatly affected by the variations in the proportions and structure of the three phases. The types of solid materials, the particle size, and the water content are probably the most significant variables that determine

TABLE 16.1 Unified Soil Classification System Used by Engineers

Major Division				Group Symbol	Soil Group Name
COARSE-GRAINED SOILS (More than half of material larger than 0.074 mm)	GRAVELS	Clean gravels	Less than 5% fines	GW	Well-graded gravel
				GP	Poorly graded gravel
		Dirty gravels	More than 12% fines	GM	Silty gravel
				GC	Clayey gravel
	SANDS	Clean sands	Less than 5% fines	SW	Well-graded sand
				SP	Poorly graded sand
		Dirty sands	More than 12% fines	SM	Silty sand
				SC	Clayey sand
FINE-GRAINED SOILS (More than half of material smaller than 0.074 mm)	SILTS, NONPLASTIC			ML	Silt
				MH	Micaceous silt
				OL	Organic silt
	CLAYS, PLASTIC			CL	Silty clay
				CH	High plastic clay
				OH	Organic clay
Predominantly organics				PT	Peat and muck

Note: The value of 0.074 mm is the boundary between sand and silt that engineers use. Geologists use 0.063 mm for the same boundary.

engineering properties of soils. Important engineering properties are strength, sensitivity, compressibility, erodibility, hydraulic conductivity, corrosion potential, ease of excavation, and shrink-swell potential.

Soil Strength

Soil strength is the ability of a soil to resist deformation. It is difficult to generalize about the strength of soils. Numerical averages of the strength of a soil can be misleading, because soils are often composed of mixtures, zones, or layers of materials with different physical and chemical properties.

The strength of a particular soil type is a function of cohesive and frictional forces. *Cohesion* is a measure of the ability of very small silt and clay soil particles to stick together. The cohesion of particles in fine-grained soils is primarily the result of electrostatic forces between particles and is a significant factor in determining the strength of a soil. In partially saturated coarse-grained, sandy soils, moisture films between the grains may cause an apparent cohesion due to surface tension caused by the attraction of water molecules to each other at the surface or between soil grains (Figure 16.5). The principle of cohesion explains the ability of damp sand, which is cohesionless when dry, to stand in vertical walls in children's

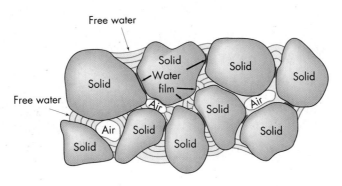

Figure 16.5 Water in soils

Partly saturated soil, showing particle-water-air relationships. Particle size is greatly magnified. The attraction between the water and the soil particles, or the surface tension, develops a stress that holds the grains together. This apparent cohesion is destroyed if the soil dries out or becomes completely saturated. *(After Pestrong, R. 1974. Slope stability. New York: American Geological Institute and McGraw-Hill)*

Figure 16.6 **Apparent cohesion**
Children's sand castle. The surface tension of water molecules in the damp sand enables the walls of the castle to stand. *(John Lei/Stock Boston)*

sand castles on the beach (Figure 16.6).[10] Friction between grains also contributes to the strength of a soil. The total frictional force is a function of the density, size, and shape of the soil particles and of the weight of overlying particles that force the grains together. Frictional forces are most significant in coarse-grained soils rich in sand and gravel. Frictional forces are the reason why you do not sink far into the sand when walking on dry sand on beaches. Most soils are a mixture of coarse and fine particles, so the strength is usually the result of both cohesion and internal friction. Although it is difficult to generalize, clay-rich soils with weak clay minerals and organic-rich soils, such as peaty soils, tend to have lower strengths than coarser soils.

Vegetation may play an important role in soil strength. For example, tree roots may provide considerable cohesion through the binding characteristics of a continuous root mat or by anchoring individual roots to bedrock beneath thin soils on steep slopes (Figure 16.7).

Figure 16.7 **Root strength** Tree roots are helping to bind the particles of this riverbank soil together. *(Edward A. Keller)*

Soil Sensitivity

Soil sensitivity measures changes in soil strength resulting from disturbances such as vibrations or excavations. Sand and gravel soils with no clay are the least sensitive. As fine material becomes abundant, soils become more and more sensitive. Some clay soils may lose 75 percent or more of their strength after a disturbance.[11]

Soil Compressibility

Soil compressibility is a measure of a soil's tendency to consolidate, or decrease in volume. Compressibility is partially a function of the elastic nature of the soil particles and is directly related to the settling of structures such as the world-famous Leaning Tower of Pisa in Italy (Figure 16.8). Excessive settling will crack foundations and walls. Coarse materials such as gravels and sands have a low compressibility, and settling will be considerably less in these materials than in highly compressible fine-grained or organic soils.

Erodibility

Soil *erodibility* refers to the ease with which soil materials can be removed by wind or water. Soils with a high erosion factor include unprotected silts and sands. Cohesive soils, which are more than 20 percent clay, naturally cemented soils, or coarse gravel-rich soils are not as easily eroded by wind or water and therefore have a low erosion factor.

Figure 16.8 Leaning Tower of Pisa, Italy The lean is a result of differential settling of the soil. If correction measures had not been taken to strengthen the tower, it would have fallen over. *(D. & J. Heaton/Stock Boston)*

Hydraulic Conductivity

Hydraulic conductivity (K) is a measure of the ease with which water moves through a material. Hydraulic conductivity is measured in units of velocity, centimeters per second or centimeters per hour (see the discussion of groundwater in Chapter 12). Saturated clean gravels or sands have the highest K values (from 5 to greater than 50 cm per hour, or 2 to greater than 20 in. per hour). As fine particles in a saturated mixture of clean gravel and sand increase, K decreases. Saturated clays generally have a very low K, less than 0.025 cm (0.01 in.) per hour.[6] Hydraulic conductivity in unsaturated soils is complex; fine-grained, unsaturated soils actually pull or suck water and hold it tightly. Hydraulic conductivity is very important in many soil environmental problems related to soil drainage, the movement of liquid pollutants in soils, and land-use potential for agriculture, waste disposal, and construction.

Corrosion Potential

Corrosion is a slow weathering or chemical decomposition that proceeds from the surface into the ground. All objects buried in the ground—pipes, cables, anchors, fence posts—are subject to corrosion. The corrosiveness of a particular soil depends on the chemistry of the soil, the buried material, and the amount of water available.[12]

Ease of Excavation

Ease of excavation pertains to the procedures, and hence the equipment, required to remove soils during construction. There are three general categories of excavation techniques. *Common excavation* is accomplished with an earth mover, backhoe, or bulldozer. This equipment essentially removes the soil without having to scrape it first; most soils can be removed by this process. *Rippable excavation* requires breaking up the soil with special ripping teeth before it can be removed. For example, a tightly compacted or cemented soil would require rippable excavation. *Blasting* or *rock cutting* is the third, and often the most expensive, category; a hard silica cemented soil might need to be cut with a jackhammer before being removed.

Shrink-Swell Potential

Shrink-swell potential refers to the tendency of a soil to gain or lose water. Soils that tend to increase or decrease in volume with water content are called **expansive soils.** The swelling is caused by the chemical attraction of water molecules to the submicroscopic flat particles, or plates, of certain clay minerals. The plates are composed primarily of silica, aluminum, and oxygen atoms, and layers of water are added between the plates as the clay expands or swells (Figure 16.9a).[13] Expansive soils tend to absorb large quantities of water and expand. Cracks in the ground form when the soil dries out and contracts (Figure 16.9c). *Montmorillonite* is the common clay mineral associated with most expansive soils. With sufficient water, pure montmorillonite may expand up to 15 times its original volume; fortunately, most soils contain limited amounts of this clay mineral, so it is unusual for an expansive soil to swell beyond 25 percent to 50 percent. However, an increase in volume of more than 3 percent is considered potentially hazardous.[13]

Expansive soils in the United States cause significant environmental problems. As one of our most costly natural hazards, expansive soils are responsible for several billions of dollars in damages annually to highways, buildings, and other structures. Every year more than 250,000 new houses are constructed on expansive soils. Of these, about 60 percent will experience some minor damage, such as

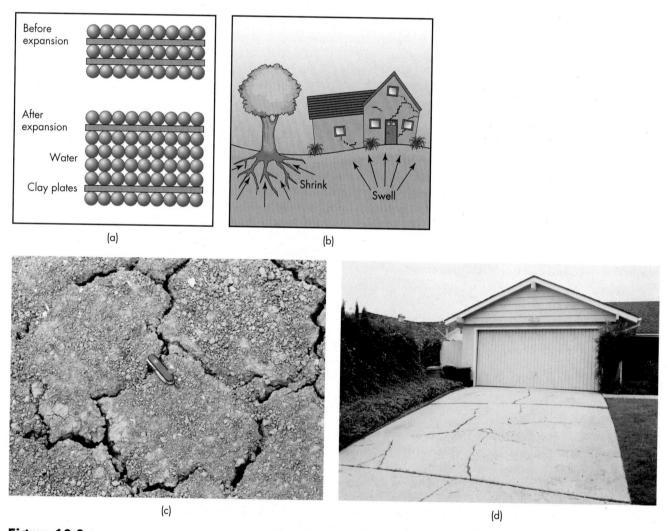

(a) (b)

(c) (d)

Figure 16.9 **Expansive soils** (a) Expansion of a clay (montmorillonite) as layers of water molecules are incorporated between clay plates. (b) Effects of soil's shrinking and swelling at a home site. *(After Mathewson, C. C., and Castleberry, J. P., II. Expansive soils: Their engineering geology. Texas A&M University)* (c) Cracks from a soil that has contracted. (d) Driveway cracked by expansion of clay soil under the foundation. *(Edward A. Keller)*

cracks in the foundation, walls, driveway, or walkway (Figure 16.9b, d); 10 percent will be seriously damaged, some beyond repair.[14,15]

Structural damage on expansive soils is caused by volume changes in the soil in response to changes in moisture content. Factors that affect the moisture content of an expansive soil include climate, vegetation, topography, drainage, and quality of construction.[14] Builders in regions that have a pronounced wet season followed by a dry season allow for a regular shrink-swell sequence. These regions, such as the southwestern United States, are more likely to experience an expansive soil problem than are regions where precipitation is more evenly distributed throughout the year. Vegetation can cause changes in the moisture content of soils. Because large trees draw and use a lot of local soil moisture, especially during a dry season, they facilitate soil shrinkage. Therefore, in areas with expansive soil, trees should not be planted close to the foundations of light structures, such as homes (Figure 16.9b).

Topography and drainage are also significant factors for evaluating expansive soils. Adverse topographic and drainage conditions cause water to form ponds around or near structures, increasing the swelling of expansive clays. However, homeowners and contractors can do a great deal to avoid this problem. Proper design of subsurface drains and rain gutters can minimize expansive soil damages by improving drainage, as may designing and constructing foundations to accommodate some shrinking and swelling of the soil.[14]

Clearly, some soils are more desirable than others for specific uses. Although planners concerned with land use will not conduct soil tests to evaluate the engineering properties of soils, they will be better prepared to design with nature and take advantage of geologic conditions if they understand the basic terminology and principles of Earth materials. Our discussion of engineering properties established two general principles. First, because of their low strength, high sensitivity, high compressibility, low permeability, and variable shrink-swell potential, clay soils should be avoided in projects involving heavy structures, structures with minimal allowable settling, or projects needing well-drained soils. Second, soils that have a high corrosive potential or that require other than common excavation should be avoided if possible. If such soils cannot be avoided, extra care, special materials and techniques, and higher-than-average initial costs—including planning, design, and construction—must be expected. The secondary costs—that is, the operation and maintenance costs of these projects—may also be greater.

16.8 Rates of Soil Erosion

Rates of soil erosion are measured as a volume, mass, or weight of soil that is removed from a location within a specified time and area, for example, kilograms per year per hectare. Soil erosion rates vary with the engineering properties of the soil, land use, topography, and climate.

There are several approaches to measuring rates of soil erosion. The most direct method is to make actual measurements on slopes over a period of at least several years and use these values as representative of what is happening over a wider area and longer time span. This approach is rarely used, however, because data from individual slopes and drainage basins are very difficult to obtain. A second approach is to use data obtained from resurveying reservoirs to calculate the change in the reservoirs' storage capacity of water; the depletion of storage capacity is equivalent to the volume of sediment eroded from upstream soils. A third approach is to use an equation to calculate rates of sediment eroded from a particular site. One of the most commonly used equations is the Universal Soil Loss Equation.[16] This equation uses data on rainfall, runoff, the size and shape of the slope, the soil cover, and erosion-control practices to predict the amount of soil moved from its original position[17] (see A Closer Look: Universal Soil Loss Equation).

Figure 16.10 **Serious soil erosion** and gully formation in central California related to diversion of runoff water. The surface was essentially ungullied several months before the photograph was taken. *(Edward A. Keller)*

16.9 Sediment Pollution

Sediment is one of our greatest pollutants. In many areas, it chokes streams; fills in lakes, reservoirs, ponds, canals, drainage ditches, and harbors; buries vegetation; and generally creates a nuisance that is difficult to remove. Natural pollutional sediment—eroded soil—is truly a resource out of place. It depletes soil at its site of origin (Figure 16.10), reduces the quality of the water it enters, and may deposit unwanted sediment on productive croplands or other useful land (Figure 16.11).[18]

Sources of the sediment include land disturbed for agriculture, land overgrazed by animals such as sheep or cattle, urban construction sites, land recently logged or burned, and land disrupted by mining. **Sediment pollution** affects rivers, streams, lakes, and even the ocean. Unfortunately, the problem promises to be with us indefinitely. One solution to sediment pollution is to implement sound conservation practices, particularly in urbanizing areas, where tremendous quantities of sediment are produced during construction (see Case History: Reduction

Figure 16.11 Soil erosion has resulted in unwanted red sediment at this site in Charlotte, North Carolina. *(Edward A. Keller)*

A CLOSER LOOK | Universal Soil Loss Equation

The Universal Soil Loss Equation is

$$A = RKLSCP$$

where

A = the long-term *average* annual soil loss for the site being considered
R = the long-term rainfall *runoff* erosion factor
K = the soil erodibility index
L = the hillslope/*length* factor
S = the hillslope/gradient, or *slope* factor
C = the soil *cover* factor
P = the erosion-control *practice* factor

The advantage of using this equation is that once the various factors have been determined and multiplied together to produce predicted soil loss, conservation practices may be applied through factors C and P to reduce the soil loss to the desired level. For slopes that are amenable to shaping, factors K, L, and S may also be manipulated to achieve desired sediment loss results. This equation is particularly valuable for evaluating construction sites and areas along corridors such as pipelines and highways. When planning construction sites, planners can use the Universal Soil Loss Equation to predict the impact of sediment loss on local streams and other resources and to develop management strategies for minimizing this impact.[16,17]

of Sediment Pollution, Maryland). Another solution is to use sediment control basins, which are designed to trap and control sediment; they must be periodically cleared out to operate effectively. A generalized cross section of a sediment control basin is shown in Figure 16.12.

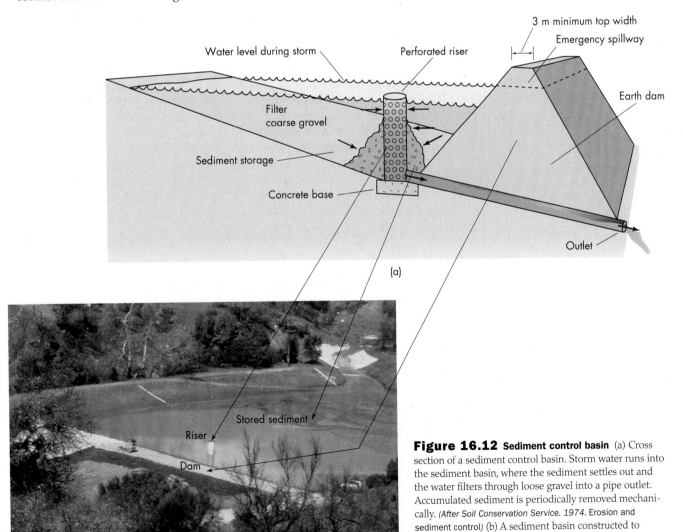

Figure 16.12 Sediment control basin (a) Cross section of a sediment control basin. Storm water runs into the sediment basin, where the sediment settles out and the water filters through loose gravel into a pipe outlet. Accumulated sediment is periodically removed mechanically. *(After Soil Conservation Service. 1974. Erosion and sediment control)* (b) A sediment basin constructed to trap sediment that eroded after a wildfire in southern California. *(Edward A. Keller)*

CASE HISTORY | Reduction of Sediment Pollution, Maryland

A study in Maryland demonstrated that sediment control measures can reduce sediment pollution in an urbanizing area.[19] The suspended sediment transported by the northwest branch of the Anacostia River near Colesville, Maryland, with a drainage area of 55 km^2 (21 mi^2), was measured over a 10-year period. During that time, construction within the basin involved about 3 percent of the area each year. The total urban land area in the basin was about 20 percent at the end of the 10-year study.

Sediment pollution was a problem in the area because the soils are highly susceptible to erosion; moreover, there is sufficient precipitation to ensure their erosion when construction has removed the vegetative cover. Most of the sediment is transported during spring and summer rainstorms.[19] A

sediment-control program was initiated, and the sediment yield was reduced by about 35 percent. The program utilized basic sediment-control principles such as tailoring development to the natural topography, exposing a minimal amount of land, providing protection for exposed soil, minimizing surface runoff from critical areas, and trapping eroded sediment on construction sites. Specific measures included scheduled grading to minimize the time of soil exposure, mulch application and temporary vegetation to protect exposed soils, sediment diversion berms, stabilized channels, and sediment basins. The Maryland study concluded that even further sediment control can be achieved by both scheduling major grading during periods of low erosion potential and designing improved sediment traps to control runoff during storms.[19]

16.10 Land Use and Environmental Problems of Soils

Human activities affect soils by influencing the pattern, amount, and intensity of surface-water runoff, erosion, and sedimentation. The most important of these human influences is the conversion of natural areas to various land uses.

The estimated and observed variation in sediment yield that accompanied changes in land use in the Piedmont region from 1800 to 2000 are summarized on Figure 16.13. Notice the sharp peak in sediment production during the construction phase of urbanization in about 1960. These data suggest that the effects of land-use change on a drainage basin, its streams, and its sediment production may be quite dramatic. Streams and naturally forested areas are assumed to be relatively stable, that is, without excessive erosion or deposition. A land-use change that converts forested land for agricultural purposes also generally increases runoff and erosion. As a result, streams become muddy and may not be able to transport all the sediment delivered to them. The channels will therefore partially fill with sediment, or *aggrade*, possibly increasing the magnitude and frequency of flooding.

Agriculture

In the past 50 years, soil erosion and overuse of soils caused by intensive agriculture have damaged about 10 percent of the world's best agricultural land. It is

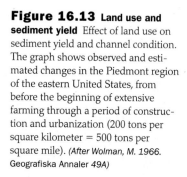

Figure 16.13 Land use and sediment yield Effect of land use on sediment yield and channel condition. The graph shows observed and estimated changes in the Piedmont region of the eastern United States, from before the beginning of extensive farming through a period of construction and urbanization (200 tons per square kilometer = 500 tons per square mile). *(After Wolman, M. 1966. Geografiska Annaler 49A)*

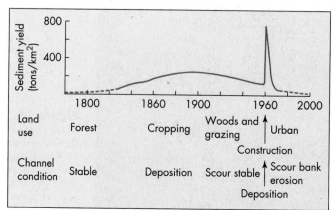

estimated that in the United States approximately one-third of the *topsoil*, the most fertile soil near the surface that supports vegetation, has been lost to erosion.[20]

Traditional agriculture, which involves plowing the soil in straight lines, or furrows, is particularly damaging. Removal of vegetation exposes the soil to erosion by wind and water. Even on gentle slopes, furrows may channel the water downslope, increasing the erosion potential. Although new soil is constantly forming, soils form very slowly, from a rate of about 1 mm (0.04 in.) in several decades to as much as 1 mm (0.04 in.) per year. The key to sustaining soil resources is to reduce erosion to a rate less than the rate that soils naturally form. Some practices to sustain soils are[20]

- *Contour plowing.* This involves plowing with the natural topography of the land. Furrows are plowed perpendicular to the slope of the land rather than in the downslope direction. This is one of the most effective ways to reduce erosion by running water and is widely used (Figure 16.14).

- *No-till agriculture.* This eliminates plowing altogether, greatly reducing soil erosion. No-till agriculture is an integrated plan to plant and harvest crops without plowing while suppressing weeds and other pests.

- *Terracing slopes.* Soil erosion on steep slopes can be managed and minimized by terracing slopes to produce flat areas for farming. Retaining walls of stone or other materials are used to form terraces and stabilize the slope. Terracing is widely used on farms around the world (Figure 16.15).

- *Planting more than one crop.* This is most effective on small farms in the tropical rain forest and other areas. The forest trees are cut in small patches of land, and some smaller trees and plants are left in place. Several crops are planted among the remaining natural vegetation (Figure 16.16). After several years, the land is allowed to recover. When the forest has grown back, the process may be repeated. This approach works if human population on the land is low. With increased population and pressure to farm more land more frequently, the practice is not sustainable.

Figure 16.14 Contour plowing can significantly reduce soil erosion. This scene is near McGregor, Iowa.
(Alex S. MacLean/Peter Arnold, Inc.)

Figure 16.15 **Terraced slopes** reduce soil erosion and produce flat land for farming. These rice paddies on terraces are in Bali, Indonesia. *(Fred Bruemmer/Peter Arnold, Inc.)*

Urbanization

The conversion of agricultural, forested, or rural land to highly urbanized land causes dramatic changes. During the construction phase of a conversion there is a tremendous increase in sediment production, which may be accompanied by a moderate increase in runoff (Figures 16.13 and 16.17). The response of streams in the area is complex and may include both channel erosion and deposition, resulting in wide, shallow channels. The combination of increased runoff and shallow channels increases the flood hazard. After the construction phase, the land is mostly covered with buildings, parking lots, and streets, so the sediment yield drops to a low level. Streams respond to the lower sediment yield and higher runoff by eroding their channels, which become deeper. However, because of the large impervious areas and the use of storm sewers, runoff increases, again increasing the risk of flooding.

The process of urbanization directly affects soils in several ways:

- Soil may be scraped off and lost. Once sensitive soils are disturbed, they may have lower strengths when they are remolded.

- Materials may be brought in from outside areas to fill a depression before construction, resulting in a much different soil than was previously there.

- Draining soils to remove water may cause *desiccation,* or drying out, and other changes in soil properties.

- Soils in urban areas are susceptible to soil pollution resulting from deliberate or inadvertent addition of chemicals to soils. This problem is particularly serious if hazardous chemicals have been applied.

Figure 16.16 **Multiple crops** Farming a small plot of land in the rain forest of Nigeria, producing several crops. *(Still Pictures/Peter Arnold, Inc.)*

Off-Road Vehicles

Urbanization is not the only land use that causes increased soil erosion and hydrologic changes. In recent years, the popularity of off-road vehicles (ORVs) has

(a)

(b)

Figure 16.17 **Soil erosion in the urban environment** (a) Urbanization and the construction of freeway on-ramps can contribute to soil erosion and increased sediment production. This eroding embankment is near the University of California, Santa Barbara, and the community of Isla Vista. (b) A few years after erosion control measures were taken by planting vegetation. *(Edward A. Keller)*

increased enormously; demand for recreational areas to pursue this interest has led to serious environmental problems as well as conflicts between users of public lands.

There are now millions of ORVs, many of which are invading the deserts, coastal dunes, and forested mountains of the United States. Problems associated with

(a)

(b)

Figure 16.18 Off-road vehicles Serious erosion problems caused by off-road vehicle use (a) in mountains and (b) on coastal dunes. *(Edward A. Keller)* (c) An ORV churning up sand on a coastal dune. *(Los Angeles Times Photo by Steve Osman)*

(c)

ORVs are common, from the shores of North Carolina and New York to sand dunes in Michigan and Indiana to deserts and beaches in the western United States. A single motorcycle need travel only 8 km (5 mi) to have an impact on an area measuring 1,000 m² (10,765 ft²), and a four-wheel-drive vehicle has an impact over the same area by traveling only 2.4 km (1.5 mi). In some desert areas, the tracks produce scars that may remain part of the landscape for hundreds of years.[21,22]

The major areas of environmental problems caused by ORVs are soil erosion, changes in hydrology, and damage to plants and animals. Off-road vehicles cause direct mechanical erosion and facilitate wind and water erosion of materials loosened by their passing (Figure 16.18). Runoff from ORV sites is as much as eight times greater than for adjacent unused areas, and sediment yields are comparable to those found on construction sites in urbanizing areas.[22] Hydrologic changes from ORV activity result primarily from near-surface soil compaction that reduces the ability of the soil to absorb water. Furthermore, water already in the soil becomes more tightly held and thus less available to plants and animals. In the Mojave Desert, tank tracks produced 50 years ago are still visible, and the compacted soils have not recovered.[22] Compaction also changes the variability of soil temperature. This effect is especially apparent near the surface, where the soil becomes hotter during the day and colder at night. Animals are killed or displaced and vegetation is damaged or destroyed by intensive ORV activity. The damage results from a combination of soil erosion, compaction, temperature change, and moisture content change.[22]

Figure 16.19 **Soil erosion by biking and walking** Deep ruts have been excavated by bicyles and walking in a sandy soil at this California site. *(Edward A. Keller)*

There is little doubt that as a management strategy some land must be set aside for ORV use. However, how much land should be involved, and how can environmental damage be minimized? Sites should be chosen in closed basins with minimal soil and vegetation variation. The possible effects of erosion by wind, must be evaluated carefully, as must the sacrifice of nonrenewable cultural, biological, and geologic resources.[21] A major problem remains: Intensive ORV use is incompatible with nearly all other land use, and it is very difficult to restrict damages to a specific site. Material removed by mechanical, water, and wind erosion will always have an impact on other areas and activities.[21,22]

In recent years the demand for self-propelled vehicles has been increasing. Use of off-road mountain bikes has grown dramatically and is having an impact on the environment. Bicycles have damaged mountain meadows, and their intensive use contributes to trail erosion (Figure 16.19). Mountain bike users are lobbying to gain entry into even more locations in the national forests, parks, and wilderness areas. Their position states that one bicycle causes less erosion than one horse. Although this is generally true, mountain bikes are cheaper and easier to maintain than horses; therefore, there are many more people riding mountain bikes than there are riding horses. Thus, the cumulative effect of bicycles on trails may be greater than that of horses. Also, hikers and other visitors may not mind seeing animals such as horses in wilderness areas but may be less receptive to bicycles, which are fast and almost silent (you can't hear them coming). Wilderness management plans will have to be developed to ensure that overenthusiastic people do not damage sensitive environments.

16.11 **Soil Pollution**

Soil pollution occurs when materials detrimental to people and other living things are inadvertently or deliberately applied to soils (see Case History: Times Beach, Missouri, at the beginning of this chapter). Many types of materials, including organic chemicals such as hydrocarbons or pesticides, or heavy metals such as selenium, cadmium, nickel, or lead, may act as soil contaminants. Soils, particularly those with clay particles, can also act to selectively attract, absorb, or bind toxins and other materials that otherwise would contaminate the environment. Soils may also contain organisms that break down certain contaminants into less harmful materials. As a result, soils offer opportunities to reduce environmental pollution. However, contaminants in soils and the products of their breakdown by soil and biochemical processes may be toxic to ecosystems and humans if they become concentrated in plants or are transported into the atmosphere or water.[5]

Problems arise when soils intended for uses other than waste disposal are contaminated, or when people discover that soils have been contaminated by previous uses. Houses and other structures, such as schools, have been built over sites where soils have been contaminated. At many sites, contamination from old waste-disposal facilities or from dumping of chemicals is now being discovered; some of these sites are being treated. However, treatment of soils to remove contaminants can be a very costly endeavor. Treatments vary from excavation and disposal to incineration or bioremediation. Often, bioremediation is done on the pollution site and does not require excavating and moving large quantities of contaminated soil (Figure 16.20).[5,23] In recent years, soil and water contaminated by leaking underground tanks have become a significant environmental concern. Businesses are now adding systems to monitor storage tanks so that leaks can be detected before significant environmental damage occurs.

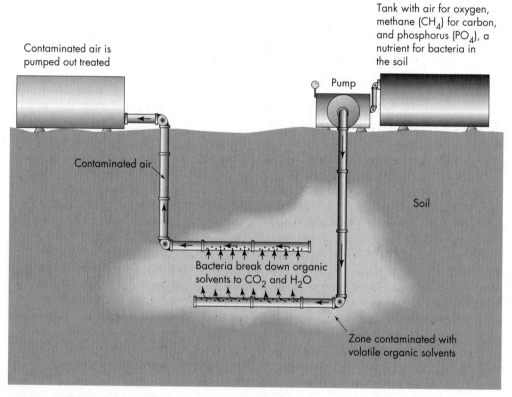

Figure 16.20 Bioremediation Idealized diagram illustrating the process of enhanced bioremediation of soil contaminated by an organic solvent. Methane (CH_4), phosphorus (PO_4), and air (with O_2) are nutrients pumped intermittently into the contaminated area and released from the lower slotted pipe. The upper pipe (also slotted) sucks contaminated air from the soil. The nutrients stimulate the growth of bacteria. The supply of methane, a carbon source, is then stopped and the carbon-hungry bacteria go after the inorganic solvents, degrading them to carbon dioxide and water as part of their life cycle. This type of process can significantly reduce the contaminated soil's treatment time and cost. (*Modified after Hazen, T. C. 1995. Savanna River site— A test bed for cleanup technologies. Environmental Protection, April: 10–16*)

16.12 Soil Surveys and Land-Use Planning

The best use of land is greatly determined by its soils; therefore, a report called a **soil survey** is an important part of planning for nearly all engineering projects. A *soil survey* should include soil descriptions; soil maps showing the horizontal and vertical extent of soils; and results of tests to determine grain size, moisture content, and strength. The purpose of the survey is to provide necessary information for identifying potential problem areas before construction.[6]

The information from detailed soil maps can be extremely helpful in land-use planning if it is used in combination with guidelines for the proper use of soils. Soils can be rated according to their limitations for a specific land use, such as housing, light industry, septic-tank systems, roads, recreation, agriculture, and forestry.

SUMMARY

Engineers define soil as Earth material that may be removed without blasting, whereas to a soil scientist, a soil is solid Earth material that can support rooted plant life. A basic understanding of soils and their properties is becoming crucial in several areas of environmental geology, including land-use planning, waste disposal, and evaluation of natural hazards such as flooding, landslides, and earthquakes.

Soils result from interactions of the rock and hydrologic cycles with biogeochemical processes. As open systems, they are affected by variables such as climate, topography, parent material, time, and biological activity. Soil-forming processes tend to produce distinctive soil layers, or horizons, defined by the processes that formed them and the type of materials present. The processes of leaching, oxidation, and accumulation of materials in various soil horizons are particularly important. Development of the argillic B horizon, for example, depends on the translocation of clay minerals from upper to lower horizons. Three important properties of soils are color, texture or particle size, and structure (the aggregation of particles).

A soil may be considered a complex ecosystem in which many types of living things convert soil nutrients into forms that plants can use. Soil fertility is the capacity of the soil to supply nutrients needed for plant growth. Soils may go through natural stages of building, sustaining, and degrading. Each stage has important implications for ecosystems.

Soil has a solid phase consisting of mineral and organic matter; a gas phase, including mostly air; and a liquid phase, which consists mostly of water. Water may flow vertically or laterally through the pores of a soil. The flow is either saturated, in which all pore space is filled with water, or, more commonly, unsaturated, in which pore spaces are partially filled with water. Soil moisture and how water moves through soils are becoming important topics in environmental geology.

Several types of soil classification exist, but none of them integrates both engineering properties and soil processes. Environmental geologists must be aware of both the soil-science classification, or soil taxonomy, and the engineering classification, known as the unified soil classification system.

A basic understanding of the engineering properties of soils is crucial in many environmental problems. These properties include soil strength, sensitivity, compressibility, erodibility, hydraulic conductivity, corrosion potential, ease of excavation, and shrink-swell potential. Shrink-swell potential is particularly important because expansive soils in the United States cause significant environmental problems and constitute one of our most costly natural hazards.

Rates of soil erosion can be determined by direct observation of soil loss from slopes, by measurement of accumulated sediment in reservoirs, or by calculation using an equation. A common method is to apply the Universal Soil Loss Equation, which uses variables that affect erosion to predict the amount of soil that will be moved from its original position. These variables can often be manipulated as part of a management strategy to minimize erosion and sediment pollution from a particular site. Sediment, both natural and human-made, may be one of our greatest pollutants. It reduces water quality and chokes streams, lakes, reservoirs, and harbors. With good conservation practice, sediment pollution can be much reduced.

Sustaining soils requires that rates of soil erosion be less than rates of soil formation. Land use and surface-water manipulation affect the pattern, amount, and intensity of surface-water runoff, soil erosion, and sediment pollution. In agricultural areas, soil erosion may be minimized most effectively by contour plowing, no-till agriculture, terracing slopes, and planting more than one crop.

Urbanization often involves loss of soil, change of soil properties, accelerated soil erosion during construction, and pollution of soils. Use of motorized and nonmotorized off-road vehicles causes soil erosion, changes in hydrology, and damage to plants and animals.

Soil pollution occurs when hazardous materials are inadvertently or deliberately added to soils. Pollution limits the usefulness of soils or even renders them hazardous to life. Processes in soils may also be useful in reducing or eliminating contaminants from soil. The deliberate use of soil processes to eliminate contamination is known as bioremediation.

Soil surveys are extremely useful in land-use planning. Soils can be rated according to their limitations for various land uses. This information can be combined with a detailed soils map to produce a simplified map that shows a soil's limitations for a specific use.

Revisiting Fundamental Concepts

Human Population Growth

Increases in human population have resulted not only in a need for more soil resources to grow food to support our growing numbers; they have also resulted in productive soils formerly used for agriculture being covered with pavement and buildings. Intensive agriculture to provide food for the world's growing population has required the use of tremendous amounts of chemical fertilizers and water for irrigation in our semiarid and arid regions. These practices have resulted in an increase in soil and water pollution from the use of chemical fertilizers.

Sustainability

Sustaining soil resources is one of the fundamental goals to ensure long-term food production in agricultural regions. Of particular importance is the development of plans to minimize soil erosion that results in loss of soil resources and produces sediment pollution.

Earth as a System

Soils are complex ecosystems involving interactions between the hydrologic and rock cycles and biogeochemical processes through time. As climate changes, so do soil-forming processes, and soils often reflect climatic changes that alter patterns of precipitation and the amount of solar energy received at the surface of Earth.

Hazardous Earth Processes, Risk Assessment, and Perception

Loss of soil fertility as a result of soil erosion and other processes is a serious hazard with respect to our ability to provide food to feed the people of the world. In addition, specific soil hazards exist, including the shrink-swell properties of soils, that may cause damage to roads, buildings, and other structures.

Scientific Knowledge and Values

The study of soils is a mature science, and we have a good understanding of how soils form, are eroded, and may be sustained. Pollution of soils is a serious problem in agricultural and urban areas. How we ultimately deal with soil pollution will reflect our values concerning the rights of people to live in an environment where soils do not produce health hazards as a result of pollution.

Key Terms

expansive soil (p. 538)

residual soil (p. 529)

sediment pollution (p. 540)

shrink-swell potential (p. 538)

soil (p. 528)

soil fertility (p. 533)

soil horizons (p. 529)

soil profile (p. 529)

soil survey (p. 549)

transported soil (p. 529)

Review Questions

1. Differentiate how we define soil from a soil scientist's perspective and that of an engineer.

2. What is the difference between a residual and a transported soil?

3. What are the major soil horizons?

4. Define soil texture.

5. What do we mean by soil fertility?

6. What are the two major ways that we classify soils?

7. What determines the strength of a soil?

8. What is the shrink-swell potential of a soil, and why is knowing this important?

9. What are some of the ways that we can evaluate rates of soil erosion?

10. Why is sediment pollution such a large environmental problem?

11. What is the role of urbanization in causing soil erosion and how can it be minimized?

12. How has the use of motorized and nonmotorized off-road vehicles caused soil erosion problems?

13. How can soil pollution occur?

14. What is a soil survey, and how can it be used in land-use planning?

Critical Thinking Questions

1. How and why could processes such as clear-cut logging, in which all trees are cut, and use of off-road vehicles lead to loss of soil fertility?

2. One of your environmentalist friends really likes to ride her mountain bike on steep terrain. She particularly likes racing downhill on ski slopes during the summer. What are some conflicts she may have in reconciling her sport with potential damage to the environment?

3. You own a consulting firm and a client hires you to evaluate several hundred hectares (acres) to start a small firm to grow organic vegetables. The land is generally flat with some rolling hills. How could you evaluate the project from a soils perspective? Outline a general plan to advise your client.

SEVENTEEN

Waste as a Resource: Waste Management

Learning Objectives

Development of management strategies to deal with our waste problems is an important environmental concern. Learning objectives for this chapter are

- Gain an appreciation for the evolution of concepts of waste management from "dilute and disperse," to integrated waste management, to materials management with the visionary goal of zero production of waste

- Know the various alternatives for solid waste disposal

- Understand important processes related to sanitary landfills, including generation of leachate, site selection, design, monitoring, and federal legislation

- Understand the principles of hazardous chemical waste management in terms of what responsible management is, alternative management strategies, and federal legislation pertaining to hazardous waste

Note: Management of radioactive waste is discussed in Chapter 15 with nuclear energy.

People Processing e-waste in China In the process they are exposed to toxic materials. *(© Basel Action Network 2007)*

CASE HISTORY | Where Does E-Waste Go?

People in the United States today discard hundreds of millions of computers and other electronic devices such as cell phones, iPods, televisions, and computer games. The average life of a computer is about 3 years. Computers are not constructed with recycling in mind, and where they often end up is an environmental problem.

When you take **e-waste** to a location where it is collected, you may naturally assume the waste will be handled properly. It is becoming clear that this is not what often happens. The United States is where the most of the e-waste is generated. E-waste includes the plastic housing for computers, TVs, and the like; when burned it produces toxic material. Electronic parts also include small amounts of heavy metals including gold, tin, copper, cadmium, and mercury that are harmful, toxic, or may cause cancer if breathed, ingested, or absorbed through the skin. At present many millions of computers are disposed of by what is sold as recycling. In the United States there is no official process administered by the Environmental Protection Agency to ensure that e-waste doesn't cause future problems. Most e-waste is exported under the label of recycling to countries such as Nigeria and China. Some of China's largest e-waste facilities are in Guiyu (see opening photograph), near Hong Kong. People in the Guiyu area process over 1 million tons

of e-waste each year with little thought to the potential toxicity of the material they are handling.

In the United States, computers cannot be recycled profitably without charging the people who dispose of them a hefty fee. Even so, many U.S. firms ship their e-waste out of the country, where greater profits are possible. The revenue to the Guiyu area is about a million dollars per year. So far the central government of China is resistant to regulating the activity. Workers at the locations where computers are disassembled by hand, sometimes with primitive tools, may not be aware of the toxic nature of some of the materials they are working with and thus have a hazardous occupation. In the Guiyu area there are over 5,000 family-run facilities that are scavenging the e-waste for raw materials, often exposing the workers to toxins and potential health problems.

The United States has not taken a proactive stance to regulate the computer industry to produce less waste. In fact, the United States did not ratify an international agreement that restricts and bans exports of hazardous e-waste.[1]

The U.S. handling of e-waste is producing environmental problems for others. If we value a quality environment, we need to ensure the safe handling and recycling of e-waste. Hopefully that is what we will do in the future.

17.1 Concepts of Waste Management: An Overview

People in the United States and throughout the world are facing a tremendous solid-waste disposal problem, particularly in growing urban areas. The problem boils down to the simple fact that urban areas are producing too much waste and there is far too little space for disposal. About half the cities in the United States are estimated to be running out of landfill space. Cost is another limiting factor— expenditures for landfill disposal have skyrocketed in recent years.[2]

All types of societies produce waste, but industrialization and urbanization have caused an ever-increasing effluence that has greatly compounded the problem of waste management. Although tremendous quantities of liquid and solid waste from municipal, industrial, and agricultural sources are being collected and recycled, treated, or disposed of, new and innovative programs remain necessary if we are to keep ahead of what might be called a waste crisis. Disposal or treatment of liquid and solid waste by federal, state, and municipal agencies costs billions of dollars every year. In fact, it is one of the most costly environmental expenditures of governments, accounting for the majority of total environmental expenditures.

A possible solution to the solid waste problem would be to develop new disposal facilities. Unfortunately, no one wants to live near a waste disposal site, be it a sanitary landfill for municipal waste, an incinerator facility that can reduce the volume of waste by 75 percent, or a disposal operation for hazardous chemical materials. This obviously creates serious siting problems even if the local geographic, geologic, and hydrologic environment is favorable. The siting problem also involves issues of *social justice*. Waste-management facilities are all too frequently located in areas where the people are of low social and economic status or belong to a minority ethnic group or race. Investigation of the issues involved in siting waste facilities to which many people object based on perceived environmental problems is part of an emerging field known as *environmental justice.*[3] The consensus seems to be that people have little confidence in the ability of government or industry to preserve and protect public health as it relates to waste disposal.[2]

The waste disposal industry in the United States, which represents a $20 billion sector of the economy, is accustomed to the relatively simple system of collection of waste and landfill disposal.[1] The rise in public consciousness concerning environmental problems and solutions is forcing the disposal industry to explore new solid waste management systems. What has emerged is the concept known as **integrated waste management (IWM),** a set of management alternatives including source reduction, recycling, landfilling, and incinerating.[2]

Earlier Views

During the first century of the Industrial Revolution, the volume of waste produced was relatively small and the concept of "dilute and disperse" was adequate. Factories were located near rivers because the water provided easy transport of materials by boat, ease of communication, sufficient water for processing and cooling, and easy disposal of waste into the river. With few factories and sparse population, "dilute and disperse" seemed to remove the waste from the environment.[4]

Unfortunately, as industrial and urban areas expanded, the concept of "dilute and disperse" became inadequate, and a new concept known as "concentrate and contain" became popular. It is now apparent, however, that containment was and is not always achieved. Containers, natural or artificial, may leak or break and allow waste to escape. As a result, another concept developed known as "resource recovery." This philosophy holds that waste can be converted to useful materials, in which case it is no longer waste but resources. However, even with our state-of-the-art technology, large volumes of waste cannot be economically converted or are essentially indestructible. Therefore, we still have waste disposal problems.[4]

Modern Trends: Integrated Waste Management

There is a growing awareness that many of our waste-management programs simply involve moving waste from one site to another and not properly disposing of it. For example, waste from urban areas may be placed in landfills, but eventually these may cause further problems because of the production of methane gas (which is a resource if managed properly) or noxious liquids that leak from the site to contaminate the surrounding areas. Disposal sites are also capable of producing significant air pollution. It is safe to assume that waste management is going to be a public concern for a long time. Of particular importance will be the development of new methods of waste management that will not endanger the public health or cause a nuisance.

Integrated waste management (IWM) emerged in the 1980s as a set of management alternatives, including resource reduction, recycling, reusing, composting, landfilling, and incinerating.[2] **Reduce, recycle, and reuse** are the three Rs of IWM. Recycling can reduce the weight of urban refuse disposed in landfills by approximately 50 percent. Some specific ways you can reduce the waste you personally produce are suggested in Table 17.1.

The recycling option of IWM, which has been seriously pursued for nearly two decades, has generated entire systems of waste management that have produced tens of thousands of jobs while reducing the amount of urban waste from homes in the United States sent to landfills from 90 percent in the 1980s to about 65 percent today (Figure 17.1). In fact, many firms have combined waste reduction with recycling to reduce the waste they deliver to landfills by over 50 percent. In spite of this success, IWM is criticized for not effectively advancing policies to prevent

---

TABLE 17.1   What Can You Do to Reduce the Waste You Generate?

- **Keep Track of the Waste You Personally Generate:** Knowing what and how much waste you produce will make you conscious of how to reduce it.

- **Recycle as Much as Possible:** Take cans, glass, and paper to a recycling center or use curbside pickup. Take hazardous materials such as batteries, cell phones, computers, paint, used oil, and solvents to a hazardous waste pickup site.

- **Reduce Packaging:** Whenever possible buy food items in bulk or concentrated form.

- **Use Durable Products:** Choose automobiles, furniture, sports equipment, toys, and tools that will last a long time.

- **Reuse Products:** Some things may be used over and over again. For example, reuse boxes and "bubble wrap" to ship packages.

- **Purchase Products Made from Recycled Material:** Many bottles, cans, cereal boxes, containers, and cartons as well as carpets, floor tiles, and other products are made from recycled material.

- **Purchase Products Designed for Easy Recycling:** Some products as large as automobiles are being designed with recycling in mind.

Source: Modified from U.S. Environmental Protection Agency. Waste. Accessed 4/21/06 at www.epa.gov

---

Figure 17.1 Recycling facility in Chicago Illinois *(Mitch Kezar/Getty Images, Inc.)*

waste production by overemphasizing recycling. In the long term, waste management policies that rely on recycling cannot be successful. If we continue with today's management of waste, in approximately 50 to 70 years when the U.S. population has doubled again, we will be producing the same volume of waste sent to landfills that we do today, given a 50 percent rate of recycling. Clearly, emphasizing recycling is not a sustainable solution to our waste problem. With this in mind, the concept of IWM needs to be rethought and expanded to include what is termed *materials management.*[5]

17.2 Materials Management

Materials management is part of IWM, but it provides a new goal. That goal is "zero production of waste," so that what is now thought of as waste will be a resource! This is a visionary goal, requiring more sustainable use of materials combined with resource conservation. It is believed that materials management as an extension of IWM can be established by[5]

- Eliminating subsidies for extraction of virgin materials such as timber, minerals, and oil.
- Establishing "green building" incentives that use recycled materials and products in new construction.
- Establishing financial penalties for production of those products that do not meet the objectives of material management practices.
- Providing financial incentives for those industrial practices and products that benefit the environment by enhancing sustainability, such as encouraging products that reduce waste production and use recycled materials.
- Providing incentives for the production of new jobs in the technology of materials management as well as incentives for practicing reducing, recycling, and reusing resources. This is the essence of materials management and sustainable resource utilization.

The concept of materials management for "zero waste" is part of what is known as **industrial ecology.** The idea is to produce urban and industrial systems that model natural ecosystems, where waste from one part of the system is a resource for another part.

With this introduction to modern trends and integrated waste management, it is advantageous to break the management treatment and disposal of waste into several categories: solid waste disposal; hazardous chemical waste management; radioactive waste management; and ocean disposal.

17.3 Solid Waste Disposal

Disposal of solid waste is primarily an urban problem.[6] Figure 17.2 summarizes major sources and types of solid waste, and Table 17.2 lists the generalized composition of solid waste (by weight, before recycling) in 1986 and 2003. We emphasize that this is only an average composition and considerable variation can be expected because of differences in such factors as land use, economic base, industrial activity, climate, and season of the year. It is no surprise that paper is by

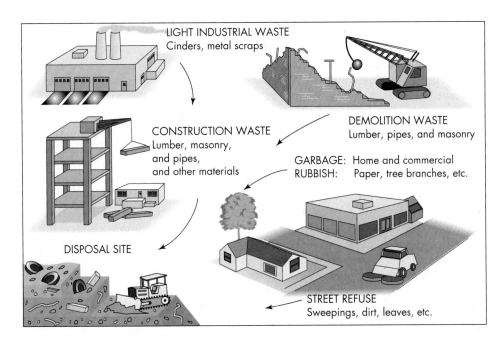

Figure 17.2 Types of materials or refuse commonly transported to a disposal site.

TABLE 17.2   Generalized Composition of
Urban Solid Waste (by Weight)
for 1986 and 2003

Material	1986 (%)	2003 (%)
Paper	36	35
Yard waste	20	12
Plastics	7	11
Metals	9	8
Food waste	9	12
Glass	8	5
Wood	4	6
Other (includes rubber, leather, textiles)	7	11

Source: Ujihara, A. M., and Gough, M. Managing ash from municipal waste incinerators, in *Resource for the future* (Center for Risk Management: 1989); U.S. Environmental Protection Agency.

far the most abundant solid waste. Plastics have increased by 60 percent since 1986 with a tremendous increase in plastic containers such as bottles. All this plastic is not going unnotice. There is an emerging industry to recycle waste plastic. Plastic recycling plants are being constructed and operated to handle many types of plastic. What was waste plastic is being recycled into new plastic as good as that made from oil.

In some areas, infectious waste from hospitals and clinics can create problems if they are not properly sterilized before disposal. Some hospitals have facilities to incinerate such waste. In large urban areas, huge quantities of toxic materials may also end up at disposal sites. Urban landfills are now being considered hazardous waste sites that will require costly monitoring and cleanup.

The common methods of solid waste disposal summarized from a U.S. Geological Survey report include on-site disposal, composting, incineration, open dumps, and sanitary landfills.[7]

On-Site Disposal

By far the most common on-site disposal method in urban areas is the mechanical grinding of kitchen food waste. Garbage disposal devices are installed in the wastewater pipe system from a kitchen sink, and the garbage is ground and flushed into the sewer system. This effectively reduces the amount of handling and quickly removes food waste, but final disposal is transferred to the sewage treatment plant where solids such as sewage sludge still must be disposed of.[7]

Hazardous liquid chemicals may be inadvertently or deliberately disposed of in sewers, requiring treatment plants to handle toxic materials. Illegal dumping in urban sewers has only recently been identified as a potential major problem.

Composting

Composting is a biochemical process in which organic materials decompose to a humuslike material. It is rapid, partial decomposition of moist, solid, organic waste by aerobic organisms. The process is generally carried out in the controlled environ-ment of mechanical digesters.[2] Although composting is not common in the United States, it is popular in Europe and Asia, where intense farming creates a demand for the compost.[4] A major drawback of composting is the necessity to separate the organic material from the other waste. Therefore, it is economically advantageous only when organic material is collected separately.[8] Nevertheless, composting is considered part of integrated waste management, and its use is growing.

Incineration

Incineration is the reduction of combustible waste to inert residue by burning at high temperatures (900 to 1,000°C). These temperatures are sufficient to consume all combustible material, leaving behind only ash and noncombustibles. Incineration ideally reduces the volume of waste that must be disposed of by 75 to 95 percent.[7] However, because of maintenance of incineration facilities and waste-supply problems, the actual reduction of waste by incineration is closer to 50 percent. As we have already mentioned, this is about the same savings that can be gained from waste reduction and recycling.[6] The advantages of incinerating urban waste are twofold:

- Incineration can effectively convert a large volume of combustible waste to a much smaller volume of ash to be disposed of at a landfill.
- Combustible waste can be used to supplement other fuels in generating electrical power.

Burning urban waste is certainly not a clean process. The burning produces air pollution and toxic ash that must be disposed of at landfills. Smokestacks from incinerators emit nitrogen and sulfur oxides, which are precursors of acid rain, as well as carbon monoxide and heavy metals such as lead, cadmium, and mercury. The smokestacks can be fitted with devices to trap some of the pollutants, but the process of pollution abatement is expensive. Furthermore, the incinerators themselves are expensive and often need government subsidies to be established. One study showed that an investment of $8 billion could construct incinerators capable of burning about 25 percent of the solid waste generated in the United States, whereas a similar investment in recycling and composting facilities could handle as much as 75 percent of the nation's urban solid waste.[6]

The economic viability of incinerators depends on revenue from the sale of energy produced by burning waste. As a result, incinerators need to run at near capacity to remain profitable. With the increase in composting and recycling, the economics are far from certain because those processes compete directly with incineration. However, it is safe to say that waste reduction and recycling can reduce the volume of waste that must be disposed of at a landfill at least as much as incineration can.[6]

Open Dumps

Open dumps are the oldest and most common way of disposing of solid waste. In many cases, open dumps are located wherever land is available without regard to safety, health hazards, and aesthetic degradation. The waste is often piled as high as equipment allows. In some instances, the refuse is ignited and allowed to burn; in others, it is periodically leveled and compacted.[7] In addition to being unsightly, open dumps generally create a health hazard by breeding pests, polluting the air, and often contaminating groundwater and surface water. In the United States, open dumps have given way to planned and managed sanitary landfills, but they are still common in many poor countries of the world.

Sanitary Landfills

A **sanitary landfill** (also called a municipal solid waste landfill) as defined by the American Society of Civil Engineering is a method of solid waste disposal that functions without creating a nuisance or hazard to public health or safety. Engineering principles are used to confine the waste to the smallest practical area, reduce it to the smallest practical volume, and cover it with a layer of compacted soil or specially designed tarps at the end of each day of operation, or more frequently if necessary. It is this covering of the waste that makes the sanitary landfill sanitary. The cover effectively denies continued access to the waste by insects, rodents, and other animals, and it also isolates the refuse from the air. The use of the cover minimizes the amount of surface water entering into the waste and gas escaping from it.[8]

The sanitary landfill as we know it today emerged in the late 1930s. Two types are used: area landfill on relatively flat sites and depression landfill in natural or artificial gullies or pits. Normally, refuse is compacted and covered at the end of each day. The finishing cover (cap) or soil (clay) is designed to minimize infiltration of surface water.[7]

Potential Hazards. One of the most significant potential hazards from a sanitary landfill is groundwater or surface-water pollution. If waste buried in a landfill comes in contact with water percolating down from the surface or with groundwater moving laterally through the refuse, **leachate**—obnoxious, mineralized liquid capable of transporting bacterial pollutants—is produced.[9] For example, two landfills dating from the 1930s and 1940s in Long Island, New York, have produced leachate plumes that are several hundred meters wide and have migrated several kilometers from the disposal site. Both the nature and the strength of leachate produced at a disposal site depend on the composition of the waste, the length of time that the infiltrated water is in contact with the refuse, and the amount of water that infiltrates or moves through the waste.[7] The concentration of pollutants in landfill leachate is much higher than in raw sewage or slaughterhouse waste. Fortunately, the amount of leachate produced from urban waste disposal is much less than the amount of raw sewage.

Another possible hazard from landfills is uncontrolled production and escape of methane gas, which is generated as organic waste decomposes. For example, gas generated in an Ohio landfill migrated several hundred meters through sandy soil to a housing area, where one home exploded and several others had to be evacuated. Properly managed, methane gas (if not polluted with toxic materials) is a resource. At new and expanded landfills, methane is often confined by barriers made of plastic liner and clay and then collected in specially constructed wells. The technology for managing methane is advancing, and landfills across the country are now producing methane and selling the gas as one way to help reduce costs associated with waste management.

Site Selection. Factors controlling the feasibility of sanitary landfills include

- Topographic relief
- Location of the groundwater table
- Amount of precipitation
- Type of soil and rock
- Location of the disposal zone in the surface-water and groundwater flow system

The best sites are those in which natural conditions ensure reasonable safety in disposal of solid waste. This means that there is little (or acceptable) pollution of ground- or surface waters, and that conditions are safe because of climatic, hydrologic, geologic, or human-induced conditions or combinations of these.[10]

The best sites for landfills are in arid regions. Disposal conditions are relatively safe there because in a dry environment, regardless of whether the burial material is permeable or impermeable, little or no leachate is produced. On the other hand, some leachate will always be produced in a humid environment, so an acceptable level of leachate production must be established to determine the most favorable sites. What is acceptable varies with local water use, local regulations, and the ability of the natural hydrologic system to disperse, dilute, and otherwise degrade the leachate to a harmless state.

The most desirable site in a humid climate is one in which the waste is buried above the water table in clay and silt soils of low hydraulic conductivity. Any

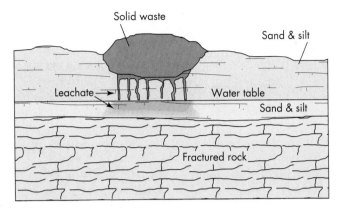

Figure 17.3 Solid waste **disposal** site where the refuse is buried above the water table over a fractured-rock aquifer. Potential for serious pollution is low to moderate because leachate is partially degraded by natural filtering as it moves down to the water table. *(After W. J. Schneider. 1970. U.S. Geological Survey Circular 601F.)*

leachate produced will remain in the vicinity of the site, where it will be degraded by natural filtering and by exchange of some ions between the clay and the leachate. This holds even if the water table is fairly high, as it often is in humid areas, provided material with low hydraulic conductivity is present.[11] For example, if the refuse is buried over a fractured rock aquifer, as shown in Figure 17.3, the potential for serious pollution is low to moderate because the leachate is partly degraded by natural filtering as it moves down to the water table. Furthermore, the dispersion of contaminants is confined to the fracture zones.[2] However, if the water table were higher or if the cover material were thinner with a moderate-to-high hydraulic conductivity, then widespread groundwater pollution of the fractured rock aquifer might result.

If a landfill site is characterized by an inclined limestone rock aquifer covered by sand and gravel with high hydraulic conductivity (Figure 17.4), considerable contamination of the groundwater could result. Leachate moves quickly through the sand and gravel soil and enters the limestone, where open fractures or cavities may transport the pollutants with little degradation other than dispersion and dilution. Of course, if the inclined rock is all shale, with low hydraulic conductivity, little pollution will result.

The following general guidelines[11] should be followed in site selection for sanitary landfills:

- Limestone or highly fractured rock quarries and most sand and gravel pits make poor landfill sites because these earth materials are good aquifers.
- Swampy areas, unless properly drained to prevent disposal into standing water, make poor sites.
- Floodplains likely to be periodically inundated by surface water should not be considered as acceptable sites for refuse disposal.
- Areas in close proximity to the coast, where trash (transported by wind or surface water) or leachate in ground- or surface water may pollute beaches and coastal marine waters, are undesirable sites.

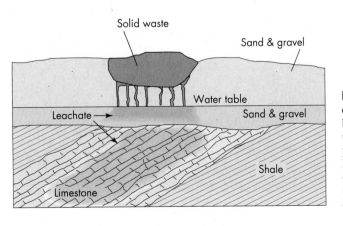

Figure 17.4 Solid-waste **disposal** site where the waste is buried. Leachate can migrate down to fractured bedrock (limestone). The potential for groundwater pollution is high because of the many open and connected fractures in the rock. *(After W. J. Schneider. 1970. U.S. Geological Survey Circular 601F.)*

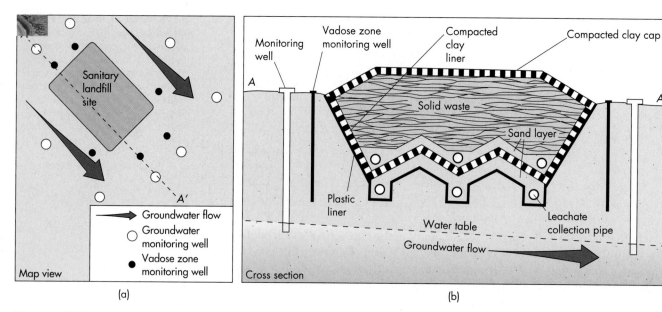

Figure 17.5 Landfill design Idealized diagrams showing map view (a) and cross section (b) of a landfill with a double liner of clay and plastic and a leachate collection system.

- Any material with high hydraulic conductivity and with a high water table is probably an unfavorable site.

- In rough topography, the best sites are near the heads of gullies where surface water is at a minimum.

- Clay pits, if kept dry, provide satisfactory sites.

- Flat areas are favorable sites, provided an adequate layer of material with low hydraulic conductivity, such as clay and silt, is present above any aquifer.

We emphasize that, although these guidelines are useful, they do not preclude the need for a hydrogeological investigation that includes drilling to obtain samples, permeability testing to determine hydraulic conductivity, and other tests to predict the movement of leachate from the buried refuse.[8]

Design of Sanitary Landfills. Design of modern sanitary landfills for municipal solid waste (MSW) is complex and employs the multiple-barrier approach. Barriers include a compacted clay liner, leachate collection systems, and a compacted clay cap. Figure 17.5 is an idealized diagram showing these features, and Figure 17.6 shows such a landfill being constructed. Depending upon local

Figure 17.6 Rock Creek municipal landfill, Calaveras County, California, under construction. The light brown slope in the central part of the photograph is a compacted clay liner. The sinuous ditch is part of the leachate collection system, and the square pond in the upper part of the photograph is the leachate evaporation pond under construction. *(Courtesy of John Kramer)*

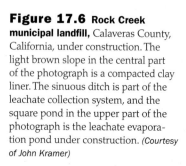

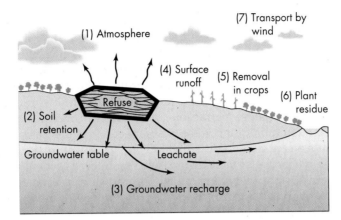

Figure 17.7 Several ways that hazardous waste pollutants from a solid waste disposal site may enter the environment.

site conditions, landfills may also have additional synthetic liners made of plastics or other materials and a system to collect natural gas that might accumulate. Finally, sanitary landfills must have a system of monitoring wells and other devices to evaluate for potential groundwater pollution. The subject of monitoring is an important one, and we will now address that issue in greater detail.

Monitoring Sanitary Landfills. Once a site is chosen for a sanitary landfill, **monitoring** the movement of groundwater should begin before filling commences. After the operation starts, continued monitoring of the movement of leachate and gases should be continued as long as there is any possibility of pollution. This is particularly important after the site is completely filled and the permanent cover material is in place because a certain amount of settlement always occurs after a landfill is completed. If small depressions form as a result of settlement, surface water can collect, infiltrate the fill material, and produce leachate. Therefore, monitoring and proper maintenance of an abandoned landfill will reduce its pollution potential.[8]

Hazardous waste pollutants from a solid waste disposal site can enter the environment[12] by as many as seven paths (Figure 17.7):

1. Gases in the soil and fill, such as methane, ammonia, hydrogen sulfide, and nitrogen, may volatilize and enter the atmosphere.

2. Heavy metals such as lead, chromium, and iron are retained in the soil.

3. Soluble substances, such as chloride, nitrate, and sulfate, readily pass through fill and soil to the groundwater system.

4. Surface runoff can pick up leachate and transport it into the surface water network.

5. Some crops and cover plants growing in the disposal area may selectively take up heavy metals and other toxic substances to be passed up the food chain as people and animals ingest them.

6. Plant residue left in the field contains toxic substances that will return to the environment through soil-forming and runoff processes.

7. Paper, plastics, and other undesirable waste may be transported off-site by wind.

A thorough monitoring program considers the seven possible paths by which pollutants enter the environment. Potential atmospheric pollution by gas from landfills is a growing concern, and a thorough monitoring program includes periodic analysis of air samples to detect toxic gas before it becomes a serious problem. Many landfills have no surface runoff; therefore, monitoring on-site surface water is not necessary. However, if surface runoff does occur, thorough monitoring is required, as well as monitoring of nearby down-gradient streams, rivers, and lakes. Monitoring of soil and plants should include periodic chemical analysis at prescribed sampling locations.

If permeable water-bearing zones exist in the soil or bedrock below a sanitary landfill, monitoring wells (see Figure 17.5) are needed for frequent sampling of groundwater quality and monitoring of the movement of any leachate that has entered the groundwater.[12] Even if the landfill is in relatively impermeable soil overlying dense permeable rock, minimal monitoring of groundwater quality through monitoring wells is still needed. In this case, leachate and groundwater movement may be less than 30 cm/yr. Water in the unsaturated (vadose) zone above the water table must also be monitored to identify potential pollution problems before they contaminate groundwater resources, where correction is very expensive. Waste transported off-site by wind is monitored, collected as necessary, and disposed of.

Sanitary Landfills and Federal Legislation. Federal legislation regulates new landfills strictly. The intent of the legislation is to strengthen and standardize the design, operation, and monitoring of sanitary landfills. Those landfills that are unable to comply with the regulations might be shut down. Specific regulations include

- Landfills are not sited in certain areas, including floodplains, wetlands, unstable land, and earthquake fault zones. They are not sited near airports because birds attracted to landfill sites present a hazard to aircraft.
- Landfill construction must include liners and a leachate collection system.
- Operators of landfills must monitor groundwater for specific toxic chemicals.
- Operators of landfills must meet financial assurance criteria. This is met through posting bonds or insurance to ensure that monitoring of the landfill continues for 30 years after closure.

17.4 Hazardous Waste Management

The creation of new chemical compounds has proliferated tremendously in recent years. In the United States alone, approximately 1,000 new chemicals are marketed annually and about 50,000 chemicals are currently on the market. Although many of these chemicals are beneficial to people, several tens of thousands of them are classified as definitely or potentially hazardous to people's health (Table 17.3).

TABLE 17.3   Examples of Materials We Use and Potentially Hazardous Waste They Generate

Materials	Potential Hazardous Waste
Plastics	Organic chlorine compounds
Pesticides	Organic chlorine compounds, organic phosphate compounds
Medicines	Organic solvents and residues, heavy metals (mercury and zinc, for example)
Paints	Heavy metals, pigments, solvents, organic residues
Oil, gasoline, and other petroleum products	Oil, phenols and other organic compounds, heavy metals, ammonia salts, acids, caustics
Metals	Heavy metals, fluorides, cyanides, acid and alkaline cleaners, solvents, pigments, abrasives, plating salts, oils, phenols
Leather	Heavy metals, organic solvents
Textiles	Heavy metals, dyes, organic chlorine compounds, solvents

Source: U.S. Environmental Protection Agency. SW-826, 1980.

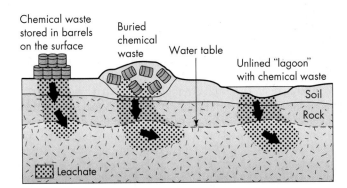

Figure 17.8 Ways that uncontrolled dumping of chemical waste may pollute soil and/or groundwater.

The United States is currently generating more than 150 million metric tons of **hazardous waste** each year. In the recent past, as much as half of the total volume of waste was being dumped indiscriminately.[13] This is now illegal, and we do not know how much illegal dumping is going on—certainly there is some, particularly in urban sewer systems. Past uncontrolled dumping of chemical waste has polluted soil and groundwater resources in several ways (Figure 17.8).

- Barrels in which chemical waste is stored, either on the surface or buried at a disposal site, eventually corroded and leaked, polluting the surface, soil, and groundwater.

- Liquid chemical waste dumped in unlined lagoons (shallow ponds for collection of waste) has percolated through the soil and rock and eventually reached the groundwater table.

- Liquid chemical waste has been illegally dumped in deserted fields or along dirt roads.

- E-waste, such as computers, TVs, and the like, is being sent to foreign countries such as China, where they are exposing workers to toxic components (see Case History at opening of the chapter).

Old abandoned hazardous landfills and other sites for the disposal of chemical waste have caused serious problems that have been very difficult to correct (see A Closer Look: Love Canal). A site near Elizabeth, New Jersey, provides an example of the casual dumping of chemicals that was once so widespread. At that site, before cleanup, there were thousands of charred drums left standing next to a brick and steel building once owned by a now-bankrupt chemical corporation. The drums and other containers, which were stacked four high in places, had been corroding for nearly a decade. Many of them were improperly labeled or burned so badly that the nature of the chemicals could not be determined from outside markings. Leaking barrels allowed unknown quantities of waste to seep into an adjacent stream that eventually flows into the Hudson River.

The New Jersey site was so polluted that cleanup efforts were very difficult. Identification of some of the materials at the site showed that there were two containers of nitroglycerine, numerous barrels of biological agents, cylinders of phosgene and gaseous phosphorus (which are extremely volatile and ignite when exposed to air), as well as a variety of heavy metals, pesticides, and solvents, some of which are very toxic. It took months of work with a large crew of people to remove most of the material from the New Jersey site. Unfortunately, it is difficult to know if all the waste has been removed; additional material may be buried at other sites that are more difficult to locate.[19] There are many such stories of terrible problems resulting from chemical waste disposal, but the best known comes from the Love Canal near Niagara Falls, New York (as for example, A Closer Look: Love Canal).

A CLOSER LOOK | Love Canal

In 1976, in a residential area near Niagara Falls, New York, trees and gardens began to die. Children found the rubber on their tennis shoes and on their bicycle tires disintegrating. Dogs sniffing in a landfill area developed sores that would not heal. Puddles of toxic, noxious substances began to ooze to the soil surface; a swimming pool popped its foundation and was found to be floating on a bath of chemicals.

A study revealed that the residential area had been built on the site of a chemical dump. The area was excavated in 1892 by William T. Love as part of a canal between the upper and lower reaches of the Niagara River. The idea was to produce inexpensive hydroelectric power for a new urban industrial center. When that plan failed because alternating current was discovered and industry could be located far from the source of power, the canal was unused (except for recreation such as swimming and ice skating) for decades. It seemed a convenient place to dump wastes. From the 1940s to the 1950s, more than 80 different substances from a chemical company were dumped there. More than 20,000 tons of chemical waste, along with urban waste from the city of Niagara Falls, was disposed of in the canal.[14] Finally, in 1953, the company dumping the chemicals donated the land to the city of Niagara Falls for one dollar. Eventually, several hundred homes adjacent to an

elementary school were built near the site (Figure 17.A). Heavy rainfall and snowfall during the winter of 1976–77 set off the events that made Love Canal a household word.

A study of the site identified a number of substances present there—including benzene, dioxin, dichlorethylene, and chloroform—that were suspected of being carcinogens. Although officials readily admitted that very little was known about the impact of these chemicals and others at the site, there was grave concern for the people living in the area. During the next few years there were allegedly higher-than-average rates of miscarriages, blood and liver abnormalities, birth defects, and chromosome damage. However, a study by the New York State health authorities suggested that no chemically caused health effects had been absolutely established.[15–17]

The cleanup of the Love Canal is an important demonstration of state-of-the-art technology in hazardous waste treatment. The objective is to contain and stop the migration of waste through the groundwater flow system and to remove and treat dioxin-contaminated soil and sediment from stream beds and storm sewers.[14] The method being used to minimize further production of contaminated water is to cover the dump site and the adjacent contaminated area with a 1-m-thick layer of compacted clay and a polyethylene plastic cover

Figure 17.A Love Canal
This is an infrared aerial photograph of the Love Canal area in New York. Healthy vegetation is bright red. This portion of Love Canal runs from the upper left corner to the lower right. It appears as a scar on the landscape. Buried chemical waste seeped to the surface to cause numerous environmental problems and concern here. The site became a household name for toxic waste.
(New York State Department of Environmental Conservation)

to reduce infiltration of surface water. Lateral movement of water is inhibited from entering or escaping the site by specially designed perforated tile drain pipe. These procedures will greatly reduce subsurface seepage of water through the site, and the water that does seep out will be collected and treated.[15–17]

The homes adjacent to Love Canal were abandoned and bought by the government. Approximately 200 of the homes had to be destroyed. During the 1980s, approximately $175 million was spent for cleanup and relocation at Love Canal. The EPA now considers some of the area clean, and some of the remaining homes (about 200) were sold. They sold despite the reputation of the area and the adverse publicity it attracted

because the price of the homes was approximately 20 percent below the market of other areas in Niagara Falls. The neighborhood is now known as Black Creek Village. In early 1995 the maintenance and operation of the area was transferred from New York State to a consulting company, which will continue long-term sampling and monitoring.[14,18] Also, in 1995, the Occidental Chemical Company agreed to pay $129 million to cover the cost of the incident.

What went wrong in Love Canal? How can we avoid such disasters in the future? The real tragedy of Love Canal is that it is probably not an isolated incident. There are many hidden "Love Canals" across the country, "time bombs" waiting to explode.[15,16]

Responsible Management

In 1976 the U.S. government moved to begin the management of hazardous waste with the passage of the Resource Conservation and Recovery Act (RCRA), which is intended to provide for "cradle-to-grave" control of hazardous waste. At the heart of the act is the identification of hazardous waste and their life cycles. Regulations call for stringent record keeping and reporting to verify that waste does not present a public nuisance or a public health problem. The act also identifies hazardous waste in terms of several categories:

- Waste that is highly toxic to people and other living things
- Waste that may explode or ignite when exposed to air
- Waste that is extremely corrosive
- Waste that is otherwise unstable

Recognizing that a great number of waste disposal sites presented hazards, Congress in 1980 passed the Comprehensive Environmental Response Compensation and Liability Act (CERCLA), which established a revolving fund (popularly called the *Superfund*) to clean up several hundred of the worst abandoned hazardous chemical waste disposal sites known to exist around the country. The EPA developed a list of Superfund sites (National Priorities List). Figure 17.9 summarizes environmental impact statistics and lists some of the pollutants encountered at Superfund sites.

Although the Superfund has experienced significant management problems and is far behind schedule, a number of sites have been treated. Unfortunately, the

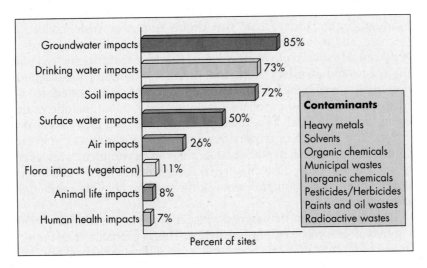

Figure 17.9 **Environmental impacts at Superfund sites** National Priorities List and some of the pollutants encountered at the sites. *(National Priorities List and U.S. Water News, November 1993)*

funds available are not sufficient to pay for decontamination of all the targeted areas. That would cost many times more, perhaps as much as $100 billion. Furthermore, because of concern that the present technology is not sufficiently advanced to treat all the abandoned waste disposal sites, the strategy may be simply to confine the waste to those areas until better disposal methods are developed. It seems apparent that the danger of abandoned disposal sites is likely to persist for some time to come.

The federal legislation also changed the way the real estate industry does business. The act has tough liability provisions, and property owners could be liable for costly cleanup of hazardous waste found on their property (even if they did not cause the problem). Banks and other lending institutions could be liable for release of hazardous materials on their property by their tenants. In 1986 the Superfund Amendment and Reauthorization Act (SARA) provided a possible defense for real estate purchasers against liability provided they completed an *environmental audit* prior to purchase. The audit is a study of past land use at the site (determined by analyzing old maps and aerial photographs, and may involve drilling and sampling of soil and groundwater) to determine if pollutants are present. Such audits now are done on a routine basis prior to purchase of property for development.

The SARA legislation required that certain industries report all releases of hazardous materials into the environment. The list of companies releasing such substances became public and was known as the "Toxic 500 list." Unwanted publicity to companies on the list is thought to have resulted in better and safer handling of hazardous waste by firms that were formerly identified as polluters of the environment. No owner wants his or her company to be the No. One (or even the twenty-fifth or hundredth) most serious polluter among U.S. firms.[20]

Management of hazardous chemical waste includes several options: recycling (on-site processing to recover by-products with commercial value), microbial breakdown, chemical stabilization, high-temperature decomposition, incineration, and disposal by secure landfill or deep-well injection. A number of technological advances have been made in the field of toxic waste management, and as land disposal becomes more and more expensive, the recent trend toward on-site treatment is likely to continue. However, on-site treatment will not eliminate all hazardous chemical waste; disposal will remain necessary. Table 17.4 compares hazardous waste reduction technology in terms of treatment and disposal. Notice that all of the technologies available will cause some environmental disruption. No one simple solution exists for all waste-management issues.

Secure Landfill

The basic idea of the **secure landfill** is to confine the waste to a particular location, control the leachate that drains from the waste, collect and treat the leachate, and detect possible leaks. Figure 17.10 demonstrates these procedures. A dike and liners (made of clay and impervious material such as plastic) confine the waste, and a system of internal drains concentrates the leachate in a collection basin from which it is pumped out and transported to a wastewater treatment plant. Designs of new facilities today must include multiple barriers consisting of several impermeable layers and filters as well as impervious covers. The function of impervious liners is to ensure that the leachate does not contaminate soil and, in particular, groundwater resources. However, this type of waste disposal procedure must have several monitoring wells to alert personnel if and when leachate migrates out of the system, possibly contaminating nearby water resources.

It has been argued that there is no such thing as a really secure landfill, implying that all landfills leak to some extent. This is probably true; impervious plastic liners, filters, and clay layers can fail, even with several backups, and drains can

TABLE 17.4 **Comparison of Hazardous Waste Reduction Technologies**

	DISPOSAL		TREATMENT		
	Landfills and Impoundments	Injection Wells	Incineration and Other Thermal Destruction	Emerging High-Temperature Decomposition[a]	Chemical Stabilization
Effectiveness: How well it contains or destroys hazardous characteristics	Low for volatiles, questionable for liquids; based on lab and field tests	High, based on theory, but limited field data available	High, based on field data, but little data on specific constituents	Very high; commercial scale tests	High for many metals; based on lab tests
Reliability issues	Siting, construction, and operation Uncertainties: long-term integrity of cells and cover, liner life less than life of toxic waste	Site history and geology; well depth, construction, and operation	Monitoring uncertainties with respect to high degree of DRE: surrogate measures, PICs, incinerability[c]	Limited experience Mobile units, on-site treatment avoids hauling risks Operational simplicity	Some inorganics still soluble Uncertain leachate test, surrogate for weathering
Environmental media most affected	Surface and groundwater	Surface and groundwater	Air	Air	Groundwater
Least compatible wastes[b]	Liner reactive, highly toxic, mobile, persistent, and bioaccumulative	Reactive; corrosive; highly toxic, mobile, and persistent	Highly toxic and refractory organics, high heavy metals concentration	Some inorganics	Organics
Relative costs: Low, Moderate, High	L-M	L	M-H	M-H	M
Resource recovery potential	None	None	Energy and some acids	Energy and some metals	Possible building material

[a]Molten salt, high-temperature fluid well, and plasma arc treatments.
[b]Relative to other technologies, this method is less effective for reducing waste exposure.
[c]DRE = destruction and removal efficiency. PIC = product of incomplete combustion.

Source: Council on Environmental Quality, 1983.

become clogged, causing overflow. Yet landfills that are carefully sited and engineered can minimize problems. Preferable sites are those with good natural barriers, such as thick clay-silt deposits, an arid climate, or a deep water table that minimizes migration of leachate. Nevertheless, land disposal should be used only for specific chemicals suitable for the method.

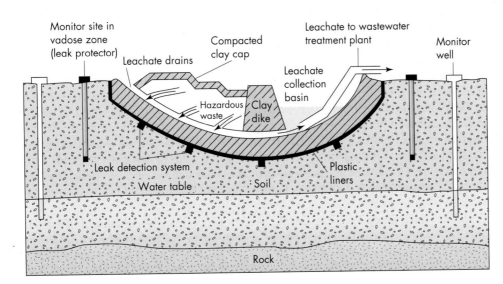

Figure 17.10 Secure landfill for hazardous chemical waste. The impervious liners and systems of drains are an integral part of the system to ensure that leachate does not escape from the disposal site. Monitoring in the vadose zone is important and involves periodic collection of soil water with a suction device.

Land Application

Application of waste materials to the surface-soil horizon is referred to as **land application,** *land spreading,* or *land farming.* Land application may be a desirable treatment method for certain biodegradable industrial wastes, including petroleum oily waste and certain organic chemical-plant waste. A good indicator of the usefulness of land application for disposal of a particular waste is the waste's **biopersistence** (the measure of how long a material remains in the biosphere). The greater or longer the biopersistence, the less suitable the waste is for land-application procedures. Land application is not an effective treatment or disposal method for inorganic substances, such as salts and heavy metals.[21]

Land application of biodegradable waste works because, when such materials are added to the soil, they are attacked by microorganisms (bacteria, molds, yeast, and other organisms) that decompose the waste material. The soil may thus be thought of as a microbial farm that constantly recycles matter by breaking it down into more fundamental forms useful to other living things in the soil. Because the upper soil zone contains the largest microbial populations, land application is restricted to the uppermost 15 to 20 cm of the soil profile.[21] As with other types of land-disposal technology, the vadose zone and groundwater near the site must be carefully monitored to ensure the disposal system is working as planned and not polluting water resources.

Surface Impoundment

Excavations and natural topographic depressions have been used to hold hazardous liquid waste. These **surface impoundments** are primarily formed of soil or other surficial materials, but they may be lined with manufactured materials such as plastic. The impoundment is designed to hold the waste; examples include aeration pits and lagoons at hazardous waste facilities. Surface impoundments have been criticized because they are especially prone to seepage, resulting in pollution of soil and groundwaters. Evaporation from surface impoundments can also produce an air-pollution problem. For these reasons, hazardous waste facilities have been prohibited from receiving noncontainerized liquid waste.

Deep-Well Disposal

Another method of hazardous waste disposal is by injection into deep wells. The term *deep* refers to rock (not soil) that is below and completely isolated from all freshwater aquifers, thereby assuring that injection of waste will not contaminate or pollute existing or potential water supplies. This generally means that the waste is injected into a permeable rock layer several hundred to several thousand meters below the surface in geologic basins that are confined above by relatively impervious, fracture-resistant rock, such as shale or salt deposits.[4]

Deep-well disposal of industrial waste should not be viewed as a quick and easy solution to industrial waste problems.[22] Even where geologic conditions are favorable for deep-well disposal, natural restrictions include the limited number of suitable sites and the limited space within these sites for disposal of waste. Possible injection zones in porous rock are usually already filled with natural fluids, mostly brackish or briny water. Therefore, to pump in waste, some of the natural fluid must be displaced by compression (even slight compression of the natural fluids in a large volume of permeable rock can provide considerable storage space) and by slight expansion of the reservoir rock as the waste is being injected.[23]

Problems with Deep-Well Disposal. Several problems associated with disposal of liquid waste in deep wells have been reported.[22,23] Perhaps the best known are the earthquakes that were caused by injecting waste from the Rocky Mountain Arsenal near Denver, Colorado (see Chapter 6). These earthquakes occurred between 1962 and 1965. The injection zone was fractured gneiss at a depth of

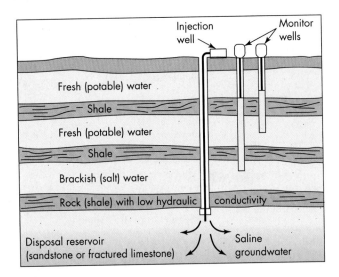

Figure 17.11 Deep-well injection system The disposal reservoir is a sandstone or fractured limestone capped by impermeable rock and isolated from all fresh water. Monitor wells are a safety precaution to ensure that there is no undesirable migration of the liquid waste into freshwater aquifers above the injection zone.

3.6 km, and the increased fluid pressure evidently initiated movement along the fractures. This is not a unique case. Similar initiations of earthquakes have been reported in oil fields in western Colorado, Texas, and Utah.[23]

Feasibility and General Site Considerations. The feasibility of deep-well injection as the best solution to a disposal problem depends on four factors: (1) the geologic and engineering suitability of the proposed site; (2) the volume and the physical and chemical properties of the waste; (3) economics; and (4) legal considerations.[24] The geologic considerations for disposal wells are twofold:[24]

- The injection zone must have sufficient porosity, thickness, hydraulic conductivity, and size to ensure safe injection. Sandstone and fractured limestone are the commonly used reservoir rocks.[23]

- The injection zone must be below the level of freshwater circulation and confined by a relatively impermeable rock with low hydraulic conductivity, such as shale or salt, as shown in Figure 17.11.

Incineration of Hazardous Chemical Waste

Hazardous waste may be destroyed through high-temperature incineration. Incineration is considered to be a waste treatment rather than a disposal method because the hazardous waste is not disposed of directly; rather, it undergoes a treatment (incineration) that produces an ash residue to be disposed of in a landfill.[21] The technology used in incineration and other high-temperature decomposition or destruction is changing rapidly. Figure 17.12 diagrams one type of high-temperature incineration system that may be used to burn toxic waste. Waste—as liquid, solid, or sludge—enters the rotating combustion chamber, where it is rolled and burned. Ash from this burning process is collected in a water tank, and the remaining gaseous materials move into a secondary combustion chamber, where the process is repeated. Remaining gas and particulates move through a scrubber system that eliminates particulates and acid-forming components. Carbon dioxide, water, and air then are emitted from the stack. As shown in Figure 17.12, ash particulates and wastewater are produced at various parts of the incineration process; these must be either treated or disposed of in a landfill.

More advanced types of incineration and thermal decomposition of waste are being developed. One of these utilizes a molten salt bed that should be useful in destroying certain organic materials. Other incineration techniques include liquid-injection incineration on land or sea, fluidized-bed systems, and multiple-hearth

Figure 17.12 **High-temperature incinerator** This system is designed to burn toxic waste.

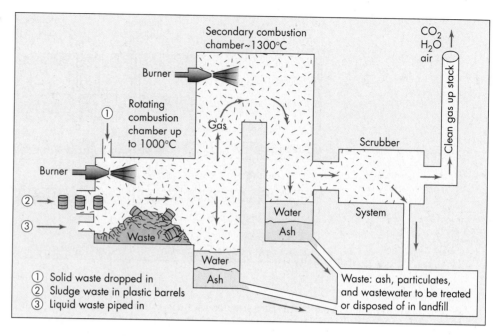

① Solid waste dropped in
② Sludge waste in plastic barrels
③ Liquid waste piped in

furnaces. Which incineration method is used for a particular waste depends upon the nature and composition of the waste and the temperature necessary to destroy the hazardous components. For example, the generalized incineration system shown in Figure 17.12 could be used to destroy PCBs.

Alternatives to Land Disposal

Direct land disposal of hazardous waste is often not the best initial alternative. Even with extensive safeguards and state-of-the-art designs, land-disposal alternatives cannot guarantee that the waste is contained and will not cause environmental disruption in the future. This holds true for all land-disposal facilities, including landfill, surface impoundments, land application, and injection wells. Pollution of air, land, surface water, and groundwater may result from failure of a land-disposal site to contain hazardous waste. Groundwater pollution is perhaps the most significant result of such failure because it provides a convenient route for pollutants to reach humans and other living things. Figure 17.13 shows some of the paths that pollutants take from land-disposal sites to contaminate the environment.

Figure 17.13 **How land disposal of waste may contaminate the environment** Examples of how land disposal/treatment methods of managing hazardous waste may contaminate the environment. *(Modified after C. B. Cox. 1985. The buried threat. No. 115-5. Sacramento, CA: California Senate Office of Research.)*

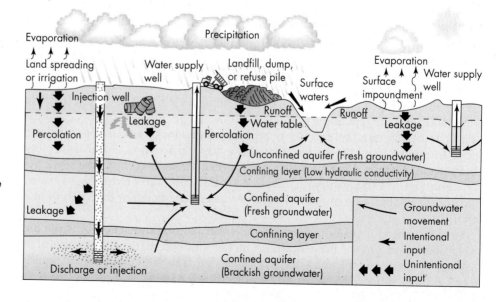

Source Reduction. Source reduction has the objective of reducing the amount of hazardous waste generated by manufacturing or other processes. For example, changes in the chemical processes, equipment, raw materials, or maintenance measures employed can successfully reduce either the amount or the toxicity of the hazardous waste produced.[25]

Recycling and Resource Recovery. Hazardous chemical waste may contain materials that can successfully be recovered for future use. For example, acids and solvents collect contaminants when they are used in manufacturing processes. These acids and solvents can be processed to remove the contaminants and can then be reused in the same or different manufacturing processes.[25]

Treatment. Hazardous chemical waste can be treated by a variety of processes to change the physical or chemical composition of the waste in such a way as to reduce its toxic or hazardous characteristics. Examples include neutralizing acids, precipitation of heavy metals, and oxidation to break up hazardous chemical compounds. Incineration, as we have pointed out, is also a type of waste treatment.

The advantages of source reduction, recycling, and treatment include

- The waste that must be later disposed of is reduced to a much smaller volume, which produces less stress on the dwindling number of acceptable landfill sites.
- Treatment of wastes may make them less toxic and therefore less likely to cause problems in landfills.
- Useful chemicals may be reclaimed and reused.

SUMMARY

Industrialization and urbanization have produced enormous amounts of waste and greatly compounded the problem of waste management. Around many large cities, space for new landfills is becoming hard to find, and few people wish to live near any waste disposal operation. We are headed toward a disposal crisis if the new methods and ideas of integrated waste management are not acted on soon.

Since the Industrial Revolution waste management practices have moved from "dilute and disperse," to "concentrate and contain," to integrated waste management (IWM), which includes alternatives such as reducing, recycling, reusing, landfilling, incinerating, and composting. The goal of many of these alternatives, which can be summarized as "reduce, recycle, and reuse," is to reduce the total amount of waste that needs to be disposed of in landfills or incinerators. More recently a new concept of materials management is emerging as part of IWM. The ultimate objective is "zero waste"—all waste will become resources.

The most common method for disposal of urban waste in the United States today is the sanitary landfill, in which the waste deposited each day is covered with a layer of compacted soil. Potential hazards from sanitary landfills are pollution of groundwater by leachate (polluted water) from the site and uncontrolled production of methane. However,

if methane is contained, it is a useful by-product of landfill operations. Under arid conditions, buried waste produces little leachate; in humid regions the most suitable sites for landfills are where waste can be buried well above the water table in clay and silt soil of low hydraulic conductivity. Modern sanitary landfills have multiple barriers to prevent leachate from infiltrating the vadose zone (above the water table). These landfills have systems to monitor the vadose zone and wells to monitor groundwater. The siting and operation of sanitary landfills is regulated by federal laws.

Hazardous waste management is a serious environmental problem in the United States. Hundreds or thousands of uncontrolled disposal sites may be time bombs that could eventually cause serious public health problems. Because we continue to produce hazardous waste, it is imperative that safe disposal methods be developed and used. Land-disposal options for the management of waste include secure landfills in which the waste is confined and the leachate controlled; land application in which suitable biodegradable materials are spread on the surface; deep-well injection; and incineration with disposal of the residue by secure landfill. Alternatives to land disposal include source reduction, on-site processing to recover by-products with commercial value, and chemical stabilization.

Revisiting Fundamental Concepts

Human Population Growth

Waste management becomes a larger potential problem as human population increases. This results because with business-as-usual waste management, the amount of waste needed to be managed increases with each additional person.

Sustainability

Sustainability is not possible without a sound waste-management program, with the objective of near zero waste production. In other words, to achieve true sustainability we need to nearly eliminate the concept of waste—waste needs to become raw materials (resources) for new uses.

Earth as a System

Taking a systems approach to waste management requires understanding of inputs and outputs and storage of materials such as paper, plastics, metals, and other materials that move through our waste stream. Some may be reused, others reconstituted, and still others transformed to new resources.

Hazardous Earth Processes

Some materials in our waste stream are clearly hazardous to people and the environment. These need to be identified and managed to reduce potential environmental damage.

Scientific Knowledge and Values

How we manage our waste is a reflection of how we regard the environment. Science can provide alternatives for waste management, but which alternatives we implement reflects our values.

Key Terms

biopersistence (p. 570)

composting (p. 558)

deep-well disposal (p. 570)

e-waste (p. 554)

hazardous waste (p. 565)

incineration (p. 559)

industrial ecology (p. 557)

integrated waste management (IWM) (p. 555)

land application (p. 570)

leachate (p. 560)

materials management (p. 556)

monitoring (p. 563)

reduce, recycle, and reuse (p. 555)

sanitary landfill (p. 559)

secure landfill (p. 568)

surface impoundments (p. 570)

Review Questions

1. What is materials management?

2. What are the three Rs?

3. Define a sanitary landfill.

4. What is leachate?

5. How are sanitary landfills monitored?

6. What are the characteristics of hazardous waste?

7. What is a secure landfill?

8. Why is Love Canal important?

9. What is an environmental audit?

10. What are the common methods of hazardous waste management?

Critical Thinking Questions

1. Complete an audit of your personal waste production and disposal where you live. How much are you presently recycling and how much do you estimate (at the high end) that you can recycle? If everyone in your neighborhood did this, what would be the impact on the local waste situation?

2. Defend or criticize the statement that management strategies consisting of recycling and incineration compete with one another and therefore we should emphasize one of these based on environmental considerations.

3. For the region in which you live, identify potential hazardous wastes that are produced by homes, businesses, and industry or agriculture. How are these wastes currently being treated and what could be done to develop a better management strategy if there are problems?

4. Where does the e-waste from your university or college go; what about your community e-waste? Was it difficult to find out the above answers?

EIGHTEEN

Air Pollution

Learning Objectives

Air pollution is one of our most serious environmental problems. It affects all of us in terms of the quality of the air we breathe and the public and private funds spent to control air pollution. In this chapter, we focus on the following learning objectives:

- Know the history, general effects, and sources of air pollution

- Know the common types of air pollution and their environmental effects

- Understand the important relationships among meteorology, topography, and air pollution

- Understand the two major types of smog and how they are produced

- Know some of the methods of controlling air pollution

- Know the basic factors that determine the total cost of air pollution to society

An officer guides a London bus at midday through thick fog on December 8, 1952. The smog killed about 4,000 people. (Bettmann/CORBIS)

CASE HISTORY | The London Smog Crisis of 1952

During the first week of December 1952, the air in London, England, became stagnant, and much of the incoming solar radiation could not penetrate the cloud cover. The humidity climbed to 80 percent, and the temperature dropped rapidly until the noontime temperature was approximately 1°C (34°F). A very thick fog developed, and the cold and dampness increased the demand for home heating. Because the primary fuel used in homes was coal, emissions of ash, sulfur oxides, and soot increased rapidly. The stagnant air became filled with pollutants, from both home heating fuels and automobile exhaust. At the height of the crisis, visibility was greatly reduced, and automobiles had to use their headlights at midday. Between December 4 and 10, an estimated 4,000 people died from the pollution. Figure 18.1 shows the increase in both sulfur dioxide and smoke and the deaths that occurred during that period. The environment, not human activities, finally solved the problem: The siege of smog ended when the weather changed and allowed the pollution to disperse.

Ever since the beginning of the Industrial Revolution, people living in London, as well as residents of other major cities, survived despite the weather and pollution. What had finally gone wrong in 1952? During the London smog crisis, the stagnant weather conditions coupled with the number of homes burning coal and cars burning gasoline exceeded the atmosphere's ability to remove or transform the pollutants; even the usually rapid natural mechanisms for removing sulfur dioxide were saturated. As a result, sulfur dioxide remained in the air and the fog became acidic, adversely affecting people and other organisms, particularly vegetation. The health effects were especially destructive because small acid droplets became fixed on larger particulates, facilitating their being drawn deep into the lungs.

The 1952 London smog crisis was a landmark event. Finally, human activities had exceeded the natural abilities of the atmosphere to serve as a sink for the waste. The crisis was due in part to a positive, or reinforcing, feedback: Burning fossil fuels added particulates to the air, increasing the formation of fog and decreasing visibility and light transmission; the dense, smoggy layer increased the dampness and cold and accelerated the use of home heating fuels. The worse the weather and the pollution became, the more people burned coal to keep warm, further worsening both the weather and the pollution.

Although London has always been well known for its fog, relatively little was known about the role of coal burning in intensifying fog conditions. Since 1952, London fogs have been greatly reduced because coal has been replaced by much cleaner gas as the primary home heating fuel.

Figure 18.1 London smog event The relationship between the number of deaths and the London fog of 1952. *(Modified from Williamson, S. J. 1973. Fundamentals of air pollution. Addison-Wesley)*

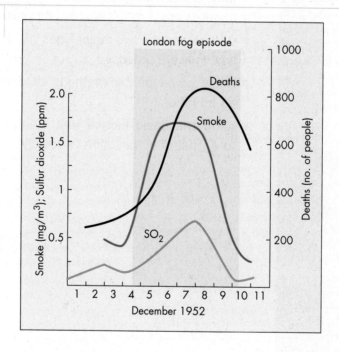

18.1 Introduction to Air Pollution

Ever since life began on Earth, the atmosphere has been both an important resource for chemical elements and a medium for depositing wastes. The earliest photosynthetic plants dumped their waste oxygen into the atmosphere, and the oxygen-rich air that eventually resulted made possible the development and survival of higher life forms.

As the fastest moving dynamic medium in the environment, the atmosphere has always been one of the most convenient places for people to dispose of unwanted materials. Ever since humans first used fire, the atmosphere has all too often been a sink for waste disposal. With the rapid urbanization and industrialization of the last two centuries, atmospheric circulation has often proved inadequate to dissipate human wastes, and urban air in particular has become increasingly polluted.

18.2 Pollution of the Atmosphere

Chemical pollutants can be thought of as compounds that are either in the wrong place or in the wrong concentrations at the wrong time. As long as a chemical is transported away or degraded rapidly relative to its rate of production, there is no pollution problem. Pollutants that enter the atmosphere through natural or artificial emissions may be degraded by natural processes within the atmosphere as well as by the hydrologic and geochemical cycles. Conversely, pollutants in the atmosphere may become pollutants in the hydrologic and geochemical cycles (as, for example, acid rain; see Chapter 15).

People have long recognized the existence of atmospheric pollutants, both naturally produced and those induced by humans. Acid rain was first described in the seventeenth century, and by the eighteenth century it was known that smog and acid rain damaged plants in London. Beginning with the Industrial Revolution in the eighteenth century, air pollution became more noticeable; by the middle of the nineteenth century, particularly following the American Civil War, air pollution became an increasing concern. A physician at a public health conference probably introduced the word *smog* in 1905 to denote poor air quality resulting from a mixture of smoke and fog.

Today, air pollution is a well-known, serious health hazard in many large cities. For example, in Los Angeles, California, millions of people are regularly exposed to unhealthy air. Although Los Angeles has the country's most serious air pollution problem, it is only one of many U.S. cities with unhealthy air. Other U.S. cities with a serious air pollution problem include Houston, Texas; Baltimore, Maryland; Charlotte, North Carolina; and Atlanta, Georgia. With the minor exception of the Pacific Northwest, no U.S. region is free from air pollution.[1] Approximately 150 million people in the United States are exposed to air pollution that may cause lung disease, contributing to more than 300,000 deaths per year. Some estimates of the health cost associated with air pollution in the United States place it at about $50 billion per year.

Effects on Human Artifacts. The effects of air pollution on buildings and monuments include discoloration, erosion, and decomposition of construction materials, as discussed in the section on acid rain in Chapter 15.

Aesthetic Effects. Air pollutants affect visual resources by discoloring the atmosphere and reducing visual range and atmospheric clarity. We cannot see as far in polluted air as in clean air, and what we do see has less color contrast. Once limited to cities, these effects now extend even to the wide open spaces of the United States. For example, near the junction of New Mexico, Arizona, Colorado, and Utah, emissions from the Four Corners fossil fuel–burning power plant are altering air clarity in an area where on a clear day visibility was once 80 km (50 mi) from a mountaintop.[2,3]

Figure 18.2 **Industrial source of air pollution** Air pollutants being emitted from a papermill smokestack in Newfoundland, Canada. *(Malcolm S. Kirk/Peter Arnold, Inc.)*

18.3 Sources of Air Pollution

The two primary types of air pollution sources are stationary and mobile. **Stationary sources** include those with a relatively fixed location, such as *point sources, fugitive sources,* and *area sources.* **Point sources** emit air pollutants from one or more discrete controllable sites, such as smokestacks of industrial power plants (Figure 18.2). **Fugitive sources** generate air pollutants from open areas exposed to wind processes. Examples include dirt roads, construction sites, farmlands, storage piles, and surface mines. Figure 18.3 shows a large dust cloud generated in an area near Kihei on the island of Maui where sugarcane fields had been burned a few days earlier. This dust from the recently burned fields is an example of a fugitive source. **Area sources** are locations that emit air pollutants from several sources within a well-defined area, as, for example, small urban communities or areas of intense industrialization within an urban complex. **Mobile sources** move from place to place while yielding emissions. Mobile sources of pollution include automobiles, aircraft, ships, and trains.[3]

18.4 Air Pollutants

The major air pollutants occur either in a gaseous form or as particulate matter (PM). The *gaseous pollutants* include sulfur dioxide (SO_2), nitrogen oxides (NO_x, where x represents a variable number of oxygen atoms, usually 1, 2, or 3), carbon monoxide (CO), ozone (O_3), volatile organic compounds (referred to as VOCs), hydrogen sulfide (H_2S), and hydrogen fluoride (HF). Particulate-matter pollutants are particles of organic or inorganic solid or liquid substances. Those that are less than 10 μm (10 millionths of a meter) in diameter are designated PM 10. Those that are less than 2.5 μm in diameter are designated PM 2.5.

Air Toxins

Air pollutants known to cause cancer or other serious health problems are classified as **air toxins.** This category of air pollutants is classified by (1) whether or not they cause cancer and (2) their tendency to cause respiratory, reproductive, neurological, or immune diseases. Toxicity is based on exposure to an air toxin by breathing the pollutant. The extent to which a specific toxin affects the health of an individual depends on the toxicity of the pollutant; frequency and duration of exposure; concentration of the pollutant a person is exposed to; and general health

(a)

(b)

Figure 18.3 **Agricultural source of air pollution** (a) Burning sugarcane fields near Kihei on the island of Maui, Hawaii, 1994. (b) Dust rising from the area a few days later. *(Edward A. Keller)*

of the person. Over 150 chemicals including ammonia, chlorine gases, and hydrogen sulfide are evaluated. The four states with the most serious exposure to air toxins are California, New York, Oregon, and New Jersey. The three cleanest states are Montana, Wyoming, and South Dakota. Evaluation of air toxins as a category of air pollutants is a relative new activity but will add increased understanding of toxic air pollution problems and hopefully lead to improved air quality.[4]

Primary and Secondary Pollutants

Air pollutants are also classified as primary and secondary according to their origin. **Primary pollutants** are emitted directly into the air and include particulate matter, sulfur oxides, carbon monoxide, nitrogen oxides, and hydrocarbons. **Secondary pollutants** are produced when primary pollutants react with normal atmospheric compounds. For example, ozone forms over urban areas through reactions among primary pollutants, sunlight, and natural atmospheric gases. Thus, ozone becomes a serious secondary pollution problem on bright, sunny days in areas with abundant primary pollution. Although particularly well documented for southern California cities such as Los Angeles, ozone becomes a secondary pollutant worldwide under appropriate conditions.

The primary pollutants that account for nearly all air pollution problems are carbon monoxide, nitrogen oxides, sulfur oxides, volatile organic compounds, and particulates. Each year, about 50 million metric tons of these materials enter the United States' atmosphere as a result of human-related processes. At first glance this quantity of pollutants appears very large. However, if uniformly distributed in the atmosphere, this would amount to only a few parts per million by weight. Unfortunately, pollutants are not uniformly distributed but tend to be released, produced, and concentrated locally or regionally. For example, large cities' weather and climatic conditions combine with urbanization and industrialization to produce local air pollution problems.

We will now discuss major air pollutants in terms of their natural and human-induced components (Table 18.1) and their effects on people, plants, and materials (Table 18.2).

Sulfur Dioxide

Sulfur dioxide (SO_2) is a colorless, odorless gas at Earth's surface. One significant aspect of SO_2 is that, once emitted into the atmosphere, it may be converted through complex reactions to fine particulate sulfate (SO_4) and removed from the atmosphere by deposition. This is the fate of about 30 percent of atmospheric SO_2.

TABLE 18.1   Major Natural and Human-Produced Components of Air Pollutants

Air Pollutant	EMISSIONS (% OF TOTAL)		Major Source of Human-Produced Component	%
	Natural	Human-Produced		
Sulfur oxides (SO_x)	50	50	Combustion of fuels (stationary sources, mostly coal)	84
			Industrial processes	9
Nitrogen dioxide (NO_2)		Nearly all	Automobiles	37
			Combustion of fuels (stationary sources, mostly natural gas and coal)	38
Carbon monoxide (CO)	91	9	Automobiles	54
Ozone (O_3)	A secondary pollutant derived from reactions with sunlight, NO_2, and oxygen (O_2)		Concentration that is present depends on reaction in lower atmosphere involving hydrocarbons and thus automobile exhaust	
Hydrocarbons (HC)	84	16	Automobiles	27
			Industrial processes	7
Particulates	85	15	Dust	85
			Industrial processes	7
			Combustion of fuels (stationary sources)	8

TABLE 18.2   **Effects of Selected Air Pollutants on People, Plants and Materials**

Pollutant	Effects on People[1,2]	Effects on Plants[1,3]	Effects on Materials[1,4]
Sulfur dioxide (SO_2)	Increase in chronic respiratory disease; shortness of breath; narrowing of airways for people with asthma	Bleaching of leaves; decay and death of tissue; younger leaves are more sensitive than older; sensitive crops and trees include alfalfa, barley, cotton, spinach, beets, white pine, white birch, and trembling aspen; if oxidized to sulfuric acid, causes damage associated with acid rain	If oxidized to sulfuric acid, damages buildings and monuments; corrodes metal; causes paper to become brittle; turns leather to red-brown dust; SO_2 fades dyes of fabrics, damages paint
Nitrogen oxides (NO_x)	Except for odor, is a mostly nonirritating gas; may aggravate respiratory infections and symptoms (sore throat, cough, nasal congestion, fever) and increase risk of chest cold, bronchitis, and pneumonia in children	No perceptible effects on many plants, but may suppress plant growth for some, and may be beneficial at low concentrations; if oxidized to nitric acid causes damage associated with acid rain	Causes fading of textile dyes; if oxidized to nitric acid may damage buildings and monuments
Carbon monoxide (CO)	Reduces the ability of the circulatory system to transport oxygen; headache, fatigue, nausea; impairs performance of tasks that require concentration; reduces endurance; may be lethal, causing asphyxiation	None perceptible	None perceptible
Ozone (O_3)	Strong irritant; aggravates asthma; injury to cells in respiratory system; decreased elasticity of lung tissue; coughing; chest discomfort; eye irritation	Flecking, stippling, spotting, and/or bleaching of plant tissue (leaves, stems, etc.); oldest leaves are most sensitive; tips of needles of conifers become brown and die; reduction of yields and damage to crops including lettuce, grapes, and corn	Cracks rubber; reduces durability and appearance of paint; causes fabric dyes to fade
Particulate matter (PM-2.5, PM-10)	Increased chronic and acute respiratory diseases; depending on chemical composition of particulates, may irritate tissue of throat, nose, lungs, and eyes[5]	Depending on chemical composition of particles, may damage trees and crops; dry deposition of SO_2, when oxidized, is a form of acid rain	Contributes to and may accelerate corrosion of metal; may contaminate electrical contacts; damages paint appearance and durability; fades textile dyes

[1]Effects are dependent upon dose (concentration of pollutant and time of exposure) and susceptibility of people, plants, and materials to a particular pollutant. For example, older people, children, and those with chronic lung diseases are more susceptible to pollutants such as O_3, SO_2, and NO_x.
[2]Annual U.S. losses exceed $50 billion.
[3]Annual U.S. losses to crops are $1 billion to $5 billion.
[4]Annual U.S. losses exceed $5 billion.
[5]Visible as dust and smoke.

Source: Modified from U.S. Environmental Protection Agency; Bunbel, R. W., Fox, D. L., Turner, D. B., and Stern, A. C. 1994. *Fundamentals of air pollution,* 3rd ed. San Diego: Academic Press; Godish, T. 1997. *Air quality,* 3rd ed. Boca Raton, FL: Lewis Publishers.

Sulfur dioxide also undergoes oxidation to form sulfuric acid (H_2SO_4), as discussed in Chapter 15.[2] The major source for the anthropogenic, or human-caused component of sulfur dioxide is fossil fuel burning, primarily from burning coal (Table 18.1). Industrial processes, ranging from refining of petroleum to production of paper, cement, and aluminum, are another major source.[3]

Destructive effects associated with sulfur dioxide include paint and metal corrosion and plant injury or death, especially in crops such as alfalfa, cotton, and barley. When present in sulfate form, it can cause severe damage to the lungs of humans and other animals (Table 18.2).[3]

Nitrogen Oxides

Nitrogen oxides (NO_x) are emitted in several forms. The most important nitrogen oxide is nitrogen dioxide (NO_2), a light yellow-brown to reddish brown gas with an irritating odor. It is toxic and corrosive.[2] Nitrogen dioxide may be converted by complex reactions in the atmosphere to fine particulate nitrate (NO_3) and then nitric acid (HNO_3). In addition, nitrogen dioxide is one of the main pollutants contributing to the development of smog. Nearly all nitrogen dioxide is emitted from anthropogenic sources; the two primary contributors are automobiles and power plants that burn fossil fuels such as coal and oil (Table 18.1).

The environmental effects of nitrogen oxides are variable but include eye, nose, and throat irritation; increased susceptibility of animals and humans to infections; suppression of plant growth and damage to leaf tissue; and impaired visibility when the oxides are converted to their nitrate form in the atmosphere. Interestingly, when nitrate is deposited on the soil, it can promote plant growth (Table 18.2).[3]

Carbon Monoxide

Carbon monoxide (CO) is a colorless, odorless gas that is extremely toxic to humans and other animals at very low concentrations. This toxicity results from a striking physiological effect: The hemoglobin in our blood readily attracts carbon monoxide. Hemoglobin absorbs carbon monoxide nearly 250 times more rapidly than it does oxygen. Therefore, if any carbon monoxide is in the vicinity, it is preferentially absorbed into our bloodstream. Many people, about 200 annually in the United States, have been accidentally asphyxiated by carbon monoxide produced from incomplete fuel combustion in campers, tents, and houses. Carbon monoxide detectors, which are similar to smoke detectors, are now available to warn people if carbon monoxide in a building becomes concentrated at a toxic level. The actual effects of carbon monoxide toxicity can range from dizziness and headaches to death. Carbon monoxide is particularly hazardous to people with known heart disease, anemia, or respiratory disease. Finally, the effects of carbon monoxide tend to be worse in higher altitudes, where oxygen levels are naturally lower.[2,3]

Approximately 90 percent of the carbon monoxide in the atmosphere comes from natural sources; the other 10 percent is generated by fires, automobiles, and other sources of incomplete burning of organic compounds (Table 18.1). Local concentrations of carbon monoxide can build up and cause serious health effects (Table 18.2).

Ozone and Other Photochemical Oxidants

Photochemical oxidants result from the atmospheric interaction of sunlight with pollutants such as nitrogen dioxide, and will be explained in Section 18.5 with smog. The most common photochemical oxidant is *ozone* (O_3), a colorless, unstable gas with a slightly sweet odor. The primary sources of photochemical oxidants, most notably ozone, include automobiles, fossil fuel burning, and industrial processes that produce nitrogen dioxide. The effects of ozone and other oxidants are well known and include damage to rubber, paint, and textiles.

Ozone is a pollutant near the surface of Earth. However, in the stratosphere, high above Earth, it is concentrated in the *ozone layer,* where it protects life on Earth from the Sun's harmful ultraviolet radiation (see Chapter 19).

Ozone's effects on plants can be subtle. At very low concentrations, ozone can reduce growth rates while not producing any visible injury. At higher concentrations, ozone kills leaf tissue, eventually killing entire leaves and, if the pollutant levels remain high, whole plants. The death of white pine trees planted along highways in New England is believed to be due in part to automobile-generated ozone. Ozone also causes harm to humans and other animals, especially to the eyes and the respiratory system (Table 18.2).[3]

Volatile Organic Compounds

Volatile organic compounds (VOCs) include a wide variety of organic compounds used as solvents in industrial processes such as dry cleaning, degreasing, graphic arts, and adhesives production. *Hydrocarbons*, a group of VOCs, are composed of hydrogen and carbon. Thousands of such compounds exist, including natural gas (methane, CH_4), butane (C_4H_{10}), and propane (C_3H_8). Other toxic examples are benzine, an industrial solvent and additive to gasoline, and arcolein, produced from manufacturing processes that include burning petroleum. Both benzine and arcolein are components of cigarette smoke.[4] Analysis of urban air has identified many different hydrocarbons, some of which react with sunlight to produce photochemical smog. There are numerous potential adverse effects of hydrocarbons because many are toxic to plants and animals; hydrocarbons can also be converted to harmful compounds through complex chemical changes that occur in the atmosphere. Globally, only 15 percent of hydrocarbon emissions are anthropogenic. However, about half of hydrocarbons entering the atmosphere in the United States are emitted from anthropogenic sources, the most important of which is the automobile (Table 18.1). Anthropogenic sources are particularly abundant in urban regions. Nevertheless, in some cities in the southeastern United States such as Atlanta, Georgia, natural emissions probably exceed those from automobiles and other human sources.[2]

Hydrogen Sulfide

Hydrogen sulfide (H_2S) is a highly toxic corrosive gas easily identified by its rotten-egg odor. Hydrogen sulfide is produced from natural sources such as geysers, swamps, and bogs and from human sources such as industrial plants that produce petroleum or that smelt metals. The potential effects of hydrogen sulfide include functional damage to plants and health problems ranging from toxicity to death for humans and other animals.[3]

Hydrogen Fluoride

Hydrogen fluoride (HF) is a gaseous pollutant released primarily by industrial activities such as the production of aluminum, coal gasification, and burning of coal in power plants. Hydrogen fluoride is very toxic, and even a small concentration (as low as 1 part per billion) may cause problems for plants and animals. It is particularly dangerous to grazing animals because some plants can become toxic upon exposure to hydrogen fluoride.[3]

Other Hazardous Gases

Rarely a month goes by without the media reporting a truck or train accident that released gaseous, toxic chemicals into the atmosphere. People are often evacuated until the leak is stopped or the gas has dispersed to a nontoxic level. Often, chlorine gases or a variety of other materials used in chemical and agricultural processes are involved.

Another source of gaseous air pollution is sewage-treatment plants. Urban areas deliver a tremendous variety of organic chemicals, including paint thinner, industrial solvents, chloroform, and methyl chloride to treatment plants through their sewer systems. These materials are not removed in the treatment plants; in fact, the treatment processes facilitate the evaporation of the chemicals into the atmosphere, where they may be inhaled. Many of the chemicals are toxic or are suspected carcinogens. It is a cruel twist of fate that the treatment plants designed to control water pollution are becoming sources of air pollution. Although some pollutants can be moved from one location to another, and transformed from liquid to gas, we really cannot get rid of them easily.

Some chemicals are so toxic that extreme care must be taken to ensure they do not enter the environment. The extreme toxicity of some chemicals was tragically demonstrated on December 3, 1984, when a toxic gas, stored in liquid form in a pesticide plant, leaked, vaporized, and formed a deadly cloud that settled over a 64 km² (25 mi²) area of Bhopal, India. The gas leak lasted less than an hour, but more than 2,000 people were killed and more than 15,000 were injured. The colorless gas, methyl isocyanate, causes severe burning on contact with eyes, nose, throat, and lungs (Figure 18.4). Breathing the gas in concentrations of only a few

Figure 18.4 Toxic gas The toxic gas leak from a pesticide plant in Bhopal, India, caused many victims to suffer eye damage. The gas released causes tissue burns on contact. *(Pablo Bartholomew/Getty Images Inc.)*

parts per million causes violent coughing, swelling of the lungs, bleeding, and death. Less exposure can cause a variety of problems, including loss of sight.

Methyl isocyanate is one of the ingredients of Sevin, a common pesticide known in the United States, and of at least two other insecticides used in India. A plant in West Virginia also makes the chemical, and small leaks have evidently occurred there both before and after the catastrophic accident in Bhopal. The Bhopal accident clearly indicates that chemicals that can cause catastrophic injuries and death should not be stored close to large population centers. Furthermore, chemical plants need to have more reliable accident-prevention equipment and personnel trained to control leaks or other problems.

Particulate Matter: PM 10 and PM 2.5

Particulate-matter pollutants are small particles of solid or liquid substances that are designated PM 10 or PM 2.5, depending on size. PM 10 particles are less than 10 μm in diameter; they are released into the atmosphere by many natural processes and human activities. About 85 percent of emissions of particulates are from natural sources (Table 18.1). Desertification, volcanic eruptions, fires, and modern farming all add large amounts of particulate matter to the air. Burning of crop residue (from wheat fields in North Dakota, bluegrass in Idaho, rice in California, and sugarcane in Hawaii) emits large amounts of PM 2.5 from smoke and soot into the atmosphere (Figure 18.3). Nearly all industrial processes release particulates into the atmosphere, as does fossil fuel burning. Much particulate matter is easily visible as smoke, soot, or dust, but some particulate matter is not readily apparent. Particulates include airborne asbestos particles and small particles of heavy metals such as arsenic, copper, lead, and zinc, which are usually emitted from industrial facilities such as smelters.

In 1997 a very serious air pollution event occurred in Indonesia. That year, the monsoonal rains came late, producing a severe drought in Indonesia. Farmers took advantage of the dry weather and burned more forest as part of the slash-and-burn farming that has been going on for centuries in that area. Subsistence farmers prefer to clear land by burning because bulldozing the land is far too expensive. In general, it is believed that limited slash-and-burn farming does not pose a major ecological threat to the rain forest environment as long as the burning is not widespread and fires are carefully controlled.[5] In 1997, however, the fires raged out of control, producing a thick toxic haze of smoke and causing a catastrophe thought to be one of the world's greatest environmental disasters.

Although much of the blame for the fires and resulting damage was placed upon small farmers, large-scale agriculture was also responsible for clear-cut logging and burning. During the dry season of 1997 at least 20,000 hectares (49,000 acres) of land burned, producing vast amounts of smoke and particulate matter. Because of the smoke and particulates caused by the fire, approximately 20 million Indonesians had to be treated for a variety of illnesses. A passenger airline crash in Sumatra, a result of the dense smoke, killed 234 people. The Air Quality Index (AQI), a measure of air quality, was as high as 800! An AQI of 500 can be responsible for the premature death of ill and elderly persons, and an index of 800 is equivalent to smoking four packs of cigarettes a day! (AQI is discussed in Section 18.8.) The burning of the land produced a dense, blinding haze that greatly reduced visibility. By the end of the summer the hazy air began to drift across the South China Sea to Singapore and Malaysia. The arrival of the monsoons extinguished many of the fires, but when the rain stopped in February of 1998 fires were again spotted. Although there is now a ban on burning for agriculture and logging, it has had mixed success. Both natural- and human-caused fires continue to occur each year. Damage was not limited to humans; wildlife, including endangered species such as orangutans and rhinoceros, and wilderness reserves were damaged.[4] Concerns linger about the immediate disastrous effects of continued cycles of fire and their cumulative environmental damage.

Very fine particle pollutants, with particles less than 2.5 μm in diameter (PM 2.5) cause the greatest damage to the lungs. Sulfates and nitrates are two of the most significant fine-particle pollutants. Both of these compounds are mostly secondary pollutants and are produced in the atmosphere through chemical reactions between sulfur dioxide or nitrogen oxides and normal atmospheric constituents. Sulfates in the form of sulfuric acid and nitric acid are formed by these reactions and are then precipitated as acid rain.[3] When measured, particulate matter is often referred to as total suspended particulates (TSP).

Particulates affect human health, ecosystems, and the biosphere (Table 18.2). Particulates entering the lungs may lodge there, producing chronic effects on respiration; asbestos is particularly dangerous in this way. Dust raised by road building and plowing is deposited on the surfaces of green plants, possibly interfering with both the plants' absorption of carbon dioxide and oxygen and their release of water; heavy dust may affect the breathing of animals. Particulates in the United States are believed to contribute to the deaths of 60,000 people per year. In cities it is estimated that up to 9 percent of all deaths are associated with particles, and the risk of mortality is 15 percent to 25 percent higher in cities with the highest levels of fine-particulate pollution.[6]

Particulates associated with large construction projects may kill organisms and damage large areas, changing species composition, altering food chains, and generally affecting ecosystems. In addition, modern industrial processes have greatly increased the total suspended particulates in the atmosphere. Particulates block sunlight and thus cause changes in climate, which can have lasting effects on the biosphere.[7]

Asbestos. *Asbestos* particles have only recently been recognized as a significant hazard to humans. In the past, asbestos was treated rather casually, and people working in asbestos plants were not protected from dust. Asbestos was used in building insulation and in brake pads for automobiles. As a result, asbestos fibers are found throughout industrialized countries, especially in European and North American urban environments. In one case, asbestos products were sold in burlap bags that were eventually reused in plant nurseries and other secondary businesses, thus further spreading the pollutant. Some asbestos particles are believed to cause cancer, so asbestos must be carefully controlled (see Chapter 3).

Lead. *Lead* is an important constituent of automobile batteries and other industrial products. When used as a gasoline additive, lead improves the performance of automobile engines. Previously, in the United States, lead emitted into the environment via automobile exhaust was widely deposited, reaching high levels in soils and waters along roadways (see Case History: Lead in the Environment, Chapter 20). Once released, lead can be transported through the air as particulates to be taken up by plants through the soil or deposited directly on plant leaves. Thus, it enters terrestrial food chains, and may be ingested by people. When lead is carried by streams and rivers, deposited in quiet waters, or transported to the ocean or lakes, it is taken up by aquatic organisms and thus enters aquatic food chains. Although lead is no longer used in gasoline in the United States, it is still used in other parts of the world. In the United States, lead emissions have decreased by 98 percent since 1970 in direct response to reductions and elimination of lead in gasoline.

Cadmium. Some of the *cadmium* in the environment comes from coal ash, which is widely dispersed from smokestacks and chimneys. Cadmium exists as a trace element in the coal, at a very low concentration of 0.05 parts per million. When the coal ash falls on plants, the cadmium is incorporated into plant tissue and concentrated. As cadmium moves up through the food chain, it undergoes *biomagnification* at each higher level. Herbivores have approximately three times the concentration of green plants, and carnivores have approximately three times the concentration of herbivores.

18.5 Urban Air Pollution

Air pollution is not distributed uniformly around the world. Much of it is concentrated in and around urban areas, where automobiles and heavy industry emit an enormous amount of waste into the environment. The visible air pollution known as smog is present in nearly all urbanized areas, although it is much worse in some regions than in others. In this section, we will consider the factors contributing to urban air pollution and discuss the composition and formation of smog.

Influence of Meteorology and Topography

The extent to which air pollution occurs in an urban area depends largely on emission rates, topography, and weather conditions; these factors determine the rate at which pollutants are concentrated and transported away from their sources or converted to harmless compounds in the air. When the rate of pollutant production exceeds the rates of transport and chemical transformation, dangerous conditions can develop.

Meteorological conditions can determine whether air pollution is only a nuisance or a major health problem. Most pollution periods in the Los Angeles Basin and other smoggy areas do not cause large numbers of deaths. However, serious pollution events can develop over a period of days and lead to an increase in deaths and illnesses.

Restricted circulation in the lower atmosphere due to the formation of an *inversion layer* may lead to a pollution event. An **atmospheric inversion** occurs when warm air overlies cool air (Figure 18.5). An inversion layer is particularly a problem when there is a stagnated air mass. Figure 18.6 shows two types of developing inversions that may worsen air pollution problems. Figure 18.6a shows descending warm air from inland warm arid areas forming a semipermanent inversion layer. Because the mountains act as a barrier to the pollution, polluted air moving in response to the sea breeze and other processes tends to move up canyons, where it is trapped. The air pollution that develops occurs primarily during the

Figure 18.5 Inversion in Los Angeles Temperature inversion over Los Angeles. Polluted air is trapped relatively close to the ground, producing a hazy or reddish-brown layer. (Stock Boston)

summer and fall, when warm inland air comes over the mountains and overlies cooler coastal air, forming the inversion. This example is representative of the situation in the Los Angeles area.

Figure 18.6b shows a valley with relatively cool air overlain by warm air, a situation that can occur in several ways. When cloud cover associated with a stagnant air mass develops over an urban area, the incoming solar radiation is blocked by the clouds, which absorb some of the energy and thus heat up. On or near the ground, the air cools. If the humidity is high, a thick fog may form as the air cools. Because the air is cool, people living in the city burn more fuel to heat their homes and factories, thus delivering more pollutants into the atmosphere. As long as the stagnant conditions exist, the pollutants continue to accumulate. This was the major cause of the 1952 London smog event, described in the case history that opened this chapter.

Figure 18.6 Temperature inversion Two causes for the development of a temperature inversion, which may aggravate air pollution problems. (a) Descending warm air from inland arid areas forms a semipermanent inversion layer. (b) A valley with relatively cool air is overlain by warm air.

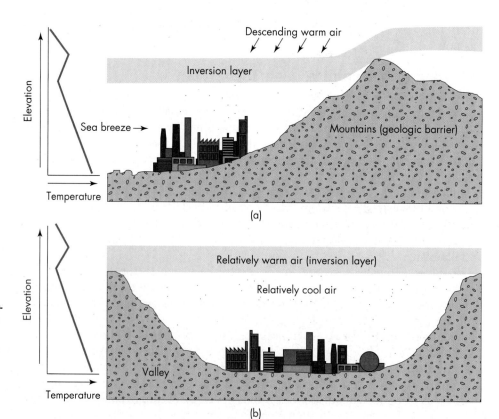

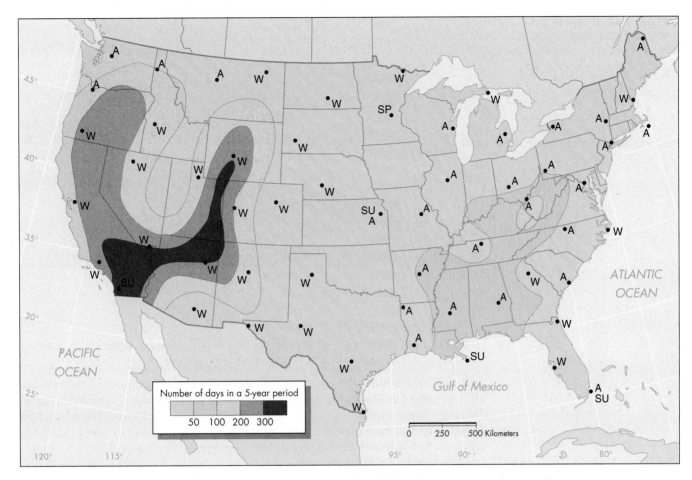

Figure 18.7 **Potential for air pollution** Number of days in a 5-year period characterized by conditions favorable for reduced dispersion and local buildup of air pollutants that existed for at least a 48-hour (2-day) period. The time of year when most of these periods occurred is shown by the season: SP (spring), A (autumn), W (winter), or SU (summer). *(Modified after Holzworth, as presented in Neiburger, Edinger, and Bonner. 1973.* Understanding our atmospheric environment. *W. H. Freeman)*

Evaluating meteorological conditions can be extremely helpful when predicting which areas will have potential smog problems. Figure 18.7 shows the number of days in a 5 year period for which conditions were favorable for reduced dispersion of air pollution for at least a 48 hour period. This illustration clearly shows that most of the problems—that is, the greatest number of days of reduced dispersion—are located in the western United States.

The meteorological and topographic conditions of the western states help explain why cities such as Los Angeles and San Diego, in California, and Phoenix, Arizona, have air quality problems from time to time. Cities situated in a topographic "bowl" surrounded by mountains are more susceptible to smog problems than are cities in open plains. Cities with weather conditions such as temperature inversions are also particularly susceptible. Both the surrounding mountains and the temperature inversions prevent the transportation, or dispersion, of pollutants by the winds and weather systems.

Potential for Urban Air Pollution

The potential for air pollution in urban areas, as illustrated in Figure 18.8, is determined by the following factors:

- Rate of pollutant emissions
- Distance that a mass of air moves through urban air pollution sources

Figure 18.8 How air pollutants may be concentrated As wind velocity increases and the mixing layer (shown here as H) becomes thicker, there is less potential for air pollution. As both the emission rate and the downwind length of the city increase, there is a greater potential for air pollution. The "chimney effect" allows polluted air to move over a mountain down into an adjacent valley.

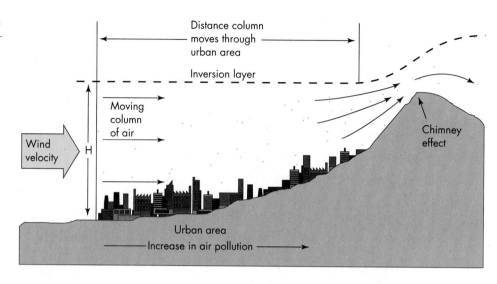

- Average speed of the wind
- Height of the mixing layer—that is, the height to which potential pollutants may be thoroughly mixed in the lower atmosphere

If we assume there is a constant rate of emission of air pollutants, the mass of air will collect more and more pollutants as it moves through the urban area. Thus, the concentration of pollutants in the air is directly proportional to the first two factors: As either the emission rate or the downwind travel distance increases, so does the concentration of pollutants. Conversely, city air pollution decreases with increased wind velocity and the height of mixing. The stronger the wind and the higher the mixing layer, the lower the pollution.

If an inversion layer is present near a geologic barrier such as a mountain, a "chimney effect" may let the pollutants spill over the top of the barrier (see Figure 18.8). This effect has been noticed in the Los Angeles Basin, where pollutants can climb several thousand meters, damaging mountain pine trees and other vegetation.

Smog Production

Wherever many sources are producing air pollutants over a wide area—as, for example, automobiles in Los Angeles—there is potential for the development of smog. The two major types of smog are *sulfurous smog*, sometimes referred to as London-type smog or gray air, and *photochemical smog*, sometimes called L.A.-type smog or brown air. **Sulfurous smog** is primarily produced by burning coal or oil at large power plants. Under certain meteorological conditions, the sulfur oxides and particulates produced by this burning combine to produce concentrated sulfurous smog (Figure 18.9).

Reactions that produce **photochemical smog** are complex, involving nitrogen oxides (NO_x), hydrocarbons, and solar radiation (Figure 18.10). Development of photochemical smog is related to automobile use. In southern California, for example, when commuter traffic begins to build up early in the morning, the concentrations of nitrogen oxide (NO) and hydrocarbons begin to increase. At the same time, the amount of nitrogen dioxide (NO_2) may decrease owing to the sunlight-driven reaction of NO_2 that produces NO plus atomic oxygen (O). The atomic oxygen is then free to combine with molecular oxygen to form ozone (O_3), increasing ozone levels after sunrise. By midmorning oxidized hydrocarbons react with NO to increase the concentration of NO_2. This reaction causes the NO concentration to decrease and allows ozone to build up, producing a midday peak in

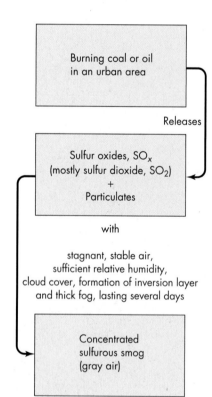

Figure 18.9 Sulfurous smog
Possible development scenario for concentrated sulfurous smog and smoke.

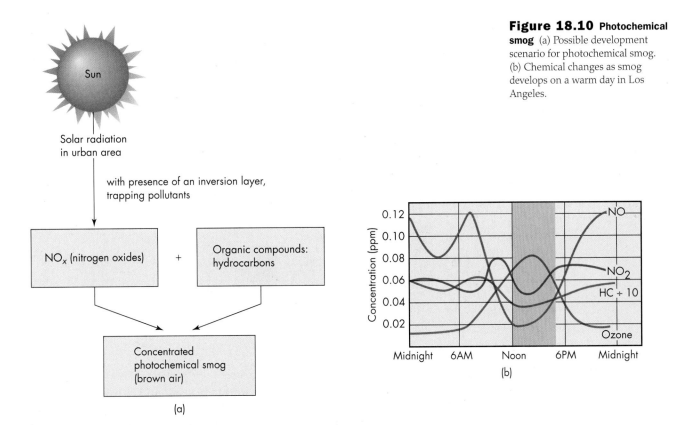

Figure 18.10 Photochemical smog (a) Possible development scenario for photochemical smog. (b) Chemical changes as smog develops on a warm day in Los Angeles.

ozone and a minimum in NO. As the smog matures, visibility is reduced by light-scattering from small particles (Figure 18.11).

Future of Air Pollution in Urban Areas

Air pollution levels in many cities in developed countries have a mixed but generally improving record. Data from major U.S. metropolitan areas in recent years show a decline in the total number of days characterized as unhealthful and very unhealthful, suggesting that the nation's air quality is improving. This suggested improvement has occurred despite the fact that there are many more motor vehicles. The improved air quality has resulted from building automobiles that burn

(a)

(b)

Figure 18.11 Smog in Los Angeles City of Los Angeles, California, on a clear day (a) and a smoggy day (b). *(Kathleen Campbell/Getty Images, Inc.)*

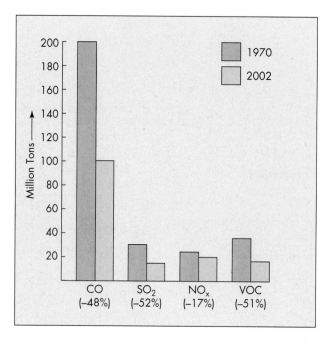

Figure 18.12 Trends in U.S. emissions of selected air pollutants (1970–2002) Emissions of carbon monoxide, sulfur dioxide, nitrogen oxide, and volatile organic compounds have decreased significantly since 1970. *(Modified from Environmental Protection Agency. 2002. Highlights accessed 3/25/04 at www.epa.gov)*

less fuel more efficiently, contain smog-control devices, and burn cleaner, improved fuel. However, most major urban areas such as New York and Los Angeles still have unhealthy air much of the time. Furthermore, many U.S. cities have poor air quality at least 30 days per year. According to this criterion, millions of Americans still live in cities that have hazardous air pollution for a significant portion of each year.[1]

Cities in countries with burgeoning populations and poverty are particularly susceptible to air pollution now and in the future. The financial base necessary to fight air pollution may not exist in these areas. People in these countries tend to be more concerned with basic survival and finding ways to house and feed their growing urban populations. For example, Mexico City is home to over 20 million people, making it one of the largest urban areas in the world. Industry and power plants in Mexico City, along with the numerous, often aging cars, trucks, and buses, emit hundreds of thousands of tons of particulates, sulfur dioxide, nitrogen oxides, and hydrocarbons into the atmosphere each year. The city is at an elevation of about 2,255 m (7,400 ft) in a natural basin surrounded by mountains—a perfect situation for a severe air pollution problem. The mountains can rarely be seen from Mexico City now. Physicians report a steady increase in respiratory diseases, and headaches, irritated eyes, and sore throats are common.

The encouraging news is that emissions of some major air pollutants in the United States are decreasing (Figure 18.12). For example, since 1970, SO_2 emissions have declined 52 percent as a result of burning less coal, using low-sulfur coal, and treating effluent gases from power plants before release into the environment. Emissions of volatile organic compounds have also decreased since 1970 to levels not recorded since the 1940s. A decrease of about 51 percent has resulted in part from the successful control of automobile emissions and substitutions of water-based components instead of VOC components in products such as asphalt.

18.6 Indoor Air Pollution

Indoor air pollution is one of the most serious environmental health hazards that people face in their homes and workplaces (Figure 18.13).[8,9] In recent years buildings have been constructed more tightly, often with windows that cannot be

1. If not properly maintained, heating, ventilation and air conditioning systems of buildings are sources of indoor air pollutants, including molds and bacteria. Gas or oil furnaces produce carbon monoxide, nitrogen dioxide, and particulate matter.

2. Poor location of fresh air intake, as for example above a loading dock, or a first-floor restaurant exhaust fan may inadvertently bring in air pollutants.

3. Remodeling, painting, and other such activities often bring a variety of chemicals and materials into a building. Fumes from these activities may enter the building's heating, ventilation, and air conditioning system, causing pollution throughout the building.

4. Furniture and carpets in buildings often contain toxic chemicals such as formaldehyde, organic solvents, and asbestos that may be released in buildings.

5. Cleaning products and solvents used in offices and other parts of buildings contain harmful chemicals. The fumes may circulate throughout a building.

6. Coffee machines, fax machines, computers, and printers may release chemicals such as ozone (O_3).

7. Insect control that uses pesticides may contaminate buildings with cancer-causing chemicals.

8. Restrooms are the source of a variety of indoor air pollutants including second-hand smoke, molds, and fungi resulting from humid conditions.

9. Dust mites and molds can exist in carpets and furniture.

10. People cause increase in carbon monoxide; may emit bioeffluents, and may spread bacterial and viral contaminants.

11. Loading docks may be the source of organic waste from garbage containers, particulates, and carbon monoxide from vehicles.

12. Radon gas can seep into buildings from soil; rising water allows the growth of molds.

13. People who smoke inside, perhaps in restrooms; people who smoke outside buildings, particularly near open or revolving doors, may cause pollution of the building as the second-hand smoke is drawn into and up through the building by the chimney effect.

14. Pollen from inside and from outside air can enter the building.

Figure 18.13 **Indoor air pollution** Common sources of indoor air pollutants.

opened, to save energy. As a result, in many buildings air is filtered. Unless filters are maintained properly, indoor air can become polluted with a variety of substances, including smoke, chemicals, disease-carrying organisms, and radon, a naturally occurring radioactive gas suspected of causing lung and other cancers (see Chapter 20). It was shown that the virus responsible for the respiratory infection known as Legionnaires' disease is transported through air filters and ventilation systems of buildings.

Modern urban structures are built of many substances, some of which release minute amounts of chemicals and other materials into the nearby air. In some buildings, asbestos fibers are slowly released from insulation and other fixtures; people exposed to a particular type of asbestos fiber may develop a rare form of lung cancer. The poisonous gases carbon monoxide and nitrogen dioxide may be released in homes from unvented or poorly vented gas stoves, furnaces, or water heaters. Formaldehyde, present in some insulation materials and wood products used in home construction, is known to irritate ears, nose, and throat. Radon is present in some building materials, such as concrete blocks and bricks manufactured from materials with a high radon concentration. Buildings that lack systems that recirculate clean air are likely to have indoor pollution problems. Improvements in indoor air quality have been developed; most of these stress improved clean air circulation and reduced pollutant emissions.

Interestingly, indoor air pollution existed centuries ago. In 1972 the body of a fourth-century Eskimo woman was discovered on St. Lawrence Island in the Bering Sea. The woman evidently was killed during an earthquake or landslide, and her body was frozen soon after death. Detailed autopsies showed that the woman suffered from black lung disease, which afflicts some coal miners today. Anthropologists and medical personnel concluded that the woman breathed very polluted air for a number of years. They speculated that the air pollution was in part due to hazardous fumes given off from lamps that burned seal and whale blubber.[10]

18.7 Control of Air Pollution

Reducing air pollution requires a variety of strategies tailored to specific sources and type of pollutants. From an environmental perspective, the best strategy is to reduce emissions through conservation and efficiency measures—that is, to burn less fossil fuel. For both stationary and mobile sources, strategies have been to collect, capture, or retain pollutants before they enter the atmosphere. Other strategies, however, may only make the problem worse, as exemplified by smelters in Sudbury, Ontario, Canada. The ores processed by the smelters contain a high percentage of sulfur. During the smelting process, smokestacks emit large amounts of sulfur dioxide as well as particulates containing nickel, copper, and other toxic metals. Attempts to minimize the local pollution by increasing the height of the smokestacks backfired; taller stacks spread the pollution over a larger area.

Pollution problems vary in different regions of the world and even within the United States. In the Los Angeles Basin, for example, nitrogen oxides and hydrocarbons are particularly troublesome because they combine in the presence of sunlight to form photochemical smog. Furthermore, in Los Angeles most of the nitrogen oxides and hydrocarbons are emitted from automobiles—a mobile source.[11] In other urban areas, such as in Ohio and the Great Lakes region, air quality problems are primarily due to sulfur dioxide and particulate emissions from industry and coal-burning power plants, which are point sources.

Because the problems vary greatly from country to country and from region to region, it is often difficult to obtain both the political consensus and the money needed for effective control. Nevertheless, effective control strategies do exist for many air pollution problems.

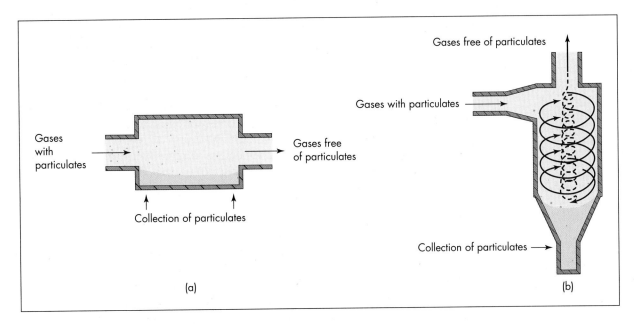

Figure 18.14 **Controlling particulates on site** Devices being used to control emissions of particulates before they enter the atmosphere. (a) Simple settling chamber that collects particulates by gravity settling. (b) Centrifugal collector, in which particulates are forced to the outside of the chamber by centrifugal force and then fall to the collection site.

Control of Particulates

Particulates emitted from stationary sources—fugitive, point, or area sources—are easier to control than are the very small particulates of primary or secondary origin released from mobile sources such as automobiles. Further research is needed to develop adequate control methods for these very small particles.

Power plants and industrial sites (area sources) utilize a variety of settling chambers or collectors to control the release of coarse particulates (Figure 18.14). These methods act to settle out particles where they can be collected for disposal in landfills.

Particulates from fugitive sources, such as a waste pile, must be controlled at the site before they are eroded by the wind and carried into the atmosphere. Measures such as protecting open areas, controlling dust, or reducing the effect of wind can help to decrease the release of particulates from fugitive sources. For example, covering waste piles with plastic or other material can protect them from wind erosion; soil piles can be protected by planting them with vegetation that inhibits wind erosion. Dust on a road can be controlled by spreading water or a combination of water and chemicals to hold it down; structures or vegetation may be used as windbreaks, that is, placed to lessen wind velocity near the ground.

Control of soot from agricultural burning that increases crop yields and reduces the need for pesticides is a difficult problem. The approach has been to advise farmers of the best days to burn and to ban burning if the concentration of PM 2.5 reaches hazardous levels.

Control of Automobile Pollution

Carbon monoxide, nitrogen oxides, and hydrocarbons in urban areas are best controlled by regulating automobile exhaust, the primary source of the anthropogenic portion of these pollutants. Furthermore, controlling these materials will also regulate the ozone in the lower atmosphere, since it forms there from reactions with nitrogen oxides and hydrocarbons in the presence of sunlight.

Nitrogen oxide from automobile exhaust is controlled by recirculating exhaust gas within the automobile's engine. Recirculation dilutes the air-to-fuel mixture being burned; reduces the temperature of combustion; decreases the oxygen concentration in the burning mixture, making it a richer fuel; and produces fewer nitrogen oxides. Unfortunately, this method increases hydrocarbon emissions. Nevertheless, exhaust recirculation to reduce nitrogen oxide emissions has been common practice in the United States for over 20 years. The most common device used to remove carbon monoxide and hydrocarbon emissions from automobiles is the *catalytic converter*, which converts carbon monoxide to carbon dioxide and hydrocarbons to carbon dioxide and water.[12]

It has been argued that the automobile emission regulation plan in the United States has not been very effective in reducing pollutants. The pollutants may be reduced while a vehicle is relatively new, but many people simply do not maintain their emission control devices over the life of the automobile. Furthermore, some people disconnect smog-control devices. It has been suggested that annual inspections of an automobile's emissions could be a powerful tool in regulating air pollution.[13] Vehicles could be tested each year for emission control, and fees would be assessed on the basis of the test results. The fees would encourage the purchase of automobiles that pollute less, and the annual inspections would ensure that pollution control devices are properly maintained.

Other options to reduce pollution from automobiles are (1) developing and using both fuel additives and reformulated fuel that produces a cleaner fuel and (2) requiring new cars to have improved fuel efficiency, which would encourage or require the use of electric cars or hybrid cars that have both an electric engine and an internal combustion engine.

Control of Sulfur Dioxide

Sulfur dioxide emissions can be reduced by abatement measures performed before, during, or after combustion. The technology is now available to purify coal, enabling it to burn cleanly; however, the cost of removing the sulfur does make the fuel more expensive.

Changing from high-sulfur coal to low-sulfur coal seems an obvious solution to reducing sulfur dioxide emissions. Unfortunately, most low-sulfur coal is located in the western part of the United States, whereas most coal burning occurs in the eastern part. Therefore, using low-sulfur coal is a solution only where it is economically feasible to transport the coal over long distances. Another possibility is cleaning relatively high-sulfur coal by washing it. Washing finely ground coal with water allows the iron sulfide, known as the mineral pyrite, to settle out because it is denser than coal. Although washing removes some of the sulfur, it is also expensive. Another option is coal gasification, which converts relatively high-sulfur coal to a gas, removing the sulfur in the process. The gas obtained from coal is cleaner than the coal and can be transported relatively easily, augmenting supplies of natural gas. The synthetic gas produced from coal is comparatively expensive but may become more competitive with other gas prices in the future.[14]

Stationary sources of SO_2 such as power plants can reduce their SO_2 emissions by removing them in the stack before they reach the atmosphere. A highly developed technology for the cleaning of gases in tall stacks is **scrubbing** (Figure 18.15). In this method, gases produced during burning, including SO_2, are treated with a slurry of lime (calcium oxide, CaO) or limestone (calcium carbonate, $CaCO_3$). The sulfur oxides react with the calcium to form insoluble calcium sulfide, which is collected. However, the calcium sulfide residue is a sludge that must be disposed of at a waste-disposal site. The sludge can cause serious water pollution if it is allowed to interact with the hydrologic cycle; therefore, it must be carefully treated.[14] Furthermore, scrubbers are expensive and add significant (10 percent to 15 percent or more) cost to producing electricity by burning coal.

An innovative approach has been taken at a large coal-burning power plant near Mannheim, Germany. The smoke from combustion is treated with liquid

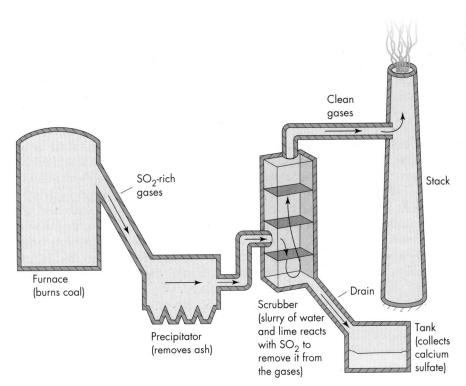

Figure 18.15 **Removing sulfur dioxide** A scrubber used to remove sulfur oxides from the gases emitted by tall smokestacks.

ammonia (NH_3), which reacts with the sulfur gas in the smoke to produce ammonium sulfate. Sulfur-contaminated gas entering the pollution control system is cooled by outgoing sulfur-depleted, or clean, gases to a temperature that favors the reaction; then outgoing clean gas is heated by incoming sulfur-rich, or dirty, gas to force it out a vent. Waste heat from the cooling towers also heats nearby buildings, and the plant sells the ammonium sulfate in solid granular form to farmers as fertilizer. The plant, finished in 1984, was built in response to tough pollution control regulations enacted to help abate acid precipitation, thought to contribute to pollution that is killing forests in Germany.

Pollution abatement has been successful in Japan. Japan had what was considered the most severe sulfur pollution problem in the world, and the health of the Japanese people was being directly affected. It was not uncommon to see residents of Yokohama and other areas wearing masks over their mouths when out on the streets. The Japanese government issued control standards in the 1970s. In the 5 year period following initiation of the program, the sulfur dioxide level was reduced by 50 percent despite doubled energy consumption. Power plants have generally met these requirements by installing scrubbers known as flue gas desulfurization systems that can remove over 95 percent of the sulfur from smokestacks. More than 1,000 of these are in use in Japan today.[15] The Japanese have also begun to control nitrogen oxides.

18.8 Air Quality Standards

Air quality standards are associated with emission standards that attempt to control air pollution. Emission standards enacted by various countries set maximum emission levels for specific pollutants. Unfortunately, these maximums vary greatly from one country to another, reflecting a lack of agreement concerning the concentrations of pollutants that cause environmental problems.[13]

Clean Air Legislation in the United States

In the United States, the Clean Air Act was passed in 1970 with the objective of improving the nation's air quality; it was amended in 1977 and 1990. The 1977 amendments define two levels or types of air quality standards. Primary standards are levels set to protect human health but may not protect against damaging effects to structures, paint, and plants. Secondary standards, although designed to help prevent other environmental degradation, are similar to the primary levels.[13]

The 1990 amendments to the Clean Air Act tighten controls on air quality. In particular, the legislation placed more stringent controls on sulfur dioxide (SO_2) emissions produced from coal-burning power plants, requiring these emissions to be reduced by approximately 50 percent by the year 2000. This goal was achieved; emissions have been reduced by about one-half. The goal was reached because utility companies introduced a variety of the pollution abatement strategies discussed above. The legislation also attempts to provide for economic incentives to reduce pollution through a system of allowances, or credits, to emit SO_2. Utility companies are issued a certain number of these allowances, which they may buy and sell on the open market. Utility companies with low emissions may sell their credits to other utilities that have higher emission rates. Of course, environmentalists may also purchase the allowances, effectively keeping the allowances from the utility companies and thereby forcing them to employ tighter pollution abatement technology to their operation.

The 1990 amendments also require reductions in nitrogen oxide emissions. Reducing nitrogen oxide emissions is more difficult to achieve than reducing sulfur dioxide emissions, since nitrogen oxide emissions are generally associated with automobiles rather than with stationary coal-burning power plants.[16] Finally, the 1990 amendments also address the emission of toxic material into the atmosphere. The amendments regulate toxins that are believed to have the most damaging effect on human health, including known carcinogens. The goal is to reduce these emissions by as much as 90 percent.

Monitoring Air Pollution

In the United States, urban areas' air quality is often reported as good, moderate, unhealthy, very unhealthy, or hazardous (Table 18.3). These levels are based on values of the Air Quality Index (AQI), which are derived from monitoring the concentration of five major pollutants: total suspended particulates, sulfur dioxide, carbon monoxide, ozone, and nitrogen dioxide. During a pollution episode in Los Angeles a first-stage smog episode begins if the AQI exceeds 100. This corresponds to unhealthy air with an Air Quality Index (AQI) between 100 and 300. A second-stage smog episode is declared if the AQI exceeds 300, a point at which the air quality is hazardous to all people. Cautionary statements, general adverse health effects, and action levels of the AQI are listed on Table 18.3. As the air quality decreases during a pollution episode, people are requested to remain indoors, minimize physical exertion, and avoid driving automobiles. Industry also may be requested to reduce emissions to a minimum during the episode.

18.9 Cost of Controlling Air Pollution

Currently, billions of dollars are spent annually in the United States in an attempt to control air pollutants. The money is dedicated to: reducing emissions from stationary sources, such as power plants and factories; reducing emissions from motor vehicles; and developing efficiency and conservation measures. The Clean Air Act Amendments of 1990 are very broad, and the impact of their pollution abatement measures is widely felt throughout many sectors of American society and industry. This impact includes increases in the cost of many industrial products and services, including higher costs of automobiles and gasoline.

TABLE 18.3 Air Quality Index (AQI) and Health Conditions

Index Value	Descriptor	Cautionary Statement	General Adverse Health Effects	Action Level (AQI)[1]
0 to 50	Good	None	None	None
51 to 100	Moderate	Unusually sensitive people should consider limiting prolonged outdoor exertion.	Very few symptoms[2] for the most susceptible people[3]	None (health advisories for susceptible individuals)
101 to 150	Unhealthy for sensitive groups	Active children and adults and people with respiratory disease, such as asthma, should limit prolonged outdoor exertion.	Mild aggravation of symptoms in susceptible people; few symptoms for healthy people	None (health advisories for all)
151 to 199	Unhealthy	Active children and adults and people with respiratory disease, such as asthma, should avoid prolonged outdoor exertion; everyone else, especially children, should limit prolonged outdoor exertion.	Mild aggravation of symptoms in susceptible people; irritation symptoms for healthy people	None (health advisories for all)
200 to 300	Very unhealthy	Active children and adults and people with respiratory disease, such as asthma, should avoid outdoor exertion; everyone else, especially children, should limit outdoor exertion.	Significant aggravation of symptoms in susceptible people; widespread symptoms in healthy people	Alert (200+) (health advisories for all; is an Alert; activities that cause pollution might be restricted)
Over 300 to 500	Hazardous	*Everyone* should avoid all outdoor exertion.	300–400: Widespread symptoms in healthy people 400–500: Premature onset of some diseases Over 500: Premature death of ill and elderly people; healthy people experience symptoms that affect normal activity	Warning (300+) (health advisories to all; triggers a Warning; probably would require power plant operations to be reduced; carpooling required) Emergency (400+) (health advisories for all; triggers an Emergency; cessation of most industrial and commercial activities, including power plants; nearly all private use of vehicles prohibited)

[1] Triggers preventive action by state or local officials
[2] Symptoms include eye, nose, and throat irritation, chest pain, breathing difficulty
[3] Susceptible people: young, old, and ill people; people with lung or heart disease

Source: U.S. Environmental Protection Agency

The cost and benefits of air pollution control are controversial subjects. Some argue that the present system of setting air quality standards is inefficient and unfair. They maintain that regulations are tougher for new sources of pollution than for existing ones. In addition, they believe that even if the benefits of pollution control exceed total costs, the cost of air pollution control varies widely from one industry to another.

Another economic consideration is that, as the degree of pollutant control increases, a point is eventually reached at which the cost of incremental control—that is, reducing additional pollution—is greater than the additional benefits. This and other economic factors have led to the position that fees and taxes for emitting pollutants might be preferable to attempting to evaluate uncertain costs and benefits and that it makes more economic sense to enforce fees rather than standards.

Some variables that must be considered in the economics of air pollution are shown in Figure 18.16. With increasing air pollution regulations, the capital cost for equipment needed to control air pollution increases. This is shown as the nearly straight blue line in Figure 18.16. As the regulations for air pollution increase, the loss from pollution damages decreases. This is shown as the gently curved red line in Figure 18.16. The total cost of air pollution is the sum of the capital cost and losses, represented in Figure 18.16 by the concave black line. Notice there is a minimum cost defined at a particular pollution level. Figure 18.16 shows that if the desired pollution level is lower than that at which the minimum total

Figure 18.16 Economics of air pollution Some of the relationships between economic cost and increasing air pollution controls. The nearly straight blue line from the lower left to upper right shows that with increasing air pollution regulations, the capital cost for equipment increases. The gently curved red line shows that as air pollution is reduced, the losses from pollution decrease. The black line is the sum of the capital costs and losses. Notice it is U-shaped. The bottom of the U is the minimum cost. *(Modified after Williamson, S. J. 1973. Fundamentals of air pollution. Addison-Wesley)*

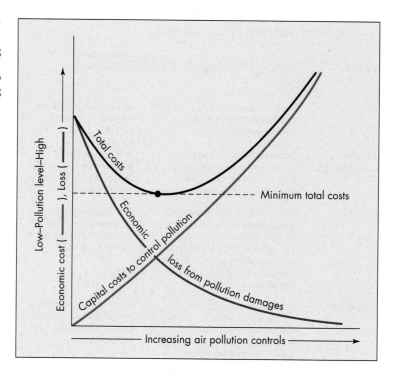

cost occurs, then additional costs will be necessary. This type of diagram, while valuable for considering some of the major variables, does not adequately consider all of the loss from pollution damages. For example, long-term exposure to air pollution may cause or aggravate chronic diseases in human beings at a very high cost. How do we quantify the portion of the cost that is attributable to air pollution? Despite these drawbacks, it seems worthwhile to reduce the air pollution level below some particular standard. Thus, in the United States, ambient air quality standards have been developed as a minimum acceptable air quality level.

SUMMARY

The atmosphere is a fast-flowing medium too often used by people for waste disposal. Every year several hundred million metric tons of pollutants enter the atmosphere above the United States from processes related to human activity. Considering the enormous volume of the atmosphere, this is a relatively small amount of material. If it were distributed uniformly, there would be fewer problems with air pollution. Unfortunately, in general, the pollutants are not evenly distributed but are concentrated in urban areas or in other areas where the air naturally lingers.

The two major types of air pollution sources are stationary and mobile. Stationary sources have a relatively fixed position and include point sources, fugitive sources, and area sources. Air pollutants are classified as gaseous and particulate or as primary and secondary. Primary pollutants are those emitted directly into the air: particulates, sulfur oxides, carbon monoxide, nitrogen oxides, and hydrocarbons. Secondary pollutants are those produced through reactions among primary pollutants and other atmospheric compounds. A good example of a secondary pollutant is ozone, which forms over urban areas through photochemical reactions among primary pollutants and natural atmospheric gases. Be careful not to confuse ozone as a pollutant in the lower atmosphere with ozone depletion in the stratosphere (see Chapter 19).

More recently the U.S. Environmental Protection Agency has been evaluating toxic air pollutants known as air toxins in terms of their impact on human health. It is expected that by analyzing air toxins, their adverse effects on health will be identified, and will leading to their control.

Meteorological and topographic conditions in a particular urban area greatly affect the potential for air pollution problems. In particular, restricted lower-atmosphere circulation associated with temperature inversion layers may lead to pollution events, especially in cities lying in a "bowl" surrounded by mountains. Air pollution over an urban area is directly proportional to the emission rate of pollution and the downwind travel distance of pollutants over the city; pollution is inversely proportional to the wind velocity and the height of the mixing layer of the atmosphere over the urban area.

The two major types of smog are sulfurous, or gray, and photochemical, or brown. Sulfurous smog is primarily produced by industrial burning of fossil fuels. Photochemical smog is produced by photochemical reactions of the nitrogen oxides and hydrocarbons in automobile exhaust. Each type causes particular environmental problems that vary with geographic region, time of year, and local urban conditions.

A growing problem in urban areas is indoor air pollution, caused by tight insulation of modern buildings, emission of toxic gases from building materials, and circulation of toxic gases, smoke, and disease-causing organisms through ventilation systems.

Methods to control air pollution are tailored to specific sources and types of pollutants. These methods vary and include settling chambers for particulates, catalytic converters to remove carbon monoxide and hydrocarbons from automobile exhaust, and scrubbers or combustion processors that use lime to remove sulfur before it enters the atmosphere.

Air quality in urban areas is usually reported in terms of whether the quality is good, moderate, unhealthy, very unhealthy, or hazardous. These levels are defined in terms of the Air Quality Index (AQI). The nation's air quality has generally improved in recent years; however, air quality in many urban areas is still unhealthful during a significant portion of the year.

The relationships between emission control and environmental cost are complex. The minimum total cost is a compromise between capital costs to control pollutants and losses or damages resulting from pollution. If additional controls are used to lower the pollution to a more acceptable level, then additional, rapidly increasing costs will be incurred.

Revisiting Fundamental Concepts

Human Population Growth

As human population has increased in numbers and density in urban regions, so have urban air pollution problems. Although in the United States and other developed countries the amount of air pollution emitted per person has generally decreased, the problem has increased as a result of the tremendous growth in the total number of people. In some developing countries, air pollution problems are more acute and are more likely to produce serious health problems.

Sustainability

One of the primary rights of people is to breathe clean air. Sustaining air quality in urban areas remains a challenge to society but one necessary for both people and other living things.

Earth as a System

The atmosphere is a dynamic, ever-changing part of the Earth system that naturally interacts with the water and land in many complex ways. For example, volcanic eruptions inject huge amounts of particles and volcanic gas into the atmosphere, causing serious natural pollution events at a regional scale. Human activities cause the emission of pollutants such as carbon dioxide, sulfur dioxide, nitrogen oxides, and particulates, which changes the lower atmosphere and thereby leads to global warming (see Chapter 19), acid rain (see Chapter 15), and pollution events.

Hazardous Earth Processes, Risk Assessment, and Perception

The possibility of another London smog crisis that might kill hundreds or thousands of people is small in developed countries today. In countries with serious air pollution problems, the possibility of a catastrophic air pollution event is much greater. On a global basis, the acid rain problem, which is damaging trees, lakes, buildings, and human health, remains a serious problem (see Chapter 15). Furthermore, since emissions of chemicals that cause acid rain have been accumulating in soils for decades, the problem will not soon go away.

Scientific Knowledge and Values

The science of air pollution and how to minimize it is a mature field of study. However, the economic costs of preventing or minimizing air pollution conflict with our desire to maintain a high-quality air environment for people and other living things. In the United States, people have repeatedly acknowledged that it is a fundamental right of everyone to breathe clean air, and they are willing to pay for the necessary technology to minimize air pollution.

Key Terms

air quality standards (p. 597)

air toxins (p. 580)

area sources (p. 580)

atmospheric inversion (p. 587)

fugitive sources (p. 580)

mobile sources (p. 580)

particulate-matter pollutants (p. 585)

photochemical smog (p. 590)

point sources (p. 580)

primary pollutants (p. 581)

scrubbing (p. 596)

secondary pollutants (p. 581)

stationary sources (p. 580)

sulfurous smog (p. 590)

Review Questions

1. What are the two major kinds of air pollution sources?

2. Differentiate between primary and secondary pollutants.

3. What is the major source for the anthropogenic component of sulfur dioxide in the atmosphere?

4. What are some of the adverse environmental effects associated with sulfur dioxide?

5. What are some of the reasons we are concerned with nitrogen oxides?

6. What is the main hazard from carbon monoxide?

7. Define ozone, and explain why we are concerned about ozone in the lower atmosphere.

8. What are volatile organic compounds?

9. What are the major concerns with hydrogen sulfide and hydrogen fluoride?

10. What was the major lesson learned from the Bhopal, India, incident?

11. Why were the 1997 fires in Indonesia an environmental disaster?

12. What is an atmospheric inversion, and how may it be produced?

13. What are the main factors that determine the potential for urban air pollution?

14. How is photochemical smog produced?

15. Why have emissions of sulfur dioxide declined in recent years?

16. Why are we so concerned about indoor air pollution?

17. What are the major methods used to control pollution from automobiles?

18. What are the major ways that we control emissions of sulfur dioxide?

19. What are air quality standards? Do you agree with the statement that as the degree of control of a pollutant increases, eventually a point is reached at which the cost of incremental control is greater than additional benefits?

20. The minimal total cost of controlling air pollution is a compromise between what two factors?

Critical Thinking Questions

1. Consider the case history of the London smog crisis in 1952. What lessons have been learned from it? Do you think they are relevant to the air pollution problems that are currently being faced in the world? Where might such events occur? Why?

2. From an environmental perspective, the best way to lower emissions of air pollutants is not to emit them in the first place. For your community, region, or state, develop a series of actions that could use principles of conservation or efficiency to help reduce real or potential air pollution problems.

3. How can identifying air toxins and their effects on human health lead to their emissions being better controlled?

4. One of the measures for helping reduce emissions from burning fossil fuels is allowing utilities to purchase allowances or credits for emissions. These may then be bought and sold on the open market. Because only a certain number of credits are available, this system limits the pollution emissions to the desired level. Some environmentalists do not like the idea because they believe it allows utilities to buy their way out of pollution problems. What do you think about this idea? List advantages and disadvantages of this system to try to help control emissions of sulfur dioxide.

PART FOUR

Environmental Management, Global Perspective, and Society

Part 4 focuses on environmental geology and its relationship to society on local, regional, and global scales. We begin in Chapter 19 with a discussion of global climate change and present important topics, including: (1) tools for studying global change, (2) Earth's atmosphere and climate change, (3) the potential effects of global climate change, and (4) the hypothesis that human activity causing emissions of greenhouse gases (primarily carbon dioxide) is resulting in global warming. There is no doubt that the global mean annual temperature has increased about 0.8°C during the twentieth century and that from the 1990s to 2005 we experienced some of the warmest years on record. We conclude that global warming is occurring, and there is a human component of the warming produced by burning fossil fuels.

The "capstone" for the fourth edition of *Introduction to Environmental Geology* is a discussion of relationships between geology, society, and the future. Most professional geologists interact with society through their work in areas such as engineering and environmental geology, petroleum geology, economic geology, groundwater geology, education, or research. Chapter 20 presents discussion of selected examples of how geology interacts with society, including environmental health, environmental impact analysis, land-use planning, and environmental law. A main conclusion is that if we are going to achieve sustainability, we need to carefully consider important links between geologic processes, the environment, and society. For example, we are just beginning to understand more fully the relationship between our geologic environment and human health and the incidence of chronic and acute diseases, including heart disease and cancer. The subjects of environmental impact analysis and land-use planning are important because our approach to these subjects reflects our commitment to our environment at all levels. The important field of environmental law is assisting us in problem solving and mediation rather than confrontation and adversarial positions when dealing with environmental matters. Finally, I present three important steps necessary to avoid a potential environmental crisis and how we might attain sustainability.

NINETEEN

Global Climate Change

Learning Objectives

It is often said that the only certainties in life are death and taxes, but change could also be added to this short list. We are all interested in changes that will affect us, our families, our society, and the world. The sorts of changes we might encounter are many—varying from gradual to accelerating, abrupt, chaotic, or surprising. Some changes will affect our local, regional, or global environment. One change that affects all of us is global climate change. In this chapter we will focus on the following learning objectives:

- Know the tools used for studying Earth system science and global change

- Understand climate change and global warming

- Know the important linkages associated with global warming

- Know some of the potential impacts of global warming and how they might be minimized

Viking Landing Artist's conception of Viking explorers discovering America near the end of the tenth century A.D. *(Library of Congress)*

CASE HISTORY | Eric the Red and Climate Change

Global climate change and in particular global warming is the hottest (pun intended) environmental issue today. People for about 700 years have experienced climate change, but the warming that is occurring today is unprecedented with respect to the rate of change.

The science concerning global warming is known, and warming as a result of human activity is occurring. Although there are very serious potential consequences, we need not be in a gloom-and-doom mind set. The science shows that we can slow down and stop global warming by taking appropriate action. With appropriate practical actions and policy we can achieve an atmosphere that is both cleaner and healthier for life on earth.[1] We will now back up a bit and discuss what happened to the Viking colonization of America as a result of climate change that was unrelated to human activity. Their story emphasizes natural change, which is also a component of global change and global warming.

The famous Viking explorer Eric the Red embarked on a voyage of exploration near the end of the tenth century A.D. His voyage coincided with a period of relatively warm climate that existed from about A.D. 1100 to 1300. During that period, known as the Medieval Warming Period, the Vikings colonized Iceland, Greenland, and northern North America. The sea temperature was probably about 4°C (7°F) warmer than it is now. The warmer seas helped warm the land, and grains, including corn, were grown for the first time in Greenland. Reportedly, a cousin of Eric the Red lived in a settlement in southwestern Greenland. While being visited by his famous cousin, he wanted to serve a feast, but he did not have a boat available to fetch an animal from a nearby island. Instead, he swam more than 3 km (1.9 mi) each way to collect a sheep and

bring it back to the settlement! Obviously, he was a very vigorous person, but he was probably not trained for long-distance swimming in cold water; it is nearly impossible to envision swimming under those conditions unless the water was at least 10°C (50°F). Today, late summer water temperatures in the fjords along that coast are probably closer to 6°C (42°F). Therefore, if Eric the Red's cousin did indeed swim to fetch the sheep, the water must have been at least 4°C (7°F) warmer than it is today. Of course, this story is not scientific evidence, but there are records reporting that generally good sailing conditions and relatively stable climate existed during the warm period. This situation changed by the early fourteenth century A.D. with the beginning of the Little Ice Age. The Little Ice Age was a period of colder, more difficult conditions for people in northern Europe. In fact, the fourteenth century was characterized by increasing climatic disturbances including severe storms, wet periods, drought, heat, and cold. The climatic change created treacherous sea conditions, severely restricting trade and commerce, and caused crop failures. During that same century, several episodes of famine and the spread of the Black Plague occurred, reaching Iceland in approximately A.D. 1400. The restricted trade, along with difficult agricultural conditions and disease, are believed to have eventually caused the abandonment of Viking settlements in North America and parts of Greenland.[2,3]

The story of the Viking settlements demonstrates that climatic change and accompanying effects can have dire consequences for people. A change in global climate that significantly affected our global food supply would be a catastrophe. Even a reduction of a few percent of the global harvest of grain would have serious implications.

19.1 Global Change and Earth System Science: An Overview

Preston Cloud, a famous Earth scientist interested in the history of life on Earth, human impact on the environment, and the use of resources, proposed two central goals of the Earth sciences.[4]

1. Understand how Earth works and how it has evolved from a landscape of barren rock to the complex landscape dominated by the life we see today

2. Apply that understanding to better manage our environment

Cloud's goals emphasize that our planet is characterized by a complex evolutionary history. Interactions between the atmosphere, the oceans, solid Earth, and the biosphere (Figure 19.1) have resulted in development of a complex and abundant diversity of landforms—continents, ocean basins, mountains, lakes, plains, and slopes—as well as the abundant and diverse life-forms that inhabit a broad spectrum of habitats.

Until recently it was thought that human activity caused only local or, at most, regional environmental change. It is now generally recognized that the effects of human activity on Earth are so extensive that we are actually involved in an unplanned planetary experiment. To recognize and perhaps modify the changes we have initiated, we need to understand how the entire Earth works as a system. The

Figure 19.1 Earth from space
Showing the blue of the oceans, the land and biosphere of Africa, and the dynamic clouds of the atmosphere, all interacting in the Earth system. *(National Aeronautics and Space Administration)*

emerging discipline, called **Earth system science,** seeks to further this understanding by learning how the various components of the system—the atmosphere, oceans, land, and biosphere—are linked on a global scale and interact to affect life on Earth.[5]

19.2 Tools for Studying Global Change

The Geologic Record

Sediments deposited on floodplains or in lakes, bogs, glaciers, or the ocean may be compared to pages of a history book. Organic material that is often deposited with sediment may be dated by a variety of methods to provide a chronology. In addition, the organic material can tell a story concerning the past climate, life-forms in the area, and environmental changes that have taken place.[6]

One of the more interesting uses of the geologic record has been the examination of glacial ice (see Section 19.4). Glacial ice contains trapped air bubbles that may be analyzed to provide information concerning atmospheric carbon dioxide (CO_2) concentrations when the ice formed. These trapped air bubbles are atmospheric time capsules from the past and have been used to analyze the carbon dioxide content of air as old as 160,000 years.[7] Glaciers also contain a record of heavy metals such as lead that settle out of the atmosphere as well as a variety of other chemicals that can be used to study recent Earth history.

Real-Time Monitoring

Monitoring is the regular collection of data for a specific purpose; real-time monitoring refers to collecting these data while a process is actually occurring. For example, we often monitor the flow of water in rivers to evaluate water resources or flood hazard. In a similar way, samples of atmospheric gases can help establish trends or changes in the composition of the atmosphere; measurements of temperature and the composition of the ocean are also used to examine changes within them. Gathering of real-time data is necessary for testing models and for calibrating the extended prehistoric record derived from geologic data.

Methods of monitoring vary with the subject being measured. For example, the impacts from mining may be monitored by evaluating remotely sensed data

Figure 19.2 **Modeling climate**
Idealized diagram illustrating the cells used in Global Circulation Models (GCMs).

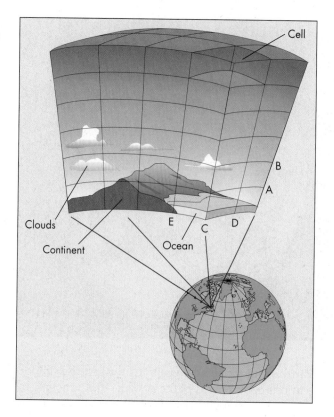

collected by satellite or high-altitude aerial photographs. However, the most reliable data are often derived from ground measurements that establish the validity of the airborne or satellite measurements.

Mathematical Models

Mathematical models use numerical means to represent real-world phenomena and the linkages and interactions between the processes involved. Mathematical models have been developed to predict the flow of surface water and groundwater, erosion and deposition of sediment in river systems, ocean circulation, and atmospheric circulation.

The global change models that have gained the most attention are the *Global Circulation Models (GCMs)*. These models predict changes in atmospheric circulation at the global scale.[8] Data used in the calculations are arranged into large cells that represent several degrees of latitude and longitude; typical cells represent an area about the size of Oregon (Figure 19.2). In addition, there are usually 6 to 20 levels of vertical cells representing the lower atmosphere. Calculations involving equations for major atmospheric processes are then used to make predictions.

Global circulation models must be viewed as a first approach used to solve very complex problems. Despite this limitation, GCMs provide information necessary for evaluating Earth as a system and point out the additional data necessary to develop better models in the future. The models do predict the regions that are likely to be relatively wetter or drier on the basis of specific changes in the atmosphere.

19.3 Earth's Atmosphere and Climate Change

To a great extent, the study of global change is the study of changes in the atmosphere and linkages between the atmosphere, lithosphere, hydrosphere, and biosphere. We define **climate** as the characteristic atmospheric conditions—that is,

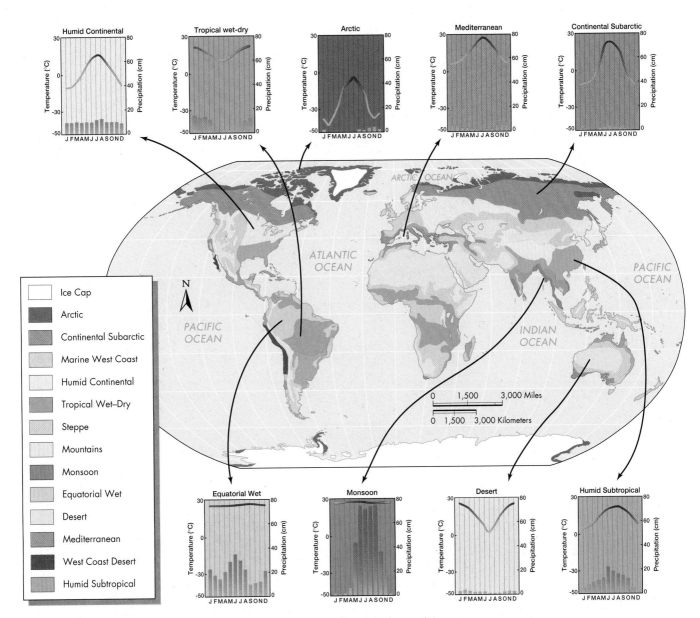

Figure 19.3 **Climates of the world** Characteristic temperature and precipitation conditions. Temperature is represented by the red line, and precipitation is shown as bars. *(Modified after Marsh, W. M., and Dozier, J. Landscapes: An introduction to physical geography. Copyright © 1981. John Wiley & Sons)*

the *weather*—at a particular place or region over time periods from seasons to years to decades. The climate at a particular location may be complex and consist of more than average precipitation and temperature. For example, it may be dependent on infrequent or extreme seasonal patterns such as rain in the monsoon season of parts of India. The major climatic zones on Earth are shown in Figure 19.3.

Global circulation and movement of air masses in the atmosphere (Figure 19.4a) produce the major climatic zones. Warm tropical air near the equator rises and moves north and south descending in the mid-latitudes (sometimes producing deserts). The air then rises again at higher latitudes and finally descends at the poles. Atmospheric conditions from the equator to the poles are idealized in Figure 19.4b.

The lower, active part of the atmosphere, where weather occurs, is the troposphere. The structure of the atmosphere is shown in Figure 19.5. Air temperature and concentration of oxygen decrease with altitude in the troposphere. At the top of the troposphere, at an altitude that varies from about 18 km at the tropics to 7 km at the poles, it is very cold and the temperature remains nearly constant for a few kilometers

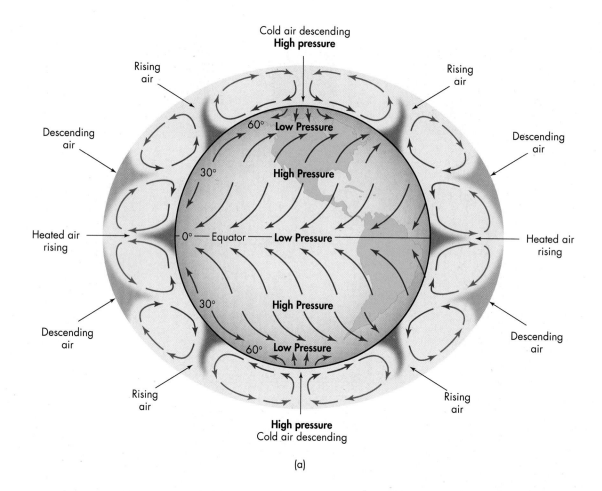

(a)

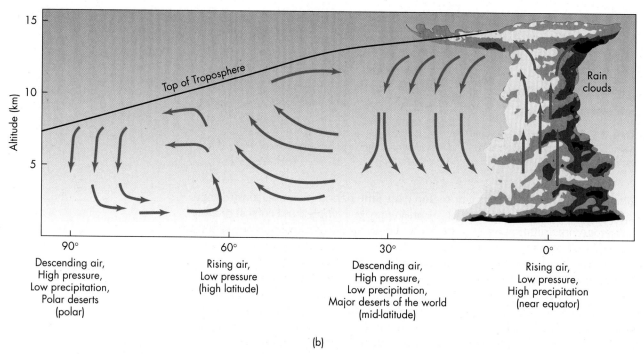

(b)

Figure 19.4 Atmospheric circulation (a) Global circulation of the lower atmosphere showing zones of rising and descending air masses. (b) Zones of rising and descending air masses from the equator to the poles. *(Keller, E. A., and Blodgett, R. H. 2006.* Natural hazards. *Upper Saddle River, NJ: Pearson Prentice Hall)*

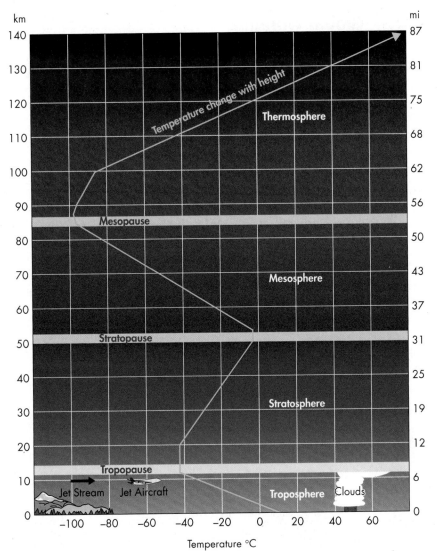

km

mi

Figure 19.5 Structure of the atmosphere Atmospheric layers and temperature profile. The weather occurs in the troposphere below about 11 km but varies from 7 km at the poles to 18 km in the tropics. *(NOAA)*

through the tropopause. The constant temperature with little air movement places a lid on the active lower atmosphere (troposphere). Temperature then increases in the stratosphere only to decline again in the mesosphere. Nearly all (99 percent) of the atmosphere by weight is below an altitude of about 30 km (20 mi).

The Atmosphere

Our atmosphere can be thought of as a complex chemical factory with many little-understood reactions taking place within it. Many of the reactions that take place are strongly influenced by both sunlight and the compounds produced by life. The air we breathe is a mixture of nitrogen (N_2) (78 percent), oxygen (O_2) (21 percent), argon (Ar) (0.9 percent), carbon dioxide (CO_2) (0.03 percent), and other trace elements (less than 0.07 percent). It also contains compounds such as methane, ozone, carbon monoxide, oxides of nitrogen and sulfur, hydrogen sulfide, hydrocarbons, and various particulates, many of which are common air pollutants. The most variable part of the atmosphere's composition is water vapor (H_2O), which can range from approximately 0 percent to 4 percent by volume in the lower atmosphere.[3]

With this brief introduction to the atmosphere and climate, we will now discuss human-induced global warming in terms of: the greenouse effect; history and process of global temperature change; and potential consequences.

19.4 **Global Warming**

A growing volume of evidence suggests that we are now in a period of global warming resulting from burning vast amounts of fossil fuels. Does this mean we are experiencing human-induced global warming? Many scientists now believe that human processes, as well as natural ones, are significantly contributing to global warming.

The Greenhouse Effect

For the most part, the temperature of Earth is determined by three factors: the amount of sunlight Earth receives, the amount of sunlight Earth reflects, and therefore does not absorb, and atmospheric retention of reradiated heat.[8] Figure 19.6 shows the basics of Earth's energy balance, which represents the equilibrium between incoming and outgoing energy. Earth receives energy from the Sun in the form of electromagnetic radiation. The various types of electromagnetic energy are shown in Figure 19.7. Radiation from the Sun is relatively short wave, mostly visible, whereas Earth radiates relatively long-wave infrared radiation (Figure 19.8). The hotter an object, whether it is the Sun, Earth, rock, or lake, the more electromagnetic energy it emits. The Sun, with a surface temperature of 5,800°C (10,472°F), radiates much more energy per unit area than Earth, with an average surface temperature of 15°C (59°F).

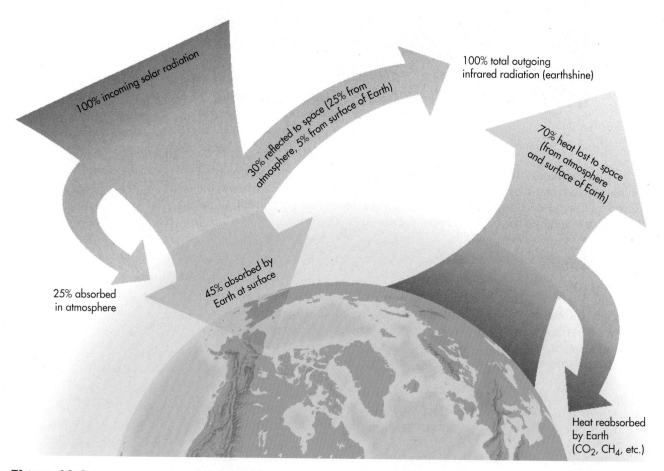

Figure 19.6 **Annual energy flow to Earth from the Sun** The relatively small component of heat from Earth's interior to the near-surface environment is also shown. *(Modified after Pruitt, N. L., Underwood, L. S., and Surver, W. 1999. BioInquiry, Learning system 1.0, making connections in biology. John Wiley & Sons, Hoboken, NJ)*

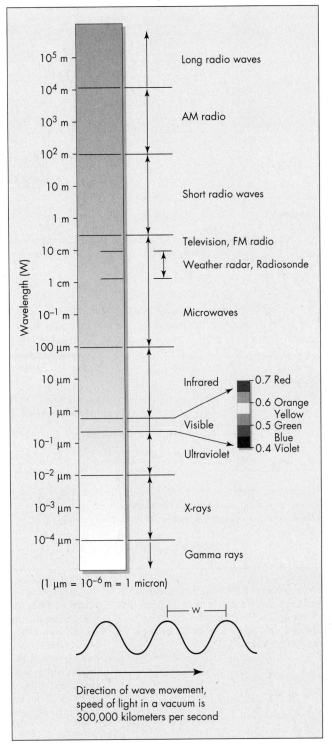

Figure 19.7 **The electromagnetic spectrum** The range in wavelength (W), distance between one wave crest to the next, is enormous, from millionths of a meter for X-rays and gamma rays to hundreds of thousands of meters for long radio waves. Waves travel at the speed of light in a vacuum—300,000 km (186,411 mi) per second.

Absorbed solar energy warms Earth's atmosphere and surface, which then reradiate the energy back into space as infrared radiation.[8] Water vapor and several other atmospheric gases—including carbon dioxide (CO_2), methane (CH_4), and chlorofluorocarbons (CFCs), human-made chemicals used in air conditioners and refrigerators—tend to trap heat. That is, they absorb some of the energy radiating from Earth's surface and are thereby warmed. As a result, the lower atmosphere of Earth is much warmer than it would be if all of its radiation escaped into space without this intermediate absorption and warming. This effect is somewhat analogous to the trapping of heat by a greenhouse and is therefore referred to as the **greenhouse effect** (Figure 19.9).

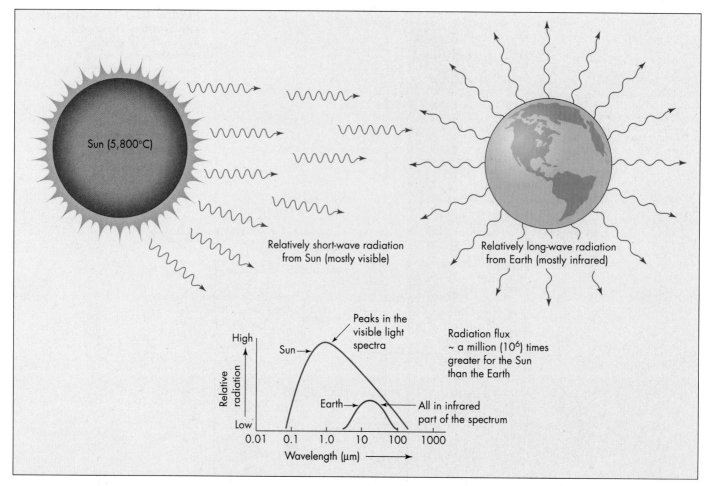

Relatively short-wave radiation from Sun (mostly visible)

Relatively long-wave radiation from Earth (mostly infrared)

Radiation flux ~ a million (10^6) times greater for the Sun than the Earth

All in infrared part of the spectrum

Figure 19.8 Earth-Sun Idealized diagram comparing the emission of energy from the Sun with that from Earth. Notice that the solar emissions have a relatively short wavelength, whereas those from Earth have a relatively long wavelength. *(Modified after Marsh, W. M., and Dozier, J. Landscapes: An introduction to physical geography. Copyright © 1981. John Wiley & Sons)*

Figure 19.9 Greenhouse effect
Idealized diagram showing the green-house effect. Incoming visible solar radiation (sunshine) is absorbed by Earth's surface, warming it. Infrared radiation is then emitted at the surface of Earth as earthshine to the atmosphere and outer space. Most of the infrared radiation emitted from Earth is absorbed by the atmosphere, heating it and maintaining the greenhouse effect. *(Developed by M. S. Manalis and E. A. Keller, 1990)*

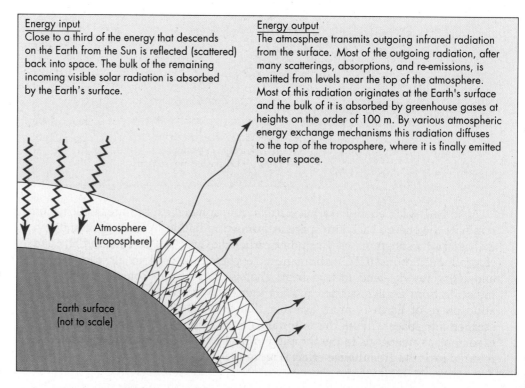

Energy input
Close to a third of the energy that descends on the Earth from the Sun is reflected (scattered) back into space. The bulk of the remaining incoming visible solar radiation is absorbed by the Earth's surface.

Energy output
The atmosphere transmits outgoing infrared radiation from the surface. Most of the outgoing radiation, after many scatterings, absorptions, and re-emissions, is emitted from levels near the top of the atmosphere. Most of this radiation originates at the Earth's surface and the bulk of it is absorbed by greenhouse gases at heights on the order of 100 m. By various atmospheric energy exchange mechanisms this radiation diffuses to the top of the troposphere, where it is finally emitted to outer space.

Atmosphere (troposphere)

Earth surface (not to scale)

TABLE 19.1   Rate of Increase and Relative Contribution of Several Gases to the Anthropogenic Greenhouse Effect

	Rate of Increase (% per year)	Relative Contribution (%)
CO_2	0.5	60
CH_4	<1	15
N_2O	0.2	5
O_3^*	0.5	8
CFC-11	4	4
CFC-12	4	8

*In the troposphere.

Source: Data from Rodhe, H. 1990. A comparison of the contribution of various gases to the greenhouse effect. *Science* 248:1218, table 2. Copyright 1990 by the AAAS.

It is important to understand that the greenhouse effect is a natural phenomenon that has been occurring for millions of years on Earth as well as other planets in our solar system. Without heat trapped in the atmosphere, Earth would be much colder than it is now, and all surface water would be frozen.[9] Most of the natural "greenhouse warming" is due to water vapor and small particles of water in the atmosphere. However, potential global warming due to human activity is related to carbon dioxide, methane, nitrogen oxides, and chlorofluorocarbons. In recent years, the atmospheric concentrations of these gases and others have been increasing because of human activities. These gases tend to absorb infrared radiation from Earth, and it has been hypothesized that Earth is warming because of the increases in the amounts of these so-called greenhouse gases. Table 19.1 shows the rate of increase of these atmospheric gases due to human-induced emissions and their relative contribution to the *anthropogenic,* or human-caused, component of the greenhouse effect. Notice that carbon dioxide produces 60 percent of the relative contribution.

Measurements of carbon dioxide trapped in air bubbles of the Antarctic ice sheet suggest that during most of the past 160,000 years the atmospheric concentration of carbon dioxide has varied from a little less than 200 ppm to about 300 ppm.[7] The highest levels are recorded during major interglacial periods that occurred approximately 125,000 years ago and at the present (Figure 19.10a). At the beginning of the Industrial Revolution, the atmospheric concentration of carbon dioxide was approximately 280 ppm. Since 1860, fossil fuel burning has contributed to the exponential growth of the concentration of carbon dioxide in the atmosphere. The change from approximately 1500 to 2000 is shown in Figure 19.10b. Data for before the mid-twentieth century are from measurements made from air bubbles trapped in glacial ice. The concentration of carbon dioxide in the atmosphere today exceeds 370 ppm, and it is predicted to reach at least 450 ppm—more than 1.5 times the preindustrial level—by the year 2050.[10] Changes in atmospheric concentration of carbon dioxide from the mid-twentieth century to today at Mauna Loa, Hawaii, are shown in Figure 19.10c. The annual cycles with a lower concentration of CO_2 during the summer are due to the summer growing season in the northern hemisphere, when plants extract more carbon dioxide from the atmosphere by photosynthesis.

Global Temperature Change

The Pleistocene ice ages began approximately 2 million years ago, and since then there have been numerous changes in Earth's mean annual temperature.[3]

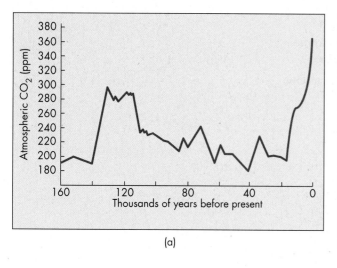

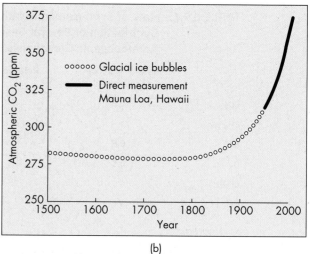

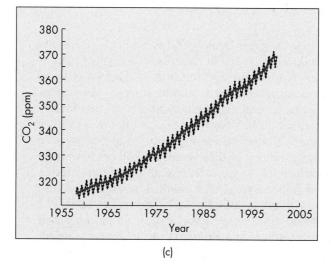

Figure 19.10 Carbon dioxide in the atmosphere (a) Concentration of atmospheric carbon dioxide for the past 160,000 years, based on evidence from Antarctica. *(Data from Schneider, S. H. 1989. The changing climate. Scientific American 261:74)* (b) Average concentration of atmospheric carbon dioxide from 1500 to 2000. *(Data in part from Post, W. M., et al. 1990, The global carbon cycle, American Scientist 78[4]:210–26)* (c) Atmospheric concentration of carbon dioxide at Mauna Loa, Hawaii. *(Source data from Scripps Institute of Oceanography, NOAA, and C. D. Keeling, accessed 1/24/04 at www.mlo.noaa.gov)*

Figure 19.11 shows the changes of approximately the past million years on several time scales. The first scale shows the entire million years (Figure 19.11a), during which there have been major climatic changes involving swings of several degrees Celsius in mean temperature. Low temperatures have coincided with major glacial events that have greatly altered the landscape; high temperatures are associated with interglacial conditions. Interglacial and glacial events become increasingly prominent in the scales showing changes over 150,000 and 18,000 years. The last major interglacial warm period, even warmer than today, was the Eemian (Figure 19.11b). During the Eemian, sea level was a few meters higher than it is today. The cold period that occurred about 11,500 years ago is known as the Younger Dryas, followed by rapid warming to the Holocene maximum, which preceded the Little Ice Age (Figure 19.11c).

A scale of 1,000 years shows several warming and cooling trends that have affected people (Figure 19.11d). For example, a major warming trend from approximately A.D. 1100–1300 allowed the Vikings to colonize Iceland, Greenland, and northern North America. When glaciers made a minor advance around A.D. 1400,

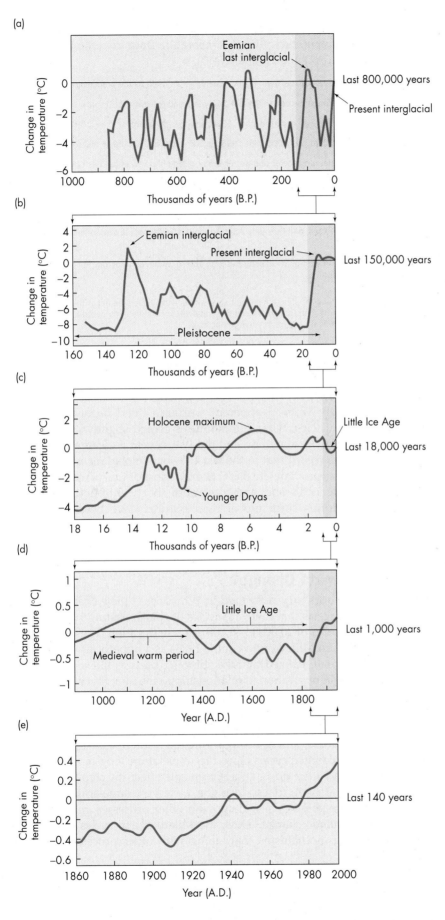

(a)

(b)

(c)

(d)

(e)

Figure 19.11 **Changes in global temperature** Change in temperature over different periods of time during the last million years. *(Modified after University Corporation for Atmospheric Research, Office for Interdisciplinary Studies. 1991. Science capsule, Changes in time in the temperature of the earth. EarthQuest 5[1]; and the UK Meteorological Office. 1997. Climate change and its impacts: A global perspective)*

TABLE 19.2   **Evidence Supporting the Late Twentieth-Century Rise in Global Temperature: Global Temperature Data from the United States (NOAA) and Europe (WMO)**

Warming since the mid-1970s has been about three times as rapid as the preceding century.

The 1990s was the warmest decade in the last 142 years and the last 1,000 years, from geologic data.

The 10 warmest years have all occurred since 1990, and the five warmest since 1997.

The warmest year on record was 2005 with 1998 second and 2002 and 2004 tied for third.

In 2003 the United States was cooler and wetter than average in much of the eastern part of the country, and warmer and drier in much of the west. Ten western states were much warmer than average; New Mexico had its warmest year. Alaska was warmer in all four seasons and had one of the 5 warmest years since Alaska began taking measurements in 1918.

Europe in 2003 experienced summer heat waves with the warmest seasonal temperatures ever recorded in Spain, France, Switzerland, and Germany. Approximately 15,000 people died in heat waves in Paris during the summer.

Warm conditions with drought in 2003 contributed to severe wildfires in Austrlia, southern California, and British Columbia, Canada.

Note: A few years of high temperatures with drought, heat waves, and wildfires are not by themselves an indication of longer-term global warming. The persistent trend of increasing temperatures over several decades is more compelling evidence that global warming is real and happening.

during a cold period known as the Little Ice Age, the Viking settlements in North America and parts of Greenland were abandoned (see Case History: Eric the Red and Climate Change, at the beginning of this chapter).

In approximately 1750 an apparent warming trend began that lasted until approximately the 1940s, when temperatures cooled slightly.[3] Over the last 140 years, more changes are apparent, and the 1940s event is clearer (Figure 19.11e). It is evident from the record that in the last 100 years global mean annual temperature increased by approximately 0.8°C (1.4°F). Most of the increase has been since the 1970s, and the 1990s and first 5 years of the twenty-first century had the warmest temperatures since global temperatures have been monitored. This period is known as the late-twentieth century increase in global temperature.[11] Table 19.2 lists evidence to support the recent warming.

Why Does Climate Change?

The question that begs to be answered is: Why does climate change? Examination of Figure 19.11 suggests that there are cycles of change lasting 100,000 years separated by shorter cycles of 20,000 to 40,000 years in duration. These cycles were first identified by Milutin Milankovitch in the 1920s as a hypothesis to explain climate change. Milankovitch realized that the spinning Earth is like a wobbling top unable to keep a constant position in relationship to the Sun; this instability partially determines the amount of sunlight reaching and warming Earth. He discovered that variability in Earth's orbit around the Sun follows a 100,000 year cycle that is correlated with the major glacial and interglacial periods of Figure 19.11a. Cycles of approximately 40,000 and 20,000 years are the result of changes in the wobble of Earth's axis. Milankovitch cycles reproduce most of the long-term cycles observed in the climate, and they do have a significant effect on climate. However, the cycles are not sufficient to produce the observed large-scale global climatic changes. Therefore, these cycles, along with other processes, must be invoked to explain global climatic change. Thus, the Milankovitch cycles can be looked at as natural forcing mechanisms that, along with other processes, may produce climatic change.[12]

We now believe that our climate system may be inherently unstable and capable of changing quickly from one state to another in as short a time as a few decades.[13] Part of what may drive the climate system and its potential to change is the *ocean conveyor belt,* a global-scale circulation of ocean waters characterized by

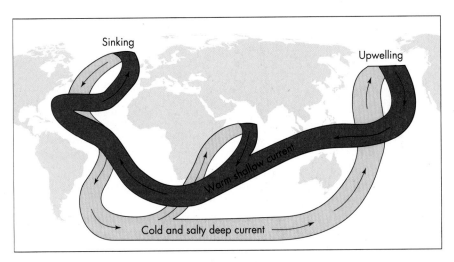

Figure 19.12 Ocean conveyor belt Idealized diagram of the ocean conveyor belt. The actual system is more complex, but, in general, warm surface water (red) is transported westward and northward (increasing in salinity owing to evaporation) to near Greenland, where it cools from contact with cold Canadian air. As the water increases in density, it sinks to the bottom and flows south, then east to the Pacific, where upwelling occurs. The masses of sinking and upwelling waters balance, and the total flow rate is about 20 million cubic meters (700 million cubic feet) per second. The heat released to the atmosphere from the warm water helps keep northern Europe warmer than it would be if the oceanic conveyor belt were not present. *(Modified after Broecker, W. 1997. Will our ride into the greenhouse future be a smooth one? Geology Today 7[5]:2–6)*

strong northward movement of 12° to 13°C (53° to 55°F) near-surface waters in the Atlantic Ocean that are cooled to 2° to 4°C (35° to 39°F) when they arrive near Greenland (Figure 19.12).[13] As the water cools it becomes saltier; the salinity increases the water's density and causes it to sink to the bottom. The current then flows southward around Africa, adjoining the global pattern of ocean currents. The flow in this conveyor belt current is huge, equal to about 100 Amazon rivers. The amount of warm water and heat released to the atmosphere along with the stronger effect of relatively warm winter air moving east and northeast across the Atlantic Ocean is sufficient to keep northern Europe 5° to 10°C (8.5° to 17°F) warmer than it would otherwise be. If the conveyor belt were to shut down it would have an effect on the climate of Europe. However, the effect would not be catastrophic to England and France by producing extreme cold and ice bound conditions.[14]

Although scientific uncertainties exist, there is sufficient evidence to state that (1) there is a discernible human influence on global climate; (2) warming is now occurring; and (3) the mean surface temperature of Earth will likely increase by from 1.5° to 4.5°C (2.6° to 7.8°F) during the twenty-first century.[1,15] The human-induced component of global warming results from increased emissions of gases that tend to trap heat in the atmosphere. There is good reason to argue that increases in carbon dioxide and other greenhouse gases are related to an increase in the mean global temperature of Earth. Over the past 160,000 years there has been a strong correlation between the concentration of atmospheric CO_2 and global temperature (Figure 19.13). When CO_2 has been high, temperature has also been high, and, conversely, low concentrations of CO_2 have been correlated with a low global temperature. However, in order to better understand global warming, we need to consider major **forcing** variables that influence global warming, including solar emission, volcanic eruption, and anthropogenic input.

Solar Forcing

Since the Sun is responsible for heating Earth, solar variation should be evaluated as a possible cause of climate change. When we examine the history of climate

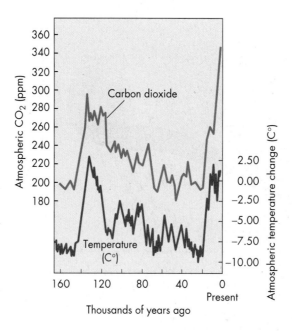

during the past 1,000 years, the variability of solar energy plays a role. Examination of the solar record reveals that the Medieval Warm Period (A.D. 1000–1300)
corresponds with a time of increased solar radiation, comparable to that which
we see today. Evaluation of the record also suggests that minimum solar activity
occurred during the fourteenth century, coincident with the beginning of the
Little Ice Age (Figure 19.11d). Therefore, it appears that variability of the input
of solar energy to Earth can partially explain climatic variability during the past
1,000 years. However, the effect is relatively small, only 0.25 percent; that is, the
difference between the solar forcing from the Medieval Warm Period to the Little
Ice Age is only a fraction of 1 percent.[11]

Volcanic Forcing

Upon eruption, volcanoes can hurl vast amounts of particulate matter, known as
aerosols, into the atmosphere. The aerosol particles reflect a significant amount of
sunlight and produce a net cooling that may offset much of the global warming
expected from the anthropogenic greenhouse effect.[16] For example, increased
atmospheric aerosols over the United States (from air pollution) have probably
reduced mean temperatures roughly 1°C (1.7°F) cooler than they would be otherwise. Aerosol particle cooling may thus help explain the disparity between model
simulations of global warming and actual recorded temperatures that are lower
than those predicted by models.[17]

Volcanic eruptions add uncertainty in predicting global temperatures. For
instance, what was the cooling effect of the 1991 Mt. Pinatubo eruption in the
Philippines (Figure 19.14)? Tremendous explosions sent volcanic ash to elevations of 30 km (19 mi) into the stratosphere, and, as with similar past events, the
aerosol cloud of ash and sulfur dioxide remained in the atmosphere, circling
Earth, for several years. The particles of ash and sulfur dioxide scattered incoming solar radiation, resulting in a slight cooling of the global climate during
1991 and 1992. Calculations suggest that aerosol additions to the atmosphere
from the Mt. Pinatubo eruption counterbalanced the warming effects of greenhouse gas additions through 1992. However, by 1994, most aerosols from the
eruption had fallen out of the atmosphere, and global temperatures returned to
previous higher levels.[18] Volcanic forcing from pulses of volcanic eruptions is
believed to have significantly contributed to the cooling of the Little Ice Age (see
Figure 19.11d).[11]

Figure 19.14 **Volcanic eruption cools Earth** The eruption of Mt. Pinatubo in the Philippines in 1991 injected vast amounts of volcanic ash and sulfur dioxide up to about 30 km (19 mi) into the atmosphere. *(T. J. Casadevall/U.S. Geological Survey)*

Anthropogenic Forcing

A recent study considered climate forcing over the past 1,000 years, allowing the late-twentieth century warming to be placed within an historical context. A mathematical model of climate forcing was developed that removed major natural forcing mechanisms, allowing anthropogenic forcing to be isolated. The study established that the level of natural variability in the climate system in the preceding 1,000 years cannot explain the warming that occurred at the end of the twentieth century and continues into the early twenty-first century. In other words, present warming greatly exceeds natural variability and closely agrees with the responses predicted from models of greenhouse gas forcing.[11] Results of the modeling in a general sense are shown in Figure 19.15. Notice that the late twentieth-century rise

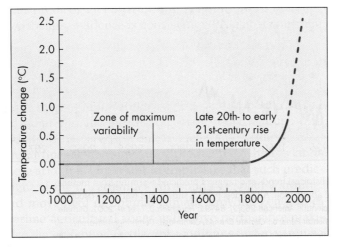

Figure 19.15 **Late twentieth-century rise in temperature** Global temperature change during the past 1,000 years. Solar and volcanic forcing have been removed, leaving anthropogenic forcing from greenhouse gases. *(Modified after Crowley, T. J. 2000. Causes of climate change over the past 1000 years.* Science *289: 270–77)*

A CLOSER LOOK | Glaciers

A **glacier** is a land-bound mass of moving ice. It is composed of glacial ice formed by transformations that take several years to several hundred years. Glacial ice begins as loosely packed snow and transforms to a more granular but still quite porous ice. Compaction from the weight of overlying snow and ice and recrystallization of snow to ice eventually form dense, blue glacial ice. Glacial ice has a density of approximately 0.9 g/cm³ (0.03 lb/in.³), which is approximately 150 times the density of new-fallen snow. Therefore, in order for glaciers to form, fallen snow must survive several years and accumulate so that the transformation to glacial ice can occur.

Glaciers that cover large tracts of land are called continental glaciers, ice sheets, or ice caps; those confined to mountain valleys at high altitudes or high latitudes are called mountain, or alpine glaciers (Figure 19.A).

Major glacial events, each lasting several million years, have occurred occasionally over the last billion years. We are now living in one of these glacial periods; during the past 1.65 million years, glaciers have repeatedly advanced (grown) and retreated (melted) across much of the landscape. Only a few thousand years ago, the most recent continental glaciers retreated from the Great Lakes region of the United States. Figure 19.B shows the maximum extent of these ice sheets. During the Ice Age of the Pleistocene epoch, glaciers covered as much as 30 percent of the land area of Earth, including the present sites of major cities such as New York, Toronto, and Chicago. During this time, maximum development of continental glaciers occurred; the most recent glacial maxima occurred about 18,000 years ago. Since large quantities of fresh water were stored in the ice, global sea level was more than 120 m (396 ft) lower than it is today. Today, glacial ice covers only about 10 percent of Earth's land. Nearly all of this ice is located in the Antarctic ice sheet, with lesser amounts in the Greenland ice sheet and in mountain glaciers of places such as Alaska, southern Norway, the Alps, and New Zealand.

We are presently in an interlude, a warm time known as an *interglacial*, when the abundance of glacial ice is relatively low. Nevertheless, we are probably still in a glacial period and ice sheets may again advance in the future.

Causes of Glaciation

The reasons that major global glacial events occurred several times during the past 1 billion years are unknown. Perhaps these events are related to the position of the continents, which significantly affects ocean circulation and global climate. On the assumption that we are still in the Pleistocene glaciation period, the amount of solar radiation reaching Earth's surface has regularly changed by a few percent. These small, cyclic variations at about 100,000, 40,000, and 20,000 years result from both regular changes in Earth's orbit and the orientation of Earth's axis of rotation.[12] These cyclic changes (known as Milankovitch cycles) are fairly well correlated with major and minor advances and retreats of glacial ice. However, correlation is not proof of causation. The cycles cause only small variations of the amount of solar radiation reaching Earth. They cannot be the main cause and certainly not the only cause that enables continental glaciers to advance or retreat.

Figure 19.A Mountain (valley) glacier Confluence of Muldrow and Traleika glaciers below Mount Denali, Alaska. The dark material on the glaciers is sediment being carried on the top of the ice. The fractures in the white glacier ice are crevasses. *(Michael Collier)*

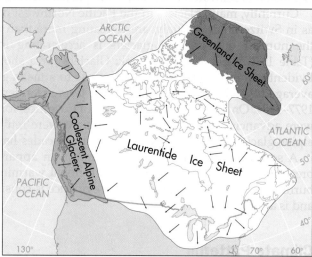

Figure 19.B Ice sheets Maximum extent of ice sheets during the Pleistocene glaciation. *(From Foster, R. J. 1983. General geology, 4th ed. Columbus, OH: Charles E. Merrill)*

Figure 19.14 Volcanic eruption cools Earth The eruption of Mt. Pinatubo in the Philippines in 1991 injected vast amounts of volcanic ash and sulfur dioxide up to about 30 km (19 mi) into the atmosphere. *(T. J. Casadevall/U.S. Geological Survey)*

Anthropogenic Forcing

A recent study considered climate forcing over the past 1,000 years, allowing the late-twentieth century warming to be placed within an historical context. A mathematical model of climate forcing was developed that removed major natural forcing mechanisms, allowing anthropogenic forcing to be isolated. The study established that the level of natural variability in the climate system in the preceding 1,000 years cannot explain the warming that occurred at the end of the twentieth century and continues into the early twenty-first century. In other words, present warming greatly exceeds natural variability and closely agrees with the responses predicted from models of greenhouse gas forcing.[11] Results of the modeling in a general sense are shown in Figure 19.15. Notice that the late twentieth-century rise

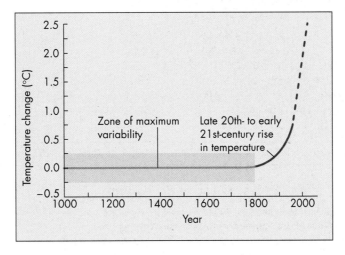

Figure 19.15 Late twentieth-century rise in temperature Global temperature change during the past 1,000 years. Solar and volcanic forcing have been removed, leaving anthropogenic forcing from greenhouse gases. *(Modified after Crowley, T. J. 2000. Causes of climate change over the past 1000 years. Science 289: 270–77)*

in temperature greatly exceeds the zone of maximum variability (change, in this case warming) that could have resulted from natural forcing mechanisms such as solar variability and volcanic activity. The projected level of twenty-first-century global warming is also shown, assuming an approximate 2.0°C (3.4°F) increase in temperature due to the doubling of carbon dioxide. Thus, the major scientific issues concerning global warming have apparently been solved. Significant global warming as a result of human activity is occurring. We will now discuss some of the effects and potential adjustments to global warming. Ultimately, the decisions made to adjust to global warming will reflect our values.

Human processes are also causing a slight cooling. Reflection from air pollution particles has reduced incoming solar energy by as much as 10 percent. This is termed **global dimming,** and may be offsetting up to 50 percent of the expected warming due to greenhouse gases.

19.5 Potential Effects of Global Climate Change

Our discussion of global warming can be summarized as follows: (1) Human activity is increasing the concentration of greenhouse gases in the atmosphere; (2) the mean temperature of Earth increased by about 0.8°C (1.4°F) in the past 100 years; and (3) a significant portion of the observed increase in mean temperature of Earth results from human activity.

Figure 19.16 shows the potential range in predicted global warming derived from computer modeling. All of the models are fairly reliable in predicting that warming will occur and possibly accelerate in the coming decades. Therefore, we need to carefully examine the potential effects of such warming.

If the greenhouse gases double in the future, it is estimated that the average global temperature will rise about 1.5° to 4.5°C (2.6° to 7.8°F), with significantly

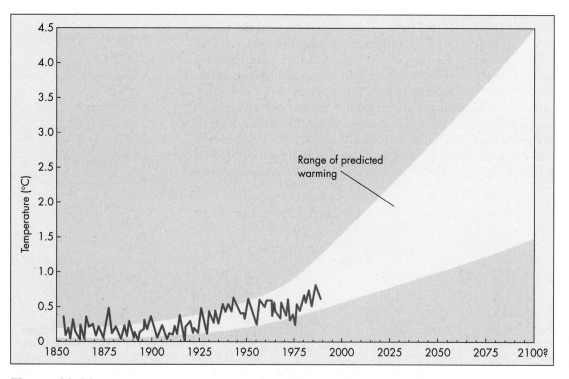

Figure 19.16 Predicted warming Historical trends and range of predicted warming from mathematical models for a doubling of CO_2 since about 1850. All models predict substantial warming. *(Modified by John Deecken from Global warming trends. Scientific American 263[2]: 84–91; Houghton, J. T., et al. 2001. Climate change 2001: The scientific basis. Intergovernmental Panel on Climate Change. Cambridge: Cambridge University Press; and Hansen, J. 2003. Can we defuse the global warming time bomb? Natural Science. www.naturalscience.com)*

TABLE 19.3 Number of Glaciers That Changed, by Type of Change 1967–1995[*]

	Advancing	Stationary	Retreating
Northern Cascades			
1967	7	8	7
1974	9	0	2
1985	5	10	32
1990	1	5	41
1995	0	0	47
Switzerland			
1967	31	14	55
1975	70	3	28
1986	42	9	13
1989	19	3	83
1993	6	0	73
Italy			
1981	25	10	10
1985	25	6	14
1988	26	13	92
1990	9	9	123
1993	6	8	127

[*]Note: The numbers of glaciers observed varied during the period of observation.

Source: Data from Pelto, M. S. 1993. Changes in water supply in alpine regions due to glacier retreat. American Institute of Physics, *Proceedings,* Vol. 277: *The World at Risk: Natural Hazards and Climate Change,* pp. 61–67; and Pelto, M. S. 1996. Recent changes in glacier and alpine runoff in the North Cascades, Washington. *Hydrological Processes* 10:1173–80.

greater warming at the polar regions.[15] Specific effects of this temperature rise in a specific region are difficult to predict, but two of the main possibilities being considered are a change in the global climate pattern and a rise in sea level due to expansion of seawater as it warms and partial melting of glacial ice.

Currently, many glaciers in the Pacific Northwest of the United States, as well as in Switzerland and Italy, are retreating more than advancing; the increase in the number of retreating glaciers is accelerating in the Cascades and Italy (Table 19.3).[19] See A Closer Look: Glaciers for a short introduction to glaciers. Evidently, this acceleration is in response to a mean global temperature that has averaged 0.4°C (0.68°F) above the long-term mean temperature during the years 1977–1994. On Mt. Baker in the Northern Cascades, for example, all eight glaciers were advancing in 1976. By 1990, all eight were retreating. In addition, 4 of 47 alpine glaciers observed in the Northern Cascades have disappeared since 1984.

A series of particularly warm years, or an apparent increase in melting or breakup of glacial ice, is not proof of global warming resulting from humans' burning of fossil fuels. Nevertheless, evidence is consistent with global warming and is a source of concern.

Climate Patterns

A global rise in temperature might significantly change rainfall patterns, soil moisture relationships, and other climatic factors important to agriculture. (See A Closer Look: Desertification.) It has been predicted that some northern areas such as Canada and Eastern Europe may become more productive, whereas lands to the south will become more arid. It is important to emphasize that such predictions are very difficult in light of the uncertainties surrounding global warming. Furthermore, the northward movement of optimal climatic growing zones does not necessarily mean that prime agricultural zones will move north, since maximum grain production is also dependent on fertile soil conditions that may not be

A CLOSER LOOK | Glaciers

A **glacier** is a land-bound mass of moving ice. It is composed of glacial ice formed by transformations that take several years to several hundred years. Glacial ice begins as loosely packed snow and transforms to a more granular but still quite porous ice. Compaction from the weight of overlying snow and ice and recrystallization of snow to ice eventually form dense, blue glacial ice. Glacial ice has a density of approximately 0.9 g/cm³ (0.03 lb/in.³), which is approximately 150 times the density of new-fallen snow. Therefore, in order for glaciers to form, fallen snow must survive several years and accumulate so that the transformation to glacial ice can occur.

Glaciers that cover large tracts of land are called continental glaciers, ice sheets, or ice caps; those confined to mountain valleys at high altitudes or high latitudes are called mountain, or alpine glaciers (Figure 19.A).

Major glacial events, each lasting several million years, have occurred occasionally over the last billion years. We are now living in one of these glacial periods; during the past 1.65 million years, glaciers have repeatedly advanced (grown) and retreated (melted) across much of the landscape. Only a few thousand years ago, the most recent continental glaciers retreated from the Great Lakes region of the United States. Figure 19.B shows the maximum extent of these ice sheets. During the Ice Age of the Pleistocene epoch, glaciers covered as much as 30 percent of the land area of Earth, including the present sites of major cities such as New York, Toronto, and Chicago. During this time, maximum development of continental glaciers occurred; the most recent glacial maxima occurred about 18,000 years ago. Since large quantities of

fresh water were stored in the ice, global sea level was more than 120 m (396 ft) lower than it is today. Today, glacial ice covers only about 10 percent of Earth's land. Nearly all of this ice is located in the Antarctic ice sheet, with lesser amounts in the Greenland ice sheet and in mountain glaciers of places such as Alaska, southern Norway, the Alps, and New Zealand.

We are presently in an interlude, a warm time known as an *interglacial*, when the abundance of glacial ice is relatively low. Nevertheless, we are probably still in a glacial period and ice sheets may again advance in the future.

Causes of Glaciation

The reasons that major global glacial events occurred several times during the past 1 billion years are unknown. Perhaps these events are related to the position of the continents, which significantly affects ocean circulation and global climate. On the assumption that we are still in the Pleistocene glaciation period, the amount of solar radiation reaching Earth's surface has regularly changed by a few percent. These small, cyclic variations at about 100,000, 40,000, and 20,000 years result from both regular changes in Earth's orbit and the orientation of Earth's axis of rotation.[12] These cyclic changes (known as Milankovitch cycles) are fairly well correlated with major and minor advances and retreats of glacial ice. However, correlation is not proof of causation. The cycles cause only small variations of the amount of solar radiation reaching Earth. They cannot be the main cause and certainly not the only cause that enables continental glaciers to advance or retreat.

Figure 19.A Mountain (valley) glacier Confluence of Muldrow and Traleika glaciers below Mount Denali, Alaska. The dark material on the glaciers is sediment being carried on the top of the ice. The fractures in the white glacier ice are crevasses. *(Michael Collier)*

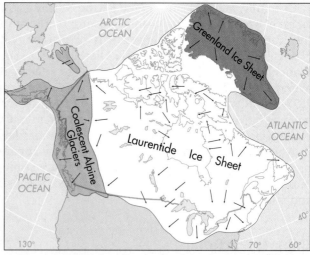

Figure 19.B Ice sheets Maximum extent of ice sheets during the Pleistocene glaciation. *(From Foster, R. J. 1983. General geology, 4th ed. Columbus, OH: Charles E. Merrill)*

A CLOSER LOOK | Desertification

Arid Lands

One of the major concerns of global warming is the change in regional and global climate patterns, especially the expansion of arid lands and accompanying desert processes.

Arid and semiarid regions of Earth, also known as the dry lands, occupy approximately 35 percent of Earth's land surface. Approximately 20 percent of the world's human population lives there; clearly, dry lands are environmentally important to people.[20]

Arid and semiarid regions of Earth are areas of relatively low to very low amounts of annual precipitation. In some very arid areas, years may pass before any measurable precipitation occurs.

Deserts owe their origin, regardless of their landscape characteristics, to their scant precipitation. The deserts of the world are primarily located in two belts located between 15° and 30° latitude north and south of the equator. These dry belts are caused by the global pattern of atmospheric circulation. At the equator, warm air rises and moves north and south toward the cooler poles. There are areas of descending air at approximately 30° latitude (see Figure 19.4). Semipermanent cells of high pressure exist in these locations, along with reduced rainfall. The high rainfall at the equator results from rising warm, moist air that cools as it rises and condenses to form rainfall. In the subtropical areas at approximately 15° to 30° latitude, the air is subsiding rather than rising, so little rain is produced.[21,22]

Global circulation of air masses explains the major deserts of the world, including those in Africa, the Middle East, parts of India, South America, Australia, and North America; however, it is not responsible for all of the deserts of the world.

Some of the semiarid lands and deserts of North America, as well as of central Asia, are dry lands because of their position far inland in continents. Dry lands may also occur in the rain shadow of mountain ranges that intercept the rainfall and block storms from moving farther inland.[23]

Desertification

At its simplest level, **desertification** may be defined as the conversion of one type of land, such as grasslands, to land that resembles a desert. This conversion does not involve the natural processes associated with climatic changes during the last 1.65 million years that are thought to cause expansion or contraction of desert areas. Rather, desertification is human-induced degradation, which introduces desertlike conditions or may transform the land from a particular state to a more desertlike landscape.[20] For example, the Sahara Desert in Africa is expanding (Figure 19.C), as are several others in North America and Asia.

Two primary characteristics of desertification are the degradation of soil, due primarily to soil erosion, and the degradation of vegetation.[22] As a result of these two characteristics, the process of desertification may significantly affect the human environment by damaging food production and thereby contributing to malnutrition and famine.[22] Damage and loss of vegetation resulting from desertification may be so extensive that the productivity of the land is essentially lost and may not be recovered in a time frame useful to humans. Therefore, in contrast to drought, which is normally a relatively short-term problem that affects land productivity, desertification involves impacts that are long term and chronic.[20]

Figure 19.C Advancing **desert** People digging out of a sand dune in the Sahara Desert, Africa, after a windstorm. The Sahara Desert continues to expand. *(Steve McCurry/Magnum Photos, Inc.)*

For example, consider the U.S. "dust bowl." Terrible dust storms (Figure 19.D) produced the "black blizzards" of the "dirty '30s," better known as the Dust Bowl, that struck the high plains of the United States, particularly Texas, Colorado, Oklahoma, and Kansas. More intensive use of tractors and trucks enabled large-scale farming; farmers cleared the land of vegetation for growing grains. Exceptionally hot and dry drought years also reduced the natural vegetation cover. The exposed topsoil was eroded by wind and covered with sand, frequently generating tremendous dust storms. Thus, both human activity and technology and natural climatic variation played a role in creating the Dust Bowl that severely degraded the farmlands and caused great hardships for the people in the region. Today, northern China is experiencing its own "dust bowl" and the reasons are similar to those of the United States in the dirty 30's—unwise agriculture and grazing practice.

The causes of desertification are well known and recognized. In a general sense, they are related to four main types of poor land use:[21] *overcultivation, overgrazing, deforestation,* and *poor irrigation practices.* These are exacerbated in some countries with abundant arid and semiarid lands when population increase is rapid. Prevention, minimization, and reversal of desertification involve the following:[24,25]

- Protection and improvement of high-quality land rather than dedicating time and money to poor land

- Application of simple, sound range management techniques to protect the land from overgrazing by livestock

- Application of sound conservation measures to agricultural lands to protect soil resources

- Use of appropriate technology to increase crop production, allowing poorer lands to be returned to less-intensive land uses other than intensive agriculture (for example, forestry, wildlife, or grazing)

- Increased land restoration efforts through vegetation management, sand-dune stabilization, and control of soil erosion

Figure 19.D U.S. "Dust Bowl" (a) Dust storm at Manteer, Kansas, in 1935. *(U.S. Department of Commerce)* (b) Farmer and sons walking into the face of a dust storm, Oklahoma, 1936. Notice that the house is partially buried by dust. *(Arthur Rothstein/AP/Wide World Photos)*

(a)

(b)

present in all new areas. For example, Canadian prairie soils tend to be thinner and less fertile than those of the Midwestern United States. To some extent, it is this uncertainty that concerns people. Stable or expanding global agricultural activities are crucial to people throughout the world who depend upon the food grown in the major grain belts. Hydrologic changes associated with climatic change resulting from global warming might seriously affect food supplies worldwide.

Global warming might also change the frequency and intensity of violent storms, and this change may be more important than the issue of which areas become wetter, drier, hotter, or cooler. Warming oceans will feed more energy into high-magnitude storms such as hurricanes. More or larger hurricanes would increase the hazard of living in low-lying coastal areas, many of which are experiencing rapid growth of human population.

That global and regional climatic changes have a significant effect on the incidence of hazardous natural events such as storm damage, landslides, drought, and fires is sometimes dramatically illustrated by El Niño. **El Niño** is a natural climatic event that occurs on an average of once every few years, most recently in 1997–1998. El Niño is both an oceanic and an atmospheric phenomenon, involving unusually high surface temperatures in the eastern equatorial Pacific Ocean and droughts and highintensity rainstorms in various places on Earth (see A Closer Look: El Niño).

Sea Level Rise

A rise of sea level is a potentially serious problem related to global warming. Estimates of the rise expected in the next century vary widely, from approximately 40 to 200 cm (16 to 80 in.), and precise estimates are not possible at this time. However, a 40 cm (16 in.) rise in sea level would have significant environmental impacts. Such a rise could cause increased coastal erosion on open beaches of up to 80 m (260 ft), rendering buildings and other structures more vulnerable to waves generated from high-magnitude storms. In areas with coastal estuaries, a 40 cm (16 in.) sea level rise would cause a landward migration of existing estuaries, again putting pressure on human-built structures in the coastal zone. A sea level rise of 1 m (3.3 ft) would have very serious consequences; significant alterations would be needed to protect investments in the coastal zone. Communities would have to choose between making very substantial investments in controlling coastal erosion or allowing beaches and estuaries to migrate landward over wide areas.[10]

Changes in the Biosphere

There is a growing body of evidence that global warming is initiating a number of changes in the biosphere, threatening both ecological systems and people. These include risk of extinction as land use changes and habitat becomes fragmented. Other changes include shifts in the range of plants and animals, with a variety of consequences: Mosquitoes carrying diseases including malaria and dengue fever in Africa, South America, Central America, and Mexico are migrating to higher elevations; butterfly species are moving northward in Europe; some bird species are moving northward in the United Kingdom; subalpine forests in the Cascade Mountains in Washington are migrating to higher meadows; alpine plants in Austria are shifting to higher elevations; sea ice melting in the Arctic is placing stress on seabirds, walruses, and polar bears; and warming of shallow water in the Florida Keys, Bermuda, Australia's Great Barrier Reef, and many other tropical ocean areas is causing the bleaching of coral reefs.[30]

Strategies for Reducing the Impact of Global Warming

Two big questions concerning the Earth climate system and people are (1) What changes have occurred? and (2) What changes could occur in the future? Answering

A CLOSER LOOK | El Niño

El Niño events probably start from random, slight reductions in the tradewinds that, in turn, cause warm water in the western equatorial Pacific Ocean to flow eastward (Figure 19.E). This change further reduces the tradewinds, causing more warm water to move eastward, until an El Niño event is established.[26] El Niños are thought to bring about an increase in some natural hazards on a nearly global scale by putting a greater amount of heat energy into the atmosphere. The heat energy in the atmosphere increases as more water evaporates from the ocean to the atmosphere. The increase of heat and water in the atmosphere produces more violent storms such as hurricanes.

Figure 19.F shows the extent of natural disasters that have been attributed in part to the El Niño event of 1997–1998, when worldwide hurricanes, floods, landslides, droughts, and fires killed people and caused billions of dollars in damages to crops, ecosystems, and human structures. Australia, Indonesia, the Americas, and Africa were particularly hard hit. There is some disagreement about how much damage

and loss of life is directly attributable to El Niño, but few disagree that it is significant.[27,28] We do not yet completely understand the cause of El Niño events. They occur every few years to a lesser or greater extent, including the large El Niño events of 1982–1983 and 1997–1998.

El Niño events can cause havoc by increasing the occurrence of hazardous natural events, and there is concern that human-induced climatic change may, through global warming, produce more and stronger El Niño events in the future. This effect would result in part because as we burn more fossil fuels and emit more greenhouse gases into the atmosphere, the oceans will continue to warm, and differential warming in various parts of the ocean may increase the frequency and intensity of El Niño events in the Pacific Ocean.

The opposite of El Niño is La Niña, in which eastern Pacific waters are cool and droughts rather than floods may result in southern California. The alternation from El Niño to La Niña is a natural Earth cycle only recently recognized.[29]

Figure 19.E El Niño

Idealized diagram contrasting (a) normal conditions and processes with (b) those of El Niño. *(Modified after National Oceanic and Atmospheric Administration. www.elnino.noaa.gov/. Accessed 3/3/99)*

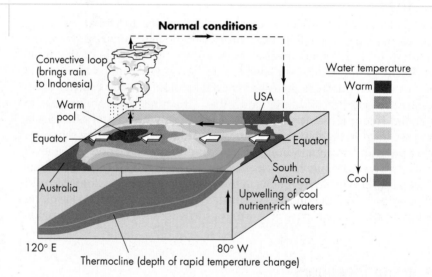

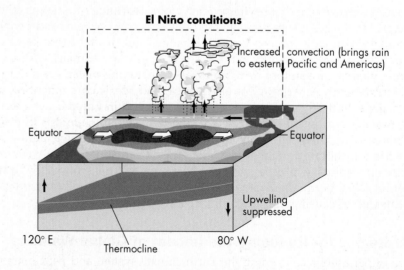

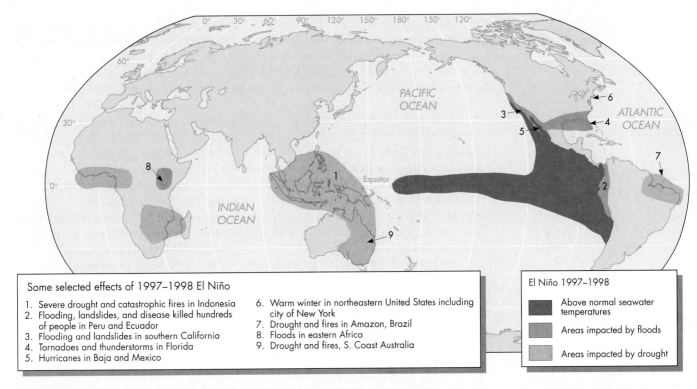

Some selected effects of 1997–1998 El Niño

1. Severe drought and catastrophic fires in Indonesia
2. Flooding, landslides, and disease killed hundreds of people in Peru and Ecuador
3. Flooding and landslides in southern California
4. Tornadoes and thunderstorms in Florida
5. Hurricanes in Baja and Mexico
6. Warm winter in northeastern United States including city of New York
7. Drought and fires in Amazon, Brazil
8. Floods in eastern Africa
9. Drought and fires, S. Coast Australia

El Niño 1997–1998

Above normal seawater temperatures

Areas impacted by floods

Areas impacted by drought

Figure 19.F The 1997–1998 El Niño event   Map shows the general extent of the El Niño effects and the regions damaged by floods, fires, or drought. *(Data from National Oceanic and Atmospheric Administration, 1998)*

these questions requires geologic evaluation of prehistoric change, and prediction through modeling and simulation of future change (Figure 19.17). Because we now know global warming is due in part to the increased concentration of greenhouse gases, reduction of these gases in the atmosphere is a primary management strategy. This was the subject of the 1997 United Nations Framework Convention on Climate Change in Kyoto, Japan. The objective of the convention was to produce an international agreement to reduce emissions of greenhouse gases,

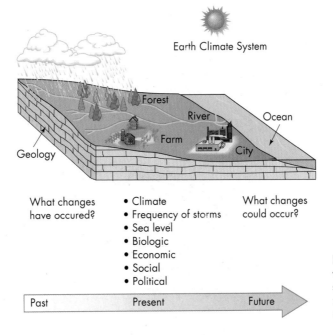

Earth Climate System

Forest

River   Ocean

Farm   City

Geology

What changes have occured?

- Climate
- Frequency of storms
- Sea level
- Biologic
- Economic
- Social
- Political

What changes could occur?

Past   Present   Future

Figure 19.17 Two big questions concerning Earth's climate system linked to people and environment. *(Modified after International Panel or Climate Change 2001, at www.ipcc)*

especially carbon dioxide. The United States originally agreed to reductions but in 2001 refused to honor the agreement, much to the disappointment of other nations, especially European allies. As a result the leadership in controlling global warming has shifted from the United States to the European Union. The Kyoto Protocol was signed by 166 nations and became a formal international treaty in February 2005.

We are faced with a dilemma with respect to burning of fossil fuels. On one hand, fossil fuels are vital to our society and are necessary for continued economic development and growth, as well as human well-being. On the other hand, scientific evidence suggests that burning fossil fuels is contributing significantly to global warming. Burning fossil fuels is linked to environmental problems that include rise in sea level, increased surface temperatures, and increase frequency and intensity of storms such as hurricanes. Impacts associated with fossil fuels depend upon how much global temperature will actually rise. Estimates of temperature rise from climate models range from about 1.5° to 4.5°C. If the increase is close to 2°C, we can probably adapt with minimal disturbance. If the increase in temperature is near the high end, then significant impacts are likely. One way to estimate potential increase in temperature that is independent of the model predictions is to examine the geologic record for past change. A recent study of the geologic record from ocean sediments deposited during the past several hundred thousand years suggests that warming in the next century will be about 5°C, consistent with models that predict warming as great as 4.5°C.[31] If global temperature rises 5°C, than we will need a strategy to reduce emissions of carbon to avoid serious environmental disruption. This results because at this level of temperature increase significant environmental impacts are much more likely. Also it is important to recognize that following peak of emissions of carbon dioxide it will probably take several hundred years or longer before temperature and sea level stabilize (Figure 19.18). Therefore, the sooner we act to reduce emissions, the lower the impacts will be and the sooner we will be on a path to stabilize climate change.

Assuming reduction of carbon emissions is necessary to reduce impacts of global warming, we need to take action (see A Closer Look: Discussion on How to Stop Global Warming Heats Up). The reduction must occur at a time when human population and energy consumption are increasing. There are several ways that we might reduce emissions of carbon dioxide into the environment: reduce emissions through improved engineering of fossil fuel-burning power plants; use those fossil fuels that release less carbon into the atmosphere such as natural gas, which on burning releases less carbon dioxide than does coal or oil (Table 19.4); conserve energy to reduce our dependence on fossil fuels; use more alternative energy sources; and store carbon in Earth's systems such as forests, soils, and in rocks below the surface of Earth.[32,33]

Storage of carbon in plants, soils, and the ocean has received considerable study. The option of sequestration of carbon in the geological or rock environment has received less attention. Sequestration is attractive because the residence time

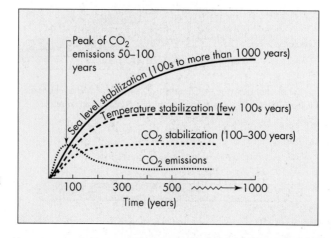

Figure 19.18 Lag times to **stabilization** of CO₂, temperature, and sea level following peak in emission. *(Modified after International Panel on Climate Change 2001, at www.ipcc)*

TABLE 19.4 Comparison of Common Fossil Fuels in Terms of Percentage of Total Carbon Released

Fossil Fuel	Carbon Emitted as % of Total Carbon from Fossil Fuels	Comparisons
Coal	36	Releases much more carbon per unit of energy than oil or gas.
Oil	43	Releases 23% less carbon per unit of energy than coal.
Natural Gas	21	Releases 28% less carbon per unit of energy than oil or gas.

Source: Data from Dunn, J. 2001. Decarbonizing the energy economy. *World Watch Institute State of the World 2001.* New York: W. W. Norton and Company. Chapters IV, V.

of carbon in the geologic environment is generally a long period of time, potentially thousands to hundreds of thousands of years.

The general principle of geologic sequestration of carbon is fairly straightforward. The idea is to capture carbon dioxide from power plants and industry and inject the carbon into the subsurface geologic environment. Two geologic environments have received considerable attention. The first is sedimentary rocks that contain salty water. These rocks, known as saline aquifers, are fairly widespread at numerous locations on Earth and have large reservoir capacity with the potential to sequester large amounts of carbon. Sedimentary rocks in these reservoirs have the potential to sequester many years of human-produced carbon dioxide emissions. They could provide the necessary time to transition to an energy economy not dependent on fossil fuels.[32] The second (but related) geologic environment for sequestration of carbon is depleted oil and gas fields. The process of placing carbon dioxide in the geologic environment involves compressing the gas to a mixture of both liquid and gases, and then injecting it underground using wells. Injecting carbon dioxide into depleted oil and gas fields has an added advantage in that the carbon dioxide is not only stored, but also serves as a way to enhance recovery of remaining oil and gas in the reservoir. The injected carbon dioxide helps move the oil toward production wells.

A demonstration project is now ongoing in Saskatchewan, Canada. The Weyburn oil field started production in the 1950s and was considered to be depleted. However, with enhanced recovery and storage of carbon dioxide, production is likely to last for several more decades. The source of carbon dioxide for the Weyburn field comes from a coal burning plant in North Dakota by way of a pipeline that delivers several thousand tons of carbon per day.[32]

Another carbon sequestration project is located beneath the North Sea. Carbon dioxide is injected into a salt-water aquifer located below a natural gas field. The project, which started in 1996, injects about 1 million tons of carbon dioxide into the subsurface environment each year. It is estimated that the entire facility can hold about the amount of carbon dioxide that is projected to be produced from all of Europe's fossil fuel plants in the next few hundred years.[34] The cost of sequestering the carbon beneath the North Sea is expensive. However, it saves the company from paying carbon dioxide taxes for emissions into the atmosphere. Finally, pilot projects to demonstrate the potential and usefulness of storing carbon in the United States have been initiated in Texas, beneath depleted oil fields. The good news is that immense salt aquifers are common beneath many areas of the United States including the Gulf Coast, Texas, and Louisiana, and the potential to store carbon is immense.[34]

Recent studies suggest that global warming is probably not an immediate emergency, and we will have a decade or so to develop alternatives to continued intensive burning of fossil fuels. If we decide that we must stabilize the concentration of atmospheric carbon dioxide in the future, it will be necessary to go through a transition from fossil fuel energy sources to alternative sources that produce much less carbon dioxide. However, speculation based on studies of glacial ice cores from Greenland indicate that significant climatic change may occur quickly, perhaps in

A CLOSER LOOK | Discussion on How to Stop Global Warming Heats Up

The science behind global warming has been solved. Earth is warming and some significant portion of that warming is a result of burning fossil fuels. About 25 percent of the greenhouse gas carbon dioxide, which is of primary concern, is emitted by the United States and one state, California, by itself is twelfth in the world in emitting the gas. U.S. policy at the national level has been criticized for lacking leadership and simply recommending voluntary controls of emissions. As a result the controls have shifted to state and local governments. Recently the state of California approved a bill to reduce emissions by 25 percent by the year 2020. The bill is a compromise that took weeks of lobbying by environmentalists and others for it to eventually be signed by the governor. The bill has been labeled by some as a "job killer," but environmentalists and those who favor the bill point out that the legislation will likely result in many new jobs for the state.

California will become the first state in the nation to directly address global warming through reduction of carbon dioxide emitted from power plants and other industrial processes. It is hoped that the California law will provide an example for other states to follow as we try to control the greenhouse gases that are causing global warming.

California is not the only state attempting to curb emissions of greenhouse gases. Several other states are working together in the northeastern part of the country to control emissions from electrical power plants that burn coal.

In California the Air Resources Board will have the responsibility of measuring the amount of carbon dioxide and other greenhouse gases coming from a variety of sources including electric power plants, oil refineries, and other industrial activities. Following the measurement of the emissions, limits will be set. Not surprisingly, some of those industries are saying that the emission controls will drive up prices of electricity in California and force large employers to move their operations to other states that have more permissive policies with respect to emission of greenhouse gases. However, other groups see opportunities from the legislation, as companies are likely to invest in pollution control technology and alternative fuels. That will likely produce thousands of new jobs for the state.

The Air Resources Board may also decide to combine the emission control strategy with market-based policies that include trading what are known as emission credits, that is, credits to emit a certain amount of greenhouse gases. Credits may initially be given to electrical power plants and industry. The total pollution allowed by all the credits is the target emission for the state. Credits may then be bought and sold on the open market. Newer power plants with low emissions would not need all their credits and so could sell them to older power plants that have higher emissions. It is hoped that the combination of emission control and market-based programs will not only reduce emissions of greenhouse gases and help reduce the threat of global warming but also herald a new era in how we approach environmental problems.

---

as short a time as a few years. A quick, natural or human-induced warming or cooling is probably unlikely, but if one does occur, the potential impacts could be fast and serious. Human civilization was born and has developed to our present highly industrialized society in only about 7,000 years. That period has been characterized by a relatively stable, warm climate that is probably not characteristic of longer periods of Earth history. It is difficult to imagine the human suffering that might result late in the twenty-first century from a quick climate change to harsher conditions when there are as many as 10 billion or more people on Earth to feed.[13]

19.6 Coupling of Global Change Processes: Ozone Depletion and Global Warming

The major global change processes discussed in this chapter have interesting linkages. For example, the chlorofluorocarbons (CFCs) that cause ozone (O_3, triatomic oxygen) depletion when they reach the stratosphere also contribute to the greenhouse effect when they are released into the lower atmosphere. The processes responsible for stratospheric **ozone depletion** and development of the Antarctic ozone hole are complex but related to chemical reactions between chlorine (Cl), and ozone (O_3) linked to the Antarctic polar vortex (where there is counterclockwise rotation of the atmosphere), and polar stratospheric clouds where ozone-depleting reactions happen. The air mass in the vortex is isolated, cools, condenses, and descends.[35] Chlorofluorocarbons, which contain chlorine, are stable in the lower atmosphere but wander up to the stratosphere. There, chlorine is split from CFCs by ultraviolet radiation from the Sun and reacts with ozone to produce chlorine oxide and oxygen: $Cl + O_3 \rightarrow ClO + O_2$. The result is destruction of the ozone.

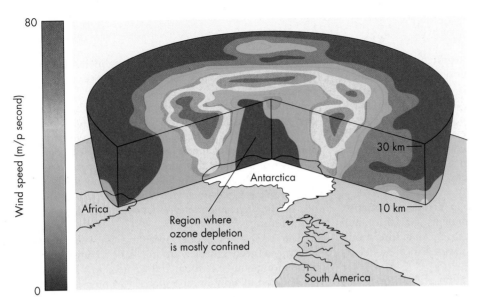

Figure 19.19 Idealized **diagrams** of the Antarctic polar vortex and the role of polar stratospheric clouds in the ozone depletion chain reaction. *(Based on Toon, O. B., and Turco, R. P. 1991. Polar stratospheric clouds and ozone depletion,* Scientific American *264(6):68–74)*

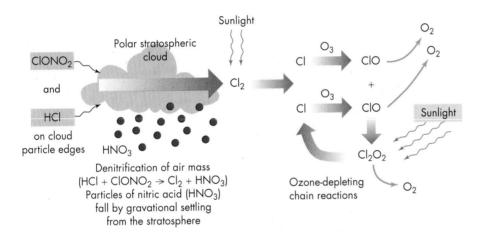

The chlorine is released from chlorine oxide (ClO) by a subsequent reaction with monoatomic oxygen (O): $ClO + O \rightarrow Cl + O_2$, releasing Cl to destroy more ozone by the first reaction (Figure 19.19). The reactions are complex but follow the general equations just mentioned. The two important sinks for chlorine in the clouds are hydrochloric acid (HCl) and chlorine nitrate. With denitrification chlorine (Cl_2) is released to enter into ozone depletion reactions (see Figure 19.19). During the Antarctic spring, these chain reactions may happen very quickly, causing up to 2 percent ozone depletion each day and ultimately producing the observed 50 percent to 70 percent annual depletion. In the lower atmosphere where they are emitted, CFCs trap more heat than carbon dioxide because they can absorb more infrared radiation. One study concluded that on a per molecule basis, chlorofluorocarbons are 10,000 times more efficient in absorbing infrared radiation than is carbon dioxide. Nevertheless, carbon dioxide contributes much more to the total anthropogenic greenhouse effect than do CFCs because so much more carbon dioxide is being emitted.[36]

The coupling of the greenhouse and ozone problems via release of materials such as CFCs is an important point to consider. Other couplings related to processes such as the burning of fossil fuels release precursors to acid rain as well as carbon dioxide and other greenhouse gases (see Chapter 15). Burning fossil fuels and volcanic eruptions, as discussed earlier, also produce particulates that result in atmospheric cooling. Carbon dioxide has the predominant effect on changing global temperature, but we see the principle of environmental unity in action. Many aspects of one problem, in this case global warming, are related to other processes or problems, for example, ozone depletion.

SUMMARY

The main goal of the emerging integrated field of study known as Earth system science is to obtain a basic understanding of how our planet works and how its various components, such as the atmosphere, oceans, and solid Earth, interact. Another important goal is to predict global changes that are likely to occur within the next several decades. The global changes involved include temperature, or climate, changes and the resulting changes in seawater and on land. Because these are short-term predictions, Earth system science is relevant to people everywhere.

Methods of studying global change include examination of the geologic record from lake sediments, glacial ice, and other Earth materials; gathering of real-time data from monitoring stations; and development of mathematical models to predict change.

Climate refers to characteristic atmospheric conditions such as precipitation and temperature over seasons, years, and decades. The atmosphere is a dynamic, complex environment where many important chemical reactions occur.

Human activity is contributing significantly to global warming. The trapping of heat by the atmosphere is generally referred to as the greenhouse effect. Water vapor and several other gases, including carbon dioxide, methane, and chlorofluorocarbons, tend to trap heat and warm Earth because they absorb some of the heat energy radiating from Earth. The effects of a global rise in temperature include a rise in global sea level and changes in rainfall patterns, high-magnitude storms, soil moisture, agriculture, and the biosphere.

Natural forcing mechanisms that may cause climatic change include Milankovitch cycles, solar variability, and volcanic activity. Anthropogenic causes include air pollution and increase in greenhouse gases, especially carbon dioxide. We now understand that global climate can change rapidly over a time period of a few decades to a few hundred years. Two important questions are: What is the nature and extent of past climate change? and What climate changes will occur in the future?

The science of global warming is well understood. Human-induced global warming is occurring. There is no reason for gloom and doom, but we need to take appropriate action soon to slow or stop global warming and associated environmental consequences.

Adjustments to global warming will vary from adapting to change, to reducing emissions of carbon dioxide, to sequestration of carbon. Several different, simultaneous adjustments are likely.

Some global changes, such as ozone depletion and the greenhouse effect, are coupled. For example, the CFCs released in the lower atmosphere add to the greenhouse effect there before diffusing to the stratosphere, where they cause ozone depletion.

Revisiting Fundamental Concepts

Human Population Growth

Increase in human population is one of the driving forces of the human-induced portion of global warming. The United States, with about 5 percent of the world's population, emits about 25 percent of the carbon dioxide, thought to contribute to warming through the greenhouse effect.

Sustainability

A relatively stable global climate is desired for future generations of people and ecosystems. Rapid or dramatic warming or cooling of the climate would be difficult for society to adjust to. Presently, our energy policies of burning vast amounts of fossil fuels are not sustainable. If we continue to emit huge amounts of greenhouse gases, we will continue to perform an unplanned global experiment with untested but predictable adverse consequences.

Earth as a System

Earth is a planetary system—with linkages between air, water, soil, rock, and living things—that changes in many ways. Understanding the Earth system and how it changes is important if we are to solve environmental problems such as global warming and acid rain.

Hazardous Earth Processes, Risk Assessment, and Perception

Global warming is linked to a variety of natural hazards, including flooding, hurricanes, and coastal erosion. The changes in risks and resulting damages from natural hazards that are likely to occur as the atmosphere warms are being studied. Our perception of hazards will change as the frequency and intensity of events increase or decrease in the future.

Scientific Knowledge and Values

Scientific understanding of global change and climate change has improved greatly in the past few years. Although details are sketchy, we have a good idea of the potential adverse consequences of our present energy policy of burning fossil fuels. Our values are being tested as never before. Reducing emissions of carbon dioxide will require a new energy policy that some fear will cause financial damage to our economy. Others believe the transition to alternative energy sources that do not pollute the environment or cause global warming will not be that difficult. Regardless of their position on energy sources, people place a high value on living in a pollution-free environment without the threat of rapid, potentially damaging climate change.

Key Terms

climate (p. 608)

desertification (p. 625)

Earth system science (p. 607)

El Niño (p. 627)

forcing (p. 619)

glacier (p. 624)

global dimming (p. 622)

greenhouse effect (p. 613)

ozone depletion (p. 632)

Review Questions

1. What are the two central goals of Earth science as defined by Preston Cloud?

2. Define Earth system science.

3. What are the three major tools to study global change?

4. Define climate.

5. Define glacier.

6. What is desertification?

7. What are the three main factors that determine the temperature of Earth?

8. Why do we think that the late twentieth-century global warming is due to human activity?

9. What is the greenhouse effect?

10. Why is the so-called ocean conveyer belt important?

11. What are the major potential effects of global warming?

12. What are the main ways to reduce emissions of carbon dioxide?

13. Why are CFCs thought to be responsible for the destruction of ozone in the stratosphere?

14. How are CFCs and carbon dioxide both involved in potential global warming?

Critical Thinking Questions

1. Have a discussion with your parents or someone of similar age and write down the major changes that have occurred in their lifetime as well as yours. Characterize these changes as gradual, abrupt, surprising, chaotic, or other description of your choice. Analyze these changes and discuss which ones were most important to you personally. Which of these affected our environment at the local, regional, or global level?

2. How do you think future change is likely to affect you, particularly considering the environmental problems we may be facing as a result of increased world population?

3. Do you think that we need additional studies to further confirm human activity is a significant component of the late twentieth-century warming before taking steps to reduce potential adverse impacts? Why? Why not?

4. Do you think that global warming from increasing CO_2 might be saving us from another glacial period? What is the role of global dimming in this question?

5. How would you respond to the assertion that we need more study before taking potentially expensive steps to slow or stop global warming? Would such steps cause financial problems or present economic opportunity? Why? Why not?

TWENTY

Geology, Society, and the Future

Learning Objectives

The study of geology and its relationship to specific functions of society, such as environmental health, environmental impact, and land-use planning relies on integrating what we have learned throughout this book. In this sense, this chapter is a capstone. At a deeper level, relationships between people and our environment provide insight into why we study environmental geology. Hopefully, we will build on past successes and learn from our mistakes. In this chapter we will focus on the following learning objectives:

- Understand geologic aspects of environmental health

- Understand the geologist's role in evaluating land for appropriate uses

- Understand environmental impact analysis, including the major components of the environmental impact statement and the processes of scoping and mitigation

- Know the processes of law, particularly the use of mediation and negotiation to resolve environmental conflicts

- Know three steps we can take to avert a potential environmental crisis due to the convergence of human population increase, reduction in water resources, and global warming, linked to our food supply

- Be able to discuss what steps are necessary to attain the goal of sustainability

Polluted river in China The rush to industrial growth in China is causing environmental problems. *(National Geographic, March 2004. Bob Sacha Photo)*

CASE HISTORY | China on the Edge of an Environmental Crisis?

China, with a human population of about 1.3 billion people, was until recently termed the "sleeping giant." The giant is stirring with industrial growth that is improving the economic conditions for millions of its people—but there is a heavy environmental cost that has recently been acknowledged and is beginning to be addressed. The Chinese people are asking questions that we in the United States asked in the 1970s. What good is economic development if you can't breath the air or drink the water? Why expand agriculture if you produce a "dust bowl" that through desertification and soil erosion lowers grain productivity that may require the import of food? The problems are clear:[1,2]

- China is currently experiencing the largest human-induced transformation the world has ever seen. Multitudes of people are migrating to cities where opportunities and rapid industrial and economic growth, often at the expense of the environment, are occurring. Environmental protection is in its infancy, but the country is aware that reducing, and eventually eliminating, environmental degradation is a necessary, if not an immediate, goal.

- China's industrial growth is fueled by coal that is contributing to air pollution and acid rain.

- Burning coal inside homes produces a serious indoor air pollution problem, contributing through respiratory diseases to the deaths of nearly one million Chinese people per year.

- More than one-quarter of a billion people in China's factories and mines are commonly in contact with toxic chemicals and dust.

- Ninety percent of China's raw sewage from cities is discharged without treatment into rivers, lakes, and the ocean. As a result, large rivers flowing into the Yellow Sea are polluted and cannot be safely used as a water supply for people.

- About one-half of China's people drink water that is contaminated by waste from people and other animals.

- Groundwater aquifers in northern China's grain belt are rapidly being depleted which, along with soil erosion and dust storms, is reducing grain production.

The Chinese government is now reacting to what is termed an "ecological meltdown"[2] by trying to develop new, non-polluting energy resources (hydropower, wind, and solar); reducing the burning of coal; returning farm land back to range lands, forest, and wetlands to reduce soil erosion; planting trees on hillslopes that were previously logged in order to stabilize soils, reduce runoff, and reduce the flood hazard; constructing water treatment plants to treat toxic wastewater from factories; and even construct a new city based on a model of sustainability. The people of China want to have larger homes and automobiles like affluent people in Europe and the United States. They also want clean water and an improved environment for their children. The big question is—can China have a clean, healthy environment and continue on its present path of rapid economic development? The answer is probably no! The present path or "business as usual" will continue to harm the environment. In the remainder of this chapter we will discuss some aspects related to this question. At the end of the chapter we will suggest an approach that may help China as well as the rest of the world.

20.1 Introduction

We live on the surface of Earth, which has developed landforms such as mountains, plains, river valleys, hillslopes, and shorelines through the interactions of many geologic and biogeologic processes. In most places, a biogeologic layer of Earth material that we call soil covers the surface. Ever since people began gathering in villages that became cities, practicing agriculture, and manufacturing products from a variety of Earth materials, geology has played an important role in society. In other parts of this book, we discussed some of the major Earth processes, including natural hazards, and resources such as minerals, water, and soil and their interactions with human behavior and processes. In this chapter, we further examine examples of relationships between geology and the needs of society. Although there are many possible relationships, we will discuss geologic aspects of environmental health, site selection, environmental-impact analysis, and environmental law. We will end this chapter and book with a short discussion of how we might avoid future human-induced environmental crises and move toward sustainability. The discussions in this chapter do not provide a complete analysis of all the aspects of societal interactions with geology; rather they serve as an introduction to the subject and future thought concerning how we might move toward a more harmonious existence with our planet Earth.

20.2 **Geology and Environmental Health**

As members of the biological community, humans have carved a niche in the biosphere—a niche highly dependent on complex interrelations among the biosphere, atmosphere, hydrosphere, and lithosphere. Yet we are only beginning to inquire into and gain a basic understanding of the total range of environmental factors that affect our health and well-being. As we continue our exploration of the geologic cycle—from minute quantities of elements in soil, rocks, and water to regional patterns of climate, geology, and topography—we are making important discoveries about how these factors may influence death rates and incidences of certain diseases. In the United States alone, the death rate varies significantly from one area to another,[3] and some of the variability is the result of the local, physical, biological, and chemical environment in which we live (see A Closer Look: Lead in the Environment).

At this point it is useful to introduce a few basic terms and definitions. For example, **toxicology** is the science involved with the study of poisons, or *toxins*, and their potential effects on people and ecosystems as well as clinical, economic, industrial, and legal problems associated with toxic materials in the environment. A special type of toxin that causes cancer is called a *carcinogen*; carcinogens are some of the most feared and heavily regulated toxins in society today. The concentration of a particular toxin or carcinogen is often reported in units such as parts per million (ppm) or parts per billion (ppb). These are incredibly small numbers. For example, perhaps you are making popcorn and would like to add some salt to it. Using your standard 737 g (1 lb 10 oz) container of iodized salt, you want to season the popcorn at a concentration of 1 ppm. To set that concentration, using the whole container of salt, you would have to pop approximately 737 metric tons of kernels of corn, more than 6 million quarter-pound bags of popcorn! The average person has a hard time understanding that such small concentrations as parts per million or parts per billion of a chemical can cause cancer or be toxic. Finally, toxins and other pollutants in the atmosphere are usually measured as micrograms

A CLOSER LOOK | Lead in the Environment

The incidence of lead poisoning is an example of geologic, cultural, political, and economic influences on patterns of disease. The effects of lead poisoning can include anemia, mental retardation, and palsy. For decades, lead was used in gasoline, resulting in polluted soil near highways. Until recently, lead was also used in many other products, including paints. Some children chewing on these painted surfaces ingested lead in toxic amounts. Lead is also found in some moonshine whiskey and has resulted in lead poisoning among adults and even unborn or nursing infants whose mothers drank it.[4]

It has been suggested that widespread lead poisoning was one of the reasons for the fall of the Roman Empire. Some historians estimate that the Romans produced about 55,000 metric tons of lead each year for several hundred years. The Romans used lead for pots, for wine cups, and in cosmetics and medicines. The ruling class also had water running through lead pipes into their homes. Historians argue that gradual lead poisoning among the upper class resulted in their eventual demise through widespread stillbirths, deformities, and brain damage. The high lead content found in the bones of ancient Romans lends support to this hypothesis.[5] Further support comes from a study of Greenland glaciers' ice cores; for the period 500 B.C. to A.D. 300, lead concentrations in the ice are approximately four times higher than either before or after the Roman Empire time period. This finding suggests that lead mining and smelting during the Roman Empire caused lead contamination in the atmosphere over the Northern Hemisphere.[6]

We have known for a long time that at high concentrations lead is an extremely toxic metal that causes serious health problems. More recently, it has been suggested that lead in urban areas can also cause social problems. In one study, children with above-average lead concentrations in their bones were found to be more likely to have attention deficit disorder or delinquency problems and to exhibit aggressive behavior than were children with lower lead concentrations.[7] On the basis of this finding, some part of the crime rate and aggressive behavior of children in urban areas may be attributed to environmental pollution as well as to socioeconomic conditions. If environmental pollution does contribute to social problems, then how we manage urban environments is enormously significant. The idea that aggressive behavior is due in part to our contaminated environment is controversial, and additional studies are necessary to verify these conclusions.

per liter (μg/L) or micrograms per cubic meter (μg/m^3) of air. Radioactive radon in air is an exception; radon gas is reported in picoCuries per liter (pCi/L), a reference to the number of radioactive disintegrations per second in a liter of air.

Disease has been described, from an environmental perspective, as an imbalance resulting from a poor adjustment between an individual and the environment.[8] It seldom has a single cause. The geologist's contribution to our understanding of its causes helps isolate aspects of the geologic environment that may influence the incidence of disease. This tremendously complex task requires sound scientific inquiry coupled with interdisciplinary research with physicians and other scientists. Although the picture is now rather vague, the possible rewards of the emerging field of medical geology are exciting and may eventually play a significant role in environmental health.

Some Geologic Factors of Environmental Health

The soil in which we cultivate plants for food, the rock on which we build our homes and industries, the water we drink, and the air we breathe all influence our chances of developing serious health problems. Conversely, these factors can also influence our chances of living a longer, more productive life. Surprisingly, many people still believe that soil, water, or air in a "natural," "pure," or "virgin" state must be "good" and that if human activities have changed or modified these resources, they have become "contaminated," or "polluted," and, therefore, "bad." This simple dichotomy is by no means the entire story.[6,9,10]

Chronic Disease and the Geologic Environment

Health can be defined as an organism's state of adjustment to its own internal environment and to its external environment. Observation over many years has suggested that some regional and local variations in human chronic diseases such as cancer and heart disease are related to the geologic environment. Although evidence continues to accumulate, the exact nature of these associations remains to be discovered. There are two reasons for the lack of conclusive results. First, hypotheses about the relationship between the geologic environment and disease have not been specific enough to be tested adequately; basic research and field verification need better coordination. Second, there are many methodological difficulties in obtaining reliable and comparable data for medical–geological studies;[11] we know much less about geologic influences on chronic disease than about the contribution of other environmental factors such as climate.

Although our evaluation of geologic contributions to disease in some cases remains an educated guess, the benefit of learning more about these relationships is obvious. The geographic variations in the incidence of heart disease in the United States may be related to the geologic environment. In addition, it has been estimated that two-thirds of the cancerous tumors in the Western Hemisphere result in part from environmental causes, while genetic factors are also important.[11]

Heart Disease and the Geochemical Environment

Our use of the term *heart disease* here includes coronary heart disease (CHD) and cardiovascular disease (CVD). Variations in heart disease mortality have generally shown interesting relationships with the chemistry of drinking water—in particular, with the hardness of drinking water. Hardness is a function of the amount of calcium, magnesium, and iron that is dissolved in water. Higher concentrations of these elements produce harder water. Water with low concentrations of these elements is termed *soft*. Studies in Japan, England, Wales, Sweden, and the United States all conclude that communities with relatively soft water have a higher rate of heart disease than communities with harder water.

Perhaps the first report of a relationship between water chemistry and cardiovascular disease came from Japan, where a prevalent cause of death is stroke, a sudden loss of body functions caused by the rupture or blockage of a blood vessel in

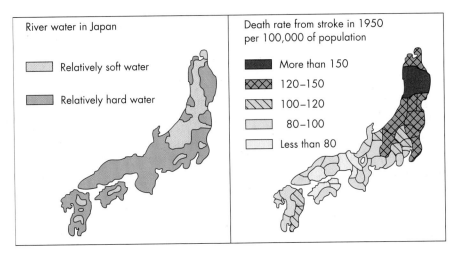

Figure 20.1 Stroke Maps of Japan comparing relatively soft and hard water in rivers with the death rate from strokes in 1950. (*Data from T. Kobayashi. 1957*)

the brain. The geographic variation of the disease in Japan is related to the relative softness or hardness of river water. Figure 20.1 shows that areas in Japan with high death rates due to stroke generally correspond to areas of relatively soft water.[12]

The relatively soft water of the northeastern part of Japan evidently stems from the sulfur-rich volcanic rock found in that region. Rivers in this region have relatively soft water. Rivers in Japan that flow through sedimentary rocks have, in contrast, relatively hard water, as do most rivers in the world.

A general inverse relationship between hard water and death rates from heart disease is also present in the United States.[13,14] For example, similar to the Japanese report, a study in Ohio suggests that hard water may influence the incidence of heart disease.[15] The Ohio study found that counties with relatively soft drinking water derived from coal-bearing rocks in the southeastern part of the state tend to have a higher death rate due to heart attack than do counties with relatively hard drinking water derived from young glacial deposits (Figure 20.2).

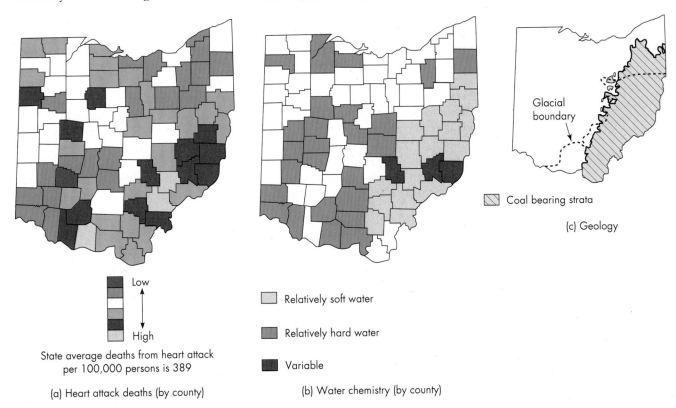

State average deaths from heart attack per 100,000 persons is 389

(a) Heart attack deaths (by county)

(b) Water chemistry (by county)

Glacial boundary

Coal bearing strata

(c) Geology

Figure 20.2 Heart attacks Occurrence of heart attack deaths by county and distribution of soft and hard water by county in Ohio. (*After Bain, R. J. 1979. Geology, 7:7–10*)

It is important to realize, however, that this generally negative correlation is not conclusive. A study in Indiana found a small positive correlation between heart disease and hardness of water, suggesting that many other variables may exert considerable influence on rates of heart disease.[16]

The correlations discussed do not necessarily show a cause-and-effect relationship between the geochemical environment and heart disease. If there is a cause-and-effect relationship, we do not know its nature, but there are several possibilities:

- Soft water is more acidic than hard water and may, through corrosion of pipes, release into the water trace elements that cause heart disease.
- Some other characteristics of soft water may contribute directly to heart disease.
- Some substances dissolved in hard water may help prevent heart disease.

Of course, some combination of these and other factors is also possible and is consistent with our observation that a disease may have several causes. Additional research is needed to prove the benefit of hard water and perhaps suggest treatment of soft water to reduce heart disease.

Cancer and the Geochemical Environment

Cancer tends to be strongly related to environmental conditions. However, as is the case for heart disease, relationships between the geochemical environment and cancer have not been proved. The causes of the various types of cancers are undoubtedly complex and involve many variables, some of which may be the presence or absence of certain Earth materials.

Cancer-causing, or *carcinogenic,* substances in the environment have two origins: Some occur naturally in Earth materials such as soil and water; others are released into the environment by human use. In recent years, attention has been focused on the many known and suspected carcinogens released by human industrial activities. This awareness has sometimes resulted in alarm concerning substances whose relationship to cancer is not well demonstrated. This does not mean that all concern about industrial carcinogens is misplaced. Recent information suggests that variable concentrations of cancer-causing substances may be found in much of our drinking water. Carcinogens may or may not be in our drinking water, but certainly water polluted with industrial waste containing toxic chemicals, some of them possible carcinogens, is being released into our surficial water supplies. The Mississippi River, particularly, has pollution problems. Ironically, present methods of water treatment that use chlorine may contribute to these problems: When combined with chlorine, some industrial waste turns into cancer-producing material. In addition, antiquated water-treatment procedures in some areas fail to remove certain carcinogens. Naturally occurring airborne carcinogenic substances are also found in some of our homes and buildings (see A Closer Look: Radon Gas).

20.3 Environmental Planning: Site Selection

Site selection is the process of evaluating an environment that will support human activities. It is a task shared by Earth scientists, engineers, landscape architects, geographers, ecologists, planners, social scientists, and economists, and thus involves a multidimensional approach to landscape evaluation.

The goal of site evaluation for a particular land use is to ensure that site development is compatible with both the possibilities and the limitations of the natural environment. Although it is obviously advantageous to know the possibilities and limitations of a site before development begins, site evaluation is often overlooked. People still purchase land for various activities without considering whether the land use they have in mind is compatible with the site they have chosen.

A CLOSER LOOK | Radon Gas

Radon is a naturally occurring radioactive gas that is colorless, odorless, and tasteless. Uranium-bearing rocks are the source of the radon gas that contaminates many homes in the United States.

Exposure to elevated concentrations of radon is associated with a greater risk of lung cancer, particularly for smokers. The risk is thought to increase as the level of radon concentration, the length of exposure to radon, and amount of smoking increase.[17,18]

In recent years, there have been approximately 140,000 lung cancer deaths each year in the United States. Although smoking is the most important factor associated with lung cancer, the Environmental Protection Agency estimates that per year, 7,000 to 30,000 of these lung cancer deaths are related to exposure to radon gas. One estimate is that exposure to both tobacco and radon is approximately 10 times as hazardous as exposure to either of these pollutants by itself. However, there have been few direct studies linking radon gas exposure in houses to increased incidence of lung cancer. Estimates of such linkage come from studies of people who have experienced high exposure to radiation through such activities as mining uranium.

In terms of the implied risk, large concentrations of radon are indeed hazardous, and it has not been shown that small concentrations are harmless. The Environmental Protection Agency (EPA) has set 4.0 picoCuries of radon gas per liter of air (4.0 pCi/L) as the radioactivity level beyond which radon gas is considered to be a hazard. The radioactivity level of 4.0 pCi/L is only an estimate of a concentration target to which indoor levels should be reduced. The average outdoor concentration of radon gas is approximately 0.4 pCi/L, and the average indoor level about 1.0 pCi/L. The comparable risk at 4.0 pCi/L, for a nonsmoker, is about the same as the risk of drowning. For a smoker, the comparable risk at this level is about 100 times the risk of dying in an airplane crash (Figure 20.A). The risk of radon exposure can also be estimated according to the number of people who might contract lung cancer, as shown on Figure 20.A. These estimations have been difficult to verify. One U.S. study of nonsmoking women reported that a significant positive relationship between radon exposure and lung cancer was not found.[19] Another study in Sweden did report a statistically significant positive relation between exposure to radon gas in homes and lung cancer.[20] The latter concluded that residential exposure to radon gas is an important contributing factor in the incidence of lung cancer in the general population of Sweden.

The Geology of Radon Gas

Both rock type and geologic structure are important in determining how much radon is likely to reach the surface of Earth. Uranium-238 concentrations in rocks and soil can vary greatly. Some rock types, such as sandstone, generally contain less than 1 ppm U-238; others, such as some dark shales and some granites, may contain more than 3 ppm U-238. The actual amount of radon that reaches the surface of Earth is related to the concentration of uranium in the rock and soil as well as the efficiency of the transfer processes from the rock or soil to soil-water and soil-gas.

Some regions of the United States contain bedrock with an above-average natural concentration of uranium. The Redding Prong that covers parts of Pennsylvania, New Jersey, and New York is famous for elevated concentrations of radon gas (Figure 20.B). Homes in this region also contain elevated concentrations of radon gas.[21] Similarly, two areas in Florida have elevated concentrations of uranium in addition to radioactive potassium from phosphate-rich rocks. Many other states, including Illinois, New Mexico, South Dakota, North Dakota, and Washington, have also identified areas with elevated indoor radon concentrations. One type of dark shale found in the Santa Barbara, California, area has been identified as a significant radon gas producer.

How Radon Gas Enters Homes

Radon gas enters homes by three major pathways (Figure 20.C):[22]

- As a gas migrating up from soil and rocks into basements and other parts of houses
- In groundwater pumped into wells
- In construction materials, such as building blocks, made of substances that emit radon gas

Most of the early interest in radon gas in homes was initiated because of the use of building materials containing high concentrations of radium, a product of the radioactive decay of uranium (see Chapter 15). For example, between 1929 and 1975, materials used to manufacture concrete in Sweden contained relatively high concentrations of radium. Similarly, in the 1960s, it was discovered that construction materials in some homes and other buildings in the Grand Junction area of Colorado were contaminated by uranium mine waste. In Florida, houses were discovered to have relatively high concentrations of radon where phosphate mining waste had been used as fill dirt. Some homes in New Jersey were built at a landfill site where wastes from a radium-processing plant had earlier been deposited.[23]

Radon is commonly found in homes because it is a gas that can move through the small openings in soils and rocks. Radon migrates up with soil gases and seeps through concrete floors and foundations, floor drains, or small cracks and pores in block walls. Radon enters homes for the same reason that smoke goes up a chimney, in a process known as the chimney, or stack, effect: Houses are generally warmer than the surrounding soil and rock, so as the gases and air rise, radon is drawn into the house. Wind is an additional factor in radon contamination because it increases air flow into and out of buildings. Radon also enters homes through the water supply, particularly if the home is supplied by a private well; however, this contribution is usually much smaller than the gas that seeps up through house foundations. Radon is released into the air when the water is used for daily functions such as showering, dishwashing, and clothes washing. In general, approximately 10,000 pCi/L of radon in water will produce about 1 pCi/L of radon in indoor air.

Scope and Perception of the Problem

Since 1985, awareness of radon gas has increased in the United States. Nevertheless, people have a hard time focusing on a problem that they cannot see, smell, hear, or touch. There is no good estimate of the number of homes in the United

RADON RISK IF YOU SMOKE

Radon level	If 1,000 people who smoked were exposed to this level over a lifetime...	The risk of cancer from radon exposure compares to...	WHAT TO DO: Stop smoking and...
20 pCi/l	About 135 people could get lung cancer	← 100 times the risk of drowning	Fix your home
10 pCi/l	About 71 people could get lung cancer	← 100 times the risk of dying in a home fire	Fix your home
8 pCi/l	About 57 people could get lung cancer		Fix your home
4 pCi/l	About 29 people could get lung cancer	←100 times the risk of dying in an airplane crash	Fix your home
2 pCi/l	About 15 people could get lung cancer	← 2 times the risk of dying in a car crash	Consider fixing between 2 and 4 pCi/l
1.3 pCi/l	About 9 people could get lung cancer	(Average indoor radon level)	
0.4 pCi/l	About 3 people could get lung cancer	(Average outdoor radon level)	(Reducing radon levels below 2 pCi/l is difficult)

Note: If you are a former smoker, your risk may be lower.

RADON RISK IF YOU'VE NEVER SMOKED

Radon level	If 1,000 people who never smoked were exposed to this level over a lifetime...	The risk of cancer from radon exposure compares to...	WHAT TO DO:
20 pCi/l	About 8 people could get lung cancer	← The risk of being killed in a violent crime	Fix your home
10 pCi/l	About 4 people could get lung cancer		Fix your home
8 pCi/l	About 3 people could get lung cancer	← 10 times the risk of dying in an airplane crash	Fix your home
4 pCi/l	About 2 people could get lung cancer	← The risk of drowning	Fix your home
2 pCi/l	About 1 person could get lung cancer	← The risk of dying in a home fire	Consider fixing between 2 and 4 pCi/l
1.3 pCi/l	Less than 1 person could get lung cancer	(Average indoor radon level)	
0.4 pCi/l	Less than 1 person could get lung cancer	(Average outdoor radon level)	(Reducing radon levels below 2 pCi/l is difficult)

Note: If you are a former smoker, your risk may be higher.

Figure 20.A Estimated risk associated with radon These estimates are calculated as long-term exposure risks for a person living about 70 years and spending about 75 percent of the time in the home with a designated level of radon. *(From U.S. Environmental Protection Agency. 1986. A citizen's guide to radon. OPA-86-004)*

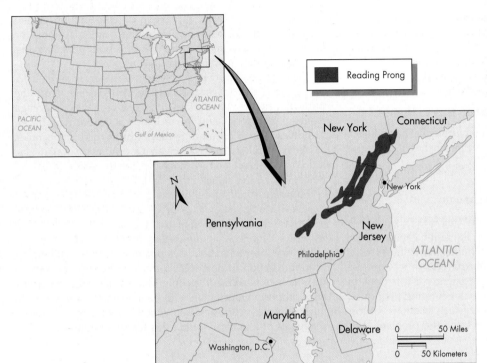

Figure 20.B The Reading **Prong area** is located in the eastern United States, where high levels of indoor radon were first discovered in the early 1980s. *(Modified after U.S. Geological Survey. 1986. U.S.G.S. Yearbook)*

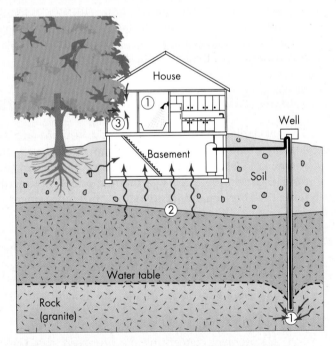

Figure 20.C **How radon may enter homes** (1) Radon in groundwater enters the well and goes to the house, where it is used as drinking water and for dishwashing, showers, and other purposes. (2) Radon gas in rock and soil migrates into the basement through cracks in the foundation and pores in construction. (3) Radon gas is emitted from construction materials used in building the house. *(From Environmental Protection Agency)*

States that may have elevated concentrations of radon gas. The tests that have been done indicate that approximately 1 in 12 homes surveyed have an indoor radon level above 4 pCi/L. If this rate is a good average, then approximately 7 million homes in the United States would have elevated rates, and millions more will need to be tested.

Reducing Concentrations of Radon Gas in Homes

A successful program to reduce or limit a potential hazard from radon gas in homes centers upon three major strategies:[24,25]

- Improve home ventilation by keeping more windows open or using fans.
- Locate and seal points of radon gas entry.
- Use construction methods that provide a venting system.

The simplest method of reducing the radon concentration is to increase the ventilation in the home. Sometimes this is sufficient to solve the problem. Sealing cracks in the foundation can also reduce radon entry, particularly if it is done in specific sites where the radon is actually entering the house. Detectors are available to help identify such locations. Unfortunately, this method often fails to solve the problem, and

new cracks may form. A variety of construction options are open to the homeowner, including venting systems in a basement or crawl space. If the house is built on a slab, then subslab ventilation systems may be installed.

There is good and bad news concerning radon gas. The bad news is that many parts of the United States and other parts of the world have relatively high emissions of radon gas from soil and rocks, and the radiation is producing a hazard in homes. The good news is that in most cases the problem can be fixed relatively easily. Even if a ventilation system is required, it costs only a few thousand dollars, generally a small percentage of the value of the home.

Future research may show that health risks from radon gas exposure are not as great as predicted by the EPA. Some scientists are skeptical and think that the hazard level of 4 pCi/L may be set too low.[26] Nonetheless, it is important for people to become informed about this potential problem. Taking appropriate action to deal with the possibility of radon gas in their homes is difficult; people seem to be more concerned with property values than with health issues. Both recognizing the whole picture and knowing that reducing radon gas is a solvable problem can reduce our fears about radon.

The Earth scientist plays a significant role in the evaluation process by providing crucial geologic information. This information includes soil and rock types, rock structure, especially fractures, drainage characteristics, groundwater characteristics, landform information, and estimates of possible hazardous Earth events and processes, such as floods, landslides, earthquakes, and volcanic activity. The engineering geologist also takes samples, makes tests, and predicts the engineering properties of the Earth materials.

20.4 Environmental Impact Analysis

The probable effects of land use by humans are generally referred to as the *environmental impact*. This term became popular in 1969 when the National Environmental Policy Act (NEPA) required that all major federal actions that could possibly affect the quality of the human environment be preceded by an evaluation of the project and its impact on the environment.

Environmental Impact Statements

To carry out both the letter and the spirit of NEPA, the Council on Environmental Quality prepared guidelines to help in preparing the **environmental impact statement (EIS).** The major components of the statement, according to the revised guidelines issued in 1979, are the following:[27]

- A summary of the EIS
- A statement concerning the proposed acts, their purpose, and need for the project
- A rigorous comparison of the reasonable alternatives
- A succinct description of the environment affected by the proposed project
- A discussion of the proposed project, its environmental consequences, and alternatives. This must include direct and indirect effects; energy requirements and conservation potential; possible depletion of resources; impact on urban quality and cultural and/or historical resources; possible conflicts with state or local land-use plans, policies, and controls; and mitigation measures.

Scoping. The environmental impact statement process was criticized during the first 10 years under NEPA because it initiated a tremendous volume of paperwork by requiring detailed reports that tended to obscure important issues. In response, the revised regulations introduced the idea of **scoping,** the process of identifying important environmental issues that require detailed evaluation early in the planning of a proposed project. As part of this process, federal, state, and local agencies, as well as citizen groups and individuals, are asked to participate by identifying issues and alternatives that should be addressed as part of the environmental analysis.

Mitigation. Another important concept in an environmental impact analysis is **mitigation.** Mitigation involves identifying actions that will avoid, lessen, or compensate for anticipated adverse environmental impacts of a particular project. For example, if a project involves filling in wetlands, a possible mitigation might be enhancement or creation of wetlands at another site.

Mitigation is becoming a common feature of many environmental impact statements and state environmental impact reports. Unfortunately, it may be overused. Sometimes mitigation is not possible for a particular environmental disruption. Furthermore, we often do not know enough about restoring and creating habitats and environments such as wetlands to do a proper job. Requiring mitigation procedures may be useful in many instances, but it must not be considered an across-the-board acceptable way to circumvent adverse environmental impacts associated with a particular project.

Negative Declarations

Negative declarations are filed when an agency determines that a particular project does not have a significant adverse impact on the environment. The negative declaration requires a statement that includes a description of the project and detailed information that supports the contention that the project will not have a significant effect on the environment. A negative declaration does not have to consider a wide variety of alternatives to the project, but it should be a complete and comprehensive statement concerning potential environmental problems.[28] Although the language and law associated with the concept of the negative declaration are different at the federal and state levels, it is an important component of environmental assessment.

An Example of Environmental Impact Analysis

The San Joaquin Valley in central California is one of the richest farm regions in the world (Figure 20.3). Farming on the west side of the valley has been made possible through extensive irrigation from deep wells and by transporting water via canals that originate in the Sierra Nevada. However, development of the land for intensive agriculture has resulted in environmental problems including soil erosion, pollution from fertilizers and pesticides, and damage to soils as a result of irrigation. In this discussion, we will explore the environmental problem of selenium toxicity in water as a result of irrigation practices and the environmental analysis completed in an attempt to solve the problem.

On the west side of the valley, soils have developed on alluvial fan deposits shed from the Coast Ranges. With an annual average precipitation of less than 25 cm (10 in.), irrigation is necessary for agriculture. Drainage problems exist because at shallow depths of 3 to 15 m (10 to 50 ft), there are often clay barriers that restrict the downward migration of water and salts. When irrigation waters are applied to fields, these clay barriers produce a local near-surface zone of saturation, an example of perched groundwater found near the surface above the regional groundwater level (see Chapter 12). As more irrigation water is applied, the saturated zone may rise to the crop-root zone and drown the plants.[29] A subsurface drainage system may be installed to remove the excess water.

Figure 20.3 **The San Joaquin Valley** The San Joaquin Valley and San Luis Drain, which terminates at Kesterson Reservoir. *(Modified from Tanji, K., Lauchli, A., and Meyer, J. 1986. Selenium in the San Joachin Valley. Environment 28[6])*

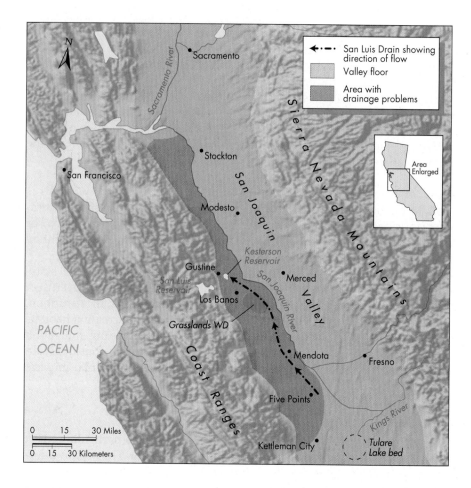

Irrigation water contains a variable amount of dissolved salts, which are left behind in the soil as soil water evaporates and plants remove water through transpiration, the evaporation of water from their leaves. Periodic leaching by applying large amounts of fresh water will remove the salts, but subsurface drains are necessary to remove the salty water (Figure 20.4). Tile drains beneath each field connect to larger drains that eventually join a master drainage system, usually a canal.

The San Luis Drain, a 135 km (85 mi) long concrete-lined canal, was constructed between 1968 and 1975, before the requirement of environmental impact analysis, to convey the agricultural drainage water northward toward the San Francisco Bay area (Figure 20.3). The drain to the bay was never completed because of limited funding and uncertainty over the potential adverse environmental effects of discharging drain water into the bay. Instead, Kesterson Reservoir, a series of 12 ponds, was used as the terminal point for the drainage water (Figure 20.5).

During the first few years following construction of Kesterson Reservoir, the flow in the drain was mostly fresh water purchased to provide water for the Kesterson Wildlife Refuge, established primarily for waterfowl. The reservoir started receiving salty agricultural drainage in 1978, and by 1981 this was its main source of water. The drain and reservoir at first seemed to be a good system to get rid of the wastewater from farming activities; unfortunately, as the water infiltrated the soil, it picked up the heavy metal selenium in addition to salts and agricultural chemicals. The selenium is derived from weathering of sedimentary rock in the Coast Ranges on the west side of the valley. The sediments, including the selenium, are carried by streams to the San Joaquin Valley and deposited there. Potential problems due to the selenium might have been anticipated had adequate environmental analysis been done prior to construction of the San Luis Drain.

In 1982, a U.S. Fish and Wildlife study revealed unusually high selenium levels in fish from Kesterson Reservoir. Dead and deformed chicks from waterbirds and

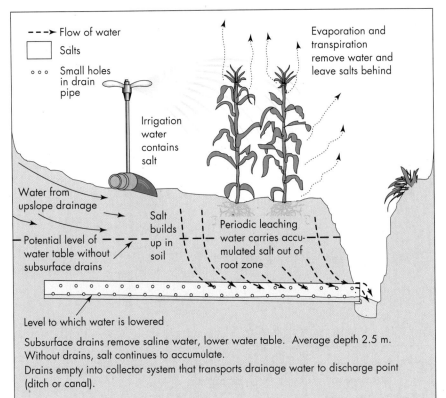

Figure 20.4 **Agricultural drainage** Diagram indicating the need for subsurface drains on the west side of the San Joaquin Valley. *(Modified after U.S. Department of the Interior, Bureau of Reclamation. 1984. Drainage and salt disposal. Information Bulletin 1. San Luis Unit, Central Valley Project, California)*

other wildfowl were reported from 1983 to 1985. Water sample analysis revealed that selenium in the ponds ranged from 60 to 390 parts per billion (ppb). Selenium concentrations as high as 4,000 ppb were measured in the agricultural drain waters. These concentrations of selenium greatly exceeded the 10 ppb drinking water standard set by the Environmental Protection Agency. As a result of the studies, it was inferred that selenium toxicity was the probable cause of the deaths and deformities of the wildfowl. The selenium was clearly shown to come from the San Luis Drain, and as the water spread out through the Kesterson Wildlife Refuge, the heavy metal was incorporated into the environment in increasingly higher concentrations due to *biomagnification* (the increase in concentrations of a substance at higher levels of the food chain).

In 1985, San Luis Drain wastewater at Kesterson was classified as hazardous and a threat to public health. At that time, the state Water Resources Control Board ordered the U.S. Department of the Interior's Bureau of Reclamation, which was running the facility, to alleviate the hazardous condition at the reservoir; they were given 5 months to develop a cleanup plan and 3 years to comply with it.[29] This order initiated the formal environmental impact analysis of Kesterson Reservoir. The immediate strategy was threefold: First, initiate a monitoring program; second, take steps (scare off the birds) to ensure that wildfowl would no longer use the wetlands; and third, reduce public exposure to the hazard.

The Kesterson case history is interesting because it provides insight into the environmental planning and review process. Primary environmental concerns and impacts were identified by the scoping process, which identified the following issues:

- Disposition of Kesterson Reservoir and the San Luis Drain after cleanup

- Environmental, social, and economic impacts of closing the Kesterson Reservoir to agricultural drainwater discharge

- Potential public health impacts of selenium and other contaminants in the drain water

Figure 20.5 Wildlife refuge
(a) Map of Kesterson Reservoir ponds. The location of Kesterson Reservoir is shown in Figure 20.8. *(After U.S. Department of the Interior. 1987. Kesterson program. Fact Sheet No. 4)* (b) Aerial photograph of the site. Pond no.4 is in the foreground and, at the time this photograph was taken in 1983, was filled with seleniferous agriculture drainage water. *(San Luis National Wildlife Refuge Complex)*

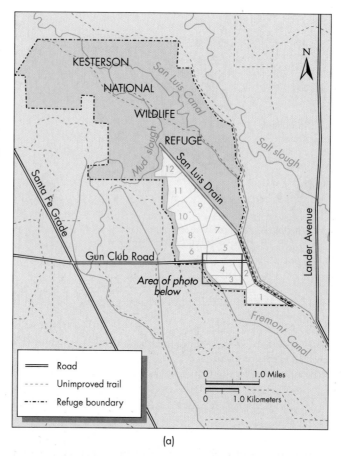

(a)

(b)

- Consideration of alternative methods to clean up the water, soils, sediment, and vegetation
- Potential for migration of contaminated groundwater away from the Kesterson area

In October 1986, the Bureau of Reclamation issued a final environmental impact statement that addressed the impacts of alternative methods to clean up Kesterson Reservoir and the San Luis Drain.[30] In the end, about one million m^3 (yd^3) of clean soil was brought in filling ponds and covering the worst areas with as much as 30 cm (20 in.) of soil. This has largely solved the threat of biological contamination. The board required the bureau to mitigate the loss of Kesterson Wildlife Refuge by providing alternative wetlands for wildlife. Although the plan may help solve the Kesterson problem, the potentially widespread selenium problem in the western San Joaquin Valley needs a more general long-term solution. People in the valley cannot afford to construct and maintain a series of toxic waste dumps for contaminated soil and vegetation. Research to evaluate the selenium problem is continuing, so the initial plans may change with time as more information is obtained.

Filling Kesterson ponds with soil has been successful in reducing impacts on wildfowl. Site monitoring has been an ongoing project since the ponds were filled. Wetlands established through the process of mitigation are providing valuable wildlife habitat in an area where, historically, wetlands have been lost to other land uses. Selenium-laden agricultural drain water from the 17,000 hectares (65 mi^2) of irrigated agriculture land in the area has been eliminated, but at a significant cost to farmers. Groundwater has risen in some fields as drainage was eliminated, threatening their productivity. Finally, the selenium problem in the broader San Joaquin Valley has not disappeared. Selenium-contaminated water from other agricultural drains still continues to cause concern.

20.5 Land Use and Planning

Land use in the contiguous United States is dominated by agriculture and forestry, with only a small portion of the land, approximately 3 percent, used for urban purposes. Currently, several thousand square kilometers of land per year are converted to uses other than agriculture. About one-half of the conversion is for wilderness areas, parks, recreation areas, and wildlife refuges. The other half is for urban development, transportation networks, and facilities. On a national scale, urbanization of rural lands may appear to be proceeding slowly. However, it often occurs in rapidly growing urban areas, where it may be viewed as destroying agricultural and natural lands and intensifying existing urban environmental problems. Urbanization in areas with scenic and recreation value is often viewed as potentially damaging to important ecosystems.

Land-Use Planning

Land-use planning is an important environmental issue. Good land-use planning is essential for sound economic development, for avoiding conflicts between land uses, and for maintaining a high quality of life in our communities. When a business manages its capital and resources in an efficient manner, we call that good business. When a city or county efficiently manages its land and resources, we call that good planning.[31] From an Earth science perspective, the basic philosophy of good land-use planning is to avoid hazards, conserve natural resources, and generally protect the environment through the use of sound ecological principles.

The land-use planning process shown in Figure 20.6 includes several steps:[32]

- Identify and define issues, problems, goals, and objectives
- Collect, analyze, and interpret data (including an inventory of environmental resources and hazards)
- Develop and test alternatives

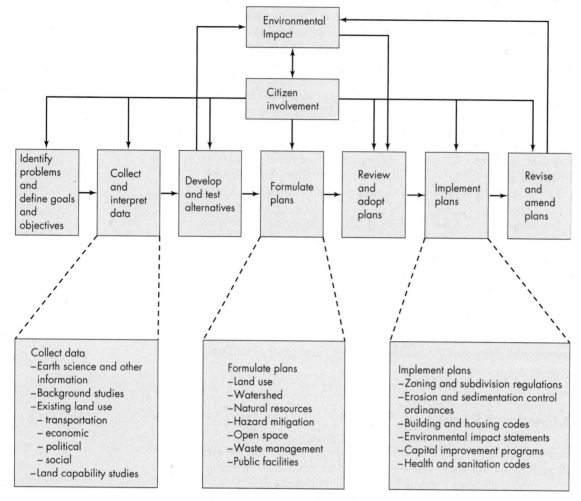

Figure 20.6 The land-use planning process *(Modified after U.S. Geological Survey Circular 721, 1976)*

- Formulate land-use plans
- Review and adopt plans
- Implement plans
- Revise and amend plans

The three most important steps, each of which comprises a complex list of factors, are data collection, formulation of plans, and implementation of plans, as illustrated on Figure 20.6 (lower part).

The role of the Earth scientist in the planning process is most significant in the data collection and analysis stage. Depending on the specific task or plan, the Earth scientist will use available Earth science information, collect necessary new data, prepare pertinent technical information such as interpretive maps and texts, and assist in the preparation of land-capability maps. Ideally, the natural capability of a land unit should match its specific potential uses.[33]

Scenic Resources

Scenery in the United States has been recognized as a natural resource since 1864 when the first state park, Yosemite Valley in California, which later became Yosemite National Park, was established. The early recognition of scenic resources was primarily concerned with outdoor recreation, focusing on preservation and management of discrete parcels of unique scenic landscapes. Public awareness of and concern for the scenic value of the "everyday" nonurban landscape is relatively recent. Society now fully recognizes scenery—even when it is not spectacular—as a valuable resource. There is now awareness that landscapes have varying degrees of scenic value, just as more tangible resources have varying degrees of economic value. Earth scientists, as members of a team evaluating the entire environment, assist in characterization of the landscape and its resources, including scenery.

Sequential Land Use

The need to use land near urban areas for a variety of human activities has, in some instances, led to application of the concept of sequential land use rather than permanent, exclusive use. The concept of sequential use of the land is consistent with the principle that the effects of land use are cumulative, and therefore we have a responsibility to future generations. The basic idea is that after a particular activity, such as mining or a landfill operation, has been completed, the land is reclaimed for another purpose.

There are several examples of sequential land use. Sanitary landfill sites have been planned so that when the site is completed, the land will be used for recreational purposes, such as a golf course. The city of Denver used abandoned sand and gravel pits once used for sanitary landfill as sites for a parking lot and the Denver Coliseum. Bay Harbor, Michigan, is a modern upscale community and a world-class resort in Little Traverse Bay, near the north end of Lake Michigan. The harbor and other development resulted from restoration of an abandoned shale quarry and cement plant. The restoration transformed a site that was a source of cement dust, an air pollutant, into a high-value landscape that people wish to visit (Figure 20.7).

Multiple Land Use

Multiple land use occurs where land may be used for more than one purpose. For example, our national forests are used for a variety of purposes including

(a)   (b)

Figure 20.7 **Sequential land use at Little Traverse Bay, Lake Michigan** Near Petoskey, Michigan, an abandoned quarry and cement plant (a) were transformed into a world-class resort (b). (Note quarry wall behind the plant and small building right of the access road.) *([a] Courtesy of Ned Tanner; [b] courtesy of Bob Fell)*

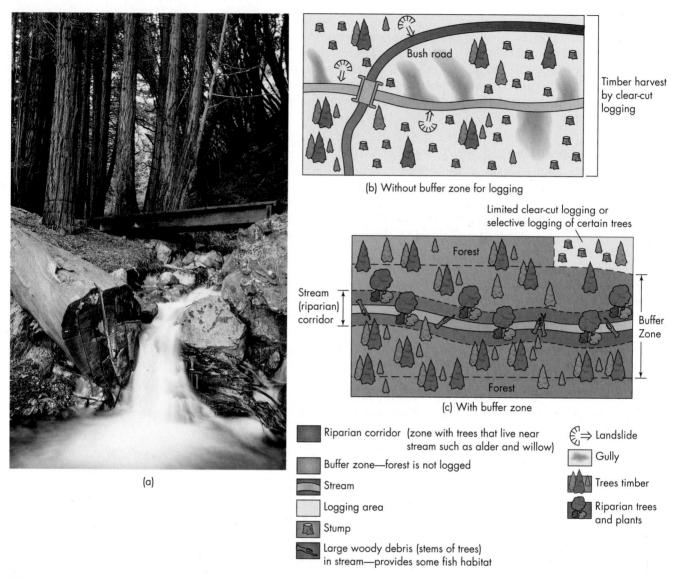

Figure 20.8 Timber harvesting and erosion (a) Stream in redwood forest of northwestern California. *(Rich Reid/Animals Animals/Earth Scenes)* (b) Landsliding and gully erosion damage a forest stream when a wide buffer zone is not part of the plan to harvest timber. This type of damage is particularly common when clear-cut logging, which removes nearly all trees, is used. (c) Timber harvest plan that leaves the stream area intact and has a wide buffer zone.

recreation and timber harvesting. Other examples include reservoirs designed for providing irrigation water, flood control, and recreation. It has often been said that it is difficult to maximize benefits for more than one purpose. This is certainly true for multiple land use. Recreation and timber harvesting are not compatible if we wish to maximize both the number of trees harvested and the number and quality of fish in a forest stream. Logging, unless done very carefully and with a wide buffer zone to protect the stream, will damage a stream environment (Figure 20.8). Damage may occur by removing trees that provide shade for the stream, keeping the water cool; through sediment pollution from increased runoff and soil erosion; and through landsliding and the formation of gullies. Likewise, a reservoir designed to provide water for summer irrigation of farmland may be full in the spring when flood protection is needed and be very low, exposing the shoreline to erosion, when people arrive for late summer recreation (Figure 20.9).

Figure 20.9 **Exposed shoreline**
Pen-y-garrea Reservoir in Wales at low
water. The exposed shoreline is not a
pleasant setting for recreational activity.
(Animals Animals/Earth Scenes)

20.6 Environmental Law

Environmental planning needs implementation and enforcement, and **environmental law** is therefore becoming an important part of our jurisprudence. Environmental law is becoming increasingly significant among lawyers and environmental law societies, and environmental courses have been established at law schools.

The Process of Law

It is beyond the scope of our discussion to consider in detail the process of law as it relates to the environment. Suffice it to say that law is a technique for the ordered accomplishment of economic, social, and political purposes, and the most desirable legal technique generally is one that most quickly allows ends to be reached. Laws related to the environment too often primarily serve the major interest that dominates the culture, and in our sophisticated culture the major concerns are wealth and power. These concerns stress society's ability to use the resource base to produce goods and services, and the legal system provides the vehicle to ensure that productivity.[34] On the other hand, in recent years many laws have been established to protect the air, water, land, and living environment at the local, regional, and global levels. This is an encouraging sign if we hope to sustain Earth's renewable resources.

Some environmental lawyers today believe that the process of law as it has generally been practiced is not working satisfactorily when environmental issues are concerned. In general, when two views conflict, adversarial confrontation occurs. Emotional levels are often high, and it may be difficult for disagreeing parties to see positions other than their own. An emerging view in environmental law is one that stresses problem solving. This may take the form of mediation through negotiation. For example, in the 1970s the Environmental Protection Agency often announced new environmental regulations that were thrust upon various sectors of society without warning. Not surprisingly, many of the people and organizations affected by these regulations ended up in lawsuits and lengthy litigation. In the 1980s, the EPA began a practice of consulting interested parties before enacting regulations. Consultation, negotiation, and *mediation* may well prove much more successful than earlier strategies that produced unproductive adversarial reactions.

One of the major problems in mediation and negotiation is getting all the important players in a given case to sit down and talk about the issues in a meaningful way. The party with the "upper hand" may try to make the negotiations particularly difficult for opposing parties. Increasingly, however, all parties in environmental issues are recognizing that it is advantageous to work together toward a solution that is satisfactory for everyone. This is basically a collaborative process that seeks solutions that favor the environment while allowing activities and projects to go forward.

It is important to recognize that collaboration is different from and broader than compromise, which often requires giving something up to get something else. Collaboration is more comprehensive in that it asks—in fact, necessitates—that the parties work together to create opportunities for mutual gain. Collaboration creates a climate of joint problem solving. What is required for negotiation and mediation to work is for all parties to clearly and honestly state their positions and then work to see where common ground might be found. Relationships built upon mutual trust are then developed. It should come as no surprise that almost all issues are negotiable, and alternatives can often be worked out that avoid or at least minimize costly litigation and lengthy delays.

The Storm King Mountain Case

The Storm King Mountain dispute is a classic example of conflict between a utility company and conservationists. In 1962 the Consolidated Edison Company of New York announced plans for a hydroelectric project approximately 64 km (40 mi) north of New York City on Storm King Mountain, part of the Hudson River Highlands, an area in the Appalachian Mountains considered by many to have unique aesthetic value. It is the only place in the eastern United States where a major river has eroded through mountains at sea level, giving the effect of a fjord.[35]

Early plans called for construction of an aboveground powerhouse, which would have required a deep cut into Storm King Mountain. The project was redesigned to site the powerhouse entirely underground, eliminating the cut on the mountain (Figure 20.10). However, conservationists continued to oppose the project, and the issues broadened to include possible damage to fisheries. The conservationists argued that the high rate of water intake from the river would draw many fish larvae into the plant, where they would be destroyed by turbulence and abrasion. The most valuable sport fish in the river is the striped bass, and one study showed that 25 percent to 75 percent of the annual bass hatch might be destroyed if the plant were operating. The fish return from the ocean to tidal water to spawn, and, since the Hudson River is the only estuary north of Chesapeake Bay where the striped bass spawn, the concern for the safety of the fisheries was justified. The problem was even more severe because the proposed plant was to be located near the striped bass spawning area.[35]

The Storm King Mountain controversy is interesting because it emphasizes the difficulty of making decisions about multidimensional issues. A utility company was trying to survive in New York City, where high peak-power demands are accompanied by high labor and maintenance costs. At the same time, conservationists were fighting to preserve a beautiful landscape and fishery resources. Both had legitimate arguments, but in light of their special interests, it was difficult to resolve the conflicts. Existing laws and procedures were sufficient to resolve the issues, but trade-offs were necessary. Ultimately, an economic and environmental price must be paid for any decision, a price that reflects our desired lifestyle and standard of living.

The first lawsuit in the Storm King dispute was filed in 1965, and, after 16 years of intense courtroom battles, the dispute was settled in 1981. The total paper trail exceeded 20,000 pages, and in the end the various parties used an outside mediator to help them settle their differences. This famous case has been cited as a major victory for environmentalists. Could the outcome have been decided much earlier if the various litigants had been able to sit down and talk openly about the issues? Perhaps through the processes of negotiation and mediation, the dispute might have been settled much sooner at lesser cost to the individual parties and society in general.[36] This case emphasizes that environmental law issues may best be approached from a problem-solving viewpoint rather than by the sides taking adversarial positions.

(a)

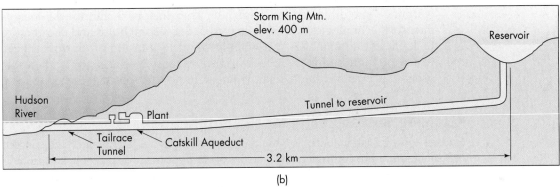

(b)

Figure 20.10 **Hudson River Highlands** (a) Storm King Mountain and the Hudson River Highlands, New York. *(Photo by Joe Deutsch)* (b) Diagram showing how the entire Storm King Mountain hydroelectric project might be placed underground. *(After Carter, L. J. 1974* Science, *184:1353–58. Copyright 1974, American Association for the Advancement of Science)*

20.7 **Geology, the Environment, and the Future**

This book started with the introduction of some philosophy and fundamental concepts associated with geology and the environment. A main message of this book is that Earth is a dynamic place and people should not build on floodplains; of course, this is an oversimplification. There are many important linkages between geologic processes, the environment, and society, and we must pay close attention to these linkages if we are to achieve sustainability.

Avoiding an Environmental Crisis: Focusing on What Can Be Done

The concept of an environmental crisis was introduced in Chapter 1. We now turn to what can be done to eliminate development of a potential crisis that could damage human society as we know it. First we will suggest short-term actions to avoid a potential food shortage in the next few years to decades. Second, we will discuss how we might achieve the long-term concept of sustainability over centuries of human use of Earth.

Lester Brown, a leader in understanding of global environmental problems and solutions, argues we should do three things (all are linked to food supply) to stop a human environmental crisis from developing that could cause tremendous social unrest and disturbance not previously seen.[1]

1. Control human population growth to the United Nations' "low" scenario of 7 to 8 billion people on Earth by 2050; we are presently on a path to the "middle" scenario of 8 to 9 billion people. Some keys to achieving this population growth reduction are the education of women, improved health care for children, and a general rise in the standard of living (see Chapter 1).

2. Conserve and sustain water resources, especially groundwater. Use of groundwater has allowed for grain production to triple in the last half century while the population has more than doubled. Water resources are simultaneously being depleted in all grain belts of the world (United States, China, India, Canada). We know how to conserve and sustain water (see Chapter 12), but need policies that include pricing water at its real value, not a subsidized lower price. We need to implement water conservation technology for agriculture and other uses sooner rather than later.

3. Control carbon emissions and stop global warming. Each incremental increase in warming results in the loss of a small portion of our grain crops to droughts and other water related hazards. Warming above a threshold temperature causes reduction of grain productivity. Solutions to stop global warming were discussed in Chapter 19. They include a shift to less energy-intensive technologies such as using small fluorescent bulbs and use of more fuel-efficient automobiles. We will also need to transition from fossil fuels, which are contributing to global warming, to those produced from renewable energy, such as wind and solar. We need to do this sooner rather than later.

Attaining Sustainability for the Future

Understanding sustainability is a main objective of this book. However, the concept of sustainability is a controversial topic. In its simplest form, to promote sustainability as applied to development or planning, we must first ensure that our renewable resources, such as water, air, soil, and living things, are maintained for future generations and ecosystems. Second, we must ensure that human processes do not degrade these resources. A well-known geologist has stated that sustainable development is impossible. This statement is accurate with respect to nonrenewable resources such as fossil fuels and minerals that are a part of our natural heritage from the geologic past; sustainable use of those resources is not possible. The best we can do is use the nonrenewable resources so that future generations have their fair share. Through conservation and recycling we can greatly extend the availability of mineral resources. With respect to renewable resources, the concept of sustainability is central to future environmental planning and our preservation of a quality environment.

The important question is, How can we develop a sustainable future? In discussing this topic I make a number of value judgments about our present development as a society. These are my opinions; your opinions may be different. This discussion is not science, based upon facts and testing hypotheses, but follows from what we learn from science. This distinction is important. Our present path of overpopulation, resource consumption, and pollution is clearly not a sustainable one. What is required is development of a new path that will transform our present way of running society, with creation of wealth as a main goal, into one that integrates social, industrial, agricultural, and environmental interest into a harmonious, sustainable system.[37] This will include[38]

- An evolution of values and lifestyles that move toward sustainability. In the United States we are already there with respect to values. Survey after survey has confirmed that the vast majority of people in the United States (90 percent) state that protecting the environment should be a top priority. About half the people believe that environmental protection does not need to conflict with economic growth.[39]

■ Recognition that sustainable development needs to include all people on Earth, rich and poor alike, and bring a higher standard of living to all without compromising the environment. That is, we should assist the disadvantaged people of the world, not take advantage of them. Working people need a living wage on which they can support their families. Exploitation of other countries' resources and labor to reduce costs of manufacturing goods or growing food undermines us all.

■ Planning for future population changes, resource use, and natural hazards in a proactive way, rather than waiting for surprises or shocks and then reacting.

World population is going to continue to increase dramatically for at least the next few decades; this increase will be most apparent in the developing world, where birth rates are the highest. As more land is needed to house our growing population, more and more people will be put at risk in terms of their proximity to hazardous processes such as hurricanes, flooding, landslides, volcanic activity, and earthquakes. Past experiences suggest that we are not particularly well prepared to deal with many of these hazards when they occur in populated areas. Therefore, a major goal of environmental geology is to identify hazardous processes and work to minimize their potential effects through sound land-use planning, a better general understanding of the processes, and a potential range of strategies that might help us learn to better live with our planet. Achieving this goal is particularly important for the future because as population increases, there will be continued demand for more and more land on which people can live. Urban areas will increase in size to become giant urban corridors, particularly along some of the coastal areas, where people tend to be most concentrated. Therefore, it behooves us to learn from our past mistakes and ensure that future development is wise development.

I hope I am "preaching to the choir" and that many of you who have studied environmental geology will have a better awareness of the world around you and the geologic processes that operate within it. As a result, you are empowered to become more in tune with the natural processes of Earth and now have the knowledge to make wiser choices concerning how we use Earth resources, where we live, and how we build a society with the objective of sustainable planning. These choices are really the heart of ensuring that our future world will be safer for people and that society will live more in harmony with the environment than in the past. This is your charge as the next generation.

SUMMARY

China is in danger of environmental collapse as it rushes into the twenty-first century. Their path is similar to that of Europe and the United States following the Industrial Revolution, in which environmental degradation was followed by attempts to improve a damaged environment. The scale of change in China is huge due to a population of 1.3 billion people seeking a more affluent lifestyle.

The death rate and incidence of specific diseases vary from one area to another; some of the variability has geologic linkages. These causes are often quite complex, and a particular disease seldom has a one-cause, one-effect relationship. Nevertheless, considerable evidence suggests that the geochemical environment is a significant factor in the incidence of some serious chronic health problems, including heart disease and cancer.

Probable effects of human use of the land are called environmental impact. The National Environmental Policy Act (NEPA) requires preparation of an environmental impact statement (EIS) for all major federal actions that could significantly affect the quality of the human environment. Scoping and mitigation are important processes in environmental impact work.

Land use is an important environmental issue. Urbanization of rural areas may destroy agricultural land or damage ecosystems with high scenic or recreational value. The limited land available for urban area expansion has led to the concept of sequential and multiple land use rather than permanent consignment of land to a single use. Land-use planning is essential for sound economic development and maintaining a high quality of life. The basic philosophy of sustainable planning, from an Earth science perspective, is to plan to avoid hazards, conserve natural resources, and generally protect the environment through the use of sound ecological principles.

The term environmental law has gained common usage and is now an important part of law. There is a move toward problem solving and mediation rather than confrontation and adversarial positions in environmental matters. When negotiation replaces inflexibility, progress in solving problems is enhanced.

Geology has an important message for our future and the environment: We live on a dynamic planet and must make careful decisions concerning where we choose to live and how we plan for sustainability. This is the essence of the linkages between geology, society, and the future. It is becoming clear that we cannot continue "business as usual" with respect to how we have treated the environment. To avoid a crisis we need to control human population growth, conserve our water resources, and stop human-induced global warming.

Revisiting Fundamental Concepts

Human Population Growth

Human population increase is an important factor in nearly all relationships between geology and society. Increasing numbers of people require the use of more resources and produce more chemicals and waste. Finding socially acceptable ways of controlling human population is a fundamental goal of environmental science and its subdiscipline, environmental geology.

Sustainability

Sustainability is part of the solution to many environmental problems, and sustaining renewable resources is an important environmental goal. How we might achieve sustainability is not entirely clear. It will require a new way of thinking that will transform our society with industrial, agricultural, and environmental interests to a system more in harmony with natural and human-modified ecosystems.

Earth as a System

Understanding relationships between society and geology requires the understanding of how physical, hydrologic, and biological systems function and change and of how they are linked to the cultures and societies of the world. Linkages between systems are related to patterns of disease and pollution that kill millions of people each year, mostly in countries where access to clean water and good sanitation is not adequate.

Hazardous Earth Processes, Risk Assessment, and Perception

Environmental analysis, regulations, and laws have the objective of reducing hazards and environmental degradation resulting from human interactions with the environment in activities such as waste management and use of land and other resources. Environmental impact analysis and assessment is an important part of perceiving and anticipating risks and changes that may result from a particular action such as constructing a dam and reservoir or disposing of toxic waste.

Scientific Knowledge and Values

Scientific understanding of geologic processes is a mature field of study. Applying this knowledge to environmental problems is an effort supported by our values. People desire a healthy environment and are willing to support legislation to protect and improve the quality of the air we breathe, the water we drink, and the land we live on. We also value other living things and ecosystems, and there are laws to protect ecosystems and endangered species.

Key Terms

environmental impact statement (EIS) (p. 646)

environmental law (p. 655)

land-use planning (p. 651)

mitigation (p. 647)

radon (p. 643)

scoping (p. 647)

site selection (p. 642)

toxicology (p. 639)

Review Questions

1. Define toxicology.

2. How can you define disease from an environmental viewpoint?

3. What are some possible explanations for the relationship between the geochemical environment and the incidence of heart disease?

4. What is radon?

5. Why do we think exposure to radon gas in homes is dangerous?

6. What are the major pathways whereby radon gas may enter homes?

7. How may concentrations of radon gas in homes be reduced?

8. What is site selection?

9. What are the major components of an environmental impact statement?

10. What is scoping?

11. Why are there potential problems with mitigation?

12. What is a negative declaration?

13. What are the major steps in land-use planning?

14. How do negotiation and mediation enter into the practice of environmental law?

Critical Thinking Questions

1. Do you see any parallels between the Industrial Revolution in Europe and the United States and what is happening in China today? Do you think China could avoid some of the environmental damages our industrial growth brought? If your answer is no, justify your response. If your answer is yes, state what China could do.

2. Consider the inverse relationship between water hardness and incidence of heart disease presented in this chapter. Four hypotheses detailing the relationship were stated, including that hardness of water has nothing to do with heart disease. Develop a strategy for testing each hypothesis.

3. Your college or university is trying to find a 10-acre site on or adjacent to the campus for future development of an academic center. Complete a survey of your campus and surrounding area and make a recommendation concerning where the new buildings should go. What criteria did you use in your decision-making process? What values are involved in your decision?

4. Is the concept of sustainability a valid one that you would support? What do you think is necessary to sustain our renewable resources? Some people say sustainability is a dead issue. Do you agree? Why or why not?

5. Do you believe the three steps stated by Lester Brown are necessary to avoid an environmental crisis in the near future? Why? Why not?

A

Minerals

Characteristic Properties of Minerals

Characteristic properties of specific minerals often have aesthetic or utility value to people. For example, Stone Age people valued natural volcanic obsidian glass and fine-grained quartz, called flint, for making tools. High-quality stone tools were collected and traded among different tribes on a regional scale. Similarly, native copper, when discovered in the Great Lakes region by Native Americans, was valued for its metallic properties of luster and malleability (discussed later) and was used for jewelry and tools. Many ancient cultures, in addition to our own, have valued gemstones such as rubies and sapphires for their beauty. Historically, one of the most valued minerals has been halite (NaCl), or common table salt, because all animals, including people, need it to live. Halite has been mined and obtained from evaporation ponds for thousands of years. Some clay minerals have been highly valued because they can be molded into useful containers, pots, and decorative statues and painted with black, red, and orange paints made from still other minerals. Today, specific physical properties of a wide variety of minerals are used for everything from ceramics to electronics, metallurgy, agriculture, and personal jewelry. Now let us take a closer look at properties of minerals.

Identifying Minerals

As discussed in Chapter 3, the identification of minerals from hand specimens is a combination of pattern recognition and testing for particular properties or characteristics of minerals. These include color, specific gravity, cleavage, crystal form, fracture, hardness, luster, and other diagnostic properties characteristic of a particular mineral or group of minerals.

Color

The *color* of a mineral can be misleading because a given mineral may have several different colors, or several different minerals may have the same color. Depending on the impurities present, quartz may be clear, pink, purple, yellow, or smoky black. The color a mineral leaves as a powder when rubbed on a porcelain plate, or "streak plate," is more reliable. The mineral hematite, Fe_2O_3, may be dull black or a shiny, dark metallic silver color, but its streak is always red.

Specific Gravity

The *specific gravity* of a mineral is the density of a mineral compared with the density of water. Water has a specific gravity of 1. The specific gravity of minerals varies from about 2.2 for halite to 19.3 for gold. Most minerals have specific gravities of about 2.5 to 4.5. In practice, we take a sample of a mineral in hand, heft it, and estimate if it is light, medium, or heavy.

Cleavage

Cleavage is the way a mineral tends to part along planes of weakness determined by the internal crystalline structure and types of chemical bonds in the mineral. Several common types of cleavage are shown in Figure A.1.

Crystal Form

Crystal form is also a useful diagnostic property for mineral identification (see Figure 3.5). For example, the elongated hexagonal pointed crystals of quartz (SiO_2), regardless of the color of the specimen, are diagnostic, as are the cube-shaped crystals of pyrite (FeS_2).

Fracture

The way a mineral commonly breaks, or *fractures*, may also help identify it. Terms to describe fracture include blocky, splintery, fibrous, or conchoidal. Conchoidal means that the material fractures like glass, usually with curved surfaces that look something like a clamshell.

Hardness

The relative *hardness* of minerals is determined by using a scale suggested by the Austrian mineralogist Friedrich Mohs. The scale ranges from 1 to 10, where 1, the hardness of the mineral talc, is the softest, and 10, the hardness of diamond, is the hardest (Table A.1). Unit

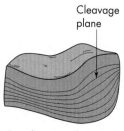

Cleavage plane

(a) One direction of cleavage. Mineral examples: muscovite and biotite micas

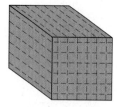

(b) Two directions of cleavage at right angles or nearly right angles. Mineral examples: feldspars, pyroxene

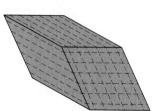

(c) Two directions of cleavage not at right angles. Mineral example: amphibole

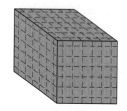

(d) Three directions of cleavage at right angles. Mineral examples: halite, galena

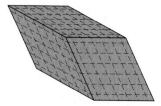

(e) Three directions of cleavage not at right angles. Mineral example: calcite

Figure A.1 **Some common types of cleavage with mineral examples** Cleavage, with the exception of mica, is shown by dashed lines. (a) Mica has one cleavage direction and breaks into sheets. (b) Feldspar and pyroxene have two cleavage directions and break at right or nearly right angles. (c) Amphibole has two directions of cleavage that do not meet at right angles. (d) Halite and galena have three cleavage directions that meet at right angles. They break into cube-shaped fragments. (e) Calcite also has three cleavage directions, but they do not meet at right angles. Calcite breaks into rhombohedrons.

increase in the scale is not equivalent to an equal increase in hardness. The hardness of a mineral is determined by a series of tests using the Mohs Hardness scale. Any mineral or substance will scratch a mineral with a lower hardness on the Mohs scale. Trying to scratch a mineral sample of unknown hardness with a knife blade (hardness about 5.5) will tell you if your unknown mineral is harder or softer than 5.5. Then, using other minerals or substances on the Mohs scale, you can determine the relative hardness of the unknown mineral sample.

Luster

Luster refers to the way light is reflected from a mineral. Some minerals have metallic luster with high reflectivity. Other minerals are translucent or transparent to light and have a nonmetallic luster. Some minerals have more than one luster. For example, some samples of graphite have metallic luster and others are nonmetallic. Terms to describe luster include pearly, greasy, and earthy.

Table A.2 contains a list of some of the common minerals and their properties. An identification key (Table A.3) may assist in the actual identification. In Table A.3 the first step is to decide if the mineral has a metallic or a nonmetallic luster. The second step is to determine the relative hardness. Third, decide whether the mineral has a cleavage. Fourth, use characteristic properties, including streak and fracture, to assist in the final identification.

It is important to note that actual identification of minerals in the field or in the laboratory from small specimens, using simple tests and perhaps a hand lens to determine cleavage and other properties, is basically a pattern recognition exercise. After you have looked at a number of minerals over a period of time and have learned how particular minerals vary, you will become more proficient at identifying a particular sample. When positive identification is necessary, mineralogists will use a variety of sophisticated equipment to determine the chemical composition and internal structure to positively identify minerals.

TABLE A.1 **Mohs Hardness Scale**

Relative Hardness		Mineral	Comment	Hardness of Common Hardness Materials
Softest	1	Talc	Softest mineral known; used to make powder	Graphite, "lead" in pencils (1–2)
	2	Gypsum	Used to make wall board	Fingernail (2.5)
	3	Calcite	Main mineral in marble, rock	Copper penny (3.0)
	4	Fluorite	Mined for fluorine, used in glass and enamel	
	5	Apatite	Enamel of your teeth	Steel knife blade, Glass (5.5–6)
	6	Orthoclase	Common rock-forming mineral	
	7	Quartz	When purple is gemstone amethyst; birthstone for February	Hard steel file (6.5)
	8	Topaz	When transparent is gemstone; birthstone for November	
Hardest	9	Corundum	When red is ruby, gemstone; birthstone for July; when blue is sapphire, gemstone; birthstone for September	
	10	Diamond	Hardest known mineral; gemstones are extremely brilliant	

Note: The scale is relative, and unit increases in hardness do not represent equal increase in hardness.

TABLE A.2 Properties and Environmental Significance of Selected Common Minerals

Mineral Group	Mineral	Chemical Formula	Color	Hardness	Other Characteristics	Comment
Silicates	Plagioclase feldspars	$(Na,Ca)Al(Si,Al)Si_2O_8$	Usually white or gray, but may be others	6	Two good cleavages at approximately 90° and may have fine striations on one of the cleavage surfaces	One of the most common of the rock-forming minerals; is a group of feldspars ranging from sodium- to calcium-rich; important industrial minerals
	Alkali feldspars	$(Na,K)AlSi_3O_8$	Gray, white to pink	6	Two good cleavages at 90°; may be translucent to opaque with vitreous luster; white streak	One of the most common of the rock-forming minerals; the potassium-rich alkali feldspar orthoclase is widely used in the porcelain industry and in a variety of industrial processes
	Quartz	SiO_2	Varies from colorless to white, gray, pink, purple, and several others, depending on impurities	7	Often has good crystal shape with six sides; conchoidal fracture	Very common rock-forming mineral; resistant to most chemical weathering; basic constituent of glasses and fluxes; commonly used as abrasive material; colored varieties such as amethyst and others provide a number of semiprecious gemstones
	Pyroxene	$(Ca,Mg,Fe)_2Si_2O_6$	Usually greenish to black	5–6	Crystals are commonly short and stout; two cleavages at about 90°	Important group of rock-forming minerals; particularly common in rocks; weathers rather quickly; an important commercial source of asbestos when they crystallize, forming a mass of easily separated thin fibers
	Amphibole	$(Na,Ca)_2(Mg,Al,Fe)_5Si_8O_{22}(OH)_2$	Generally light green to black	5–6	Distinguished from pyroxene by the cleavage angle being 120° rather than 90°; amphibole also generally has a better cleavage with higher luster than does pyroxene	Important rock-forming mineral; particularly for and metamorphic rocks; relatively nonresistant to weathering
	Olivine	$(Mg,Fe)_2SiO_4$	Generally green, but also may be yellowish	6.5–7	Generally has a glassy luster, no cleavage	Important rock-forming mineral, particularly for and metamorphic rocks; relatively nonresistant to chemical weathering

TABLE A.2 (Continued)

Mineral Group	Mineral	Chemical Formula	Color	Hardness	Other Characteristics	Comment
	Clay	Various hydrous aluminum silicates with elements such as Ca, Na, Fe, Mg	Generally white but may vary due to impurities	1–2	Generally found as soft, earthy masses composed of very fine grains; may have an earthy odor when moist; often difficult to identify the particular clay mineral from hand specimen	Clay minerals are very important from an environmental viewpoint; many uses in society today; clay-rich soils often have many engineering geology problems
	Biotite (black mica)	$K(Mg,Fe)_3AlSi_3O_{10}(OH)_2$	Black to dark brown	2.5–3	Breaks apart in parallel sheets as a result of excellent cleavage in one direction	Important rock-forming mineral common in igneous and metamorphic rocks
	Muscovite (white mica)	$KAl_2(AlSi_3)O_{10}(OH)_2$	Colorless to light gray or green or even brown if the specimen is several millimeters thick	2–3	Breaks apart into thin sheets due to excellent cleavage in one direction	Important rock-forming mineral, particularly for igneous and metamorphic; used for several purposes, including industrial roofing materials, paint, and rubber
Carbonates	Calcite	$CaCO_3$	Colorless, but may have a variety of colors resulting from impurities	3	Effervesces strongly in dilute hydrochloric acid; often breaks apart to characteristic rhombohedral pieces as a result of two good cleavages at 78°; transparent varieties display double refraction, when a single dot on a piece of white paper looks like two dots if viewed through the mineral	Main constituent of the important sedimentary rock limestone and the metamorphic rock marble; associated with a variety of environmental problems with these rocks, including the development of sinkholes; chemically weathers rapidly; used in a variety of industrial processes including asphalts, fertilizers, insecticides, and plastics
	Dolomite	$(Ca,Mg)CO_3$	Generally white but may be a variety of other colors, including light brown and pink	3.5–4	When powdered, will slowly effervesce in dilute hydrochloric acid; two cleavages at 78°; may be transparent to translucent and have a vitreous to pearly luster	Common mineral in the important sedimentary rock type dolomite and dolomitic limestones
	Malachite	$CuCO_3 \cdot Cu(OH)_2$	Bright to emerald green, but may be dark green	3.5–4	Effervesces slightly in dilute hydrochloric acid and turns the acid solution green; often displays a swirling banded structure of light and darker green bands	Valued as a decorative stone; used in jewelry making and is a copper ore

TABLE A.2 **Properties and Environmental Significance of Selected Common Minerals** *(Continued)*

Mineral Group	Mineral	Chemical Formula	Color	Hardness	Other Characteristics	Comment
Oxides	Hematite	Fe_2O_3	Usually various shades of reddish brown to red to dark gray	5.5–6.5	Streak is dark red	A common mineral found in small amounts in many igneous rocks, particularly basaltic rocks
	Magnetite	Fe_3O_4	Black	6	Magnetic	The most important ore of iron
	Bauxite	Hydrous aluminum oxides	Variable from yellow to brown, gray, and white	1–3	Usually has an earthy luster composed of dull, earthy masses with a colorless streak	Primary ore of aluminum; results from intense weathering of rocks in tropical and subtropical environments
	Limonite	Hydrous iron oxide with small amounts of other elements	Generally yellow to yellow-brown or black	1–5.5	Often present as earthy masses or in the form of a crust	Limonite is a term often used to describe a variety of very fine-grained hydrous oxides. Often forms from the chemical weathering of iron minerals and is present as "rust"
Sulfides	Pyrite	FeS_2	Generally a brassy yellow tarnishing to brown	6–6.5	Often present as well-formed cubic crystals with striations on crystal faces	Has been used in the production of sulfuric acid, but mostly known for adding sulfur content to coal and contributing to the formation of acid-rich waters that result from the weathering of the mineral
	Chalcopyrite	$CuFeS_2$	Dark to brassy yellow, tarnishing to iridescent films that are reddish to blue-purple	3.5–4	Easily disintegrates; greenish black streaks; lacks cleavage	An important ore of copper; often associated with gold and silver ores
	Galena	PbS	Silver gray	2.5	Gray to black streak; high specific gravity; general metallic luster	Primary ore of lead
Sulfates	Gypsum	$CaSO_4 \cdot 2H_2O$	Generally colorless to white, but may have a variety of colors due to impurities	2	Often is transparent to opaque with one perfect cleavage; may form fibrous crystals but often is an earthy moss	Several industrial uses in making of plaster of Paris for construction material, fertilizer, and flux for pottery
	Anhydrite	$CaSO_4$	Commonly white but may be colorless	3–3.5	Commonly observed as massive fine aggregates; may be translucent to transparent; colorless streak	Used for the production of sulfuric acid, as a filler material in paper, and occasionally as an ornamental stone

TABLE A.2 (Continued)

Mineral Group	Mineral	Chemical Formula	Color	Hardness	Other Characteristics	Comment
Halides	Fluorite	CaF_2	Variable colors, often purple, yellow, white, or green	4	Often observed as cubic crystals but may also be massive; four perfect cleavages	A number of industrial uses, including a flux in the metal industry, certain lenses, and in the production of hydrofluoric acid
	Halite	$NaCl$	Generally colorless to white, but may be a variety of colors due to impurities	2–2.5	Salty taste; often forms a cubic form due to three perfect cleavages	Common table salt; application to snowy roads in winter may pollute water; suggested as a host rock for nuclear waste
Native Elements	Gold	Au	Gold yellow	2.5–3	Crystals are rare; extremely high specific gravity; very ductile and malleable	Primary use as a monetary standard; used in jewelry and in the manufacture of computer chips and other electronics
	Diamond	C	Generally colorless but also occurs in various shades of yellow and blue	10	Commonly cut to display brilliant luster	Used in industrial abrasives and, of course, jewelry
	Graphite	C	Black to gray	1–2	Often occurs as foliated masses; black streak and and marks paper as well; is the "lead" in pencils	A variety of industrial uses including lubricants, dyes, and the "lead" in pencils
	Sulfur	S	Usually yellow when pure buy may be brown to black from impurities	1.5–2.5	Characteristic sulfurous smell	Often a by-product of the oil industry; may be used to manufacture sulfuric acid

Note: The scale is relative, and unit increases in hardness do not represent equal increase in hardness.

Source: Modified after Davidson, J. P. Reed, W. E., and Davis, P. M. 1997. *Exploring Earth*. Upper Saddle River, NJ: Prentice Hall; and Birchfield, B. C., Foster, R. J., Keller, E. A., Melhom, W. N., Brookins, D. G., Mintt, L. W., and Thurman, H. V. 1982. *Physical geology*, Columbus, OH: Charles E. Merrill.

TABLE A.3   **Key to Assist Mineral Identification. To use this table, (1) make a decision as to whether the luster of the mineral to be identified is nonmetallic or metallic. If it is nonmetallic, decide whether it is a light-colored or a dark-colored mineral; (2) use a knife blade to determine if the mineral specimen is harder or softer than the knife blade; (3) look for evidence of cleavage; and (4) compare your samples with the other properties shown here and use Table A.2 for final identification.**

Light-colored nonmetallic luster	Hard—not scratched by knife	Shows cleavage	White or flesh colored; 2 cleavage planes at nearly right angles; hardness, 6. Large crystals that show irregular veining are perthite.	Orthoclase (potassium feldspar)
			White or green-gray; 2 cleavage planes at nearly right angles; hardness, 6; striations on cleavage.	Plagioclaseone
		No cleavage	White, clear, or any color; glassy luster; transparent to translucent; hexagonal (six-sided) crystals; hardness, 7; conchoidal fracture.	Quartz
			Various shades of green and yellow; glassy luster; granular masses and crystals in rocks; hardness, 6.5–7 (apparent hardness may be much less).	Olivine
	Soft—scratched by knife	Shows cleavage	Colorless to white; salty taste; 3 perfect cleavages forming cubic fragments; hardness, 2–2.5.	Halite
			White, yellow to colorless; 3 perfect cleavages forming rhombohedral fragments; hardness, 3; effervesces with dilute hydrochloric acid.	Calcite
			Pink, colorless, white, or dark; rhombohedral cleavage; hardness, 3.5–4; effervesces with dilute hydrochloric acid only if powdered.	Dolomite
			White to transparent; 1 perfect cleavage; hardness, 2.	Gypsum
			Green to white; feels soapy; 1 cleavage; hardness, 1.	Talc
			Colorless to light yellow or green; transparent in thin sheets that are very elastic; 1 perfect cleavage; hardness, 2–3 (white mica).	Muscovite
			Green to white; fibrous cleavage; may form veins.	Asbestos
		No cleavage	Green to white; feels soapy; hardness, 1.	Talc
			White to transparent; hardness, 2.	Gypsum
			Yellow to greenish; resinous luster; hardness, 1.5–2.5.	Sulfur

TABLE A.3 *(Continued)*

Dark-colored nonmetallic luster	**Hard—not scratched by knife**	**Shows cleavage**	Black to dark green; 2 cleavage planes at nearly 90°; hardness, 5–6.	Pyroxine
			Black to dark green; 2 cleavage planes about 120°; hardness, 5–6.	Amphibole
		No cleavage	Various shades of green and yellow; glassy luster; granular masses and crystals in rocks; hardness, 6.5–7 (apparent hardness may be much less).	Olivine
			White, clear, or any color; glassy luster; transparent to translucent; hexagonal (six-sided) cyrstals; hardness, 7; conchoidal fracture.	Quartz
			Red to brown; red streak; earthly appearance; hardness, 5.5–6.5 (apparent hardness may be much less).	Hematite
	Soft—scratched by knife	**Shows cleavage**	Brown to black; cleavage, 1 direction; hardness, 2.5–3 (black mica).	Biotite
			Various shades of green; cleavage, 1 directon; hardness, 2–2.5 (green "mica").	Chlorite
		No cleavage	Red to brown; red streak; earthy appearance; hardness, 5.5–6.5 (apparent hardness may be less).	Hematite
			Lead-pencil black; smudges fingers; hardness, 1–2; 1 cleavage that is apparent only in large crystals.	Graphite
			Yellow-brown to dark brown; may be almost black; streak yellow-brown; earthy; hardness, 1–5.5 (usually soft).	Limonite
Metallic luster			Black; strongly magnetic; hardness, 6.	Magnetite
			Lead-pencil black; smudges fingers; hardness, 1–2; 1 cleavage that is apparent only in large crystals.	Graphite
			Brass yellow; black streak; cubic crystals, commonly with striations; hardness, 6–6.5.	Pyrite
			Brass yellow; may be tarnished; black streak; hardness, 3.5–4; massive.	Chalcopyrite
			Shiny gray; black streak; very heavy; cubic cleavage; hardness, 2.5.	Galena

Source: Modified after Birchfield, B. C., Foster, R. J., Keller, E. A. Melhom, W. N., Brookins, D. G., Mintt, L. W., and Thurman, H. V. 1982. *Physical geology.* Columbus, OH: Charles E. Merrill.

B

Rocks

In Chapter 3 we stated that a rock is an aggregate of one or more minerals. In traditional geologic investigations, this is the most commonly used definition. The terminology of environmental geology is somewhat different; in environmental geology, we are primarily concerned with properties that affect engineering design and environmental problems. In environmental and engineering geology, the term *rock* is reserved for Earth materials that cannot be removed without blasting using explosives such as dynamite; Earth materials that can be excavated with normal earth-moving equipment, for example, a shovel or a bulldozer, are called soil. Thus, a very friable, or loosely compacted, poorly cemented sandstone may be considered a soil, whereas a well-compacted clay may be called a rock. This pragmatic terminology conveys more useful information to planners and designers than do the conventional terms. Clay, for example, is generally unconsolidated and easily removed. A contractor, assuming that a material described as clay is a soil, might bid low for an excavation job, thinking that the material can be removed without blasting. If the clay turns out to be well compacted, however, it may have to be blasted, which is a much more expensive process. Performing a preliminary investigation and forewarning the contractor to consider the clay a rock can avoid this kind of error. (Soils and their properties are considered in detail in Chapter 16.)

Identifying Rocks

Rocks are composed of minerals. Whereas some rocks contain only one mineral, most contain several. As it is for minerals, the identification of rocks from a hand specimen is primarily a pattern recognition exercise. However, there are some useful hints that will assist you. The first task is to decide if the rock is igneous, sedimentary, or metamorphic. Sometimes this decision is not as easy as you might think. It is particularly difficult to identify rocks that are fine grained. A specimen is considered fine grained if you cannot see individual mineral grains with your naked eye. It is advisable to use a magnifying glass or hand lens to assist you in your evaluation of the properties of minerals and rocks. Some general rules of thumb will assist you in getting

started: (1) If the sample is composed of bits and pieces of other rocks, it is most likely sedimentary; (2) if the rock specimen is foliated (a feature sometimes called rock cleavage, produced by the parallel alignment of mineral grains), then it is metamorphic; (3) if individual mineral crystals are relatively coarse grained, meaning you can see them with the naked eye or easily with a hand lens, and the crystals are interlocking and composed of minerals such as quartz and feldspar, it is probably a plutonic igneous rock such as granite; (4) if the rock is mostly fine grained but contains some larger crystals (phenocrysts), it is probably an extrusive igneous rock; (5) if the rock is relatively soft and effervesces when diluted hydrochloric acid is applied, it is probably a carbonate rock, most likely limestone or, if metamorphic, marble; (6) if it's a really grungy, soft, highly weathered, or altered rock, as many in nature seem to be, you are going to have some problems identifying it!

Strength of Common Rock Types

The strength of a rock is commonly reported in terms of its compressive strength, which is the compression, as mechanically applied in a viselike machine, necessary to break or fracture the specimen. The strength of a rock is reported as a force per unit area, which in the metric system is reported in Newtons per meter squared (N/m^2). Ranges of compressive strength for some of the common rock types are shown on Table B.1.

Physical Properties of Rocks

Physical properties of rocks include color, specific gravity, relative hardness, porosity, permeability, texture, and strength.

Color. The color of a rock varies depending upon the minerals present and the amount of weathering that has occurred. Rocks encountered in their natural environment are various shades of light gray to brown to black. They have black to orange linear stains frequently, at the surface, produced by iron oxides.

TABLE B.1   Strength of Common Rock Types

	Rock Type	Range of Compressive Strength (10^6 N/m^2)	Comments
Igneous	Granite	100 to 280	Finer-grained granites with few fractures are the strongest. Granite is generally suitable for most engineering purposes.
	Basalt	50 to greater than 280	Brecciated zones, open tubes, or fractures reduce the strength.
Metamorphic	Marble	100 to 125	Solutional openings and fractures weaken the rock.
	Gneiss	160 to 190	Generally suitable for most engineering purposes.
	Quartzite	150 to 600	Very strong rock.
Sedimentary	Shale	Less than 2 to 215	May be a very weak rock for engineering purposes; careful evaluation is necessary.
	Limestone	50 to 60	May have clay partings, solution openings, or fractures that weaken the rock.
	Sandstone	40 to 110	Strength varies with degree and type of cementing material, mineralogy, and nature and extent of fractures.

Source: Data primarily from Bolz, R. E., and Tuve, G. L., eds. 1973. *Handbook of tables for applied engineering science.* Cleveland, OH: CRC Press.

Specific Gravity and Relative Hardness. The same definition of specific gravity we gave for minerals applies to rocks and refers to the weight of the rock relative to the weight of water. Some rocks are lighter than water and will float. For example, pumice, a light-colored, volcanic rock with many vesicles, or cavities, is lighter than water. Rocks become heavier as iron- and magnesium-bearing minerals become more abundant. There is no scale for relative rock hardness similar to the Mohs scale for minerals. Soft rocks can be broken with your fingers, and hard rocks require a sledgehammer to break apart.

Porosity and Permeability. *Porosity* is the percentage of a rock's volume that contains empty space between grains, or open space in fractures. *Permeability* is the capacity of a porous rock to transmit a fluid, usually oil or water (see Chapter 11). The coefficient of permeability is known as hydraulic conductivity with units of velocity (length per unit time). Values of porosity and hydraulic conductivity for several types of rocks and sediments are shown on Table 3.4. The properties of porosity and permeability are very important in understanding many environmental problems and geologic hazards, including waste disposal, water pollution, landsliding, and even earthquakes.

Texture. The texture of a rock refers to the size, shape, and arrangement of the crystals or grains within it. In general, a rock is fine grained if you cannot see mineral crystals or grains with your naked eye; a rock is considered coarse grained if crystals or grains are visible. The shape of crystals and grains varies from regular and rounded to interlocking and angular. Angular grains are irregular with sharp corners. A rock may have lots of openings between grains or in fractures with a relatively high porosity. If these openings are partially filled with smaller grains or cementing materials, the porosity is reduced. Porosity decreases from about 30 percent to 5 percent as pore spaces are filled with smaller grains or cementing material such as calcite ($CaCO_3$), quartz (SiO_2), or gypsum ($CaSO_4 \cdot 2H_2O$). Coarse grains of sand (1 to 2 mm, or 0.04 to 0.08 in., in diameter) are shown on Figure B.1a. If the range of sizes of grains is large, the distribution of grains is *poorly sorted* (Figure B.1b). Conversely, if the grains of sand are all approximately the same size, the distribution of grains is described as *well sorted* (Figure B.1a, c).

Other rock textures are characterized by vesicles, or cavities, formed by gas expansion during their formation. For example, vesicular rocks commonly occur near the surface of a lava flow where gas is escaping.

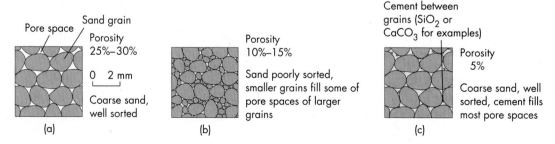

Figure B.1 (a) **Porosity** of a well-sorted coarse sand. (b) Poorly sorted sand. (c) Well-sorted sand with grains cemented together.

C

Maps and Related Topics

Topographic Maps

As the name suggests, topographic maps illustrate the topography at the surface of Earth. The topographic maps show an area's natural and human-made features graphically. Elevations and some other natural features are shown by contour lines, which are lines of equal elevation. Individual contour lines on a topographic map are a fixed interval of elevation apart, and this interval is known as the *contour interval.* Common contour intervals are 5, 10, 20, 40, 80, or 100 meters or feet. The actual contour interval of a particular map depends upon the topography being represented as well as the scale of the map. If the topography has a relatively low relief (difference in elevation between the highest point and lowest point of a map), then the contour interval may be relatively small, but if the relief is large, then larger contour intervals are necessary or the lines will become so bunched up as to be impossible to read. This brings up an important point: Where contour lines are close together, the surface of Earth at that point has a relatively steep slope, compared to other areas where the contour lines are farther apart.

The scale of a topographic map (or of any map, for that matter) may be delineated in several ways. First, a scale may be stated as a ratio, say 1 to 24,000 (1:24,000), which means that 1 inch on a map is equal to 24,000 inches or 2,000 feet on the ground. Second, a topographic map may have a bar scale often found in the lower margin of the map useful for measuring distances. Finally, scales of some maps are stated in terms of specific units of length on the map; for example, it might be stated on the map that 1 inch equals 200 feet. This means that 1 inch on the map is equivalent to 200 feet (which is 2,400 inches) on the ground. In this example we could also state that the scale is 1 to 2,400. The most common scale used by the U.S. Geological Survey Topographic Maps is 1:24,000, but scales of 1:125,000 or smaller are also used. Remember $1 \div 24,000$ is a larger number than $1 \div 125,000$, so $1 \div 125,000$ is the smaller scale. The smaller the scale of maps of similar physical map size, the more area is shown.

In addition to contour lines, topographic maps also show a number of cultural features, such as roads, houses, and other buildings. Features such as streams and rivers are often shown in blue. In fact, a whole series of symbols are commonly used on topographic maps. These symbols are shown in Figure C.1.

Reading Topographic Maps

Reading or interpreting topographic maps is as much an art as it is a science. After you have looked at many topographic maps that represent the variety of landforms and features found at the surface of Earth, you begin to recognize these forms by the shapes of the contours. This is a process that takes a fair amount of time and experience in looking at a variety of maps. However, there are some general rules that do help in reading topographic maps.

- Valleys containing rivers or even small streams have contours that form *V*s that point in the upstream direction. This is sometimes known as the *rule of Vs.* Thus, if you are trying to draw the drainage pattern that shows all the streams, you should continue the stream in the upstream direction as long as the contours are still forming a V pattern. Near a drainage divide, the Vs will no longer be noticeable.

- Where contour lines are spaced close together, the slope is relatively steep, and where contour lines are spaced relatively far apart, the slope is relatively low. Where contour lines that are spaced relatively far apart change to become closer together, we say a "break in slope" has occurred. This is commonly observed at the foot of a mountain or where valley side slopes change to a floodplain environment.

- Contours near the upper parts of hills or mountains may show a closure. These may be relatively oval or round for a conical peak or longer and narrower for a ridge. Remember, the actual elevation of a peak is higher than the last contour shown and may be estimated by taking half the contour interval. That is, if the highest contour on a peak were 1,000 m with a contour interval of 50 m, then you would assume the top of the mountain is 1,025 m.

- Topographic depressions that form closed contours have hachure marks on the contours and are useful in indicating the presence of such a depression.

Control data and monuments

Vertical control

Third order or better, with tablet	$^{BM}\times_{16.3}$
Third order or better, recoverable mark	$\times_{120.0}$
Bench mark at found section corner	$^{BM!}_{18.6}$
Spot elevation	$\times_{5.3}$

Contours

Topographic

Intermediate	
Index	
Supplementary	
Depression	
Cut; fill	

Bathymetric

Intermediate	
Index	
Primary	
Index primary	
Supplementary	

Boundaries

National	
State or territorial	
County or equivalent	
Civil township or equivalent	
Incorporated city or equivalent	
Park, reservation, or monument	

Surface features

Levee	Levee
Sand or mud area, dunes, or shifting sand	(Sand)
Intricate surface area	(Strip mine)
Gravel beach or glacial moraine	(Gravel)
Tailings pond	(Tailings pond)

Mines and caves

Quarry or open pit mine	
Gravel, sand, clay, or borrow pit	
Mine dump	(Mine dump)
Tailings	(Tailings)

Vegetation

Woods	
Scrub	
Orchard	
Vineyard	
Mangrove	(Mangrove)

Glaciers and permanent snowfields

Contours and limits	
Form lines	

Marine shoreline

Topographic maps

Approximate mean high water	
Indefinite or unsurveyed	

Topographic-bathymetric maps

Mean high water	
Apparent (edge of vegetation)	

Coastal features

Foreshore flat	
Rock or coral reef	
Rock bare or awash	
Group of rocks bare or awash	
Exposed wreck	
Depth curve; sounding	
Breakwater, pier, jetty, or wharf	
Seawall	

Rivers, lakes, and canals

Intermittent stream	
Intermittent river	
Disappearing stream	
Perennial stream	
Perennial river	
Small falls; small rapids	
Large falls; large rapids	
Masonry dam	
Dam with lock	
Dam carrying road	
Perennial lake; Intermittent lake or pond	
Dry lake	(Dry lake)
Narrow wash	
Wide wash	Wide wash
Canal, flume, or aquaduct with lock	
Well or spring; spring or seep	

Submerged areas and bogs

Marsh or swamp	
Submerged marsh or swamp	
Wooded marsh or swamp	
Submerged wooded marsh or swamp	
Rice field	(Rice)
Land subject to inundation	Max pool 431

Buildings and related features

Building	
School; church	
Built-up area	
Racetrack	
Airport	
Landing strip	
Well (other than water); windmill	
Tanks	
Covered reservoir	
Gaging station	
Landmark object (feature as labeled)	
Campground; picnic area	
Cemetery: small; large	(Cem)

Roads and related features

Roads on Provisional edition maps are not classified as primary, secondary, or light duty. They are all symbolized as light duty roads.

Primary highway	
Secondary highway	
Light duty road	
Unimproved road	
Trail	
Dual highway	
Dual highway with median strip	

Railroads and related features

Standard gauge single track; station	
Standard gauge multiple track	
Abandoned	

Transmission lines and pipelines

Power transmission line; pole; tower	
Telephone line	Telephone
Aboveground oil or gas pipeline	
Underground oil or gas pipeline	Pipeline

Figure C.1 **Some of the common symbols** used on topographic maps prepared by the U.S. Geological Survey.
(From U.S. Geological Survey)

- Sometimes the topography on a slope is hummocky and anomalous to the general topography of the area. Such hummocky topography is often suggestive of the existence of mass wasting processes and landslide deposits.

In summary, after observing a variety of topographic maps and working with them for some time, you will begin to see the pattern of contours as an actual landscape consisting of hills and valleys and other features.

Locating Yourself on a Map

The first time you take a topographic map into the field with you, you may have some difficulty locating where you are. Determining where you are is crucial in trying to prepare maps to show particular features—for example, the locations of floodplains, landslides, or other features. One way to locate yourself on a map is to recognize certain features such as mountain peaks, intersections of roads, a prominent bend in a road or river, or some other readily recognized feature. Then work it out from there with a compass. That is, you might take a bearing (compass direction) to several prominent features and draw these bearings on the map; your location is where they intersect. Today, however, we have more modern technology for locating ourselves and working with maps. Global Positioning Systems (GPS) are readily available at a very modest price and assist in locating a position on the surface of Earth. Hand-held GPS systems are readily available today and can generally locate your position on the ground with an accuracy of about 30 m. The GPS receivers work by receiving signals from three or four satellites and measuring the distance from the satellite to your location. This is done by measuring the time it takes for the signal from your receiver to reach the satellite and return. The accuracy of defining your position can be reduced to approximately 1 m by utilizing a reference receiver on the ground that communicates with the satellites and then working out your position relative to the reference receiver, as shown in Figure C.2.

Global Positioning Systems are commonly linked to computers, so when a position is known, it may be

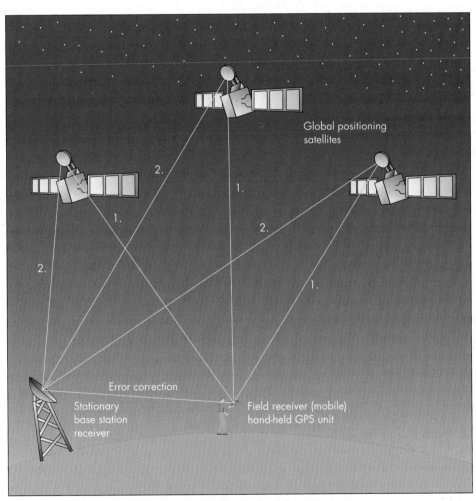

1. Field receiver and satellites — accuracy of position ~ 100 m of true position
2. Field receiver with stationary base station (differential GPS)
   — accuracy of position ~ 6 m of true position

Figure C.2 **Idealized diagram** showing how GPS systems work.

plotted directly on a map viewed on a computer screen. GPS technology is revolutionizing the way we do our mapping, is becoming widely available, and is a valuable research tool in the field.

A word of information here concerning the term *in the field.* When geologists use the term in the field, we mean "outside on the surface on Earth" and are not referring to the "field of geology." For example, I am now about to go out and study the coastline of California after the El Niño storms. I would say to my colleagues, "I am going into the field."

An Example from a Coastal Landscape

A coastal landscape is shown in Figure C.3. As illustrated in (a), which is an oblique view of the topography, the area is characterized by two hills with an intervening valley. A coastal seacliff is shown on the east (right) part of the diagram, and a sandspit produced by longshore coastal processes (see Chapter 10) with a hook on the end suggests that the direction of sand transport in the surf zone and beach is from the east (right) to the west (left) along this coastal area. A topographic map for the area is shown in (b). Notice that the contour interval (CI) is 20 ft. The elevation at the top of the highest hill on the east side of the diagram would be approximately 290 ft; because the last contour is 280, and the hill cannot be higher than 300, we split the difference. Several streams flow into the main valley; notice that the contours V in the upstream direction, particularly toward the peak of elevation of approximately 270 ft. Other information that may be "read" from the topographic map includes the following.

- The landform on the western portion of the map (left side) is a hill with elevation of about 275 ft and a gentle slope to the west, and a steep slope to the east (right) and toward the ocean. The eastern slope is particularly steep near the top of the mountain where the contours are very close together.

- The center part of the map shows a river flowing into a bay protected by a hooked sand bar. The relatively flat land in the vicinity of the river is a narrow floodplain with a light-duty road (see Figure C.3) delineated by two closely spaced lines extending along the western side. A second unimproved road crosses the river and extends out to the head of the sandspit, providing access to a church and two other buildings. The floodplain is delineated on both sides of the river by the 20-ft contour, above which to the 40-ft contour is a break in slope at the edge of the valley.

- The eastern and southern slopes of the hill on the west (left) side of the map with elevation of about 275 ft have a number of small streams flowing mostly southward into the ocean. These streams may be defined as relatively steep gullies that are dissecting (eroding) the hill.

- In contrast to the hill on the western part of the map, the hill on the eastern part with elevation of approximately 290 ft has gentler slopes on the eastern flank, with streams spaced relatively far apart flowing toward the river and the ocean. Thus, we could state that the streams flowing to the main river valley from the east have dissected (eroded) the landscape more (to a lower slope) than those streams flowing into the main valley from the west.

- In particular, note the stream channel just south of the hill with elevation of about 275 ft on the west side of the map. That stream is actively eroding headward, forming a narrow, steep gully denoted by the pattern of closely spaced contours that form a concave indentation into the more gentle topography that forms the surface that is sloping to the west from the top of the mountain.

To continue a study of this area, we might construct more topographic profiles across the area from east to west and also construct profiles of some of the streams flowing into the main valley. The next task might be to obtain aerial photographs and geologic maps to draw more conclusions concerning the topography and geology.

Geologic Maps

Geologists are interested in the types of rocks found in a particular location and in their spatial distribution. Producing a geologic map is a very basic step in understanding the geology of an area. Geologic maps are the fundamental database from which to interpret the geology of an area. The first step in preparing a geologic map is to obtain good base maps (usually topographic) or aerial photographs on which the geologic information may be transferred. The geologist then goes into the field and makes observations at outcrops where the rocks may be observed. Different rock types, units, or formations are mapped and the contacts between varying rock types are mapped as lines on the map. The attitude of sedimentary rock units is designated by a T-shaped strike and dip symbol. The strike is the compass direction of the intersection of the rock layer with a horizontal plain, and the dip is the maximum angle the rock layer makes with the horizontal (Figure C.4). Figure C.5a shows a very simple geologic map with three rock types—sandstone, shale, and limestone—over an area of approximately 1,350 km^2. The arrangement of the strike and dip symbols suggests that the major structure is an anticline. A geologic cross section constructed across the profile along the line E–E' is shown in Figure C.5b. Geologists often make a series of cross sections to try

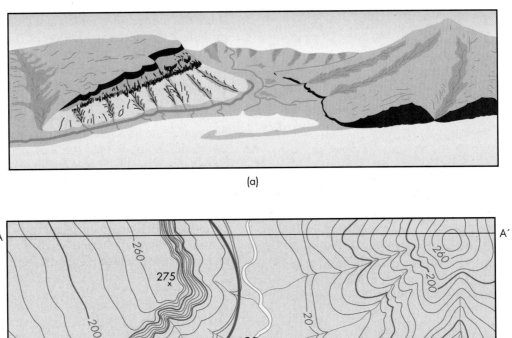

(a)

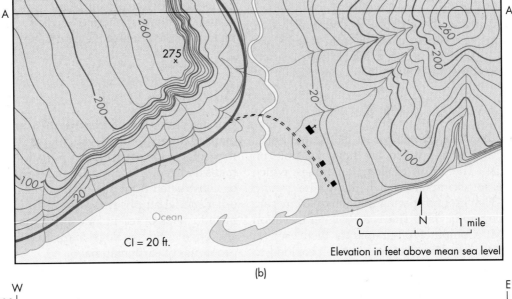

(b)

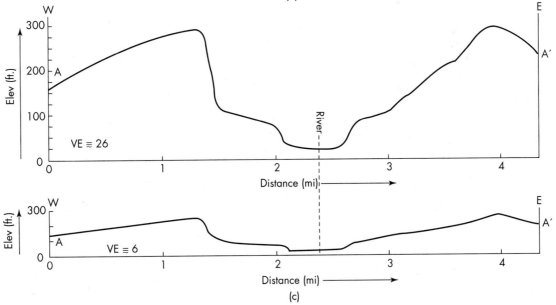

(c)

Figure C.3 Drawing of a landscape along a coastal area (a) Set of diagrams illustrating topographic maps. (b) Topographic map for the same area with a contour interval of 20 ft. (c) Topographic profiles along line T of the topographic map shown with vertical exaggeration of approximately 26 times and 8 times. The vertical exaggeration is the ratio of the vertical to horizontal scale of the topographic profile. The ratio for the upper profile is approximately 26, so we say that the vertical exaggeration is approximately 26 times. For the lower profile, the vertical exaggeration is about 8 times. In the real world, of course, there is no vertical exaggeration (the vertical and horizontal scales are the same). As an experiment, you might try to make a topographic profile along line T with no vertical exaggeration. What do you conclude? *(From U.S. Geological Survey)*

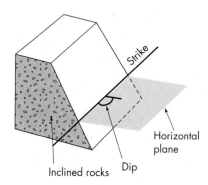

Figure C.4 **Idealized block diagram** showing the strike and dip of inclined sedimentary rocks.

to better understand the geology of a particular area. Geologic maps at a variety of scales from 1 to 250,000 down to 1 to 24,000 are generally available from a variety of sources, including the U.S. Geological Survey.

Digital Elevation Models

Topographic data for many areas in the United States and other parts of the world are now available on computer disk. The arrays of elevation values, which may include, for example, the elevation of the ground on a 30 m × 30 m ground spacing grid (area, 900 m^2) are the basic topographic data. This is the Digital Elevation Model (DEM). Computer programs are then used to view the data, and color shading may be provided to

show the topography, which is a DEM product. DEMs provide a visual representation of the surface of Earth and can be "viewed" from a variety of angles. That is, you may decide you would like to view the topography obliquely from the south, north, east, or west. The vertical dimension may also be exaggerated so that minor topographic differences may become more apparent. Figure C.6 shows a relief map constructed from a digital elevation model for the Los Angeles Basin. This clearly illustrates that Los Angeles is nearly surrounded by mountains and hills, which are uplifted by recent tectonic activity. DEMs are becoming important research tools in evaluating the topography of an area. They may be used to delineate and map such features as fault scarps, drainages, marine terraces, floodplains, and many other features of the landscape.

Summary

Our discussion establishes that there are several types of maps that are useful in evaluating a particular area and its geology. Of particular importance are topographic maps and geologic maps. Digital elevation models may be constructed from topographic data, and a variety of other special purpose maps are also available. Examples include maps of recent landslides, maps of floodplains, and engineering geology maps that show engineering properties of earth materials.

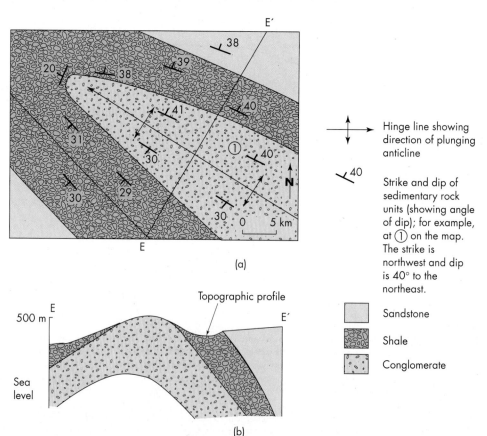

Figure C.5 (a) **Idealized diagram** showing a very simple geologic map with three rock types and (b) geologic cross section showing the topography and structure of the anticline.

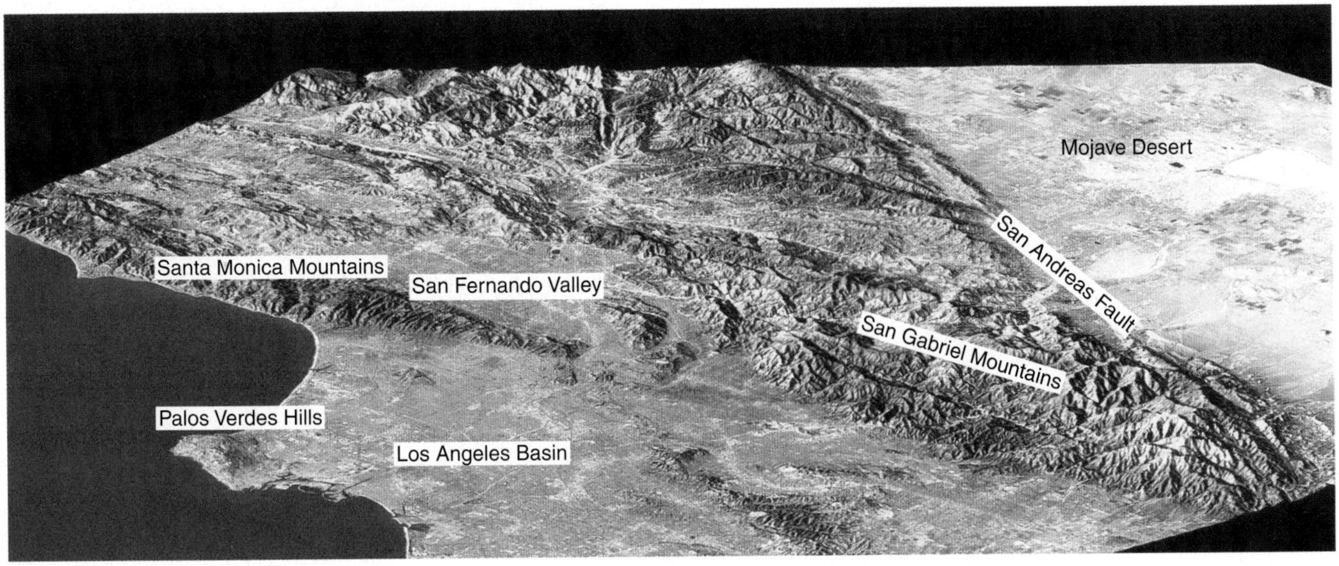

Figure C.6 **Color relief map made from a Digital Elevation Model of the Los Angeles area** with a fair amount of vertical exaggeration. The flat area in the central part of the image is the Los Angeles Basin. The smaller basin just to the left is the San Fernando Valley, which is separated from the Los Angeles basin by the Santa Monica Mountains. *(Courtesy of Robert Crippin, NASA, Jet Propulsion Laboratory)*

D

How Geologists Determine Geologic Time

In order to understand the history of Earth, we have to determine the actual age of rocks and sediments. The science of dating rocks and sediments is known as **geochronology.** Although it is possible to establish relative geochronology based upon fossil evidence using the law of faunal assemblages and geologic relationships such as the laws of superposition and crosscutting relationships (see Chapters 1 and 3), it is absolute dating, (numerical dating) that provides the geochronology necessary to establish rates of geologic processes and ages of rocks. The science of geochronology is expanding rapidly and includes a number of techniques that provide numerical ages from a few years to a few billion years (the age of Earth).

From an environmental perspective, it is important to be able to establish rates of geologic processes and when geologic events such as volcanic eruptions, earthquakes, floods, and landslides occurred in the past. The chronology of natural hazard events is critical in establishing their return period as part of better understanding hazards and predicting when they are likely to occur in the future. That is, we wish to establish rates of geologic processes of significance to society.

Geologic time is much different, in a way, than our normal time framework. Although geologic time and "normal" time use the same units of measure—years—they differ vastly in duration and in the instruments we use to measure duration.[1] Normal time is counted in hours, days, seasons, or decades, and the instrument used to measure time is a clock. For "deep time," (geologic time), measured in millions to hundreds of millions to several billion years, geologists take advantage of naturally occurring isotopes such as uranium-235, uranium-238, potassium-40, or carbon-14 to date the rocks in which these isotopes occur. Dating is possible because the natural decay of each type of unstable radioactive isotope occurs at a constant rate that can be used to determine the age of rocks in which that isotope occurs. The process of radioactive decay is a spontaneous process in which the nucleus of a particular isotope undergoes a change while emitting one or more forms of radiation (see Chapter 15, A Closer Look: Radioactivity). The three major kinds of radiation that are emitted during radioactive decay are called alpha particles, beta particles, and gamma radiation. Alpha decay is particularly significant because the decay consists of two protons and two neutrons. The isotope present before alpha decay is known as the parent, and the new isotope present after decay is known as the daughter product, which is an isotope of a different element. Radioactive isotopes, particularly those of very heavy elements, undergo a series of radioactive decay steps that finally ends when a stable, nonradioactive isotope is produced. For example, uranium-238 undergoes 14 nuclear transformations emitting alpha, beta, and gamma radiation in different steps to finally decay to stable lead-206, which is not radioactive. An important characteristic of a radioisotope such as U-238 is its half-life, which is the time required for one-half of a given amount of the isotope to decay to another form. Every radioisotope has a unique and characteristic half-life. Table D.1 shows four parent radioactive isotopes, their daughter stable isotopes, and characteristic

TABLE D.1 Parent and Daughter Isotopes with Half-Lives for Four Elements Commonly Used to Absolutely (Numerically) Date Earth Materials Such as Rock or Organic Material

Parent radioactive isotopes	Daughter stable isotope	Half-Life
Uranium-238	Lead-206	4.5 billion yr
Uranium-235	Lead-207	700 million yr
Potassium-40	Argon-40	1.3 billion yr
Carbon-14	Nitrogen-14	5,730 yr

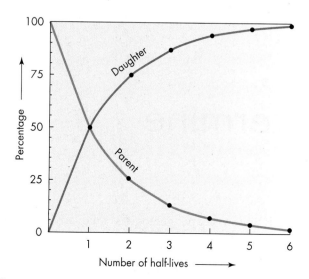

Figure D.1 **Reduction of a parent radioactive isotope** to a daughter isotope with increasing half-lives.

half-lives. Figure D.1 shows in diagrammatic form the reduction of a parent radioactive isotope and concurrent increase in the daughter product. For example, after one half-life, 50 percent of the original parent isotope remains; this is reduced to 25 percent after two half-lives, and to 12.5 percent after three half-lives. By the time six half-lives have nearly passed, all the parent material has been lost through radioactive decay. As the parent material decreases, the daughter product increases to be nearly 100 percent by six half-lives. Because the process of radioactive decay is irreversible and occurs at a constant rate, it serves as a clock that may be used for absolute (numerical) age dating in years. Because radioisotopes such as uranium-238, uranium-235, and potassium-40 have relatively long half-lives (Table D.1), their decay is useful in dating rocks on the order of millions to billions of years. For example, radioactive uranium has been used to date the oldest rocks on Earth at about 4.6 billion years before present. The actual methods of measuring amounts of parent and daughter isotopes and calculating numeric dates are complex and tedious, although the concept is easy to grasp. These methods have been used successfully to develop the geochronology necessary to place numerical dates in the geologic time table and to delineate important Earth history events such as mountain building, ice ages, and the appearance of life forms.

Archeologists and anthropologists concerned with human history are interested in developing a geochronology with a duration of a few million years or less, that is the period from which human and ancestral human fossils are found. From an environmental geology perspective, we are often interested in establishing the geochronology of the past few hundred to few hundred thousand years of Earth's history, and for this we have several potential methods. For example, the radioisotope uranium-234 undergoes radioactive decay at a known rate to thorium-230, and the ratio of these two isotopes is useful in numerically dating a variety of materials, such as coral, back to several hundred thousand years. For sediments younger than about 40,000 years, carbon-14 is used extensively for numerical dating. The most common form of carbon is stable carbon-12, but carbon-14, which is radioactive, occurs in small quantities and undergoes radioactive decay to the stable daughter isotope nitrogen-14. The half-life of the decay of carbon-14 to nitrogen-14 is 5,730 years. The carbon-14 method works because organic material incorporates carbon-14 into living tissue, and when the organism dies radioactive decay begins. Common materials dated by carbon-14 include wood, bone, charcoal, and other types of buried organic material. Carbon-14 has a relatively short half-life, and, as a result, after about 40,000 years the amount of carbon-14 remaining is very small and difficult to measure. Consequently, use of this technique is limited.

The field of geochronology is expanding rapidly, and new techniques are being developed as I write this. For example, it is now possible to directly date the duration that a landform has been exposed at Earth's surface. This is known as exposure dating. The basic idea is that certain isotopes—for example, beryllium-10, aluminum-26, and chlorine-36—that are produced when cosmic rays interact with Earth's atmosphere, accumulate in measurable quantities in surface materials such as soils, alluvial deposits, and exposed rock surfaces. Thus, the amount of accumulation of these isotopes is a measure of the minimum time of exposure to the surface environment.

Other innovative methods of developing the geochronology of an area are also being used. For example, lichens, which are mosslike plants that grow on rock surfaces, are known to have given growth rates for particular species, and they tend to grow in circular patches. Careful measurements of the size of lichen patches, coupled with known growth rates, provide minimum numerical ages for the rocks they are found on. The method has been successfully used in California and New Zealand to date regional occurrences of rockfalls generated by large earthquakes, thus dating past seismic activity.[2,3] Lichenometry can provide dates to about 1,000 years before present. Sediments may also be accurately dated through the use of dendrochronology, which is the analysis of annual growth rings to measure the age of woody material found in sediments. The method has been used extensively by archeologists to date prehistoric sites of human habitation and also by climatologists as a method of reconstructing past climates in terms of wet and dry years. Growth rings record climate because during dry years tree rings are narrow relative to wet years. Very accurate numerical geochronology may also be obtained from counting varves. A varve is a layer of sediment representing one year of deposition, usually in a lake or an ocean. Careful counting of varves may extend the chronology back several thousand years.

The most accurate chronology is the historical record, but in many parts of the world the historical record is only a few hundred years (as for example, in the United States). In other areas, such as China, the historical record extends several thousand years. However, given the brevity of human history compared with the great length of geologic history, we resort to numerical dating methods to establish geochronology. There are more than 20 methods useful in establishing geochronology that yield numerical dates. This discussion presents only a few of these at a very elementary level. The science

of geochronology is complex but extremely crucial to understanding rates of geologic processes and placing geologic events in chronological order.

References

1. **Ausich, W. I., and Lane, G. N.** 1999. *Life of the past,* 4th ed. Upper Saddle River, NJ: Prentice Hall.
2. **Bull, W. B., and Brandon, M. T.** 1998. Lichen dating of earthquake-generated regional rock fall events. Southern Alps, New Zealand. *Geological Society of America Bulletin* 110(1):608–84.
3. **Bull, W. B.** 1996. Dating San Andreas fault earthquakes with lichenometry. *Geology* 24: 111–14.

E

Darcy's Law

In 1856 an engineer named Henry Darcy was working on the water supply for Dijon, France. He performed a series of important experiments that demonstrated that the discharge (Q) of groundwater may be defined as the product of the cross-sectional area of flow (A), the hydraulic gradient (I), and the hydraulic conductivity (K). Thus,

$$Q = KIA$$

The unit on each side of the equation is a volumetric flow rate (such as cubic meters per day), and the relationship is known as **Darcy's law.** The quantity $Q/A = KI$ is the **Darcy flux** (v). We may say that

$$v = Q/A \quad \text{or} \quad Q = vA$$

Although v has the unit of a velocity, the Darcy flux is only an apparent velocity. To determine the actual velocity of groundwater in an aquifer (vx) we must remember that the water moves through pore spaces, so its velocity is affected by the porosity of the aquifer material. If we let n represent the porosity, then the actual cross-sectional area of flow is An, and it follows from $Q = vA$ that

$$vx = Q/An = v/n \quad \text{or} \quad vx = KI/n$$

The actual velocity vx is about three times the Darcy flux (assuming an average value of $n = 0.33$).

The driving force for groundwater flow is called the **fluid potential** or **hydraulic head,** which at the point of measurement is the sum of the elevation of the water (elevation head) and the ratio of the fluid pressure to the unit weight of water (pressure head). The difference in hydraulic heads between two points (h) divided by the flow length (L) gives us the hydraulic gradient (I).

Groundwater always moves from an area of higher hydraulic head to an area of lower hydraulic head and may therefore move down, laterally, or upward, depending upon local conditions.

Darcy's law has many important applications to groundwater problems. For example, consider an area underlain by sedimentary rocks with a semiarid climate. The area is dissected by a river system in a valley approximately 4 km wide. Alluvial deposits in the valley form an aquifer confined by a clay layer, and two wells have been drilled approximately 1 km apart in the down-valley direction (Figure E.1, part a). A cross-valley section between the wells (Figure E.1, part b) shows that the saturated zone is 25 m thick, consists of sand and gravel, and has a hydraulic conductivity of 100 m/day (1.2×10^{-3} m/sec). Porosity (n) of the aquifer materials is 30 percent (0.3). A down-valley section is shown in Figure E.1, part c. The wells are separated by 1,000 m and the elevation of the water in wells 1 and 2 are, respectively, 98 and 97 m. Two questions we might ask concerning the conditions shown in Figure E.1 are

1. What is the discharge Q (m^3/sec or gallons per day) of water moving through the aquifer in the down-valley direction?

2. What is the travel time (T) of the groundwater between wells 1 and 2? This question is particularly interesting from an environmental standpoint if a water pollution event is detected at well 1 and we want to know when the pollution will reach well 2.

Answering these two questions requires us to apply Darcy's law to the situation outlined above. To answer the first question, which asks how much water is moving through the aquifer, recall that $Q = KIA$. We will solve for Q. The hydraulic gradient is the ratio of the difference in elevation of the water (pressure head) between the two wells to the length of the groundwater flow between the wells. The difference in elevation of the water in wells 1 and 2 is 1 m and the flow length is 1,000 m. Thus, the hydraulic gradient (I) is 0.001 (1×10^{-3}). The hydraulic conductivity is given as 1.2×10^{-3} m/sec. The cross-sectional area of the aquifer (A) is 25 m $\times$ 4,000 m, or 100,000 m^2 (1×10^5 m^2). Multiplying these numbers, we find that Q is equal to 0.12 m^3/sec. This is equivalent to 10,368 m^3/day, which is approximately 2.7 million gallons per day. Of course, all of this water could not be pumped from the aquifer. Pump tests of the wells would be necessary to determine how much of the 2.7 million gallons per day could be pumped without depleting the resource.

Turning now to the second question, which concerns the travel time of the water from one well to the other, we again apply Darcy's law. In this case we calculate the Darcy flux (v), which is

$$v = Q/A = KI$$

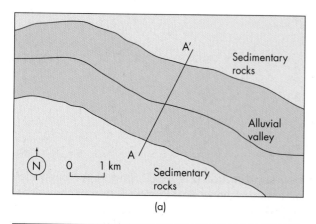

(a)

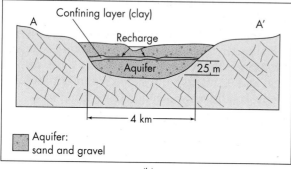

(b)

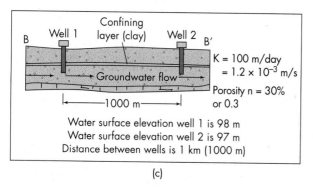

(c)

Figure E.1 **Hypothetical map of an alluvial valley** (a) Cross-valley profile. (b) Profile down-valley. (c) Showing groundwater conditions.

Remember that the Darcy flux is only an apparent velocity and does not reflect the fact that the actual movement of the groundwater is through the pore spaces between the grains of sand and gravel in the aquifer. The actual velocity (vx) is the ratio of the product of KI to the porosity.

$$vx = KI/n$$
$$= (1.2 \times 10^{-3} \text{ m/sec}) (1 \times 10^{-3})/0.3$$
$$= 4.0 \times 10^{-6} \text{ m/sec}$$

Travel time (T) then is the ratio of the length of flow (L) to the velocity of the water moving through the pore spaces (vx). This follows from the fact that distance L is the product of velocity vx and time T, ($L = vxT$). Thus, $T = 1{,}000 \text{ m}/4.0 \times 10^{-6} \text{ m/sec} = 2.5 \times 10^{8}$ sec. This is approximately 7.9 years.

GLOSSARY

A soil horizon   Uppermost soil horizon, sometimes referred to as the zone of leaching.

Abrasion   With respect to wind processes, refers to the erosion of rock surfaces as the result of collision of sand or silt grains being transported by the wind colliding with rock surfaces to cause pitting or polishing.

Absorption   The process of taking up, incorporating, or assimilating a material.

Acid mine drainage   An environmental problem related to the discharge of acidic waters resulting from the weathering of sulfide minerals, such as iron pyrite, associated with coal and sulfide mineralization of important metals, such as copper, silver, and zinc.

Acid rain   Rain made artificially acidic by pollutants, particularly oxides of sulfur and nitrogen; natural rainwater is slightly acidic owing to the effect of carbon dioxide dissolved in the water.

Active fault   There are a variety of definitions, but one is displacement along a fault in the past 10,000 years; another definition is multiple displacement in the past 35,000 years.

Active method   With respect to permafrost, refers to the method of thawing the frozen ground before constructing buildings and other structures; generally used when permafrost is thin or discontinuous.

Adsorption   Process in which molecules of gas or molecules in solution attach to the surface of solid materials with which they come into contact.

Advanced treatment   With respect to wastewater treatment, includes a variety of processes, such as use of chemicals, sand filters, or carbon filters to further purify and remove contaminants from wastewater.

Aerobic   Characterized by the presence of free oxygen.

Aesthetics   Originally a branch of philosophy, defined today by artists and art critics.

Aftershocks   Earthquakes that occur a few minutes to a few months to a year or so after the main event.

Agglomerate volcanic breccia   See Volcanic breccia, agglomerate.

Aggregate   Any hard material, such as crushed rock, sand, gravel, or other material that is added to cement to make concrete.

Air Quality Index   Method by which air quality in urban areas is usually reported, defining levels or stages of air pollution in terms of the National Ambient Air Quality Standards.

Air Quality Standards   Levels of air pollutants that define acceptable levels of pollution generally over a particular time period.

Air toxin   Air pollutant known to cause cancer or other serious health conditions.

Albedo   A measure of reflectivity or the amount (by decimal or percent) of electromagnetic radiation reflected by a material or surface.

Alkaline soil   Soil found in arid regions that contains a large amount of soluble mineral salts (primarily sodium) that in the dry season may appear on the surface as a crust or powder.

Alluvial fan   A fan-shaped deposit that is a segment of a cone that forms where a stream emerges from a mountain front; sediments deposited include stream flow deposits and mudflow or debris deposits.

Alluvium   Unconsolidated sediments, including sand, gravel, and silt, deposited by streams.

Alpha particles   Type of nuclear radiation consisting of two protons and two neutrons emitted during radioactive decay.

Alternative energy   Refers to energy sources that are alternative to the commonly used fossil fuels.

Anaerobic   Characterized by the absence of free oxygen.

Andesite   A type of volcanic rock consisting of feldspar and other silicate minerals rich in iron and magnesium with an intermediate silica content of about 60 percent; an example is Mount Fuji in Japan.

Angle of repose   The maximum angle that loose material will sustain.

Anhydrite   Evaporite mineral ($CaSO_4$) calcium sulfate.

Anthracite   A type of coal characterized by a high percentage of carbon and low percentage of volatiles, providing a high heat value; often forms as a result of metamorphism of bituminous coal.

Anticline   Type of fold in rock characterized by an upfold or arch; the oldest rocks are found in the center of the fold.

Appropriation doctrine, water law   Holds that prior usage of water is a significant factor; the first to use the water for beneficial purposes is prior in right.

Aquiclude   Also known as aquitard; earth material that will hold groundwater but cannot transmit it at a rate sufficient to be pumped from a well. See Impermeable.

Aquifer   Earth material containing sufficient groundwater that the water can be pumped out; highly fractured rocks and unconsolidated sands and gravels make good aquifers.

Aquitard   Earth material that retards the flow of groundwater.

Area source   Type of nonpoint air or water pollution that is diffused, such as runoff from an urban area or automobile exhaust.

Area (strip) mining   Type of strip mining practiced on relatively flat areas.

Argillic B soil horizon   Designated as Bt, a soil horizon enriched in clay minerals that have been translocated downward by soil-forming processes.

Arid   Refers to regions with an annual precipitation less than approximately 25 cm/yr.

Artesian   Refers to a groundwater system in which the groundwater is isolated from the surface by a confining layer and the water is under pressure. Groundwater that is under sufficient pressure will flow freely at the surface from a spring or well.

Asbestos Fibrous mineral material used as insulation. It is suspected of being either a true carcinogen or a carrier of carcinogenic trace elements.

Ash, volcanic Unconsolidated volcanic debris, less than 4 mm in diameter, physically blown out of a volcano during an eruption.

Ash fall Volcanic ash eruption that blows up into the atmosphere and then rains down on the landscape.

Ash flow Mixture of volcanic ash, hot gases, and fragments of rock and glass that flows rapidly down the flank of a volcano; may be an extremely hazardous event.

Asteroid Rock or metallic particle in space from about 10 m to 1,000 km in diameter.

Asthenosphere The upper zone of Earth's mantle, located directly below the lithosphere; a hot, slowly flowing layer of relatively weak rock that allows for the movement of the tectonic plates.

Atmosphere Layer of gases surrounding Earth.

Atmospheric inversion When warm air overlies cool air in the lower atmosphere.

Atom The smallest part of a chemical element that can take part in a chemical reaction or combine with another element.

Avalanche A type of landslide involving a large mass of snow, ice, and rock debris that slides, flows, or falls rapidly down a mountainside.

Average residence time The amount of time it takes for the total stock or supply of material in a system to be cycled through the system.

B soil horizon Intermediate soil horizon; sometimes known as the *zone of accumulation*.

Balance of nature The idea that nature undisturbed by human activity will reach a balance or state of equilibrium; is thought by some to be an antiquated idea that probably has never existed in nature.

Barrier island Island separated from the mainland by a salt marsh; generally consists of a multiple system of beach ridges and is separated from other barrier islands by inlets that allow exchange of seawater with lagoon water.

Basalt A type of volcanic rock consisting of feldspar and other silicate minerals rich in iron and magnesium; has a relatively low silica content of about 50 percent; the most common volcanic rock-forms of shield volcanoes.

Basaltic Engineering geology term for fine-grained igneous rocks.

Base flow The low flow discharge of a stream or river, which is produced by groundwater seeping into the channel.

Base level The theoretical lowest elevation to which a river may erode; generally is at or about sea level.

Batholith A large intrusion of igneous rock that may exceed thousands of cubic kilometers generally with surface expression greater than 100 km^2; commonly composed of several intrusions known as *plutons*.

Bauxite A rock composed almost entirely of hydrous aluminum oxides; a common ore of aluminum.

Beach Accumulation of whatever loose material (commonly sand, gravel, or bits of shell and so forth) that accumulate on a shoreline as the result of wave action.

Beach budget Inventory of sources and losses of sediment to a particular stretch of coastline.

Beach face That part of the beach environment that is a nearly planar part of the beach profile below the berm and normally exposed to the swash action from waves.

Beach nourishment Artificial process of adding sediment (sand) to a beach for recreational and aesthetic purposes as well as to provide a buffer to coastal erosion.

Bedding plane The plane that delineates the layers of sedimentary rocks.

Bed load That portion of the total load that a stream carries that moves along the bottom of the channel by rolling, sliding, or skipping.

Bed material Sediment transported and deposited along the bed of a stream channel.

Bentonite A type of clay that is extremely unstable; when wet, it expands to many times its original volume.

Berm The relatively flat part of a beach profile produced by deposition from waves; the part of the beach where people sunbathe.

Beta particles Type of nuclear radiation consisting of electrons emitted during radioactive decay.

Biochemical oxygen demand (BOD) A measure of the amount of oxygen necessary to decompose organic materials in a unit volume of water. As the amount of organic waste in water increases, more oxygen is used, resulting in a higher BOD.

Biodiversity In general, refers to the variety of life in an area, region, or Earth. Also refers to the total number of species (richness) or the main species encountered (dominance).

Biofuel A number of fuels including ethanol (alcohol), derived from plants.

Biogeochemical cycle Movement of a chemical element or compound through the various Earth systems including atmosphere, lithosphere, biosphere, and hydrosphere.

Bioleaching Use of microorganisms to recover metals, as for example release of finely disseminated gold that may then be treated by cyanide leaching.

Biomagnification The process whereby chemicals may accumulate at higher and higher concentrations in the food chain; also referred to as biological concentration.

Biomass Organic matter. As a fuel, biomass can be burned directly (as wood) or converted to a more convenient form (such as charcoal or alcohol) and then burned.

Biopersistence A measure of how long a particular material will remain in the biosphere.

Bioremediation With respect to soil pollution, refers to technology that utilizes natural or enhanced microbial action in the soil to degrade organic contaminants *in situ* (at the site), not requiring excavation of the soil.

Biosphere The zone adjacent to the surface of Earth that includes all living organisms.

Biotechnology With respect to resource management, refers to use of organisms to assist in mining of ores or cleaning up of waste from mining activities.

Biotite A common ferromagnesian mineral, a member of the mica family.

Bituminous coal A common type of coal characterized by relatively high carbon content and low volatiles; sometimes called *soft coal*.

Bk soil horizon Soil horizon characterized by accumulation of calcium carbonate that may coat individual soil particles and fill some pore spaces but does not dominate the morphology of the horizon.

Blowout Failure of an oil, gas, or disposal well resulting from adverse pressures that can physically blow part of the well casing upward; may be associated with leaks of oil, gas, or, in the case of disposal wells, harmful chemicals.

Braided river A river channel characterized by an abundance of islands that continually divide and subdivide the flow of the river.

Breaker zone That part of the beach and nearshore environment where incoming waves peak up, become unstable, and break.

Breakwater A structure (as a wall), which may be attached to a beach or located offshore, designed to protect a beach or harbor from the force of the waves.

Breccia A rock or zone within a rock composed of angular fragments; sedimentary, volcanic, and tectonic breccias are recognized.

Breeder reactor A type of nuclear reactor that produces more fissionable fuel than it consumes.

Brine Water that has a high concentration of salt.

British thermal unit (Btu) A unit of heat defined as the heat required to raise the temperature of 1 lb of water 1°C.

Brittle Material that ruptures before any plastic deformation.

Burner reactor Type of nuclear reactor used to generate electricity that consumes more fissionable material than it produces.

C soil horizon Lowest soil horizon, sometimes known as the zone of partially altered parent material.

Calcite Calcium carbonate ($CaCO_3$); common carbonate mineral that is the major constituent of the rock limestone; weathers readily by solutional processes; large cavities and open weathered fractures are common in rocks that contain calcite.

Caldera Giant volcanic crater produced by very rare but extremely violent volcanic eruption or by collapse of the summit area of a shield volcano after eruption.

Caldera eruption Relatively infrequent large volcanic eruption that is associated with a catastrophic explosion that may produce a very large volcanic crater 30 or more kilometers in diameter.

Caliche A white-to-gray irregular accumulation of calcium carbonate in soils of arid regions.

Calorie The quantity of heat required to raise the temperature of 1 g of water from 14.5° to 15.5°C.

Cambic soil horizon Soil horizon diagnostic of incipient soil profile development, characterized by slightly redder color than the other horizons.

Capacity A measure of the total load a river may carry.

Capillary action The rise of water along narrow passages, facilitated and caused by surface tension.

Capillary fringe The zone or layer above the water table in which water is drawn up by capillary action.

Capillary water Water that is held in the soil through capillarity.

Carbon-14 Radioactive isotope of carbon with a half-life of approximately 5,740 years; used in radiocarbon dating of materials back to approximately 40,000 years.

Carbonate A compound or mineral containing the radical CO_3. The common carbonate is calcite.

Carbon cycle One of Earth's major biogeochemical cycles that involves the movement of carbon in the atmosphere, biosphere, lithosphere, and hydrosphere.

Carbon monoxide (CO) A colorless, odorless gas that at low concentration is very toxic to humans and other animals; anthropogenic component of emissions is only about 10 percent and comes mainly from fires, automobiles, and other sources of incomplete burning of organic compounds; many people have accidentally been asphyxiated by carbon monoxide from poorly ventilated campers, tents, and homes.

Carcinogen Any material known to produce cancer in humans or other animals.

Carrying capacity The maximum number of a population of a species that may be maintained within a particular environment without degrading the ability of that environment to maintain that population in the future.

Catastrophe An event or situation causing sufficient damage to people, property, or society in general from which recovery and/or rehabilitation is long and involved; natural processes most likely to produce a catastrophe include floods, hurricanes, tornadoes, tsunamis, volcanoes, and large fires.

Cave A natural subterranean void that often consists of a series of chambers that are large enough for a person to enter, most commonly formed in the rock types limestone or marble.

Chain reaction With respect to nuclear fission, refers to the splitting of atomic nuclei by neutron bombardment that results in splitting of uranium and release of ever-more neutrons; such an action uncontrolled is the kind observed in nuclear explosions, whereas sustained stable reactions are those that occur in nuclear fission reactors that produce electricity.

Channelization An engineering technique to straighten, widen, deepen, or otherwise modify a natural stream channel.

Channel pattern The shape of a river channel as viewed from above ("bird's-eye view"); patterns include straight, meandering, and braided.

Channel restoration The process of restoring stream channels and adjacent areas to a more natural state.

Chemical bonding The holding together of atoms by attractive forces between atoms and/or sharing of electrons.

Cinder cone A volcanic, conical hill formed by the accumulation of volcanic ash and other pyroclastic deposits.

Circum-Pacific belt One of the three major zones where earthquakes occur; essentially the border of the Pacific plate; also known as the *ring of fire*, as many active volcanoes are found on the edge of the Pacific plate.

Clay May refer to a mineral family or to a very fine-grained sediment; associated with many environmental problems, such as shrinking and swelling of soils and sediment pollution.

Clay skins Oriented plates of clay minerals surrounding soil grains and filling pore spaces between grains.

Climate The characteristic atmospheric condition (weather) at a particular place or region over time periods of seasons, years, or decades.

Closed system A system with boundaries that restrict the flow of energy and matter; for example, with respect to mineral resources, Earth can be considered a closed system.

Coal A sedimentary rock formed from plant material that has been buried, compressed, and changed.

Coal-bed methane Methane stored on surfaces of organic matter in coal.

Coastal erosion   Erosion of a coastline caused by a number of processes, including wave action, landsliding, wind, and runoff from the land.

Cogeneration   With respect to energy resources, the recycling of waste heat to increase the efficiency of a typical power plant or factory; may involve production of electricity as a by-product from industrial processes.

Cohesion   With respect to engineering properties of soils, refers to electrostatic forces that hold fine soil grains together and constitute a part of the shear strength of a soil.

Colluvium   Mixture of weathered rock, soil, and other, usually angular, material on a slope.

Columnar jointing   System of fractures (joints) that break rock into polygons of typically five or six sides; the polygons form columns; common in basalt and most likely caused by shrinking during cooling of the lava.

Comet   Particle in space composed of a rocky core surrounded by ice from a few m to a few hundred km in diameter.

Common excavation   Excavation that can be accomplished with an earthmover, backhoe, or dragline.

Competency   A measure of the largest sized particle that a river may transport.

Complex response   Mechanism of operation of a system in which changes occur at a variety of scales and times without input of multiple perturbations from outside the system.

Composite volcano   Steep-sided volcanic cone produced by alternating layers of pyroclastic debris and lava flows.

Composting   A biochemical process in which organic materials are decomposed to a humuslike material by aerobic organisms.

Compound   In chemistry, any substance that contains more than one element of definite proportions by weight.

Comprehensive plan   Planning document adopted by local governments that states general and long-range policies for how the community plans to deal with future development.

Compressibility (soil)   Measure of a soil's tendency to decrease in volume.

Concentration factor   With respect to mining of resources, the ratio of a metal's necessary concentration for profitable mining to its average concentration in Earth's crust.

Conchoidal fracture   Fracture in rock or mineral that is smoothly curved, like a seashell.

Cone of depression   A cone-shaped depression in the water table caused by withdrawal of water at rates greater than those at which the water can be replenished by natural groundwater flow.

Confined aquifer   An aquifer that is overlain by a confining layer (aquitard).

Conglomerate   A detrital sedimentary rock composed of rounded fragments, 10 percent of which are larger than 2 mm in diameter.

Conservation   Policy for resources such as water and energy that moderates or adjusts demands in order to minimize expenditure of the resource; may mean getting by with less through improved technology to provide just the amount of the resource necessary for a given task.

Consumptive use   A type of offstream use in which the water does not return to the stream or groundwater resource after use; the water evaporates, is incorporated into crops or products, or is consumed by animals or humans.

Contact metamorphism   Type of metamorphism produced when country rocks are in close contact with a cooling body of magma below the surface of Earth.

Continental drift   Movement of continents in response to seafloor spreading; the most recent episode of continental drift supposedly began about 200 million years ago with the breakup of the supercontinent Pangaea.

Continental shelf   Relatively shallow ocean area between the shoreline and the continental slope that extends to approximately a 600-ft water depth surrounding a continent.

Continuity equation   With respect to hydrology of rivers, refers to the equation that the discharge of flow is equal to the product of the cross-sectional area of flow times the velocity of the flow.

Contour (strip) mining   Type of strip mining used in hilly terrain.

Contour plowing   Practice of plowing the land along natural contours perpendicular to the downslope direction to reduce erosion.

Convection   Transfer of heat involving movement of particles; for example, the boiling of water, in which hot water rises to the surface and displaces cooler water that moves toward the bottom.

Convergence of wave normals   With respect to coastal processes, refers to areas where wave heights are higher and potential erosive power of waves are greater due to offshore topography or other features that concentrate the wave energy.

Convergent boundary   Boundary between two lithospheric plates in which one plate descends below the other (subduction).

Core   With respect to the interior of Earth, the central part of Earth below the mantle, divided into a solid inner core with a radius of approximately 1,300 km and a molten outer core with a thickness of about 2,000 km; the core is thought to be metallic and composed mostly of iron.

Corrosion   A slow chemical weathering or chemical decomposition that proceeds inward from the surface; objects such as pipes experience corrosion when buried in soil.

Corrosion potential   Potential of a particular soil to cause corrosion of buried iron pipes as a result of soil chemistry.

Cost–benefits analysis   A type of site selection in which benefits and costs of a particular project are compared; the most desirable projects are those for which the benefits-to-cost ratio is greater than 1.

Creep   A type of downslope movement characterized by slow flowing, sliding, or slipping of soil and other earth materials.

Crust   The outermost layer of the solid Earth, embedded in the top of the lithosphere, that varies in thickness from 6 to 7 km below the oceans to as much as 70 km beneath continental mountain ranges.

Crystalline   A material with a definite internal structure such that the atoms are in an orderly, repeating arrangement.

Crystallization   Processes of crystal formation.

Crystal settling   A sinking of previously formed crystals to the bottom of a magma chamber.

Cultural eutrophication   Rapid increase in the abundance of plant life, particularly algae, in freshwater or marine environments resulting from input of nutrients from human sources to the water.

Cutoff trench A trench that is excavated across a slope to intercept surface and subsurface waters and provide drainage, thus decreasing the likelihood of a landslide.

Darcy flux The product of the hydraulic conductivity and hydraulic gradient, which has the units of a velocity.

Darcy's law Empirical relationship that states that the volumetric flow rate such as cubic meters per day is a product of hydraulic conductivity, hydraulic gradient, and cross-sectional area of flow; developed by Henry Darcy in 1856.

Debris avalanche A type of mass wasting characterized by a rapid downslope movement of soil and/or other rock debris on steep slopes; may result from saturation after a heavy rainfall or other processes such as volcanic eruptions that cause the failure on the side of a volcano.

Debris flow (mudflow/lahar) Rapid downslope movement of Earth material often involving saturated, unconsolidated material that has become unstable because of torrential rainfall.

Deep-well disposal Method of waste disposal that involves pumping waste into subsurface disposal sites such as fractured or otherwise porous rocks.

Deflation The process of removing loose sand or dust by the action of wind.

Delta A deposit of sediments that forms where a river flows into a lake or the ocean and commonly has the shape of the Greek letter delta (Δ); actual shape may be highly variable depending on the relative importance of river processes and coastal processes. It is a landform of very low relief with distributary channels that spread out the flow through a system of channels.

Dendrochronology Study of tree rings to establish a chronology.

Desalination Engineering processes and technology that reduce salinity of water to such a level that it may be consumed by people or used in agriculture.

Desert A difficult term to define scientifically, but classification is mostly based on climatic data, type of vegetation, type of soil, and the general look of a landscape, commonly in an arid or semiarid region.

Desertification Conversion of land from a more productive state to one more nearly resembling a desert.

Detrital Mineral and rock fragments derived from preexisting rocks.

Diagenesis Physical and chemical processes and changes in sediments following deposition which produce sedimentary rocks.

Diamond Very hard mineral composed of the element carbon.

Dike Generally, a relatively long, narrow igneous intrusion.

Dioxin An organic compound composed of oxygen, hydrogen, carbon, and chlorine; a by-product resulting from chemical reactions in the production of other chemicals such as herbicides; may be extremely toxic to mammals and damage ecosystems.

Directivity With respect to earthquake hazards, refers to the fact that during some moderate to large earthquakes the rupture of the fault is in a particular direction and the intensity of seismic shaking is greater in that direction.

Disaster preparedness With respect to natural hazards, refers to the actions of individuals, families, cities, states, or entire nations taken before a hazardous event to plan for that event and to minimize losses.

Discharge The quantity of water flowing past a particular point on a stream, usually measured in cubic feet per second (cfs) or cubic meters per second (cms).

Disease From an environmental viewpoint, disease may be considered as an imbalance that results in part from a poor adjustment between an individual and his or her environment.

Disseminated mineral deposit Mineral deposit in which ore is scattered throughout the rocks; examples are diamonds in kimberlite and many copper deposits.

Dissolved load That part of the load that a river carries that results from chemical dissolution of rocks in the drainage basin.

Disturbance From an ecological viewpoint, refers to an event that disrupts a system; examples include wildfires or hurricanes that cause considerable environmental change.

Divergence of wave normals With respect to coastal processes, refers to areas where wave heights and erosive capability of waves are relatively low compared to other locations; often found in bays and other areas where sand is more likely to be deposited on beaches.

Divergent boundary Boundary between lithospheric plates characterized by production of new lithosphere; found along oceanic ridges.

Dose dependency Refers to the effects of a certain trace element on a particular organism being dependent on the dose or concentration of the element.

Dose-response curves A graph showing relationship between response and dose of a particular trace element on a particular population of organisms.

Doubling time The time necessary for a quantity of whatever is being measured to double.

Downstream floods Floods produced by storms of long duration that saturate the soil and produce increased runoff over a relatively wide area. Often are of regional extent.

Drainage basin Area that contributes surface water to a particular steam network.

Drainage control Refers to the development of surface and subsurface drains to increase the stability of a slope.

Drainage net System of stream channels that coalesce to form a stream system.

Dredge spoils Solid material, such as sand, silt, clay, or rock deposited from industrial and municipal discharges, that is removed from the bottom of a water body to improve navigation.

Driving forces Those forces that tend to make Earth material slide.

Ductile Material that ruptures after elastic and plastic deformation.

E soil horizon A light-colored horizon underlying the *A* horizon that is leached of iron-bearing compounds.

E-waste Waste from electronic devices such as computers, cell phones, iPods, and the like.

E-zone With respect to coastal erosion, refers to the zone that is expected to erode within a particular time period.

Earth flow Type of mass wasting or landslide characterized by water-saturated Earth materials moving downslope, often with an upper slumping and lower flow type of deformation.

Earthquake   Natural shaking or vibrating of Earth in response to the breaking of rocks along faults. The earthquake zones of Earth generally are correlated with lithospheric plate boundaries.

Earthquake cycle   A hypothesis to explain periodic occurrence of earthquakes based on drop in elastic strain after an earthquake and reaccumulation of strain before the next event.

Earth's energy balance   Refers to the balance between incoming solar radiation and outgoing radiation from Earth; involves consideration of changes in the energy's form as it moves through the atmosphere, oceans, and land, as well as living things, before being radiated back into space.

Earth systems science   The study of Earth as a system.

Ease of excavation (soil)   Measure of how easily a soil may be removed by human operators and equipment.

Ecology   Branch of biology that treats relationships between organisms and their environments.

Ecological restoration   Application of ecology to restore ecosystems such as rivers, wetlands, beaches, or sand dunes that have been degraded. Also the restoration of land following activities such as mining and timber harvesting.

Economic geology   Application of geology to locating and evaluating mineral materials.

Ecosystem   A community of organisms and its nonliving environment in which chemical elements cycle and energy flows.

Efficiency   With respect to energy resources, refers to designing and using equipment that yields more power from a given amount of energy, resulting in wasting less of the energy to the environment as heat.

Effluent   Any material that flows outward from something; examples include wastewater from hydroelectric plants and water discharged into streams from waste-disposal sites.

Effluent stream   Stream in which flow is maintained during the dry season by groundwater seepage into the channel.

Elastic deformation   Type of deformation in which material returns to its original shape after the stress is removed.

Element   A chemical substance composed of identical atoms that cannot be separated by ordinary chemical means into different substances.

El Niño   An event during which trade winds weaken or even reverse and the eastern equatorial Pacific Ocean becomes anomalously warm; the westward moving equatorial current weakens or reverses.

Emergency planning   Planning for projects after catastrophic events such as hurricanes, floods, or other events.

Engineering geology   Application of geologic information to engineering problems.

Environment   Both physical and cultural surroundings that surround an individual or a community; also sometimes denotes a certain set of circumstances surrounding a particular occurrence, for example, environments of deposition.

Environmental audit   The study of past land use at a particular site, often determined from analyzing old maps and aerial photographs but may involve drilling and sampling of soil and groundwater.

Environmental crisis   Refers to the hypothesis that environmental degradation has reached a crisis point as a result of human use of the environment.

Environmental geology   Application of geologic information to environmental problems.

Environmental geology map   A map that combines geologic and hydrologic data expressed in nontechnical terms to facilitate general understanding by a large audience.

Environmental impact statement (EIS)   A written statement that assesses and explores the possible impacts of a particular project that may affect the human environment; required by the National Environmental Policy Act of 1969.

Environmental law   A field of law concerning the conservation and use of natural resources and the control of pollution.

Environmental resource unit (ERU)   A portion of the environment with a similar set of physical and biological characteristics, a supposedly natural division characterized by specific patterns or assemblages of structural components (such as rocks, soils, and vegetation) and natural processes (such as erosion, runoff, and soil processes).

Environmental unity   A principle of environmental studies that states that everything is connected to everything else.

Ephemeral   Temporary or very short lived; characteristic of beaches, lakes, and some stream channels that change rapidly (geologically).

Epicenter   The point on the surface of Earth directly above the hypocenter (area of first motion) of an earthquake.

Erodibility (soil)   Measure of how easily a soil may erode.

Evaporite deposits   Sediments deposited from water as a result of extensive evaporation of seawater or lake water; dissolved materials left behind following evaporation.

Expansive soil   With respect to engineering properties of soils, refers to soils that, upon wetting and drying, will alternately expand and contract, causing problems for foundations of buildings and other structures.

Exponential growth   A type of compound growth in which a total amount or number increases at a certain percentage each year, and each year's rate of growth is added to the total from the previous year; characteristically stated in terms of a particular doubling time, that is, the time in years it will take the original number to double; commonly used in reference to population growth.

Extrusive igneous rocks   Igneous rock that forms when magma reaches the surface of Earth; a volcanic rock.

Facies   With respect to sediments, in a sedimentary rock refers to characteristics that most often reflect the process or conditions of original deposition of the sediment with respect to its composition and grain size.

Falling   With respect to mass wasting and landslides, refers to Earth materials such as rocks that fall from steep slopes.

Fault   A fracture or fracture system that has experienced movement along opposite sides of the fracture.

Fault gouge   A clay zone formed by pulverized rock during an earthquake, which may create a groundwater barrier.

Fault scarp   A steep slope that is formed by a fault rupture at the surface of Earth.

Fault segmentation   A concept recognizing that faults may be divided into specific segments depending upon their geometry, structure, and earthquake history.

Fecal coliform bacteria   A type of bacteria commonly found in the gut of humans and other animals; usually harmless, but can cause some diseases; commonly used as a measure of biological pollution.

Feedback   The response of a system by which output from the system serves as input back into the system, causing change.

Feldspar   The most abundant family of minerals in the crust of Earth; silicates of calcium, sodium, and potassium.

Ferromagnesian mineral   Silicate minerals containing iron and magnesium, characteristically dark.

Fertile material   Material such as uranium-238, which is not naturally fissionable but upon bombardment by neutrons is converted to plutonium-239, which is fissionable.

Fetch   The distance in which wind blows over a body of water; one of the factors significant in determining the height of windblown waves.

Fission   The splitting of an atom into smaller fragments with the release of energy.

Flash flood   A type of upstream flood in which the floodwaters rise quickly.

Flashy discharge   Stream flow characterized by a short lag time or response time between when precipitation falls and peak discharge of a stream occurs.

Flood-hazard mapping   The mapping of the floodplain and levels of water inundation from floods of a particular magnitude for the purpose of delineating the flood hazard.

Flooding   From an environmental perspective, refers to overbank flow of rivers causing potential damage to human facilities; as a natural process, refers to overbank flows that may lead to the construction of floodplains adjacent to the river channel.

Floodplain   Flat topography adjacent to a stream in a river valley, produced by the combination of overbank flow and lateral migration of meander bends.

Floodplain regulation   A process of delineating floodplains and regulating land uses on them.

Floodplain zoning   Designating appropriate land uses in areas where flooding has occurred or is likely to occur in the future.

Flood-proofing   With respect to flood hazards, refers to the construction and modification of buildings and other structures so that they are not inundated by floodwaters.

Flowage (flow)   With respect to mass wasting and landsliding, refers to downslope movement of Earth materials that deform as a fluid.

Fluid potential   The primary driving force of moving water both in the surface and subsurface environments; in general, refers to the elevation or height of a mass of water above a particular reference.

Fluorine   Important trace element, essential for nutrition.

Fluvial   Concerning or pertaining to rivers.

Fly ash   Very fine particles (ash) resulting from the burning of fuels such as coal.

Focus   The point or location in Earth where earthquake energy is first released; during an earthquake event, seismic energy radiates out from the focus.

Fold   Bend that develops in stratified rocks because of tectonic forces.

Foliation   Property of metamorphic rock characterized by parallel alignment of the platy or elongated mineral grains; environmentally important because it can affect the strength and hydrologic properties of rock.

Forcing   With respect to global warming, a factor or variable that contributes to global warming, as, for example, anthropogenic forcing from burning fossil fuels.

Forecast   With respect to natural hazards, refers to an announcement that states that a particular event such as a flood is likely to occur at a particular time, often with some probability as to how likely the event is.

Foreshocks   Small to moderate earthquakes occurring before the main event.

Formation   Any rock unit that can be mapped.

Fossil   The remains or evidence of past life, including bones, shells, impressions, and trails, preserved naturally in the geologic record.

Fossil fuels   Fuels such as coal, oil, and gas formed by the alteration and decomposition of plants and animals from a previous geologic time.

Fracture zone   A fracture system that may or may not be active and may or may not have an alteration zone along the fracture planes; environmentally important because fracture zones greatly affect the strength of rocks.

Frequency   The number of waves passing a point of reference per second (units are cycles per second or hertz, Hz); the inverse of the wave period.

Friction   With respect to deformation of Earth materials, for example, faulting or landsliding, refers to forces that resist motion, usually defined along a plane, such as a fracture or slip plane of a landslide.

Fuel cell   A device that produces electricity directly from a chemical reaction; commonly uses hydrogen as a fuel to which an oxidant is supplied.

Fugitive sources   Stationary air pollution sources that generate pollutants from open spaces exposed to wind processes.

Fumarole   A natural vent from which fumes or vapors are emitted, such as the geysers and hot springs characteristic of volcanic areas.

Fusion, nuclear   Combining of light elements to form heavy elements with the release of energy.

Gabbro   A dark, coarse-grained igneous rock with minerals such as calcium-rich feldspar, olivine, and pyroxene.

Gaging station   Location at a stream channel where discharge of water is measured.

Gaia hypothesis   A series of hypotheses that explain how Earth as a system may operate with respect to life. Metaphorically, Earth is viewed as a giant organism consisting of various interactive systems with distinct feedback and thresholds that result in producing an environment beneficial to the many life forms on Earth. Furthermore, life is an important ingredient in producing that environment.

Gamma radiation   Type of nuclear radiation consisting of energetic and penetrating rays similar to X-rays emitted during radioactive decay.

Gasification   Method of producing gas from coal.

Geochemical cycle   Migratory paths of elements during geologic changes and processes.

Geochemistry   Earth chemistry; the study of the abundance and distribution of chemical elements within soil, rock, and water.

Geochronology   Chronology of Earth events from numeric dating methods.

Geographic Information System (GIS) Technology capable of storing, retrieving, transforming, and displaying spatial environmental data.

Geologic cycle A group of interrelated cycles known as the hydrologic, rock, tectonic, and geochemical cycles.

Geologic time Time extending from the beginning of Earth to the present; determined in part from Earth's history as recorded in the rocks and sediments that have been deposited and formed at various times; the geologic time scale is the chronological arrangement of rocks of various ages, generally from the oldest event to the youngest.

Geology The science of Earth, including its structure, composition, and history.

Geomorphology The study of landforms and surface processes.

Geopressured system Type of geothermal energy system resulting from trapping the normal heat flow from Earth by impermeable layers such as shale rock.

Geothermal energy The useful conversion of natural heat from the interior of Earth.

Geothermal gradient The rate of increase of temperature beneath the surface of Earth with depth; the average increase is approximately 25 °C per kilometer.

Geyser A particular type of hot spring that ejects hot water and steam above the surface of Earth; perhaps the most famous is Old Faithful in Yellowstone National Park.

Glacial surge A sudden or quick advance of a glacier.

Glacier A landbound mass of moving ice.

Global circulation models Refers to computer models used to predict global change, such as increase in mean temperature, precipitation, or some other climatic variable.

Global dimming Slight cooling caused by human release of air pollution particles that reflect incoming solar radiation back to space.

Global warming Refers to the hypothesis that the mean annual temperature of the lower atmosphere is increasing as a result of burning fossil fuels and emitting greenhouse gases into the atmosphere.

Gneiss A coarse-grained, foliated metamorphic rock in which there is banding of light and dark minerals.

Grading of slopes Cut-and-fill activities designed to increase the stability of a slope.

Granite Coarse-grained intrusive igneous rock with minerals such as potassium-rich feldspar, quartz, and mica.

Gravel Unconsolidated, generally rounded fragments of rocks and minerals greater than 2 mm in diameter.

Gravitational water Water that occurs in pore spaces of a soil and is free to drain from the soil mass under the influence of gravity.

Greenhouse effect Trapping of heat in the atmosphere by water vapor, carbon dioxide, methane, and chlorofluorocarbons (CFCs).

Groin A structure designed to protect shorelines and trap sediment in the zone of littoral drift, generally constructed perpendicular to the shoreline.

Groin field With respect to coastal processes, refers to a group of groins.

Groundwater Water found beneath the surface of Earth within the zone of saturation.

Groundwater discharge Refers to the outflow of groundwater, as from a well, spring, or seepage into stream channels.

Groundwater flow Movement of water in the subsurface below the groundwater table.

Groundwater recharge Refers to the process whereby surface waters infiltrate the soil and vadose zone to eventually augment groundwater resources.

Groundwater system With respect to geothermal energy, refers to the use of groundwater resources at normal temperatures to provide energy for heating and cooling.

Groundwater treatment Refers to a variety of physical, chemical, and biological processes utilized to remove pollutants from groundwater.

Grout A mixture of cement and sediment that is sufficiently fluid to be pumped into open fissures or cracks in rocks, thereby increasing the strength of a foundation for an engineering structure.

Growth rate A rate usually measured as a percentage by which something is changing; for example, if you earn 5 percent interest in a bank account per year, then the growth rate is 5 percent per year.

Gypsum An evaporite mineral, $CaSO_4 \cdot 2H_2O$.

Half-life The amount of time necessary for one-half of the atoms of a particular radioactive element to decay.

Halite A common mineral, NaCl (salt).

Hardpan soil horizon Hard, compacted, or cemented soil horizon, most often composed of clay but sometimes cemented with calcium carbonate, iron oxide, or silica; nearly impermeable and often restricts the downward movement of soil water.

Hard path From an environmental energy perspective, refers to use of large centralized power plants; may be coupled with energy conservation and cogeneration.

Hazardous chemicals Chemicals that are harmful, carcinogenic, or otherwise toxic to humans or other living things or ecosystems; most are produced by our industrial/agricultural industries, but some have natural sources as well.

Hazardous waste Waste materials determined to be toxic or otherwise harmful to people and the environment.

Heavy metals With respect to geochemistry and environmental health, refers to metals such as lead, selenium, zinc, and others that may be harmful in small concentrations in the environment.

Hematite An important ore of iron, a mineral (Fe_2O_3).

High-level waste Commercial and military spent nuclear fuel consisting in part of uranium and plutonium that is extremely toxic.

High-value resource Materials such as diamonds, copper, gold, and aluminum; these materials are extracted wherever they are found and transported around the world to numerous markets.

Hot igneous system Type of geothermal energy system in which heat is supplied by the presence of magma.

Hot spot Assumed stationary heat source located below the lithosphere that feeds volcanic processes near Earth's surface.

Hot spring A natural discharge of groundwater at a temperature higher than the human body.

Hot springs and geysers   Features at the surface of Earth where hot water and steam are quietly released or may be explosively erupted.

Humus   Black organic material in soil.

Hurricane   Tropical cyclone characterized by circulating winds of 100 km/hr or greater generated over an area of about 160 km in diameter; known as typhoons in the Pacific Ocean.

Hydraulic conductivity   Measure of the ability of a particular material to allow water to move through it; units are length per time, such as meters per day.

Hydraulic gradient   With respect to movement of groundwater, refers to the slope of the groundwater surface or the piezometric surface that is important in the movement of groundwater.

Hydraulic head   With respect to groundwater movement, refers to the height of the groundwater above a datum such as sea level.

Hydrocarbon   Organic compounds consisting of carbon and hydrogen.

Hydroconsolidation   Consolidation of Earth materials when wet.

Hydroelectric power   *See* Water power.

Hydrofracturing   Pumping of water under high pressure into subsurface rocks to fracture the rocks and thereby increase their permeability.

Hydrogen fluoride   An extremely toxic gas, dangerous at very small concentrations.

Hydrogen sulfide   A toxic gas that is flammable and has the smell of rotten eggs.

Hydrogeology   Discipline that studies relationships between groundwater, surface water, and geology.

Hydrograph   A graph of the discharge of a stream over time.

Hydrologic cycle   Circulation of water from the oceans to the atmosphere and back to the oceans by way of precipitation, evaporation, runoff from streams and rivers, and groundwater flow.

Hydrologic gradient   The driving force for both saturated and unsaturated flow of groundwater; quantitatively, it is the slope or rate of change of the hydraulic head, which at the point of measurement is the algebraic sum of the elevation head and pressure head.

Hydrology   The study of surface and subsurface water.

Hydrosphere   The water environment in and on Earth and in the atmosphere.

Hydrothermal convection system   Geothermal energy system characterized by the circulation of hot water; may be dominated by water vapor or hot water.

Hydrothermal ore deposit   A mineral deposit derived from hot-water solutions of magmatic origin.

Hygroscopic water   Refers to water absorbed and retained on fine-grained soil particles; may be held tenaciously.

Hypocenter   The point in Earth where an earthquake originates; also known as the *focus*.

Hypothesis   A statement intended to be a possible answer to a scientific question. The best hypothesis may be tested. Often multiple hypotheses are developed to answer a particular question.

Icebergs   Large blocks of glacial ice that break off from the front of a glacier, dropping into the ocean by a process called calving; the block of ice, 90 percent of which is below the surface of the water, then moves with the currents of the ocean.

Igneous rocks   Rocks formed from solidification of magma; extrusive if they crystallize on the surface of Earth, and intrusive if they crystallize beneath the surface.

Impermeable   Earth materials that greatly retard or prevent movement of fluids through them.

Impervious cover   With respect to urban hydrology, represents the surface of Earth that is covered with concrete, roofs, or other structures that impede the infiltration of water into the soil; in general, as urbanization proceeds, the percentage of the land that is under impervious cover increases.

Incineration   Reduction of combustible waste to inert residue (ash) by burning at high temperatures.

Indoor air pollution   Refers to those pollutants that are concentrated within buildings we live and work in.

Industrial ecology   Design of industrial systems to be similar to ecosystems where waste from one part of the system is a resource for another.

Infiltration   Movement of surface water into rocks or soil.

Influent stream   Stream that is everywhere above the groundwater table and flows in direct response to precipitation; water from the channel moves down to the water table, forming a recharge mound.

Input-output analysis   A type of systems analysis in which rates of input and output are calculated and compared.

Instream use   Water that is used but not withdrawn from its source; for example, water used to generate hydroelectric power.

Instrumental intensity   The intensity of shaking from an earthquake obtained from data recorded from a dense network of high-quality seismographs.

Integrated waste management (IWM)   A complex set of management alternatives for waste management, including source reduction, recycling, composting, landfill, and incineration.

Intraplate earthquake   Earthquakes that occur in the interior of a lithospheric plate, far away from any plate boundary.

Intrusive igneous rock   Igneous rock that forms when magma solidifies below the surface of Earth; a volcanic rock.

Island arc   A curved group of volcanic islands associated with a deep oceanic trench and subduction zone (convergent plate boundary).

Isostasy   The principle stating that thicker, more buoyant crust is topographically higher than crust that is thinner and denser. Also, with respect to mountains, the weight of rocks of the upper crust is compensated by buoyancy of the mass of deeper crystal rocks; that is, mountains have "roots" of lighter crustal rocks extending down into the denser mantle rocks, like icebergs in the ocean.

Isotopes   Atoms of the same element having the same number of protons in the nucleus, but differing number of neutrons.

Itai itai disease   Extremely painful disease that attacks bones, causing them to become very brittle so that they break easily; associated with the consumption of or exposure to heavy metals, especially cadmium, in concentrations of a few parts per million in the soil or in food.

Jetty   Often constructed in pairs at the mouth of a river or inlet to a lagoon, estuary, or bay, designed to stabilize a channel, control deposition of sediment, and deflect large waves.

Joint A rock fracture along which there has been no displacement. Parallel joints form a joint set.

Juvenile water Water derived from the interior of Earth that has not previously existed as atmospheric or surface water.

K soil horizon A calcium carbonate–rich horizon in which the carbonate often forms laminar layers parallel to the surface; carbonate completely fills the pore spaces between soil particles.

K-T boundary Geologic time boundary between the Cretaceous and Tertiary periods, approximately 65 million years ago.

Karst topography A type of topography characterized by the presence of sinkholes, caverns, and diversion of surface water to subterranean routes.

Kimberlite pipe An igneous intrusive body that may contain diamond crystals disseminated (scattered) throughout the rock.

Lag time The time period between when the main mass of precipitation falls and a peak discharge in a stream occurs. With urbanization, the lag time generally decreases.

Land application Alternative for disposal of certain types of hazardous chemical waste in which the waste is applied to the soil and degraded by natural biological activity in the soil.

Land ethic Ethic that affirms the right of all resources, including plants, animals, and earth materials, to continued existence and, at least in some locations, continued existence in a natural state.

Landslide Specifically, rapid downslope movement of rock and/or soil; also a general term for all types of downslope movement.

Land-use planning Complex process involving development of a land-use plan to include a statement of land-use issues, goals, and objectives; summary of data collection and analysis; land-classification map; and report describing and indicating appropriate development in areas of special environmental concern. An extremely controversial issue.

Lateral blast Type of volcanic eruption characterized by explosive activity that is more or less parallel to the surface of Earth. Lateral blast may occur when catastrophic failure of the side of a volcano occurs.

Laterite Soil formed from intense chemical weathering in tropical or savanna regions.

Lava Molten material produced from a volcanic eruption, or rock that forms from solidification of molten material.

Lava flow Eruption of magma at the surface of Earth that generally flows downslope from volcanic vents.

Lava tube A natural conduit or tunnel through which magma moves from a volcanic event downslope (sometimes many kilometers) to where the magma may again emerge at the surface; after the volcanic eruption, the tubes are often left as open voids and are a type of cave.

Law of cross-cutting relationships Fundamental law of geology that states a rock is younger than any rock it cuts across; application of this law assists in determining the relative ages of rocks.

Law of faunal assemblages Also known as the *law of faunal succession*, this is a general law of the geological sciences that states that the fossils or organisms succeed one another in an order that may be recognized. In other words, the fossil content of sedimentary rocks suggests the rocks' relative ages.

Law of original horizontality A principal law of the geological sciences that states that a sedimentary stratum, at the time it was deposited, was nearly horizontal; this does not mean that individual grains in a stratum are deposited horizontally, but that the sedimentary bed itself is essentially horizontal.

Law of superposition A fundamental law of the geological sciences that states that for any sequence of sedimentary strata that has not been overturned, the youngest rocks are at the top and the oldest at the bottom or base of the sedimentary sequence. Another way of stating it is that for any given strata or sedimentary unit, rocks above are younger and those below are older.

Leachate Noxious liquid material capable of carrying bacteria, produced when surface water or groundwater comes into contact with solid waste.

Leaching Process of dissolving, washing, or draining Earth materials by percolation of groundwater or other liquids.

Lead A toxic heavy metal that has been heavily utilized by people for thousands of years, most recently in paints and gasoline.

Levee Natural levees result from overbanked flows of rivers; human-constructed levees are earthen embankments along a river channel to protect land adjacent to the river from flooding.

Lignite A type of low-grade coal.

Limestone A sedimentary rock composed almost entirely of the mineral calcite.

Limonite Rust, hydrated iron oxide.

Liquefaction Transformation of water-saturated granular material from the solid state to a liquid state.

Lithosphere Outer layer of Earth, approximately 100 km thick, that comprises the plates that contain the ocean basins and continents.

Littoral Pertaining to the nearshore and beach environments.

Littoral cell Segment of coastline that includes an entire cycle of sediment delivery to the coast, longshore littoral transport, and eventual loss of sediment from the nearshore environment.

Loess Deposits of windblown silt.

Longitudinal profile With respect to the study of rivers, refers to the profile of a stream channel that is generally concave and represents a graph of the relationship between elevation and distance downstream from a reference point.

Longshore bar and trough Elongated depression and adjacent ridge of sand roughly parallel to shore produced by wave action.

Longshore current A current of water and moving sediment that develops in the surf zone as the result of waves that strike the land at an angle.

Longshore sediment transport With respect to coastal processes, refers to the transport of sediment in the nearshore environment by wave activity.

Low-level radioactive waste Materials that contain only small amounts of radioactive substances.

Low-value resource Resources such as sand and gravel that have primarily a place value, economically extracted because they are located close to where they are to be used.

Magma A naturally occurring molten rock material, much of which is in a liquid state.

Magma tap   Attempt to recover geothermal heat directly from magma; feasibility of such heat extraction is unknown.

Magmatic differentiation   Physical and chemical processes that result in different chemical composition and thus mineralogy of igneous rock from a common magma source.

Magnetic reversal   Involves the change of Earth's magnetic field between normal polarity and reverse polarity; also sometimes known as *geomagnetic reversal*.

Magnetite   A mineral and important ore of iron, Fe_3O_4.

Magnitude-frequency concept   The concept that states that the magnitude of an event is inversely proportional to its frequency.

Manganese oxide nodules   Nodules of manganese, iron with secondary copper, nickel, and cobalt, which cover vast areas of the deep-ocean floor.

Mantle   An internal layer of Earth approximately 3,000 km thick composed of rocks that are primarily iron and magnesium-rich silicates. The lower boundary of the mantle is with the core, and the upper boundary is with the crust. The boundary is known as the *Mohorovičić discontinuity* (also called the *Moho*).

Marble   Metamorphosed limestone.

Marl   Unconsolidated clays, silts, sands, or mixtures of these materials that contain a variable content of calcareous material.

Mass extinction   Sudden loss of large numbers of plant and animal species relative to new species being added.

Mass wasting   A comprehensive term for any type of downslope movement of earth materials.

Material amplification   Refers to the phenomenon that some Earth materials will cause the amplitude of seismic shaking to increase. This is generally associated with soft sediment, such as silt and clay deposits.

Materials management   Making better use of materials to reduce the waste we produce.

Maximum credible earthquake   The largest earthquake that may reasonably be assumed to occur at a particular area in light of the tectonic environment, historic earthquakes, and paleoseismicity.

Meandering   A type of channel pattern characterized by a sinuous channel with a series of gentle bends that migrate back and forth across a floodplain.

Meanders   Bends in a stream channel that migrate back and forth across the floodplain, depositing sediment on the inside of the bends, forming point bars, and eroding outsides of bends.

Mediation   Process of working toward a solution to a conflict concerning the environment that is advantageous for everyone; a collaborative process that seeks a solution that favors the environment while allowing activities and projects to go forward.

Meltdown   An accident at a nuclear power station in which the reactor core overheats and fuel rods melt.

Metamorphic rock   A rock formed from preexisting rock by the effects of heat, pressure, and chemically active fluids beneath Earth's surface. In foliated metamorphic rocks, the mineral grains have a preferential parallel alignment or segregation of minerals; nonfoliated metamorphic rocks have neither.

Meteor   Particle from dust to centimeters in size that is destroyed in the Earth's atmosphere (shooting star).

Meteoric water   Water derived from the atmosphere.

Meteorite   Particle from dust to asteroid size that impacts Earth's surface.

Meteoroid   Particle in space from less than 10 m to larger than dust size—may form from breakup of asteroids.

Methane   A gas, CH_4; the major constituent of natural gas.

Methane hydrate   Icelike compound made of molecules of methane gas that is trapped within cages of frozen water beneath the seafloor at water depths of about 1,000 m.

Mica   A common rock-forming silicate mineral.

Mid-oceanic ridge   A topographic high commonly found in the central part of oceans characterized by seafloor spreading. An example is the Mid-Atlantic Ridge.

Milankovitch cycles   Natural cycles of variation of solar radiation that reach Earth's surface at approximately 20,000, 40,000, and 100,000 years.

Mineral   An element or chemical compound that is normally crystalline and is formed as the result of geologic processes.

Mining reclamation   See Reclamation, mining.

Mining spoils   See Spoils, mining.

Mitigated negative declaration   Environmental statement filed when the initial study of a project suggests that any significant environmental problems that will result from or occur during a project could be modified to mitigate those problems.

Mitigation   The identification of actions that will avoid, lessen, or compensate for anticipated adverse environmental impacts.

Mobile sources   Moving sources of air pollution such as automobiles.

Modified Mercalli Scale   A scale with 12 divisions that subdivide the amount and severity of shaking and damage from an earthquake.

Moho   The boundary between the crust and mantle, also known as the *Mohorovičić discontinuity*; distinguished by compositional differences between the rocks of the crust and the mantle.

Moment magnitude   The magnitude of an earthquake based on its seismic moment, which is the product of the average amount of slip on the fault that produced the earthquake, the area that actually ruptured, and the shear modulus of the rocks that failed.

Monitoring   With respect to waste management, refers to periodic or continuous gathering of samples of soil, vegetation, vadose zone water, and groundwaters in and near waste management facilities, such as landfills or hazardous waste disposal facilities.

Mudflow   A mixture of unconsolidated materials and water that flows rapidly downslope or down a channel.

Multiple land use   A principle of land use that involves multiple uses at the same time, as, for example, a dam and reservoir designed for flood control, water supply, and recreation.

Myth of superabundance   The myth that land and water resources are inexhaustible and management of resources is therefore unnecessary.

National Environmental Policy Act of 1969 (NEPA)   Act declaring a national policy that harmony between humans and their physical environment be encouraged; established the Council on Environmental Quality and requirements that an environmental impact statement be completed before major federal actions that significantly affect the quality of the human environment.

Natural gas Sometimes also referred to as *natural energy gas* or *hydrocarbons* that include ethane, propane, butane, and hydrogen.

Near earth object Asteroids that reside and orbit between the Earth and the Sun or have orbits that intersect Earth's orbit.

Natural hazards Refers to processes, such as earthquakes, floods, and volcanic eruptions, that produce a hazard to people and property.

Negative declaration The filing of a statement that declares that the environment will experience no significant effects from a particular project or plan.

Negative feedback A type of feedback in which the outcome moderates or decreases the process, often leading to a steady-state system or a system in quasi-equilibrium (is self-regulating).

Negotiation With respect to environmental law, refers to processes whereby parties that differ on a particular issue sit down and talk and try to work out an agreement.

Neutron A subatomic particle having no electric charge, found in the nuclei of atoms; crucial in sustaining nuclear fission in a reactor.

Nitrogen oxides (NOₓ) Group of gases emitted as a result of burning fossil fuels in automobiles and at power plants; includes compounds such as nitrogen dioxide (NO_2), which is a light yellow-brown to reddish brown gas that is a main pollutant contributing to the development of photochemical smog.

Nonpoint sources Diffused and intermittent sources of air or water pollutants.

Nonrenewable resource A resource cycled so slowly by natural Earth processes that, once used, it will be essentially unavailable during any useful time frame.

No-till agriculture Combination of farming practices that does not include plowing of the land.

Normal fault A generally steep fault with vertical displacement (dip-slip) in which the hanging-wall has moved down relative to the foot-wall.

Nuclear energy Generation of electricity using a nuclear reactor.

Nuclear fusion *See* Fusion, nuclear.

Nuclear reactor Device in which controlled nuclear fission is maintained; the major component of a nuclear power plant.

O soil horizon Soil horizon that contains plant litter and other organic material; found above the *A* soil horizon.

Ocean pollution Pollution of the oceans of the world due to direct or indirect interjection of contaminants to the marine environment, whether deliberate or not; often results from the process of ocean dumping, but there are many sources of ocean pollution.

Offstream use Water removed or diverted from its primary source for a particular use.

Oil When referring to energy resources, may also be known as *petroleum* or *crude oil*; a liquid hydrocarbon generally extracted from wells.

Oil shale Organic-rich shale containing substantial quantities of oil that can be extracted by conventional methods of destructive distillation.

Open system A type of system in which there is a constant flow of energy and matter across the borders of the system.

Ore Earth material from which a useful commodity can be extracted profitably.

Outcrop A naturally occurring or human-caused exposure of rock at the surface of Earth.

Overburden Earth materials (spoil) that overlie an ore deposit, particularly material overlying or extracted from a surface (strip) mine.

Overland flow Flow of water on the surface of Earth not confined to channels; results because the intensity of precipitation is greater than the rate at which rainwater infiltrates into the ground.

Oxidation Chemical process of combining with oxygen.

Oxides With respect to mineral types, refers to mineral compounds that link oxygen with one or more metallic elements, as for example the mineral hematite (Fe_2O_3).

Ozone Triatomic oxygen (O_3).

Ozone depletion Refers to stratospheric loss of ozone, generally at the South Pole, related to release of chlorofluorocarbons (CFCs) into the atmosphere.

P wave One of the seismic waves produced by an earthquake; the fastest of the seismic waves, it can move through liquid and solid materials.

Paleomagnetic Also known as *paleomagnetism;* refers to the study of magnetism of rocks and the intensity and direction of the magnetic field of Earth in the geologic past.

Particulate matter With respect to air pollutants, refers to small particles of solid or liquid substances that are released into the atmosphere by natural processes and human activities. Examples include smoke, soot, or dust, as well as particles of heavy metals such as copper, lead, and zinc.

Passive method With respect to permafrost, refers to one of the common methods of trying to minimize problems associated with permafrost that involve keeping the ground frozen and not upsetting the natural balance of environmental factors.

Pathogen Any material that can cause disease; for example, microorganisms, including bacteria and fungi.

Peak oil The time when half of all oil on Earth will have been extracted.

Pebble A rock fragment between 4 and 64 mm in diameter.

Ped An aggregate of soil particles; classified by shape as spheroidal, blocky, prismatic, and so on.

Pedology The study of soils.

Pegmatite A coarse-grained igneous rock that may contain rare minerals rich in elements such as lithium, boron, fluorine, uranium, and others.

Perched water table Existence of a water table of relatively limited extent that is found at a higher elevation than the more regional water table.

Percolation test A standard test for determining rate at which water will infiltrate into the soil; primarily used to determine feasibility of a septic-tank disposal system.

Permafrost Permanently frozen ground.

Permeability A measure of the ability of an Earth material to transmit fluids such as water or oil. *See* Hydraulic conductivity.

Petrology Study of rocks and minerals.

Phenocrysts Large crystals in an igneous rock with a porphyritic texture.

Photochemical smog Usually referred to as LA smog or brown air; forms as a result of interactions between solar radiation and automobile exhaust.

Photovoltaics Type of solar technology that converts sunlight directly to electricity.

Physiographic determinism Site selection based on the philosophy of designing with nature.

Physiographic province Region characterized by a particular assemblage of landforms, climate, and geomorphic history.

Placer deposit Ore deposit found in material transported and deposited by such agents as running water, ice, or wind; for example, gold and diamonds found in stream deposits.

Plastic deformation Deformation that involves a permanent change of shape without rupture.

Plate tectonics A model of global tectonics that suggests that the outer layer of Earth, known as the *lithosphere*, is composed of several large plates that move relative to one another; continents and ocean basins are passive riders on these plates.

Plume With respect to groundwater and groundwater pollution, refers to an often elongated three-dimensional mass of polluted or contaminant groundwater that is generally moving away from a contaminant source.

Plunging breaker A type of wave or breaker from a storm that strikes a shoreline with a relatively steep beach profile; tends to be associated with beach erosion.

Pluton Any of several types of igneous intrusions that are variable in size, including dikes and sills; generally, batholiths are composed of a number of plutons.

Plutonium-239 A radioactive element produced in a nuclear reactor; has a half-life of approximately 24,000 years.

PM-2.5 With respect to air pollution, refers to particulate matter less than 2.5 millionths of a meter in size.

PM-10 With respect to air pollution, refers to particulate matter less than 10 millionths of a meter in size.

Point bar Accumulation of sand and other sediments on the inside of meander bends in stream channels.

Point sources Usually discrete and confined sources of air or water pollutants such as pipes that enter into a stream or river or stacks emitting waste from factories or other facilities into the atmosphere.

Pollutant Any substance in the environment, that in excess is known to be harmful to people or other desirable living organisms.

Pollution Any substance, biological or chemical, of which an identified excess is known to be detrimental to desirable living organisms.

Pool Common bed form produced by scour in meandering and straight stream channels with relatively low channel slope; characterized at low flow by slow-moving, deep water; generally, but not exclusively, found on the outside of meander bends.

Porosity The percentage of void (empty space) in earth material such as soil or rock.

Porphyritic Refers to a specific texture of igneous rocks characterized by relatively few, often earlier-formed, coarse-grained crystals (phenocrysts) surrounded by a mass of fine-grained, later-formed crystals.

Positive feedback A type of system in which the output amplifies the input, leading to what some call a vicious cycle.

Another way of looking at positive feedback is the more you have, the more you get.

Potable water Water that may be safe to drink.

Precautionary principle An environmental planning tool that advocates taking cost-effective, proactive steps to eliminate or reduce the consequences of an environmental problem even if the science is not completely worked out. In simple words, better safe than sorry.

Precursor events With respect to natural hazards, refers to physical, chemical, or biological events that occur before an event such as a flood, earthquake, or volcanic eruption.

Prediction A statement that an event with specified magnitude, such as a tsunami or flood, will happen during a particular time interval. Contrast with forecast, which provides a percent chance of something happening.

Primary pollutants With respect to air pollution, refers to those pollutants emitted directly into the atmosphere, including particulates, sulfur oxides, carbon monoxide, nitrogen oxides, and hydrocarbons.

Primary treatment With respect to a wastewater treatment plant, includes screening and removal of grit and sedimentation of larger particles from the waste stream.

Pyrite Iron sulfide, a mineral commonly known as *fool's gold*; environmentally important because, in contact with oxygen-rich water, it produces a weak acid that can pollute water or dissolve other minerals.

Pyroclastic activity Type of volcanic activity characterized by eruptive or explosive activity in which all types of volcanic debris, from ash to very large particles, are physically blown from a volcanic vent.

Pyroclastic deposits Refers to particles forcefully ejected from a volcanic vent, explosive in origin, containing volcanic ash particles and those of larger size such as blocks and bombs.

Pyroclastic flow Rapid subaerial flowage of eruptive material consisting of volcanic gases, ash, and other materials that move rapidly down the flank of a volcano; often form as the result of the collapse of an eruption column; may also be known as *ash flows, fiery clouds,* or *nueé ardentes*.

Quartz Silicon oxide, a common rock-forming mineral.

Quartzite Metamorphosed sandstone composed of quartz grains.

Quick clay Type of clay that, when disturbed, such as by seismic shaking, may experience spontaneous liquefaction and lose all shear strength.

R soil horizon Consolidated bedrock that underlies the soil.

Radioactive waste Type of waste produced in the nuclear fuel cycle, generally classified as high level or low level.

Radioactive waste management Refers to waste-management policies and procedures related to radioactive waste.

Radioisotope A form of a chemical element that spontaneously undergoes radioactive decay, changing from one isotope to another and emitting radiation in the process.

Radon A colorless, radioactive, gaseous element.

Rapid draw-down Rapid decrease in the elevation of the water table at a particular location due to a variety of processes, including receding flood waters or lowering of the water in a reservoir.

Reclaimed water Water that has been treated by wastewater handling facilities and may be used for other purposes on discharge, such as irrigation of golf courses or croplands.

Reclamation, mining Restoring land used for mining to other useful purposes, such as agriculture or recreation, after mining operations are concluded.

Record of decision A concise statement by an agency planning a proposed project as to which alternatives were considered and, specifically, which alternatives are environmentally preferable; becoming an important part of environmental impact work.

Recurrence interval The time between events, such as floods or earthquakes or other natural processes. Often we are interested in the average recurrence interval, which is determined by finding the mean of a series of recurrence intervals between events.

Recycling The reuse of resources reclaimed from waste.

Reduce, recycle, and reuse The three R's of integrated waste management that describe the objective of reducing the amount of waste that must be disposed of in landfills or other facilities.

Refraction With respect to coastal processes, refers to the bending of surface waves as they enter shallow water and "feel bottom"; that part of a wave that first "feels bottom" slows down, bending the wave.

Regional metamorphism Wide-scale metamorphism of deeply buried rocks by regional stress accompanied by elevated temperatures and pressures.

Relative profile development Refers to soils that may be weakly, moderately, or strongly developed, depending on specific soil properties.

Renewable energy Energy sources that are replenished quickly enough to maintain a constant supply if they are not overused; examples include solar energy, water power, and wind energy.

Renewable resource A resource such as timber, water, or air that is naturally recycled or recycled by human-induced processes within a useful time frame.

Reserves Known and identified deposits of Earth materials from which useful materials can be extracted profitably with existing technology under present economic and legal conditions.

Residual deposits With respect to mineral resources, refers to those deposits that form as a result of the mechanical concentration of a particular mineral, as for example gold nuggets in a stream or decomposition in weathered rocks.

Residual soil Refers to soils that develop on bedrock.

Resisting forces Forces that tend to oppose downslope movement of Earth materials.

Resistivity A measure of an Earth material's ability to retard the flow of electricity; the opposite of conductivity.

Retaining wall A structure constructed to buttress the toe area of a slope to minimize the likelihood of a landslide.

Resources Includes reserves plus other deposits of useful Earth materials that may eventually become available.

Reverse fault A fault with vertical displacement in which the hanging-wall has moved up relative to the foot-wall; a low-angle reverse fault is called a thrust fault.

Rhyolite A type of volcanic rock consisting of feldspar, ferromagnesian, and quartz minerals with a relatively high silica content of about 70 percent; associated with volcanic events that may be very explosive.

Richter magnitude A measure of the amount of energy released by an earthquake, determined by converting the largest amplitude of the shear wave to a logarithmic scale in which, for example, 2 indicates the smallest earthquake that can be felt and 8.5 indicates a devastating earthquake.

Ridge push A gravitational push similar to a gigantic landslide responsible for pushing the plates apart at divergent plate boundaries.

Riffle A section of stream channel characterized at low flow by fast, shallow flow; generally contains relatively coarse bed load particles.

Riparian doctrine Part of our prevailing water law restricted for the most part to owners of land adjoining a stream or body of standing water.

Riparian rights, water law Right of the landowner to make reasonable use of water on his or her land, provided the water is returned to the natural stream channel before it leaves the property; the property owner has the right to receive the full flow of the stream undiminished in quantity and quality.

Rip current A seaward flow of water in a confined narrow zone from a beach to beyond the breaker zone.

Rippable excavation Type of excavation that requires breaking up soil before it can be removed.

Riprap Layer or assemblage of broken stones placed to protect an embankment against erosion by running water or breaking waves.

Risk From an environmental viewpoint, risk may be considered as the product of the probability of an event times the consequences.

Risk assessment In terms of toxicology, the process of determining potential adverse environmental health effects following exposure to a particular toxic material.

Risk management With respect to toxicology, the process of integrating risk assessment with legal, social, political, economic, and technical issues to develop a plan of action for a particular toxin.

Riverine environment Land area adjacent to and influenced by a river.

Rock Geologic: An aggregate of a mineral or minerals. Engineering: Any Earth material that must be blasted to be removed.

Rock cycle Group of processes that produce igneous, metamorphic, and sedimentary rocks.

Rock salt Rock composed of the mineral halite.

Rock texture See Texture, rock.

Rotational landslide Type of landslide that develops in homogeneous material; movement is likely to be rotational along a potential slide plane.

Runoff Water moving over the surface of Earth as overland flow on slopes or stream flow; that part of the hydrologic cycle represented by precipitation or snowmelt that results in stream flow.

S wave Secondary wave, one of the waves produced by earthquakes.

Safety factor (SF) With respect to landsliding, refers to the ratio of resisting to driving forces; a safety factor of greater than 1 suggests a slope is stable.

Saline Salty; characterized by high salinity.

Salinity   A measure of the total amount of dissolved solids in water.

Salt dome   A structure produced by upward movement of a mass of salt; frequently associated with oil and gas deposits on the flanks of a dome.

Saltwater intrusion   Process whereby fresh groundwater may be displaced by saltwater, often as a result of pumping of groundwater resources.

Sand   Grains of sediment with a size between 1/16 and 2 mm in diameter; often, sediment composed of quartz particles of this size.

Sand dune   Ridge or hill of sand formed by wind action.

Sandstone   Detrital sedimentary rock composed of sand grains that have been cemented together.

Sanitary landfill   Method of solid-waste disposal that does not produce a public health problem or nuisance; confines and compresses waste and covers it at the end of each day with a layer of compacted, relatively impermeable material, such as clay.

Saturated flow   A type of subsurface or groundwater flow in which all the pore spaces are filled with water.

Scarp   Steep slope or cliff commonly associated with landslides or earthquakes.

Scenic resources   The visual portion of an aesthetic experience; scenery is now recognized as a natural resource with varying values.

Schist   Coarse-grained metamorphic rock characterized by foliated texture of the platy or elongated mineral grains.

Schistosomiasis   Snail fever, a debilitating and sometimes fatal tropical disease.

Scientific method   The method by which scientists work, starting with the asking of a question concerning a particular problem, followed by development and testing of hypotheses.

Scoping   Process of identifying important environmental issues that require detailed evaluation early in the planning of a proposed project; an important part of environmental impact analysis.

Scrubbing   Process of removing sulfur dioxide from gases emitted from burning coal in power plants producing electricity.

Seacliff   Steep (commonly near-vertical) bluff adjacent to and adjoining a beach or coastal environment; produced by a combination of erosional processes including wave activity and subaerial processes such as weathering, landsliding, and runoff of surface water from the land.

Seafloor spreading   The plate tectonics concept that new crust is continuously added to the edges of lithospheric plates at divergent plate boundaries as a result of upwelling of magma along mid-oceanic ridges.

Seawall   Engineering structure constructed at the water's edge to minimize coastal erosion by wave activity.

Secondary enrichment   Weathering process of sulfide ore deposits that may concentrate the desired minerals.

Secondary pollutants   With respect to air pollution, refers to pollutants produced when primary pollutants react with normal atmospheric compounds; an example is ozone, which forms through reactions between primary pollutants, sunlight, and natural atmospheric gases.

Secondary treatment   With respect to wastewater treatment, includes aerobic and anaerobic digestion of waste in the wastewater stream, primarily by bacterial breakdown; final stage is disinfection of treated water, usually with chlorine.

Secure landfill   Type of landfill designed to contain and dispose of hazardous chemical waste; many of these facilities have been shut down because containment of the hazardous waste has been impossible to maintain.

Sedimentary environment   Environments conducive to the deposition of sediments including lakes, floodplains, sand dunes, and glacial deposits.

Sedimentary rock   A rock formed when sediments are transported, deposited, and then lithified by natural cement, compression, or other mechanism; detrital sedimentary rock is formed from broken parts of previously existing rock; chemical sedimentary rock is formed by chemical or biochemical processes removing material carried in chemical solution.

Sedimentology   Study of environments of deposition of sediments.

Sediment pollution   Pollution of some part of the environment either on land or in a body of water by sediment that has been transported into that environment by wind or water; an example is turbidity of a water supply (muddy water).

Sediment yield   Volume or mass of sediment per unit time produced from a particular area.

Seismic   Refers to vibrations in Earth produced by earthquakes.

Seismic gaps   Areas along active fault zones that are capable of producing large earthquakes but have not produced one recently.

Seismic risk map   Map that depicts the seismic risk of a particular area or region; often based on past earthquake activity or on the probability of a specified intensity of a shaking occurring over a specified period of time.

Seismograph   Instrument that records earthquakes.

Seismology   The study of earthquakes as well as the structure of Earth through the evaluation of both natural and artificially generated seismic waves.

Selenium   Important nonmetallic trace element with an atomic number of 34.

Semiarid   Those lands characterized with an annual precipitation between 25 and 50 cm/yr.

Sensitivity (soil)   Measure of loss of soil strength due to disturbances such as human excavation and remolding.

Septic tank   Tank that receives and temporarily holds solid and liquid waste. Anaerobic bacterial activity breaks down the waste, solid wastes are separated out, and liquid waste from the tank overflows into a drainage system.

Sequential land use   Development of land previously used as a site for the burial of waste; the specific reuse must be carefully selected.

Serpentine   A family of ferromagnesian minerals; environmentally important because they form very weak rocks.

Sewage sludge   Solid material that remains after municipal wastewater treatment.

Shake map   Map showing pattern and extent of seismic shaking from an earthquake.

Shale   Sedimentary rock composed of silt- and clay-sized particles; the most common sedimentary rock.

Shield volcano   A broad, convex volcano built up by successive lava flows; the largest of the volcanoes.

Shrink-swell potential (soil) Measure of a soil's tendency to increase and decrease in volume as water content changes.

Silicate minerals The most important group of rock-forming minerals.

Sill A generally planar, tabular igneous intrusion.

Silt Sediment between 1/16 mm and 1/256 mm in diameter.

Sinkhole Surface depression formed by solution of limestone or collapse over a subterranean void such as a cave.

Sinuous channel Type of stream channel (not braided).

Site selection A method of environmental analysis, the purpose of which is to select a site or series of sites for a particular activity.

Slab pull The plate tectonics concept that dictates that as a plate moves farther from a ridge access, it cools and gradually becomes more dense than the asthenosphere beneath it. At the same time, at a subduction zone the heavy plate falls through the lighter mantle and the weight of the descending slab pulls the entire plate.

Slate A fine-grained, foliated metamorphic rock.

Sliding With respect to mass wasting and landsliding, refers to the deformation or downslope movement of a nearly intact block of earth materials along a slip plain.

Slip plane Refers to inclined features such as rock fractures and bedding planes along which landsliding occurs.

Slip rate Long-term rate of slip (displacement) along a fault; usually measured in millimeters or centimeters per year.

Slow earthquake Earthquake produced by fault rupture that can take days to months to complete.

Slump Type of landslide characterized by downward slip of a mass of rock, generally along a curved slide plane.

Smog A general term to refer to visible air pollution, probably first coined in the early part of the twentieth century as a combination of smoke and fog.

Snow avalanche Rapid downslope movement of snow, ice, and rock.

Soft path With respect to energy resources, refers to development of an energy policy that involves alternatives that are renewable, flexible, decentralized, and (to some people) more benign from an environmental viewpoint than the hard path.

Soil Soil science: Earth material so modified by biological, chemical, and physical processes that the material will support rooted plants. Engineering: Earth material that can be removed without blasting.

Soil chronosequence A series of soils arranged in terms of relative soil profile development from youngest to oldest.

Soil fertility Capacity of a soil to supply nutrients (such as nitrogen, phosphorus, and potassium) needed for plant growth when other factors are favorable.

Soil horizons Layers in soil (A, B, C, etc.) that differ from one another in chemical, physical, and biological properties.

Soil profile Weathering of earth materials that, along with biological activity and time, produces a soil that contains several horizons distinct from the parent material from which the soil formed.

Soil sensitivity A relative estimate of a soil's ability to maintain its original strength when disturbed and remolded; soils that keep only a portion of their original strength are said to be sensitive.

Soil slip A type of mass wasting event that generally is narrow relative to width and is linear in form; develops during precipitation events on steep slopes; in California, shallow soil slips are commonly referred to as *mud slides*.

Soil strength The shear strength of a soil in terms of a soil's cohesive and frictional forces.

Soil survey A survey consisting of a detailed soil map and descriptions of soils and land-use limitations; usually prepared in cooperation with local government.

Soil taxonomy With respect to soil science, refers to a method of classifying soils developed by the U.S. Department of Agriculture.

Soil texture Refers to the relative proportions of sand-, silt-, and clay-sized particles in a soil.

Solar collector Any of a variety of active and passive devices that collect solar energy; the most common is the flat-plate collector used to heat water.

Solar energy Energy that is collected from the Sun.

Solid waste Material such as refuse, garbage, and trash.

Specific risk The product of the elements at risk, the probability that a specific event will occur, and the vulnerability defined as the proportion of elements at risk.

Spilling breaker A type of wave associated with a shoreline of relatively low slope; spilling breakers tend to be associated with deposition of sand on a beach.

Spoils, mining Banks or piles that are accumulations of overburden removed during mining processes and discarded on the surface.

Spreading center Synonymous with mid-oceanic ridges where new crust is continuously added to the edges of lithospheric plates.

Spring With respect to groundwater processes, refers to the natural discharge of groundwater where the groundwater system intersects the surface of Earth.

Stationary sources With respect to air pollutants, refers to those sources that are relatively fixed in location.

Steady-state system A system in which the input is approximately equal to the output, so a rough equilibrium is established.

Storm surge Wind-driven oceanic waves, usually accompanying a hurricane.

Strain Change in shape or size of a material as a result of applied stress.

Stream power The product of the discharge of a river and its energy slope.

Strength (soil) Ability of a soil to resist deformation; results from cohesive and frictional forces in the soil.

Stress Force per unit area; may be compressive, tensile, or shear.

Strike-slip A fault across which displacement is predominantly horizontal.

Strip mining A method of surface mining.

Subduction Process in which one lithospheric plate descends beneath another.

Subduction zone Convergence of tectonic plates where one plate dives beneath another and is consumed in the mantle.

Submarine trench A relatively narrow, long (often several 1000 km), deep (often several km) depression on the ocean floor that forms as a result of convergence of two tectonic plates with subduction of one.

Subsidence Sinking, settling, or other lowering of parts of the crust of Earth.

Subsurface water All of the waters within the lithosphere.

Sulfur dioxide (SO₂) Colorless and odorless gas whose anthropogenic component in the atmosphere results primarily from the burning of fossil fuels.

Sulfurous smog Sometimes referred to as *London-type smog* or *gray air,* produced primarily by burning coal or oil at large power plants where sulfur oxides and particulates produced by the burning produce a concentrated smog.

Surface impoundment Excavated or natural topographic depressions used to hold hazardous liquid waste. Although impoundments are often lined, they have been criticized because they are especially prone to seepage and pollution of soil and groundwaters.

Surface water Waters above the solid surface of Earth.

Surface wave One type of wave produced by earthquakes; generally cause most of the damage to structures on the surface of Earth.

Surf zone That part of the beach and nearshore environment characterized by borelike waves of translation after waves break.

Suspended load Sediment in a stream or river carried off the bottom by the fluid.

Sustainability A difficult term to define but generally refers to development or use of resources in such a way that future generations will have a fair share of Earth's resources and inherit a quality environment. In other words, sustainability refers to types of development that are economically viable, do not damage the environment, and are socially just.

Sustainable energy policy Development of energy policy that finds useful sources of energy that do not have adverse environmental effects or minimizes those effects in such a way that future generations will have access to energy resources and a quality environment.

Sustainable global economy A global economic development that will not harm the environment, will provide for future generations, and is socially just.

Swash zone That part of the coastal environment where waves run up on the beach face and then back again into the ocean where the land meets the water; the runup or wave swash covers part of the beach with shallow water and then the backwash exposes it again.

Swell With respect to coastal processes, refers to the sorting out of waves by period from a storm into groups of waves having more or less uniform heights and lengths, allowing groups of waves to move long distances from storms to coastal areas with relatively little loss of energy.

Syncline Fold in which younger rocks are found in the core of the fold; rocks in the limbs of the fold dip inward toward a common axis.

System Any part of the universe that is isolated in thought or in fact for the purpose of studying or observing changes that occur under various imposed conditions.

Tar sand Naturally occurring sand, sandstone, or limestone that contains an extremely viscous petroleum.

Tectonic Referring to rock deformation.

Tectonic creep Slow, more or less continuous movement along a fault.

Tectonic cycle Part of the geologic cycle; at the global scale, it is the cycle of plate tectonics that produces ocean basins and mountain ranges.

Tephra Any material ejected and physically blown out of a volcano; mostly ash.

Texture, rock The size, shape, and arrangement of mineral grains in rocks.

Theory A strong scientific statement. A hypothesis may become a theory after it has been tested many times and has not been rejected.

Thermal pollution With respect to water pollution, refers to water of elevated temperature, often from the disposal of water used to cool industrial processes or to produce electricity into a body of water such as a lake, river, or ocean.

Threshold A point of change where something happens; for example, a stream bank may erode when the water has sufficient force to dislodge particles; the point at which the erosion starts is the threshold.

Throughflow Downslope shallow subsurface flow of water above the groundwater table.

Thrust fault A low-angle reverse fault.

Tidal energy Electricity generated by tidal power.

Tidal flood A type of flood that occurs in estuaries or coastal rivers as the result of interactions between high tides and storm waves.

Tidal power With respect to energy resources, refers to the useful conversion of tidal currents to produce electrical power.

Till Unstratified, heterogeneous material deposited directly by glacial ice.

Total load With respect to stream processes, refers to the sum of the dissolved, suspended, and bed load that a stream or river carries.

Toxic Harmful, deadly, or poisonous.

Toxicology The science of the study of toxins and their effects on people and other living organisms and ecosystems; also concerned with relationships between toxic materials and resulting clinical and industrial processes.

Transform boundary Synonymous with transform faults, occurring where edges of two plates slide past one another; most are boundaries within oceanic crust; an example on land is the San Andreas fault in California.

Transform fault Type of fault associated with oceanic ridges; may form a plate boundary, such as the San Andreas fault in California.

Translation (slab) landslide Type of landslide in which movement takes place along a definite fracture plane, such as a weak clay layer or bedding plane.

Transported soil Does not refer to any transport process of a soil but rather reflects that the parent material of a soil is material that has been transported to a particular location, such as alluvium to a floodplain or glacial deposit.

Transuranic waste Nuclear waste composed of human-made radioactive elements heavier than uranium.

Trench With respect to plate tectonics, refers to an elongated depression of the seafloor associated with convergent boundaries and subduction zones; often referred to as submarine trenches, which are the sites of some of the deepest oceanic waters on Earth; often located seaward of the subduction zone.

Triple junction   Areas where three tectonic plates and their boundaries join.

Tropical cyclone (typhoon)   Severe storm generated from a tropical disturbance; called *typhoons* in most of the Pacific Ocean and *hurricanes* in the Western Hemisphere.

Tsunami   Seismic sea wave generated mostly by submarine earthquake, but also by submarine volcanic eruption, landslide, or impact of an asteroid; characteristically has very long wave length and moves rapidly in the open sea; incorrectly referred to as tidal wave.

Tuff   Volcanic ash that is compacted, cemented, or welded together.

Typhoon   A tropical cyclone that occurs over the Indian Ocean or western Pacific; the counterpart of a hurricane that occurs in the Atlantic.

Unconfined aquifer   Aquifer in which there is no impermeable layer restricting the upper surface of the zone of saturation.

Unconformity   A buried surface of erosion representing a time of nondeposition; a gap in the geologic record.

Unified soil classification system   Classification of soils, widely used in engineering practice, based on amount of coarse particles, fine particles, or organic material.

Uniformitarianism   Concept that the present is the key to the past; that is, we can read the geologic record by studying present processes.

Unsaturated flow   Type of groundwater flow that occurs when only a portion of the pores is filled with water.

Urban ore   Refers to the fact that in some communities the sewage sludge from waste-disposal facilities contains sufficient metal deposits to be considered an ore.

Vadose zone   Zone or layer above the water table in which some water may be suspended or moving in a downward migration toward the water table or laterally toward a discharge point.

Volatile organic compound   Usually abbreviated as VOC when used to describe water or air pollutants; generally are hydrocarbon compounds such as gasoline, benzene, and propane.

Volcanic ash   *See* Ash, volcanic.

Volcanic breccia, agglomerate   Large rock fragments mixed with ash and other volcanic materials cemented together.

Volcanic crisis   A condition in which a volcanic eruption or prospect of an eruption produces a crisis situation for society.

Volcanic dome   Type of volcano characterized by very viscous magma with high silica content; activity is generally explosive.

Wadati-Benioff Zone   Inclined zone of earthquakes produced as a tectonic plate is subducted.

Warning   With respect to natural hazards, the announcement of a possible disaster such as a large earthquake or flood that could occur in the near future.

Wastewater renovation and conservation cycle   A process of recycling liquid waste that includes return of treated wastewater to crops or irrigation and continued renovation through recharge of groundwater; the reused part involves pumping out of the groundwater for municipal, industrial, or other purposes.

Water budget   Analysis of sources, sinks, and storage sites for water in a particular area.

Water conservation   Practices taken to use water more efficiently and to reduce withdrawal and consumption of water.

Water cycle   *See* Hydrologic cycle.

Water management   Practice of managing our water resources.

Water pollution   Degradation of water quality as measured by biological, chemical, or physical criteria.

Water power   Use of flowing water such as in a reservoir to produce electrical power.

Watershed   Land area that contributes water to a particular stream system. *See* Drainage basin.

Water quality standards   In the United States, refers to Environmental Protection Agency minimum standards for drinking water and water for other uses.

Water table   Surface that divides the vadose zone from the zone of saturation; the surface below which all the pore space in rocks is saturated with water.

Wave climate   Statistical characterization on an annual basis of wave height period and direction for a particular site.

Wave height   Refers to the difference in elevation between the trough and the crest of a wave.

Wave period   Refers to the time in seconds for successive wave crests to pass a reference point; the inverse of the frequency of the wave.

Wave length   Refers to the horizontal length between successive crests of waves.

Weathering   Changes that take place in rocks and minerals at or near the surface of Earth in response to physical, chemical, and biological changes; the physical, chemical, and biological breakdown of rocks and minerals.

Wetlands   Landscape features such as swamps, marshes, bogs, or prairie potholes that are frequently or continuously inundated by water.

Wind power   Technology (mostly windmills) used to extract electrical energy from the wind.

Zero waste   The concept related to waste management that asserts that there is no such thing as waste, but only resources out of place.

Zone of saturation   Zone or layer below the water table in which all the pore space of rock or soil is saturated.

REFERENCES

Chapter 1

1. **Diamond, J.** 2005. *Collapse*. London, England: Penguin Books.
2. **Hunt, T. L.** 2006. Rethinking the fall of Easter Island. *American Scientist* 94(5):412–19.
3. **Cloud, P.** 1978. *Cosmos, Earth, and man*. New Haven, CT: Yale University Press.
4. **Ermann, M.** 1927. *Desiderata*. Terre Haute, IN.
5. **Davidson, J. P., Reed, W. E., and Davis, P. M.** 1997. *Exploring Earth*. Upper Saddle River, NJ: Prentice Hall.
6. **Population Reference Bureau.** 2000. World Population Data Sheet. Washington, DC.
7. **Brown, L. R., Flavin, C., and Postel, S.** 1991. *Saving the planet*. New York: W. W. Norton & Co.
8. **Smil, V.** 1999. How many billions to go? *Nature* 401: 429.
9. **Hooke, LeB.** 1994. On the efficiency of humans as geomorphic agents. *GSA Today* 4(9):217, 224–25.
10. **Moncrief, L. W.** 1970. The cultural basis for our environmental crisis. *Science* 170: 508–12.
11. **National Research Council.** 1971. *The Earth and human affairs*. San Francisco: Canfield Press.
12. **Barnhardt, W.** 1987. The death of Ducktown. *Discover*, October, 35–43.
13. **Ellis, W. S.** 1990. A Soviet sea lies dying. *National Geographic* 177(2):73–92.
14. **Earth Systems Science Committee.** 1988. *Earth systems science*. Washington, DC: National Aeronautics and Space Administration.
15. **Lovelock, J.** 1988. *The ages of Gaia*. New York: W. W. Norton & Co.
16. **Leopold, A.** 1949. *A Sand County almanac*. New York: Oxford University Press.
17. **Foster, K. R., Vecchia, P., and Repacholi, M. H.** 2000. Science and the precautionary principle. *Science* 5(288):979–81.
18. **Easton, T. A., and Goldfarb, T. D.,** eds. 2003. *Taking sides, environmental issues*, 10th ed. Issue 5. Is the precautionary principle a sound basis for international policy? pp. 76–101. Guilford, CT: McGraw-Hill/Dushkin.
19. **Shepard, P.** 1998. *Coming home to the Pleistocene*. Washington, DC: Island Press.

Chapter 2

1. **Wysession, M.** 1995. The inner workings of Earth. *American Scientist* 83: 134–47.
2. **Glatzmaier, G. A.** 2001. The geodynamo. www.es.ucsc.edu/~glatz/geodynamo.html. Accessed 2/21/01.
3. **Fowler, C. M. R.** 1990. *The solid Earth*. Cambridge: Cambridge University Press.
4. **Le Pichon, X.** 1968. Sea-floor spreading and continental drift. *Journal of Geophysical Research* 73: 3661–97.
5. **Isacks, B. L., Oliver, J., and Sykes, L. R.** 1968. Seismology and the new global tectonics. *Journal of Geophysical Research* 73: 5855–99.
6. **Cox, A., and Hart, R. B.** 1986. *Plate tectonics*. Boston: Blackwell Scientific Publications.
7. **Keller, E. A., and Pinter, N.** 1996. *Active tectonics*. Upper Saddle River, NJ: Prentice Hall.
8. **Pinter, N., and Brandon, N. T.** 1997. How erosion builds mountains. *Scientific American* 276(4):60–65.
9. **Dewey, J. F.** 1972. Plate tectonics. *Scientific American* 22: 56–68.
10. **Heirtzler, J. R., Le Pichon, X., and Baron, J. G.** 1966. Magnetic anomalies over the Reykjanes Ridge. *Deep Sea Research* 13: 427–43.
11. **Cox, A., Dalrymple, G. B., and Doell, R. R.** 1967. Reversals of Earth's magnetic field. *Scientific American* 216(2):44–54.
12. **Claque, D. A., Dalrymple, G. B., and Moberly, R.** 1975. Petrography and K-Ar ages of dredged volcanic rocks from the western Hawaiian Ridge and southern Emperor Seamount chain. *Geological Society of America Bulletin* 86: 991–98.

Chapter 3

1. **Ross, M.** 1990. Hazards associated with asbestos minerals. In *Proceedings of a U.S. Geological Survey workshop on environmental geochemistry*, ed. B. R. Doe, pp. 175–76. U.S. Geological Survey Circular 1033.
2. **Skinner, H. C. W., and Ross, M.** 1994. Minerals and cancer. *Geotimes* 39(1):13–15.
3. **Gribble, C. D.,** ed. 1988. *Rutley's elements of mineralogy*, 27th ed. Boston: Unwin Hyman.
4. **Nickel, E. H.** 1995. Definition of a mineral. *Mineralogical Magazine* 59: 767–68.
5. **Davidson, J. P., Reed, W. E., and Davis, P. M.** 1997. *Exploring Earth*. Upper Saddle River, NJ: Prentice Hall.
6. **Krynine, D. P., and Judd, W. R.** 1957. *Principles of engineering geology and geotechnics*. New York: McGraw-Hill.
7. **Schultz, J. R., and Cleaves, A. B.** 1955. *Geology in engineering*. New York: John Wiley.
8. **Rogers, J. D.** 1992. Reassessment of the St. Francis Dam failure. In *Engineering geology practice in Southern California*, ed. R. Proctor and B. Pipkin, pp. 639–66. Association of Engineering Geologists, Special Publication No. 4.

Chapter 4

1. **Botkin, D. B., and Keller, E. A.** 2005. *Environmental science*, 5th ed. Hoboken, NJ: John Wiley, p. 664.
2. **Tallis, J. H.** 1991. *Plant community history*. London: Chapman and Hall, p. 398.
3. **Ripple, J. W., and Beschta, Robert L.** 2004. Wolves and the ecology of fear: Can predation risk structure ecosystems? *BioScience* 54(8):755–66.
4. **Dugan, J. E., and Hubbard, D. M.** 2006. Ecological responses to coastal armoring on exposed sandy beaches. *Shore and Beach* 74(1):10–16

5. **Gould, S. J.** 1993. *The golden rule: A proper scale for our environmental crisis in Eight Little Piggies: Reflections in natural history.* New York: W. W. Norton.

6. **Vitousek, P. M., Mooney, H. A., Lubchenco, J., and Melillo, J. M.** 1997. Human dominator of ecosystems. *Science* 277(5325):494–499.

7. **Riley, A. L.** 1998. *Restoring streams in cities.* Washington, DC: Island Press.

8. **Society for Ecological Restoration.** 2004. The SER international primer on ecological restoration. www.ser.org. Accessed 3/11/06.

9. **South Florida Water Management District.** Kissimmee River restoration. www.sfwmd.gov. Accessed 3/11/06.

10. **Comprehensive Everglades Restoration Plan.** www.evergladesplan.org. Accessed 3/11/06.

Chapter 5

1. **Dokka, R. K.** 2006. Modern-day tectonic subsidence in coastal Louisiana. *Geology* 34: 281–84.

2. **U.S. Army Corps of Engineers.** 2006. *Performance evaluation of the New Orleans and southeast Louisiana hurricane protection system,* vol. 1. Executive summary and overview. Washington DC.

3. **Advisory Committee on the International Decade for Natural Hazard Reduction.** 1989. *Reducing disaster's toll.* Washington, DC: National Academy Press.

4. **White, G. F., and Haas, J. E.** 1975. *Assessment of research on natural hazards.* Cambridge, MA: MIT Press.

5. **Peterson, D. W.** 1986. Volcanoes—Tectonic setting and impact on society. In *Studies in geophysics: Active tectonics,* pp. 231–46. Washington, DC: National Academy Press.

6. **Crowe, B. W.** 1986. Volcanic hazard assessment for disposal of high-level radioactive waste. In *Studies in geophysics: Active tectonics,* pp. 247–60. Washington, DC: National Academy Press.

7. **Kates, R. W., and Pijawka, D.** 1977. Reconstruction following disaster. In *From rubble to monument: The pace of reconstruction,* ed. J. E. Haas, R. W. Kates, and M. J. Bowden. Cambridge, MA: MIT Press.

8. **Costa, J. E., and Baker, V. R.** 1981. *Surficial geology: Building with the Earth.* New York: John Wiley.

9. **Abramovitz, J. N., and Dunn, S.** 1998. *Record year for weather-related disasters.* Worldwatch Institute, Vital Signs Brief, 98–5.

10. **Magnuson, E.** 1985. A noise like thunder. *Time* 126(13):35–43.

11. **Abramovitz, J. N.** 2001. Averting unnatural disasters. In L. R. Brown, et al. *State of the world 2001.* Worldwatch Institute. New York: W. W. Norton, pp. 123–42.

12. **Russell, G.** 1985. Colombia's mortal agony. *Time* 126(21):46–52.

13. **Herd, D. G.** 1986. The 1985 Ruiz Volcano disaster. *EOS, Transactions of the American Geophysical Union,* May 13, 457–60.

14. **IAVCEE Subcommittee on Decade Volcanoes.** 1994. Research on decade volcanoes aimed at disaster prevention. *EOS, Transactions of the American Geophysical Union* 75(30):340, 350.

Chapter 6

1. **U.S. Geological Survey.** 1996. *USGS response to an urban earthquake, Northridge '94.* U.S. Geological Survey Open File Report 96–263.

2. **U.S. Geological Survey.** 2005. *Magnitude 7.6—Pakistan.* Earthquake Hazards Program. earthquake.usgs.gov. Accessed 5/2/06.

3. **Achenbach, J.** 2006. The next big one. *National Geographic* 209(4):120–47.

4. **U.S. Geological Survey.** 2003. *Shake map—A tool for earthquake response.* USGS Fact Sheet FS-087-03.

5. **Hamilton, R. M.** 1980. Quakes along the Mississippi. *Natural History* 89: 70–75.

6. **Mueller, K., Champion, J., Guccione, M., and Kelson, K.** 1999. Fault slip rates in the modern New Madrid Seismic Zone. *Science* 286: 1135–38.

7. **Melbourne, T. I., and Webb, F. H.** 2003. Slow but not quite silent. *Science* 300: 1886.

8. **Bolt, B. A.** 2004. *Earthquakes,* 5th ed. San Francisco: W. H. Freeman.

9. **Jones, R. A.** 1986. New lessons from quake in Mexico. *Los Angeles Times,* September 26.

10. **Hough, S. E., Friberg, P. A., Busby, R., Field, E. F., Jacob, K. H., and Borcherdt, R. D.** 1989. Did mud cause freeway collapse? *EOS, Transactions of the American Geophysical Union* 70(47):1497, 1504.

11. **Hart, E. W., Bryant, W. A., and Treiman, J. A.** 1993. Surface faulting associated with the June 1992 Landers earthquake, California. *California Geology,* January, February, 10–16.

12. **Hanks, T. C.** 1985. *The national earthquake hazards reduction program: Scientific status.* U.S. Geological Survey Bulletin 1659.

13. **Evans, D. M.** 1966. Man-made earthquakes in Denver. *Geotimes* 10: 11–18.

14. **Youd, T. L., Nichols, D. R., Helley, E. J., and Lajoie, K. R.** 1975. Liquefaction potential. In *Studies for seismic zonation of the San Francisco Bay region,* ed. R. D. Borcherdt, pp. 68–74. U.S. Geological Survey Professional Paper 941A.

15. **Hansen, W. R.** 1965. *The Alaskan earthquake, March 27, 1964: Effects on communities.* U.S. Geological Survey Professional Paper 542A.

16. **Oppenheimer, D., Beroza, G., Carver, G., Dengler, L., Eaton, J., Gee, L., Gonzales, F., Jayko, A., Li, W. H., Lisowski, M., Magee, M., Marshall, G., Murray, M., McPherson, R., Romanowicz, B., Sataker, K., Simpson, R., Somerville, P., Stein, R., and Valentine, D.** 1993. The Cape Mendocino, California, earthquakes of April, 1992: Subduction at the triple junction. *Science* 262: 433–38.

17. **U.S. Geological Survey.** 2005. Magnitude 9.0 Sumatra-Andaman Island earthquake SAIE. Earthquake hazards program EHP. earthquake.usgs.gov. Accessed 1/13/05.

18. **Chapman, C.** 2005. The Asian tsunami in Sri Lanka: A personal experieence. *EOS, Transactions, American Geophysical Union* 86(2):13–14.

19. **Reuters, M. B.** Elephants saved tourists from tsunami. savetheelephants.org. Accessed 6/13/06.

20. **Danielsen, F., and 11 others.** 2005. The Asian tsunami: A protective role for coastal vegetation. *Science* 310: 643.

21. **U.S. Geological Survey.** 2004. *Is a powerful quake likely to strike in the next 30 years?* USGS Fact Sheet 039-03, revised in 2004.

22. **Scholz, C.** 1997. Whatever happened to earthquake prediction? *Geotimes* 42(3):16–19.

23. **Raleigh, B., et al.** 1977. Prediction of the Haicheng earthquake. *EOS, Transactions of the American Geophysical Union* 58(5):236–72.
24. **Press, F.** 1975. Earthquake prediction. *Scientific American* 232: 14–23.
25. **Scholz, C. H.** 1990. *The mechanics of earthquakes and faulting.* New York: Cambridge University Press.
26. **Silver, P. G., and Wakita, H.** 1996. A search for earthquake precursors. *Science* 273: 77–78.
27. **Rikitakr, T.** 1983. *Earthquake forecasting and warning.* London: D. Reidel.
28. **Page, R. A., Boore, D. M., Bucknam, R. C., and Thatcher, W. R.** 1992. *Goals, opportunities, and priorities for the USGS Earthquake Hazards Reduction Program.* U.S. Geological Survey Circular 1079.
29. **Hait, M. H.** 1978. Holocene faulting, Lost River Range, Idaho. *Geological Society of America Abstracts with Programs* 10(5):217.
30. **Reilinger, R., Toksot, N., McClusky, S., and Barka, A.** 2000. 1999 Izmit, Turkey earthquake was no surprise. *GSA Today* 10(1):1–5.
31. **Stein, R. S.** 1999. The role of stress transfer in earthquake occurrence. *Nature* 402(6762):605–609.
32. **Eberhart-Phillips, D., and 28 others.** 2003. The 2002 Denali fault earthquake, Alaska: A large magnitude, slip-partitioned event. *Science* 300: 1113–18.
33. **Hendley, J. W., II, and Stauffer, P.H.,** eds. 2003. *Rupture in south-central Alaska—The Denali earthquake of 2002.* U.S. Geological Survey Fact Sheet 014-03.
34. **Holden, R., Lee, R., and Reichle, M.** 1989. *Technical and economic feasibility of an earthquake warning system in California.* California Division of Mines and Geology Special Publication 101.
35. **Southern California Earthquake Center.** 1995. *Putting down roots in earthquake country.* Los Angeles: University of Southern California.

Chapter 7

1. **Wright, T. L., and Pierson, T. C.** 1992. *Living with volcanoes.* U.S. Geological Survey Circular 1073.
2. **Pendick, D.** 1994. Under the volcano. *Earth* 3(3):34–39.
3. **IAVCEE Subcommittee on Decade Volcanoes.** 1994. Research at decade volcanoes aimed at disaster prevention. *EOS, Transactions of the American Geophysical Union* 75(30):340, 350.
4. **Decker, R., and Decker, B.** 2006. *Volcanoes,* 4th ed. New York: W. H. Freeman.
5. **Fisher, R. V., Heiken, G., and Hulen, J. B.** 1997. *Volcanoes.* Princeton, NJ: Princeton University Press.
6. **Francis, P.** 1983. Giant volcanic calderas. *Scientific American* 248(6):60–70.
7. **Office of Emergency Preparedness.** 1972. *Disaster preparedness.* 1, 3. Washington, DC.
8. **Crandell, D. R., and Waldron, H. H.** 1969. Volcanic hazards in the Cascade Range. In *Geologic hazards and public problems, conference proceedings,* ed. R. Olsen and M. Wallace, pp. 5–18. Office of Emergency Preparedness Region 7, Washington, DC.
9. **Williams, R. S., Jr., and Moore, J. G.** 1973. Iceland chills a lava flow. *Geotimes* 18: 14–18.
10. **Tilling, R. I.** 2000. Volcano notes. *Geotimes* 45(5):19.
11. **U.S. Geological Survey.** 1997. *Volcanic air pollution.* U.S. Geological Survey Fact Sheet 169-97.
12. **U.S. Geological Survey.** 1999. Pilot project Mount Rainier volcano lahar warning system. volcanoes. usgs.gov. Accessed 8/29/06.
13. **Ancochea, E., Fuster, J. M., Ibarrola, E., Cendrero, A., Hernan, F., Cantagrel, J. M., and Jamond, C.** 1990. The volcanic evolution of the island of Tenerife (Canary Islands) in the light of new K-Ar data. *Journal of Volcanology and Geothermal Research* 44(3–4):231–49.
14. **Cantagrel, J. M., Arnaud, N. O., Ancochea, E., Fuster, J. M., and Huertas, M. J.** 1999. Repeated debris avalanches on Tenerife and genesis of Las Cañadas caldera wall (Canary Islands). *Geology* 27(8):739–42.
15. **Watts, A. B., and Masson, D. G.** 1995. A giant landslide on the north flank of Tenerife, Canary Islands. *Journal of Geophysical Research* 100(12):24487–98.
16. **American Geophysical Union.** 1991. Pinatubo cloud measured. *EOS, Transactions of the American Geophysical Union* 72(29):305–06.
17. **Tilling, R. I.** 2000. Mount St. Helens 20 years later. *Geotimes* 45(5):14–18.
18. **Hammond, P. E.** 1980. Mt. St. Helens blasts 400 meters off its peak. *Geotimes* 25: 14–15.
19. **Brantley, S., and Topinka, L.** 1984. *Earthquake Information Bulletin* 16(2).
20. **Pendick, D.** 1995. Return to Mount St. Helens. *Earth* 4(2):24–33.
21. **Francis, P.** 1976. *Volcanoes.* London: Pelican Books.
22. **Richter, D. H., Eaton, J. P., Murata, K. J., Ault, W. U., and Krivoy, H. L.** 1970. Chronological narrative of the 1959–60 eruption of Kilauea Volcano, Hawaii. U.S. Geological Survey Professional Paper 537E.
23. **Murton, B. J., and Shimabukuro, S.** 1974. Human response to volcanic hazard in Puna District, Hawaii. In *Natural hazards,* ed. G. F. White, pp. 151–59. New York: Oxford University Press.

Chapter 8

1. **Rahn, P. H.** 1984. Flood-plain management program in Rapid City, South Dakota. *Geological Society of America Bulletin* 95: 838–43.
2. **U.S. Department of Commerce.** 1973. *Climatological data, national summary* 24(13).
3. **Anonymous.** 1993. The flood of '93. *Earth Observation Magazine,* September: 22–23.
4. **Mairson, A.** 1994. The great flood of '93. *National Geographic* 185(1):42–81.
5. **Bell, G. D.** 1993. The great midwestern flood of 1993. *EOS, Transactions of the American Geophysical Union* 74(43):60–61.
6. **Anonymous.** 1993. Flood rebuilding prompts new wetlands debate. *U.S. Water News,* November, 10.
7. **Committee on Alluvial Fan Flooding.** 1996. *Alluvial fan flooding.* Washington, DC: National Academy Press.
8. **Edelen, G. W., Jr.** 1981. Hazards from floods. In *Facing geological and hydrologic hazards, Earth-science considera-tions,* ed. W. W. Hays, pp. 39–52. U.S. Geological Survey Professional Paper 1240–B.
9. **Keller, E. A., and Capelli, M. H.** 1992. Ventura River flood of February, 1992: A lesson ignored? *Water Resources Bulletin* 28(5):813–31.
10. **Mackin, J. H.** 1948. Concept of the graded river. *Geological Society of America Bulletin* 59: 463–512.

11. **Keller, E. A., and Florsheim, J. L.** 1993. Velocity reversal hypothesis: A model approach. *Earth Surface Processes and Landforms* 18: 733–48.

12. **Beyer, J. L.** 1974. Global response to natural hazards: Floods. In *Natural hazards*, ed. G. F. White, pp. 265–74. New York: Oxford University Press.

13. **Linsley, R. K., Jr., Kohler, M. A., and Paulhus, J. L.** 1958. *Hydrology for engineers.* New York: McGraw-Hill.

14. **Leopold, L. B.** 1968. *Hydrology for urban land planning.* U.S. Geological Survey Circular 559.

15. **Seaburn, G. E.** 1969. *Effects of urban development on direct runoff to East Meadow Brook, Nassau County, Long Island, New York.* U.S. Geological Survey Professional Paper 627B.

16. **McCain, J. F., Hoxit, L. R., Maddox, R. A., Chappell, C. F., and Caracena, F.** 1979. *Storm and flood of July 31–August 1, 1976, in the Big Thompson River and Cache la Poudre River Basins, Larimer and Weld Counties, Colorado.* U.S. Geological Survey Professional Paper 1115A.

17. **Shroba, R. R., Schmidt, P. W., Crosby, E. J., and Hansen, W. R.** 1979. *Storm and flood of July 31–August 1, 1976, in the Big Thompson River and Cache la Poudre River Basins, Larimer and Weld Counties, Colorado.* U.S. Geological Survey Professional Paper 1115B.

18. **Bradley, W. C., and Mears, A. I.** 1980. Calculations of flows needed to transport coarse fraction of Boulder Creek alluvium at Boulder, Colorado. *Geological Society of America Bulletin*, Part II, 91: 1057–90.

19. **Agricultural Research Service.** 1969. *Water intake by soils.* Miscellaneous Publication no. 925.

20. **Strahler, A. N., and Strahler, A. H.** 1973. *Environmental geoscience.* Santa Barbara, CA: Hamilton Publishing.

21. **Terstriep, M. L., Voorhees, M. L., and Bender, G. M.** 1976. *Conventional urbanization and its effect on storm runoff.* Illinois State Water Survey Publication.

22. **Office of Emergency Preparedness.** 1972. *Disaster preparedness,* 1, 3. Washington, DC.

23. **Mount, J. F.** 1997. *California rivers and streams.* Berkeley: University of California Press.

24. **Baker, V. R.** 1984. Questions raised by the Tucson flood of 1983. In *Proceedings of the 1984 meetings of the American Water Resources Association and the Hydrology Section of the Arizona–Nevada Academy of Science,* pp. 211–19.

25. **Baker, V. R.** 1994. Geologic understanding and the changing environment. *Transactions of the Gulf Coast Association of Geological Societies* 44: 1–8.

26. **Pinter, N., Thomas, R., and Wlosinski, J. H.** 2001. Assessing flood hazard on dynamic rivers. *Transactions, American Geophysical Union* 82(31):333, 38–39.

27. **U.S. Congress.** 1973. *Stream channelization: What federally financed draglines and bulldozers do to our nation's streams.* House Report No. 93–530. Washington, DC: U.S. Government Printing Office.

28. **Rosgen, D.** 1996. *Applied river morphology.* Lakewood, CO: Wildland Hydrology.

29. **Pilkey, O. H., and Dixon, K. L.** 1996. *The Corps and the Shore.* Washington, DC: Island Press.

30. **Smith, K., and Ward, R.** 1998. *Floods.* New York: John Wiley.

31. **Bue, C. D.** 1967. *Flood information for floodplain planning.* U.S. Geological Survey Circular 539.

32. **Schaeffer, J. R., Ellis, D. W., and Spieker, A. M.** 1970. *Flood-hazard mapping in metropolitan Chicago.* U.S. Geological Survey Circular 601C.

33. **Baker, V. R.** 1976. Hydrogeomorphic methods for the regional evaluation of flood hazards. *Environmental Geology* 1: 261–81.

Chapter 9

1. **Gurrola, L. D., and Keller, E. A.** 2005. Prehistoric landslides complexes in the landscape and associated hazards: La Conchita, California. *Geological Society of America Abstracts with Programs* 37(7):519.

2. **Jibson, R.W.** 2005. Landslide hazards at the La Conchita, California. U.S. Geological Survey Open file report 2005-1067.

3. **U.S. Geological Survey.** 2004. *Landslide types and processes.* Fact Sheet 2004-3072.

4. **Pestrong, R.** 1974. *Slope stability.* American Geological Institute. New York: McGraw-Hill.

5. **Nilsen, T. H., Taylor, F. A., and Dean, R. M.** 1976. *Natural conditions that control landsliding in the San Francisco Bay region.* U.S. Geological Survey Bulletin 1424.

6. **Campbell, R. H.** 1975. *Soil slips, debris flows, and rainstorms in the Santa Monica Mountains and vicinity, southern California.* U.S. Geological Survey Professional Paper 851.

7. **Terzaghi, K.** 1950. *Mechanisms of landslides.* Geological Society of America: Application of Geology to Engineering Practice, Berkey Vol.: 83–123. Boulder, CO: Geological Society of America.

8. **Leggett, R. F.** 1973. *Cities and geology.* New York: McGraw-Hill.

9. **Kiersch, G. A.** 1964. Vaiont Reservoir disaster. *Civil Engineering* 34: 32–39.

10. **Swanson, F. J., and Dryness, C. T.** 1975. Impact of clear-cutting and road construction on soil erosion by landslides in the Western Cascade Range, Oregon. *Geology* 7: 393–96.

11. **Jones, F. O.** 1973. *Landslides of Rio de Janeiro and the Sierra das Araras Escarpment, Brazil.* U.S. Geological Survey Professional Paper 697.

12. **Leighton, F. B.** 1966. Landslides and urban development. In *Engineering geology in southern California,* ed. R. Lung and R. Proctor, pp. 149–97. Los Angeles: Los Angeles Section of the Association of Engineering Geology.

13. **Briggs, R. P., Pomeroy, J. S., and Davies, W. E.** 1975. *Landsliding in Allegheny County, Pennsylvania.* U.S. Geological Survey Circular 728.

14. **Jones, D. K. C.** 1992. Landslide hazard assessment in the context of development. In *Geohazards,* ed. G. J. McCall, D. J. Laming, and S. C. Scott, pp. 117–41. New York: Chapman and Hall.

15. **Slosson, J. E., Yoakum, D. E., and Shuiran, G.** 1986. Thistle, Utah, landslide: Could it have been prevented? In *Proceedings of the 22nd symposium on engineering geology and soils engineering,* pp. 281–303.

16. **Piteau, D. R., and Peckover, F. L.** 1978. Engineering of rock slopes. In *Landslides,* ed. R. Schuster and R. J. Krizek. Transportation Research Board, Special Report 176: 192–228.

17. **Poland, J. F., and Davis, G. H.** 1969. Land subsidence due to withdrawal of fluids. In *Reviews in engineering geology*, ed. D. J. Varnes and G. Kiersch, pp. 187–269. Boulder, CO: Geological Society of America.
18. **Bull, W. B.** 1974. Geologic factors affecting compaction of deposits in a land subsidence area. *Geological Society of America Bulletin* 84: 3783–3802.
19. **Kenny, R.** 1992. Fissures. *Earth* 1(3):34–41.
20. **Cornell, J., ed.** 1974. *It happened last year—Earth events—1973.* New York: Macmillan.
21. **Dougherty, P. H., and Perlow, M., Jr.** 1987. The Macungie sinkhole, Lehigh Valley, Pennsylvania: Cause and repair. *Environmental Geology and Water Science* 12(2):89–98.
22. **Rahn, P. H.** 1996. *Engineering geology*, 2nd ed. Upper Saddle River, NJ: PrenticeHall.
23. **Craig, J. R., Vaughan, D. J., and Skinner, B. J.** 1996. *Resources of the Earth*, 2nd ed. Upper Saddle River, NJ: Prentice Hall.

Chapter 10

1. **McDonald, K. A.** 1993. A geology professor's fervent battle with coastal developers and residents. *Chronicle of Higher Education*, 40(7):A8–89, A12.
2. **Coates, D. R.,** ed. 1973. *Coastal geomorphology*. Binghamton, NY: Publications in Geomorphology, State University of New York.
3. **Davis, R. E., and Dolan, R.** 1993. Nor'easters. *American Scientist* 81: 428–39.
4. **Komar, P. D.** 1998. *Beach processes and sedimentation*, 2nd ed. Upper Saddle River, NJ: Prentice Hall.
5. **El-Ashry, M. T.** 1971. Causes of recent increased erosion along United States shorelines. *Geological Society of America Bulletin* 82: 2033–38.
6. **Norris, R. M.** 1977. Erosion of sea cliffs. In *Geologic hazards in San Diego*, ed. P. L. Abbott and J. K. Victoris. San Diego, CA: San Diego Society of Natural History.
7. **Flanagan, R.** 1993. Beaches on the brink. *Earth* 2(6):24–33.
8. **Carter, R. W. G., and Oxford, J. D.** 1982. When hurricanes sweep Miami Beach. *Geographical Magazine* 54(8):442–48.
9. **U.S. Department of Commerce.** 1978. *State of Maryland coastal management program and final environmental impact statement.* Washington, DC: U.S. Department of Commerce.
10. **Leatherman, S. P.** 1984. Shoreline evolution of North Assateague Island, Maryland. *Shore and Beach*, July, 3–10.
11. **Wilkinson, B. H., and McGowen, J. H.** 1977. Geologic approaches to the determination of long-term coastal recession rates, Matagordo Peninsula, Texas. *Environmental Geology* 1: 359–65.
12. **Larsen, J. I.** 1973. *Geology for planning in Lake County, Illinois.* Illinois State Geological Survey Circular 481.
13. **Buckler, W. R., and Winters, H. A.** 1983. Lake Michigan bluff recession. *Annals of the Association of American Geographers* 73(1):89–110.
14. **White, A. U.** 1974. Global summary of human response to natural hazards: Tropical cyclones. In *Natural hazards: Local, national, global,* ed. G. F. White, pp. 255–65. New York: Oxford University Press.
15. **Office of Emergency Preparedness**. 1972. *Disaster preparedness*, 1, 2. Washington, DC.

16. **Lipkin, R.** 1994. Weather's fury. In *Nature on the rampage*, pp. 20–79. Washington DC: Smithsonian Institution.
17. **Rowntree, R. A.** 1974. Coastal erosion: The meaning of a natural hazard in the cultural and ecological context. In *Natural hazards: Local, national, global,* ed. G. F. White, pp. 70–79. New York: Oxford University Press.
18. **Baumann, D. D., and Sims, J. H.** 1974. Human response to the hurricane. In *Natural hazards: Local, national, global,* ed. G. F. White, pp. 25–30. New York: Oxford University Press.
19. **National Research Council.** 1990. *Managing coastal erosion.* Washington, DC: National Academy Press.
20. **Neal, W. J., Blakeney, W. C., Jr., Pilkey, O. H., Jr., and Pilkey, O. H.** 1984. *Living with the South Carolina shore.* Durham, NC: Duke University Press.
21. **Pilkey, O. H., and Dixon, K. L.** 1996. *The Corps and the shore.* Washington, DC: Island Press.

Chapter 11

1. **Lewis, J. S.** 1996. Rain of iron and ice. Reading, MA: Addison-Wesley; Rubin, A. E. 2002. *Disturbing the solar system.* Princeton, NJ: Princeton University Press.
2. **Rubin, A. F.** 2002. *Disturbing the solar system.* Princeton, NJ: Princeton University Press.
3. **Brown, P., Spalding, R. E., ReVelle, D. O., Tagliaferri, E. and Worden, S. P.,** 2002. The flux of small near-Earth objects colliding with the Earth. *Nature* 420: 294–96
4. **Cloud, P.** 1978. *Cosmos, Earth and man.* New Haven, CT: Yale University Press.
5. **Davidson, J. P., Reed, W. E., and Davis, P. M.** 1997. *Exploring Earth.* Upper Saddle River, NJ: Prentice Hall.
6. **Grieve, R., and Cintala, M.** 1999. Planetary impacts. In *Encyclopedia of the solar system*, ed. P. R. Weissman, L. McFadden, and T. V. Johnson. San Diego, CA: Academic Press.
7. **Williams, S. J., Barnes, P., and Prager, E. J.** 2000. *U.S. Geological Survey coastal and marine geology research—recent highlights and achievements.* U.S. Geological Survey Circular 1199, p. 28.
8. **Weissman, P. R., McFadden, L., and Johnson, T. V.,** eds. 1999. *Encyclopedia of the solar system.* San Diego, CA: Academic Press.
9. **Dott, R. H., Jr., and Prothero, D. R.** 1994. *Evolution of the Earth,* 5th ed. New York: McGraw-Hill.
10. **Alvarez, W.** 1997. *T. rex and the crater of doom.* New York: Vintage Books. Random House.
11. **Alvarez, L.W., Alvarez, W., Asaro, F., and Michel, H. V.** 1980. Extraterrestrial cause for Cretaceous-Tertiary extinction. *Science* 208(4448):1095–1108.
12. **Pope, K. O., Ocampo, A. C., and Duller, C. E.** 1991. Mexican site for the K/T impact crater? *Nature* 351: 105.
13. **Swisher, C. C., III, Grajales-Nishimura, J. N., Montanari, A., Margolis, S. V., Claeys, P., Alvarez, W., Ranne, P., Cedillo-Pardo, E., Maurrasse, F. J.-N. R., Curtis, G. H., Smit, J., and McWilliams, M. O.** 1992. Ages of 65.0 million years ago from Chicxulub crater melt rocks and Cretaceous-Tertiary boundary tektites. *Science* 257: 954–58.

14. **Hildebrand, A. R., Penfield, G. T., Kring, D. A., Pilkington, N., Camargo, Z.A., Jacobsen, S. B., and Boynton, W. V.** 1991. Chicxulub crater: A possible Cretaceous/Tertiary boundary impact crater on the Yucatan peninsula, Mexico. *Geology* 19: 867–71.

Chapter 12

1. **Foxworthy, G. L.** 1978. Nassau County, Long Island, New York—Water problems in humid country. In *Nature to be commanded*, ed. G. D. Robinson and A. M. Spieker, pp. 555–68. U.S. Geological Survey Professional Paper 950.
2. **Alley, W. M., Reilly, T. E., and Franke, O. L.** 1999. *Sustainability of ground-water resources*. U.S. Geological Survey Circular 1186.
3. **Water Resources Council.** 1978. *The nation's water resources, 1975–2000*, vol. 1. Washington, DC: Water Resources Council.
4. **Gleick, P. H.** 1993. An introduction to global fresh water issues. In *Water in crisis*, ed. P. H. Gleick, pp. 3–12. New York: Oxford University Press.
5. **Winter, T. C., Harvey, J. W., Franke, O. L., and Alley, W. M.** 1998. *Ground water and surface water. A single resource*. U.S. Geological Survey Circular 1139.
6. **Solley, W. B., Pierce, R. R., and Perlman, H. A.** 1993. *Estimated use of water in the United States in 1990*. U.S. Geological Survey Circular 1200.
7. **Sharp, J. M., Jr., and Banner, J. L.** 1997. The Edwards aquifer: A resource in conflict. *GSA Today* 7(8):1–8.
8. **Loaiciga, H. A., Maidment, D. R., and Valdes, J. B.** 1999. Climate-change impacts in a regional karst aquifer, Texas, U.S.A. *Journal of Hydrology* 227: 173–94.
9. **Leopold, L. B.** 1977. A reverence for rivers. *Geology* 5: 429–30.
10. **Graf, W. L.** 1985. *The Colorado River*. Association of American Geographers.
11. **Nash, R.** 1986. Wilderness values and the Colorado River. In *New courses for the Colorado River*, ed. G. D. Weatherford and F. L. Brown. Albuquerque: University of New Mexico Press.
12. **Hundley, N., Jr.** 1986. The West against itself: The Colorado River—An institutional history. In *New courses for the Colorado River*, ed. G. D. Weatherford and F. L. Brown. Albuquerque: University of New Mexico Press.
13. **Dolan, R., Howard, A., and Gallenson, A.** 1974. Man's impact on the Colorado River and the Grand Canyon. *American Scientist* 62: 392–401.
14. **Lavender, D.** 1984. Great News from the Grand Canyon. *Arizona Highways Magazine*, January, 33–38.
15. **Hecht, J.** 1996. Grand Canyon flood a roaring success. *New Scientist* 151: 8.
16. **Lucchitta, I., and Leopold, L. B.** 1999. Floods and sandbars in the Grand Canyon. *Geology Today* 9: 1–7.
17. **Covich, A. P.** 1993. Water and ecosystems. In *Water in crisis*, ed. P. H. Gleick, pp. 40–55. New York: Oxford University Press.
18. **Gleick, P. H.,** ed. 1993. *Water in crisis*, Table F.1. New York: Oxford University Press.
19. **Levinson, M.** 1984. Nurseries of life. *National Wildlife*. Special Report, February/March, 18–21.
20. **Holloway, M.** 1991. High and dry. *Scientific American* 265(6):16–20.
21. **Brown, L. R.** 2003. *Plan B. Rescuing a planet under stress and a civilization in trouble*. New York: W. W. Norton.

Chapter 13

1. **Bowie, P.** 2000. No act of God. *The Amicus Journal* 21(4):16–21.
2. **Mallin, M. A.** 2000. Impacts of industrial animal production on rivers and estuaries. *American Scientist* 88(1):26–37.
3. **Schwarzenbach, R. P., and six others.** 2006. The challenge of micropollutants in aquatic systems. *Science* 313: 1072–77.
4. **Mitch, W. J., Day, J. W., Jr., Gilliam, J. W., Groffman, P. M., Hey, D. L., Randall, G. W., and Wang, N.** 2001. The Gulf of Mexico hypoxia—Approaches to reducing nitrate in the Mississippi River or reducing a persistent large-scale ecological problem. *BioScience* (in press).
5. **Oil Spill Issue.** 1989. *Alaska Fish and Game* 21(4).
6. **McGinn, A. P.** 2000. POPs culture. *World Watch*, April 1, 26–36.
7. **Delzer, G. C., Zogorski, J. S., Lopes, T. J., and Basshart, R. L.** 1996. *Occurrence of gasoline oxygenate MTBE and BTEX compounds in urban storm water in the United States, 1991–1995*. U.S. Geological Survey Water Resources Investigations Report 96–4145.
8. **Waldbott, G. L.** 1978. *Health effects of environmental pollutants*, 2nd ed. Saint Louis. MO: C. V. Moseby.
9. **U.S. Geological Survey.** 1995. *Mercury contamination of aquatic ecosystems*. U.S. Geological Survey FS-216-95.
10. **Author unknown.** Arsenic exposure. sos-arsenic.net. Accessed 9/22/06.
11. **Parfit, M.** 1993. Troubled waters run deep. *National Geographic* 184(5A):78–89.
12. **Environmental Protection Agency.** 2002. American Heritage Rivers, Cuyahoga River. www.epa.gov/rivers/98river/fscuya.html. Accessed 1/9/02.
13. **Carey, J.** 1984. Is it safe to drink? *National Wildlife*, Special Report, February/March, 19–21.
14. **U.S. Geological Survey.** 1999. *National water-quality assessment program, Delaware River basin*. U.S. Geological Survey Fact Sheet FS-056-99.
15. **Moss, M. E., and Lins, H. S.** 1989. *Water resources in the 21st century*. U.S. Geological Survey Circular 1030.
16. **Environmental Protection Agency.** 1991. *Is your drinking water safe?* EPA 570-9-91-0005.
17. **Jewell, W. J.** 1994. Resource-recovery wastewater treatment. *American Scientist* 82(4):366–75.
18. **Bedient, P. B., Rifai, H. S., and Newell, C. J.** 1994. *Groundwater contamination*. Englewood Cliffs, NJ: Prentice Hall.
19. **Leeden, F., Troise, F. L., and Todd, D. K.** 1990. *The water encyclopedia*, 2nd ed. Chelsea, MI: Lewis Publishers.
20. **U.S. Geological Survey.** 1997. *Predicting the impact of relocating Boston's sewage outfall*. U.S.G.S. Fact Sheet 185-97.
21. **American Chemical Society.** 1969. *Clean our environment: The chemical basis for action*. Washington, DC: U.S. Government Printing Office.
22. **Breaux, A., Fuber, S., and Day, J.** 1995. Using natural coastal wetland systems: An economic benefit analysis. *Journal of Environmental Management* 44: 285–91.

23. **Parizek, R. R., and Myers, E. A.** 1968. Recharge of ground water from renovated sewage effluent by spray irrigation. *Proceedings of the Fourth American Water Resources Conference*, pp. 425–43.
24. **Bastian, R. K., and Benforado, J.** 1983. Waste treatment: Doing what comes naturally. *Technology Review*, February/March, 59–66.
25. **Hileman, B.** 1995. Rewrite of Clean Water Act draws praise, fire. *Chemical & Engineering News* 73: 8.

Chapter 14

1. **Prospectors & Developers Association of Canada.** Minerals & use. www.pdac.ca. Accessed 9/5/06.
2. **U.S. Department of the Interior, Bureau of Mines.** 1991. *Minerals in 1991* I 28.156/3: 991.
3. **Barsotti, A. F.** 1992. Wake up and smell the coffee. *Minerals Today*, October, 12–17.
4. **U.S. Geological Survey.** 1997. The role of non-fuel minerals in the U.S. economy. Accessed 1/15/07 @minerals .usgs.gov
5. **Brobst, D. A., Pratt, W. P., and McKelvey, V. E.** 1973. *Summary of United States mineral resources.* U.S. Geological Survey Circular 682.
6. **Kesler, S. F.** 1994. *Mineral resources, economics and the environment.* Upper Saddle River, NJ: Prentice Hall.
7. **National Oceanic and Atmospheric Administration.** 1977. Earth's crustal plate boundaries: Energy and mineral resources. *California Geology*, May, 108–09.
8. **Craig, J. R., Vaughan, D. J., and Skinner, B. J.** 1996. *Resources of the Earth*, 2nd ed. Upper Saddle River, NJ: Prentice Hall.
9. **U.S. Department of the Interior, Bureau of Mines.** 1991. *Research 92. Biotechnology—Using nature to clean up wastes* I 28.115(992):16–21.
10. **Silva, M. A.** 1988. Cyanide heap leaching in California. *California Geology* 41(7):147–56.
11. **Woodbury, R.** 1998. The giant cup of poison. *Time* 151(12):4.
12. **Pettyjohn, W. A.** 1972. Nothing is without poison. In *Man and his physical environment*, ed. G. D. McKenzie and R. O. Utgard, pp. 109–10. Minneapolis: Burgess Publishing.
13. **Takahisa, H.** 1971. *Discussion on environmental geochemistry in health and disease*, ed. H. L. Cannon and H. C. Hupps, pp. 221–22. Geological Society of America Memoir 123.
14. **Jeffers, T. H.** 1991. Using microorganisms to recover metals. *Minerals Today*, June, 14–18.
15. **Haynes, B. W.** 1990. Environmental technology research. *Minerals Today*, May, 13–17.
16. **Sullivan, P. M., Stanczyk, M. H., and Spendbue, M. J.** 1973. *Resource recovery from raw urban refuse.* Report of Investigations 7760. Washington, DC: U.S. Bureau of Mines.
17. **Davis, F. F.** 1972. Urban ore. *California Geology*, May, 99–112.
18. **U.S. Geological Survey.** 2005. Minerals yearbook 2004— Recycling metals. Accessed 1/15/07 @minerals.usgs.gov.
19. **Brown, L., Lenssen, N., and Kane, H.** 1995. Steel recycling rising. In *Vital signs 1995*. Washington, DC: Worldwatch Institute.
20. **Wellmar, F. W., and Kosinowoski, M.** 2003. Sustainable development and the use of non-renewable sources. *Geotimes* 48(12):14–17.

Chapter 15

1. **Alekett, K.** 2006. Oil: A bumpy road ahead. *World Watch* 19: 1, 10–12.
2. **Cavanay, R.** 2006. Global oil about to peak? A recurring myth. *World Watch* 19: 1, 13–15.
3. **Butti, K., and Perlin, J.** 1980. *The golden thread: 2500 years of solar architecture and technology.* Palo Alto, CA: Cheshire Books.
4. **Craig, J. R., Vaughan, D. J., and Skinner, B. J.** 1996. *Resources of the Earth.* Upper Saddle River, NJ: Prentice Hall.
5. **Rahn, P. H.** 1982. *Engineering geology: An environmental approach.* New York: Elsevier.
6. **Garbini, S., and Schweinfurth, S. P.,** eds. 1986. U.S. Geological Survey Circular 979.
7. **U.S. Environmental Protection Agency.** 1973. *Processes, procedures and methods to control pollution from mining activities.* EPA-430/9-73-001.
8. **Committee on Environment and Public Planning.** 1974. Environmental impact of conversion from gas or oil to coal for fuel. *Geologist*, Supplement 9(4).
9. **Vendetti, J.** 2001. Storing coal slurry. *Geotimes* 46(12):7.
10. **Kesler, S. E.** 1994. *Mineral resources, economics and the environment.* New York: Macmillan.
11. **McCabe, P. J., Gautier, D. L., Lewan, M. D., and Turner, C.** 1993. *The future of energy gases.* U.S. Geological Survey Circular 1115.
12. **McCulloh, T. H.** 1973. Oil and gas. In *United States mineral resources*, ed. D. A. Brobst and W. P. Pratt, pp. 477–96. U.S. Geological Survey Professional Paper 820.
13. **Nuccio, V.** 1997. *Coal-bed methane—An untapped energy resource and an environmental concern.* U.S. Geological Survey Fact Sheet FS-019-97.
14. **Suess, E., Bohrmann, G., Greinert, J., and Lauch, E.** 1999. Flammable ice. *Scientific American* 28(5):76–83.
15. **Culbertson, W. C., and Pitman, J. K.** 1973. Oil shale. In *United States mineral resources*, ed. D. A. Brobst and W. P. Pratt, pp. 497–503. U.S. Geological Survey Professional Paper 820.
16. **Office of Oil and Gas, U.S. Department of the Interior.** 1968. *United States petroleum through 1980.* Washington, DC: U.S. Government Printing Office.
17. **Allen, A. R.** 1975. Coping with oil sands. In *Perspectives on energy*, ed. L. C. Ruedisili and M. W. Firebaugh, pp. 386–96. New York: Oxford University Press.
18. **Kerr, R. A.** 2000. USGS optimistic on world oil prospects. *Science* 289: 237.
19. **Youngquist, W.** 1998. Spending our great inheritance. Then what? *Geotimes* 43(7):24–27.
20. **Edwards, J. D.** 1997. Crude oil and alternative energy production forecast for the twenty-first century: The end of the hydrocarbon era. *American Association of Petroleum Geologists Bulletin* 81(8):1292–1305.
21. **Maugeri, L.** 2004. Oil: Never cry wolf—When the petroleum age is far from over. *Science* 304: 1114–15.
22. **Krajick, K.** 2001. Long-term data show lingering effects from acid rain. *Science* 292: 195–96.
23. **Finch, W. I., et al.** 1973. Nuclear fuels. In *United States mineral resources*, ed. D. A. Brobst and W. P. Pratt, pp. 455–76. U.S. Geological Survey Professional Paper 820.

24. **U.S. Geological Survey.** 1973. *Nuclear energy resources: A geologic perspective.* U.S. Geological Survey, INF-73-14.

25. **Duderstadt, J. J.** 1977. Nuclear power generation. In *Perspectives on energy,* ed. L. C. Ruedisili and M. W. Firebaugh, pp. 249–73. New York: Oxford University Press.

26. **Brenner, D. J.** 1989. *Radon: Risk and remedy.* New York: W. H. Freeman.

27. **Till, C. E.** 1989. Advanced reactor development. *Annals of Nuclear Energy* 16: 301–05.

28. **Cohen, B. L.** 1990. *The nuclear energy option: An alternative for the '90s.* New York: Plenum.

29. **Lake, J. A., Bennett, R. G., and Koten, J. F.** 2002. Next-genereation nuclear power. *Scientific American,* January, 73–81.

30. **MacLeod, G. K.** 1981. Some public health lessons from Three Mile Island: A case study in chaos. *Ambio* 10: 18–23.

31. **Anspaugh, L. R., Catlin, R. J., and Goldman, M.** 1988. The global impact of the Chernobyl reactor accident. *Science* 242: 1513–18.

32. **Balter, M.** 1995. Chernobyl's thyroid cancer toll. *Science* 270: 1758.

33. **Fletcher, M.** 2000. The last days of Chernobyl. *Times 2,* November 14, 3–5. London.

34. **Office of Industry Relations.** 1974. *The nuclear industry, 1974.* Washington, DC: U.S. Government Printing Office.

35. **Fischer, J. N.** 1986. *Hydrologic factors in the selection of shallow land burial for the disposal of low-level radioactive waste.* U.S. Geological Survey Circular 973.

36. **Weart, W. D., Rempe, M. T., and Powers, D. W.** 1998. The waste isolation plant. *Geotimes,* October, 14–19.

37. **U.S. Department of Energy.** 1999. Waste isolation pilot plant, Carlsbad, New Mexico. www.wipp.carlsbad .nm.us. Accessed 6/20/01.

38. **U.S. Department of Energy.** 1992. *DOE's Yucca Mountain studies.* DOE/RW-0345P.

39. **Hunt, C. B.** 1983. How safe are nuclear waste sites? *Geotimes* 28(7):21–22.

40. **Heiken, G.** 1979. Pyroclastic flow deposits. *American Scientist* 67: 564–71.

41. **U.S. Department of Energy.** 1990. *Yucca Mountain project: Technical status report.* DE90015030.

42. **Bredehoeft, J. D., England, A. W., Stewart, D. B., Trask, J. J., and Winograd, I. J.** 1978. *Geologic disposal of high-level radioactive wastes—Earth science perspectives.* U.S. Geological Survey Circular 779.

43. **Duffield, W. A., Sass, J. H., and Sorey, M. L.** 1994. *Tapping the Earth's natural heat.* U.S. Geological Survey Circular 1125.

44. **Wright, P.** 2000. Geothermal energy. *Geotimes* 45(7):16–18.

45. **Muffler, L. J. P.** 1973. Geothermal resources. In *United States mineral resources,* ed. D. A. Brobst and W. P. Pratt, pp. 251–61. U.S. Geological Survey Professional Paper 820.

46. **Tenenbaum, D.** 1994. Deep heat. *Earth* 3(1):58–63.

47. **Worthington, J. D.** 1975. Geothermal development. *Status Report—Energy resources and technology, a report of the Ad Hoc Committee on Energy Resources and Technology.* Atomic Industrial Forum.

48. **Flavin, C., and Dunn, S.** 1999. Reinventing the energy system. In *State of the world 1999: A Worldwatch Institute report on progress toward a sustainable society,* ed. L. R. Brown and others. New York: W. W. Norton.

49. **Berger, J. J.** 2000. *Beating the heat.* Berkeley, CA: Berkeley Hills Books.

50. **Eaton, W. W.** 1978. Solar energy. In *Perspectives on energy,* 2nd ed., ed. L. C. Ruedisili and M. W. Firebaugh, pp. 418–36. New York: Oxford University Press.

51. **Miller, E. W.** 1993. *Energy and American society. A reference handbook.* Santa Barbara, CA: ABC-CLIO.

52. **Mayur, R., and Daviss, B.** 1998. The technology of hope. *The Futurist,* October, 46–51.

53. **Brown, L. R.** 1999. Crossing the threshold. *Worldwatch,* March–April, 12–22.

54. **Quinn, R.** 1997. Sunlight brightens our energy future. *The World and I,* March, 156–63.

55. **Hunt, S. C., Sawin, J. L., and Stair, P.** 2006. *Cultivating renewable alternatives to oil.* In *State of the world 2006,* ed. W. Stark, pp. 61–77. New York: W. W. Norton.

56. **Seth, D.** 2001. *Hydrogen futures: Toward a sustainable energy system.* Worldwatch Paper 157. Washington, DC: Worldwatch Institute.

57. **Kartha, S., and Grimes, P.** 1994. Fuel cells: energy conversion for the next century. *Physics Today* 47: 54–61.

58. **Haggin, J.** 1995. Fuel-cell development reaches demonstration stage. *Chemical & Engineering News* 73: 28–30.

59. **Alward, R., Eisenbart, S., and Volkman, J.** 1979. *Micro-hydro power: Reviewing an old concept.* Washington, DC: U.S. Department of Energy, National Center for Appropriate Technology.

60. **Committee on Resources and Man, National Academy of Sciences.** 1969. *Resources and man.* San Francisco: W. H. Freeman.

61. **Zich, R.** 1997. China's Three Gorges: Before the flood. *National Geographic* 192(3):2–33.

62. **Nova Scotia Department of Mines and Energy.** 1981. *Wind Power.*

63. **Flavin, C.** 1999. The bull market in energy. *Worldwatch,* March–April, 24–27.

64. **U.S. Congress, Office of Technology Assessment.** 1993. *Potential environmental impacts of bioenergy crop production—Background paper.* Washington, DC: U.S. Government Printing Office.

65. **Wihersaari, M.** 1996. Energy consumption and greenhouse gas emissions from biomass production chains. *Energy Conversion and Management* 37: 1217.

66. **Darmstadter, J., Landsberg, H. H., Morton, H. C., with Coda, M. J.** 1983. *Energy today and tomorrow.* Englewood Cliffs, NJ: Prentice Hall.

67. **Steinhart, J. S., Hanson, M. E., Gates, R. W., Dewinkel, C. C., Briody, K., Thornsjo, M., and Kabala, S.** 1978. A low energy scenario for the United States: 1975–2050. In *Perspectives on energy,* 2nd ed., ed. L. C. Ruedisili and M. W. Firebaugh, pp. 553–88. New York: Oxford University Press.

68. **Lovins, A. B.** 1979. *Soft energy paths: Towards a durable peace.* New York: Harper & Row.

69. **Davis, G. R.** 1990. Energy for planet Earth. *Scientific American* 263(3):55–74.

Chapter 16

1. **Grady, D.** 1983. The dioxin dilemma. *Discover,* May, 78–83.

2. **Roberts, L.** 1991. More pieces of the dioxin puzzle. *Science* 254: 377.

3. **Vitousek, P. M., Chadwick, O. A., Crews, T. E., Fownes, J. H., Hendricks, D. M., and Herbert, D.** 1997. Soil and ecosystem development across the Hawaiian Islands. *GSA Today* 7(9):1–10.

4. **Birkland, P. W.** 1984. *Soils and geomorphology.* New York: Oxford University Press.

5. **Brady, N. C., and Weil, R. R.** 1996. *The nature and properties of soils,* 11th ed. Upper Saddle River, NJ: Prentice Hall.

6. **Miller, R. W., and Gardiner, D. T.** 1998. *Soils in our environment,* 8th ed. Upper Saddle River, NJ: Prentice Hall.

7. **Anonymous.** 1979. *Environmentally sound small-scale agricultural projects.* Mt. Rainier, MD: Mohonk Trust, Vita Publications.

8. **Olson, G. W.** 1981. *Soils and the environment.* New York: Chapman and Hall.

9. **Singer, M. J., and Munns, D. N.** 1996. *Soils,* 3rd ed. Upper Saddle River, NJ: Prentice Hall.

10. **Krynine, D. P., and Judd, W. R.** 1957. *Principles of engineering geology and geotechnics.* New York: McGraw-Hill.

11. **Pestrong, R.** 1974. *Slope stability.* New York: American Geological Institute and McGraw-Hill.

12. **Flawn, P. T.** 1970. *Environmental geology.* New York: Harper & Row.

13. **Hart, S. S.** 1974. Potentially swelling soil and rock in the Front Range urban corridor. *Environmental Geology* 7. Denver: Colorado Geological Survey.

14. **Mathewson, C. C., Castleberry, J. P., II, and Lytton, R. L.** 1975. Analysis and modeling of the performance of home foundations on expansive soils in central Texas. *Bulletin of the Association of Engineering Geologists* 17(4):275–302.

15. **Jones, D. E., Jr., and Holtz, W. G.** 1973. Expansive soils: The hidden disaster. *Civil Engineering,* August, 49–51.

16. **Wischmeier, W. H., and Meyer, L. D.** 1973. Soil erodibility on construction areas. In *Soil erosion: Causes, mechanisms, prevention and control,* pp. 20–29. Highway Research Board Special Report 135. Washington, DC: Highway Research Board.

17. **Dunne, T., and Leopold, L. B.** 1978. *Water in environmental planning.* San Francisco: W. H. Freeman.

18. **Robinson, A. R.** 1973. Sediment: Our greatest pollutant? In *Focus on environmental geology,* ed. R. W. Tank, pp. 186–92. New York: Oxford University Press.

19. **Yorke, T. H.** 1975. Effects of sediment control on sediment transport in the northwest branch Anacostia River Basin, Montgomery County, Maryland. *U.S. Geological Survey Journal of Research* 3: 487–94.

20. **Botkin, D. B., and Keller, E. A.** 2000. *Environmental science.* New York: John Wiley.

21. **Wilshire, H. G., and Nakata, J. K.** 1976. Off-road vehicle effects on California's Mojave Desert. *California Geology* 29: 123–32.

22. **Wilshire, H. G., et al.** 1977. *Impacts and management of off-road vehicles.* Report to the Committee on Environment and Public Policy. Washington, DC: Geological Society of America.

23. **Hazen, T. C.** 1995. Savanna River site—A test bed for cleanup technologies. *Environmental Protection,* April, 10–16.

Chapter 17

1. **Harder, B.** 2005. Toxic e-waste is coached in poor nations. *National Geographic News,* November 8.

2. **Relis, P., and Dominski, A.** 1987. *Beyond the crisis: Integrated waste management.* Santa Barbara, CA: Community Environmental Council.

3. **Bullard, R. D.** 1990. *Dumping in Dixie: Race, class and environmental quality.* Boulder, CO: Westview Press.

4. **Galley, J. E.** 1968. Economic and industrial potential of geologic basins and reservoir strata. In *Subsurface disposal in geologic basins: A study of reservoir strata,* ed. J. E. Galley, pp. 1–19. American Association of Petroleum Geologists, Memoir 10.

5. **Relis, P., and Levenson, H.** 1998. *Discarding solid waste as we know it: Managing materials in the 21st century.* Santa Barbara, CA: Community Environmental Council.

6. **Young, J. E.** 1991. Reducing waste, saving materials. In *State of the world,* ed. L. R. Brown, pp. 39–55. New York: World Watch Institute, W. W. Norton & Co.

7. **Schneider, W. J.** 1970. *Hydraulic implications of solid-waste disposal.* U.S. Geological Survey Circular 601F.

8. **Turk, L. J.** 1970. Disposal of solid wastes—Acceptable practice or geological nightmare? In *Environmental geology,* pp. 1–42. Washington, DC: American Geological Institute.

9. **Hughes, G. M.** 1972. Hydrologic considerations in the siting and design of landfills. *Environmental Geology Notes, No. 51.* Illinois State Geological Survey.

10. **Bergstrom, R. E.** 1968. Disposal of wastes: Scientific and administrative considerations. *Environmental Geology Notes, No. 20.* Illinois State Geological Survey.

11. **Cartwright, K., and Sherman, F. B.** 1969. Evaluating sanitary landfill sites in Illinois. *Environmental Geology Notes, No. 27.* Illinois State Geological Survey.

12. **Walker, W. H.** 1974. Monitoring toxic chemical pollution from land-disposal sites in humid regions. *Ground Water* 12: 213–18.

13. **Environmental Protection Agency.** 1980. *Everybody's problem: Hazardous waste.* SW-826. Washington, DC: U.S. Government Printing Office.

14. **New York State Department of Environmental Conservation.** 1994. *Remedial chronology: The Love Canal hazardous waste site.* Albany, NY: New York State.

15. **Elliot, J.** 1980. Lessons from Love Canal. *Journal of the American Medical Association* 240: 2033–34, 2040.

16. **Kufs, C., and Twedwell, C.** 1980. Cleaning up hazardous landfills. *Geotimes* 25: 18–19.

17. **Albeson, P. H.** 1983. Waste management. *Science* 220: 1003.

18. **Return to Love Canal.** 1990. *Time* 135(22):27.

19. **Magnuson, E.** 1980. The poisoning of America. *Time* 116(12):58–69.

20. **Bedient, P. B., Rifai, H. S., and Newell, C. J.** 1994. *Ground water contamination.* Englewood Cliffs, NJ: Prentice Hall.

21. **Huddleston, R. L.** 1979. Solid-waste disposal: Landfarming. *Chemical Engineering* 86(5):119–24.

22. **Committee of Geological Sciences.** 1972. *The Earth and human affairs.* San Francisco: Canfield Press.

23. **Piper, A. M.** 1970. *Disposal of liquid wastes by injection underground: Neither myth nor millennium.* U.S. Geological Survey Circular 631.

24. **Warner, D. L.** 1968. Subsurface disposal of liquid industrial wastes by deep-well injection. In *Subsurface disposal in geologic basins: A study of reservoir strata,* ed. J. E. Galley, pp. 11–20. American Association of Petroleum Geologists Memoir 10.

25. **Cecelia, C.** 1985. *The buried threat.* No. 115-5. Sacramento: California Senate Office of Research.

Chapter 18

1. **American Lung Association.** 2001. *State of the Air 2000.*

2. **Godish, T.** 1991. *Air quality,* 2nd ed. Chelsea, MI: Lewis Publishers.

3. **National Park Service.** 1984. *Air resources management manual.*

4. **U.S. Environmental Protection Agency.** 2006. National-scale air toxics assessment for 1999: Estimated emissions, concentrations and risks. www.epa.gov. Accessed 4/10/06.

5. **Simons, L. M.** 1998. Plague of fire. *National Geographic* 194(2):100–19.

6. **Pope, C. A., III, Bates, D. V., and Raizenne, M. E.** 1995. Health effects of particulate air pollution: Time for reassessment? *Environmental Health Perpsectives* 103: 472–80.

7. **Pittock, A. B., Frakes, L. A., Jenssen, D., Peterson, J. A., and Zillman, J. W.,** eds. 1978. Climatic change and variability: A southern perspective. (Based on a conference at Monash University, Australia, December 7–12, 1975.) New York: Cambridge University Press.

8. **Colin, M.** 2000. Is your office killing you? *BusinessWeek,* June 5, 114–24.

9. **Zummo, S. M., and Karol, M. H.** 1996. Indoor air pollution: Acute adverse health effects and host susceptibility. *Environmental Health* 58: 25–29.

10. **Zimmerman, M. R.** 1985. Pathology in Alaskan mummies. *American Scientist* 73: 20–25.

11. **Gates, D. M.** 1972. *Man and his environment: Climate.* New York: Harper & Row.

12. **Stoker, H. S., and Seager, S. L.** 1976. *Environmental chemistry: Air and water pollution,* 2nd ed. Glenview, IL: Scott, Foresman.

13. **Stern, A. C., Boubel, R. T., Turner, D. B., and Fox, D. L.** 1984. *Fundamentals of air pollution,* 2nd ed. Orlando, FL: Academic Press.

14. **Anthes, R. A., Cahir, J. J., Fraser, A. B., and Panofsky, H. A.** 1981. *The atmosphere,* 3rd ed. Columbus, OH: Charles E. Merrill.

15. **Anonymous.** 1981. How many more lakes have to die? *Canada Today* 12(2):1–11.

16. **Molina, B. F.** 1991. Washington report. *GSA Today* 1(2):33.

Chapter 19

1. **Hansen, J.** 2003. Can we defuse the global warming time bomb? *Natural Science.* www.naturalscience.com.

2. **Lamb, H. H.** 1977. Climate: Present, past and future, vol. 2. *Climatic history and the future.* New York: Barnes & Noble Books.

3. **Marsh, W. M., and Dozier, J.** 1981. *Landscape.* Reading, MA: Addison-Wesley.

4. **Cloud, P.** 1990. Personal written communication.

5. **National Aeronautics and Space Administration (NASA).** 1990. *EOS: A mission to planet Earth.* Washington, DC: NASA.

6. **Chumbley, C. A., Baker, R. G., and Bettis, E. A., III.** 1990. Midwestern Holocene paleoenvironments revealed by floodplain deposits in northeastern Iowa. *Science* 249: 272–74.

7. **Post, W. M., Peng, T., Emanuel, W. R., King, A. W., Dale, V. H., and De Angelis, D. L.** 1990. The global carbon cycle. *American Scientist* 78(4):310–26.

8. **Moss, M. E., and Lins, H. F.** 1989. *Water resources in the twenty-first century.* U.S. Geological Survey Circular 1030.

9. **Titus, J. G., and Seidel, S. R.** 1986. In *Effects of changes in the stratospheric ozone and global climate,* vol. 1. ed. by J. G. Titus, pp. 3–19. Washington, DC: U.S. Environmental Protection Agency.

10. **Titus, J. G., Leatherman, S. P., Everts, C. H., Moffatt and Nichol Engineers, Kriebel, D. L., and Dean, R. G.** 1985. *Potential impacts of sea level rise on the beach at Ocean City, Maryland.* Washington, DC: U.S. Environmental Protection Agency.

11. **Crowley, T. J.** 2000. Causes of climate change over the past 1000 years. *Science* 289: 270–277.

12. **Kennett, J.** 1982. *Marine geology.* Englewood Cliffs, NJ: Prentice Hall.

13. **Broecker, W.** 1997. Will our ride into the greenhouse future be a smooth one? *GSA Today* 7(5):1–7.

14. **Seager, R.** 2006. The source of Europe's mild climate. *American Scientist* 94: 334–41.

15. **Alley, et al.** 2007. *Climate Change: The Physical Science Basis.* Intergovernmental Panel on Climate Change (IPCC). Summary For Policy Makers at www.ipcc.ch accessed 2/8/07.

16. **Charlson, R. J., Schwartz, S. E., Hales, J. M., Cess, R. D., Coakley, J. A. J., Hansen, J. E., and Hofmann, D. J.** 1992. Climate forcing by anthropogenic aerosols. *Science* 255: 423–30.

17. **Kerr, R. A.** 1995. Study unveils climate cooling caused by pollutant haze. *Science* 268: 802.

18. **McCormick, P. P., Thomason, L. W., and Trepte, C. R.** 1995. Atmospheric effects of the Mt. Pinatubo eruption. *Nature* 373: 399–436.

19. **Pelto, M. S.** 1996. Recent changes in glacier and alpine runoff in the North Cascades, Washington. *Hydrological Processes* 10: 1173–80.

20. **Mainguet, M.** 1994. *Desertification,* 2nd ed. Berlin: Springer-Verlag.

21. **Goudie, A.** 1984. *The nature of the environment,* 3rd ed. Oxford: Blackwell Scientific.

22. **Grainger, A.** 1990. *The threatening desert.* London: Earthscan Publications.

23. **Oberlander, T. M.** 1994. Global deserts: a geomorphic comparison. In *Geomorphology of desert environments,* ed. A. D. Abrahams and A. J. Parsons, pp. 13–35. London: Chapman and Hall.

24. **Dregne, H. E.** 1983. *Desertification of arid lands.* Advances in desert and arid land technology and development, vol. 3. Chur, Switzerland: Harwood Academic Publishers.

25. **Sheridan, D.** 1981. *Desertification of the United States.* Washington, DC: Council on Environmental Quality.

26. **University Corporation for Atmospheric Research.** 1994. *El Niño and climate prediction.* Washington, DC: NOAA Office of Global Programs.

27. **Dennis, R. E.** 1984. A revised assessment of worldwide economic impacts: 1982–1984 El Niño/southern oscillation event. *EOS, Transactions of the American Geophysical Union* 65(45):910.

28. **Canby, T. Y.** 1984. El Niño's ill winds. *National Geographic* 165: 144–81.

29. **Philander, S. G.** 1998. Who is El Niño? *EOS, Transactions of the American Geophysical Union* 79(13):170.

30. **Holmes, N.** 2000. Has anyone checked the weather (map)? *Amicus Journal* 21(4):50–51.

31. **Lea, D. W.** 2004. The 100,000 year cycle in tropical SST, greenhouse forcing, and climate sensitivity. *Journal of Climate* 17(11), 2170–79.

32. **Friedman, S. J.** 2003. Storing carbon in Earth. *Geotimes* 48(3):16–20.

33. **Nameroff, T.** 1997. The climate change debate is heating up. *GSA Today* 7(12):11–13.

34. **Bartlett, K.** 2003. Demonstrating carbon sequestration. *Geotimes* 48(3):22–23.

35. **Toon, O. B., and Turco, R. P.** 1991. Polar stratospheric clouds and ozone depletion. *Scientific American,* 264(6):68–74.

36. **Molina, M. J., and Rowland, F. S.** 1974. Stratospheric sink for chlorofluoromethanes: Chlorine atom-catalyzed destruction of ozone. *Nature* 249: 810–12.

Chapter 20

1. **Brown, L. R.** 2003. *Plan B: Rescuing a planet under stress and a civilization in trouble.* New York: W. W. Norton & Co.

2. **Becker, J.** 2004. China's growing pains. *National Geographic* 205(3):68–95.

3. **Sauer, H. I., and Brand, F. R.** 1971. Geographic patterns in the risk of dying. In *Environmental geochemistry in health,* ed. H. L. Cannon and H. C. Hopps, pp. 131–50. Boulder, CO: Geological Society of America Memoir 123.

4. **Bylinsky, G.** 1972. Metallic menaces. In *Man, health and environment,* ed. B. Hafen, pp. 174–85. Minneapolis: Burgess Publishing.

5. **Hong, S., Candelone, J.-P., Patterson, C. C., and Boutron, C. F.** 1994. Greenland ice evidence of hemispheric lead pollution two millennia ago by Greek and Roman civilizations. *Science* 265: 1841–43.

6. **Warren, H. V., and Delavault, R. E.** 1967. A geologist looks at pollution: Mineral variety. *Western Mines* 40: 23–32.

7. **Needleman, H. L., Riess, J. A., Tobin, M. J., Biesecker, G. E., and Greenhouse, J. B.** 1996. Bone lead levels and delinquent behavior. *Journal of the American Medical Association* 275: 363–69.

8. **Hopps, H. C.** 1971. Geographic pathology and the medical implications of environmental geochemistry. In *Environmental geochemistry in health,* ed. H. L. Cannon and H. C. Hopps, pp. 1–11. Boulder, CO: Geological Society of America Memoir 123.

9. **Pettyjohn, W. A.** 1972. No thing is without poison. In *Man and his physical environment,* ed. G. D. McKenzie and R. O. Utgard, pp. 109–10. Minneapolis: Burgess Publishing.

10. **Takahisa, H.** 1971. *Environmental geochemistry in health and disease,* ed. H. L. Cannon and H. C. Hopps, pp. 221–22. Boulder, CO: Geological Society of America Memoir 123.

11. **Armstrong, R. W.** 1971. Medical geography and its geologic substrate. In *Environmental geochemistry in health and disease,* ed. H. L. Cannon and H. C. Hopps, pp. 211–19. Boulder, CO: Geological Society of America Memoir 123.

12. **Kobayashi, J.** 1957. On the geographical relationship between the chemical nature of river water and death-rate of apoplexy. *Berichte des Ohara Institute für landwirtschaftliche biologie* 11: 12–21.

13. **Winton, E. F., and McCabe, L. J.** 1970. Studies relating to water mineralization and health. *Journal of the American Water Works Association* 62: 26–30.

14. **Schroeder, H. A.** 1966. Municipal drinking water and cardiovascular death-rates. *Journal of the American Medical Association* 195: 125–29.

15. **Bain, R. J.** 1979. Heart disease and geologic setting in Ohio. *Geology* 7: 7–10.

16. **Klusman, R. W., and Sauer, H. I.** 1975. *Some possible relationships of water and soil chemistry to cardiovascular diseases in Indiana.* Boulder, CO: Geological Society of America Special Paper 155.

17. **U.S. Environmental Protection Agency.** 1989. *Radon measurements in schools.* Office of Radiation Programs. EPA 520/1 89–010.

18. **U.S. Environmental Protection Agency.** 1986. *A citizen's guide to radon.* OPA-86-004.

19. **Alavanja, M. C., Brownson, R. C., Lubin, J. H., Berger, E., Chang, J. C., and Boice, J. D., Jr.** 1994. Residential radon exposure and lung cancer among non-smoking women. *Journal of the National Cancer Institute* 80(24):1829–37.

20. **Pershagen, G., Akerblom, G., Axelson, O., Clavensjo, B., Damber, L., Desai, G., Enflo, A., Lagarde, F., Mellander, H., Svartengren, M., and Swedjemark, G. A.** 1994. Residential radon exposure and lung cancer in Sweden. *New England Journal of Medicine* 330(3):159–64.

21. **Brenner, D. J.** 1989. *Radon: Risk and remedy.* New York: W. H. Freeman.

22. **University of Maine and Maine Department of Human Services.** 1983. Radon in water and air. *Resource Highlights,* February.

23. **Hurlburt, S.** 1989. Radon: A real killer or just an unsolved mystery? *Water Well Journal,* June, 34–41.

24. **U.S. Environmental Protection Agency.** 1986. *Radon reduction techniques for detached houses.* EPA 625/5-86-019.

25. **U.S. Environmental Protection Agency.** 1988. *Radon resistant residential new construction.* EPA 600/8-88/087.

26. **Store, R.** 1993. Radon risk up in the air. *Science* 261: 1515.

27. **Council on Environmental Quality.** 1979. *Environmental quality.* Annual Report.

28. **Remy, M. H., Thomas, T. A., and Moose, J. G.** 1991. *Guide to the California Environmental Quality Act,* 5th ed. Point Arena, CA: Solano Press Books.

29. **Tanji, K., Lauchli, A., and Meyer, J.** 1986. Selenium in the San Joaquin Valley. *Environment* 28(6):6–11, 34–39.

30. **U.S. Department of the Interior, Bureau of Reclamation.** 1987. *Kesterson Program.* Fact Sheet No. 4.

31. **Rohse, M.** 1987. *Land-use planning in Oregon.* Corvallis: Oregon State University Press.

32. **Curtin, D. J., Jr.** 1991. *California land-use and planning—Law,* 11th ed. Point Arena, CA: Solano Press Books.

33. **William Spangle and Associates, F. Beach Leighton and Associates, and Baxter, McDonald and Company.** 1976. *Earth-science information in land-use planning—Guidelines for Earth scientists and planners.* U.S. Geological Survey Circular 721.

34. **Murphy, E. F.** 1971. *Man and his environment: Law.* New York: Harper & Row.

35. **Carter, L. J.** 1974. Con Edison: Endless Storm King dispute adds to its troubles. *Science* 184: 1353–58.

36. **Bacow, L. S., and Wheeler, M.** 1984. *Environmental dispute resolution.* New York: Plenum Press.

37. **Hawken, P., Lovins, A., and Lovins, L. H.** 1999. *Natural capitalism.* Boston: Little, Brown and Co.

38. **Hubbard, B. M.** 1998. *Conscious evolution.* Novato, CA: New World Library.

39. *Los Angeles Times.* 2001. *Los Angeles Times Poll Alert.* Study No.458.

INDEX

INDEX

System Requirements

Windows System Requirements:
- Windows NT/2000/XP
- Pentium III 450 processor
- In addition to the minimum RAM required by the operating system you are running, this CD-ROM requires 64 MB RAM
- 800 × 600 or 1024 × 768 pixel screen resolution
- Color monitor running "Thousands of Colors"
- Audio capable system is recommended
- CD-ROM drive
- Speakers
- Internet Explorer 5.5 or 6.0 or Netscape Communicator 7.0 or Firefox
- Flash 6 or 7 browser plug-in installed
- Internet connection for associated websites

Macintosh System Requirements:
- OS X
- G3
- In addition to the minimum RAM required by the operating system you are running, this CD-ROM requires 64 MB RAM
- 800 × 600 or 1024 × 768 pixel screen resolution
- Color monitor running "Thousands of Colors"
- Audio capable system is recommended
- CD-ROM drive
- Speakers
- Internet Explorer 5.2 or Netscape Communicator 7.0 or Safari or Firefox
- Flash 6 or 7 browser plug-in installed
- Internet connection for associated websites

Support Information

If you are having problems with this software, call (800) 677-6337 between 8:00 a.m. and 8:00 p.m. EST, Monday through Friday, and 5:00 p.m. through Midnight EST on Sundays. You can also get support by filling out the web form located at: http://247.prenhall.com/mediaform.

Our technical staff will need to know certain things about your system in order to help us solve your problems more quickly and efficiently. If possible, please be at your computer when you call for support. You should have the following information ready:
- textbook ISBN
- CD-ROM ISBN
- corresponding product and title
- computer make and model
- operating system (Windows or Macintosh) and version
- RAM available
- hard disk space available
- Sound card? Yes or No
- printer make and model
- network connection
- detailed description of the problem, including the exact wording of any error messages.

NOTE: Pearson does not support and/or assist with the following:
- third-party software (i.e. Microsoft including Microsoft Office suite, Apple, Borland, etc.)
- homework assistance
- Textbooks and CD-ROMs purchased used are not supported and are non-replaceable. To purchase a new CD-ROM, contact Pearson Individual Order Copies at 1-800-282-0693.

Hobart M. King/B. M. Carpenter/Nicole D. Wilson
Hazard City: Assignments in Applied Geology, 3e
0-13-156682-2 / 978-0-13-156682-8
© 2008 Pearson Education, Inc.
Pearson Prentice Hall
Pearson Education, Inc.
Upper Saddle River, NJ 07458
Pearson Prentice Hall™ is a trademark
of Pearson Education, Inc.